AF334460

International Winterschool on
Electronic Properties of Novel Materials

PROGRESS IN FULLERENE RESEARCH

International Winterschool on
Electronic Properties of Novel Materials

PROGRESS IN FULLERENE RESEARCH

Kirchberg, Tyrol, Austria 5 –12 March 1994

Editors

Hans Kuzmany
Universität Wien

Jörg Fink
ITW, DRESDEN

Michael Mehring
Universität Stuttgart

Siegmar Roth
Max-Planck-Institut, Stuttgart

World Scientific
Singapore • New Jersey • London • Hong Kong

Published by

World Scientific Publishing Co. Pte. Ltd.

P O Box 128, Farrer Road, Singapore 9128

USA office: Suite 1B, 1060 Main Street, River Edge, NJ 07661

UK office: 73 Lynton Mead, Totteridge, London N20 8DH

**ELECTRONIC PROPERTIES OF NOVEL MATERIALS:
PROGRESS IN FULLERENE RESEARCH**

ISBN: 981-02-1887-7

Printed in Singapore by Uto-Print

PREFACE

The *Second International Winter School on Electronic Properties of Novel Materials: Progress in Fullerene Research* took place between March 5-12, 1994, in Kirchberg (Tyrol) Austria. It is part of a series of meetings devoted to Electronic Properties of Novel Materials, which was started in March 1985. Consistent with the ideology of the previous meetings, the goal was to provide an opportunity for experienced scientists from universities and industry to discuss their most recent results and for students and young scientists to become familiar with the present status of research and applications in this new but rapidly developing field.

Meanwhile, for three years, research in the field of organic carbon materials has developed at a rapid pace due to the advent of the fullerenes and related materials. These forms of carbon are considered as a missing link between the previously discussed electroactive polymers and the oxidic superconductors. It was therefore challenging to select this topic for the Kirchberg Winter Schools. The success of this first Kirchberg meeting on fullerenes in 1993 was so large and interest in this field continued to such an extend that the organisers decided to devote the 1994 winter school to fullerenes also.

Since the onset of research, the physics and chemistry of fullerenes and related compounds has already led to a wealth of results, which is reflected in the wide range of topics covered and the numerous discussions which emerged at the meeting. For pure C_{60} as well as for intercalated compounds, preparation methods and crystal growth techniques continue to evolve, while the understanding of the observed phenomena, as there are structural phase transitions, glass transitions, metallic phases, phototransformation and more, continues to pose challenges to experimental and theoretical physicists. In particular, demand still exists for a theory to describe the superconducting phase in intercalated fullerenes.

For synthetic chemistry, the endohedral form of metallofullerenes is expected to have new exciting aspects, in addition to offering a new range of synthetic compounds. The prospect of a periodic table of endohedral fullerene complexes has already been discussed earlier. Now availability of these compounds has attracted the attention of physicists, and research with the full range of spectroscopic techniques is under progress. Similar research is underway with higher fullerenes and related compounds, such as nanotubes and onions

The success of this winter school, the second to focus on fullerenes and related materials, resulted from the range of the topics of interest, as well as from the merging of the many disciplines represented by the participants. As pointed out already in earlier discussions, the fullerene is a unique material, not only due to its molecular structure and symmetry, but more so due to the diversity of phenomena observable in its solid state. This diversity requires an interdisciplinary approach to the scientific research of these materials. We have met this challenge by providing an interdisciplinary discussion at this meeting.

This book summarises the tutorial and research lectures as well as the poster displays presented at the winter school in Kirchberg. We acknowledge all the authors for their contributions and all those who took part in the discussions for their stimulating remarks, which were essential ingredients to this exciting and informative event.

We are grateful to the "Bundesministerium für Wissenschaft und Forschung" in Austria, the "Commission of the European Communities" in Brussels and to the "US-Army Research, Development and Standardization Group", as well as to the many sponsors from industry, for their financial support. This support was more than just a great help; it was, in fact, indispensable in attaining the goal of the meeting. The sponsorship by the Commission of the European Communities is of particular importance at the eve of Austria joining the European Union and it reflects the European dimension which the Kirchberg Winterschools have obtained.

Finally we are indebted, as ever, to the manager of the Hotel Sonnalp, Herr J.R. Jurgeit, and his staff for their continuous support and for their patience with the many special arrangements required during the meeting.

<table>
<tr><td>Vienna</td><td>H. Kuzmany</td></tr>
<tr><td>Dresden</td><td>J. Fink</td></tr>
<tr><td>Stuttgart</td><td>M. Mehring</td></tr>
<tr><td></td><td>S. Roth</td></tr>
</table>

May 1994

CONTENTS

Part Two: METALLOFULLERENES; NANOTUBES AND NANOCLUSTERS

Part Three: STRUCTURE AND STRUCTURAL PHASE TRANSITION

Part Five: ELECTRONIC STRUCTURE AND ELECTRON SPECTROSCOPY

Part Seven: MAGNETIC RESONANCE

FULLERENE BALLS, TUBES, AND ONIONS

Wolfgang Krätschmer
Max-Planck-Institut für Kernphysik
D-69029 Heidelberg
P.O. Box 10 39 80
Germany

ABSTRACT

Fullerenes are carbon molecules with closed-cage structures, some of which, like e.g. the soccer ball shaped C_{60} Buckminsterfullerene can be prepared in bulk quantities. The fullerenes smaller than C_{60} are unstable and may exist only as derivatives e.g. with hydrogen. Most of the known stable larger fullerenes are spheroidal in shape, like balls. They form crystals in which the molecules tend to assume close-packed arrangements. Giant, multi-shelled fullerenes form a class by itself. They predominantly consist either of nested tubular or nested spherical structures - the bucky tubes and onions. All types of fullerenes can enclose foreign atoms, molecules or clusters of material, a feature which may be of future importance in science and technology.

INTRODUCTION

Until a decade ago, not very much was known about the structure and properties of carbon molecules larger than C_3. Early workers noticed that C_3 is the most abundant species in carbon vapour at about 2500°C and that there are also larger molecules occurring in sparks, discharges, laser ablation, or other processes involving carbon vapour or plasma (for a review, see ref. [1]). The molecule C_3 has a linear structure and its spectrum was first observed in comets. Theory predicted linear arrangements not only for C_3 but also for carbon molecules up to about C_{10}, and monocyclic ring structures for the larger species.

The situation changed when C_{60} Buckminsterfullerene was discovered and studied in bulk quantities. In order to understand the surprisingly high yields in the process of fullerene formation, one had to study the reaction pathways which involve smaller fullerene-precursor molecules C_n (with n < 30). These attempts considerably revived the research in this field. Surprisingly rich spectra of different carbon molecular structures were discovered. Besides linear and monocyclic rings, still more complicated ring structures seem to occur, and larger molecules may exhibit three or more isomeric forms[2]. However, from a certain molecular size on, the spatially extended carbon ring-structures tend to collapse into the much more compact fullerene arrangement[3]. A study of the smaller fullerenes C_n (with n < 60) may thus be valuable to better understand this crucial re-arrangement process of carbon atoms.

Whereas the "classical" (i.e. single shelled) fullerenes form rapidly by condensation out of the vapour phase, the known giant fullerene structures - the bucky tubes [4] and onions [5] - grow steadily, probably in an epitaxial-like fashion. The tubes grow on graphitic surfaces and the onions by re-arrangement of carbon atoms within pre-existing carbon grains. All giant fullerenes may also form molecular solids. In advance of their synthesis, solids consisting of bucky-onions have already been named as "onionite". The carbon network of fullerenes mainly consists of 6-membered rings and some lower- and/or higher-membered rings to achieve closure. Suitable ring combinations also allow variable curvature, so that in principle very complex cage structures are possible [6]. Carbon thus seems to be the key element for a future nano-architecture.

The closed cage structure is common to all fullerenes. Chemistry on the fullerene outside has already given rise to substances with surprising properties, e.g. superconductivity. Since the fullerene structures are usually hollow, the cages can be filled with, or can enclose almost anything. The classical fullerenes can host one or a few atoms and form endohedral compounds [7], whereas bucky tubes can be filled with metals to form nano-wires [8]. These features make fullerenes very attractive for future electronic and materials research.

SMALL FULLERENES

The first mass spectra obtained by Kaldor and co-workers in which C_{60} and C_{70} were detected [9] also exhibited a number of fullerene species of smaller mass (see figure 1).

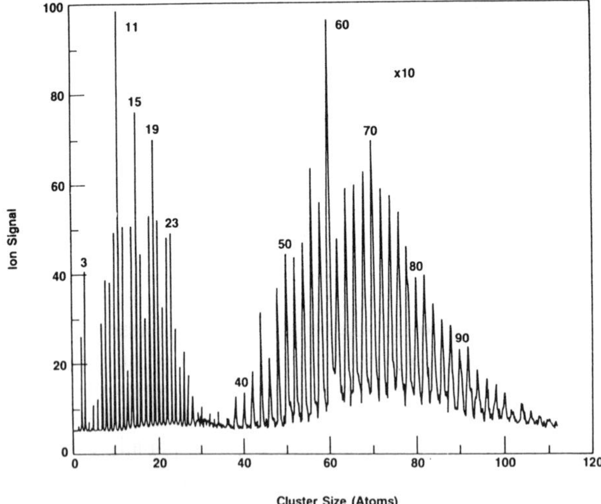

Figure 1: One of the first mass spectra obtained on fullerene molecules (from ref. [9]). Later experiments showed that the sequence of fullerenes starts at about C_{24}. Only the heavier fullerenes, like e.g. C_{60}, C_{70} etc. are chemically stable. The smaller fullerenes may become important e.g. as cage-forming hydrocarbons [12].

Additional molecular beam experiments indicated fullerenes in the size range C_{24} and larger [10]. These small species vanish from the mass spectra under intense clustering conditions, and they are not detectable in fullerene soot. Apparently, small fullerenes react chemically with other fullerenes or excess carbon, disintegrate, and end their existence in forming soot. A plausible explanation for this reactivity is the mechanical strain, i.e. the curvature produced by the abutting pentagons of small fullerenes. It may nevertheless be possible to isolate the most stable isomers of these fullerenes, e.g. by removing them fast from the carbon-vapour reaction zone and trapping then in cryogenic matrices. There are various indications that the small fullerenes can be also stabilised chemically in the form of either endohedral (e.g. in $U@C_{28}$, see ref. [11]) or exohedral compounds. An example of the latter is the hydrocarbon dodecahedran $C_{20}H_{20}$, which may be considered as the hyrogenated form of the smallest possible fullerene C_{20}. The hydrogen bonds allow the carbon cage to relieve part of the strain since the stressed sp^2 bonds can relax into less strained sp^3 arrangement (see, e.g. ref. [12]). Recently, a more controllable method of fullerene synthesis has been found which is based on the pyrolysis of naphthalene $C_{10}H_8$ [13]. In the course of such a process, intermediates may occur which resemble the hydrogenated small fullerenes.

FULLERENES AND ENDOHEDRALS

Remarkable applications of C_{60} and its exohedral chemical derivatives in materials science and technology are still not in sight. However, there seems to be a potential importance of water-soluble derivatives in pharmacy [14]. Research in superconductivity is in progress, although the transition temperatures have not increased above the level achieved two years ago. Work on C_{70} and C_{84} fullerites has started. It appears that, like in the case of C_{60} fullerite, the crystal structure follows the closest packing of spheres and the phase transitions are determined by the choreography of molecular rotations, which for the higher fullerenes is of course more complex due to the non-spherical molecular geometry and the presence of different isomers (e.g. for C_{84}). STM studies of extractable fullerenes up to about C_{330} show only spheroidal and no elongated or needle-like shapes [15]. Solids of such fullerenes very likely also crystallize in a closed packed fashion. Such carbon-balls may become interesting since the transition temperatures of fullerene-based superconductors seems to increase with the fullerene cage size, i.e. crystal lattice spacing. However, no other superconducting fullerene compound has yet been reported besides those based on the highly symmetric C_{60}.

Significant progress has been achieved in the characterisation of endohedral La and Sc compounds such as $La@C_{82}$ (see, e.g. ref. [16]). The evidence that the metal is in fact encaged inside the fullerene appears to be overwhelming, but a definite proof and an elucidation of the detailed endohedral structure is still lacking. Major difficulties in this research arise from the low efficiency in producing metal endohedrals by the arc process (yields are typically less than 10^{-3} of the soot), and in the difficulties of separating endohedrals from the empty fullerenes. Recent molecular beam experiments indicate that endohedrals should form with very high efficiency in the molecular vapour of carbon. The results suggest a kind of nucleation process in which the metal acts as the seed

4

around which the carbon cage condenses [17]. Consequently, much more efficient production processes for endohedrals should be feasible.

The formation of metal endo-fullerenes requires the presence of the metal atoms within the reaction zone where the fullerenes are created. Surprisingly, it also possible to produce endohedrals at a much later stage, after the fullerenes have been formed. By applying elevated temperatures and high pressures, rare gas atoms can be pushed from the outside into the fullerene cage [18]. The reported concentrations obtained range to about 10^{-3} $He@C_{60}$ relative to empty C_{60}. Apparently, a transient "window opening" process is operative where at elevated temperatures some C-C bonds of the fullerene re-arrange to form higher membered rings [19]. When this happens, the cage opens for a short time, allowing the rare gas atoms to slip inside. Unfortunately, the same procedure applied using more reactive gasses (like e.g. H_2) leads predominantly to a disintegration of the fullerene cage rather than to endohedral compounds [20].

BUCKY TUBES

The classical single shelled fullerenes soon decay when exposed to the beam of an electron-microscope. The bucky-tubes discovered by S. Iijima [4] are much more stable and thus can be studied in greater detail and at high resolution. The tubes are usually growing at the negative graphite electrode of fullerene generators and form nested cylinders of a few microns length. (see figure 2).

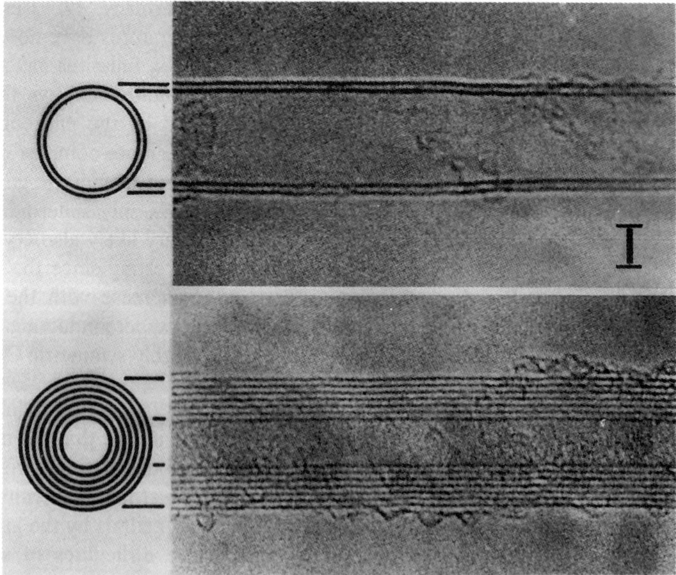

Figure 2: Electron-microscope picture of the middle portion of a bucky-tube. Notice the nested tubular structure, which is also sketched in cross section. The distance of the adjacent layers is similar to that in graphite (see ref.[4]). The scale bar has a length of 2 nm.

Each cylinder has closed ends - the tubes thus are in fact fullerenes. The narrowest tube detected has a diameter of about 0.7 nm which suggests that the closing caps are C_{60} hemispheres. A more detailed study of the different shapes of the closing caps reveals many interesting features of the nano-architecture of the tubes, such as occasionally occurring negative curvature which indicate the presence of seven- or even higher membered rings. Usually, the arrangement of hexagons of the straight graphitic cylinders is twisted with respect to the tube axis. Thus the tubes exhibit helicity (see figure 3).

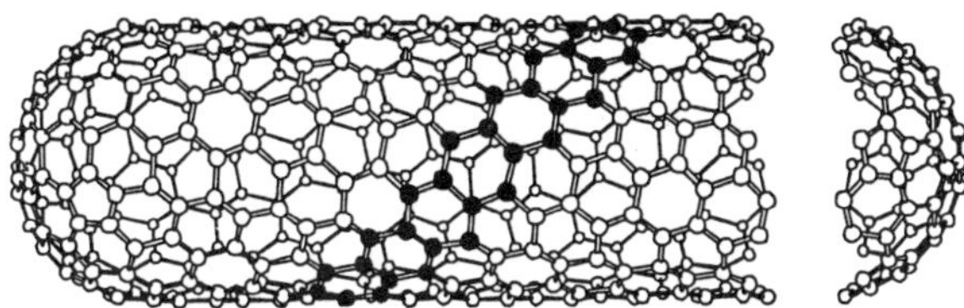

Figure 3: Sketch of the structure of a single tublet of a bucky-tube. Notice the closing caps and the helical arrangement of the hexagonal graphite (adapted from ref.[30]).

When it was discovered that tubes can be produced in bulk quantities, these kind of fullerenes became very attractive, e.g. for materials research [21]. Since the C-C bond in graphite is stronger than that in diamond, bucky-tubes should show very high mechanical strength. When grown to much larger lengths than possible at present, the tubes may replace the conventional carbon fibres. The electronic properties of bucky tubes are also remarkable and the expected band structure has been calculated. It turns out that the tube diameter and helicity determines the bandgap, i.e. whether an individual tube is metallic or semiconducting [22]. In a conventional multi-layered bucky tube, each of the individual tubelets may exhibit different helicities. Thus the electronic properties of such bucky-tubes are rather difficult to control. For this reason, it would be of advantage to grow single walled (and also single diameter tubes). First successful attempts into this direction have recently been reported [23,24].

The closing caps of the bucky-tubes contain five-membered carbon rings and are thus the sites of enhanced chemical reactivity. The bucky-tube chemistry is not yet so advanced as that of the classical fullerenes. A first step into this new domain is the controlled oxidation of the tubes which allowed the removal of carbon at the caps and the opening of the tubes at their ends. The open tubes provide strong capillary forces by which foreign materials can easily be sucked in or absorbed. The nano-wires obtained by this technique may become one of the most exciting fullerene products [8].

6

BUCKY ONIONS

When a bucky-tube or any other small carbon particle type is exposed to large doses of electrons in the beam of a transmission electron microscope, these units gradually transform into spherical bucky-onions. D. Ugarte, the discoverer of this effect, documented the transformation with many breath-takingly beautiful pictures (see figure 4 and ref. [5]).

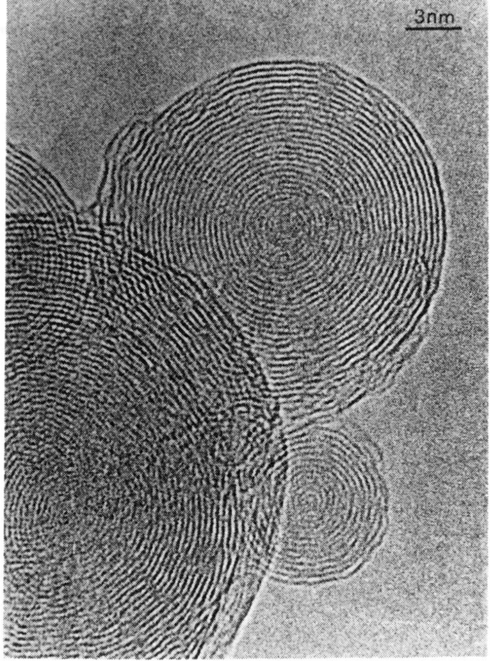

Figure 4: Electron-microscope picture of a bucky-onion. The structure is spherical. The innermost cage has about 1 nm diameter, i.e. has approximately the dimensions of C_{60}. The individual balls may rotate with respect to each other (see ref. [5]).

He interprets the rearrangement of the carbon structure by the concept of "structural fluidity", according to which the combination of both the steady bond breaking and the thermal energy input by the incoming electrons allows the carbon atoms to migrate into the energetically most favourable configurations. Ugarte believes that the onions are the most stable form which can be assumed by small carbon particles. It should be emphasized that the spherical onions are really something new and unexpected. In order to stay spherical, especially the outer shells must allow several local distortions in the carbon network: for each carbon atom to maintain a three-fold coordination with bond angles of about $120°$ and equal bond lengths, additional larger ring types, e.g. five- and seven-membered rings must occur [25]. So far, the larger carbon clusters were assumed to be non-spherical and it was believed that these units show simple icosahedral structures, respectively exhibiting more facetted shapes with sections of almost plane graphitic networks (see, e.g. ref. [26]). It remains open whether the onions or the facetted carbon clusters are the more stable configurations (a contribution to this controversy can be found in ref. [27]).

Ugarte observed that small, i.e. two and more shelled onions are already much more stable under the electron beam than the classical single-shelled fullerenes [28]. This finding may indicate that mini-onions may be much more durable compared to C_{60}. Naturally, such a feature would be highly appreciated for many applications, and the solid "onionite" may have more spectacular properties than fullerite. So far, however, no method has been found to produce spherical onions in bulk quantities. It remains a challenging task.

OUTLOOK

C_{60} and fullerenes have changed from being a rarity to becoming almost common place materials. It is therefore recommended to play with this substance on an even larger scale than has been done so far, e.g. to alloy it with metals or other compounds of interest. New applications may also emerge when fullerenes are alloyed with carbon, e.g. intercalated with graphite [29] or polymerised with each other. In the latter case, mixed sp^2 - sp^3 bonded solids may emerge, like e.g. fullerene-diamonds. It has already been suggested that the endohedral $U@C_{28}$ may be a building block for such solid [11]. The fullerene C_{28} has four bonds on the cage-outside which form a tetrahedral arrangement, i.e. has the same bond-symmetry as a single sp^3 carbon atom. Therefore, one occasionally finds remarks in the literature that fullerenes (including endohedrals) should be regarded as pseudo-atoms, which may mimic a new periodic table [12,16]. Whether these comparisons are fruitful and will lead further remains to be seen. In conclusion, it may be stated that the work with fullerenes remains extremely inspiring.

REFERENCES

1.	W. Weltner and R.J. van Zee, *Chem. Rev.* **89** (1989) 1713.

2.	G. von Helden, Ph. D. Thesis, *Investigation of the structure and energetics of gas phase cluster ions using ion chromatography* (University of California, Santa Barbara, 1993).

3.	G. von Helden, N.G. Gotts, M.T. Bowers, *Nature* **363** (1993) 60.

4.	S. Iijima, *Nature* **354** (1991) 56.

5.	D. Ugarte, *Nature* **359** (1992) 707.

8

6.   S. Iijima, T. Ichihashi, Y. Ando, *Nature* **356** (1992) 776.

7.   D.S. Bethune, R.D. Johnson, J.R. Salem, M.S. de Vries, C.S. Yannoni, *Nature* **366** (1993) 123.

8.   P.M. Ajayan, S. Iijima, *Nature* **361** (1993) 333.

9.   E.A. Rohlfing, D.M. Cox, A.J. Kaldor, *Chem. Phys.* **81** (1984) 3322.

10.   D.M. Cox, K.C. Reichmann, A. Kaldor, *J. Am. Chem. Soc.* **110** (1988) 1588.

11.   T. Guo, M.D. Diener, Y. Chai, M.J. Alford, R.E. Haufler, S.M. McClure, T. Ohno, J.H. Weaver, G.E. Scuseria, R.E. Smalley, *Science* **257** (1992) 1661.

12.   H.W. Kroto and D.R.M. Walton, *Chem. Phys. Letters* **214** (1993) 353.

13.   R. Taylor, G. J. Langley, H. W. Kroto, D. R.M. Walton, *Nature* **366** (1993) 728.

14.   S. H. Friedman, D.L. DeCamp, R.P. Sijbesma, G. Srdanov, F. Wudl, G.L. Kenyon, *J. Am. Chem. Soc.* **115** (1993) 6505; R.M. Baum, *C&EN* (November 22, 1993) 8.

15.   L.D. Lamb, D.R. Huffman, R.K. Workman, S. Howells, T. Chen, D. Sarid, R.F. Ziolo, *Science* **255** (1992) 1413.

16.   K. Kikuchi, S. Suzuki, Y. Nakao, N. Nakahara, T. Wakabayashi, H. Shiromaru, K. Saito, I. Ikemoto, Y. Achiba, *Chem. Phys. Letters* **216** (1993) 67.

17.   D.E. Clemmer, K.B. Shelimov, M.F. Jarrold, *Nature* **367** (1994) 718.

18.   M. Saunders, H.A. Jimenez-Vazquez, R.J. Cross, S. Mroczkowski, D.I. Freedberg, F.A.L. Anet, *Nature* **367** (1994) 256.

19.   R.L. Murry and G.E. Scuseria, *Science* **263** (1994) 791.

20.   Ch. Jin, R. Hettich, R. Compton, D. Joyce, J. Blencoe, T. Burch, *Direct solid phase hydrogenation of fullerenes* (1993) Preprint.

21.   T.W. Ebbesen and P.M. Ajayan, *Nature* **358** (1992) 220.

22.   J.W. Mintmire, B. I. Dunlap, C.T. White, *Phys. Rev. Letters* **68** (1992) 631.

23.   S. Iijima, T. Ichihashi, *Nature* **363** (1993) 603.

24. D.S. Bethune, C.H. Kiang, M.S. de Vries, G. Gorman, *Nature* **363** (1993) 605.

25. T. Tarnai, *Phil. Trans. R. Soc. Lond.* **A, 343** (1993) 145.

26. M. Yoshida and E. Osawa, *Fullerene Science and Technology*, **1** (1993) 55.

27. J. P. Lu and W. Yang, *The shape of bucky onions* (1993), preprint.

28. D. Ugarte, *Europhysics Letters* **22** (1993) 45.

29. S. Saito and A. Oshiyama, *Designing C_{60}-graphite co-intercalation compounds* (1993) preprint.

30. M.S. Dresselhaus, *Nature* **358** (1992) 195.

Part One

CHEMISTRY AND PREPARATION OF SIMPLE FULLERENES

Fullerenes:
Purification, Stability and Functional Covalent Derivatives

H. L. ANDERSON, F. CARDULLO, F. DIEDERICH, R. FAUST,
A. HERRMANN, L. ISAACS, D. PHILP, C. THILGEN*, A. WEHRSIG
*Laboratorium für Organische Chemie, ETH Zürich, Universitätstrasse 16, CH-8092
Zürich, Switzerland*

and
G. CALZAFERRI, D. HERREN
*Institut für Anorganische, Analytische und Physikalische Chemie, Universität Bern,
Freiestrasse 3, CH-3000 Bern 9, Switzerland*

and
U. JONAS, H. RINGSDORF
*Institut für Organische Chemie, Johannes-Gutenberg-Universität, J.-J.-Becher-Weg
18-20, D-55099 Mainz, Germany*

and
H.-D. BECKHAUS, C. RÜCHARDT, S. VEREVKIN
*Institut für Organische Chemie und Biochemie, Albert-Ludwigs-Universität Freiburg,
Albertstraße 21, D-79104 Freiburg, Germany*

and
H.-U. TER MEER, H. MOHN, W. MÜLLER
Angewandte Physik G 864, Hoechst AG, D-65926 Frankfurt am Main, Germany

ABSTRACT

A simple plug filtration method that allows the purification of gram amounts of C_{60} on activated charcoal within half an hour is described. For the higher fullerenes C_{76}, C_{78} and C_{84}, baseline separation can be achieved by HPLC on a TCPP (Tetrachlorophthal-imidopropyl) - modified silica gel stationary phase. Structural investigations on a number of methanofullerene[60] derivatives show the singular behaviour of the parent compound $C_{61}H_2$. A simple rationale is presented for the preferred occurrence of the bridging at a [6,6]-junction with a closed transannular bond as the thermodynamical product. The results of PM3 calculations are in agreement with the experimental findings and lead to a suggestion for a possible rearrangement pathway from [6,5]- to [6,6]-bridged methanofullerenes. Synthesis and first investigations of some bucky-ionophores, a class of fullerene derivatives that combines the metal complexing abilities of an ionophore moiety with the electron-accepting properties of C_{60}, are presented. Finally, efforts towards the synthesis of new carbon allotropes led to the preparation of alkynyl derivatives of C_{60}, the chemistry of which is currently being explored.

1. Improved Purification of C_{60} and of Higher Fullerenes

The major problem common to all of the early chromatographic fullerene purification methods is the relative insolubility of the fullerenes in the mobile phase, a fact which severely limits the amount of material that can be processed. Best solvents for fulle-renes are CS_2 and benzene, as well as alkylated or halogenated aromatics. With these solvents, however, fullerenes show virtually no retention on most stationary phases.

High pressure gel permeation chromatography was the first method which used pure toluene as the mobile phase and, as a result, was amenable to larger scale purifications of C_{60}.[1] Some time later, *Tour et al.* reported a separation of C_{60} from C_{70} and the higher fullerenes by flash chromatography on a mixture of silica gel/*Norit-A* charcoal (2:1) with toluene as the eluent.[2] This method can be further simplified. We found that by using the highly effective activated charcoal *Darco G60* (*Fluka*), mixed with silica gel (*Merck*, 0.040 - 0.063 mm), flash chromatography can be replaced by a simple plug filtration.[3,4] In our method, a fritted funnel (7 cm diameter) is covered with a slurry of 63 g of charcoal and 125 g of silica gel to give a plug of 9 cm height. A concentrated solution of soluble fullerene soot extract in toluene is loaded and eluted by the application of a slight vacuum at the filter flask. Only 30 min are necessary for the elution of 3.9 g of C_{60} ($\approx 90\%$ of the C_{60} input) starting from 5.0 g of fullerene soot extract. HPLC analysis of the eluate shows no contamination ($< 0.05\%$) by C_{70} or higher fullerenes.

C_{70} on the other hand is retained so strongly on this stationary phase that it can not be eluted with toluene. Sonication of the charcoal in *o*-dichlorobenzene leads to extraction of about one third of the total amount of C_{70} present in addition to some C_{60}.

An excellent separation of the higher fullerenes C_{76}, C_{2v}-C_{78}, D_3-C_{78} and C_{84} (mixture of D_2 and D_{2d}-isomers) can be achieved by HPLC using a Vydac 201TP54 C18 reversed phase and toluene/acetonitrile 1:1 as the eluent.[5] However, due to the low solubility of the carbon spheres in this solvent mixture, 50 µg of material only can be separated per injection on a 250 x 20 mm I.D. column. A twentyfold quantitative increase (1 mg per injection) with no loss in the quality of the separation could be achieved by switching to a silica gel stationary phase modified by the attachment of the strong π-electron accepting tetrachlorophthalimidopropyl (TCPP) groups in combination with the mixture CH_2Cl_2/hexane (1:2) as the eluent.[6] This π-acidic phase which had been developed for the separation of polycyclic aromatic hydrocarbons, a class of compounds having in common the lack of functional groups with fullerenes, leads to strong interactions with the carbon molecules and hence to their separation. The purification of fullerenes on the Trident-Tri-DNP (Buckyclutcher I) phase developed by *Welch et al.* is based on the same principle.[7]

2. The Stability of Fullerenes: Experimental Determination of the Heat of Formation of C_{60} and C_{70}

The time lapse between the first isolation of macroscopic amounts of fullerenes and the first determination of the heat of formation of C_{60} was due to problems with both the production of highly pure C_{60} in sufficient quantities and the accuracy of calorimetric measurements on small samples - a procedure which is still far from being trivial. Both problems are reflected by large differences among the values published in the literature[8] and called for a re-determination of the heats of formation.[9]

"Gold grade" C_{60} and C_{70} (*Hoechst AG*) were both subjected to fractional sublimation using He as the transport gas at 600 and 650°C / atmospheric pressure, respectively, and stored under argon prior to the combustion experiments which were performed in an isoperibolic aneroid microcalorimeter[10] as well as in a water-stirred isoperibolic macrocalorimer.[11] Analytical HPLC (Vydac 201TP54 RP-C18, toluene/acetonitrile 1:1) was used to determine the composition of the samples (C_{60}: 99.82% C_{60} and 0.18% $C_{60}O$; C_{70}: 0.44% C_{60} and 99.56% C_{70}) and to check their stability against oxygen under the experimental conditions preceding the ignition.

Reduction to the isothermal bomb process with $c_p = 0.71$ (C_{60}) and 0.79 $JK^{-1}g^{-1}$ (C_{70}) and to standard conditions was done according to the usual procedure[12] and afforded the results shown in Table 1.

Table 1. Experimental and Calculated Heats of Formation ΔH_f^o(g,C) per C-atom of C_{60} and C_{70}.[13]

Method	$\boxed{C_{60}}$ ΔH_f (g,C)	± σ	$\boxed{C_{70}}$ ΔH_f (g,C)	± σ	Ref.
exp.	10.16	0.06	9.65	0.03	this work
exp.	10.01	0.02			[8a]
exp.	10.58	0.06			[8b]
exp.	9.98	0.06	9.03		[8c]
exp.	10.01	0.06			[8d]
MM3	9.49		9.10		this work
Incremental	10.57		9.27		[14]
MNDO	14.5		13.4		[15]
ab initio			$\Delta = 0.91$, $C_{60} = 0.0$		[16]

The redetermination of the heat of formation per C-atom of C_{60} using sublimed samples led to a value (10.16 kcal mol^{-1}) that is only slightly higher compared to the one previously determined by us (10.01 kcal mol^{-1}).[8a] Furthermore and as predicted,[17] C_{70} is more stable per C-atom (9.65 kcal mol^{-1}) than C_{60} (10.16 kcal mol^{-1}), a fact that can be rationalized in terms of a decrease in strain due to the incorporation of additional hexagons into a framework with the same number (12) of pentagons. Finally, among the computational methods considered, MM3 reproduces the experimental results best.

3. Structural Investigations on Methanofullerenes

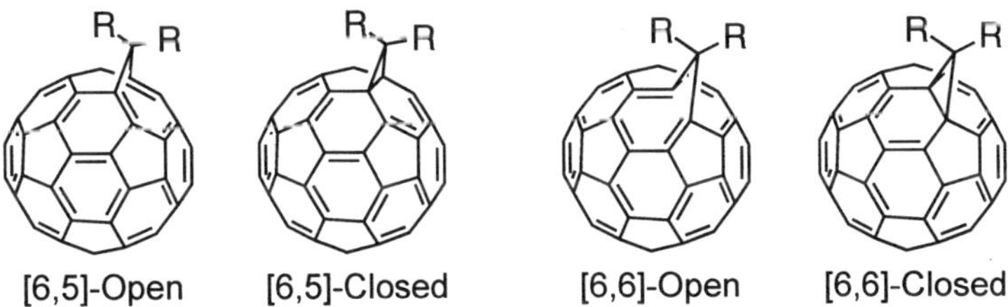

The question of the structure of methanofullerene derivatives that had been the object of controversy for some time could be settled in favor of a bridged [6,6]-junction with a closed transannular bond as the thermodynamically most stable form in most cases.[3,18,19] Preparation of the parent hydrocarbon $C_{61}H_2$ (R = H) in a thermal reaction however leads to the [6,5]-open structure,[20] which in other cases is the kinetic product. No [6,5]-closed or [6,6]-open structures have been observed so far.

For the methanofullerenes with R = H (**1**),[20,21] OMe (**2**),[18] CO$_2$Et (**3**),[21] CO$_2t$-Bu (**4**),[21] Ph (**5**)[22], the experimental studies have been complemented by PM3 calculations.[21] In all cases except **1**, the [6,6]-closed structure is more stable by 4.7-7.3 kcal

16

mol^{-1} compared to the [6,5]-open structure. Only in the case of the parent methanofullerene (**1**) is the stability order reversed, and the difference between the two isomers (0.34 kcal mol^{-1}) is very small. The preferred occurrence of the [6,6]-closed structure is rationalized by the preservation of the [5]radialene type bonding pattern found in C_{60}. The [6,5]-open structure is less stable due to a violation of Bredt's rule leading to the pyramidalization of the sp^2-hybridized bridgehead carbon atoms. The PM3 method locates a [6,5]-closed structure for methanofullerenes **2-5** and the absence of a similar structure for **1**. Experimentally, a thermal rearrangement [6,5]-open to [6,6]-closed is observed for **3-5**, but not for **1** (the preparation method of **2** is different and leads directly to the thermodynamic product). The combined computational and experimental findings suggest that the thermal interconversion between the [6,5]-open and [6,6]-closed isomers proceeds *via* a stepwise mechanism ([6,5]-open to [6,5]-closed valence isomerization followed by a 1,5-shift) and occurs only in those cases where the corresponding [6,5]-closed structure is located in a local energy minimum.

4. C_{60}-Ionophores

In a program aimed at the preparation of buckminsterfullerene derivatives with potentially interesting biological and materials properties, the crown ether derivative **6**[4] and the benzo[2.2.2]cryptand derivative **7**[23] of C_{60} were prepared. The interest in compounds of this type arises from their metal complexing abilities, their amphiphilic character, and their potential to form new conducting solid state phases through doping with electropositive metals. Of particular interest should be electride-type structures which **6** and **7** could form by dissolving potassium under complexation of K$^+$ in the ionophore moiety and delocalization of the electrons in the carbon sphere and possibly over the entire crystal lattice.

The bucky-crown **6** was obtained by Diels-Alder addition of the *o*-quinodimethane intermediate resulting from bromine elimination by Bu$_4$NI from 4,5-bis(bromomethyl)benzo[18]crown-6.[4] Condensation of methanofullerene[60] carboxylic acid, a versatile intermediate for further derivatizations of C_{60},[18] with 4-aminobenzo[2.2.2]cryptand under peptide coupling conditions afforded the C_{60}-cryptand **7**.

Solid-liquid saturation extraction experiments monitored by UV/VIS spectroscopy showed that the complexation of potassium ions strongly increases the solubility of **6** in protic solvents ($\approx$20 times in methanol). Using *Langmuir-Blodgett* techniques, monolayers of the highly amphiphilic fullerene crown and its potassium ion complex were prepared.

Pressure/area isotherms measured with a *Langmuir* pressure pickup system[4] point to an intensive interaction of **6** with the water subphase and show a surface area of 80 Å^2/molecule corresponding to the theoretical value for C$_{60}$ of 86.6 Å^2/molecule. Attempts to transfer the film of **6** to mica at low pressure (1 mN m^{-1}) via the *Langmuir-Blodgett* technique were successful for the monolayer, but multilayers could be obtained neither on hydrophilic nor on hydrophobic substrates.

Using a subphase containing K$^+$-ions with a concentration of 1 M, a significant increase in the surface area requirement was observed in the isotherm, leading to the ideal value as expected for C$_{60}$. The obvious stabilization of the monolayer upon complexation of K$^+$-ions is due to the increased hydrophilic character of the crown moiety, leading to stronger interactions with the subphase.

The properties of the bucky-cryptand **7** as well as of a cyclam derivative of C$_{60}$ currently being prepared are the subject of further investigations.

5. Fullerene-Acetylene Hybrids: *En Route* towards New Carbon Allotropes

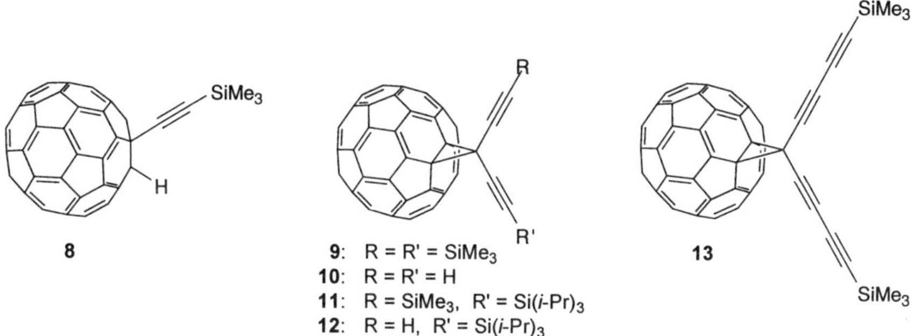

Our interest in carbon-rich materials[24] has prompted us to explore routes to molecular carbon allotropes which combine fullerene and acetylene chemistry.[25]

As opposed to chloromagnesium acetylide, lithium trimethylsilylacetylide, like a host of other nucleophiles, adds smoothly and regioselectively in a 1,2-fashion to C$_{60}$ to yield **8** upon quenching with acetic acid. The structure of **8** is evident from its spectral data, notably the ^{1}H NMR resonance at 6.25 ppm corresponding to the proton attached directly to the fullerene and the 32 ^{13}C NMR resonances indicative of C_s-symmetry.

Reaction of C$_{60}$ with carbenes derived from 1,4-pentadiyne-3-one tosylhydra-zone[26] or with silyl derivatives of 3-bromo-1,4-pentadiyne[27] in the presence of a base[28] gives access to diethynylmethanofullerenes **9** and **11** which, upon removal of the trimethylsilyl protecting groups by K$_2$CO$_3$ in MeOH/THF, yield the species **10** and **12** bearing terminal acetylenic groups.

Unsymmetrical Hay coupling (CuCl, TMEDA, O$_2$) of **10** with an excess of trimethylsilylacetylene furnished bis(butadiynyl)methanofullerene **13**, whereas symmetrical coupling of **12** afforded a butadiynyl-linked dimeric dumbbell-shaped molecule (**14**) containing two fullerene moieties.

Current investigations are focused on homocoupling reactions of dialkynyl-methanofullerenes with two terminal acetylenic groups like **10** and desilylated **13** or **14** which, upon cyclooligomerization or polymerization, may lead to new synthetic molecular or polymeric carbon allotropes.

References

1. (a) A. Gügel, M. Becker, D. Hammel, L. Mindach, J. Räder, T. Simon, M. Wagner, K. Müllen, *Angew. Chem.* **104** (1992) 666-667; *Angew. Chem. Int. Ed. Engl.* **31** (1992) 644; (b) A. Gügel, K. Müllen, *J. Chromatogr.* **628** (1993) 23-29; (c) M. S. Meier, J. P. Selegue, *J. Org. Chem.* **57** (1992) 1924-1926.
2. W. A. Scrivens, P. V. Bedworth, J. M. Tour, *J. Am. Chem. Soc.* **114** (1992) 7917-7919.
3. L. Isaacs, A. Wehrsig, F. Diederich, *Helv. Chim. Acta* **76** (1993) 1231-1250.
4. F. Diederich, U. Jonas, V. Gramlich, A. Herrmann, H. Ringsdorf, C. Thilgen, *Helv. Chim. Acta* **76**, (1993) 2445-2453.
5. (a) F. Diederich, R. L. Whetten, C. Thilgen, R. Ettl, I. Chao, M. M. Alvarez, *Science (Washington)* **254** (1992) 1768-1770; (b) C. Thilgen, F. Diederich, R. L. Whetten in *Buckminsterfullerenes*, eds. W. E. Billups, M. A. Ciufolini (VCH, New York, 1993).
6. D. Herren, C. Thilgen, G. Calzaferri, F. Diederich, *J. Chromatogr.* **644** (1993) 188-192.
7. C. J. Welch, W. H. Pirkle, *J. Chromatogr.* **609** (1992) 89-101.
8. (a) H.-D. Beckhaus, C. Rüchardt, M. Kao, F. Diederich, C. S. Foote, *Angew. Chem.* **104** (1992) 69-70; *Angew. Chem. Int. Ed. Engl.* **31** (1992) 63-64; (b) W. V. Steele, R. D. Chirico, N. K. Smith, W. E. Billups, P. R. Elmore, A. E. Wheeler, *J. Phys. Chem.* **96** (1992) 4731-4733; (c) T. Kiyobaiashi, M. Sakiyama, *Annual Report of Microcalorimetry Research Center*, Faculty of Science, Osaka University, **13** (1992) 58; *Fullerene Sci. Tech.* **1** (1993) 269-273; (d) H. P. Diogo, M. E. Minas da Pielade, T. J. S. Dennis, J. P. Hara, H. W. Kroto, R. Taylor, D. R. M. Walton, *J. Chem. Soc., Faraday Trans.* **89** (1993) 3541-3544.
9. H.-D. Beckhaus, S. Verevkin, C. Rüchardt, F. Diederich, C. Thilgen, H.-U. ter Meer, H. Mohn, W. Müller, *Angew. Chem.* in press.
10. H.-D. Beckhaus, C. Rüchardt, M. Smisek, *Thermochimica Acta* **79** (1984) 149-159.
11. S. Sunner in *Combustion Calorimetry Vol. 1*, eds. S. Sunner, M. Månsson (Pergamon Press, Oxford, 1979, ch. 2).
12. W. N. Hubbard, D. W. Scott, G. Waddington in *Experimental Thermochemistry Vol 1.*, ed. F. D. Rossini (Interscience, New York, 1956, ch. 6).
13. ΔH_{sub}(298K, C) = 0.93 kcal/g-atom, extrapolated from ΔH_{sub}(707K, C) = 0.67 kcal/g-atom measured for C_{60}[8b] with c_p(c) measured and c_p(g) estimated.[8b]
14. D. A. Armitage, C. W. Bird, *Tetrahedron Lett.* **34** (1993) 5811-5812.
15. D. Bakowies, W. Thiel, *J. Am. Chem. Soc.* **113** (1991) 3704-3714.
16. J. Cioslowski, *Chem. Phys. Lett.* **216** (1993) 389-393.
17. (a) J. Tersoff, *Physical Review B* **46** (1992) 15546-15549; (b) B. L. Zhang, C. H. Xu, C. Z. Wang, C. T. Chan, K. M. Ho, *Phys. Rev. B* **46** (1992) 7333-7336.
18. L. Isaacs, F. Diederich, *Helv. Chim. Acta* **76** (1993) 2454-2464.
19. M. Prato, V. Lucchini, M. Maggini, E. Stimpfl, G. Scorrano, M. Eiermann, T. Suzuki, F. Wudl, *J. Am. Chem. Soc.* **115** (1993) 8479-8480.
20. (a) T. Suzuki, Q. Li, K. C. Khemani, F. Wudl, *J. Am. Chem. Soc.* **114** (1992) 7301-7302; (b) A. B. Smith III, R. M. Strongin, L. Brard, G. T. Furst, W. J. Romanow, *J. Am. Chem. Soc.* **115** (1993) 5829-5830.
21. F. Diederich, L. Isaacs, D. Philp, *J. Chem. Soc., Perkin Trans. 2* **1994** 391-394.
22. T. Suzuki, Q. Li, K. C. Khemani, F. Wudl, O. Almarsson, *Science (Washington D.C.)* **254** (1992) 1186-1188.
23. F. Cardullo, F. Diederich, C. Thilgen, *unpublished results*.
24. (a) F. Diederich, Y. Rubin, *Angew. Chem.* **104** (1992) 1123-1146; *Angew. Chem. Int. Ed. Engl.* **31**, (1992) 1101-1123; (b) F. Diederich, *Nature (London)*, in press.
25. H. L. Anderson, R. Faust, Y. Rubin, F. Diederich, *submitted*.
26. H. Hauptmann, *Tetrahedron* **32** (1976) 1293-1297.
27. H. Hauptmann, *Tetrahedron Lett.* **1974** 3587-3588.
28. C. Bingel, *Chem. Ber.* **126** (1993) 1957-1959.

THE REGIOCHEMISTRY OF NUCLEOPHILIC ADDITIONS TO C_{60}

Andreas Hirsch*, Iris Lamparth
Institute for Organic Chemistry
Auf der Morgenstelle 18, University of Tübingen
72076 Tübingen, Germany

Heinrich R. Karfunkel
Ciba Geigy AG, Basel, Switzerland

Abstract

Buckminsterfullerene (C_{60}) contains thirty electrophilic double bonds being located at the junctions of the hexangons. As a consequence, for higher adducts of C_{60} a multitude of regioisomers (2 to 6 addents) are possible. Using nucleophilic cyclopropylation reactions we systematically studied the regioselectivity of addition reactions to the fullerene core with chromatographic, spectroscopic and theoretic methods. In this way the seven stable of the eight possible bisadducts of C_{60} with diethoxycarbonylmethylene have been isolated and identified for the first time. Subsequent cyclopropylations with these bisadducts lead to definined regioisomeric trisadducts. A remarkable regioselectivity of additions in *equatorial-* and *trans-3* positions to addents already bound to C_{60} was observed.

Introduction

The exohedral chemistry of C_{60} has been dominated thus far mainly with the investigation of its reactivity and the isolation and characterization of monoaddition products [1]. The regiochemistry of polyadducts has hardly been examined up this point [2]. The synthesis of polyadducts with defined three-dimensional structure is also topical in view of fullerene derivatives with interesting biological [3] or materials [4] properties. We describe here the systematic investigation of the regiochemistry of C_{60} derivatives with two and three methano-bridges [5].

Results

The cyclopropanation of C_{60} with diethyl bromomalonate was chosen as a model reaction [6]. This reaction proceeds very uniformly, and only bonds common to two six-membered rings in the fullerene framework (6-6 bonds) are attacked. We first isolated the monoaddition product $C_{61}(COOEt)_2$. Bisadducts $C_{62}(COOEt)_4$ can be synthesized from the monoadduct using the identical experimental conditions [Scheme 1]. In principle 8 regioisomeric bisadducts are possible. The seven stable out of the eight possible regioisomers of $C_{62}(COOEt)_4$ were isolated from the reaction mixture by high performance liquid chromatography (HPLC) [5].

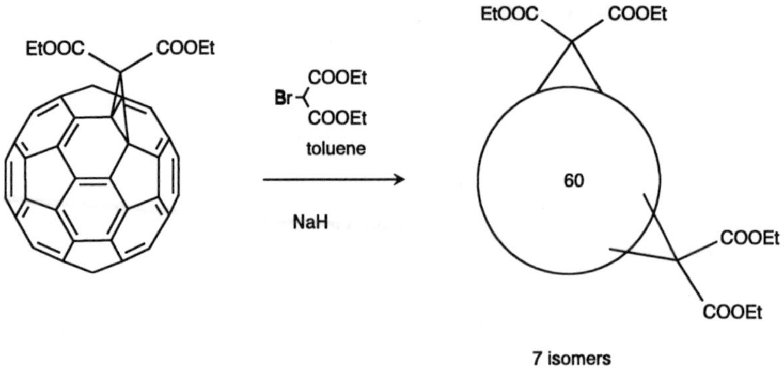

20

Scheme 1

Relative to a given position (6-6 ring bond) in C_{60}, a second addition to a 6-6 ring bond can take place in three areas of the molecule namely, in the same hemisphere (*cis*) at the equator (*e*) or in the opposite hemisphere (*trans*) [Fig. 1]. The three-dimensional relationship between two 6-6 bonds can be *cis*, *e* or *trans* and the *cis* and *trans* cases are further subdivided into *cis*-1, *cis*-2, *cis*-3 as well as *trans*-1, *trans*-2, *trans*-3 and *trans*-4.

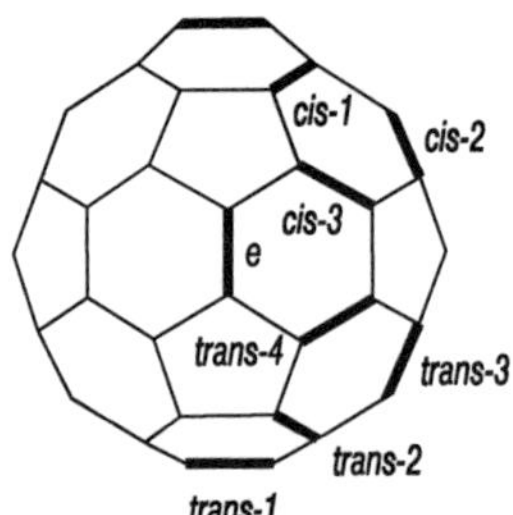

Figure 1   Positions of the ligand carrying bonds in the 8 possible regioisomers of C_{60} with symmetrical additions to 6-6 bonds.

The three bisadducts in which the attacked 6-6 bonds are in *cis*-3-, *trans*-2- and *trans*-3 positions are C_2-symmetric and therefore chiral. These regioisomers of $C_{62}(COOEt)_4$ were isolated as pairs of enantiomers using achiral stationary phases (HPLC). According to theoretical calculations (AM1), the *cis*-1 isomer is considerably destabilized with respect to the others [Tab. 1] [5]. This is mainly due to steric repulsion of the methano addents which would be located in the direct neighborhood. Indeed, this regioisomer has not been found in the reaction mixture. The other seven isomers of $C_{62}(COOEt)_4$ could be identified by NMR spectroscopy (symmetry) as well as on the basis of their elution order during HPLC separation (silica gel), which is in line with the calculated AM1-dipole moments of the different regioisomers [Tab. 1]. The formation of the bisadducts $C_{62}(COOEt)_4$ shows a remarkable regioselectivity. The *e*- as well as the *trans*-3 isomers are significantly preferred [Fig. 2]. The *cis*- and the *trans*-1 adducts of $C_{62}(COOEt)_4$ in general are less abundant in the reaction mixture compared to the other isomers. It is very likely that not only thermodynamic but also kinetic factors govern the product distribution, since the heats of formation especially of the *e*- and *trans*- isomers seem to be very similar [Tab. 1] [5]. Responsible for the low yield of the D_{2h}-symmetric *trans*-1 isomer is also the reduced statistical probability of its formation. Whereas starting from a given monoadduct the attack of only one 6–6 bond leads to a *trans*-1 bisadduct, the attack of each four bonds can lead to the other regioisomers. The regioisomers of $C_{62}(COOEt)_4$ have slightly different colors ranging from chestnut brown to red-orange. This is reflected by their electronic absorption spectra. Each regioisomer exhibits a characteristic feature in the visible part with the main absorptions at 430 and 470 nm. The relative intensity, the splitting and the width of these absorptions differ from isomer to isomer. Except for the most symmetrical *trans*-1 isomer, the bisadducts $C_{62}(COOEt)_4$ are very soluble in organic solvents like chloroform or methylene chloride.

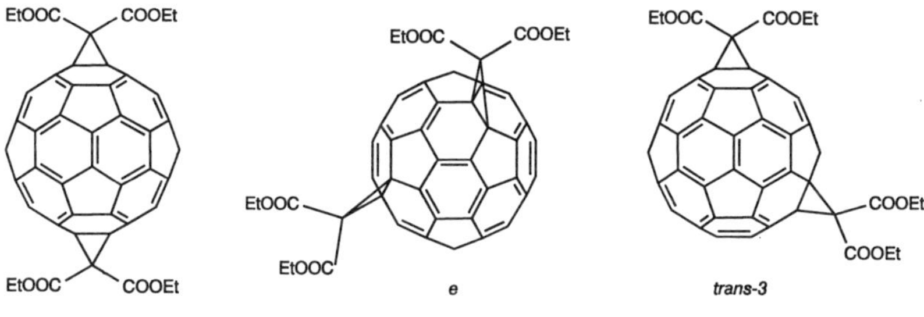

Figure 2   Front view representations of the most symmetric and least soluble *trans*-1 isomer as well as of the most abundant *e*- and *trans*-3 isomers of $C_{62}(COOEt)_2$ formed by cyclopropylation of $C_{61}(COOEt)_2$.

Table. 1   Symmetry, order of elution, relative yield, calculated stabilities and dipole moments of the different regioisomers of $C_{62}(COOEt)_2$ formed by cyclopropylation of $C_{61}(COOEt)_2$ [5].

Symmetry	Order of Elution	Relative yield (%)	Relative AM1 ΔH_f^0 (kcal/mol)	AM1- Dipole moment (Debye)
trans-1	1	2	0.2	0.1
trans-2	2	13	0.2	1.2
trans-3	3	30	0.1	2.0
trans-4	4	9	0.0	2.3
e	5	38	0.0	2.6
cis-3	6	6	1.3	3.0
cis-2	7	2	1.8	3.5
cis-1	8	0	17.7	3.8

To study the regiochemistry of the trisadducts (46 possible isomers) and isolate single isomers in reasonable yields, defined bisadducts of $C_{62}(COOEt)_4$ have been used as starting materials and treated with diethylbromo malonate in toluene in the presence of NaH to give the trisadducts $C_{63}(COOEt)_6$. Some trisadducts, for example the regioisomers with all ligands in the *e* position (trisadduct *e,e,e*) to each other, are only accessible out of one particular bisadduct (bisadduct *e* with the ligands in *e* position). Others can be formed out of two different bisadducts, such as the trisadduct *trans-4,trans-4,trans-2* from the bisadducts *trans-4* and *trans-2*, or out of three different bisadducts such as *e,trans-4,trans-3* out of *e*, *trans-4* and *trans-3*. These considerations are useful in identifying the trisadducts by their formation pathway as well as by their symmetry, determined by NMR spectroscopy. All the trisadducts with three identical positional relations, for example *e,e,e*, have a threefold symmetry (D_3, C_3 or C_{3v}), those with two different relations, for example *trans-4,trans-4,trans-2*, have a twofold symmetry (C_2 or C_s) and those with three different relations, for example *e,trans-4,trans-3*, are unsymmetric (C_1). The number of possible regioisomers that can be expected from the cyclopropylation of a certain bisadduct depends on the structure of the bisadduct. For example, 14 regioisomeric trisadducts can in principle be formed if an addition reaction with *e*-$C_{62}(COOEt)_4$ is carried out. However, the number of preferably formed trisadducts can be expected to be much lower, because trisadducts with addents bound in *cis-* especially *cis-1* positions will be considerably destabilized. In the higher adducts (tetra, penta, hexa etc.), the number of preferably formed isomers will further decrease, since many of the possible regioisomers would have the addents bound in *cis-*positions. For example, among the 43 possible structures for the hexaadduct $C_{66}(COOEt)_{12}$ without the unfavorable *cis-1*-positions, which can be formed out of *e,e,e*-$C_{63}(COOEt)_6$, one regioisomer with T_h symmetry and all addents in *e* positions ("octahedral sites") is significantly more stable than all the others. The calculated heat of formation of the next most stable isomer is 5 kcal/mol higher. Indeed, the cyclopropylation of $C_{62}(COOEt)_4$ shows a remarkable regioselectivity. The chiral orange-red colored chiral C_3-symmetric *e,e,e*-isomer of $C_{63}(COOEt)_6$ was isolated in 40% yield [Scheme 2]. In this regioisomer, the addents are bound in "octahedral sites" of the C_{60} molecule and occupy one trigon of the trigonal antiprism (octahedron).

40 %

chiral C_3 *(e,e,e)*

Scheme 2

In a similarly pronounced regioselectivity, the D_3-symmetric *trans-3,trans-3,trans-3* isomer can be obtained from the corresponding *trans*-3-$C_{62}(COOEt)_4$. In this compound, which is also chiral, the addents are bound in a zigzag fashion at an equatorial belt around the C_3-axis of symmetry. The structural characterization of these highly symmetric adducts can be achieved by NMR spectroscopy. For example, the ^{13}C NMR spectrum of *trans-3,trans-3,trans-3*- $C_{63}(COOEt)_3$ shows only 10 signals for the fullerene C-atoms.

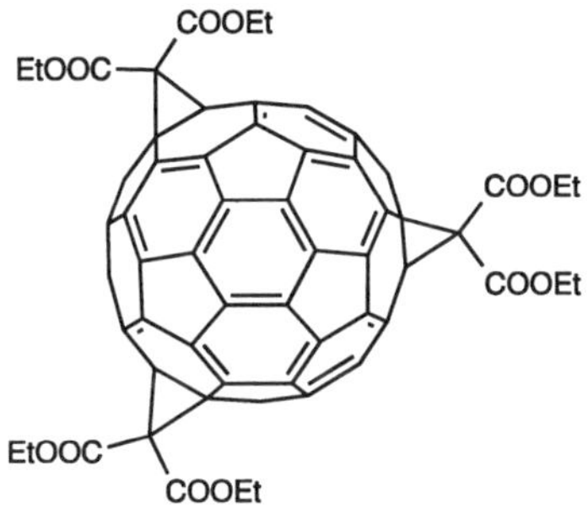

chiral D_3 *(trans-3, trans-3, trans-3)*

In analogy to the formation of the bisadducts, preferred additions into *e* and *trans-3* positions relative to addents already bound to C_{60} are observed. The addition to other regioisomers of $C_{62}(COOEt)_4$ also proceeds preferably in an *e* or *trans-3* mode. This regioselectivity can be used to systematically synthesize the T_h-symmetric hexaadduct by stepwise additions into "octahedral sites" (*e*-additions) starting from *trans-1*- or *e*-$C_{62}(COOEt)_4$.

References

[1]   Reviews: a) R. Taylor, D. R. M. Walton, *Nature* 363 (1993) 685; b) A. Hirsch, *Angew. Chem.* 105 (1993) 1189; *Angew. Chem. Int. Ed. Engl.* 32 (1993) 1138.

[2]   J. M. Hawkins, A. Meyer, T. A. Lewis, U. Bunz, R. Nunlist, G. E. Ball, T. W. Ebbesen, K. Tanigaki, *J. Am. Chem. Soc.* 114 (1992) 7954.

[3]   S. H. Friedman, D. L. DeCamp, R. P. Sijbesma, G. Srdanov, F. Wudl, G. L. Kenyon, *J. Am. Chem. Soc.* 115 (1993) 6506.

[4]   A. Hirsch, *Adv. Mater.* 5 (1993) 859.

[5]   A. Hirsch, I. Lamparth, H. R. Karfunkel, *Angew. Chem.* 106 (1994) 453; *Angew. Chem. Int. Ed. Engl.* 33 (1994) 437.

[6]   C. Bingel, *Chem. Ber.* 126 (1993) 1957.

THERMALLY AND PHOTOCHEMICALLY INDUCED CYCLOADDITIONS TO FULLERENE C_{60}: ADDITION OF 6b,10a-DIHYDROFLUORANTHENE AND 2,3-DIPHENYL-2*H*-AZIRINE TO C_{60}

Johannes Averdung and Jochen Mattay*
Organisch-Chemisches Institut, Westfälische Wilhelms-Universität Münster,
Orléansring 23, 48149 Münster, Germany

and

Holger Mohn, Wolfgang H. Müller and Hans-Ulrich ter Meer
Angewandte Physik, Hoechst AG, Postfach 8003200,
65926 Frankfurt/Main, Germany

ABSTRACT

The reaction of C_{60} **4** with in situ generated 6b,10a-dihydrofluoranthene **3** affords the stable [4+2] cycloadduct 1,9-(7,10-etheno-6b,7,8,9,10,10a-hexahydrofluorantheno)-fulleren-60 **5**. Upon irradiation 2,3-diphenyl-2*H*-azirine **6** is added to C_{60} **4** with formation of 1,9-(3,4-dihydro-2,5-diphenyl-2H-pyrrolo)fullerene-60 **7**. Mechanistic studies revealed two reaction paths leading to **7**, i.e. the classic 1,3-dipolar cyclo-addition via the nitrile ylide **8** (direct irradiation) or a route via 2-azaallenyl radical cations **9** (sensitization by photoinduced electron transfer).

1. Introduction

Since its discovery[1] and its production in gram quantities[2], the derivatization of fullerene C_{60} continues to yield fascinating results. Cycloadditions, in particular, lead to easily separable, well-defined monoaddition adducts, due to its strong dienophilic and dipolarophilic character[3]. We now report the succesful Diels-Alder reaction of 6b,10a-di-hydrofluoranthene **3** with C_{60} **4** and the results of our investigations of photochemically induced cycloadditions of 2,3-diphenyl-2*H*-azirine **6** to C_{60}, which are part of our general studies of cycloadditions via electron and energy transfer[4].

2. [4+2] Cycloaddition of 6b,10a-dihydrofluoranthene 3 to C_{60}

The stable [4 + 2] adduct 1,9-(7,10-etheno-6b,7,8,9,10,10a-hexahydrofluoran-theno)fullerene-60 **5** was obtained by reaction of C_{60} with 6b,10a-dihydrofluoranthene **3** in 27% yield (64% based on reacted C_{60}) after purification through high-performance liquid chromatography (HPLC) with a preparative RP-18 column[5]. The diene **3** was syn-thesized in situ by addition of 1,8-dehydronaphthalene **2** to benzene[6]. The reactive inter-mediate **2** was generated by oxidation of 1-aminonaphtho[1,8-de]triazine **1** with lead tetraacetate[6].

1. Synthesis of 1,9-(7,10-etheno-6b,7,8,9,10,10a-hexahydrofluorantheno)fulleren-60 **5**.

The structure of the 6-6-ring fused dihydrofullerene **5** has been identified by standard spectroscopic methods. The FD mass spectrum indicates that **5** is a 1:1 adduct. The proton decoupled ^{13}C NMR spectrum shows only 38 signals indicating C_s symmetry of the molecule. 3 signals are observed in the aliphatic region of the ^{13}C NMR spectrum, including the sp^3 hybridized carbons of C_{60} appearing at 71.06 ppm. A total of 35 signals out of the 37 expected signals for the fullerene and aromatic/alkenyl carbons are observed in the sp^2 region. The signals at 4.71, 5.37, 6.76 in the ^{1}H NMR spectrum display on a spin decoupling experiment a AA'MM'XX' splitting for the six non aromatic protons, the six aromatic protons appear at 7.56 and 7.68 ppm. The weak long-wavelength absorption at 710 nm in the UV-Vis spectrum is diagnostic for the dihydrofullerene structure[7]. The IR bands at 527, 576, 1182 and 1428 cm^{-1} are attributed to the fullerene moiety[8].

The present reaction confirms the dienophilic character of C_{60} and demonstrates the usefulness of [4 + 2] cycloaddition for C_{60} functionalization, and, thus, further studies on the direct addition of 1,8-dehydronaphthalene **2** to C_{60} are currently under way.

3. Synthesis of 1,9-(3,4-dihydro-2,5-diphenyl-2*H*-pyrrolo)fullerene-60 7

The direct irradiation of 2,3-diphenyl-2*H*-azirine **6** with light of 300 nm wavelength leads to the opening of the three-membered ring and to the formation of a nitrile ylide **8** [9]. The 1,3 dipolar nitrile ylide adds via [3+2] cyloaddition to C_{60} **4** with the formation of 1,9-(3,4-dihydro-2,5-diphenyl-2*H*-pyrrolo)fullerene-60 **7**[10]. The dihydrofullerene **7** has been isolated in 31% yield (55% based on reacted C_{60}) and identified spectroscopically. It is stable at room temperature and can be sublimated at a pressure below 0.01 mbar without decomposition. In contrast, **7** decomposes in toluene solution at a temperature above 70° C.

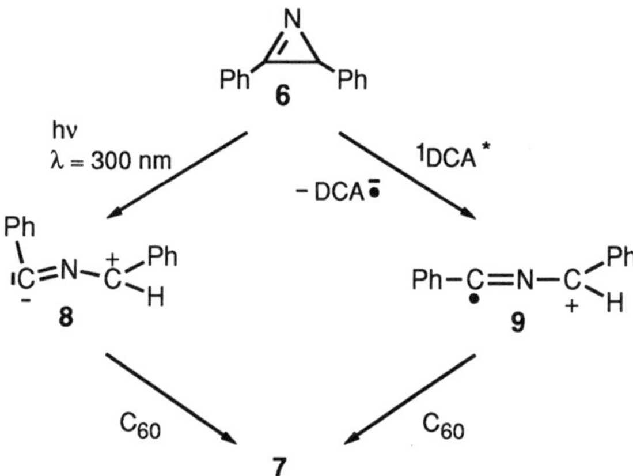

2. Photoinduced [3 + 2]-Cycloaddition of 2,3-Diphenyl-2*H*-azirne **6** to C$_{60}$ **4**

Mechanistic studies exposed a second reaction path leading to **7**. Upon irradiation under photoinduced electron transfer (PET) conditions with 9,10-dicyanoanthracene (DCA) as a PET sensitizer and with light above 400 nm wavelength the reaction obviously occurs via 2-azaallenyl radical cations **9**. However, there is no reaction by irraditation with light above 400 nm wavelength without sensitization; therefore, the reaction does not occur via excited C$_{60}$.

3. Photoreaction of 2,3-Diphenyl-2*H*-azirine **6** with C$_{60}$ **4** under direct irradiation and PET-conditions.

4. Experimental

1,9-(7,10-etheno-6b,7,8,9,10,10a-hexahydrofluorantheno)fullerene-60 **5**.
100 mg (0.54 mmol) 1-Aminonaphtho[1,8-de]triazine, in 40 ml of benzene was added in drops to a stirred solution of 96 mg (0.13 mmol) C$_{60}$ (Hoechst goldgrade, C$_{60}$ > 99.4%) and 380 mg (0.86 mmol) lead tetraacetate in benzene (170 ml). Nitrogen was immediately

evolved and stirring was continued for 75 min. Lead salts were separated by column filtration (toluene, neutral alumina) and the solvent was distilled off under reduced pressure. The adduct **5** was finally separated by HPLC (column: 250 x 20 mm, Merck LiChrosorb RP-18 7 μm, eluent: toluene/acetonitrile 1.75:1, detection wavelength: 310 nm) in 27% (33 mg) yield (64% based on reacted C_{60}).

FD-MS (m/e (%)): 924 (74; M^+), 720 (100; $C_{60}^+ = M^+ - C_{16}H_{12}$), 204 (44; $C_{16}H_{12}^+$). ^{1}H-NMR (360 MHz, $CS_2/(D_6)$acetone 10:1, υ/υ): 4.71 (s, 2 H), 5.37 (m, 2 H), 6.76 (m, 2 H), 7.56 (m, 4 H), 7.68 (m, 2 H). ^{13}C-NMR (90.5 MHz; $CS_2/(D_6)$acetone 10:1, υ/υ): 49.69, 49.91, 71.06, 119.89, 123.89, 128.95, 132.28, 133.91, 137.18, 137.91, 140.65, 140.81, 141.62, 142.24, 142.35, 142.53, 142.57, 142.73, 143.02, 143.16, 143.20, 143.67, 143.78, 145.28, 145.30, 146.04, 146.08, 146.14, 146.33, 146.73, 146.80, 146.82, 147.00, 147.15, 147.38, 148.12, 156.44, 157.06. FT-IR (KBr): 3053w, 2922s, 2850m, 1735, 1716, 1699, 1684m, 1653s, 1634s, 1617s, 1558w, 1540w, 1505w 1457m, 1428m, 1364w, 1324w, 1260w, 1182w, 1029s, br, 827m, 778ss, 744m, 691w, 668s, 599m, 576m, 563m, 527ss. UV/Vis (λ_{max} (ε) in toluene): 311 (47000), 415 (sh, 5100), 438 (sh, 3600), 710 (400-450).

5. Acknowledgments

Support provided by the Deutsche Forschungsgemeinschaft, the Minister für Wissenschaft und Forschung NRW, the Fonds der Chemischen Industrie, and Hoechst AG is gratefully acknowledged. We thank Priv.-Doz. Dr. J. Lauterwein and Dr. H. Luftmann for NMR and MS analysis.

6. References

1. H. Kroto, J. R. Heath, S. C. O'Brian, R. F. Curl and R. E. Smalley, *Nature* **318** (1985) 162-163.
2. W. Krätschmer, L. D. Lamb, K. Fostiropoulos and D. R. Huffmann, *Nature* **347** (1990) 354-358.
3. A. Hirsch, *Angew. Chem.* **105** (1993) 1189-1192, *Angew. Chem. Int. Ed. Engl.* **32** (1993) 1138 and references cited therein.
4. F. Müller and J. Mattay,. *Chem. Rev.* **93** (1993) 99-117.
5. J. Averdung, *Diploma Thesis* (Münster, 1994).
6. C. W. Rees and R. C. Storr, *J. Chem. Soc. Org. Chem. C.* (1969) 756-760, 760-764.
7. Y. Z. An, J. L. Anderson and Y. Rubin, *J. Org. Chem.* **58** (1993) 4799-4801.
8. A. Hirsch, T. Grösser, A. Skiebe and A. Soi, *Chem. Ber.* **126** (1993) 1061-1067.
9. F. Müller and J. Mattay, *Chem. Ber.* **126** (1993) 543-549.
10. J. Averdung, E. Albrecht, J. Lauterwein, H. Luftmann, J. Mattay, H. Mohn, H. W. Müller and H.-U. ter Meer, *Chem. Ber.* **127** (1994) 787-789.

ADDITION CHEMISTRY OF C_{60}: SYNTHESIS, CHARACTERIZATION AND FUNCTIONALIZATION OF FULLEROPYRROLIDINES

MAURIZIO PRATO
Dipartimento di Scienze Farmaceutiche, Università di Trieste
Piazzale Europa 1, 34127 Trieste, Italy

MICHELE MAGGINI AND GIANFRANCO SCORRANO
Centro Meccanismi di Reazioni Organiche del CNR
Dipartimento di Chimica Organica, Via Marzolo 1, 35131 Padova, Italy

ABSTRACT

1,3-Dipolar cycloaddition of azomethine ylides to fullerene C_{60} gives fulleropyrrolidines as products of addition across a 6,6-ring fusion of C_{60}. The scope of the reaction appears very wide, and many derivatives with different substituents can be synthesized. The parent compound, N-H fulleropyrrolidine, can also be prepared as a versatile intermediate for further transformations. Several examples, with potential applications ranging from materials science to medicinal chemistry, are reported.

Fulleropyrrolidines are C_{60} derivatives in which the 3,4-bond of a pyrrolidine ring is fused with a 6,6 bond of the fullerene.[1] They are conveniently prepared by cycloaddition of C_{60} with azomethine ylides, one of the most powerful classes of 1,3-dipoles.[2] Among the large number of methods used to generate the highly reactive intermediates, the *"decarboxylation route"* seems a particularly easy way to substituted azomethine ylides. Heating a mixture of N-methylglycine (sarcosine), paraformaldehyde and C_{60} in toluene gives rise to 41% of the N-methylpyrrolidine **1** after chromatography.

$$CH_3\text{-}NH\text{-}CH_2\text{-}COOH + CH_2O + C_{60} \xrightarrow[-H_2O]{\Delta \\ -CO_2} $$

1

Fulleropyrrolidine **1** is a highly symmetrical compound. Owing to the fast nitrogen inversion (as compared to the NMR time scale) and to the expected attack on the 6,6 ring fusion of C_{60}, the four protons of the two methylene groups are identical. Thus, fulleropyrrolidine **1** has C_{2v} symmetry. Accordingly, the ^{1}H-NMR spectrum of **1** shows a singlet (integration 4) for the methylene protons and a singlet (integration 3) for the methyl group. The ^{13}C-NMR spectrum shows 17 signals for the fullerene carbons, 13 of which, in the aromatic region, integrate 4 carbons each. Of the other 4 signals (which integrate 2 carbons each), 3 lie in the aromatic region and 1 in the sp^3 fullerene zone (71.1 ppm). The methyl and methylene carbons of the N-methylpyrrolidine ring are responsible for the remaining 2 signals in the aliphatic region. The UV-vis spectrum of **1** is typical of all the monoaddition fullerene derivatives, with the weak but characteristic peaks at around 430 and 700 nm.

Fulleropyrrolidines functionalized at nitrogen can be obtained via reaction of the parent N-H derivative. This latter compound is prepared by reaction of C_{60} with the N-trityl protected oxazolidinone **2**.

2   **3**

Upon heating, **2** looses CO_2 and generates the corresponding ylide, which reacts promptly with C_{60} affording excellent yields of the N-trityl pyrrolidine **3**. The action of trifluoromethanesulfonic acid on **3** causes the immediate precipitation of the ammonium salt **4**, which can be centrifuged and easily purified by repetitive washings. Addition of excess pyridine to a suspension of **4** in dichloromethane generates the free amine **5**.

3 $\xrightarrow{CF_3SO_3H}$ **4** $\xrightarrow{Py}$ **5**

According to the known reactivity of C_{60} with amines,[3] **5** is unstable upon concentration. However, if kept in dilute solutions, **5** is stable for days, even at room temperature. Therefore, amine **5** can survive long enough to react with a variety of electrophiles. For instance, addition of dansyl chloride to a solution of **5** leads to the formation of the sulfonamide **6**. It is interesting to note that the dansyl group is a widely used fluorescing label and that compound **6** can be an interesting substrate for photophysical studies involving fullerenes.

Alternatively, amine **5** can be treated with a wide range of acylating agents. Amides **7a-d** are easily obtained by reaction with the corresponding acyl chlorides. The presence of the polar amide group in compounds **7** is important for conferring amphiphilic properties to these fullerene derivatives.[4] The possibility that compounds **7a-d** give monomolecular layers under Langmuir-Blodgett conditions is currently under investigation.

NMe$_2$

SO$_2$

6

H

N

5

DNSCl

RCOCl

O

R

7a, R = CH$_3$
7b, R = (CH$_2$)$_{16}$CH$_3$
7c, R = (CH$_2$)$_{20}$CH$_3$
7d, R = (CF)$_6$CF$_3$

Another valuable approach to functionalized fulleropyrrolidines is based on the thermal ring-opening of aziridines. For instance, heating the aziridine **8** in the presence of C$_{60}$ leads to the formation in good yields of the N-benzylated fulleroproline **9**.

Recent work by Wudl[5] and Nakamura[6] groups has demonstrated that water soluble fullerene derivatives can play a significant role as biologically active compounds. An increase of their potential in pharmacological applications is also expected with specifically designed C_{60} derivatives. Fulleroproline **9** is a C_{60}-based aminoacid, and can be considered a building block in the preparation of C_{60}-containing peptides.[7]

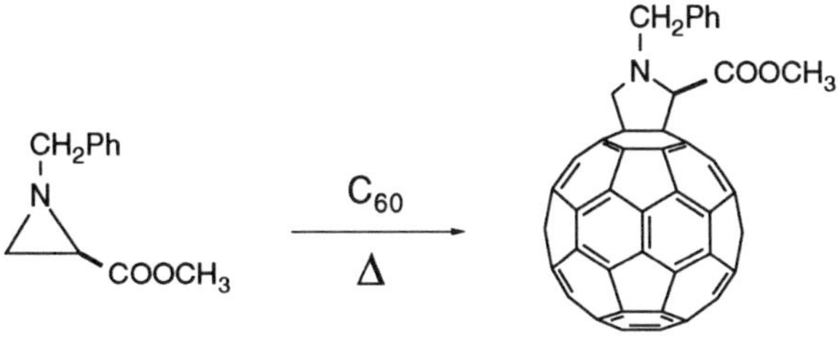

8   **9**

However, **9** cannot be considered a valid candidate for the unsubstituted fulleroproline, as the N-benzyl group appears to be difficult to remove without interfering with the reactivity of the fullerene spheroid.

The best way to the parent fulleroproline employs another azomethine ylide generation method, the so called "tautomerization route". Glycine t-butyl ester was refluxed with formaldehyde in toluene in the presence of C_{60}. The t-butyl fulleroproline **10** was obtained in good yield after chromatography.[8]

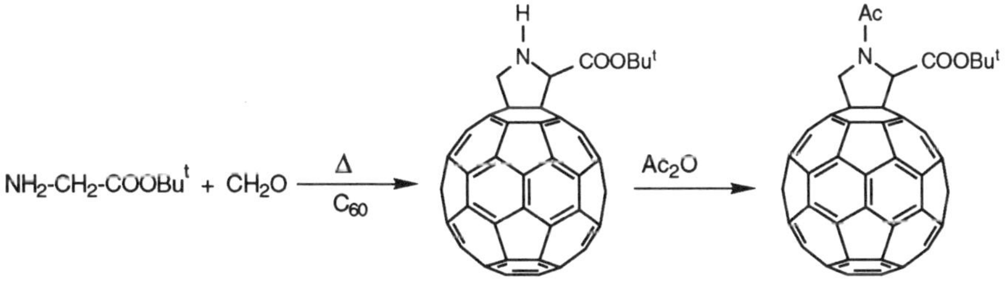

1 0   **1 1**

However, analogously to **5**, also **10** is unstable upon concentration and must be functionalized in dilute solutions. For instance, addition of acetic anhydride affords the corresponding N-acetyl fulleroproline t-butyl ester **11**. In $CDCl_3$ solution, acetamido fulleroproline **11** is present as mixture of two rotational isomers. The influence of the C_{60} sphere on the E-Z equilibrium about the tertiary amide bond of the proline residue[9] is currently under investigation.

The t-butyl ester of **10** can be hydrolyzed by reaction with triflic acid, which affords the parent N-protonated fulleroproline acid **12**.

Functionalization of the carboxylic acid on one side, and of the amine on the other side, can be achieved in **12** by means of standard procedures usually employed in peptide chemistry and is currently being studied.

References

[1] M. Maggini, G. Scorrano and M. Prato, *J. Am. Chem. Soc.* **115** (1993) 9798.

[2] (a) J. W. Lown, in *1,3-Dipolar Cycloaddition Chemistry*, ed. A. Padwa (Wiley, New York, 1984). (b) O. Tsuge and S. Kanemasa, *Adv. Heterocyclic Chem.* **45** (1989) 231.

[3] A. Hirsch, Q. Li and F. Wudl, *Angew. Chem., Int. Ed. Engl.* **30** (1991) 1309.

[4] M. Maggini, A. Karlsson, L. Pasimeni, M. Prato, G. Scorrano and L. Valli, *Tetrahedron Lett.* (1994) in press.

[5] (a) S. H. Friedman, D. L. DeCamp, R. P. Sijbesma, G. Srdanov, F. Wudl and G. L. Kenyon, *J. Am. Chem. Soc.* **115** (1993) 6506. (b) R. Sijbesma, G. Srdanov, F. Wudl, J. A. Castoro, C. Wilkins, S. H. Friedman, D. L. DeCamp and G. L. Kenyon, *J. Am. Chem. Soc.* **115** (1993) 6510. (c) R. F. Schinazi, R. Sijbesma, G. Srdanov, C. L. Hill and F. Wudl, *Antimicr. Ag. Chemother.* **37** (1993) 1707.

[6] H. Tokuiama, S. Yamago, E. Nakamura, T. Shiraki, and Y. Sugiura, *J. Am. Chem. Soc.* **115** (1993) 7918.

[7] M. Prato, A. Bianco, M. Maggini, G. Scorrano, C. Toniolo, F. Wudl, *J. Org. Chem.* **58** (1993) 5578.

[8] M. Maggini, G. Scorrano, A. Bianco, C. Toniolo, R. P. Sijbesma, F. Wudl and M. Prato, *J. Chem. Soc., Chem. Comm.* (1994) 305.

[9] (a) W. A. Thomas and M. K. Williams, *J. Chem. Soc., Chem. Comm.* (1972) 788. (b) H. Nishihara, K. Nishihara, T. Uefuji and N. Sakota, *Bull. Chem. Soc. Jpn.* **48** (1977) 553.

SYNTHESIS AND ELECTROCHEMICAL PROPERTIES OF FERROCENYL FULLEROPYRROLIDINES

MICHELE MAGGINI and GIANFRANCO SCORRANO
Centro Meccanismi di Reazioni Organiche del CNR
Dipartimento di Chimica Organica, Via Marzolo 1, 35131 Padova, Italy

MAURIZIO PRATO
Dipartimento di Scienze Farmaceutiche, Università di Trieste
Piazzale Europa 1, 34127 Trieste, Italy

ABSTRACT

1,3-Dipolar cycloaddition of azomethine ylides to fullerene-C_{60} has been employed to synthesise fulleropyrrolidines where a ferrocene group is covalently bound to C_{60}. The cyclic voltammetry shows that these compounds possess interesting properties.

An important target in molecular electronics is the design of assemblies where an electron donor and an electron acceptor moieties are part of the same molecule. These should efficiently interact and, if an electron transfer is operative in the ground or in the excited state, in principle, this feature could be used for the construction of molecular electronic devices.[1]

Due to its high electron affinity, fullerene-C_{60} can be considered a promising candidate as an acceptor unit.[2,3] In fact, in solution C_{60} has been found to accept reversibly up to six electrons.[4,5]

As a preliminary investigation, we selected the ferrocene moiety as donor unit. The methodology used to link together donor and acceptor has been recently devised in our laboratory and relies on an efficient route to fulleropyrrolidines; C_{60} derivatives in which the 3,4 bond of a pyrrolidine ring is fused with a 6,6 ring junction of C_{60}.[6]

1

34

Their preparation is based on the 1,3-dipolar cycloaddition of azomethine ylides to C_{60}. These highly reactive 1,3-dipoles can be generated by means of several different and general approaches.[7] One of these methods, synthetically very easy, is the "decarboxylation route". An α-aminoacid is heated with an aldehyde: decarboxylation and cycloaddition to C_{60} gives rise to good yields of fulleropyrrolidines. As a model compound, we prepared the N-methyl pyrrolidine **1**. N-methyl glycine, paraformaldehyde and C_{60} in toluene were heated to reflux. The N-methyl derivative **1** was isolated in 41% yield (82% based on C_{60} conversion). The procedure is experimentally very simple and the scope of the reaction is very wide. Change in the aldehyde structure allows the preparation of differently substituted fulleropyrrolidines.

Two fullerene derivatives have been prepared with the ferrocene moiety in different positions of the pyrrolidine ring. Compound **2** was obtained in a one pot synthesis by allowing ferrocene carboxaldehyde (FcCHO), N-methyl glycine and C_{60} to react in refluxing toluene. The reaction proceeded smoothly in a few hours, affording the N-methyl pyrrolidine ferrocene derivative in 40% yield after column chromatography (80% based on C_{60} conversion).

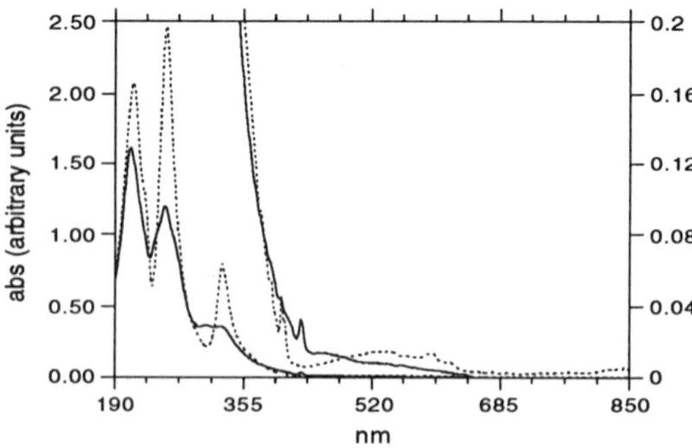

2

The UV-vis spectrum of **2** shows that this fullerene derivative retains most of the electronic properties of C_{60} (Figure 1).

Figure 1. UV-vis spectrum of **2** (solid line) and C_{60} (dotted line) in cyclohexane

Fulleropyrrolidine **3** has been prepared by treating ferrocenyl carboxylic acid chloride with the NH parent compound **6**. The synthesis began with the preparation of the N-protected fulleropyrrolidine **5** accomplished by reacting the N-triphenylmethyl oxazolidinone **4** and C_{60} in refluxing toluene. Treatment of **5** with trifluoromethane sulfonic acid afforded the triflate salt of the amine. The free amine could be liberated in dichloromethane solution with pyridine just before the coupling with the acid chloride in order to avoid undesired side reactions of the amine with the fullerene cage.[8]

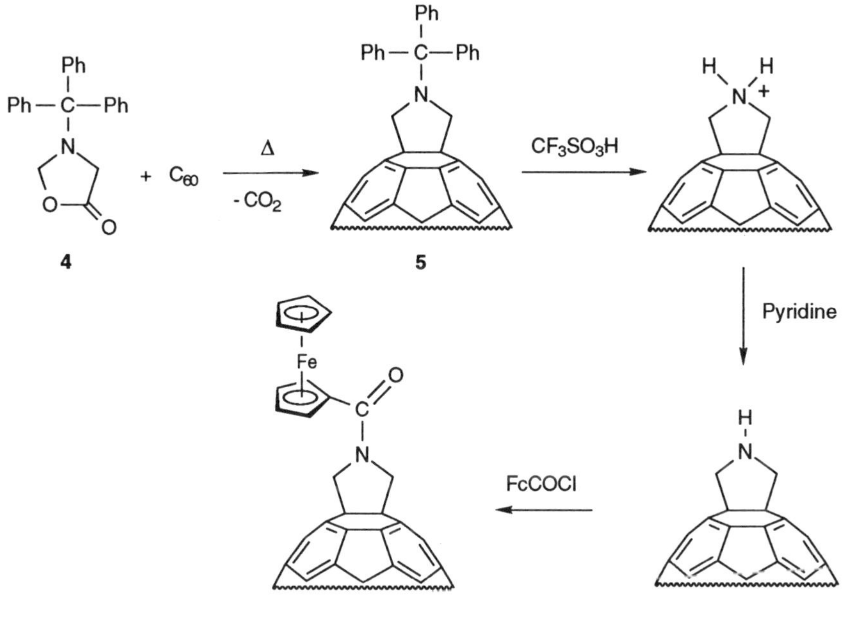

The NMR data are in agreement with structure **3** and the UV-vis spectrum is superimposable to that of **2**.

Cyclic voltammetry (CV) was employed to study the electrochemical properties of compounds **2** and **3** in solution together with those of C_{60} and the N-methyl pyrrolidine derivative **1** for comparison.[9] $E_{1/2}$ values ($E_{1/2} = (E^{an}_{peak} + E^{cat}_{peak})/2$) for C_{60} and **1-3** are reported in Table 1. As illustrated in the table, the observed reduction potentials of **1-3** are shifted to more negative values when compared to those of unsubstituted C_{60}. As a result, only five reduction peaks for the C_{60} moiety are detected in the accessible potential range (the ferrocenyl amide group is responsible for the sixth peak observed at $E_{1/2} = -3.22$ V in the CV of **3**). In the fulleropyrrolidines containing ferrocene, the relative reversible oxidation peak for the ferrocene moiety is also observed.

	$E^{0/1-}$	$E^{1-/2-}$	$E^{2-/3-}$	$E^{3-/4-}$	$E^{4-/5-}$	$E^{5-/6-}$	$E^{1+/0}$
C_{60}	-0.94	-1.33	-1.83	-2.28	-2.74	-3.14	
1	-1.05	-1.44	-2.01	-2.42	-3.12		
2	-1.08	-1.47	-2.03	-2.44	-3.14		+0.05
3	-1.00	-1.38	-1.95	-2.36	-3.05	-3.22	+0.16

Table 1. $E_{1/2}$ values (V *vs* Fc$^+$/Fc) of the redox couples of C_{60} and compounds **1-3** detected by cyclic voltammetry (sweep rate 0.1 V s^{-1}) in 3:1 toluene/acetonitrile solutions (0.1 M TBAP), at -45 °C. Errors are estimated at ± 5 mV.

From the $E_{1/2}$ values of **1-3** reported in the Table, it can be concluded that the interactions between the ferrocene donor and the spheroid acceptor are extremely labile. This would mean that charge transfer in the ground state is not operative and that the excited state of these ferrocene-fullerene substrates may deserve careful examination.

Acknowledgments

We thank Prof. G. Farnia and Dr. G. Sandonà for CV experiments.

References

1. J.-M. Lehn, *Angew. Chem., Int. Ed. Engl.* **29** (1990) 1304.
2. N. S. Sariciftci, L. Smilowitz, A. J. Heeger and F. Wudl, *Science* **258** (1992) 1474.
3. S. I. Khan, A. M. Oliver, M. N. Paddon-Row and Y. Rubin, *J. Am. Chem. Soc.* **115** (1993) 4919.
4. Q. Xie, E. Perez-Cordero and L. Echegoyen, *J. Am. Chem. Soc.* **114** (1992) 3978.
5. Recently, a one-electron, reversible oxidation process has been reported: Q. Xie, F. Arias and L. Echegoyen, *J. Am. Chem. Soc.* **115** (1993) 9818.
6. M. Maggini, G. Scorrano and M. Prato, *J. Am. Chem. Soc* **115** (1993) 9798.
7. (a) J. W. Lown, In *1,3-Dipolar Cycloaddition Chemistry*, Padwa, A., Ed.: Wiley; New York, 1984. (b) O. Tsuge and S. Kanemasa, *Adv. Heterocyclic Chem.* **45** (1989) 231.
8. A. Hirsch, Q. Li and F. Wudl, *Angew. Chem. Int. Ed. Engl.* **30** (1991) 1309.
9. M. Maggini, A. Karlsson, G. Scorrano, G. Sandonà, G. Farnia and M. Prato, *J. Chem. Soc. Chem. Comm.* (1994) 589.

Electron Transfer Reactions with C_{60}

Dirk M. Guldi,[a,b] Pedatsur Neta[a] and Klaus-Dieter Asmus[b]

[a]Chemical Kinetics and Thermodynamics Division, National Institute
  of Standards and Technology, Gaithersburg, Maryland 20899, USA
[b]Hahn-Meitner-Institut Berlin, Bereich S, Abteilung Strahlenchemie,
  Glienicker Straße 100, 14109 Berlin, Germany

Abstract. Electron transfer reactions with C_{60} leading to the singly reduced and oxidized radical ions, $C_{60}^{\bullet-}$ and $C_{60}^{\bullet+}$, have been studied by pulse radiolysis. π-Radical anions of several metalloporphyrins reduced C_{60} with rate constants of $(1-3) \times 10^9$ L mol^{-1} s^{-1}, whereas metalloporphyrins which are reduced at the metal center reacted with the fullerene somewhat more slowly, with rate constants of $(0.7-2.3) \times 10^8$ L mol^{-1} s^{-1}. $Sn^{IV}(Ph)_3(Py)P^{\bullet-}$ reacted in an equilibrium process ($K = 14$). Electron transfer from C_{60} to several aromatic radical cations also took place rapidly ($k = 2.5-7.9 \times 10^9$ L mol^{-1} s^{-1}) to produce $C_{60}^{\bullet+}$.

C_{60} undergoes a facile one-electron reduction to yield $C_{60}^{\bullet-}$ radical anions.[1] It also undergoes one-electron oxidation to the radical cation, $C_{60}^{\bullet+}$.[3-5] One-electron reduction and oxidation of C_{60} have been achieved by electrochemical techniques[6] and also by radiolytic methods in solution, where C_{60} is known to react with strongly reducing and oxidizing species.[2,7] Pulse radiolysis studies[2] permitted recording of the optical absorption spectra of $C_{60}^{\bullet-}$ and $C_{60}^{\bullet+}$ at short times after the pulse and determination of the rate constants leading to the formation of these species. The respective radical ions exhibit strong absorption in the near IR with peaks at 1080 ($C_{60}^{\bullet-}$) and 980 nm ($C_{60}^{\bullet+}$). In the present study we examined electron transfer reactions of C_{60} that have a lower driving force than those reported earlier, and which may therefore provide a quantifiable correlation between the rate constant and the free energy of the reaction. We chose to study the reduction of C_{60} by several metalloporphyrin π-radical anions and metal-reduced metalloporphyrins. In addition, we studied the oxidation of C_{60} by several aromatic π-radical cations. The capacity of C_{60} to serve as a multielectron acceptor, in general, and the moderate redox potential for the first reduction step, in particular, suggest that C_{60} might be successfully employed in solar energy conversion schemes.

The porphyrins were obtained from Midcentury Chemicals, Posen IL, with purity >99%. C_{60} was purchased from Kaesdorf (Geräte für Forschung und Industrie, München, Germany). The pulse experiments were performed with the apparatus described before, which utilizes 50-ns pulses of 2-MeV electrons from a Febetron Model 705 pulser.[8] The dose per pulse was varied between 7 to 40 Gy, which, in aqueous solutions, corresponds to a 4 to 24 µmol L^{-1} radical concentration. All experiments were carried out at ambient temperature.

Electron transfer from metalloporphyrin π-radical anions to C_{60}. Metalloporhyrins were reduced by radiolysis in a solvent mixture containing 65 vol% 2-propanol, 25 vol% toluene and 10 vol% acetone. The reducing species in this solvent mixture is the radical derived from 2-PrOH by H-abstraction, i.e. $(CH_3)_2C^{\bullet}(OH)$. This radical is known to reduce a number of metalloporphyrins quite rapidly.[9]

Pulse irradiation of deaerated solutions containing $(1-5) \times 10^{-4}$ mol L^{-1} of certain metalloporphyrins, MP, resulted in the formation of broad absorptions in the 600-800 nm range. This absorption is ascribed to the π-radical anion, the product of reduction of the porphyrin ligand, as discussed in a number of previous studies.[9,10]

$$(CH_3)_2C^{\bullet}(OH) + MP \longrightarrow (CH_3)_2CO + H^+ + MP^{\bullet-} \qquad (1)$$

In the absence of other electron acceptors the π-radical anion decays via disproportionation and protonation. The lifetime of the π-radical anion has been shown to depend strongly on the medium and on the reduction potential of the metalloporphyrin.[9,11]

Addition of various concentrations of C_{60} resulted in an accelerated decay of the π-radical anion. The observed rate ($k_{obs} = \ln 2 / t_{1/2}$) was linearly dependent on the C_{60} concentration, indicating that the underlying process has to be attributed to a reaction with C_{60}, most probably an electron transfer.

$$C_{60} + MP^{\bullet-} \longrightarrow C_{60}^{\bullet-} + MP \tag{2}$$

To confirm this reaction, we also monitored the formation of the characteristic $C_{60}^{\bullet-}$ absorption in the IR ($\lambda_{max} = 1080$ nm). The rate of formation of the absorption at various wavelengths in the 980-1060 nm range was similar to the rate of decay of the $MP^{\bullet-}$ absorption at 650-750 nm. For example, in the case of $ZnTPP^{\bullet-}$, we derived a rate constant of $(2.5\pm1.0) \times 10^9$ L mol^{-1} s^{-1} from the decay at 720 nm and $(1.4\pm1.0) \times 10^9$ L mol^{-1} s^{-1} from the formation at 970 nm. These two values are in reasonable agreement considering the large error margins in the kinetic measurements, particularly for the weak IR signal.

The rate constants for electron transfer from various metalloporphyrin π-radical anions to C_{60} (Table) are found to be in the range of $(1-3) \times 10^9$ L mol^{-1} s^{-1}, despite the fact that the one-electron reduction potentials for the metalloporphyrins examined cover a range of 0.55 V ($E_{ZnP/ZnP^{\bullet-}} = -1.35$ V; $E_{SnP/SnP^{\bullet-}} = -0.8$ V vs. SCE [12]). This lack of dependence of the rate constant on the driving force for the reaction probably reflects the fact that the reduction potentials for these porphyrins are considerably more negative than that of C_{60} and all these reactions are nearly diffusion-controlled (in the solvent mixture the calculated value is $k_{diff} = 3.5 \times 10^9$ L mol^{-1} s^{-1}).

Electron transfer between Sn-porphyrin and C_{60}. Sn^{IV}-porphyrins are very easily reduced to give long-lived π-radical anions.[11] Because their reduction potential is only slightly more negative than that of C_{60}, it was expected that the electron transfer between these two species, if sufficiently rapid as compared with the decay of the radicals, may lead to observation of equilibrium conditions.

$$Sn^{IV}P^{\bullet-} + C_{60} \rightleftharpoons Sn^{IV}P + C_{60}^{\bullet-} \tag{3}$$

Indeed, we found such an equilibrium with $Sn^{IV}(Ph)_3(Py)P$. Reduction of this porphyrin resulted in formation of the π-radical anion with absorption in the 700-800 nm range. Upon addition of various C_{60} concentrations, the rate of decay of this π-radical anion increased linearly with $[C_{60}]$. However, the rate of decay was also found to increase upon raising the porphyrin concentration at constant $[C_{60}]$. Such a dependence of the rate of reaction upon both concentrations is indicative of an equilibrium process as formulated in reaction 3. To confirm this equilibrium process and to determine the equilibrium constant, we measured the rate constant with a series of different $Sn^{IV}P$ and C_{60} concentrations. The results show the linear dependence as expected from eq. 4 and 5.

$$k_{obs}/[Sn^{IV}P] = k_3[C_{60}]/[Sn^{IV}P] + k_{-3} \tag{4}$$

$$K_3 = [Sn^{IV}P]/[C_{60}] \times [C_{60}^{\bullet-}]/[Sn^{IV}P^{\bullet-}] \tag{5}$$

The kinetic plot gives $k_3 = 3.2 \times 10^9$, $k_{-3} = 2.1 \times 10^8$ L mol^{-1} s^{-1}, and $K_3 = 15$. In good agreement with this figure, the absorbance plot gives $K_3 = 13$. Thus the average equilibrium constant from these plots is $K_3 = 14 \pm 3$ and the reduction potential of this tin porphyrin in this solvent mixture is 64 mV more negative than that of C_{60}.

Electron transfer from metalloporphyrins at low oxidation states to C_{60}. One-electron reduction of $Cr^{III}MSP$, $Ni^{II}TPP$, and $Cu^{II}TPP$ is known to occur at the metal center to yield $Cr^{II}MSP$, $Ni^{I}TPP$, and $Cu^{I}TPP$.[13] This reduction path results in only minor spectral changes and lacks, in particular, the intense absorption in the 600-800 nm range. The decay of these species upon reaction with C_{60} was monitored in the 500 nm range, where the reduced metalloporphyrin absorbs more intensely than the parent compound. The rate constants for these reactions (Table) were derived from the linear dependence of the decay rate upon C_{60} concentration.

$$Cr^{II}MSP \; + \; C_{60} \; \longrightarrow \; Cr^{III}MSP \; + \; C_{60}^{\bullet -} \tag{6}$$

These rate constants are substantially lower than those measured for the reduction by the π-radical anions (reaction 2) despite the similarity in the reduction potentials ($E_{Cr^{III}P/Cr^{II}P}$ = -1.14 V; $E_{Ni^{II}P/Ni^{I}P}$ = -1.18 V; $E_{Cu^{II}P/Cu^{I}P}$ = -1.2 V vs. SCE[12]). Possibly, the small size of the reactive metal center, vs. the large size of that in the porphyrin π-radical anions, i.e. geometric constrains, may be the cause for the decreased rate constants.

Table: Rate Constants for Reduction of C_{60} by Reduced Metalloporphyrins[a]

Reducing species	λ, nm	k, L mol^{-1} s^{-1}
$Zn^{II}TPP^{\bullet -}$ [b]	720	$(2.6\pm1.0) \times 10^{9}$
$Zn^{II}TPyP^{\bullet -}$	700	$(1.1\pm1.0) \times 10^{9}$
$In^{III}TPP^{\bullet -}$	700	$(2.6\pm1.0) \times 10^{9}$
$Ge^{IV}TPP^{\bullet -}$	700	$(2.2\pm1.0) \times 10^{9}$
$Al^{III}TPyP^{\bullet -}$	700	$(2.0\pm1.0) \times 10^{9}$
$Ga^{III}TPP^{\bullet -}$	720	$(2.0\pm1.0) \times 10^{9}$
$Sn^{IV}(Ph)_3(Py)P^{\bullet -}$	700	$(3.2\pm1.0) \times 10^{9}$
$Ni^{I}TPP$ [b]	500	$(1.3\pm1.0) \times 10^{8}$
$Cu^{I}TPP$ [b]	490	$(2.4\pm1.0) \times 10^{8}$
$Cr^{II}MSP$ [b]	490	$(6.8\pm2.0) \times 10^{7}$

[a] Determined by following the rate of decay of the reducing species at the wavelength indicated as a function of C_{60} concentration (0.1-2 x 10^{-5} mol L^{-1}). [b] In the presence of 4 mmol L^{-1} $(CH_3)_2CHO^-$.

One-electron oxidation of C_{60} by aromatic π-radical cations. Oxidative electron transfer from C_{60} to various organic radical cations was studied in CH_2Cl_2 as solvent. Radiolysis of this solvent, and other halogenated solvents, is known to lead to oxidation of many organic compounds.[14,15]

$$CH_2Cl_2^{\bullet +} \; + \; terphenyl \; \longrightarrow \; CH_2Cl_2 \; + \; terphenyl^{\bullet +} \tag{7}$$

$$terphenyl^{\bullet +} \; + \; C_{60} \; \longrightarrow \; terphenyl \; + \; C_{60}^{\bullet +} \tag{8}$$

To study the electron transfer from C_{60} to the arene radical cation we irradiated CH_2Cl_2 solutions containing 10^{-2} mol L^{-1} of the arene in the presence of various C_{60} concentrations. The various arene radical cations exhibited quite different lifetimes; some of them decayed within several microseconds, which restricted the ability to measure their reaction with C_{60}. However, the addition of C_{60} resulted in a measurable increase of the first order decay rate, showing a linear dependence on the C_{60} concentration. To confirm the formation of $C_{60}^{\bullet +}$, we measured its absorption in the near IR region (λ_{max} = 980 nm [2,4]). The differential absorption spectrum recorded with a CH_2Cl_2 solution containing 10^{-2} mol L^{-1} m-terphenyl and 2×10^{-5} mol L^{-1} C_{60} is in good agreement with that found upon direct oxidation of C_{60} in various halogenated hydrocarbons.[5] The kinetic traces in the IR region did not permit determination of exact rate constants for the formation of $C_{60}^{\bullet +}$ because the signals were too weak. It was clear, however, that the time scale for the formation of $C_{60}^{\bullet +}$ monitored in the IR region was similar to that of the decay of the (arene)$^{\bullet +}$ monitored in the visible region.

Among the compounds studied, only biphenyl has been shown to oxidize C_{60} in an earlier photooxidation study.[4] The rate constant for this reaction, 7.9×10^9 L mol^{-1} s^{-1}, is close to the diffusion-controlled limit (in CH_2Cl_2 the calculated value is k_{diff} = 1.3×10^{10} L mol^{-1} s^{-1}). The rate constants for t-stilbene, m-terphenyl, and naphthalene are 6.8×10^9, 3.8×10^9, and 2.5×10^9 L mol^{-1} s^{-1}, respectively and show little correlation with the ionization potentials of these compounds.

In conclusion, C_{60} undergoes one-electron reduction and oxidation by π-radical anions and cations, respectively, with rate constants that are very close to the diffusion-controlled limit, even when the driving force for the reaction (the difference in reduction potentials) is relatively small. This suggests a very high self-exchange rate between C_{60} and its radical ions, probably due to the high degree of delocalization of the unpaired electron and the unchanged geometry upon electron transfer. Reducing species which are not delocalized π-radical anions, such as the low oxidation state metalloporphyrins, reacted somewhat more slowly with C_{60}.

Acknowledgment. This research was supported by the Office of Basic Energy Sciences of the US Department of Energy.

References.
(1) Special issue on fullerenes *Acc. Chem. Res.* **1992**, *25*, 98.
(2) Guldi, D.M.; Hungerbühler, H.; Janata, E.; Asmus, K.-D. *J. Chem. Soc., Chem. Commun.* **1993**, 84 and *J. Phys. Chem.* **1993**, *97*, 11258.
(3) Kato, T.; Kodama, T.; Shida, T.; Nakagawa, T.; Matsui, Y.; Suzuki, S.; Shiromaru, H.; Yamauchi, K.; Achiba, Y. *Chem. Phys. Lett.*, **1991**, *180*, 446.
(4) Nonell, S.; Arbogast, J.W.; Foote, C.S. *J. Phys. Chem.* **1992**, *96*, 4169.
(5) Lichtenberger, D.L.; Nebesny, K.W.; Ray, C.D.; Huffman, D.R.; Lamb, L.D. *Chem Phys. Lett.* **1991**, *176*, 203.
(6) Jehoulet, C.; Bard, A.J.; Wudl, F.; *J. Am Chem. Soc.* **1991**, *113*, 5456. Dubois, D.; Kadish, K.M.; Flanagan, S.; Wilson, L.J. *J. Am. Chem. Soc.* **1991**, *113*, 7773. Xie, Q.; Perez-Cordero, E.; Echegoyen, L. *J. Am. Chem. Soc.* **1992**, *114*, 3978. Dubois, D.; Kadish, K.M.; Flanagan, S.;Haufler, R.E.; Chibante, L.P.F.; Wilson, L.J. *J. Am. Chem. Soc.* **1991**, *113*, 4364.
(7) Dimitrijevic, N.M. *Chem. Phys. Lett.* **1992**, *194*. Hou, H.; Luo, C.; Liu, Z.; Mao, D.; Qin, Q.; Lian, Z.; Yao, S.; Wang, W.; Zhang, J.; Lin, N. *Chem. Phys. Lett.* **1993**, *203*, 555.
(8) Nahor, G.S.; Neta, P.; Hambright, P.; Robinson, L.R. *J. Phys. Chem.* **1991**, *95*, 4415
(9) Neta, P.; Scherz, A.; Levanon, H. *J. Am. Chem. Soc.* **1979**, *101*, 3624. Neta, P. *J. Phys. Chem.* **1981**, *85*, 3678.
(10) Guldi, D.M.; Kumar, M.; Neta, P.; Hambright, P. *J. Phys. Chem.* **1992**, *96*, 9576.
(11) Richoux, M.-C.; Neta, P.; Harriman, A.; Baral, S.; Hambright, P. *J. Phys. Chem.* **1986**, *90*, 2462 and refrences therein.
(12) Felton, R.H. in "The Porphyrins", Vol. 5, chapter 3; Dolphin, D., ed.; Academic Press: New York, 1978, p. 53.
(13) Guldi, D.M.; Hambright, P.; Lexa, D.; Neta, P.; Saveant, J.-M. *J. Phys. Chem.* **1992**, *96*, 4459.
(14) Dorfman, L. M.; Wang, H.-J.; Sujdak, R.J. *Faraday Disc. Chem. Soc.* **1977**, *63*, 149. Shank, N.E.; Dorfman, L.M. *J. Chem. Phys.* **1970**, *52*, 4441.
(15) Alfassi, Z.B.; Mosseri, S.; Neta, P. *J. Phys. Chem.* **1989**, *93*, 1380.

ELECTROCHEMICAL PROPERTIES OF FULLEROL AQUEOUS SOLUTIONS

Masaki Ozawa
Department of Industrial Chemistry, The University of Tokyo, 7-3-1, Hongo, Bunkyo-ku, Tokyo, 113/ Japan

*Atsushi Tadano, *Ryohichi Aogaki, Jing Li and Koichi Kitazawa
Department of Industrial Chemistry, The University of Tokyo, 7-3-1, Hongo, Bunkyo-ku, Tokyo, 113/ Japan
**Department of Industial Design, The Polytechnic University, 4-1-1, Hashimoto-dai, Sagamihara-shi,*
Kanagawa/ Japan

ABSTRACT

Alcohol of C_{60} i.e. fullerol with 26.5 hydroxyl groups in average was synthesized via. the reaction of C_{60} in NaOH aqueous solution in the presence of tetrabutylammonium hydroxide as a catalyst. Its electrochemical properties using platinum micro-electrode were investigated by means of cyclic voltammetry and impedance measurements in aqueous solutions containing various supporting electrolytes. The fullerol did not show any electrochemical activities of redox reactions, while C_{60} easily receives stepwise reduction to $C_{60}^{n-}(n=1\sim6)$ in organic solvents. However, the fullerol was adsorbed strongly onto the platinum electrode. Moreover, it was found that this specific adsorption of fullerol gave rise to promotion of hydrogen adsorption in cathodic polarization.

1. Introduction

It is well known that fullerene can be dissolved in nonpolar solvent, e.g. benzene and toluene[1], but not in polar solvents like water. Water soluble fullerene derivative was first reported by Chiang et al. in 1992[2, 3, 4]; mixture of C_{60} and C_{70} with sulfuric acid and nitric acid near 100°C in aqueous solution turned yellowish brown colour to form the fullerol. According to Mass spectrometric, IR and XPS measurements, it has been observed that at least 13 hydroxyl groups were introduced in the form of $C_{60}(OH)_n$. Bis(phenethylamino-succinate)C_{60} was also synthesized as one of water soluble derivatives[5, 6], which has been reported about its medical effects to the IIIV protease. On the other hand, the electrochemistry of fullerene in organic solvents[7], its derivatives[8], and films[9] has been investigated, and it has been known that C_{60} exhibits up to six reversible redox peaks in various organic electrolyte solutions[10]. However, the electrochemical properties of fullerene derivatives in aqueous solutions have not been reported so far. The fullerol is expected to form polymerized film by electrochemical reactions, as for electrochemically synthesized polyphenylene-type films[11, 12] In the present study, fullerol with a higher n value, which turned out to be highly soluble in water, was synthesized[13] by a different method from previous one. Then electrochemical activities of fullerol and its behaviours against hydrogen adsorption in aqueous solutions on platinum electrodes were investigated.

2. Synthesis and characterization of fullerol

A deep violet benzene solution of C_{60} (80 mg in 60 ml) and aqueous NaOH (2 g in 2 ml water) were vigorously stirred in addition with 3 drops of 40% aqueous tetrabutylammonium hydroxide (TBAH) as catalyst at room temperature in the air, becoming with colourless, and sludge precipitated in a few hours. After removal of benzene, the sludge was again stirred with an additional 10 ml of water for 10 h. Then, at the end of reaction, more 20ml water was added. The clear reddish brown solution obtained was filtered to remove a trace of water-insoluble residue, which was condensed

42

to about 5 ml. Adding MeOH (50 ml), a brown precipitate was obtained. The precipitation was repeated three times to remove TBAH and NaOH completely. Then the products were finally obtained as powder after drying the precipitate.

A broad hydroxyl absorption centered at 3430cm^{-1}, a C-O stretching absorption centered at 1070cm^{-1} and a C=C absorption centered at 1600cm^{-1} were observed in IR spectrum of the product shown in fig.1. It is quite similar to the IR spectrum of the fullerol reported by Chiang et al.[2, 3, 4] The ^{1}H NMR spectrum of the fullerene derivative in dimethylsulfoxide(DMSO) showed a broad OH peak at δ 3.35 and a weak peak due to water in the solvent at δ 3.1. The ^{13}C NMR spectrum showed quite a broad peak, indicating the presence of unequivalent carbon atoms. The elemental analysis exhibited an average composition of $C_{60}(OH)_n$ (n = 26.5, Calc.: C, 61.5; H, 2.25%). The product obtained was highly soluble in water, because ca. 26.5 hydroxyl groups can be introduced.

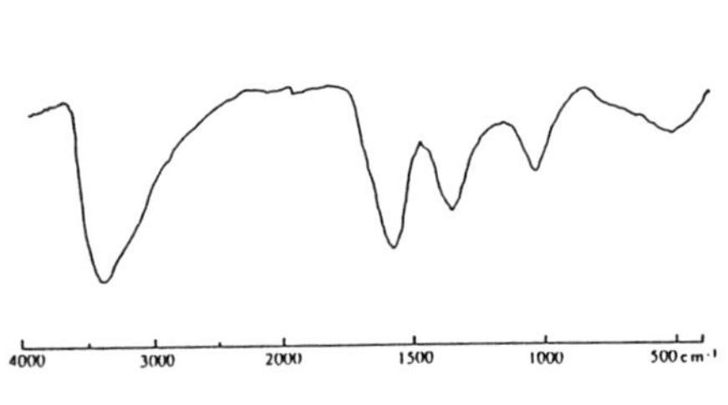

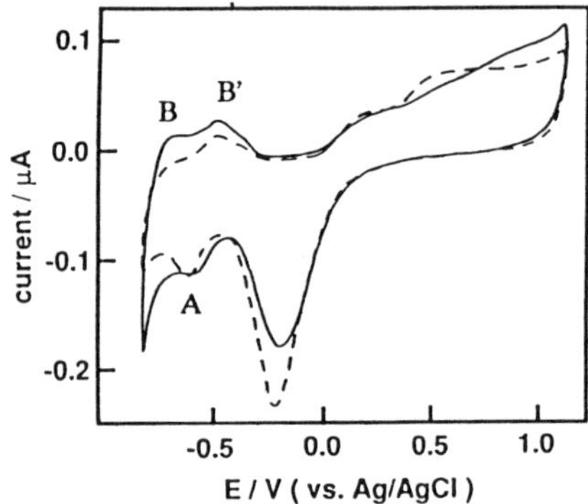

Fig.1 IR spectrum of the synthesized fullerol

Fig.2 Cyclic voltammograms in various electrolyte solutions. (- - -): 0.25M NaF aqueous solution. (——): 0.25M NaF aqueous solution containing 2.5mM fullerol. Scannimg rate is 0.5V/s.

3. Electrochemistry of fullerol aqueous solutions

The electrolysis of fullerol-containing aqueous solutions was carried out by the cyclic voltammetry and impedance measurements using a standard three-electrode cell. The electrode system consisted of a platinum microelectrodes (80μm in diameter) as a working electrode, a platinum spiral wire as a counter electrode, and Ag/AgCl electrode as a reference electrode. 0.25M aqueous solutions of $LiClO_4$, KCl, KF, KNO_3, NaCl, NaF, $NaNO_3$ and $NaClO_4$ were used as supporting electrolytes. The temperature of solutions were maintained at 25°C, and dissolved oxygen was removed by bubbling of N_2 gas prior to each measurement.

The activity of fullerol to electrochemical reactions was investigated by comparing cyclic voltammograms in solutions of various supporting electrolytes with and without fullerol. Typical cyclic voltammograms are shown in fig.2. No new peaks corresponding

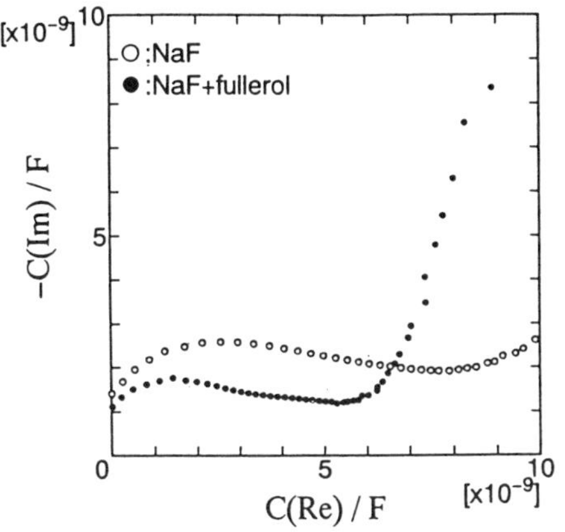

Fig.3 Capacitance plots for fullerol-free and fullerol-containing solutions measured at 0.12V.

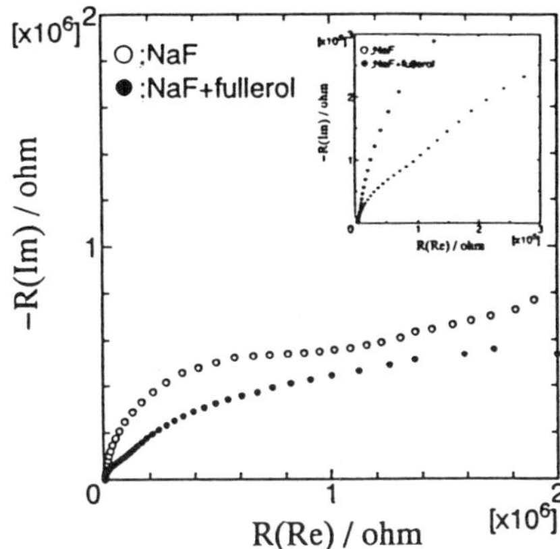

Fig.4 Impedance plots for fullerol-free and fullerol-containing solutions at the hydrogen adsorption voltage -0.8V.

to reduction or oxidation of fullerol were observed in the solid line. This suggests that fullerol molecules have no electrochemical activity under these conditions. However, comparison shows that both the adsorption (A) and desorption (B, B') peaks of hydrogen are enhanced by adding fullerol.

According to the results of capacitance measurements in electrochemical inert region (fig.3), capacitance of electric double layer, diameter of the semicircle in the fullerol-containing solutions was smaller than that in the fullerol-free counter parts. Likely cause of this phenomenon is that fullerol molecules adsorbed on the platinum electrode decreases effective active area of the electrode surface, because these are insulated. Fig.4 shows impedance plots in electrolyte solutions with and without fullerols at -0.8V, at which hydrogen adsorption occurs. and those corresponding to various kinds of electrolytes containing fullerols, respectively. As shown in fig.4, fullerol-containing solutions gave lower resistance of charge transfer, diameter of the first semicircle. This implies that the addition of fullerols promoted the hydrogen adsorption. Moreover, the resistance of fullerol-containing solutions in fig.5 was nearly constant for various supporting electrolyte. This is supposed to be an evidence of the specific adsorption of fullerol molecules on the platinum surface. This property is considered to a be helpful factor, when to attempt synthesizing fullerol-polymerized film electrochemically.

The promotion of hydrogen adsorption by specifically adsorbed fullerol as mentioned above, can be explained as fig.5 and follows: when fullerol adsorbs, oxygen atoms in OH groups make contact with the electrode. In this case, large void spaces arise for hydrogen between the cage and surface of the electrode, because ionic radius of oxygen is much larger than that of hydrogen. Hydrogen atoms are considered to be structurally stabilized, being adsorbed in this space, hence hydrogen atoms may be stacked multi-layer

4. Conclusion

Highly water soluble fullerol, $C_{60}(OH)_n$ with n ca. 26.5, was synthesized under basic condition. It was shown to be electrochemically inert in various aqueous solutions in the window between hydrogen oxidation and oxygen reduction potentials on platinum electrode. However, fullerol molecules specifically adsorbed on the platinum surface in various electrolyte aqueous solutions. Even though adsorbed fullerols was insulated, these molecules.propoted hydrogen adsorption rather than suppressed it.

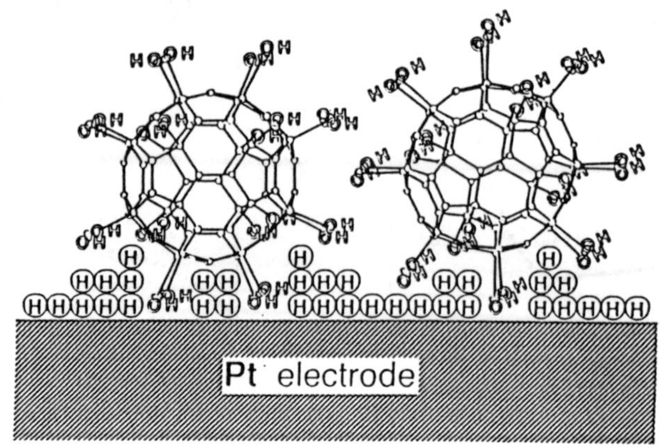

Fig.5 Schematic configulation of specifically adsorbed fullerol and hydrogen on electrode.

References

1. W. Krätschmer, L. D. Lamb, K. Fostiropoulos ans D. R. Huffman, *Nature*, **347** (1990) 354
2. L. Y. Chiang, R. B. Upasani and J. W. Swirczewski, *J. Am. Chem. Soc.*, **114** (1992) 10154
3. L. Y. Chiang, R. B. Upasani, J. W. Swirczewski and K. Creegan, *Mater. Res. Soc. Symp. Proc.*, **247** (1992) 285
4. L. Y. Chiang, J. W. Swirczewski, C. S. Hsu, S. K. Chowdhury, S. Cameron and K. Creegan, *J. Chem. Soc., Chem. Commun.*, (1992) 1791
5. S. H. Friedman, D. L. Decamp, R. P. Sijbesma, G. Srdanov, F. Wudl and G. L. Kenyon, *J. Am. Chem. Soc.*, **115** (1993) 6505
6. R. Sijbesma, G. Srdanov, F. Wudl, J. A. Castoro, C. Wilkins, S. H. Friedman, D. L. Decamp and G. L. Kenyon, *J. Am. Chem. Soc.*, **115** (1993) 6510
7. R. E. Haufler, J. Conceicao, L. P. F. Chibante, Y. Chai, N. E. Byrne, S. Flanagan, M. M. Haley, S. C. O'Brien, C. Pan, Z. Xiao, W. E. Billups, M. A. Ciufolini, R. H. Hauge, J. L. Margrave, L. J. Wilson, R. F. Curl and R. E. Smalley, *J. Phys. Chem.*, **94** (1990) 8634
8. R. S. Koefod, C. J. Xu, W. Y. Lu, J. R. Shapley, M. G. Hill and K. R. Mann, *J. Phys. Chem.*, **96** (1992) 2928
9. C. Jehoulet, Y. S. Obeng, Y. T. Kim, F. Zhou and A. J. Bard, *J. Am. Chem. Soc.*, **114** (1992) 4237
10. Q.Xie, E. Perez-Cordero and L. Echegoyen; *J. Am. Chem. Soc.*, **114** (1992) 3978
11. M. C. Pham, P. C. Lacaze and J. E. Doubois, *J. Electroanal. Chem.*, **86** (1978) 147
12. Y. Ohnuki, H. Matsuda, T. Ohsaka and N. Oyama, *J. Electroanal. Chem.*, **158** (1983) 55
13. J. Li, A. Takeuchi, M. Ozawa, X. Li, K. Saigo and K. Kitazawa, *J. Chem. Soc., Chem. Commun.*, (1993) 1784

Cluster origin of fullerene solubility

V.N. Bezmelnitsin, A.V. Eletskii and E.V. Stepanov

Russian Research Center "Kurchatov Institute", Kurchatov Square,
Moscow 123182, Russia

The solubility of fullerenes in organic solvents is explored taking into account the formation of clusters containing a number of fullerene molecules. On this basis, the extraordinary temperature dependence of the fullerene solubility obtained recently is described. The role of a disordering phase transition for the solubility behavior is shown. Comparison of the calculated data with the experimental ones permits to obtain energetic parameters characterizing the interaction of a fullerene molecule with its surrounding in a solid phase and solution.

The experiments performed at the recent years have shown a possibility for fullerenes to form clusters consisting of a number of fullerene molecules [1–5]. This work is devoted to the analysis of the behavior of fullerenes in solutions taking into account such a possibility.

To obtain the distribution function of the clusters by size we will use a droplet model for a cluster, based on assumption that the number of fullerenes per cluster $n \gg 1$. Thus the thermodynamic energy of a cluster in solution is combined by two terms, the first of which is a volume term being proportional to the number n of molecules in a cluster, and the second one is a surface term determined by the surface square proportionally $n^{2/3}$. Using a standard thermodynamic approach [7,8], we represent the distribution functions of clusters in solution over the number n of the constituent fullerene molecules in the next form:

$$f_n = g_n \exp\left[\left(-An + Bn^{\frac{2}{3}}\right)/T\right] \tag{1}$$

Here the parameter A refers to the difference of thermodynamic specific energies of the interaction of a fullerene molecule with its surrounding in the bulk of a solid phase and in the body of a cluster, and the parameter B is the similar difference of energies for molecules placed on the cluster surface. In general, these parameters can be of any sign, but for the distribution function to be normalized the condition $A > 0$ should be fulfilled. The statistical weight of a cluster g_n also, generally, depends on n as well as on temperature, however, we neglect those in comparison with strong exponential dependency (1). It corresponds to the assumption $n \gg 1$ being valid in the scope of a droplet model.

The net solubility of a fullerene is determined by the sum:

$$C(T) \sim \sum_n n f_n(n, T) \tag{2}$$

46

which, with regard for $n \gg 1$, can be replaced by integration:

$$C(T) \sim \bar{g} \int_1^\infty n \exp\left(\frac{-An + Bn^{\frac{2}{3}}}{T}\right) dn \qquad (3)$$

Here, in accordance with the above mentioned assumption, the averaged statistical weight $\bar{g}$ of clusters has taken out from integration. This coefficient comprising an entropy factor can generally take different values for various solvents. The parameters A and B, the relation of which determines the temperature dependence of solubility, vary with a kind of solvents too.

Let us try to use the expression (3) for the description of the observed temperature dependencies of C_{60} solubility presented in Fig 1 in relative units. It is easy to understand that the non-monotone dependence $C(T)$ can take place only if $B > 0$, $A > 0$. In this case, the integral (3) as a function of temperature can have only one extremum being a minimum. Indeed, the second derivation of $C(T)$ in the point of extremum is an integral of a positive expression which is apparently positive. Hence the expression (3) has no maximum in terms of temperature. However, change in the character of the temperature dependency from ascending to descending can be caused by a phase transition in a solid C_{60} occurring at a critical temperature $T \simeq 260\ K$ [6]. This transition is endothermic and characterized by specific enthalpy $\Delta h = 7.1\ kJ/mol$ [9,10]. This fact implies a considerably different role of cluster at temperatures above and below the critical one.

Taking into account the phase transition and the connection the expression (3) takes the next form:

$$C = C_0 \int_1^\infty n \exp\left(\frac{Bn^{\frac{2}{3}} - [A_{fcc} + (\Delta h - T\Delta s)\theta(T_c - T)]}{T}\right) dn \qquad (4)$$

which is applicable throughout the temperature range involved (Δs is the specific entropy of the transition, $\theta(T - T_c)$ is a step function, indices sc and fcc refer to respective lattices and temperature regions below and above T_c).

As is shown by numerical computing, the expression (4) qualitatively describes the data obtained in [6] and presented in Fig. 1. The best accordance is reached at the following values of the parameters: $B = 970\ K$, $A_{fcc} = 320\ K$, $C_0 = 5 \cdot 10^{-8}$ [C_{60} mole fraction]. The dependency $C(T)$ computed at these values of the parameters is shown by a solid curve in Fig. 1. The calculated dependence proved to be virtually insensitive to the width δT of the phase transition if $\delta T < 10\ K$. Besides, we have tracked how a characteristic size n^* of clusters corresponding the maximum of the integrated expression in (4) is changed with temperature at the obtained values of the parameters. At $T < T_c$ this size monotonically increases from $n^*(190\ K) = 3$ to the value $n^*(260\ K) = 11$ which remains practically the same up to $T = 380\ K$. Thereby, at the temperature region $T > T_c$, the value $n \gg 1$, which proves the feasibility of a droplet model and justifies the above assumption that the main contribution to the fullerene solubility at temperatures within the region of descending the dependency $C(T)$ is caused by clusters of a large size.

Thus the conjecture that the solubility of fullerenes is of a cluster nature, being indirectly suggested by the results of the experiments [2,3], enables to explain the non–monotone temperature dependency of the solubility of C_{60} observed experimentally [6] for several solvents. In fact, the decrease of solubility on warming is caused by the thermal dissociation of clusters containing the great number of fullerene molecules. Further increase of temperature is to ultimately lead to a rise of solubility, however, numerical computations show that the minimum of solubility at the above obtained values of the parameters A and B is expected at $T \simeq 4000\ K$ wherein neither a solvent nor a fullerene exists. At low temperatures $T < 260\ K$, the role of clusters is likely to be insignificant, which is connected with the change of a crystalline structure as well as with the increase of the energy of the interaction of fullerene molecules with their surrounding in a solid phase.

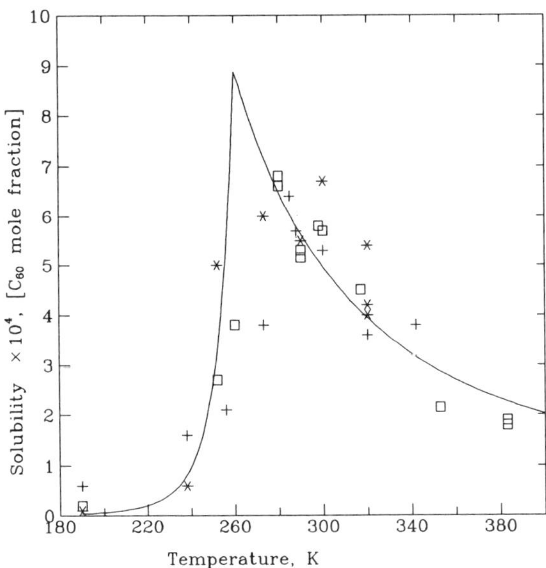

Fig. 1. The solubility of C_{60} in hexane (+, multiplied by 55.), toluene ($\square$, multiplied by 1.4) and CS_2 ($*$) [6]. The solid curve represents our calculated data.

In closing, it should be noted that for a cluster nature of the fullerene solubility to be established completely, direct experimental exploration is necessary. As a possible scheme for such experiments (see also [5]), the measurements of infrared absorption spectra of the fullerene solution in dependence on concentration at various temperatures being to be performed for an optical path–length held to be constant can be conceived. The fact of such a dependency will indicate to the presence of fullerene clusters in solution. Another way, having been partially realized in [3], is based on applying Raoult's law. According to this law, the saturation vapor pressure of a solvent above a solution differs from that above a pure solvent by a value being proportional to the concentration of solute particles. The measurement of the flow of the solvent vapor being determined by the difference of the pressures will enable to obtain the dependence of the solute particle concentration on the concentration of the solute. Provided this dependence is non–linear, it will be a basis to draw a quantitative conclusion concerning the existence of the fullerene clusters in solution. At present, the mentioned design of experiments are under development.

The authors are grateful to B.M. Smirnov for helpful advises. This work was partially supported by the Russian Foundation of Fundamental Studies.

References

1. Eletskii A.V., Smirnov B.M., *Soviet Physics – Uspekhi* **163** (1993) 33.
2. Sun Y.-P., Bunker C.E., *Nature* **365** (1993) 398.
3. Honeychuck R.V., Cruger T.W., Milliken J., *J. Am. Chem. Soc.* **115** (1993) 3034.
4. Wang Y.-M., Kamat P.V., Patterson L.K., *J. Phys. Chem.* **97** (1994) 8793.
5. Blau W.J. et. al., *Phys. Rev. Lett.* **67**(1991) 1493.
6. Ruoff R.S. et. al., *Nature* **361**(1993) 140.
7. Landau L.D., Lifshitz E.M., *Statistical Physics* (Nauka Publishing, Moscow, 1990).
8. Smirnov B.M., *Soviet Physics – Uspekhi* **162** (1993), No. 1, p. 119; No. 12, p. 97; **163** (1993), No. 10, p. 29.
9. Heiney P.A. et. al., *Phys. Rev. B* **45** (1992) 4544.
10. Samara G.A. et. al., *Phys. Rev. B* **47** (1993) 4756.

ION-FULLERENE INTERACTION: EFFECTS OF ENERGY TRANSFER

D.FINK, J.ERXMEIER, R.KLETT, H.MELZER, H.J.MÖCKEL, M.MÜLLER, D.NAGENGAST, and P.SZIMKOVIAK
Hahn-Meitner-Inst. Berlin GmbH, Glienicker Str.100, D-14109 Berlin, Germany
M.BEHAR, P.GRANDE, and J.KASCHNY
Instituto de Fisica, Universidade Federal do Rio Grande do Sul, 91500 Porto Alegre, RS, Brazil
L.T.CHADDERTON
Inst. for Advanced Studies, Australian Natl. Univ., P.O.Box 4, Canberra, ACT 2601, Australia
J.KASTNER
Inst. für Festkörperphysik, Universität Wien, Strudlhofgasse 4, A-1090 Wien, Austria
L.PALMETSHOFER
Inst. für Experimentalphysik, J.Kepler Universität, A-4040 Linz, Austria
and
L.WANG
Gesellschaft für Schwerionenforschung, Planckstr.1, D-64220 Darmstadt, Germany

ABSTRACT

This paper summarizes our recent results in the field of ion-fullerene interaction. We compare measured ranges of ions in fullerite with those in amorphous carbon and with theoretical predictions. Fullerite destruction takes place upon single-ion hit of heavy projectiles, but is a cumulative effect for light ion irradiation. Finally, we succeeded for the first time to present reliable production yields for the ion-induced creation of fullerene in carbonaceous matter at very high transferred energy densities.

1. Introduction

Properties of ion-fullerene interaction such as implantation profiles and energy transfer by nuclear and electronic processes have been studied by our groups in a joint effort since recently. The aim is to gain better understanding of the behavior of this unusual material, for future applications.

For this sake fullerite layers of 300 nm to 1.5 lm thickness were prepared by evaporation onto polished Si wafers held at room temperature in a vacuum of 2×10^{-4} Pa. Subsequently 100 keV H^+ and B^+, 2 MeV He^+, 20 to 200 keV Cs^+ and Pb^+, 50 keV Ar^+, Kr^+, and Xe^+, and 500 MeV I^{17+} ions have been implanted at room temperature into the samples at fluxes of a few nA/cm² and fluences between 3×10^{11} and 10^{17} ions/cm². Depth distributions of the implants have been determined by means of standard RBS, respectively NRA (^{15}N(H,Gamma)) techniques. The ion-induced C_{60} damage was examined by Raman scattering.

2. Range Distributions of Ions Implanted into Fullerite

In Fig.1, ranges of Cs^+ and Pb^+ ions in C_{60} and a-C are shown together with theoretical predictions. The measured ranges are party smaller (H^+ results, not shown here), equal (Cs^+), or larger (Pb^+) than theoretical predictions. Within $\pm 10\%$, ranges in fullerite and amorphous carbon agree with each other. The remaining differences and the deviation from theory might be attributed to different cross terms between electronic and nuclear energy transfer at low energies, i.e. to electronic excitations during nuclear collisions [1]. This peculiarity has been observed earlier for a number of heavy ions in several light target materials [2]. Specifically for ranges in different carbon allotropes, this would mean that the *different states of hybridization* of their outer electrons have a small but non-negligible influence on the ions' slowing down.

Fig. 1: Ranges R_p and range straggling δ of 20-200 keV Cs and Pb in fullerite and amorphous carbon. The ranges are given here in [ug/cm^2] to allow for direct comparison between different carbon allotropes. For conversion to the nm scale, please use the densities 1.65 g/cm^3 for fullerite, respectively 2.266 g/cm^3 for a-C. Range straggling values δ not deconvoluted.

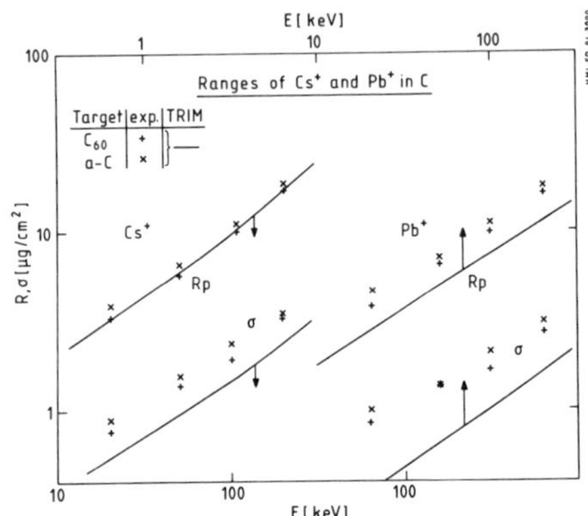

3. Ion-Induced Fullerite Destruction and Creation

We preferentially analyzed the fate of the most pronounced 1469 cm^{-1} Raman line, which is characteristic for C_{60} [4]. This line decreases after implantation, giving rise to a broad, asymetric line between 1000 and 1700 cm^{-1} which stems from amorphous carbon. (In case of very low ion ranges, the signal of the residual undestroyed fullerite at greater depths has, of course, to be subtracted in the analysis.) Additionally to our own results on fullerite destruction, we used values from literature[5],[6] to expand the still poor data basis for C_{60} damage.

The characteristic parameters of ion-induced fullerene destruction are summarized in Table 1. For light ions, fullerite destruction sets in only at fluences when collision cascades overlap, i.e. we can describe the destruction process in this case as a *cumulative effect* of many impinging ions. For contrast, heavy ions at any energy are capable to destroy fullerene molecules along their ion tracks upon *individual impact*, i.e. at fluences well below the onset of ion track overlapping.

One of our future aims will be to decide whether *electronic* (leading to preferential destruction via bond breaking) or *nuclear* processes (leading to knock-out of atoms from the C_{60} molecules) are responsible for fullerite destruction. The presently accumulated data base is not yet useful for this purpose, as the fullerite destruction yield correlates as well with the nuclear stopping (see Ref.6) as with the density of transferred electronic energy (see Fig.2), due to simple mathematical relationship between both paramters. For such a decision, either the energy dependence of fullerite destruction near the nuclear or electronic stopping power maxima, or the shape of the depth profile of fullerite destruction have to be determined.

There is a difference in Raman spectra between low and high energy irradiated fullerite insofar as only in the high energy irradiated samples an additional line at 1575 cm^{-1} shows up. It might point at the creation of carbon nanotubes, onion-like, or graphite-like

Table 1: Average values derived for fullerene destruction yield Y_d, damage cross section r_d, and mean radius r_d of the zone of fullerene destruction, for different types of ions - light ions (H, He, B, C), heavy ions (Cs, Pb, I) - , and different energies - low energies (20-2000 keV), high energies (2-500 MeV).

type of projectiles	Y_d [C_{60}/ion]	r_d [cm^2]	$r_{d_\circ}$ [Å]
light ions, low energies	12 ...100	$(1..3)x10^{-16}$	0.5...1
heavy ions, low energies	150...1200	$(2..6)x10^{-14}$	5...20
heavy ions, high energies	$2 \cdot 10^5...3 \cdot 10^6$	$(0.6..3)x10^{-12}$	40...100

Fig. 2: Threshold fluence to achieve the onset of (encircled numbers), respectively to complete (numbers in rectangles) the fullerite destruction, as a function of the transferred electronic (enclosed numbers) and nuclear (free standing numbers) energy densities. Included are values derived from literature [5],[6] for the sake of completeness. Meaning of the numbers: 1:100 keV H, 2:160 keV H [5], 3:300 keV He [5], 4:2 MeV He, 5:300 keV C [5], 6:100 keV Ar, 7:220 keV Ar [5], 8:100 keV Kr, 9:55 MeV I [6], 10:500 MeV I, 11:100 keV Cs, 12:200 keV Cs, 13:200 keV Pb, 14:100 keV B. Please note that the data band describing the complete destruction has been shifted by one order of magnitude to the right, for the sake of better clearness.

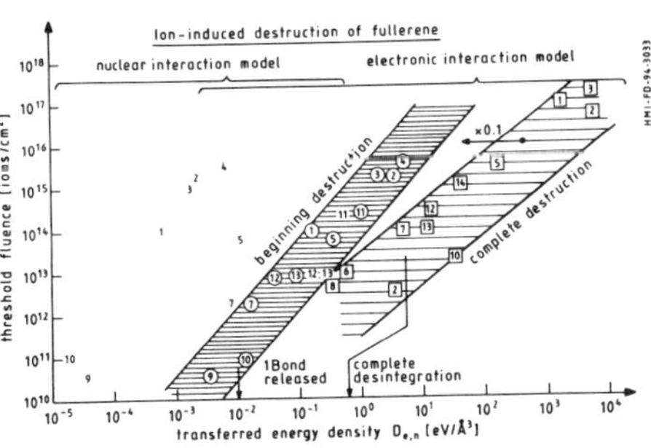

structures. In fact, the creation of fullerenes, nanotubes and carbon clusters above some threshold energy density of roughly 1 eV/Å^3 has been recently observed in irradiated pyrolytic graphite and polymers by HPLC [7], TEM [8], SANS [9], SAXS [10,11] and UV/vis. [12] spectrometry. A theory has been established [3] to describe this process in terms of complete destruction of all molecular bonds along the ion track core, i.e. of creation of a zone of highly excited but neutral target atoms which can react with each other, provided that the transferred energy density is high enough to allow for a sufficiently long chemical reaction time.

In order to quantify this effect, we have irradiated sacharose with 3 GeV U^{20+} ions at a fluence of 10^{12} ions/cm^2, and examined the product by ultrasensitive HPLC for fullerene. It turned out that 150 C_{60} and a few C_{70} molecules are created by each ion. Taking into account the high degree of dilution of carbon in the examined material, this is a large yield which cannot be explained by simple collision statistics alone: we therefore have to conclude that CC bonding is favored in comparison to CH and CO bonding during the buildup process of the fullerenes.

4. Conclusions

Implantation profiles of low energy light and heavy ions in fullerite and amorphous carbon show only little difference from each other. Their deviations from theoretical predictions might be explained by the influence of the specific state of hybridization of the outer target atoms' electrons on the magnitude of the electronic energy transfer.

Fullerene destruction results from single-ion hits in case of heavy projectiles, but is a cumulative effect for light ion impact. Some basic parameters such as the destruction yield, the damage cross section, and the effective track radius for destruction have been derived. Above a few eV/Å^3 transferred, *fullerene creation* in carbonaceous material is initiated. Its production yield is typically 150 C_{60} molecules per ion in case of 3 GeV U ion impact onto sacharose.

5. References

1. P.F.P.Fichtner, M.Behar, D.Fink, P.Goppelt, and P.L.Grande, *Nucl. Instr. Meth.* B64 (1992) 668.
2. P.L.Grande, M.Behar, D.Fink, and F.C.Zawislak, *Nucl. Instr. Meth.* B61 (1991) 282.
3. L.T.Chadderton, D.Fink, Y.Gamaly, H.Möckel, L.Wang, H.Omichi, and F.Hosoi, *Nucl. Instr. Meth.* B, in print (1994)
4. J.Kastner, H.Kuzmany, L.Palmetshofer, P.Bauer, and G.Stingeder, *Nucl. Instr. Meth.* B80/81 (1993) 1456.
5. J.Kastner, H.Kuzmany, and L.Palmetshofer, submitted to *Appl.Phys.Lett.* (1994)
6. R,M.Papaleo, S.Hallen, J.Erikson, G.Brinkmalm, P.Demirev, P.Hakansson, and B.U.R.Sundqvist, *Nucl. Instr. Meth.*, in print (1994)
7. L.T.Chadderton, D.Fink, H.J.Möckel, K.K.Dwivedi, and A.Hammoudi, *Radiation Effects and Defects in Solids* 127 (1993) 163.
8. A.Dunlop and L.T.Chadderton, pers. comm. (1993)
9. D.Fink, K.Ibel, P.Goppelt, J.P.Biersack, L.Wang, and M.Behar, *Nucl Instr. Meth.* B46 (1990)342
10. R.Montiel, pers. comm. (1993)
11. Y.Ayal, pers. comm. (1993)
12. R.M.Papaleo, M.A.de Araujo, and R.P.Livi, *Nucl. Instr. Meth.* B65 (1992) 442.

INTERACTIONS OF FREE ELECTRONS WITH C_{60}: UNUSUAL ATTACHMENT PROPERTIES

D. SMITH, P. ŠPANEL, M. LEZIUS, P. SCHEIER AND T.D.MÄRK
Institut für Ionenphysik, Universität Innsbruck,
Technikerstraße 25, A-6020, Innsbruck, Austria

ABSTRACT

We have studied electron attachment to C_{60} using our two experimental techniques: (i) the crossed electron/neutral beam, double-focusing mass spectrometer system and (ii) the flowing afterglow/Langmuir probe (FALP) apparatus. Using the former, electron attachment was observed from near-zero energy to about 15 eV, whereas using the FALP apparatus, absolute attachment rate coefficients were determined in the electron temperature range from 500 K to 4500 K from which absolute mean thermal attachment cross sections have been derived. In the light of recent theoretical work and the FALP data, it is clear that electron capture to C_{60} in the low energy regime occurs via the p-wave (not the s-wave) process, the height of the centrifugal barrier (the "activation energy" for the process) being 0.26 eV, and the maximum cross section being about 100 Å^2. Using these FALP data as a calibration, the absolute attachment cross sections for electron attachment to C_{60} have been obtained over the wide energy range explored in our beam experiment, and they are seen to remain very high up to an electron energy of about 10 eV. The "peaks" (structure) on the cross section versus electron energy curve are interpreted in terms of the electron energy loss processes which occur as the electrons interact with the C_{60} molecules. The structure on this cross section curve is mirrored closely by the structure on the electron energy loss spectra for C_{60} that have recently been obtained.

1. Introduction

It is now well known that the fullerene molecule C_{60} readily accepts electrons both in the solution and solid phases [1] and free electrons in the gas phase [2]. In the solid phase the intercalation of C_{60} with alkali metals can result in materials that range in electrical properties from superconductors to insulators depending on the filling of the electronic states in the C_{60} molecule [3]. It has been shown from gas phase studies [4] that C_{60} has a relatively large electron affinity of 2.65 eV, but this in itself does not mean that it will be an efficient gas phase electron scavenger. The most commonly used gas phase electron scavenger SF_6 has a relatively small electron affinity of only 1.06 eV but the cross section for electron capture at near-zero electron energy to produce the parent negative ion SF_6^- is very large, close to that predicted by *s-wave* scattering theory [5], but decreases rapidly with increasing energy. We show in this paper that single electron attachment to C_{60} exhibits a very different energy variation from that for SF_6, and that C_{60} very efficiently attaches a single electron over a much wider energy range. Furthermore, it is interesting to note that a free C_{60} molecule can attach two electrons [6].

Gas phase electron attachment has been researched for decades [7] and it is known that there are two basic mechanisms: (i) direct attachment in which the parent negative ion of the attaching molecule is formed; low energy attachment to SF_6 is the best-known example and fluorinated hydrocarbons such as C_4F_6, C_4F_8, C_6F_6 and C_7F_{14} also efficiently form parent anions, (ii) dissociative attachment in which a fragment neutral radical and negative ion are formed; a good example is attachment to CCl_4 in which CCl_3 and Cl^- are produced very efficiently at low energies. Often attachment reactions, both direct and dissociative, occur via 'resonances' in which the cross section is appreciable over a small electron energy range (or ranges), like the famous 'zero energy resonance' in SF_6 in which the cross section falls rapidly from its maximum according to *s-wave* theory to near zero at an electron energy of only 100 meV [5]. Resonances often occur at higher energies also as is exemplified by the dissociative attachment resonance for O_2 at about 7 eV [8]. However, many attachment reactions, notably but not exclusively dissociative attachment reactions, require 'activation energy', E_a, to promote them, i.e. they are slow at low energies/temperatures and become increasingly efficient as the energy/temperature is increased [9]. We have shown by our experimental work that electron attachment to the C_{60} molecule combines all of these features and that it is the most remarkably efficient electron scavenger.

2. Experimental

Our experimental investigations of electron attachment processes involve the use of two major apparatuses, a crossed electron/neutral beam double focusing mass spectrometer system with which we can study attachment processes over a wide electron energy range [10] but not with great accuracy at low energies (below about 0.5eV) because of the energy spread in the electron beam, and the flowing afterglow/Langmuir probe (FALP) apparatus using which we can study attachment and other plasma reaction processes at electron temperatures up to about 5000 K, i.e. mean electron energies up to about 0.5 eV [11]. Thus in co-ordinated beam and FALP experiments we can study electron attachment to selected species from near-zero electron energy to energies in excess of 100 eV. Very importantly, in the FALP measurements the *absolute* rate coefficients, β, (and hence *absolute* mean thermal cross sections) for the attachment reactions are obtained, and this allows the calibration of the cross sections, σ, obtained using our beam apparatus in the overlapping energy regime and hence provides *absolute* σ at the higher energies. The results of such co-ordinated experiments on electron attachment to C_{60} and their comparison with other related work are now presented.

3. Results

Our first experiments were carried out using the beam apparatus [10]. The C_{60} molecular beam (which emerged from a small hole in an ohmically-heated oven containing C_{60} powder) is crossed with the electron beam and the currents of the C_{60}^- ions are detected (following mass selection) as a function of the electron energy. Since the flux of C_{60} molecules in the beam is not known, only relative cross sections can be obtained but nevertheless the results obtained are remarkable in that the σ remain approximately constant over the very wide energy range from near-zero to more than 10 eV ! As can be seen in Fig. 1a, the σ apparently rises from a low value at near-zero energy to a first peak, A, before falling into an obvious trough, and with further increase in energy it rises and falls again to two other obvious peaks before finally falling towards zero at an energy of about 14 eV.

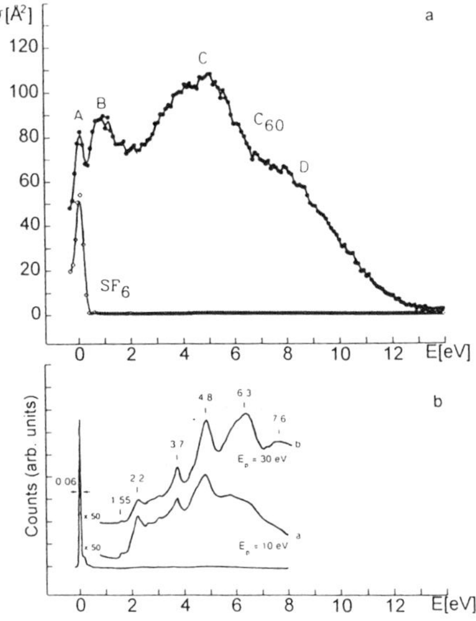

Fig. 1: a) The data for the electron attachment to C_{60} obtained from our beam experiment over a wide range of electron energy, E [10]. The absolute cross sections, σ, have been deduced using the FALP data as a calibration. The SF_6 data are included for comparison and to indicate the energy spread of the electron beam.
b) The data obtained from the high resolution electron energy loss spectroscopy (HREELS) experiment [12] at two electron energies Ep. The sharp peak near zero energy indicates the energy width of the electron beam (in eV). Note the correlation of the peaks at 3.7, 4.8 and 6.3 eV with the broad peak marked D in a).

What are the reasons for this 'structure' on this σ curve? Immediately it should be stated that peak **A** may not be real because a similar peak is apparent in the SF_6 data taken under identical conditions which is not real but is rather due to the spread in energy of the electron beam (about 0.3 eV). Therefore from these data it cannot be excluded that the cross section for C_{60}, like SF_6, is high at zero energy and then falls of rapidly at an energy of a few tenths of an eV, as is clearly suggested by the fall from peak **A** (which is surely real). However, the subsequent FALP studies do not support this conclusion so we return to the discussion of this below. The other peaks labelled **B** and **C**, and the weaker feature labelled **D** on Fig. 1a and can be explained by the energy loss processes that occur as the electrons interact with the C_{60} molecule and which result in trapping (attachment). Thus peak **B** at an energy of about 1.5 eV is due to excitation of the molecules across the fundamental HOMO-LUMO gap by the incoming electrons and the subsequent trapping of the resulting slow electrons. The width of the peak is presumably a manifestation of the variable number of collisions that must occur at the different electron energies prior to the trapping, including electron/phonon interactions [4]. The very broad peak at **C** can be interpreted in the light of results from other experiments, specifically the high resolution electron energy loss spectra (HREELS) of thin films of C_{60} [12] and, more recently, the electron energy loss spectra (EELS) of free C_{60} molecules [13]. The data from the

HREELS experiment are reproduced as Fig.1b with the same energy coordinate as our beam data shown in Fig.1a. It can clearly be seen that the broad peak at C is the net result of three distinct electron energy loss processes which have been assigned as specific one-electron or collective excitations [12, 13]. The weak feature labelled **D** is presumably due to a further excitation mechanism .

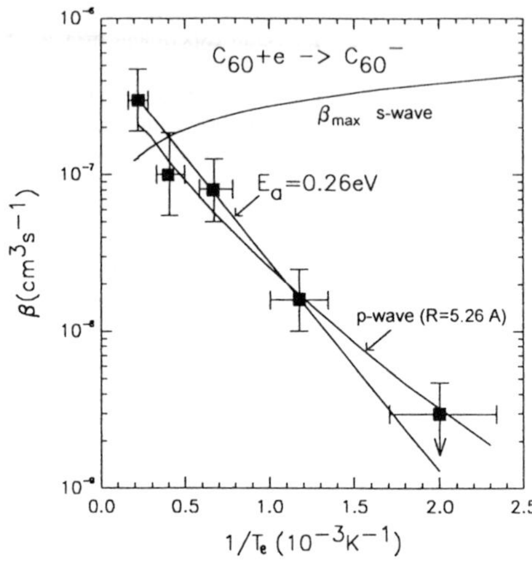

Fig. 2: The measured values of the electron attachment coefficient, β, for C_{60} obtained from our FALP experiment as a function of the inverse electron temperature, T_e^{-1}. The slope of this Arrhenius plot provides a value for the activation energy E_a (see [14] and the text). Also shown is the predicted maximum value of the rate coefficient, β_{max}, according to *s-wave* capture theory [5] and the predicted β from *p-wave* capture theory, assuming a value of R=5.26 Å (see the text and [15]).

Now returning to the consideration of the very low energy electron attachment. To accurately define the form of the variation with electron energy of the attachment cross section below an electron energy of 1 eV, an energy spread on the electron beam of <0.1 eV is required. The alternative approach is to determine the rate coefficient as a function of the electron temperature, T_e, and this we have done in the FALP experiment [14]. The results are shown in Fig. 2 in the form of an Arrhenius plot of log β against T_e^{-1}, where β is the measured rate coefficient for electron attachment to C_{60}. As can be seen, β is very small at low T_e (immeasurably so at 300 K) but increases rapidly with T_e to reach a value at 5000 K that is greater than the value predicted by *s-wave* scattering theory [5]. From the slope of this Arrhenius plot an 'activation energy', E_a, of 0.26 eV is obtained. In our original paper we incorrectly attributed the very high rate coefficient at 4500 K, and hence the high mean thermal cross section of 80 Å^2, as a summation of the upper-limit *s-wave* capture cross section (~30 Å^2) and the cross section of the C_{60} molecule (~50 Å^2). Now very recent theoretical work [15] has shown that electron attachment to C_{60} occurs preferentially in the *p-wave* not the *s-wave* channel, and further, by adopting a value of 5.26 Å for the trapping radius (close to the equilibrium half distance between the C_{60} molecules in the solid), the centrifugal barrier to the incoming *p-wave* electron is calculated to be 0.27 eV, in close agreement with our experimental E_a. Further, on the assumption that all the electrons that overcome this barrier attach to the C_{60}, the calculated σ and the derived β agree very well with our measured values over the

T_e range explored, as can be seen from Fig.2. This good agreement is a remarkable vindication of both the experimental and the theoretical work. It also demonstrates the great value of the FALP technique in the determination of absolute attachment rate coefficients (and hence absolute mean thermal attachment rate coefficients) [9, 14]. The theory also predicts [15] that the cross section should begin to fall at an electron energy of about 1 eV, and this can indeed be seen on our beam data shown in Fig.1a.

So now the features of electron attachment to C_{60} can be explained from zero electron energy out to about 10 eV. At zero energy, σ is essentially zero but increases rapidly with increasing electron energy towards a peak value of about 100 Å^2 at an electron energy of about 0.5 eV as the p-wave centrifugal barrier is overcome. Then σ falls again, increases again and remains at a high value up to an energy of about 10 eV or so (see Fig. 2). We are now able to give absolute cross sections over the entire energy regime (Fig. 1a) because of the calibration of our beam data using our FALP data. A final point worthy of note is that the apparent rapid fall-off in the attachment efficiency around an electron energy of 10 eV is probably due to thermionic emission (detachment) of the captured electrons from the hot C_{60}^- ions as their temperature is increased by the deposition of this large amount of energy into each molecule [16].

4. Acknowledgments

We thank Professor E. Tosatti for providing us with his theoretical results prior to their publication. This work was partially supported by grants from the Bundesministerium für Wissenschaft und Forschung, Wien and from the Fonds zur Förderung der Wissenschaftlichen Foschung, Wien.

5. References

1.   R.C.Haddon, *Phil. Trans. Roy. Soc. Lond.* A, **343** (1993) 53.
2.   R.F.Curl and R.E.Smalley, *Science, Wash.*, **242** (1988) 1017.
3.   R.C.Haddon, *et al.*, *Nature, Lond.*, **350** (1991) 320.
4.   L.Wang, C.L.Pettiette, J.Conceivao, O.Chesnovsky and R.E.Smalley, *Chem. Phys. Letters*, **182** (1991) 5.
5.   C.E.Klots, *Chem.Phys.Letters*, **38** (1976) 61.
6.   R.L.Hettich, R.N.Compton and R.H.Ritchie, *Phys.Rev.Letters*, **67** (1991) 1242.
7.   H.S.W.Massey, *Negative Ions* (Cambridge University Press, Cambridge, 1976).
8.   D.Rapp and D.D.Briglia, *J.Chem.Phys.*, **43** (1965) 1480.
9.   D.Smith and P.Spanel, *Adv.Atom.Mol.Opt.Phys.*, **32** (1994) 307.
10. M.Lezius, P.Scheier and T.D.Märk, *Chem. Phys. Letters*, **203** (1993) 232.
11. P.Spanel and D.Smith, *Intern. J. Mass Spectrom. Ion Processes*, **129** (1993) 193.
12. G.Genstenblum, J.J.Piraux, P.A.Thiry, R.Caudano, J.P.Vigneron, P.Lambin, A.A.Lucas and W.Krätschmer, *Phys.Rev.Letters*, **67** (1991) 2171.
13. A.W.Burose, T.Dresch and A.M.G.Ding, *Z.Phys.*, **D26** (1993) S294.
14. D.Smith, P.Spanel and T.D.Märk, *Chem. Phys. Letters*, **213** (1993) 202.
15. E. Tosatti and N Manini, *Chem. Phys. Letters*, (1994), in press.
16. C. Yeretzian, K.Hansen and R.Whetten, *Science*, **260** (1993) 652.

THE USE OF DIFFERENT IONIZATION METHODS IN MASS SPECTROMETRIC STUDIES OF FULLERENES

UWE KIRBACH
and
LOTHAR DUNSCH

Institut für Festkörperforschung, IFW Dresden e.V., Helmholtzstraße 20,
01069 Dresden, Germany

KLAUS KLOSTERMANN

Institut für Analytische Chemie der TU Dresden, Postfach,
01062 Dresden, Germany

ABSTRACT

The influence of the ionization method on the mass spectra of fullerenes was studied using a sector field mass spectrometer MAT 95 (FINNIGAN) and electron impact ionization (EI), chemical ionization (CI), fast atom bombardment ionization (FAB) as well as field desorption/field ionization (FD/FI) as the ion generation methods. The aim of the study was the comparison of a known mixture of C_{60}/C_{70} with the C_{60}/C_{70} ratios detected by mass spectrometry.

The mass spectra, the yield of ions, the molecule fragmentation and multiple ionization were shown to differ markedly.

The ionization by negative CI is the best method for the detection of fullerenes in the most cases. The advantages are high sensitivity and a good reflection of sample composition.

1 Introduction

Mass spectrometry is one of the most important and most promising method for the characterization of fullerenes and their reaction products in the pure state (for example HPLC fractions) as well as in mixtures (soot extract). Using time of flight (TOF), quadrupole or sector field mass spectrometry positive or negative fullerene ions can be detected. Often quantitative estimations are derived from these mass spectra. Furthermore the chemical ionization of fullerenes has been studied qualitatively by McElvany et.al.[1].

In this paper we follow the influence of the ionization method of the fullerene mass spectra. Mass spectra produced by electron impact ionization (EI), chemical ionization (CI), field desorption (FD) and fast atom bombardment (FAB) were made to find the best kind of ionization technique regarding the composition of the sample, sensitivity, multiple ionization and fragmentation of fullerenes.

2 Experimental

The mass spectra were recorded with a sector field mass spectrometer MAT 95 (Finnigan). The spectrometer is suitable for four kinds of ionization methods (EI, CI, FAB (liquid SIMS), FD/FI) and the detection of the generated positive or negative ions[2, 3].

All measurements were carried out with the same mixture of HPLC pure C_{60} and C_{70} (ratio 1 : 1) dissolved in toluene[4, 5]. In the case of the EI and CI the samples can be evaporated from a wire (heated up to 1500°C) and ionized directly (EI, 70 eV or 15 eV, 1 mA) or indirectly (CI, 150 eV, 100 µA) with an electron beam. Methane was used as ionization gas in CI. The samples studied by FD were dropped onto the FD whisker and ionized by an electric field of 6 kV. In contrast to these ionization methods the FAB samples were mixed with 4-nitrobenzyl alcohol (m-NBA) and ionized with a cesium-ion beam (10 kV). The positive and the negative ions were measured by CI and FAB but only the positive ions by EI and FD.

For each sample the average of all spectra during the evaporation of the sample was taken. The result is a spectrum with the true sample composition.

3 Results

As shown in the Figures 1 to 7 the mass spectra of the same sample differ very strongly. Especially the differences in the C_{60} : C_{70} ratios are considerable.

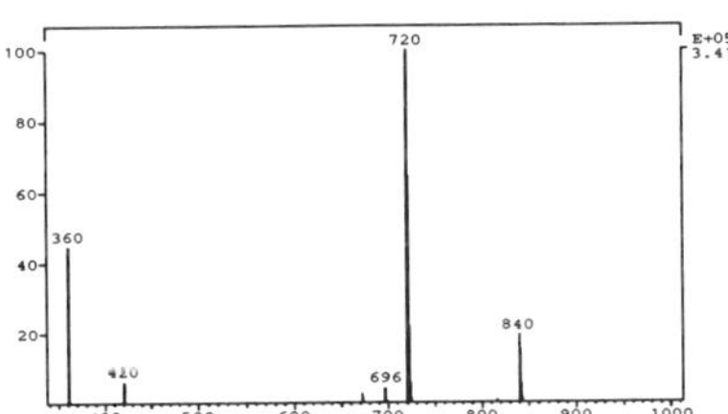

Fig.1: positive EI, 70 eV

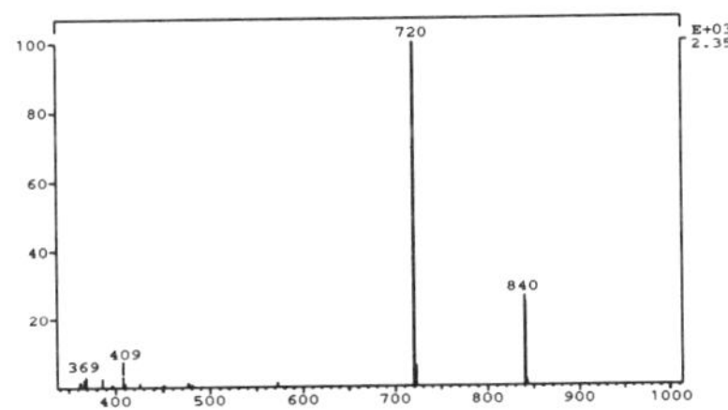

Fig.2: positive EI, 15 eV

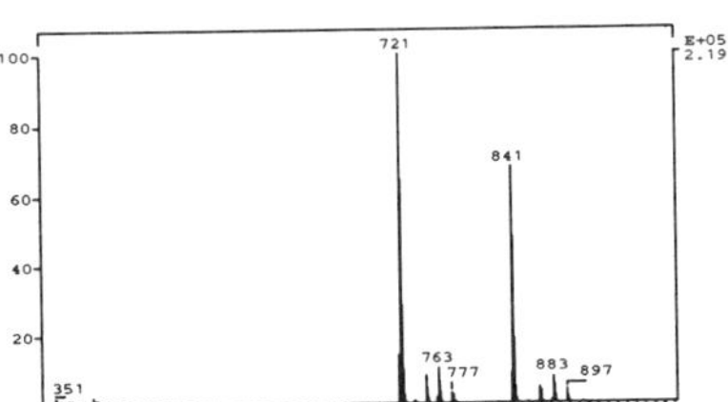

Fig.3: positive CI, methane

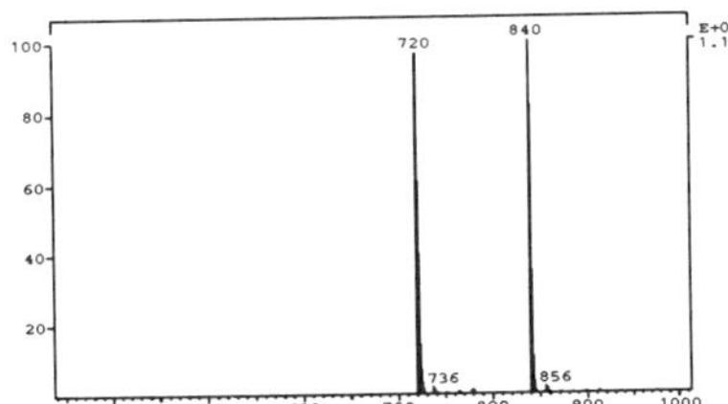

Fig.4: negative CI, methane

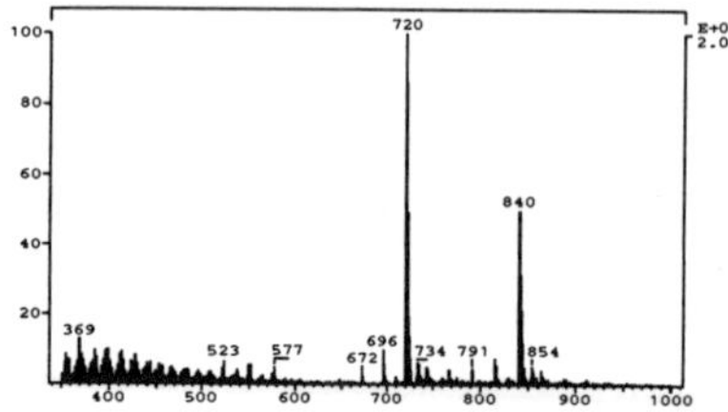

Fig.5: positive FAB, m-NBA

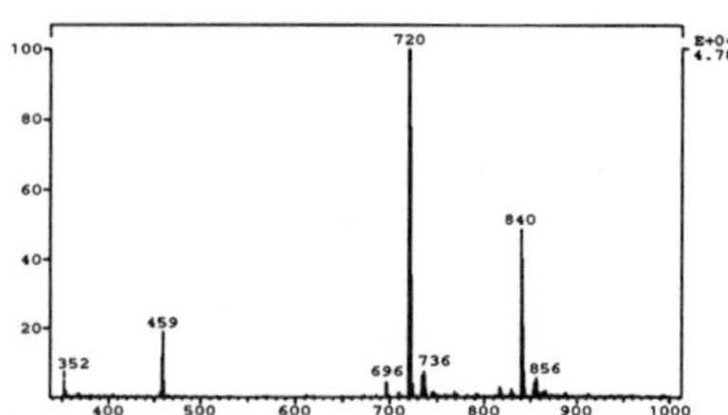

Fig.6: negative FAB, m-NBA

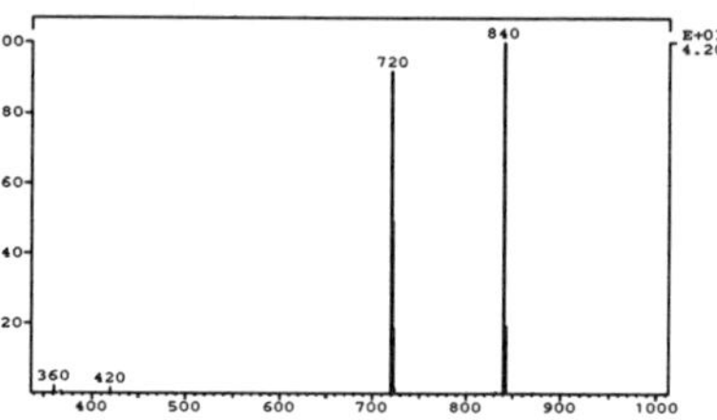

Fig.7: positive FD, 6 kV

In case of the positive-ion EI (PEI) with 70 eV we found high fragmentation and multiple ionization as well as the ratio of C_{60} : C_{70} of about 100 : 20. With a reduced electron energy of 15 eV the fragmentation and multiple ionization disappeared, but the sensitivity decreased strongly (see also Table 1).

Table 1:   Survey of the fullerene mass spectra

ionization	amount	ion current	ratio C_{60} : C_{70} (± 5)	remarks *
PEI 70 eV	1 µl	$3.41 \cdot 10^5$	100 : 20	f, m
PEI 15 eV	10 µl	$2.35 \cdot 10^3$	100 : 30	
PCI	1 µl	$2.19 \cdot 10^5$	100 : 70	a
NCI	**1 µl**	**$1.14 \cdot 10^8$**	**97 : 100**	**a**
PFAB	-	$2.02 \cdot 10^4$	100 : 50	f, a
PFAB	-	$4.78 \cdot 10^4$	100 : 50	f, a
PFD	2 µl	$4.20 \cdot 10^3$	90 : 100	m

* f - fragmentation
   m - multiple ionization
   a - adducts

Positive-ion CI (PCI), positive- and negative-ion FAB (PFAB, NFAB) spectra give the sample composition in a wrong manner. The adduct formation with the CI-gas (M+14, M+16, etc.) or the embedding matrix in the case of FAB as well as additional matrix signals below m/z = 600 can make the analysis of the fullerenes more difficult and may cause wrong interpretation of the mass spectra. For example the fullerene-oxygen adduct peak (M+15.999) can only distinguished from the CI-gas adduct peak (M+16.032) by high resolution mass spectrometry.

The negative-ion CI (NCI) and positive-ion FD (PFD) spectra reflect the real ratio of C_{60} and C_{70} in the best way. All other ionization methods give a lower C_{70} content in the mixture.

It was found that NCI is the best ionization method for fullerenes in comparison to PCI, PEI, PFD and FAB. The sensitivity of the detection by NCI was about a thousand times better than of all other ionization methods. The NCI can be used to detect traces of fullerenes (endohedral or higher fullerenes), too. A disadvantage of the CI is the formation of adducts with the ionization gas. Furthermore high sample introduction rates may lead to saturation effects. In this case quantitative measurements are impossible.

4 Conclusions

Quantitative measurements of fullerenes by mass spectrometry require the determination of the response factors in dependence on the kind of ionization. By the different ionization techniques fragments and adducts of fullerenes can be formed and a misinterpretation of the spectra can occur.

Ionization by negative-ion CI is the best method for the detection of fullerenes in the most cases. The advantages are high sensitivity and a good reflection of the sample composition.

5 Acknowledgments

This work was supported by the Sächsisches Ministerium für Wissenschaft und Kunst (SMWK).

6 References

1. S. W. McElvany, J. H. Callahan, *J. Phys. Chem.* **95** (1991) 6186
2. H. Budzikiewicz, *Massenspektrometrie - Eine Einführung*, 3. Aufl., VCH, Weinheim, 1992
3. E. Schröder, *Massenspektrometrie - Begriffe und Definitionen*, (Heidelberger Taschenbücher, Band **260**), Springer Verlag, Berlin, 1991
4. W. Krätschmer, L. D. Lamb, K. Fostiropoulos, and D. R. Huffman, *Nature* **347** (1990) 354
5. L. Dunsch, F. Ziegs, J. Fröhner, U. Kirbach, K. Klostermann, A. Bartl, and U. Feist in: Springer Series in *Solid-State Sciences* **117**, *Electronic Properties of Fullerenes*, Eds.: H. Kuzmany, J. Fink, M. Mehring, and S. Roth, Springer-Verlag, Berlin, 1993, p. 39

FULLERENES PRODUCTION BY LASER ABLATION

M. FIEBER-ERDMANN, R. STÜBNER, A. DING

Optisches Institut, Technische Universität Berlin,
Str. des 17. Juni 135, D-10623 Berlin, Germany

and

T. BECK, H.-G. EBERLE

Festkörper-Laser-Institut, Str. des 17. Juni 135, D-10623 Berlin, Germany

ABSTRACT

An apparatus was designed to investigate the fullerene production by laser ablation. Graphite blocks kept at a diluted rare gas atmosphere were irradiated by a pulsed high power Nd:YAG laser (λ = 1064 nm, pulse energy up to 45 J). The produced soot was characterized using High Performance Liquid Chromatography (HPLC) and time-of-flight mass spectrometry.

1. Introduction

The first observation of carbon molecules [1,2] and the following discovery [3] of stable fullerenes and their production technique [4] started an interesting field of research. The original technique, laser evaporation from graphite rods, did not yield any quantitative amounts of these molecules. Then Meijer and Bethune [5] found, that laser produced fullerenes could be trapped, stored and accumulated on a surface. After that Millon et al. [6] reported a method to produce larger quantities of fullerenes, which they were able to analyse by a chromatographic method. Until now fullerene production by laser ablation techniques was done using Q-switch lasers.

We report the application of lasers to produce fullerenes in macroscopic quantities, which are detectable in High Performance Liquid Chromatography (HPLC). We will discuss the production efficiency of fullerenes as a function of the generation conditions (gas pressure, laser energy, laser power density).

2. Experimental

The apparatus designed to produce fullerenes by laser ablation is schematically shown in Fig. 1. The pure graphite was evaporated using pulses of 1064 nm light from a Nd:YAG laser. Typical pulse energy was about 4 J, the repetition rate was 40 Hz. A turbo molecular pumped chamber provided a background free from hydrocarbons. The evaporation was carried out in an argon atmosphere of 300 mbar. This apparatus is able to produce 500 mg graphite soot per hour. The soot was deposited on the wall of the surrounding quartz tube. It was removed mechanically from the tube and then analysed by HPLC [7] and time-of-flight mass spectrometry. For the time-of-flight mass spectrometry analysis two different techniques were employed. In one type of experiment the soot was heated up to 450°C, the evaporated molecules were ionized by electron impact [8] and then mass analysed. In another type of experiments the soot was desorbed by laser irradiation, the molecules were ionized by a second F_2 excimer laser[9] ($\lambda = 157$ nm) and analysed in a mass spectrometer.

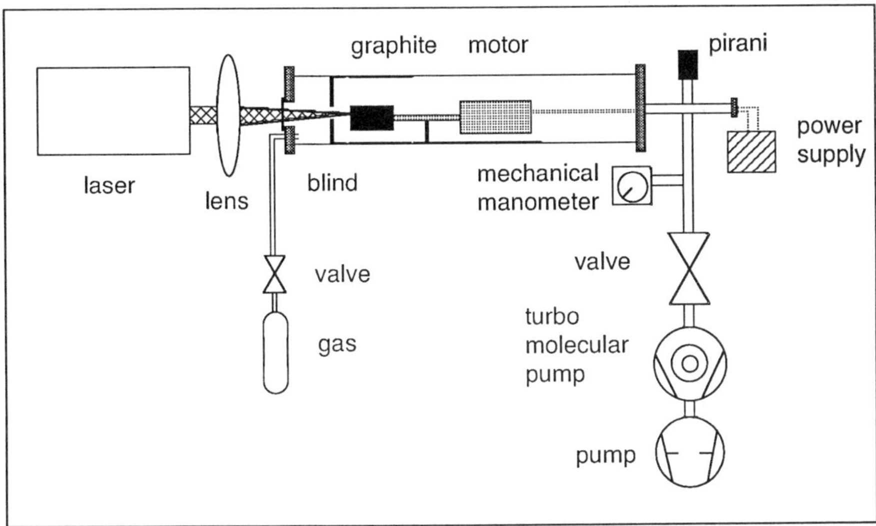

Fig. 1: The experimental set-up

3. Results and Discussion

With this experimental set-up it was possible to produce soot which contained fullerenes in macroscopic quantities. The average energy of the Nd:YAG laser determines the amount of soot, which is larger than the yield obtained by using Q-switch lasers[5,6,10]. The optimized process parameters were: *laser pulse energy* 4 J, *laser pulse width* 0.6 ms, *repetition rate* 40 Hz, *static argon pressure* 300 mbar. At these conditions the best efficiency for the C_{60} production was 2 % of the total amount of soot. The ratio of C_{70} was 70 % of that of C_{60}, which agrees with other results[5,6], see Fig. 2. In Fig. 3 a time-of-flight mass spectrum by laser ionization of the soot is shown. C_{70} is nearly as intense as C_{60} and higher fullerenes up to C_{90} are also clearly visible. In spite of larger laser energies long laser pulses (*energy* > 40 J, *pulse width* > 10 ms) are very inefficient for the production of fullerenes.

At lower argon pressure the production rate of C_{70} is higher than that of C_{60}, but the total efficiency for the fullerenes production was decreased. At an argon pressure of 150 mbar the production rate of C_{70} is 2.4 times higher than that of C_{60} (Fig. 2). Helium was also used as buffer gas, but generated negligible yields, which may be due to the rapid cooling process. From these experimental results we are able to estimate that a laser intensity larger than 1 MW / cm^2 is necessary for optimal fullerene production.

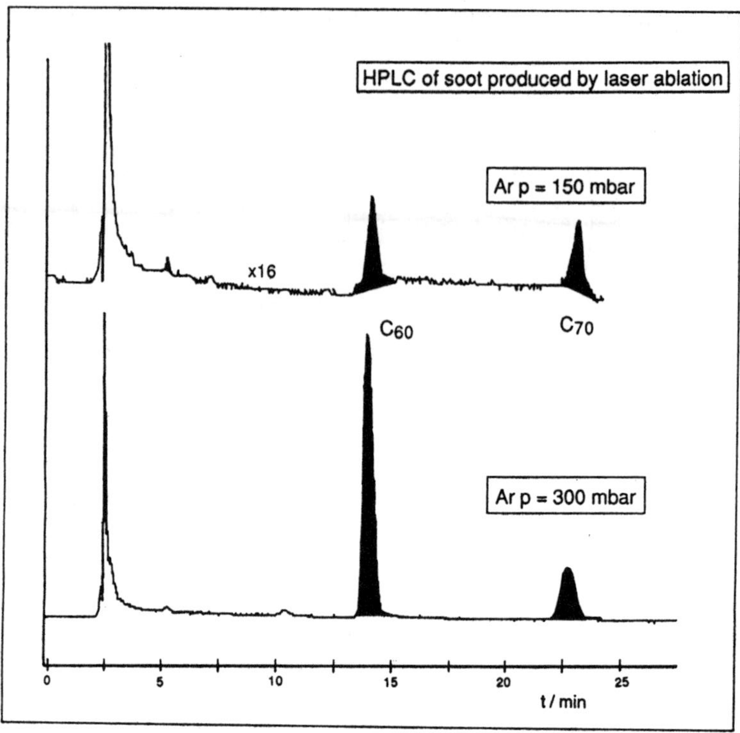

Fig. 2: The HPLC signal for C_{60} and C_{70} for two different static argon pressures (*Column*: ODS optilab, Gunther Tek, reversed phase C18. *Eluant*: 60 % CH_2Cl_2 and 40 % Methanol. *Detection*: UV 240 nm).

4. Conclusion

We have produced fullerenes in macroscopic quantities by a laser without Q-switch. The best efficiency for the C_{60} production was 2 % of the total amount of soot. Higher fullerenes were also produced with this set-up. Furthermore, we observed that an argon pressure of 300 mbar is optimal for fullerene production. Only negligible amounts of fullerenes was found when helium was used as a buffer gas.

5. Acknowledgement

The authors wish to thank Dr. D. Guldi, Dr. H. Hungerbühler, Dr. R. Fliout and Dr. T. Drewello, *Hahn-Meitner-Institut*, for the HPLC measurements, H. Delfs, and R. Henninger for the electron bombardment time-of-flight mass spectra, Dr. G. Phillipps, *Festkörper-Laser-Institut Berlin*, for help with by the laser experiments, and Prof. G. Meijer, and I. Holleman, *University of Nijmegen*, for the laser ionized time-of-flight mass spectra. Financial support by the German Ministry of Research and Technology (BMFT) is gratefully acknowledged.

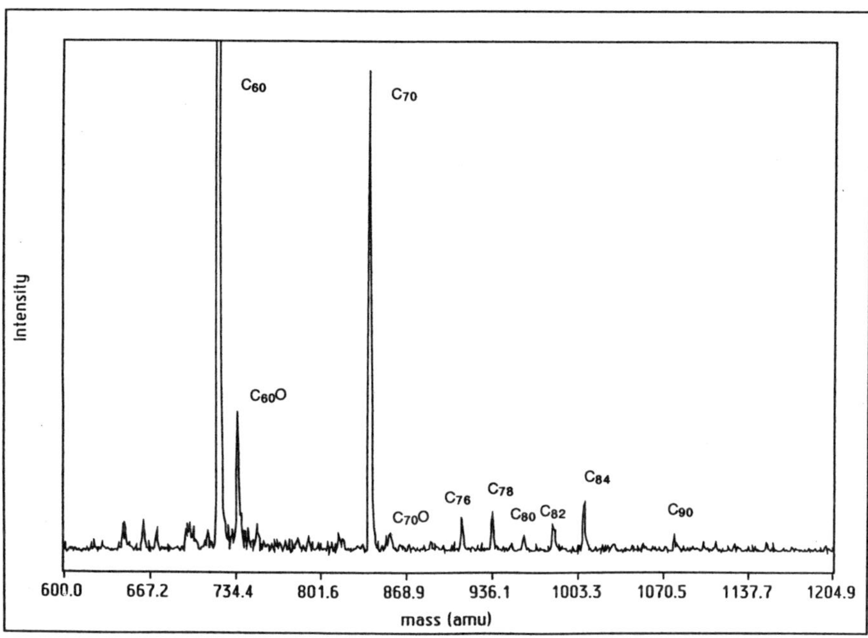

Fig. 3: The time-of-flight mass spectra by laser ionization of produced fullerenes. C_{60}, C_{70} and higher fullerenes are clearly visible.

6. References:

1. O. Hahn, F. Strassmann, J. Mattauch and H. Ewald, *Naturwiss.* **30** (1942) 541.
2. E. A. Rohlfing, D. M. Cox and A. Klador, *J. Chem. Phys.* **81** (1984) 3322.
3. H. W. Kroto, J. R. Heath, S. C. O'Brien, R. E. Curl and R. E. Smalley, *Nature* **318** (1985) 162.
4. W. Krätschmer, L. D. Lamb, K. Fostiropoulos and D. R. Huffman, *Nature* **347** (1990) 354.
5. G. Meijer and D. S. Bethune, *J. Chem. Phys.* **93(11)** (1990) 7800.
6. E. Millon, J. V. Weber, J. Theobald and J. F. Muller, *C. R. Acad. Sci. Paris* **315(II)** (1992) 947.
7. J. M. Hawkins, T. A. Lewis, S. D. Loren, A. Meyer, J. R. Heath, Y. Shibato and R. J. Saykally, *J. Org. Chem.* **55** (1991) 6251.
8. A. W. Burose, T. Dresch and A. M. G. Ding, *Z. Phys. D* **26(2)** (1993) 295.
9. G. Meijer, G. Berden, W. L. Meerts, H. E. Hunziker, M. S. de Vries and H. R. Wendt, *Chem. Phys.* **163** (1992) 209.
10. R. E. Haufler, Y. Chai, L. P. F. Chibante, J. Conceicao, C. Jin, L.-S. Wang, S. Maruyama and R. E. Smalley, *Mat. Res. Soc. Symp. Proc.* **206** (1991) 627.

GROWTH MODE AND INTERFACE STUDY OF C_{60} ON GeS(001)

K. Hevesi, G. Gensterblum, B.-Y. Han, L.M. Yu, J.-J. Pireaux,
P.A. Thiry, R. Caudano
Laboratoire Interdisciplinaire de Spectroscopie Electronique
Facultés Universitaires Notre-Dame de la Paix, B-5000 Namur, Belgium

ABSTRACT

The heteroepitaxial growth of C_{60} on GeS(001) has been studied using low energy electron diffraction, X-ray photoelectron spectroscopy and X-ray diffraction. The Van der Waals epitaxy is demonstrated by the observation of LEED patterns characteristic of the (111) face of the fcc C_{60} structure. For low coverages, the substrate spots and the $C_{60}(111)$ spots allow to specify the geometry of the epitaxy. The continuous decrease in intensity of the substrate diffraction spots with increasing coverage and the shape of the intensity curves of the C1s and Ge3d photoemission lines strongly suggest a layer by layer growth mode. This is confirmed by the observation in the X-ray diffraction spectrum of finite size oscillations on the (111) Bragg reflection peak of a thin $C_{60}(111)$ film. From a theoretical simulation of the C1s and Ge3d line intensity curves the mean free path of a C1s photoelectron in solid C_{60} is estimated to be about 17 Å. The photoemission lines behaviour also show a small difference in sticking coefficient between the first monolayer and the upper ones.

1. Introduction and experimentals

For the study of the fullerenes crystalline properties, the ordered thin films provide the best alternative to single crystals. What we will discuss here is the growth mode and the interface of C_{60} on the GeS(001) surface.

The commercial C_{60} (99.9%) was sublimed from a Knudsen cell at 400°C. The GeS(001) substrate (7x7x1 mm^3) was cleaved in UHV and held at about 250°C during the fullerene deposition. The XPS photoemission experiments were carried out in Namur on a ESCA-300 SCIENTA spectrometer (hv=1486.6 eV) equipped with a rear-view OMICRON LEED diffractometer. The X-ray diffraction was performed on the BW2 wiggler beamline at the Hamburg Synchrotron Radiation Laboratory (HASYLAB).

2. Growth study of C_{60} on GeS(001)

2.1 From the LEED patterns

The Low Energy Electron Diffraction patterns of C_{60} submonolayer coverages (fig. 1) suggest an interfacial geometry which aligns the GeS(001) surface lattice axes with the $<\bar{1}01>$ and $<1\bar{2}1>$ C_{60} axes (fig.2a) leading to the molecular arrangement shown in (fig. 2b), where C_{60} is sitting in the grooves of the GeS surface with a systematic skip of one groove to reach a nearly perfect lattice match along $<1\bar{2}1>$.

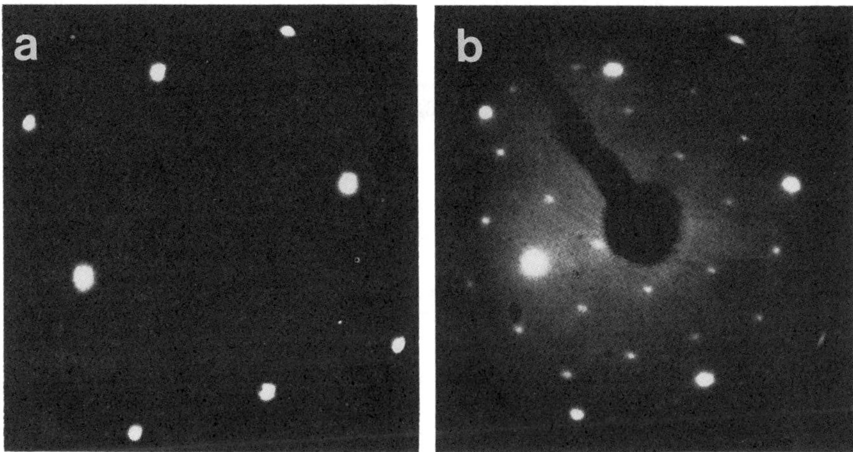

Fig. 1: Diffraction patterns of (a) the GeS(001) substrate (Ep=35.6 eV) and (b), about 0.3 monolayer of C_{60} on GeS(001) (Ep=35.1 eV).

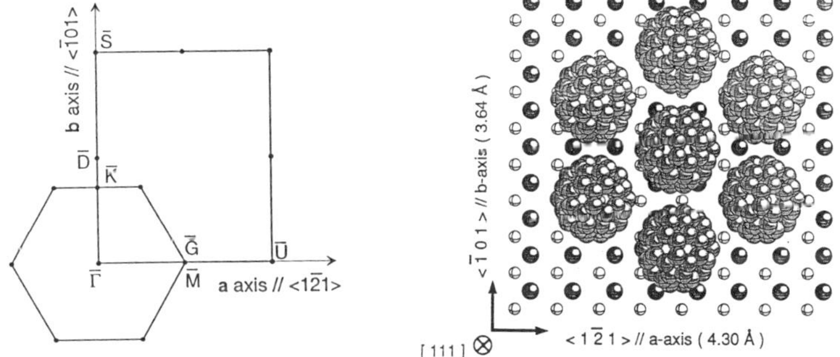

Fig. 2: a) Surface Brillouin zones of GeS(001) and C_{60}(111) corresponding to the observed LEED patterns. **b)** The structure of the C_{60}/GeS interface as deduced from LEED. The lattice parameter for the small white (grey) spheres of sulphur (germanium) along $<1\bar{2}1>$ is 4.3 Å whereas the parameter of bulk C_{60} along the same axis is 8.68 Å.

The clear LEED patterns observed on thicker (>1000Å) C_{60} films as well as the continuous intensity evolution of the spots for submonolayer coverages (fig. 1) support a layer by layer growth mode for C_{60} on GeS(001).

2.2 From X-ray diffraction

The θ-2θ scan of a 150Å-thick C_{60} film grown on GeS(001) shows two main peaks at 0.603 Å^{-1} and 0.771 Å^{-1}, corresponding respectively to the GeS(001) and the C_{60}(111) Bragg reflections(fig. 3).

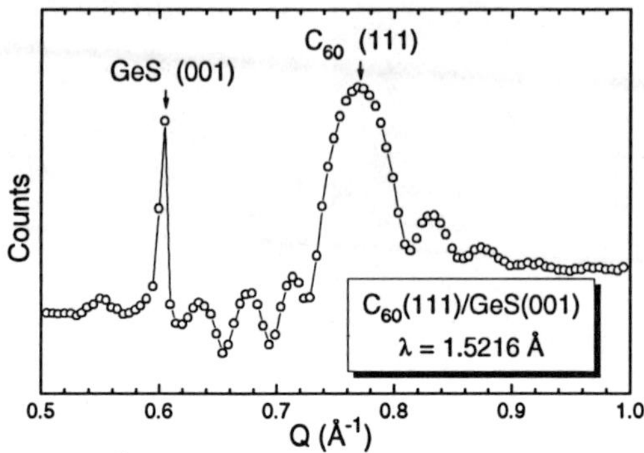

Fig. 3: θ-2θ scan of a 150Å-thick C_{60}/GeS film. Straightforward calculation yields the interlayer parameter c = 14.42±0.02 Å for the GeS(001) layers and the interlayer distance d_{111} = 8.14±0.02 Å for C_{60}.

The finite size oscillations on both sides of the C_{60}(111) peak arise from the interference between the beams diffracted by the limited number of (111) planes as confirmed by the corresponding calculated structure factor of the film. The observation of such features confirms the layer by layer growth and the high crystalline quality of the sample.

2.3 From the XPS data

During the growth of a C_{60} film on GeS(001), the intensity behaviour of the photo-emission lines from both the substrate and the adsorbate is typical of a layer by layer Franck-Van der Merwe growth mode (fig. 4a). The simulation of this growth leads to an Inelastic Mean Free Path λ=17Å for the C1s photoelectrons (E_k=1200 eV) in solid C_{60}.

3. About the C_{60} on GeS(001) interface

A drastic change of the sticking coefficient is observed between the first and the following C_{60} monolayers (fig. 4a). This change (s_1/s_2=1.4 and s_1/s_3=1.5) is indicative of a stronger interaction between the substrate and the first monolayer.

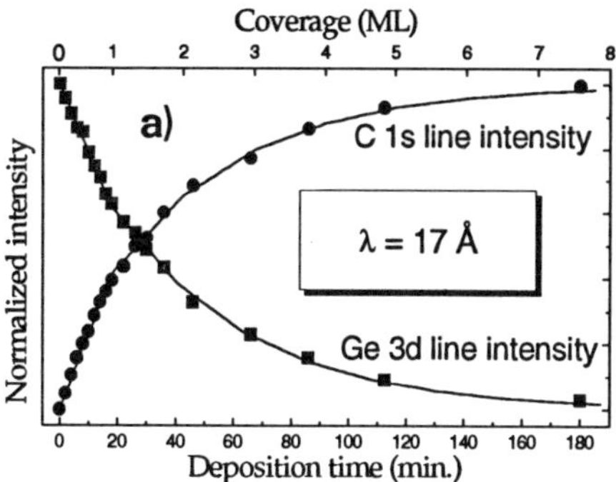

fig. 4: a) Intensity plot of the C1s and the Ge3d lines during the growth of a C_{60} film on GeS showing an F-M type of growth. The best coincidence of the simulations with the C1s data is obtained for $\lambda=17$Å. From the simulation, one can estimate the sticking coefficient ratios for the first, second and third monolayers: $s_1/s_2=1.4$ and $s_1/s_3=1.5$.

4. Conclusions

We have thus studied the Van der Waals epitaxy of C_{60} on GeS(001) which leads to the layer by layer growth of an outstanding crystalline C_{60} film. Our growth simulation yielded an IMFP of 17 Å in C_{60} for electrons of $E_k = 1200$ eV. The interface study have shown stronger interaction for C_{60} than in the bulk fullerene. An hypothetical electron transfer interaction would lead to 0.02 electron transferred to each C_{60} molecule. This does not explain the interface interaction which would rather involve induced dipole effects of the GeS ionic substrate on the C_{60} molecules. More details about this work will be found in an article submitted to Phys. Rev. B in April 1994 by G. Gensterblum, K. Hevesi et al.

5. Acknowledgments

IRSIA (*Institut pour l'encouragement de la Recherche Scientifique dans l'Industrie et l'Agriculture*) is gratefully acknowledged for pre-doctoral support to K. Hevesi.

For their active role in the X-ray diffraction experiments, we would like to thank G. Bendele, T. Buslaps, R.L. Johnson (II. Insitut für Experimentalphysik, Universität Hamburg, Germany), M. Foss (Institute of Physics and Astronomy, Aarhus University, Denmark), R. Feidenhans'l (Department of Solid State Physics, Riso National Laboratory, Denmark) and G. LeLay (Centre de Recherche sur les Mécanismes de la croissance Cristalline, Campus de Luminy, Marseille, France).

This work is supported by the Belgian National Program of Interuniversity Attraction Poles sponsored by the Prime Minister office (Science Policy Programming) and the Wallonia Region.

AN ANALYSIS OF ANNEALED C_{60} FILMS

L. AKSELROD, H.J. BYRNE, and S. ROTH
*Max-Planck-Institut für Festkörperforschung, Heisenbergstr1,
70569-Stuttgart, Germany.*

Abstract

C_{60} films are annealed under vacuum at 200°C. UV/VIS absorption studies on non-oxygenated annealed films show a decrease in the intensities of allowed transitions and a significant red-shift. IR analyses reveal no chemical changes. X-ray diffraction shows the presence of mixed phases identified as cubic and hexagonal, a closer packed structure, in the annealed nonoxygenated films. Time of illumination dependent, low input power, room temperature Raman studies of the pentagonal pinch mode of the non-oxygenated annealed films reveal the presence of two spectral components. The first, identified as the $1468cm^{-1}$ Raman line is seen to phototransform. The second, identified by the $1464cm^{-1}$ Raman line remains stable with prolonged low intensity illumination, exhibiting no photochemical changes. The results suggest that the effect of film annealing is to cause structural changes resulting in a closer packing of the lattice.

1. Introduction

Distinct samples of solid state C_{60} are observed to show a somewhat variable optical response[1,2,3], which brings up the question as to whether these different processes are linked to structural variability. The effect of crystallinity on the photochemical and photophysical properties manifested by the material have not been thoroughly studied. One method of altering sample crystallinity may lie in post sublimation annealing under inert conditions. For example, annealing of C_{60} films has been shown to result in a large increase in the photoconductive response[4]. Here, we report on the changes in film crystallinity as a function of annealing by observing changes in X-ray diffraction and the optical properties of the material.

2. Experimental, Results and Discussion

C_{60} powder and films were prepared as previously discussed[5]. Films were annealed by heating under vacuum (10^{-5} mbar) for given time increments at 200°C after sublimation with no exposure to oxygen during annealing or measurement unless specified. All Raman ($50W/cm^2$ prolonged low level irradiation with 514.5 Ar+ laser) and uv/vis measurements were made at room temperature under argon atmosphere.

Solubility studies of annealed C_{60} films show that they are easily dissolved in toluene, indicating that the chemical character of the material is preserved. Figure 1 shows a comparison of the ultra violet/visible absorption spectra taken on a film unexposed to oxygen taken with increasing annealing time. The unheated film shows the well accepted C_{60} spectrum. With increasing annealing, a red shifting of allowed electronic transitions and of the feature observed in the solid state is observed. A strong intensity decrease of the allowed electronic transitions is also evident. The former appears similar to effects seen in luminescence spectra of organic molecular crystals under high pressure, where the individual molecules are forced closer together and the latter to hypochromic effects seen in e.g. highly stacked DNA, where the alignment of individual dipole moments leads to a screening effect and the decrease in the $\pi-\pi^*$ transition, where the absorption of the system as a whole is less than the sum of individual absorptions of the monomers[6]. The effect is hindered in oxygenated films(fig. 2) subject to annealing, where some loss of features distinct from that in fig.1 is observed only after several days heating, suggesting the processes in heated oxygenated and nonoxygenated films are different in nature. This observation is supported by infrared studies on annealed and unannealed films with no and significant exposure to oxygen (fig. 3). No spectral changes between annealed and unannealed film are seen, indicating no apparent thermo-chemistry or film degradation. In contrast, oxygenated annealed films show a broad feature ~1700cm^{-1}, suggesting a thermal oxidation process, consistent with observation of Vassalo et al.[7]. Both uv/vis and IR measurements indicate that the effects of annealing are not chemical in nature, but relate to changes in structure.

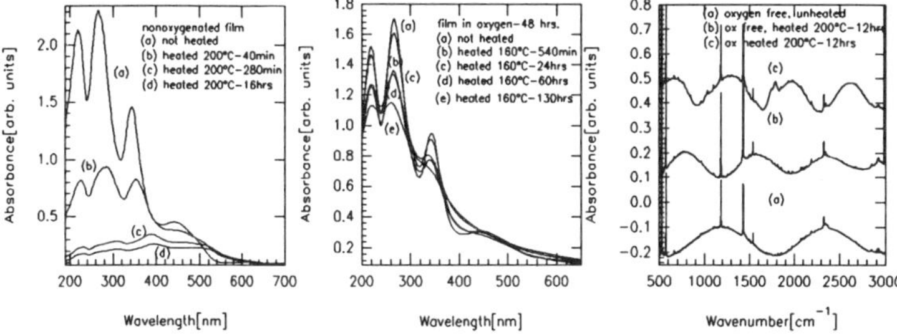

Fig.1: UV/VIS spectra of a nonoxygenated film at different annealing times. Fig.2: UV/VIS spectra of an oxygenated film at different annealing times. Fig.3: Infrared spectra of (a)unannealed nonoxygenated, (b)annealed nonoxygenated and (c)annealed oxygenated C_{60} films.

X-ray diffraction studies on unannealed films(table1) reveal cubic packing. In contrast, x-ray diffraction patterns of annealed films show a sharpening consistent with increased crystallinity and a splitting of peaks, indicative of mixed phases, namely that of the cubic phase and new features consistent with hexagonal packing similar to that identified by Zhennan et al.[8]. Hexagonal packing is associated with a more closely packed lattice, which is in agreement with the earlier observations of the changes in the uv/vis spectra following annealing in a nonoxygenated environment. Resonant Raman scattering can be employed to analyse the effects of annealing on the photo-driven processes, specifically the previously observed photochemistry in unannealed films and some crystals[5,9] in the material. The nonoxygenated sample is illuminated at low laser intensities($50W/cm^2$) and observed as a function of time(fig.4-6). Figure 4 shows the Raman spectrum after 30 sec of illumination. The $1468cm^{-1}$ line, observed in unannealed samples and a new feature at $1464cm^{-1}$ are seen to strongly dominate the spectrum. Figs. 5 and 6 and a plot of Raman intensities as a function of time for the 1468, 1458, and $1464cm^{-1}$ lines(fig.7) show the $1468cm^{-1}$ Raman

d	2θ	intensity	width	d	2θ	intensity	width
8.7299	10.124	21.3	.286	2.9034	30.77	25.3	.286
8.2939	10.658	22.0	.286	2.7370	32.692	24.5	.286
5.0603	17.511	52.2	.286	2.5408	35.296	20.7	.286
4.2715	20.778	79.5	.286	2.4941	35.98	22.0	.286
4.1616	21.333	92.6	.286	2.2782	39.524	22.0	.286
3.8463	23.105	49.6	.286	2.2399	40.228	21.3	.286
3.7573	23.660	57.0	.286	2.1567	41.851	24.5	.286
				2.0701	43.691	17.4	.600

Table 1: X-ray diffraction data of annealed film showing mixed phases.

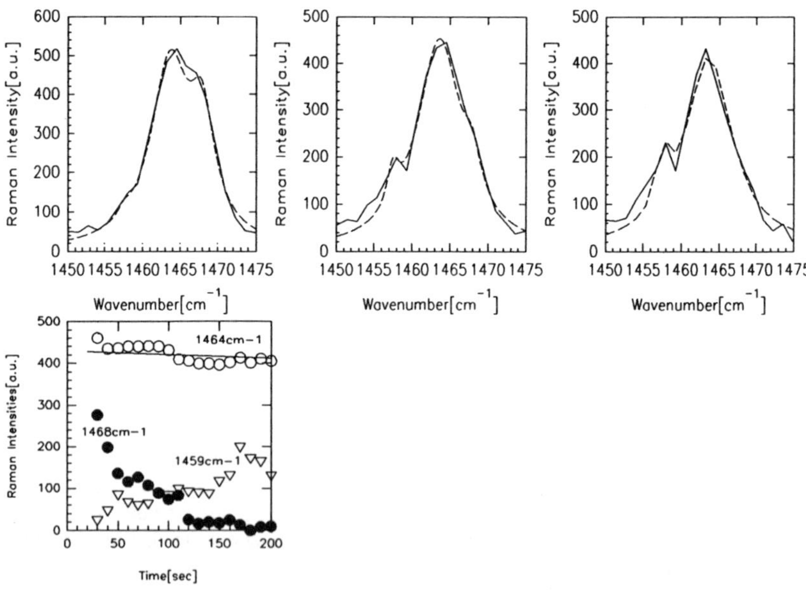

Fig.4-6(left to right at top): Raman spectra of an annealed nonoxygenated film after 30sec(fig.4), 120sec(fig.5) and 200sec(fig.6) illumination. Dashed lines are Lorentzian fits of the spectra. Fig.7(bottom left): Raman intensity as a function of illumination time for the $1468cm^{-1}$, $1459cm^{-1}$ and $1464cm^{-1}$ lines.

line decays with time and the $1458cm^{-1}$ increases with illumination time to a stable value, consistent with the phototransformation of the C_{60} phase associated with the $1468cm^{-1}$ mode. Contrary, the $1464cm^{-1}$ Raman line seen to be strongly present only in the annealed films at lower laser intensities does not decay and is seen to be stable with prolonged illumination. Annealed films thus seem to posses two phases, namely one that polymerises and one in which polymerisation is blocked.

3. Conclusions

UV/VIS and IR spectra show that annealing of oxygen-free films causes structural, not chemical changes. X-ray data reveals the presence of mixed phases in annealed films, namely the cubic seen in unannealed films and the hexagonal phase associated with a closer packing. In resonant Raman, the presence of two spectral features is observed, which appears to correspond to the two phases seen in X-ray diffraction. A new feature is observed which is not subject to polymerisation under prolonged illumination. The data suggests annealing creates a distinctly closer packed lattice of intact C_{60} balls.

References

1. T.W. Ebbesen, Y. Mochizuke, K. Tanagaki, H. Hiura, to be published in *Europhys. Lett.* (1994).
2. L. Akselrod, H.J. Byrne, J. Callaghan, A. Mittelbach and S. Roth, *Proceedings of the "International Winterschool on the Electronic Properties of Novel Materials: Electronic Properties of Fullerenes"*, Kirchberg, Austria, March (1993), Springer Series in Solid-State Sciences, Vol. 117.
3. J. Feldman. W. Guss, E.O. Göbel, C. Taliani, H. Mohn, P. Häussler, and H.U. Ter Meer, to be published in the *Proceedings of the "International Winterschool on the Electronic Properties of Novel Materials: Progress in Fullerene Research"*, Kirchberg, Austria, March (1994).
4. M. Kaiser, W.K. Maser, H.J. Byrne, A, Mittelbach, S. Roth, *Solid State Commun.*, **57**, 81 (1993).
5. L. Akselrod, H.J. Byrne, C. Thompsen, A. Mittelbach and S. Roth, *Chem. Phys. Lett.*, **212**, 384 (1993).
6. C.R. Cantor, P.R. Schimmel, *Biophysical Chemistry Part II, Techniques for the Study of Biological Structure and Function*, W.H. Freeman and Company, 385.
7. A.M. Vassallo, L.S.K. Pang, P.A. Cole-Clark, and M.A. Wilson, *J. Am. Chem. Soc.*, **113**, 7820 (1991).
8. G. Zhennan, Q. Jiuxin, Z. Xihuang, W. Yongqing, Z. Xing, F. Sungi, and G. Zizhao, *J. of Phys. Chem.*, <u>95</u>, 9615, (1991).
9. M. Rao, Ping Zhou, Kai-An Wang, G.T. Hager, J.M. Holden, Ying Wang, W.-T. Lee, Xiang-Xin Bi, P.C. Eklund, D.S. Cornett, M.A. Duncan, I.J. Amster, *Science*, **259**, 955 (1993).

INFLUENCE OF HAFNIUM CARBIDE ON FULLERENE FORMATION

G. Aced, T. Almeida Murphy, J. Erxmeyer, B. Mertesacker,
H. J. Möckel, D. Nagengast, B. Pietzack, C. Rau, A. Weidinger
Hahn-Meitner-Institut Berlin GmbH, Glienicker Str. 100,
D - 14109 Berlin, Germany

ABSTRACT

Fullerenes were prepared by co-vaporizing carbon and hafnium carbide in an arc discharge. A new extraction apparatus was developed in which the soot can be extracted in a very short time. The toluene extract was analysed by HPLC. The yield of fullerenes, eluted after C_{70}, i.e. in the retention range of higher fullerenes, was exceptionally high (60 area %). The composition of the extract was determined by FAB-mass spectrometry.

1 Introduction

The increasing success in the production and separation of metallofullerenes such as La@C_{82}[1], Y@C_{82}[2], and the observation of Hf@C_{28}[3] in a mass spectrometric experiment, prompted us to produce hafnium fullerene complexes. For this purpose we used spectrographic graphite rods which contained hafnium carbide. The yield of higher fullerenes (C_n, n>70) was significantly increased with this procedure.

2 Experimental

2.1 Preparation of Fullerene Soot

The soot was produced in a carbon arc discharge in a fullerene reactor according to the Krätschmer/Huffmann method[4]. Helium with a pressure of 200 mbar was used as buffer gas. DC current for the arc discharge was 140A. Electrodes were made out of spectrographic-grade graphite and filled with pellets containing hafnium carbide. The fullerene reactor was placed in a glove box. The resulting soot was collected under nitrogen atmosphere in order to avoid oxidation.

2.2 High Pressure/Temperature Extraction

Extraction is the first step in the isolation and purification of fullerenes from carbon soot. A new extraction apparatus was developed in which carbon soot can be extracted in a very short time. In this setup, the carbon soot is mixed with quartz and packed into an empty stainless steel extraction column. The solvent is pumped through the high pressure extractor (25 MPa). The advantage of this high pressure extraction method are time savings of up to 95% as compared to Soxhlet extraction[5] with toluene.
Fullerenes can be extracted exhaustilvely with toluene from carbon soot in about one hour with a yield of ~10%.

R. S. Ruoff et al[6] investigated the temperature dependence of solubility of C_{60} in toluene, hexane and CS_2 and found the solubility maximum at ambient. We compared high pressure extractions (25 MPa) at 250°C and at room temperature and also observed the solubility maximum near room temperature. The following experiments were therefore carried out at room temperature.

3   Results

3.1 High Performance Liquid Chromatography/Mass Spectrometry

Figure 1a shows a HPLC chromatogram of fullerene extract which was produced by the standard method, i.e. with Hf-free electrodes. In this case the usual mixture of about 78 area-% C_{60}, 19 area-% C_{70} and 2 area-% higher fullerenes was obtained.
A quite different result was obtained from the vaporization of graphite with added hafnium carbide. The yield of compounds eluting after C_{70}, i.e. in the retention range of higher fullerenes, was significantly higher (60 area-%) (figure 1b). For a better resolution the chromatographic conditions for the sample of figure 1b were changed (figure 2). The chromatographic results and HPLC conditions are summarized in table 1 for the experiments with and without hafnium carbide in the electrodes.

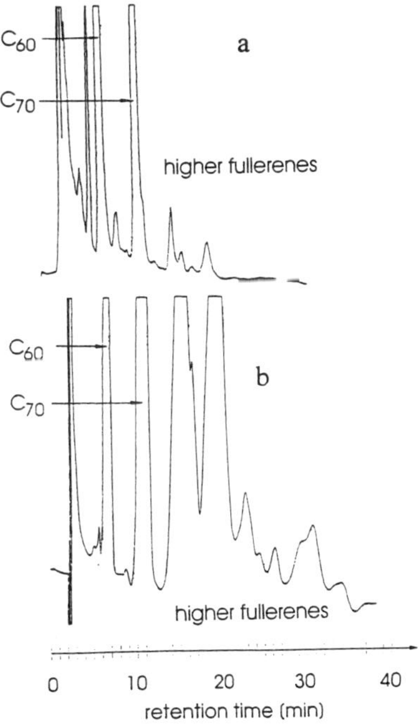

Fig. 1a, b: Comparison of fullerene extracts by HPLC which were produced by vaporization of graphite (a) and graphite with added HfC (b) (chromatographic conditions see table 1).

Table 1: Chromatographic conditions and results for HPLC-separations of soot extracts.
(Column: Nucleosil 100 C18 5 µm, 250 x 4 mm, flow: 1 mL/min)

Electrodes	Area % C_{60}	Area % C_{70}	Area % "higher fullerenes"	Eluent/ Wavelength
Graphite (fig. 1a)	78.5	18.8	1.8	Dichloromethane/ Isopropanol = 50/50 (v/v) 290 nm
Graphite/ **Hafnium carbide** (fig. 1b)	9.8	22.9	67.1	Dichloromethane/ Isopropanol = 50/50 (v/v) 290 nm
Graphite/ **Hafnium carbide** (fig. 2)	25.2	14.7	64.1	Dichloromethane/ Methanol = 60/40 (v/v) 240 nm

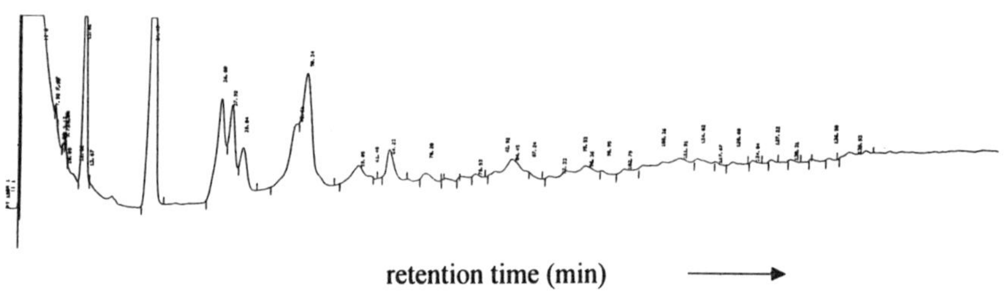

retention time (min) ⟶

Fig. 2: Chromatogram of fullerene extract resulting from the vaporization of graphite and HfC. (chromatographic conditions see table 1)

Figure 3 shows the negative FAB mass spectrum of the toluene extract, which was produced by decomposition of graphite with HfC. The important features of the MS are the peaks at m/z = 720 (C_{60}) and 840 (C_{70}), and the presence of higher fullerenes C_{76-96}. LC-MS measurements of the extract are currently in preparation.

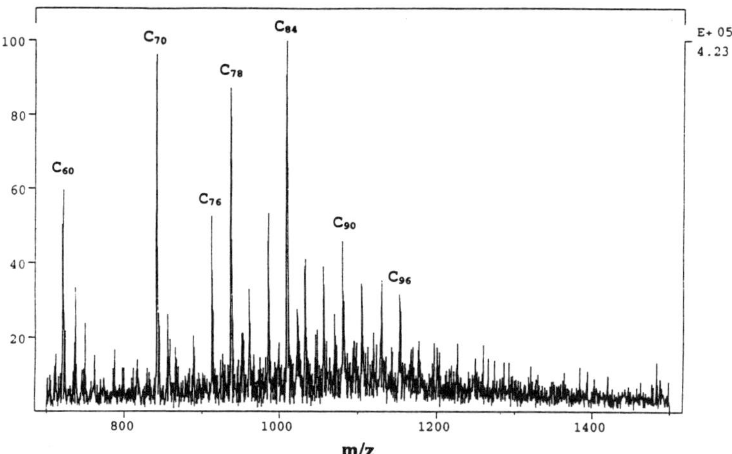

Fig. 3: Negative FAB mass spectrum of fullerene-rich sample in the mass spectral range m/z = 700-1500 (matrix: MNBA).

4 Conclusions

Unusually high yields of higher fullerenes were obtained using HfC-doped electrodes. We assume that HfC has a catalytic effect on the formation of higher fullerenes. The higher fullerenes are stable in toluene. The stability of the dried extract is currently being investigated.

Acknowledgment

The authors thank Dr. Andreas Lehmann (Bundesanstalt für Materialforschung und -prüfung, Berlin-Adlershof) for the MS measurements.

5 References

1. Y. Chai, T. Guo, C. Jin, R.E. Haufler, L.P. Chibante, J. Fure, L. Wang, J.M. Alford, R.E. Smalley, *J. Phys. Chem.* **95** (1991) 7564
K. Kikuchi, S. Suzuki, Y. Nakao, N. Nakahara, T. Wakabayashi, H. Shiromaru, K. Saito, I. Ikemoto, Y. Achiba, *Chem. Phys. Letters* **216** (1993) 177
2. J.H. Weaver, Y. Chai, G.H. Kroll, C. Jin, T.R. Ohno, R.E. Haufler, T. Guo, J.M. Alford, J. Conceicao, L.P.F. Chibante, A. Jain, G. Palmer, R.E. Smalley, *Chem. Phys. Letters* **190**, 460 (1992)
3. T. Guo, M. Diener, Y. Cai, M.J. Alford, R.E. Haufler, S.M. McClure, T. Ohno, J.H. Weaver, G.E. Scuseira, and R.E. Smalley, *Science* **257** (1992) 1661
4. W.Krätschmer, L.D. Lamb, K. Fostiropoulos, D.R. Huffman, *Nature* **347** (1990) 354
5. K.C. Khemani, M. Prato, F. Wudl, *J. Org. Chem.* **57** (1992) 3254
6. R.S. Ruoff, R. Malhotra, D.L. Huestis, D.S. Tse, D.C. Lorents, *Nature* **362** (1993) 140

THERMAL PROPERTIES, STRUCTURES AND PREPARATION OF IRON CONTAINING FULLERIDES

P.Byszewski[1,2], R.Diduszko[1], E.Kowalska[1], Z. Kucharski[3]
1. Institute of Vacuum Technology, ul.Długa 44/50, 00-241 Warsaw;
2. Institute of Physics, Polish Academy of Sciences;
al.Lotników 32/46, 02-668 Warsaw, Poland;
3. Institute of Atomic Energy, Świerk, Warsaw, Poland;

ABSTRACT

Iron in solid fullerides may bind fullerene molecules by a ferrocene like bonds, at the composition $C_{60}Fe_2$ the solid crystallizes in the monoclinic structure. This state is stable at low temperatures, annealing breaks irreversibly these bonds. Then the lattice undergoes the structural transformation to fcc cubic and iron ionization state changes from +2 to +3, it seems that after this process iron ions are bound to single fullerenes.

1. Introduction

Transition metals as dopants of fullerides are of special interest because they form several coordination compounds with cyclic hydrocarbons. In case of fullerenes the pentagons or hexagons might be expected to play the role of the hydrocarbon. The trivalent intercalants on the other hand, were considered as multi electron donors in possible C_{60} based superconductors[1] in the fcc-derived structures. The ability of these metals of entering the coordination compounds and filling C_{60} LUMO open the class of conducting material in other crystalline structure. The possibility attracted attention, and metaloorganic compounds based on fullerenes were searched for and obtained e.g. $C_{60}Pd_6$ [2] or $C_{60}Fe_2$ [3].

One of the most stable metal-hydrocarbon complexes is formed by iron, which binds two cyclopentadien rings C_5H_5 (Cp) in the ferrocene $FeCp_2$ molecule, and it may be expected, that bonds $Fe-C_{60}$ might also be the strongest. It was shown in [4] that a reaction between $FeCl_2$ and $C_{60}Na_x$ intercalate results in a mixture of $C_{60}Fe_y$ and NaCl compounds.

The other method for obtaining $C_{60}Fe$ compounds could consist in substitution of Cp ligands in ferrocene molecules by fullerenes to obtain $C_{60}FeC_{60}$ complexes[3]. The reactions between ferrocene and fullerenes were performed at various experimental conditions and the products were investigated by means of x-ray diffraction and Moessbauer spectroscopy [5]. The crystalline powder obtained in some of the reactions lead us to the suggestion that in fact there exists the thermally unstable phase $C_{60}Fe_2$ of monoclinic lattice, which at elevated temperatures undergoes transformation to a cubic one.

2. Thermal stability and structure of iron containing fullerides

In order to determine the conditions required for the ligands exchange reaction in ferrocene, experiments were started with known $C_{60}(FeCp_2)_2$ compound [6]. It was prepared by cocystallization of fullerenes and ferrocene from toluene solution. Then ferrocene molecules were built in between the hexagonal (111) planes of pure fulleride. This host-guest compound was found to be stable at atmospheric pressure to 120°C when it started to decompose, fullerenes reconstructed the fcc lattice and ferrocene sublimed. The decomposition process of the $C_{60}(FeCp_2)_2$ crystallites was analyzed by the DSC (Derivative Scanning Calorimetry) method (Fig.1) and x-ray diffraction. It was observed in other experiments, that the decomposition temperature rose with pressure and the $C_{60}(FeCp_2)_2$ crystallites could be prepared by thermal diffusion of ferrocene into fullerides.

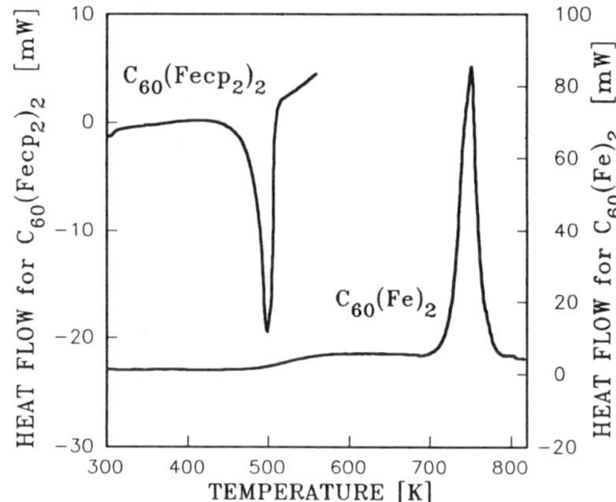

Fig.1. The decomposition of $C_{60}(FeCp_2)_2$ compound and transition of $C_{60}Fe_2$ from the monoclinic phase to the cubic phase observed by DSC.

2.1. Preparation of $C_{60}Fe_2$ samples

Those experiments indicated that bonds within ferrocene molecules are much stronger than C_{60}-$FeCp_2$ complexes interaction and the ligands substitution reaction required temperatures above the decomposition of $C_{60}(FeCp_2)_2$. Such conditions might be reached if the starting compound or a mixture of ferrocene and appropriate amount of fullerenes was annealed in sealed ampoules above decomposition of $C_{60}(FeCp_2)_2$ but below decomposition of ferrocene at normal pressure. Experimentally determined temperature ranged from 250° to 400°C with the best results at 370°C. Since the processes were performed in sealed ampoules the pressure during reactions was determined by the reagents and products partial pressures. It was observed that when the ampoules were placed in the oven in a temperature gradient, then at the cold end of the ampoules condensed the hydrocarbon complexes what considerably increased efficiency of the reaction. Our reaction vessel was therefore further complicated in order to remove the volatile reaction products during reaction from the reaction region.

The chemical reaction employed for the samples preparation might lead to fullerenes "contamination" by hydrocarbons, especially Cp rings might react with C_{60} as described by Wudl [7] for the case of C_5H_5. The crystalline powder prepared in this process prior to other investigation was annealed in vacuum to remove nonreacted ferrocene or hydrocarbons, then it contained only crystallites of C_{60} and crystallites of the new lattice structure and other thermal properties than either reagents or $C_{60}(FeCp_2)_2$.

2.2. The structure of the reaction product

The powder x-ray diffraction of the new crystalline form is shown in the Fig.2. It is compared with diffraction of C_{60} and of $C_{60}(FeCp_2)_2$. To interpret the x-ray diffraction of Fig.2 we first verified whether the structure was indeed based on fullerenes.

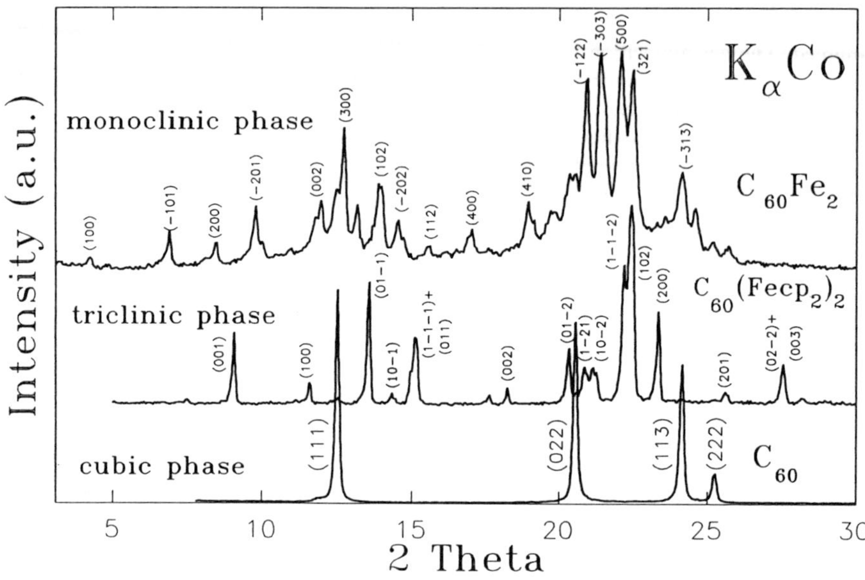

Fig.2. The comparison of the x-ray powder diffraction pattern of the product of the ligands exchange reaction with pure C_{60} in fcc and $C_{60}(FeCp_2)_2$ in the triclinic structure.

As an obvious candidate for construction of the new crystalline form, we searched for possible arrangement of $C_{60}FeC_{60}$ clusters in the three dimensional lattice. It proved possible if each of the fullerenes was bound to only four of its nearest neighbors by iron ions which occupied only some of the C_{60} planes. The distance between iron bound fullerenes had to be increased by 0.065 nm as compared to pure C_{60} and $Fe-C_{60}-Fe$ bonds angle was assumed to be 72.8°. The distances correspond well to Cp-Cp distance in ferrocene but the angle is larger than between pentagons in fullerenes. The three dimensional lattice consists of stacks of the two dimensional $C_{60}Fe_2$ networks shifted in [100] direction to form the monoclinic lattice with the following unit cell dimensions a=2.43 nm, b=1.26 nm, c=1.71 nm and β=96.9°. This choice allowed to index all of the observed diffraction peaks.

In order to determine precisely the transition temperature of the $C_{60}Fe_2$ compound from the monoclinic to cubic phase, DSC method was employed. The irreversible transition with the energy release of $\Delta E=606$ kJ/mole took place at 750 K (Fig.1). It means that the monoclinic structure with the ferrocene like bonds is unstable against transition to the cubic one where cations are randomly distributed in the lattice.

These samples were investigated by Moessbauer spectroscopy [5] which proved that cations, in the monoclinic phase were in the state Fe^{+2} as in the ferrocene molecules with similar isomer shift and quadruple splitting, and changed to Fe^{+3} as a result of ferrocene-like bonds breaking during annealing. The cations ionization change from +2 to +3 with lattice transformation suggests localization of Fe cations on C_{60} cage in the cubic structure. Such complexes were already oberved in the gas phase [8].

3. Summary

There are at least two states of iron in the fullerene based lattice. The ferrocene like bonds and lattice distortion to the monoclinic structure is unstable, though the cubic lattice with randomly distributed cations seems also difficult to prepare because of possible iron grains precipitate. The ESR experiments [9] indicate that in the $C_{60}FeC_{60}$ complexes iron do not donate free electrons to fullerene. Cobalt which also form $CoCp_2$ complexes and in its molecular orbitals scheme one electron occupies the anti bonding level could be more suitable as a donor, unfortunately it immediately forms the charge transfer complexes with fullerene instead of $C_{60}CoC_{60}$ complexes and the monoclinic structures.

Acknowledgement - This work was supported by the State Committee for Scientific Research (KBN) under grant No. P303 25703 .

4. References

1. R.C.Haddon, A.F.Hebard, M.J.Rosseinsky, D.W.Murphy, S.H.Glarum,T.T.M.Palstra, A.P.Ramirez, S.J.Duclos, R.M.Fleming, T.Siegrist, R.Tycko in *ACS Symposium Series 481, Fullerenes. Synthesis Properties and Chemistry of Large Carbon Clusters*, edt. G.S.Hammond, V.J.Kuck, American Chemical Society, Washington DC , pp. 71-89, (1992).

2. H.Nagashima, A.Nakaoka, Y.Saito, M.Kato, T.Kawanishi, K.Itoh, *J.Chem.Soc., Chem.Commun.*, pp. 377-379,(1992).

3. P.Byszewski, R.Diduszko, M.Baran, *Acta Phys.Pol.* 85,2, pp. 297-302, (1992).

4. T.Pradeep, G.U.Kulkarni, K.R.Kannan, T.N.Guru Row, C.N.R.Rao, *J.Am.Chem. Soc.*1992, pp. 114, (1992).

5. P.Byszewski, R.Diduszko, Z.Kucharski, J.Suwalski submitted to *Sol.State Comm.* for publication; P.Byszewski, Z.Kucharski, J.Suwalski this Conference.

6. J.D.Crane, P.B.Hitchcock, H.W.Kroto, R.Taylor, D.R.M.Walton, *J.Chem.Soc., Chem.Commun.*, 1992, pp. 1764-1765, (1992).

7. F.Wudl, A.Hirsch, K.C.Khemani, T.Suzuki, P.-M.Allemand, A.Koch, H.Eckert, G.Srdanov, H.M.Webb, same as [1] pp. 160-175.

8. L.M.Roth, Y.Huang, J.T.Schwedler, C.J.Cassady, D.Ben-Amotz, B.Kahr, B.S.Freiser, *J.Am.Chem.Soc.1991*, **113**, pp. 6298-6299, (1991).

9. P.Byszewski, R.Jablonski, S.Piechota, Proc. XV Conf. RAMIS'93, Poznan (Poland) 1993, published in *Molecular Physics Reports*, **5**, pp. 149-153, (1994).

IRON IN FULLERITES

P. Byszewski
Institute of Vacuum Technology, ul.Długa 44/50, 00-241 Warsaw

Z. Kucharski, J. Suwalski
Institute of Atomic Energy, 05-400, Otwock, Świerk

ABSTRACT

Mössbauer effect measurements have been used to characterise the solids fullerites containing iron. The experimental data proved that in the case of $C_{60}Fe_2$ lattice the fullerenes are bound by the ferrocene-like bonds to iron ions. The Mössbauer parameters, isomer shift and quadrupole splitting are similar to that found for pure ferrocene and $C_{60}(ferrocene)_2$. From the temperature dependence of the absorption area, the Mössbauer lattice temperature has been calculated in the high temperature limit and within the thin absorber approximation. The value obtained 103 K is markedly lower than in the case of ferrocene molecules. The quadropole coupling hyperfine parameter temperature dependence are discussed.

1. Introduction

Buckminsterfullerene (C_{60}) have been subject of much interest during last few years. It can be used for the synthesis of the novel materials. Particularly interesting seems to be modification of the fullerenes by intercalation process. Some fullerenes intercalated by alkali metals are superconducting[1] where doped by or TDAE $(C_2N_4(CH_3)_8)$ exhibit ferromagnetic properties[2].

In this communication we present results of Mössbauer investigation on fullerites-iron systems. In our previous study we proved that Mössbauer Spectroscopy (MS) is very useful techniques to characterise properties of some organic systems (e.g. conducting polymers[3]). Therefore in the present work we try to use the MS to determine the nature of iron build in fulletites.

2. Experimental

The crystalline complex $C_{60}(ferrocene)_2$ was prepared by chemical reaction described in papers[4,5] and then vacuum dried. Such materials are unstable in temperatures above approx. 100 °C, and can be transformed, by the ligands exchange reaction to $C_{60}Fe_2$[5].

The spectra were recorded in a standard transmission geometry using constant acceleration spectrometer coupled to a $^{57}Co/Rh$ source (Amersham, Buchler). For statistical reliability the spectra were recorded with several million counts per channel. A thin absorber layer was used, typically the samples contained less than 0.1 mg of the ^{57}Fe isotope per 1 cm^2. Isomer shift are given relative to metallic iron (α-Fe) at room temperature. The Mössbauer cryostat was a helium bath cryostat (MD-306, Oxford

Instruments) coupled to the temperature controller (ITC-4, Oxford Instruments). The spectra were computed with a least-squares routine using Lorentzian lines.

3. Results and Discussion

The $C_{60}(ferrocene)_2$ and $C_{60}Fe_2$ systems gives a clear MS spectra which corresponds to one Fe position with Mössbauer parameters very close to those characteristic of pure ferrocene[6]. The representative Mössbauer spectra are shown in fig.1. The characteristic feature of the room temperature ^{57}Fe spectrum in ferrocene is a doublet with isomer shift IS= 0.442 mm/s and quadrupole splitting QS= 2.38 mm/s. These

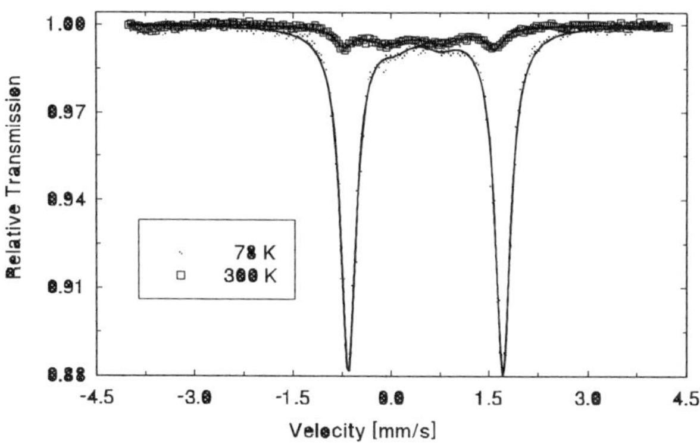

Fig.1 Mössbauer spectra of the $C_{60}Fe_2$ sample recorded at temperature 78 K and 300 K.

parameters are consistent with double oxidation of the iron and a low symmetry of the electric field gradient at the nuclei. The strong temperature dependence of the absorption amplitude points to the low Mössbauer lattice (Θ_M) temperature i.e. weak coupling of the iron to the lattice.

The spectrum of iron in the triclinic lattice of $C_{60}(ferrocene)_2$, when the ferrocene molecules are dispersed, is almost identical with spectrum of pure ferrocene. That is because in the structure of ferrocene molecule, iron is so well screened from the lattice that its influence on the spectrum is small.

However, It is not the case when iron bridges two C_{60} molecules in $C_{60}Fe_2$. The predominant feature is again similar doublet with parameters close to previously found in ferrocene molecules. The temperature dependence of the absorption intensity and quadrupole splitting are compared, for the two lattices, in Fig.2 and 3. The IS in both cases is identical within the experimental accuracy. It indicates that in both lattices iron is at the same oxidation state - Fe^{2+} with similar symmetry of the electric field at the ion site.

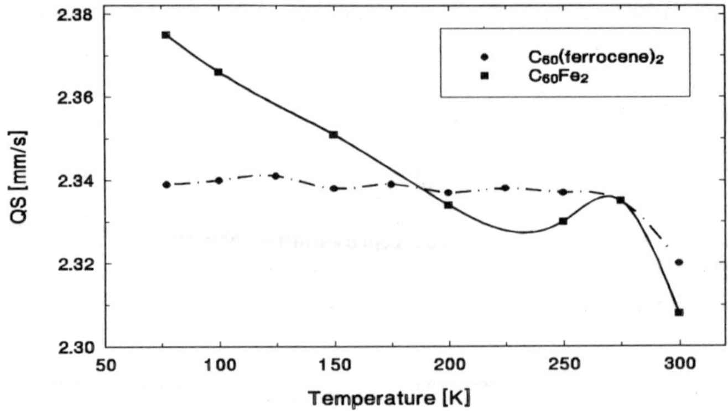

Fig.2. The quadrupole splitting (QS) temperature dependence in $C_{60}(ferrocene)_2$ and $C_{60}Fe_2$ sample. The lines show the averaged temperature dependence of QS.

There is another doublet of smaller QS and much lower intensity, at low temperatures, in $C_{60}Fe_2$ spectrum (Fig. 1). It is clearly observed only at room temperature, and corresponds only to about 1% of total iron concentration. We are inclined to ascribe it to defects arising from the imperfect technology.

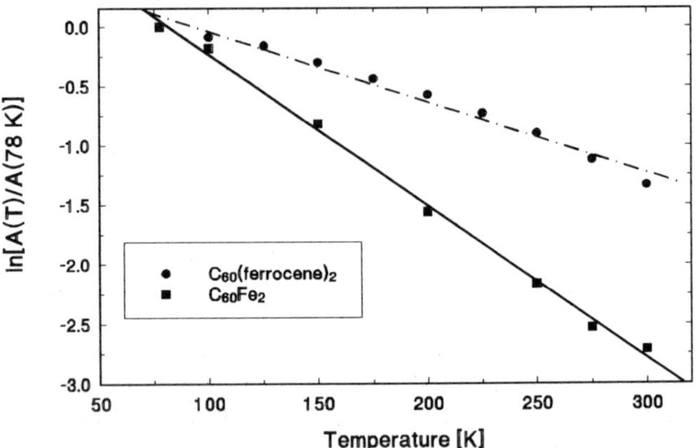

Fig.3   Temperature dependence of the ln of normalised absorption for $C_{60}(ferrocene)_2$ and $C_{60}Fe_2$ sample.

.All performed experiments strongly support the ferrocene type bonding between iron and pentagons of fullerenes in the $C_{60}Fe_2$ solid. They also show the differences in the lattice dynamic and different temperature dependence of the Mössbauer parameters: IS and QS in $C_{60}Fe_2$ and $C_{60}(ferrocene)_2$.

The Mössbauer lattice temperature can be calculated, using the high temperature limit of the Debye model, from the equitation[7]:

$$\Theta_M = \frac{E_\gamma}{c} \sqrt{-\frac{3}{M_{eff} k \, d(\ln A)/dT}}$$

where E_γ, c, k, M_{eff}, A are the energy of γ radiation, the light velocity, the Boltzman constant, the Mössbauer nucleus atomic effective mass and the normalised absorption, respectively.

For the $C_{60}Fe_2$ compound the standard fitting procedure applied to the data in Fig. 2 give $\Theta_M = 103$ K in contrast to $\Theta_M = 151$ K in $C_{60}(ferrocene)_2$ (value close to that found in pure ferrocene). The variation of the linewidth Γ versus temperature in $C_{60}(ferrocene)_2$ and in $C_{60}Fe_2$ even better represent differences in the elastic constants of the lattice. The narrow lines at room temperature in ferrocene proves that Fe occupy the same sites in all molecules while at low temperature the sites are slightly different, it reflects the ability of Fe to adjust positions of the pentadiens by the bonding forces at elevated temperatures, while in $C_{60}Fe_2$ bonding forces, being of the same order of magnitude, do not determine high angles librations and vibration of the lattice consisting of heavy molecules. In result, at high temperature, the pentagons do not face iron ions stabile. The two-dimensional network of $C_{60}Fe_2$ further contributes to the instability at iron sites. The weak bonding forces are also reflected in the thermal expansion coefficient and its anisotropy[5].

4. Conclusion

In the crystalline $C_{60}Fe_2$ compound, the position of Fe ions is determined by the interaction of individual pentagons from the nearest C_{60} molecules with Fe^{2+} ions. The imperfect alignment and orientation of the C_{60} pentagons which face the Fe^{2+} ions, the lattice vibration and the molecules librations further modulate the symmetry at the Fe^{2+} sites and the overlap of the iron atomic orbital with C_{60} π electrons contribute to the observed spectra.

The ferrocene like bonding implies that Fe $4s^2$ electrons transferred to fullcrenes are localised on the pentagons facing iron ions, it further means that in this case the molecular orbital calculated for C_{60} are inadequate to discuss the electronic energy levels, instead molecular orbital developed for ferrocene can be used for energy levels of the complex $pentagon^{-1} Fe^{2+} pentagon^{-1}$.

4. Acknowledgments

This work was partly supported by the State Committee for Scientific Research (KBN) under grant No. P302 25703 and partly by Institute of Atomic Energy under contract No. 88-5008.

5. References

1.   A.R. Kortan, N. Kopylev, S. Glarum, E.M.Gyorgy, A.P. Ramirez, R.M. Fleming,

O.Zhou, F.A. Thiel, P.L. Trevor and R.C. Haddon, *Nature* **352**, (1992) 230.

2. P.M. Allemand, K.C. Khemani, A. Koch, F.Wudl, K. Holczer, S. Donovan, G. Gruner and J.D.Thompson, *Science* **253** (1991) 301.

3. Z. Kucharski, H. Winkler, A.X. Trautwein and C. Budrowski, in *Electrinic Properties of Polymers,* ed. H. Kuzmany, M. Mehring and S. Roth, (Springer, Berlin, 1992) p. 315.

4. J.D. Crane, P.B. Hitchcock, H.W. Kroto, R.Taylor and D.R.M. Walton, *J.Chem.Soc.Chem.Comm,.* (1992) 1764.

5. P. Byszewski, R. Diduszko, E. Kowalska and Z. Kucharski, the same book.

6. T.C. Gibb, *J.Chem.Soc., Dalton Trans.,* (1976) 1237.

7. R.H. Herber and Y. Maeda, *Physica* **B 99** (1980) 352.

Implantation of ^{57}Fe into fullerites

K. Misof, G. Vogl*, L. Wende[+] and R. Sielemann[+]*

* Institut für Festkörperphysik der Universität Wien, Strudlhofgasse 4, A-1090 Wien, Austria
[+] Hahn-Meitner-Institut Berlin GmbH, Bereich Schwerionenphysik, Glienicker Str. 100, D-14109 Berlin 39

Abstract. C_{60} and C_{70} were irradiated with high-energy heavy ions at temperatures between 10K to 473K. The landing positions of the iron ions were studied by means of in-beam Mössbauer spectroscopy. For both materials the Mössbauer data indicate that the iron atoms landed on interstitial sites between two fullerene molecules with two pentagons facing each other. A second site is as yet unidentified.

1   Introduction

Solid C_{60} is a Van der Waals-bonded molecular crystal [1]. In a number of experiments phase transitions have been studied: Above 260K the C_{60} molecules form a fcc lattice with the soccer-ball shaped molecules rotating completely freely [2], whereas below 260K their rotation is partially hindered and the structure changes to simple cubic with four molecules per unit cell [3]. Below about 90 K the motion of the molecules is completely hindered [2]. Because of the rugby-ball shape of C_{70} the phase diagramme of this material is more complex than that one of C_{60} [4,5,6]. Another interesting feature of C_{60} and C_{70} is the large inner space in the molecules. Some work was done to study the endohedral chemistry [7].

The aim of our work was to irradiate fullerenes C_{60} and C_{70} with high-energy ions in order to study the positions of well separated iron atoms between the bucky-balls. In addition to earlier measurements on C_{60} at low temperature [8] we performed further investigations on C_{60} at higher temperatures and on C_{70}.

2   Experiments and Results

C_{60} and C_{70} in powder form were purchased from Hoechst. Layers of 30 μm and 20 μm thickness, respectively were evaporated at about 450°C in a vacuum of $5 \cdot 10^{-5}$ torr onto Al backings and Si backings. The structure of the samples was checked by means of X-ray diffraction measurements [9]. The C_{60} and C_{70} samples were then irradiated at temperatures between 10K and 473K at the heavy-ion accelerator VICKSI of the Hahn-Meitner-Institut Berlin with iron ions (energy spectrum up to 50 MeV). According to calculations with the well-known TRIM programme the average penetration depth of 50 MeV iron ions into fullerenes is about $12 \mu m$. The accumulated density of iron ions in the sample was about 10^{12} particles/cm^2, therefore the concentration of iron atoms in the fullerenes was tiny (about 10^{-6} per one C_{70} molecule). The landing positions of the ^{57}Fe ions were studied by means of in-beam Mössbauer spectroscopy in an analogous way as for Fe implanted into metals [10]. Fig.1a shows the spectra of C_{60} and Fig.1b the spectra of C_{70} at different temperatures, respectively. The data of C_{60} and C_{70} as

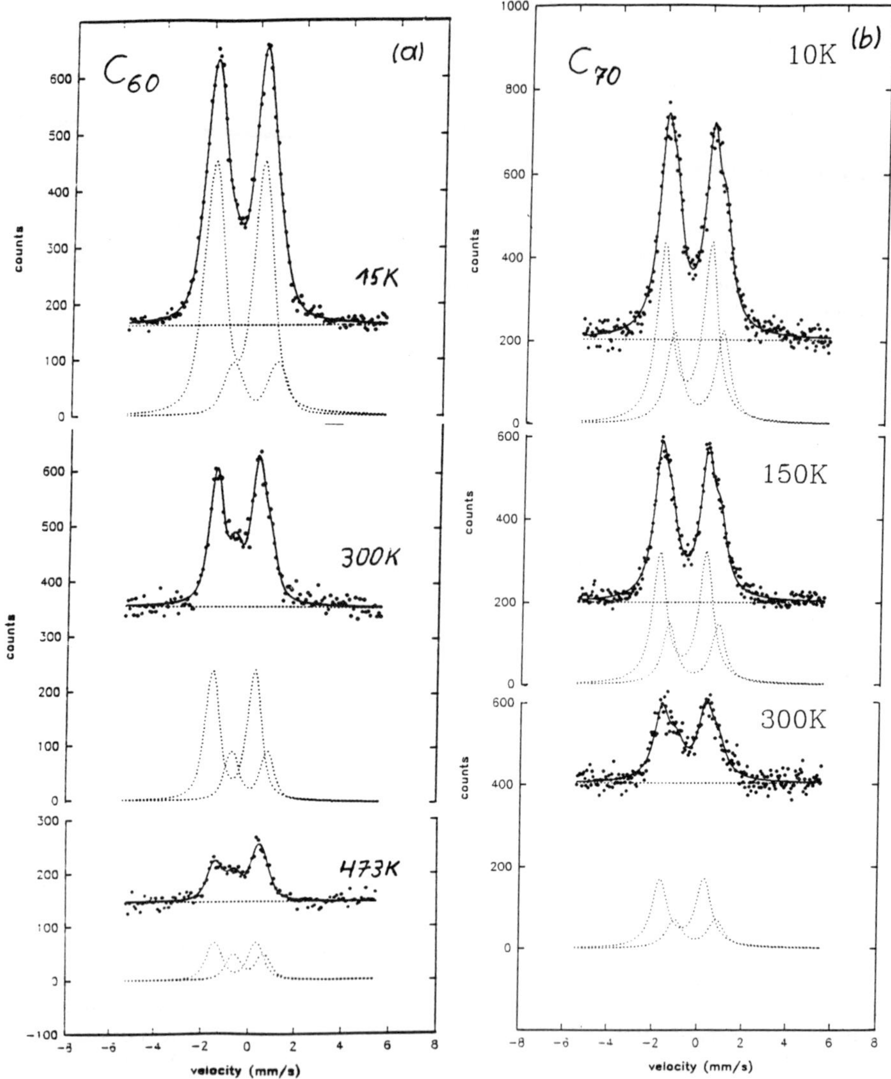

Figure 1: Mössbauer spectra of (a) C_{60} and (b) C_{70} at different temperatures. The solid line is a fit with two doublets. The fit indicates two different sites for the iron atoms.

well can be fitted with two doublets indicating two different well defined sites of the iron atoms. The isomer shift for the larger doublet in both materials indicates a high-spin Fe(II) complex. The second doublet has an isomer shift which indicates an Fe(0) or a low-spin Fe(III) compound. The temperature dependences of the splittings of the larger doublet for C_{60} indicates the well known phase transition at 260K as will be discussed below [2,3]. The fractions of the two doublets of C_{60} (proportional to Debye-Waller factor) show no anomalous behaviour. In contrast the data of C_{70} indicate a phase transition between 100K and 200K (Fig.2).

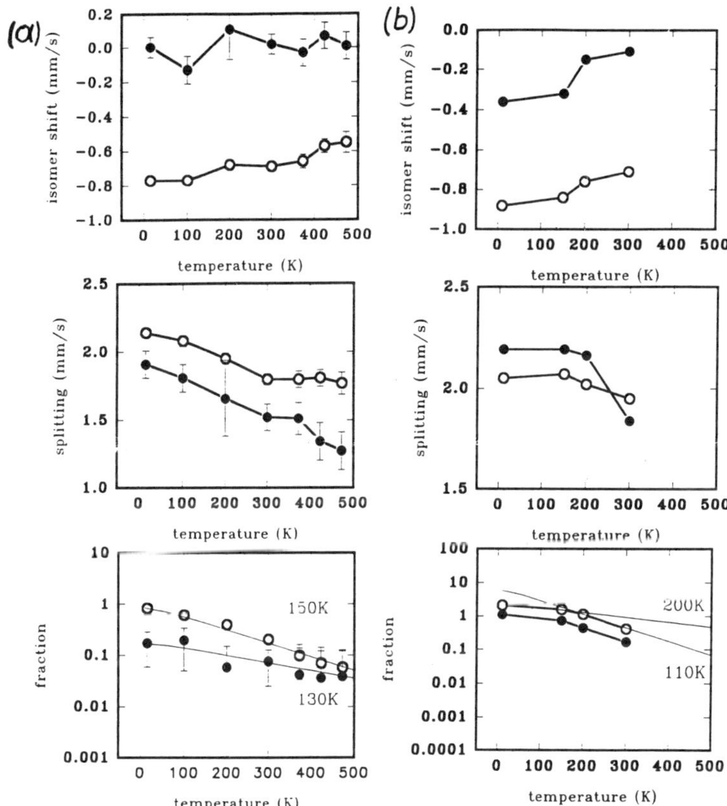

Figure 2: Isomer shift, quadrupole splitting and normalized fraction (a) of C_{60}, (b) of C_{70}; empty dots, larger doublet, full dots, smaller doublet.

3 Discussion

It is a remarkable fact that at temperatures above 260K the splitting of the larger component of C_{60} is constant. This can be interpreted as follows: The iron atoms after having slowed down through a number of collisions landed on interstitial sites between two bucky-balls with two pentagons facing each other without damaging their near surrounding. At higher temperatures this complex which is similar to ferrocene [11] hinders the free rotation of the bucky-balls bound to iron atoms. Rotation is only possible around the axis defined by the two bucky-balls and the iron atom. The observed averaging of the quadrupole splitting up to 260K (Fig.2a) indicates that the rotation of the bucky-balls increases gradually with higher temperatures. Above this temperature the rotation is too fast for the time scale of the Mössbauer atom. Thus the splitting of the larger doublet remains constant above the phase transition. The assignment of the second doublet whose splitting decreases further with higher temperature will be discussed in a future paper.

The data of C_{70} show interesting behaviour at about 160K: At this temperature we observe a jump in the isomer shift and a bent in the fraction (the fraction is proportional to the Debye-Waller factor) (Fig.2b). We interprete this as follows: Below 160K the C_{70} molecules rotate only around the long axis, whereas above this temperature reorientational disorder of the long molecular axis sets in. These data are in good agreement with the μSR data of C. Christides et al [12] and indicate a phase transition at 160K. Further investigations on C_{70} at higher temperatures are planned in order to see if the phase transition from the rhombohedral phases to the fcc phase is visible in the Mössbauer data, too.

We thank S. Klaumünzer for valuable discussions. We are particularly indebted to P. Szimkowiak for his skillful preparation of the C_{60} and C_{70} foils. This work was supported by the Austrian Fonds zur Förderung der wissenschaftlichen Forschung, project S5601.

References

[1] J.E. Fischer, P.A. Heiney, A.R. McGhie, W.J. Romanow, A.M. Denenstein, Jr.J.P. McCauley and A.B. Smith III: Science **252**, 1288 (1991).

[2] R. Tycko , G. Dabbagh, R.M. Fleming, R.C. Haddon, A.V. Makhija, and S.M. Zahurka: Phys.Rev.Lett. **67**, 1886 (1991).

[3] P.A. Heiney, J.E. Fischer,A.R. McGhie, W.J. Romanow, A.M. Denenstein, Jr.J.P. McCauley, A.B. Smith III and D.E. Cox: Phys.Rev.Lett. **66**, 2911 (1991).

[4] M.A. Verheijen, H. Meekes, G. Meijer, P. Bennema, J.L. de Boer, S. Van Smaalen, G. Van Tendeloo, S. Amelinckx, S. Muto and J. Van Landuyt: Chem.Phys. **166**, 287 (1992).

[5] G. Van Tendeloo, S. Amelincks, J.L. De Boer, S. Van Smaalen, M. A. Verheijen, H. Meekes and G. Meijer: Europhys.Lett. **21**, 329 (1993).

[6] G.B.M. Vaughan, P.A. Heiney, D.E. Cox, J.E. Fischer, A.R. McGhie, A.L. Smith, R.M. Strongin, M.A. Cichy and A.B. Smith III: submitted to Chem. Phys.

[7] D.S. Bethune, R.D. Johnson, J.R. Salem, M.S. de Vries and C.S. Yannoni: Nature **366**, 123 (1993).

[8] K. Misof, G.Vogl, P. Fratzl, R. Sielemann, B. Keck and Y. Yoshida: Springer Series in Solid State Sciences **117**, 44 (1993).

[9] K. Misof, P. Fratzl and G. Vogl: Europhy.Lett. **22**, 585 (1993).

[10] Y.Yoshida, M. Menningen, R. Sielemann, G. Vogl, G. Weyer and K. Schroeder: Phys.Rev.Lett. **61**, 195 (1988).

[11] R.L. Collins: Journal of Chem. Physics **42**, 1072 (1964).

[12] C. Christides, T. J. S. Dennis and K. Prassides: Phys.Rev.B **49**, 2897 (1994).

Part Two

METALLOFULLERENES; NANOTUBES AND NANOCLUSTERS

TRENDS IN THE STRUCTURAL AND ELECTRONIC PROPERTIES OF METALLOFULLERENES

Wanda Andreoni[1] and Alessandro Curioni[1,2]
[1]IBM Research Division, Zurich Research Laboratory
Säumerstrasse 4, CH-8803 Rüschlikon, Switzerland
[2]Scuola Normale Superiore, piazza dei Cavalieri 7, I-56126 Pisa, Italy

ABSTRACT

We discuss the results of local-density-functional based Car-Parrinello calculations on metallofullerenes with emphasis on comparison and trends.

1 Introduction

Given their special topology, fullerenes have long been thought of as possible ideal cages for atoms of any kind. In fact, the discovery of C_{60} [1] was soon followed by that of C_{60} with metals encaged, the metal atom being either La or an alkaline earth atom [2]. Significant progress has been made in the mean time on the art of encapsulating atoms in fullerenes (see e.g. [3-7]), especially after the synthesis of solid C_{60} [8]. The reader is referred especially to D.S. Bethune's lecture in this volume. In particular, the name "metallofullerenes" has now become common to denote fullerenes with metal atoms inside. Indeed, only for few cases have encaging processes been successful. Even fewer – for the moment – are the cases of metallofullerenes which have been synthesized in macroscopic quantities: La@C_{82} [4], Y@C_{82} [9], Sc$_2$@C_{84} [10, 11], and more recently La$_2$@C_{80} [9]. It is fair to recognize that there is so far no clue to the understanding of such a situation. While the search for efficient methods to form metallofullerenes is intense, there is now a strong need for theoretical research to provide insight into their physical and chemical behavior.

While one will eventually be concerned primarily with the electronic and magnetic properties of these new materials, one is currently confronted with the basic problem of their structure. In the case of La@C_{82}, for instance, the crystal structure of the solid phase has not yet been satisfactorily resolved [12], due to the presence of the solvent in the aggregate. Moreover, a direct observation of the structure of the individual component molecules has not yet been possible. Preserving a "constructive" point of view, before worrying about the solid-state effects, one would like to answer all the nontrivial questions concerning the molecular unit itself. These are primarily about its structure, e.g.: Which are the energetically favorable locations of the encaged metal atom? Which are the favorable cage structures?; about its stability as an individual cluster, e.g.: Why has K@C_{60} been observed but not Na@C_{60}?; about its fundamental electronic and chemical properties, e.g.: What type of bonding exists between the metal atom and the carbon cage? To what extent is the bond pattern of the cage affected? Which is the formal charge state of the metal atom? When two or three atoms are encaged, does a chemical bond form between them or not?; about its dynamical properties, e.g.: To what extent is the encaged atom "free" to move inside the molecule? How does its motion couple with the vibrations of the cage? It is still difficult to obtain answers from experimental observations. Moreover, little progress has been made so far on theoretical grounds, in spite of the existence of several calculations made at various levels of sophistication (see e.g. [13-22]). One of the reasons is the general difficulty of determining the geometrical structure of the metallofullerenes. In principle, *ab initio* molecular dynamics calculations [23, 24] are ideal for this search. In practice, however, one cannot map out the entire potential energy surface of such a complex molecule. Moreover, establishing the relative stability of isomers with diverse cage structures by using nontrivial optimization procedures such as simulated annealing is an extremely complicated task even for the simpler case of the undoped molecules. The barriers involved in the bond-breaking and re-bonding processes are too high to allow one to observe isomerization on the time scale of a few picoseconds typical of our *ab initio* molecular dynamics simulations. One has to resort to separate calculations for each possible (guessed) configuration of the cage. Determining the energetically favorable positions of the encaged atoms, on the other hand, is more easily and suitably feasible with our calculations, because we leave the atom "free" to explore a large portion of the interior of the molecule and allow all the atoms to relax simultaneously without imposing any symmetry constraint. Although even this complex search for the energetically favorable endohedral locations cannot be considered exhaustive, we have already obtained several useful pieces of information on the metal-cage chemistry as well as on the energetics and the electronic character [17, 21, 24, 25]. In the following sections, we shall describe some of our results to date. We shall focus our attention on comparing the behavior of different atoms in C_{60} (Na, K, Al, La, Y) as well as that of different fullerenes (C_{60} and C_{82}), and that of isomers with a different cage structure in the case of the La@C_{82} molecule. La@C_{82} and Y@C_{82} are of particular interest, since several experimental data have recently become available, such as extended X-ray absorption fine-structure (EXAFS) and photoemission (PES) data. For a detailed discussion of our calculations and results, compared with other methods and with experiment, we refer the reader to a forthcoming publication [21] .

2   Electron donors in C_{60}

Earlier *ab initio* (both Hartree-Fock [15] and LDA [13]) calculations on C_{60} with metal atoms inside were limited to high-symmetry configurations with the atom fixed at the center of the molecule. Owing to the major complexity of calculations free of symmetry constraints, investigations are still often made [19] in this way. Such an approach, however, is not appropriate in general, because, as we shall discuss below, the electronic and chemical properties are often severely dependent on the atom location inside the molecule. For all electron donors, the center of the molecule

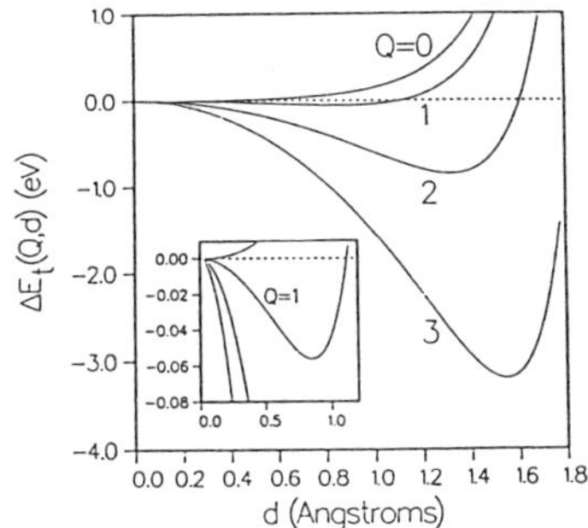

Fig. 1. Me@C_{60}: Energy change associated with the displacement from the center (from Ref. [26]).

Me	d(Å)	ΔE(eV)
Na	0.58	0.07
K	< 0.05	0.01
Al	0.96	0.24
Y	1.47	3.5
La	1.22	2.9

Tab. I. Me@C_{60}: Calculated values of the displacement of the metal atom from the center of the molecule and of the corresponding energy gain.

is **not** a minimum of the potential energy surface. Two years ago, Erwin proposed an elegant model [26] for the energetics of C_{60} endohedral complexes that explains why all the electron donors prefer to lie off-center. It is mainly the polarization of the electronic shell delocalized on the cage which drives the ion away from the center, thus determining the off-center displacement. The key result is contained in Fig. 1, which illustrates the energy change associated with the off-center displacement for several different values of the formal charge of the ionized atom, but for an ionic size fixed to that of Ne. Qualitatively, our *ab initio* calculations reproduce such behavior in all cases considered so far. As expected, the value of the off-center displacement and the energy gain associated with it depend not only on the charge, but also on the size of the metal atom and on the type of chemical bond it forms with the carbon atoms. Table I summarizes the calculated values for Na, K, Al, La, Y. The effect of the ionic size is evident from the comparison of isovalent impurities, such as Na and K, or La and Y.

In the case of Na and K, which act as one-electron donors, the potential energy surface is radially rather flat and the distortions induced on the molecular structure are minor. In fact, the splitting of both single and double bonds does not exceed 0.01 Å. In these cases, therefore, a model that disregards the discrete nature of the cage, its detailed shape and *a fortiori* its relaxation to the endohedral doping is a very reasonable approximation. It is also true that the chemistry of the complex does not change if one considers these atoms to be fixed at the center of the molecule. As explained in the work by Dunlap et al. [16] for a Na^+ ion, a manifold of equivalent as well as of quasi-degenerate minima also exists in these cases, so as to make the dynamics of these atoms inside C_{60} extremely rich.

We find that Al also acts as a one-electron donor. Although forming a stronger bond, the modification induced on the carbon molecule is still minor, the largest changes of the bond lengths being only 0.02 Å. An interesting feature of Al@C_{60} is that the $3s$ level lies within the energy interval of the occupied states of C_{60}. We find that, in the low-energy configurations, it is located about 3.3 eV below the singly occupied molecular orbital (SOMO) level, and that it is quite sensitive to the off-center displacement. In a hypothetical solid, this would correspond to a resonance in the valence band of the fullerite, and could be used not only as a fingerprint of the presence of Al, but to some extent also as a probe of its location. The $3s$ state only very weakly hybridizes with the cage orbitals. This is illustrated in Fig. 2.

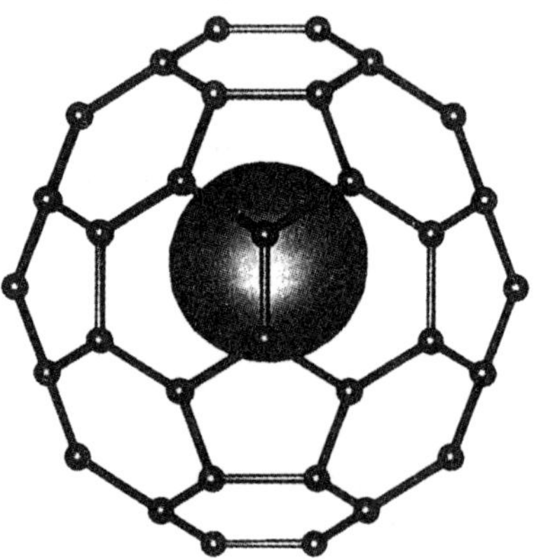

Fig. 2 . Al@C_{60}: Constant electron density surface of the "encaged" atomic $3s$ state of Al ($\rho = 0.008e/(\text{a.u.})^3$).

The effect of the interaction of Na, K and Al with C_{60} simply results in the transfer of one electron ($3s$, $4s$, $3p$, respectively) to the three-fold degenerate t_{1u} LUMO of the molecule. In all the calculated low-energy configurations, the degeneracy lifting of this level is $\lesssim 0.15$ eV.

Also in the case of La and Y, we find [25] that the cage accepts electrons (this time essentially three) and that both the SOMO and the SOMO-1 can be classified as cage states. Figure 3(A) shows the schematic diagram of the Kohn-Sham levels in the HOMO-LUMO gap region of the undoped molecule (in (a)). The (c) diagram corresponds to the lowest-energy La endohedral configuration calculated, and is representative of all low-energy geometries with La on off-center positions. The three split levels at about the position of the t_{1u} LUMOs of C_{60} correspond to orbitals which, in spite of the doping-induced distortions, largely keep the character they have in the undoped molecule. The actual occupation in the calculations is 2, 1, 0, respectively.

However, in order to illustrate their nature and the comparison with C_{60} more clearly, in Fig. 3(B) we also plot the superposition of the probability densities of the three molecular orbitals, taken with equal weights. The binding of La and Y to the carbons is stronger than in the case of Al. Owing to the higher charge transfer, the relaxation in the geometry of the fullerene is far from being negligible in these two cases, but is quite localized on the environment of the metal atom. The antibonding nature of the molecular acceptor states causes the carbon-carbon double-bonds to be more sensitive to the charge transfer and to tend to elongate. Close to the La or Y ion, in fact, they change character, since the bond length increases by as much as 0.07–0.08 Å.

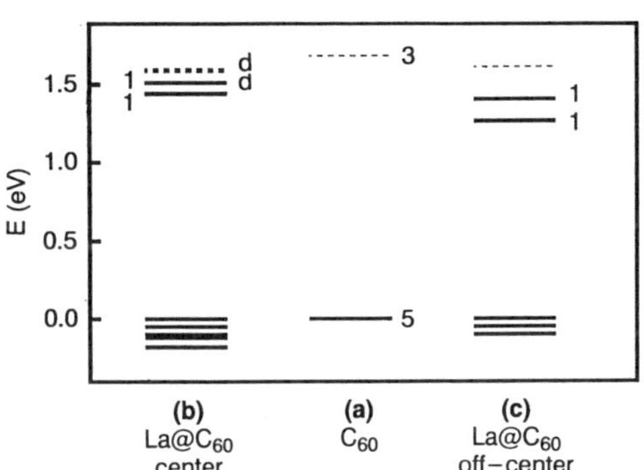

Fig. 3(A). La@C_{60}: Energy level diagram at the HOMO-LUMO gap region of C_{60} (see text).

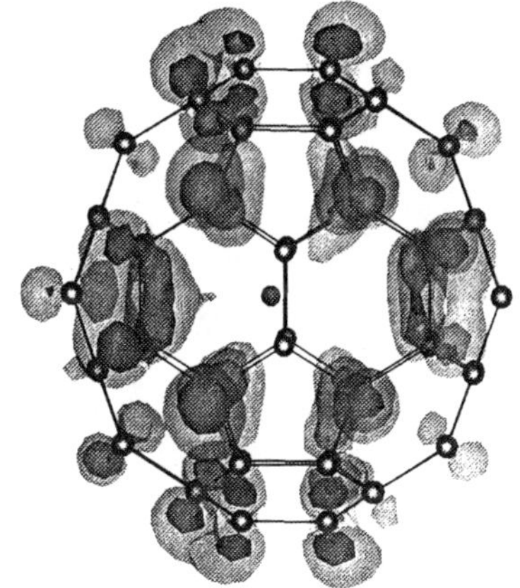

Fig. 3(B). La@C_{60} : Probability density associated with the t_{1u}-derived orbitals, taken with uniform occupation for the sake of comparison with C_{60} (see text). The plotted surfaces correspond to 0.001 and 0.003 $e/(\text{a.u.})^3$.

Figure 3(A) also reports the level diagram (b) corresponding to the configuration with La at the center of C_{60}. Here, the $5d$ level of La turns out to be close to the LUMO of C_{60}. This results in a formal charge on La close to 2+, and agrees with earlier Hartree-Fock calculations [15], performed on several electronic configurations. We find that such a geometry, however, is highly improbable. Its energy is about 3 eV higher than that of the low-energy configurations with La off-center. It is therefore obvious that in this case the correct determination of the structure is crucial for the understanding of the chemistry and the electronic properties, in general, of the metallofullerene. The same is true for Y@C_{60} [25].

98

3   (La,Y)@C$_{82}$

C_{82} is an especially interesting case. When undoped, it is not a particularly stable fullerene. Once doped with one atom of either La, Y or Sc, in spite of the relatively modest amount generally found in the arc-generated soot, it turns out to be relatively more abundant than the other fullerenes after separation. This is true both when extracted from the solvent, be it toluene, CS_2 or other organic solvents, and when separated by gradient sublimation [27]. This behavior is at odds with what happens for the cases of C_{60} and C_{70}, and has found no explanation so far.

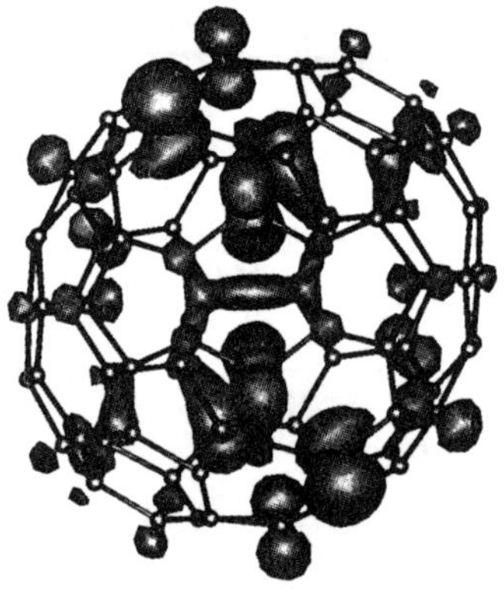

Fig. 4(A). La@C$_{82}$-C_2 isomer: Probability density of the SOMO. The contour level plotted is $0.002e/(\text{a.u.})^3$.

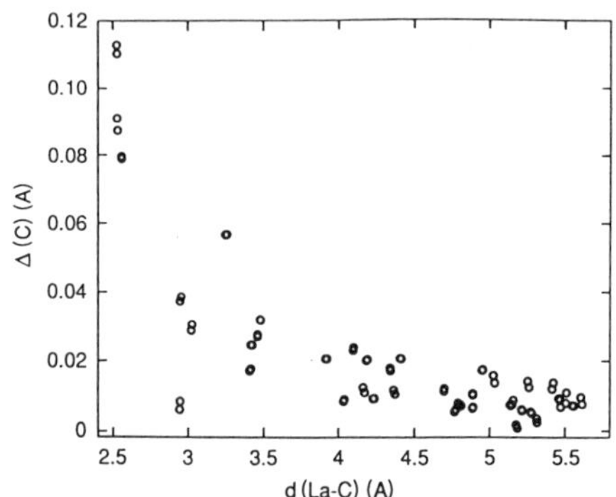

Fig. 4(B). Displacement of the carbon atoms from C_{82} to La@C$_{82}$, as a function of their distance from La (from Ref. [21]).

Macroscopic quantities of La@C$_{82}$ have recently been obtained, which have allowed several properties to be measured. Theoretical results are obtained on the isolated molecule. A detailed comparison of theory and experiment to date is contained in Ref. [21]. The main results can be summarized as follows: i) The molecule is a radical with one unpaired electron. This was demonstrated by electro-paramagnetic-resonance (EPR) measurements [28] at the very beginning of the investigation of La@C$_{82}$, and readily supported by X-ray photoemission (XPS) [29]. This finding invalidated current models of this molecule that predicted a closed-shell electronic configuration [30] and consequent speculations on its special stability. The experimental result was fully explained by our earlier calculations [17] on a C_{3v}-like cage and is confirmed by more recent calculations on a C_2 isomer [21]. Figure 4(A) shows the SOMO in the latter case. The fact that the fullerene becomes a radical, once it is endohedrally doped with La, is in fact independent of the specific cage and of the particular fullerene. In fact, in analogy with the case of C_{60} discussed above, also in C_{82} is La located at largely off-center positions, i.e. at distances compatible with a strong bonding with the carbon atoms, and donates three electrons to the cage. Preliminary results on La@C_{70} [25] further confirm this picture. Further support of the fact that the higher occupied states have a vanishing La character comes from the analysis of more recent photoemission spectra [21]. ii) Photoemission spectra on the solid phase [21] reveal the presence of a gap at the Fermi level. As the individual molecules are radicals, this observation suggests that a naive one-electron picture is inappropriate but does not exclude the fact that in the aggregated phase the coupling of the molecules is strong and may result, for instance, in the formation of dimers. However, the solid state structure is still unknown. iii) The $5p$ core states of La are not passive to doping, but hybridize with the lowest σ states of C_{82}, thus giving rise to a multiplet of states spanning a width of about 1.2 eV. Experimentally, the XPS feature with $5p$ character extends over a larger interval (about 2.5 eV), probably due to the spin-orbit splitting as well as to lifetime broadening. iv) The relaxation of the cage is far from being negligible but is still quite confined to the environment of La. This can be seen in Fig. 4(B), where the displacement of the carbon atoms relative to the undoped molecule is reported as a function of their distance from La in the case of the C_2 isomer.

As concerns trends, we find that the basic chemistry of the La-fullerene endohedral complexes does not change on passing from C_{60} to C_{82} or from one isomer to the other with different cages. However, our results on the isomers derived from the C_2 and the C_{3v} structures of C_{82} show that in the latter the relaxation of the cage and of the electronic structure is stronger. The main result is that, while the C_{3v} isomer is higher in energy than the C_2 one by as much as 0.9 eV, when endohedrally doped with La, at least two isomers exist with the two different cages that are close in energy. In fact, the C_{3v}-derived one is higher than the C_2 one by only 0.1 eV.

Preliminary results for Y@C_{82} [25] give a picture very similar to that of La@C_{82}. In this case, our results for the C_2 isomer can be compared with EXAFS data [31] for the Y–C coordination shells and distances. Experimentally there are two carbon shells (with 6 ± 1 atoms each) at distances of 2.40 ± 0.05 Å and 2.85 ± 0.05 Å, respectively. In the C_2 isomer, by definition, there are two shells of six carbon atoms. The calculated values for the average Y–C distances are 2.38 Å and 2.86 Å, respectively. In spite of the ionic size difference which allows Y to approach the cage more closely than La (by about 0.16 Å), the doping induced cage relaxation is essentially the same in the two molecules. The main difference in the electronic structure of the two endohedral complexes lies in the spectra of the unoccupied states and in the core p-states of the metal atom. In contrast with the case of La, the Y $4p$ states do not hybridize with the lowest σ cage states, which are however strongly polarized.

4 Conclusions

Research in the field of metallofullerenes is in its infancy, but is rapidly growing and hopefully will soon provide interesting new materials with well-characterized solid phases. For the moment, this appears to be possible for systems such as $Sc_2@C_{84}$ and $La_2@C_{80}$, where the most probable closed-shell nature of the electronic configuration helps stability and the structure is close-packed. However, from the materials point of view, the most interesting cases are clearly those for which the molecular unit is a radical. There is still a lot to be understood about the electronic properties, already at the molecular level. The interplay between geometrical structure and electronic properties in not trivial. Standard calculations neglect the effects of structural relaxation due to the presence of doping. We have shown that these are in fact sizable for the cases of major interest, such as those of La and Y. They will be crucial for a correct description of the dynamics of the encaged atom, which is not as free to move as in the case of the one-electron donors, and will affect the cage in a different way, depending on its location.

Unfortunately, in contrast to the case of C_{82}, current state-of-the-art attempts to produce $La@C_{60}$ in quantities that are sufficiently large to be amenable to experimental observation have failed [4, 27]. This system is particularly interesting, being isoelectronic with the superconducting alkali-metal fullerides K_3C_{60} and Rb_3C_{60}.

In conclusion, in spite of the fact that a very limited data base has allowed us to begin to recognize certain trends, it is clear that the chemistry and physics of metallofullerenes is rich, and that specific calculations are needed for each system. A common and intriguing characteristic is the presence of several isomers which are energetically competitive, and which differ either concerning the position of the encaged atom or the geometry of the host fullerene.

References

[1] H.W. Kroto, J.R. Heath, S.C. O'Brien, R.F. Curl, and R.E. Smalley, *Nature*, **318**, 162 (1995)

[2] J.R. Heath, S.C. O'Brien, Q. Zhang, Y. Liu, R.F. Curl, H.W. Kroto, F.K. Tittl, and R.E. Smalley, *J. Am. Chem. Soc.*, **107**, 779 (1985)

[3] T. Weiske, D.K. Bohme, H. Schwarz, *J. Chem. Phys.*, **95**, 8451 (1991)

[4] K. Kikuchi, S. Suzuki, Y. Nakao, N. Nakahara, T. Wakabayashi, H. Shiromaru, K. Saito, I. Ikemoto and Y. Achiba, Chem. Phys. Lett. **216**, 67 (1993); T. Suzuki, Y. Maruyama, T. Kato, K. Kikuchi and Y. Achiba, *J. Am. Chem. Soc.*, **115**, 11006 (1993); S. Hino, H. Takahashi, K. Iwasaki, K. Matsumoto, T. Miyazaki, S. Hasegawa, K. Kikuchi, and Y. Achiba, *Phys. Rev. Lett.*, **71**, 4261 (1993)

[5] E.G. Gillan, C. Yeretzian, K.S. Min, M.M. Alvarez, R.L. Whetten and R.B. Kaner, *J. Phys. Chem.*, **96**, 6869 (1992)

[6] D.S. Bethune, R.D. Johnson, J.R. Salem, M.S. deVries, and C.S. Yannoni, *Nature*, **366**, 123 (1993)

[7] M. Saunders, H.A. Jimenezvazquez, R.J. Cross and S. Mroczkowski, *J. Am. Chem. Soc.*, **116**, 2193 (1994)

[8] W. Krätschmer, L.D. Lamb, K. Fostiropoulos, D.R. Huffman, *Nature*, **347**, 354 (1990)

[9] Y. Achiba et al. (to be published)

[10] H. Shinohara, N. Hayashi, H. Sato, Y. Saito, X.-D. Wang, T. Hashizume, and T. Sakurai, J. Phys. Chem. (in press)

[11] R. Beyers, C-H. Klang, R.D. Johnson, J.R. Salem, M.S. de Vries, C.S. Yannoni, D.S. Bethune, H.C. Dorn, P. Burbanik, K. Harich and S. Stevenson (preprint)

[12] H. Suematsu, Y. Murakami, T. Arai, H. Kawata, Y. Fujii, N. Hamaya, O. Shimomura, K. Kikuchi, Y. Achiba, and I. Ikemoto, in *Novel Forms of Carbon II*, MRS Symp. Proc. Series 349 (1994) in press

[13] A. Rosen and B. Wastberg, *J. Am. Chem. Soc.*, **110**, 8701 (1988); *Z.Phys.*, **D 12**, 387 (1989)

[14] J. Ciolowski and E.D. Fleischmann, *J. Chem. Phys.*, **94**, 3730 (1991)

[15] A.H.H. Chang, W.C. Ermler, and R.M. Pitzer, *J. Chem. Phys.*, **94**, 5004 (1991)

[16] P.P. Schmidt, B.I. Dunlap and C.T. White, *J. Phys. Chem.* **95**, 10537 (1991); J.L. Ballester and B.I. Dunlap, *Phys. Rev.* **A45**, 7985 (1992)

[17] K.E. Laasonen, W. Andreoni and M. Parrinello, *Science*, **258**, 1916 (1992)

[18] S. Nagase, K. Kobayashi, T. Kato and Y. Achiba, *Chem. Phys. Lett.*, **201**, 475 (1993); *ibidem*, **214** (1993)

[19] A. Rosen, *Z. Phys.*, **D25**, 175 (1993)

[20] L.S. Wang, J.M. Alford, Y. Chai, M.D. Diener, J.L. Zhang, S.M. McClure, T. Guo, R.E. Smalley and G.E. Scuseria *Chem. Phys. Lett.*, **207**, 354 (1994)

[21] D.M. Poirier, M. Knupfer, J.H. Weaver, W. Andreoni, K. Laasonen, M. Parrinello, D.S. Bethune, K. Kikuchi and Y. Achiba, *Phys. Rev. B*, **in press**

[22] Y.S. Li and D. Tomanek, *Chem. Phys. Lett.*, **in press**

[23] R. Car and M. Parrinello, *Phys. Rev. Lett.*, **55**, 2471 (1985)

[24] For a review of the application of the Car-Parrinello method to fullerenes, see W. Andreoni, in *Physics and Chemistry of Fullerenes*, Ed. K. Prassides (Kluwer, Dordrecht, 1994)

[25] A. Curioni and W. Andreoni (to be published)

[26] S.C. Erwin, in: *Buckminsterfullerenes*, W.E. Billups and M.A. Ciufolini (eds.), (VCH Publishers, New York, 1993), p. 217ff

[27] C. Yeretzian, J.B. Wiley, K. Holczer, T. Su, S. Nguyen, R.B. Kaner, and R.L. Whetten, *J. Phys. Chem.*, **97**, 10097 (1993)

[28] R.D. Johnson, M.S. deVries, J.R. Salem, D.S. Bethune, and C.S. Yannoni, *Nature*, **355**, 239 (1992)

[29] J.H. Weaver, Y. Chan, G.H. Kroll, C.M. Jin, T.R. Ohno, R.E. Haufler, T. Guo, J.M. Alford, J.J. Conceicao, L.P.F. Chibante, A. Jain, G. Palmer and R.E. Smalley, Chem. Phys. Lett. **190**, 460 (1993)

[30] P.W. Fowler and D.E. Manolopoulos, *Nature*, **355**, 428 (1992)

[31] C.H. Park, B.O. Wells, J. DiCarlo, Z.-X. Shen, J.R. Salem, D.S. Bethune, C.S. Yannoni, R.D. Johnson, M.S. deVries, C. Booth, F. Bridges, and P. Pianetta, *Chem. Phys. Lett.*, **213**, 196 (1993)

PRODUCTION AND PROPERTIES OF FULLERENE RELATIVES: METALLOFULLERENES AND SINGLE-LAYER CARBON NANOTUBES

D.S. Bethune, C-H. Kiang*, R. Beyers, P.H.M. van Loosdrecht,
M.S. de Vries, J.R. Salem, C.S. Yannoni and R.D. Johnson

IBM Almaden Research Center, 650 Harry Rd., San Jose, CA 95120, USA

P. Burbank, J. Haynes, T. Glass, S. Stevenson and H.C. Dorn

Department of Chemistry, Virginia Tech, Blacksburg, VA 24061, USA

Abstract.
Co-vaporizing metals with carbon in a fullerene generator leads to a diverse collection of structures related to fullerenes. On the one hand, Group IIIB metals such as La, Y, and Sc form complexes with fullerene cages of the form M_nC_{2m} where the metal atoms are believed to be encapsulated in the fullerene cage. Techniques to purify such metallofullerenes have been developed, and milligram quantities are now being produced and characterized. On the other hand, the addition of transition metals such as Co, Fe and Ni to the carbon vapor lead to the catalytic production of carbon nanotubes with diameters on the order of 1 nm and with walls a single layer of atoms thick. These tubules can be several microns in length. The discovery of such tubules provides a new form of carbon with extended covalent bonding in 1 dimension, complementing the family of carbon forms with extended bonding in 0, 2, and 3 dimensions (fullerenes, graphite and diamond respectively).

1. Introduction

A vapor of carbon atoms, produced for example by a high current arc in a fullerene generator, is in a highly reactive and energetic state. This means that a vast number of reaction channels are open, and a great variety of condensation products are possible. Which species are actually produced depends on whether additional elements are present and on kinetic factors such as the presence of surfaces or particles, which can serve as nucleation sites and growth catalysts. In the absence of surfaces with only inert buffer gas present, it is now well known that carbon condenses into fullerenes: closed shells of atoms bonded in networks consisting of hexagons and pentagons.[1, 2] Adding metal atoms to this active environment gives rise to new possibilities. Here we discuss some very different

---

*Affiliated with Materials and Molecular Simulation Center, Beckman Institute, Division of Chemistry and Chemical Engineering, Caltech, Pasadena, CA 91125, USA

types of structures that have been discovered in the course of research involving co-condensation of metal and carbon vapors.

1. Metallofullerenes

Atoms can be encapsulated in fullerene shells.[3] Recent research has provided methods to produce metallofullerene (MFs) — fullerenes with metal atoms inside the cage — in macroscopic quantities; experimental and theoretical efforts to characterize these novel species are beginning to yield detailed information about them.[4,5] In two early experiments $La@C_{82}$ was found to be the only prominent metallofullerene in a toluene extract of soot produced by co-vaporizing La and C using a laser[6] or an arc.[7] Additional work has shown that increasing the concentration of metal leads to the formation of MFs with 1, 2 or more metal atoms, in cages ranging from 28 to more than 140 carbons. Trivalent metals such as the lanthanides (and Y and Sc) seem particularly favorable, but evidence that other metals can be encapsulated has been reported.[4] Figure 1 shows mass and EPR spectra of CS_2 extracts of soots produced in arc burns of La packed bars.[8] The mass spectrum (Fig. 1a) shows that with relatively high La loading, dimetal species $(La_2@C_{2n})$ are the dominant MFs, with prominent peaks for 2n = 72, 74, 76, and 88, and a smooth series of peaks for 2n running from 90 to about 140, cresting at $La_2@C_{100}$. Note that the $La_2@C_{72}$ peak is larger than that for $La@C_{82}$ (as usual, the most abundant monometal MF!).

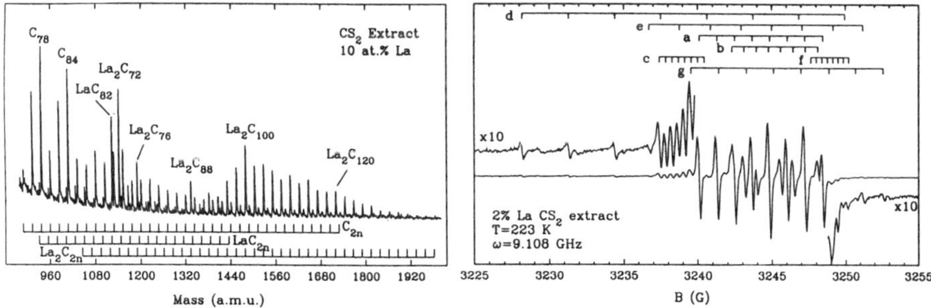

Figure 1. a) Mass spectrum of CS_2 extract of 10% La soot; b) EPR spectrum of CS_2 extract (in toluene) of 2% La soot. The 7 octets correspond to various $La@C_{2n}$ species. Percentages are wt % La relative to graphite in the mixture packed into graphite tubes (4 mm id, 6 mm od).

The EPR spectrum in Figure 1b shows the presence of seven octets associated with different monometal species. Some of these spectra have previously been assigned: *a* and *b* to $La@C_{82}$, and *c* (tentatively) to $La@C_{76}$.[7,9-11]

Such complex mixtures make separation and purification of these species difficult. Significant progress has been made using multistage high-performance liquid chromatography (HPLC). Coarse separation of MFs from small empty fullerenes is achieved using polystyrene columns. Then, different MFs are separated using a Buckyclutcher (Regis Chemical Co.). Thus, pure samples of $Sc_2@C_{2n}$, with $2n = 74$, 82, and 84,[12-15] and $La@C_{82}$ [16] have been obtained, and characterized by UV/Vis spectroscopy,[13, 16] IR,[16] and STM imaging on surfaces.[17-19] We used multistage HPLC to isolate two $Sc_2@C_{84}$ isomers.[14, 15] From $Sc_m@C_n$ samples containing mainly discandium species with cages of 74, 76, or even numbers of carbons between 80-104 Crystals of the more abundant $Sc_2@C_{84}$ species were grown from CS_2 solution. Figures 2 and 3 show TEM results obtained from these crystals.[20] Both electron diffraction and high-resolution TEM images show that the $Sc_2@C_{84}$ molecules pack in a hexagonal close-packed (HCP) structure with $a = 11.2 \pm 0.2$ Å and $c = 18.3 \pm 0.2$ Å, $c/a = 1.63 \sim \sqrt{8/3}$,

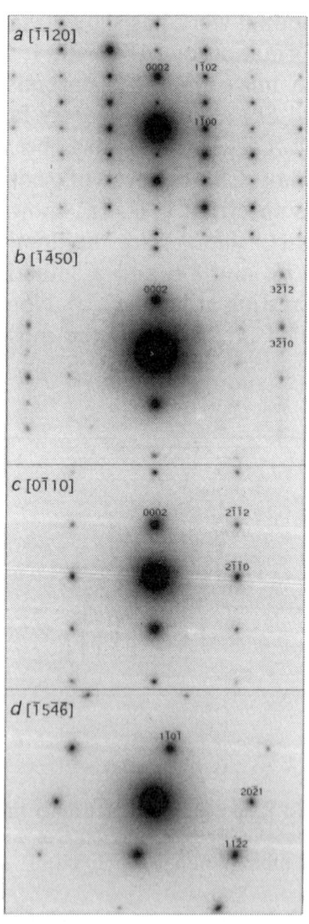

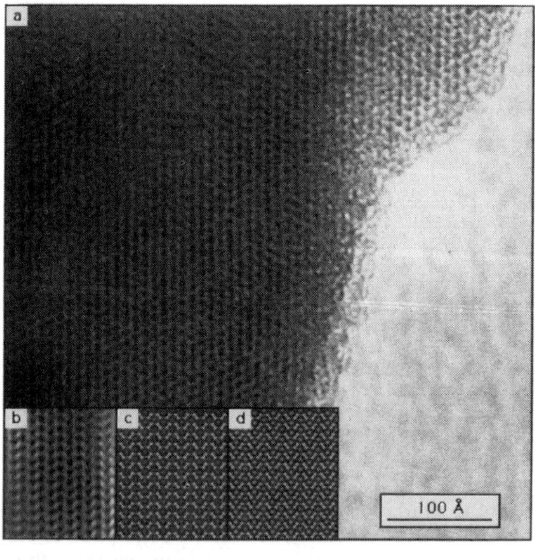

Figure 3. (a) High-resolution TEM image of $Sc_2@C_{84}$ crystal, along the $[\bar{1}\,\bar{1}\,2\,0]$ direction. Insets: (b) Fourier filtered image (to emphasize periodicities); (c), (d): simulated images of 22 Å thick $Sc_2@C_{84}$ and C_{84} crystals, respectively, at -500 Å defocus.

Figure 2. Electron diffraction zone-axis patterns of $Sc_2@C_{84}$, taken along the zone axes noted.

the value expected for ideal-sphere packing. The 11.2 Å molecular spacing in $Sc_2@C_{84}$ is the same as that found earlier in crystalline C_{84}[21-23], which has a face-centered cubic (FCC) structure. In the TEM image in Figure 3, there is higher projected charge density within columns of MF molecules than expected for empty fullerenes (based on simulations as in Fig. 3c,d and empty cage images). This leads us to conclude that the metals are within the columns. The matching intermolecular spacings for $Sc_2@C_{84}$ and C_{84} (11.2 Å) rules out the possibility that the metal atoms are between the balls, so together the TEM images and diffraction data confirm that the metal atoms are indeed inside the cages. Additionally, the observation of the c/a ratio expected for ideal sphere packing indicates molecular orientational disorder, static or dynamic.[20]

2. Single-layer Carbon Nanotubes

In the course of investigating the possibility of encapsulating transition metals in fullerenes, we discovered that when cobalt is co-vaporized with carbon, an entirely different structural possibility is realized. The nanometer scale Co particles catalyze the growth, in the gas phase, of tubules of carbon with walls a single atomic layer thick, diameters ~1 nm, and lengths up to several μm.[24] Figure 4a shows a TEM image of Co-catalyzed nanotubes.

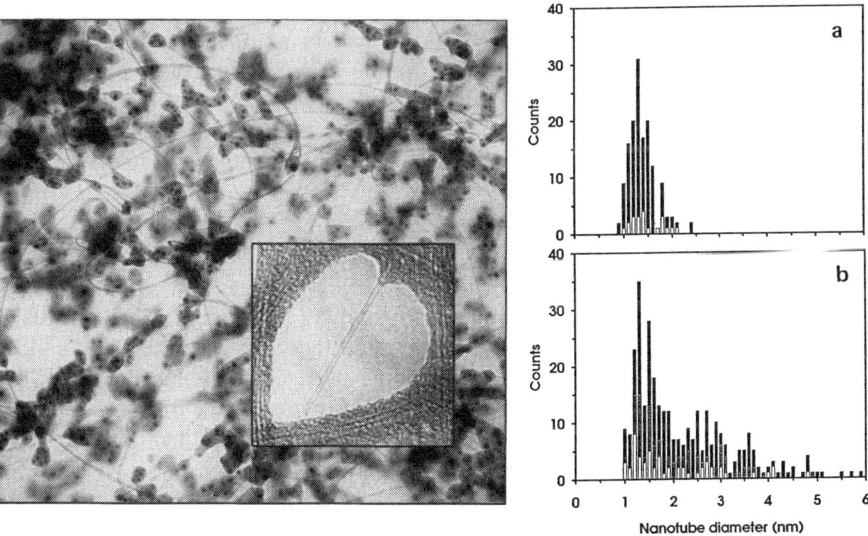

Figure 4. TEM image of carbon soot produced with cobalt (1.7x1.6 μm). Inset (62x60 nm) shows section of bare nanotube, 1.3 nm in diameter.

Figure 5. Nanotube diameter distributions. Produced (a) using Co metal or cobalt oxide; (b) with sulfur added, elemental or from CoS.

Work at the same time by Iijima and Ichihashi showed that single-layer nanotubes could also be grown with iron as a catalyst in an Ar/CH_4 atmosphere.[25, 26] Recent work has shown that nickel or Ni/Co and Ni/Fe mixtures can also catalyze single-layer nanotube growth.[27] Figure 5a shows that the nanotubes produced with cobalt have diameters of 1.3 $\pm$ 0.3 nm. For some applications, such as gas storage and chemical separation, single-layer tubes with diameters larger than 1 nm, or with a broad size distribution, could be advantageous. Figure 5b shows that the size distribution of the Co-catalyzed nanotubes can be greatly extended by using sulfur as a promoter.[28] With sulfur present, tubes with diameters up to 6 nm are found. In addition, sulfur enhances the production of nanotubes. Such effects have been noted before in connection with production of larger vapor grown carbon fibers (VGCF),[29, 30] but are not well understood. Diverse possible applications for nanotubes have been suggested: they could alter mechanical, electronic and electromagnetic properties of composite materials, or be useful as catalysts, catalyst supports or molecular-scale electrical conductors.[31-33] Work on this novel 1-D form of pure carbon has just begun.

3. Metallo-Carbohedrenes

Co-vaporizing carbon and metals which form strongly bound carbides, such as titanium and vanadium, apparently can lead to another class of structures: the metallocarbohedrenes or metcars.[34-39] The best known example, Ti_8C_{12}, is proposed to have a dodecahedral mixed metal/carbon cage structure, with each carbon bonded to one carbon and two titanium atoms.[34] In addition to such metcars, cubic TiC nanocrystals form in a mixed Ti/C vapor,[40] particularly if the Ti/C ratio is high.[38]

4. Conclusions

The production of metallofullerenes, nanotubes, metcars and carbide clusters demonstrates the richness of possibilities for making novel structures in a mixed carbon/metal environment. Many of these nanostructures are stable and can be produced in bulk, suggesting that they may be useful as building blocks for the construction of new materials. Investigating this diverse set of fullerene relatives and assessing their utility will keep researchers busy for some time to come.

5. Acknowledgments

We thank W. A. Goddard III for useful discussions. CHK acknowledges support by the NSF and the Materials and Molecular Simulation Center. HCD acknowledges support from the IBM Research Division and the Virginia Center for Innovative Technology. PvL acknowledges support from the Netherlands Organization for Scientific Research (NWO) and from the IBM Corporation.

References

1. H. W. Kroto et al., *Nature* **318**, 162-163 (1985).
2. W. Krätschmer et al., *Nature* **347**, 354-358 (1990).
3. J. R. Heath et al., *J. Am. Chem. Soc.* **107**, 7779-7780 (1985).
4. D. S. Bethune et al., *Nature* **366**, 123-128 (1993).
5. H. Schwarz et al., *Buckminsterfullerenes,* eds. W. E. Billups and M. A. Ciufolini, 257-282 (VCH Publishers, New York, 1993).
6. Y. Chai et al., *J. Phys. Chem.* **95**, 7564-7568 (1991).
7. R. D. Johnson et al., *Nature* **355**, 239-240 (1992).
8. P.H.M. van Loosdrecht et al., *to be published* (1994).
9. M. Hoinkis et al., *Chem. Phys. Lett.* **198**, 461-465 (1992).
10. S. Suzuki et al., *J. Phys. Chem.* **96**, 7159-61 (1992).
11. S. Bandow et al., *J. Phys. Chem.* **96**, 9609-9612 (1992).
12. H. Shinohara et al., *Mat. Sci. and Eng. B* **19**, 25-30 (1993).
13. H. Shinohara et al., *J. Phys. Chem.* **97**, 4259-4261 (1993).
14. H. C. Dorn et al., *Analytical Chem.* **submitted for publication**, (1993).
15. S. Stevenson et al., *Analytical Chem.* **submitted for publication**, (1993).
16. K. Kikuchi et al., *Chem. Phys. Lett.* **216**, 67-71 (1993).
17. X-D. Wang et al., *Jpn. J. Appl. Phys.* **32**, L866-8 (1993).
18. X-D. Wang et al., *Phys. Rev. B* **48**, 15492-15495 (1993).
19. H. Shinohara et al., *J. Phys. Chem.* **97**, 13438-13440 (1993).
20. R. Beyers et al., *Nature* **in press**, (1994).
21. Y. Saito et al., *Phys. Rev. B* **48**, 9182-9185 (1993).
22. S. Muto et al., *Phil. Mag. B* **67**, 443-463 (1993).
23. J. F. Armbruster et al., *Phys. Rev. B* **submitted**, (1993).
24. D. S. Bethune et al., *Nature* **363**, 605-7 (1993).
25. S. Iijima and T. Ichihashi, *Nature* **363**, 603-5 (1993).
26. S. Iijima and T. Ichihashi, *Nature* **364**, 737 (1993).
27. S. Seraphin and D. Zhou, *Appl. Phys. Lett. (USA)* **in press**, (1994).
28. C-H. Kiang et al., W. A. Goddard III, R. Beyers, J. R. Salem, and D. S. Bethune, *submitted* (1994).
29. H. Katsuki et al., *Carbon* **19**, 148-150 (1981).
30. T. Kato et al., *J. Mater. Sci. Lett (UK)* **11**, 674-677 (1992).
31. M. R. Pederson and J. Q. Broughton, *Phys. Rev. Lett.* **69**, 2689-2692 (1992).
32. R. Saito et al., *Phys. Rev.* **B46**, 1804-1811 (1992).
33. R. A. Jishi, M. S. Dresselhaus, and G. Dresselhaus, *Phys. Rev.* **B48**, 11385-11389 (1993).
34. B. C. Guo et al., *Science* **255**, 1411-1412 (1992).
35. S. Wei et al., *J. Phys. Chem.* **96**, 4166-4168 (1992).
36. Z. Y. Chen et al., *Chem. Phys. Lett.* **198**, 118-122 (1992).
37. S. F. Cartier et al., *Science* **260**, 195-6 (1993).
38. B. V. Reddy, S. N. Khanna, and P. Jena, *Science* **258**, 1640-1643 (1992).
39. M. Methfessel et al., *Phys. Rev. Lett.* **70**, 29-32 (1993).
40. S. Seraphin et al., *Appl. Phys. Lett. (USA)* **63**, 2073-2075 (1993).

PREPARATION AND SEPARATION OF LANTHANUM METALLOFULLERENES

D. FUCHS

Kernforschungszentrum Karlsruhe, Institut für Nukleare Festkörperphysik, Postfach 3640, D-76021 Karlsruhe, FRG

and

P. ADELMANN, R. MICHEL*

**Institut für Physikaliche Chemie, Universität Karlsruhe, Kaiserstraße 12, D-76128 Karlsruhe, FRG*

ABSTRACT

Lanthanum metallofullerenes were produced by arc-burning carbon electrodes packed with La_2O_3 and graphite pitch. The soot was collected anaerobically and extracted with toluene and CS_2. The extract was analyzed by laser desorption post ionization mass spectrometry. Numerous endohedral fullerenes are detected, among them $La@C_{82}$ as the most abundant metallofullerene. The separation of $La@C_{82}$ was performed by a two step HPLC (high performance liquid chromatography).

1. Introduction

Since the breakthrough in synthesizing and purifying fullerenes[1-4], the chemistry of fullerenes has gained great interest. In particular chemical compounds by including metall atoms inside the fullerene cage was discussed very early on. Such metallofullerenes could be novel in many ways.

Indeed many groups have succeeded in synthesizing metallofullerenes[5-7], but only a few have been equally successful in isolating them[8-9]. Especially for $La@C_{82}$ only a few milligrams are available and measurements of physical properties of this new kind of material are therefore hard to perform. The challenge is therefore to synthesize larger amounts of pure samples. In this paper we report on high yield preparation and separation of $La@C_{82}$ providing the base for a number of investigations on this new class of materials.

2. Preparation

$La@C_{82}$ was prepared by electric arc burning of lanthanum/graphite rods. The preparation procedure of lanthanum/graphite rods was basically the same as used by Chai et al.[8]. Graphite rods (Schunk Kohlenstofftechnik GmbH), 300 mm long by 15 mm diameter, were drilled out to 270 mm by 10 mm and filled with a composite of La_2O_3 powder (Alfa Products, ultrapure purity) and high-strength graphite pitch (Dylon inc., Cerastil GC). The overall lanthanum/carbon ratio was about 1.6 at%.

To harden the mixture, the prepared rods were at first heated in a tube furnace at $100°C$ for 30 h and then slowly heated up (30°/h) to $900°C$ for another 10 h under flowing argon.

To heat out organic residues from the pitch and to reduce the oxygen-content of the composite rods, an additional heat treatment at $950°C$ under vacuum (10^{-5} mbar) was carried out for 30 h. After this procedure, the rods were strongly hygroscopic and sensitive to air and had to be used imediately to get a   high yield of metallofullerenes.

The so obtained lanthanum-graphite rods were used as positive electrodes and were burned down in a dc discharge oven. Details of the oven-setup are shown in Fig. 1. The arc between the pure graphite rod (negative electrode) and the lanthanum-graphite composite rod was sparked at 250 A in a helium (500 mbar) circulating atmosphere. The circulation was realized with a big blower. By this method it is furthermore possible to transport the soot away from the arc-chamber and to collect the soot anaerobically by passing it through a filter. After burning, the filter-system was coupled off the oven and the soot was extracted with hot toluene and CS_2 under helium (1 atm) for 24 h.

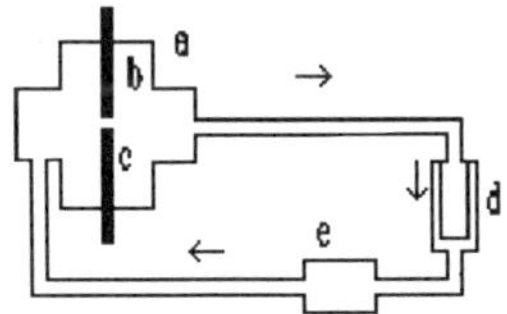

Fig. 1: The major parts of the oven-setup are the arc-chamber (a) with a pure carbon rod as a negative (b) and a composite rod as a positive electrode, the filter-system (d) and the blower (e). The helium circulates in the direction indicated by the arrows.

3. Separation

The separation of La@C_{82} was performed by a two step HPLC. In order to analyze the isolation products, mass spectroscopic measurements on extracts and HPLC-fractions were carried out by laser desorption post ionization mass spectroscopy. For desorption an excimer laser was used with a repetition rate of 20 Hz and an energy of about 1 mJ per shot. In Fig. 2 an anion mass-spectrum after the CS_2-extraction is shown. It is evident that La@C_{82} is the most abundant metallofullerene.

In the first step of our separation process the fullerenes from the crude extract were dissolved in toluene and given on a C_{18}-column (Vydac 201 TP 510 C_{18}-phase 22 mm x 250 mm) with a 1:1 mixture of acetonitrile/toluene as eluent. In the second purification step, the fraction containing La@C_{82} was given on a buckyclutcher-column (Buckyclutcher I 110 mm x 250 mm) with a 6:4:1 mixture of toluene/hexane/dichlormethane respectively as eluent.

In Fig. 3 we show an anion mass-spectrum from a HPLC-fraction after the first run on a buckyclutcher-column. After this run we achieve a La@C_{82}-concentration of about 50%. The second step has to be repeated in order to achieve a high purification of the lanthanum-fullerenes. The complete isolation of larger amounts of La@C_{82} is in progress, the results will be published elsewhere.

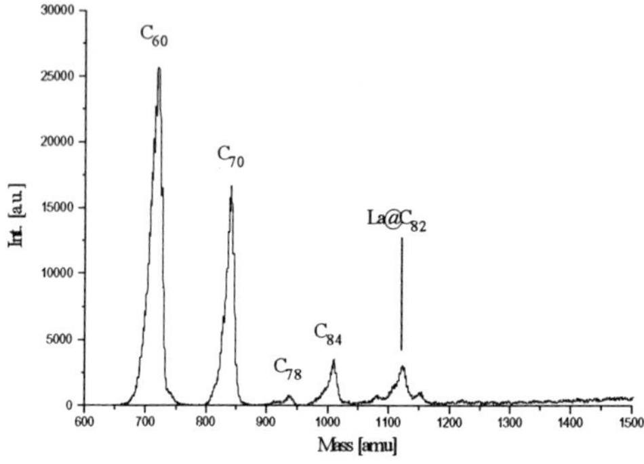

Fig.2: Anion mass-spectrum after crude extraction (CS_2). La@C_{82} is the most abundant metallofullerene.

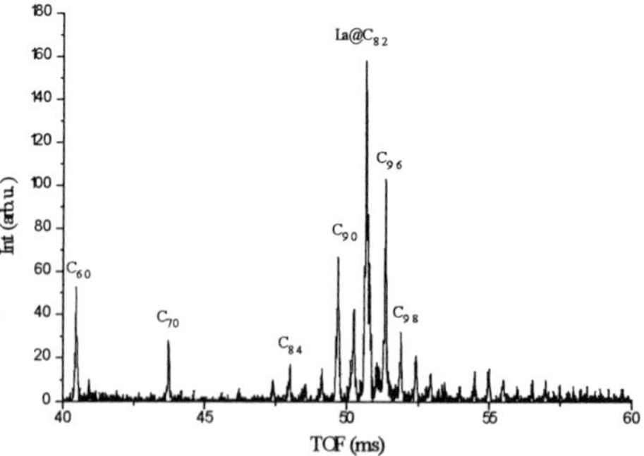

Fig.3: Anion mass-spectrum of a HPLC-fraction (first run). The concentration of La@C_{82} is about 50%.

In conclusion, we have shown that we are able to produce and to separate metallofullerenes. During the time of separation no remarkable decay of La@C$_{82}$ was observed so that material can be collected at least over a few weeks. Therefore measurements on physical properties of these new molecules are possible in near future.

5. References

1. H. R. Kroto, J. R. Heath, S. C. O'Brien, R. F. Curl, and R. E. Smalley, *Nature* **318** (1985) 162.
   R. F. Curl and R. E. Smalley, *Sci. Am.* **265** (1991) 32.
2. W. Krätschmer, K. Fostiropoulus and D. R. Huffman, *Chem. Phys. Lett.* **170** (1990) 167;
   W. Krätschmer, L. D. Lamb, K. Fostiropoulus and D. R. Huffman, *Nature* **347** (1990) 354.
3. K. Kikuchi, N. Nakahara, T. Wakabayashi, M. Honda, H. Matsumiya, T. Moriwaki, S. Suzuki, H. Shiromaru, K. Saito, K. Yamakuchi, I. Ikemoto and Y. Achiba, *Chem. Phys. Lett.* **188** (1992) 177.
4. F. Diederich, R. L. Whetten, C. Thilgen, R. Ettl, I. Chao and M.M. Alvarez, *Science* **254** (1991) 1768.
5. Y. Chai, T. Guo, C. Jin, R. E. Haufler, L. P. F. Chibante, J. Fure, L. Wang, J. M. Alford and R. S. Smalley, *J. Phys. Chem.* **95** (1991) 7564.
6. J. H. Weaver, Y. Chai, G. H. Groll, C. Jin, T. R. Ohno, R. E. Haufler, T. Guo, J. M. Alford, J. Conceicao, L. P. F. Chibante, A. Jain, G. Palmer and R. S. Smalley, *Chem. Phys. Lett.* **190** (1992) 460.
7. H. Shinohara, H. Sato, M. Ohkohchi and Y. Saito, *Nature* **357** (1992) 52.
8. H. Shinohara, H. Yamaguchi, N. Hayachi, H. Sato, M. Ohkohchi, Y. Ando and Y. Saito, *J.Phys. Chem.* **97** (1993) 4259.

PARAMAGNETIC STATES IN LANTHANUM, YTTRIUM, SCANDIUM AND HOLMIUM CONTAINING FULLERENES

A. Bartl, U. Kirbach, L. Dunsch, B. Schandert and J. Fröhner
Institut für Festkörperforschung im
Institut für Festkörper- und Werkstofforschung e.V. Dresden,
Helmholtzstr. 20, D-01069 Dresden

ABSTRACT

Electron spin resonance spectroscopy and mass spectrometry are used to characterize the electronic state of endohedral systems. Fullerenes produced by arc vaporization in presence of lanthanum, yttrium, scandium and holmium and their oxides show in the solid soot extract powder low resolved and in solution well resolved ESR spectra. The splitting is interpreted as isotropic hyperfine coupling of an unpaired electron to the nuclear magnetic moment of a metal ion inside a fullerene molecule. It is concluded that La, Y, Sc and Ho exist in ionic form in endohedral fullerenes both in the solid and liquid state of the fullerene. Furthermore it is shown that there is more than one stable position of the metal ion inside the fullerene molecule. Mass spectrometric measurements demonstrated that Me@C82 structures are the most stable endohedral fullerenes which were therefore studied in the present paper.

1. Introduction

The first type of metallofullerene extracted from fullerene soot was lanthanum-fullerene La@C$_{82}$ [1-4] followed a short time later by the detection of scandium fullerene Sc@C$_{82}$ [5,6] and yttrium fullerene Y@C$_{82}$ [5,12]. In general these new endohedral fullerenes have been identified by mass spectrometry and characterized in their electronic state by ESR spectroscopy.

In this paper we report on the preparation as well as the mass spectrometric and electron spin resonance spectroscopic measurements at La, Y, Sc and Ho containing fullerenes. Because of the small quantities of the produced material these two methods are most promising for characterization of the samples which consists of a few micrograms of metallofullerene only.

2. Experimental

The arc vaporization method by Krätschmer et al [9] was used for the preparation of metallofullerenes. Pure metal or metal oxide powders were mixed with graphite powder and pressed into a concentric hole inside the pure graphite rods handled in a dry box. With these electrodes the fullerene soot was produced by a dc arc vaporization using the apparatus developed by our group and described in [10]. This apparatus is characterized by a strong water cooling of the burning rods and the metallic mantle of the burning chamber.

The soot deposited onto the wall of the burning chamber was extracted in a Soxhlet by toluene. The resulting solid fullerene mixture was dried under vacuum. The samples were investigated by mass spectrometry (MS) using a MAT 95 (FINNIGAN) spectrometer capable of the detection of fullerenes by electron impact, fast atom bombardment, field desorption and chemical ionisation. The mass spectrometric characterization of the metallofullerenes were performed by chemical ionization for all samples.

Electron spin resonance spectroscopy (ESR) was done with the ERS 300 X-band spectrometer (ZWG Berlin) with 100 kHz field modulation and 1 mW microwave power at room temperature.

3. Results

Correlative mass spectrometric and ESR spectroscopic studies were done to identify metallofullerenes with respect to their chemical and their electronic state. First of all the composition of the soot extract was studied by mass spectometry to get the information of all species which might be seen in the ESR spectroscopic measurements (Fig. 1). Concerning the lanthanum, scandium and holmium containing fullerenes we obtained the ESR spectrum of $Sc@C_{82}$ already in the solid soot extract. Additional to the intense single-line signal of the C-60/C-70 fullerenes eight weak lines with a coupling constant of 0.36 mT could be found (Fig. 1). The ESR spectrum of the dried powder samples of yttrium containing fullerenes measured on air shows a symmetric single line signal having a linewidth of ΔB = 0.13 mT without any hint of hyperfine splitting.

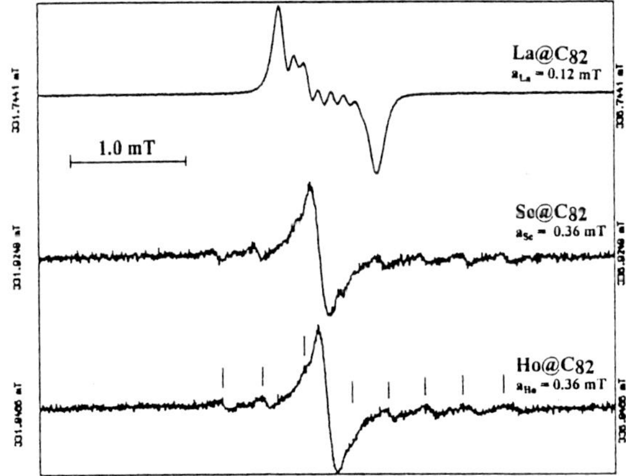

Fig. 1: ESR spectra of metallofullerenes in soot extract powder

The X-band ESR experiments of lanthanum containing fullerenes at room temperature in solution resulted in one octet and in other samples in two overlapping eight line spectra. Both these octets arise from electron hyperfine coupling of an unpaired electron of the fullerene molecule with the lanthanum nuclear spin of $I = 7/2$. The observation of a single peak in the mass spectrum but of two octets in the ESR spectrum suggests two different conformations of the same metallofullerene molecule. Theoretical calculations [7,8] led to the conclusion that the electronic state of lanthanum in $La@C_{82}$ is $La^{3+}@C_{82}^{3-}$ at the energetically most stable position and this is in agreement with the interpretation of the ESR results.

114

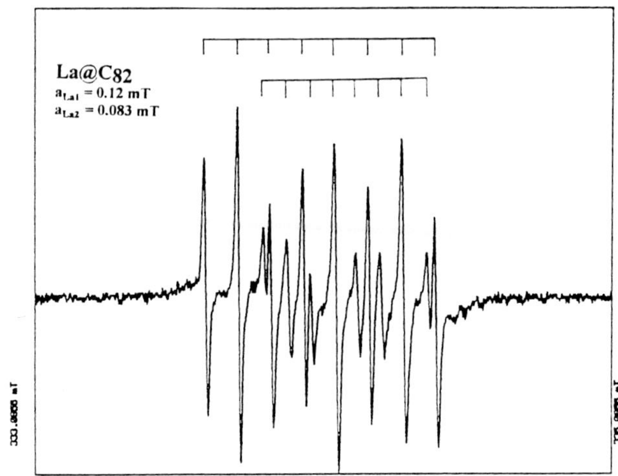

Fig. 2: ESR spectra of lanthanum containing C_{82}

The yttrium containing C_{82} spectrum in Fig. 3 consists in all cases of two doublets. The intense doublet has a hyperfine coupling constant of a_{Y1} = 0.048 mT, the weaker doublet has a coupling constant of a_{Y2} = 0.032 mT. The doublets arise from the hyperfine interaction of an unpaired electron at the fullerene molecule with the nuclear spin of the yttrium atom $I = 1/2$. The intensity of the weaker doublet is about four to five times lower then that of the main doublet. The weaker doublet is shifted to higher magnetic field values and is attributed to a second electronic state with a lower g-factor. The spectra of $Y@C_{82}$ were accompanied by weak satellite lines. These lines were attributed to natural abundant ^{13}C hyperfine coupling of the fullerene carbons with the unpaired electron.

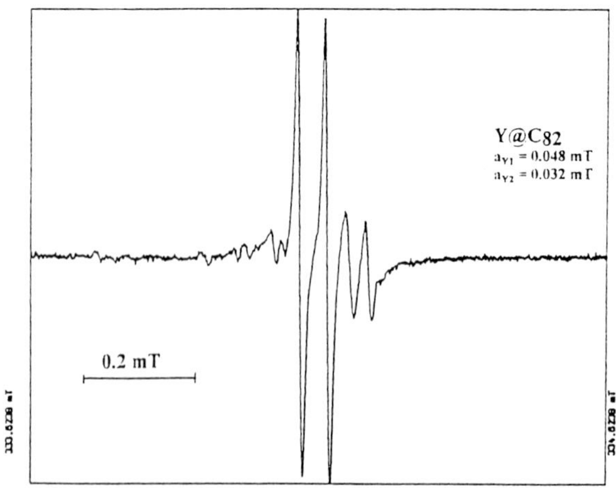

Fig. 3: ESR spectrum of yttrium containing C_{82} in solution; two conformations can be seen

The explanation of the second doublet is given by the existence of two different conformations of the $Y@C_{82}$ molecule. One conformation site contains the Y atom in the centre of the C_{82} molecule. A second conformation of $Y@C_{82}$ contains the Y atom in an asymmetric site in the C_{82} molecule. An analogous model was calculated by Laarsonen et al for La containing C_{82} [11]. Since only one peak of $Y@C_{82}$ is found in the mass spectrum the two doublets in the ESR spectrum suggest that two different conformations of the same $Y@C_{82}$ molecule are stable.

4. Acknowledgment

Financial support of this work by a grant of the **Sächsisches Ministerium für Wissenschaft und Kunst** is gratefully acknowledged.

5. References

1. R. D. Johnson, H. S. de Vries, J. Salem, D. S. Bethune and C. S. Yannoni, Nature 355 (1992) 239.
2. R. D. Johnson, D. S. Bethune and C. S. Yannoni, Acc. Chem. Res. 25 (1992) 169.
3. S. Bandow, H. Shinohara, Y. Saito, M. Ohkohchi and Y. Ando, J. Phys. Chem. 97 (1993) 6101.
4. C. S.Yannoni, H. R. Wendt, M. S. de Vries, R. L. Siemens, J. R. Salem, J. Lyerla, R. D. Johnson, M. Hoinkis, M S. Crowder, C. A. Brown, D. S. Bethune, L. Taylor, D. Nguyen, P. Jedrzejewski and H. C. Dorn, Synthetic Metals 59 (1993) 279.
5. D. S. Bethune, R. D. Johnson, J. R. Salem, M. S. de Vries and C. S. Yannoni, Nature 366 (1992) 123.
6. T. Kato, S. Suzuki, K. Kikuchi and Y. Achiba, J. Phys. Chem. 97 (1993) 13425.
7. K. Lassonen, W. Andreoni and M. Parrinello, Science 258 (1992) 1916.
8. S. Nagase, K. Kobayashi, T. Kato and Y. Achiba, Chem. Phys. Lett. 201 (1993) 475.
9. W. Krätschmer, L. D. Lamb, K. Fostiropoulos and D. R. Huffmann, Nature 347 (1990) 354.
10. L. Dunsch, F. Ziegs, J. Fröhner, U. Kirbach, K. Klostermann, A. Bartl and U. Feist, in H. Kuzmany, J. Fink, M. Mehring and S. Roth "Electronic Properties of Fullerenes"; Springer Series in Solid-State Sciences 117 (1993).
11. K. Laarsonen, W. Andreoni and M. Parrinello, Science 258 (1993) 1916.
12. M. Hoinkis, C. S. Yannoni, D. S. Bethune, J. R. Salem, R. D. Johnson, M. S. Crowder and M. S. de Vries; Chem. Phys. Lett. 198 (1992) 461.

NANOTUBE PROPERTIES & GRAPHITE ORIGAMI

Thomas W. Ebbesen
Fundamental Research Laboratories
NEC Corporation, 34 Miyukigaoka
Tsukuba 305, Japan

ABSTRACT

Some of the most important parameters controlling the yield and quality of nanotube samples during production are discussed together with their purification by oxidation. Notions related to the capillary action of nanotubes are then presented in view of using the inner hollow cavity for further experiments. Finally, with the eventual aim of creating novel 3 dimensional carbon structures, the rules governing the folding and tearing of graphitic sheets are discussed.

1. Introduction

The discovery of fullerenes has lead to the understanding that a variety of closed carbon structures can potentially be made. Some of these form spontaneously under the right conditions such as nanotubes which I will discuss in the first parts of this presentation. There are many other structures of which we can only dream for the moment. If we were able to design and construct any carbon structure then we would have the means of producing, at will, materials with a pre-chosen set of properties. Such possibility is of great importance and merits further investigation. It will require among other things the complete understanding of the rules that govern the formation of three dimensional carbon structures. We already know that Eulers rule plays a critical role since we need for instance 12 pentagons to close a hexagonal carbon network such as C_{60} or carbon nanotubes. However, in addition, there is another feature which is critical: the ability of carbon to rehybridize between sp^2 and sp^3. This gives graphitic sheets tremendous out-of-plane flexibility to form three dimensional structures. The rules that govern this property will be discussed in the last part.

2. Carbon Nanotubes: Production and Purification

There are basically three methods to produce nanotubes: carbon arc [1,2], carbon vapor deposition [3,4] and catalytic formation [5,6]. I will only the discuss the parameters related to the first method.

The carbon arc method has been described in details elsewhere [1,2]. I will just remind the reader that in this approach, a plasma is generated between two carbon rods in an inert He atmosphere by passing a DC current. As the positive electrode is consumed, a deposit forms on the negative electrode. This deposit contains nanotubes in various amounts depending on on the following parameters:

- inert gas pressure
- current between electrodes
- spacing between electrodes
- cooling of reaction chamber

The best inert gas appears to be He and its pressure should be about 500 torr. The gas must be flowing continuously. If for example the local pressure of He at the electrodes fluctuates due to bursts (instability) in the plasma, the yield might be significantly affected. The current must be as stable as possible and as low as possible in order to obtain the highest yields. Of course a minimum current is necessary to form the plasma but it should be kept as close as possible to this minimum. Not only does the yield drop with increasing current [7] but a sudden excess current (even for a few seconds) will harden an otherwise nice deposit of nanotubes into a useless solid [2,8]. The stability of the current can be maintained by keeping a small spacing between the electrodes and monitoring it visually.

Another parameter appears to be the cooling of the electrodes and the chamber which seems to affect not only the yield but also the shape and structure of the deposit. Typically two types of deposits are reported as shown in Fig. 1. One is a layered

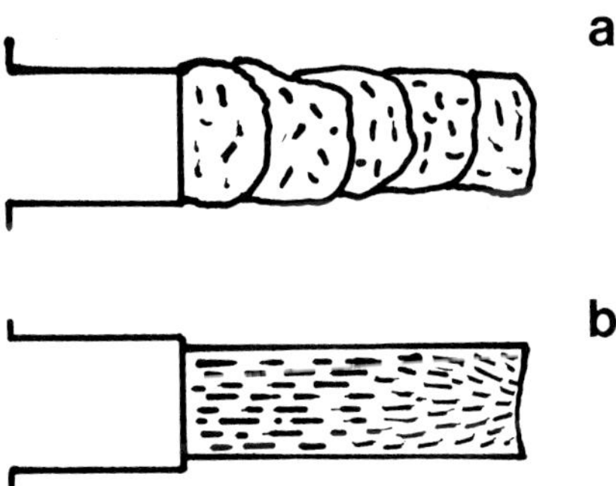

Figure 1: Typical deposits formed in the carbon arc: a) low yield deposit and b) high yield deposit.

structure perpendicular to the direction of the current in which the overall yields are small (Fig. 1a). The nanotubes are growing in all directions and are found in small pockets inside the deposit. It seems that such deposits are formed in poorly cooled reaction chambers. In a well cooled system, as shown in Fig. 2, in which the electrodes are cooled a few centimeters from the actual plasma, a cylindrical deposit will form (if all the other conditions are met). In such deposits (Fig. 1b), the cylindrical structure and the nanotubes are aligned with the flow of the current [2]. The outer hard shell can be

118

cut open and the core will contain fibers of aligned nanotubes and nanotube bundles. If the conditions are right, the fibers are clearly visible to the eye and they will be very soft to the touch of tweezers which are used to remove them from the deposit.

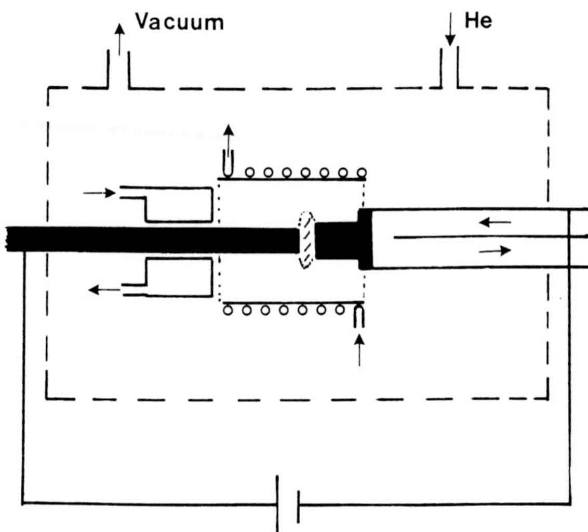

Figure 2: Diagram of the carbon arc apparatus where the nanotubes are formed from the plasma between the two black carbon rods. Notice the extensive cooling system indicated by the arrows [8].

Even with all these precautions, the nanotube yield varies from one group to another. Some places obtain mostly nanoparticles and only 5 to 10% nanotubes. It is still not clear why there such a difference and we are still trying to investigate this point. Understanding why this occurs would allows to go in the opposite direction: improve the ratio of nanotubes to nanoparticles which at present does not appear to exceed about 2 to 1.

This brings us to the subject of purity of the sample. Ideally if we could make only nanotubes in one step it would be, of course, the best. However until then we need a purification method to eliminate the nanoparticles. We first tried many traditional methods used in chemistry and biology such as chromatography and centrifugation but none of them worked well. Finally we found that nanotubes could be purified through oxidation [9]. It was already known that nanotubes could be opened through the preferential oxidation of the tips [10,11]. The tips react faster (due to high strain) then the cylindrical part of the tube which is consumed gradually. The nanoparticles are oxidized at an intermediate rate. However the tubes being much longer than the particles, there is a chance they will be left even when the nanoparticles have been entirely consumed. This is actually observed and results in pure samples as shown in Fig. 3. Although the yield of this purification method is very small, at best 1%, it is enough to obtain pure nanotubes for accurate evaluation of their properties.

The yield of the purification depends on the relative size of nanoparticles to the nanotubes and also on the ratio of nanoparticles to nanotubes in the sample. We have also found that the best results are obtained when the sample is introduced into a cool oven and then gradually bring the temperature up to 750 °C.

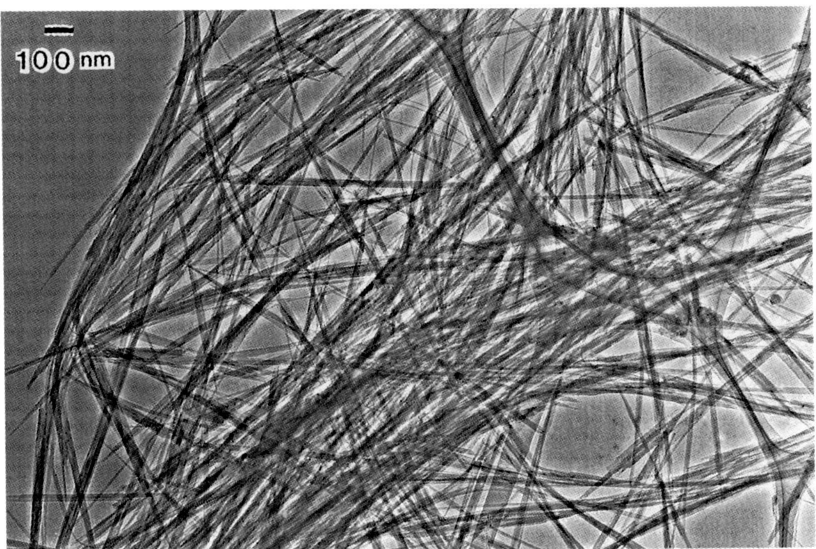

Figure 3: Purified carbon nanotubes

3. Nanotube Capillarity

Introducing materials inside nanotubes at will should open up many possibilities for doing interesting experiments in chemistry, physics and material science. The first example of such possibility came with the filling of nanotube with lead compounds by capillarity [12]. It seemed then that nanotubes might be super-straws sucking up materials as suggested by theoretical predictions [13]. However further studies revealed that filling nanotubes with pure metals was very difficult [11]. In fact nanotubes are not wet by metals such as lead [14]. And wetting (even the outside) is a pre-condition for capillarity to occur. Furthermore, even in the case of wetting, the solid- liquid contact angle θ between a molten droplet and the nanotube surface must be less than $90°$ as it is clear from the Young-Laplace equation:

$$P\,r = 2\,\gamma\cos\theta$$

where P is the pressure, r the radius of the capillary and γ the surface tension of the liquid.

The above equation is known to hold for very small capillaries and so it should

120

be applicable to nanotubes. It is still difficult to predict the capillary action without knowing the contact angles of a molten element. In general, it can be said that the polarizability of the solid should be higher than that of the liquid for good wetting. In practice, it is possible to screen potential candidates by simply seeing if they will wet the outside of the nanotubes in the bulk. We took this approach and were able to narrow down a number of potential candidates. It turns out that good wetting is typically obtained for materials with lower surface tension than lead [14]. Such materials are also seen to be sucked in by capillarity when starting with open nanotubes [14]. Furthermore, it is clear that most solvents, including water should be drawn into nanotubes through capillary action. This is important if one wishes to carry out chemical reactions inside the nanotubes which is one of our aims.

4. Graphite Origami

As already mentioned in the Introduction, it should be possible to tailor the properties of graphitic sheets if we can control their three dimensional structure [15]. We are already aware of the fact that, according to Eulers rule, the introduction of rings with a number of atoms different from 6 (pentagons, heptagons, octagons, etc.) in the hexagonal carbon network will create all kinds of geometries. However we must also understand how graphitic sheets fold and tear to create the novel structures. In a final phase, the technology to actually construct and assemble these novel nano-structures must be developped.

Here I will only discuss the rules governing the folding and tearing of graphitic sheets which we recently clarified [15]. Using atomic force microscopy, AFM, and scanning tunneling microscopy, STM, it is observed that the single sheets of graphite of HOPG (highly oriented pyrolitic graphite) fold and tear along axes displaced by multiples of 30°, i.e. n x 30° where n is an integer. This can be understood by the fact that when one folds or tears a graphitic sheet, it involves the rehybridization of the sp^2 carbon atoms along a line since pure sp^2 is only possible when the graphitic sheet remains flat and continuous. The change from sp^2 towards sp^3 (and eventual bond breaking) must always involve a pair of carbon atoms since a double bond is being disrupted. Then if one consider the possible carbon pairs that are in the graphitic sheet, one finds that there are 4 distinct pairs, 2 along the [100] symmetry axis and 2 along the [210] symmetry axis [15]. Since the symmetry axes are separated from each other by 30°, it is then immediately clear why the folding and tearing occurs at axes separated by n x 30°. These directions are those that are energetically most favored. Only along these axes can the sp^3 type line defects occur un-interrupted from one end of the fold to the other. These considerations allows us to predict the most likely structures that will form and that are actually observed. This is beyond the scope of this proceeding and will be explained in more detail elsewhere [16]. However to illustrate the possibilities, Fig. 4 shows a single graphitic ribbon which has been folded 4 times along very precise angles. This structure occured due to the friction of the AFM tip on the graphite. If we could control this process, we could no doubt make many novel structures at will.

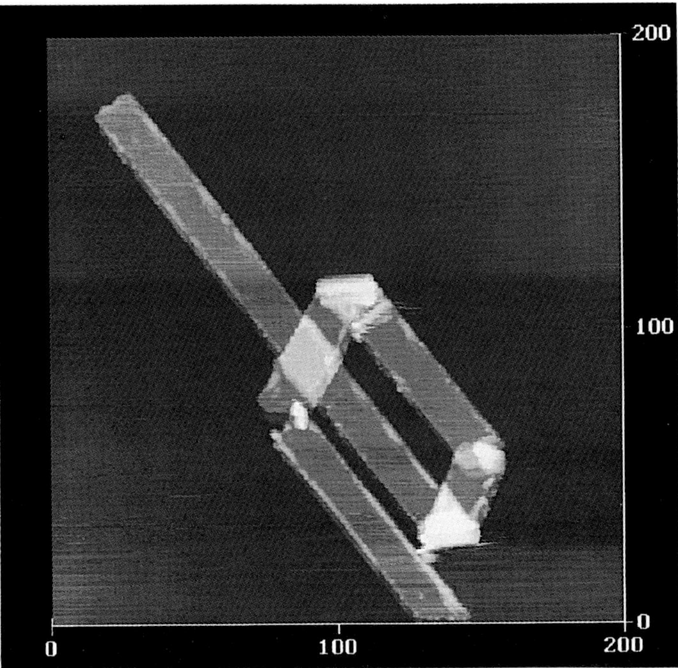

Fig. 4: An example of graphite origami as seen by AFM (courtesy of H. Hiura). Scale is in nanometers.

5. References

1. T.W. Ebbesen and P.M. Ajayan, Nature **358** (1992) 220.
2. T.W. Ebbesen et al, Chem. Phys. Lett. **209** (1993) 83.
3. M. Ge and K. Sattler, Science **260** (1993) 515.
4. Z.Y.A. Kosakovskaya, L.A. Chernozatonskii and E.A. Federov, JETP Lett. **56** (1992) 26.
5. D.S. Bethune et al, Nature **363** (1993) 605.
6. S. Iijima and T. Ichihashi, Nature **363** (1993) 603.
7. S. Seraphin et al, Carbon **31** (1993) 685.
8. T.W. Ebbesen, Annu. Rev. Mat. Sci. **24** (1994) 235.
9. T.W. Ebbesen, P.M. Ajayan, H. Hiura and K. Tanigaki, Nature **367** (1994) 519.
10. S.C. Tsang, P.F.J. Harris and M.L.H. Green, Nature **362** (1993) 520.
11. P.M. Ajayan et al, Nature **362** (1993) 522.
12. P.M. Ajayan and S. Iijima, Nature **361** (1993) 333.
13. M.R. Pederson and J.Q. Broughton, Phys. Rev. Lett. **69** (1992) 2689.
14. E. Dujardin, T.W. Ebbesen, H. Hiura and K. Tanigaki, submitted
15. H. Hiura et al, Nature **367** (1994) 148.
16. T.W. Ebbesen and H. Hiura, submitted.

122

COILED CARBON NANOTUBES

A. A. Lucas, Ph. Lambin, V. Ivanov and J.B. Nagy,
ISIS, University of Namur, rue de Bruxelles 61, B-5000 Namur, Belgium.

D. Bernaerts, X.B. Zhang, X.F. Zhang, S. Amelinckx, G. Van Tendeloo and
J. Van Landuyt,
EMAT, University of Antwerp (RUCA), Groenenborgerlaan 171, B-2020 Antwerp,
Belgium.

Abstract.

The paper presents Transmission Electron Micrographs (TEM) of helical nanotubes
produced by the catalytic breaking of acetylene. A plausible structural model for the
coiling is proposed based on the introduction of pentagon-heptagon pairs in the hexagonal
carbon network.

Carbon tubes have been grown via two distinct routes : by the carbon-arc method for
the production of fullerenes [1] and by the pyrolitic decomposition of hydrocarbons on a
finely dispersed metal catalyst [2].

We have prepared carbon nanotubes by thermal craking of acetylene on a Co/Silica
catalyst. By adjusting the synthesis conditions, we otained very thin filaments mostly
free of amorphous carbon [3]. Electron microscopy revealed that a fraction ($\approx$10 %) of
the tubes have a helix shape similar to but much thinner than those prepared catalytically
by other groups, e.g. by Motojima *et al* [4].

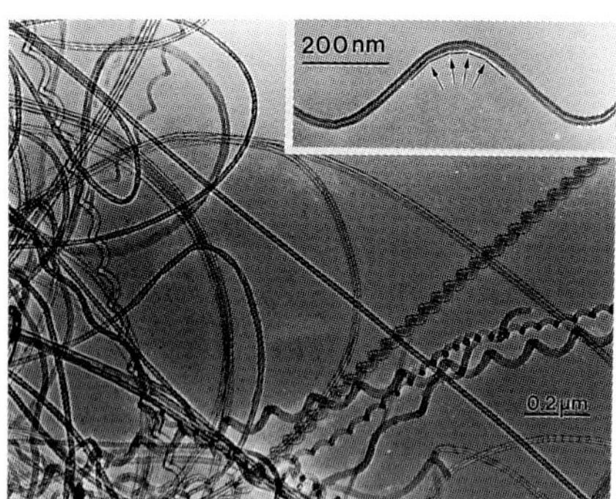

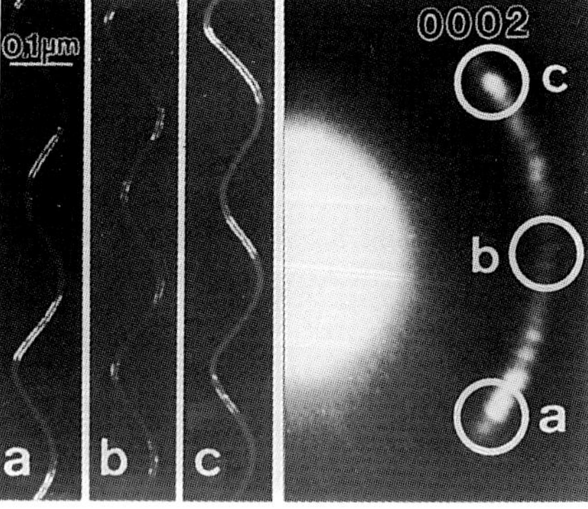

Figure1 (left). TEM micrograph of straight, bent or regularly coiled nanotubes of various radius and
pitch. Some tubes are in bundles of two or more. The inset suggests that the helical tubes are
polygonized.

Figure 2 (right). Selected area diffraction (right) and imaging (left) of a polygonized, coiled tube seen
perpendicular to its axis. Note the spotty appearance of the diffraction arc of the coil. The selected circles
a, b, c serve as apertures to produce the corresponding images of the left panels.

Fig.1 is a TEM picture of several very thin tubes. Some are straight or bent, isolated or in bundles, while others are coiled with various radii and pitches. All tubes are hollow. The inner and outer diameters range from 2 to 7 nm and 15 to 50 nm, respectively. The inner sizes reflect the diameters of the Co catalytic particles [5]. The inset of Fig.1 shows that the coiled tubes do not curve smoothly but proceed by sharp bends of about 30° (seen as smaller angles in the planar projection). Straight cylindrical segments are somehow connected at each bend. There are about 12 such segments to complete each turn of this polygonized helix.

The right part of Fig.2 is the diffraction pattern obtained from a selected area of a coiled tube comprising a couple of helix turns and whose axis is normal to the electron beam. Next to the strong specular spot, there is an arc of discrete spots labelled 0002. These spotty arcs are produced by the diffraction of the straight cylindrical segments just discussed. To understand this, one should visualize one segment as made of several nested cylindrical graphene layers arranged coaxially like tree rings around the hollow core. The interlayer distance is the characteristic graphite spacing $c = 0.34$ nm. The layers seen tangentially by the electron beam act as a diffraction grating of period c and produce pairs of $(000\pm2n)$ diffracted beams of order 2n [1] in a plane perpendicular to the axis of the cylinder segment (the factor 2 stems from the 2c lattice periodicity in hexagonal graphite). Since in a regularly polygonized coil the axis orientations of the successive segments form a cone, their perpendicular must generate the arc of spots shown in Fig.2. Thus the diffraction pattern confirms the polygonized appearance of the TEM images. The left part of Fig.2 further strengthens this interpretation. These are images obtained by placing the electron imaging aperture at positions a, b, c in the spotty arc. The brighter segments in the a, b, c images clearly indicate those parts of the helix which cause the corresponding diffraction spots. The maximum intensity towards the extremities of the arc arises from the extremum property of the planar projection angle of the corresponding segments [5].

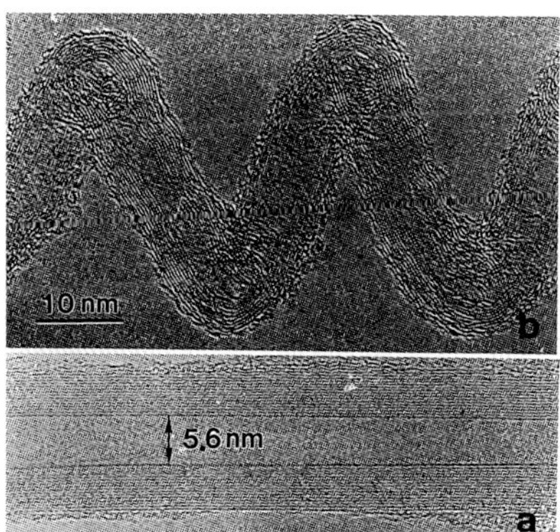

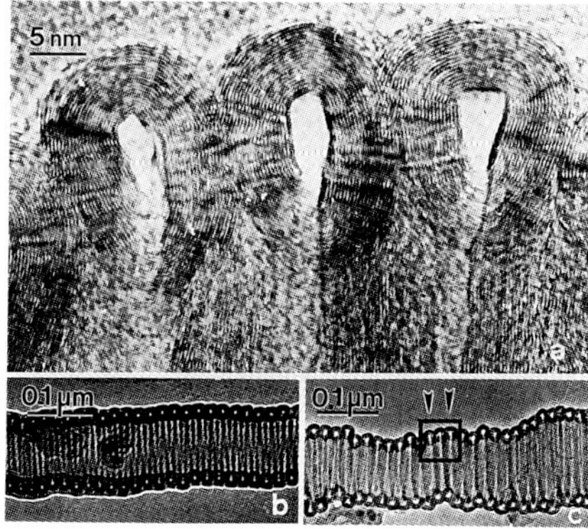

Figure 3 (left). High resolution images of nanotubes. Note the continuity of the lattice fringes in **a** (straight tube) and in **b** (coiled tube).

Figure 4 (right). Tightly wound coiled tubes. In **a** the circular sections are seen to be deformed by adhesion forces. **b** and **c** show the squeezed coil in both edge-on and transverse directions in two different focusing conditions.

124

Figs. 3 and 4 are high resolution images of tightly wound coils seen in projection perpendicular to the helix axis. Notwithstanding the apparent disorder and the overlap of layer fringes, one can clearly count in Fig.3b a dozen coaxial graphene layers around the hollow core of 2 or 3 nm diameter. A straight tube is shown in Fig.3a, having the same 0.34 nm interlayer spacing. One revealing feature is the *longitudinal continuity* of the layer fringes along the coil of Fig.3b. The *circular continuity* of the fringes in the tree ring structure is also evident from Fig.4a. Here the helix is so tightly wound that the successive turns touch each other and are deformed under their mutual van der Waals attraction.

We now discuss a model atomic structure which can explain the observed polygonized coiling. From the continuity of the graphene layers just discussed, we surmise that point defects could be responsible rather than extended ones such as grain boundaries. We also remember from the structure of globular fullerenes that pentagon rings cause positive curvatures in a hexagonal network while heptagon rings cause saddle points. Both positive and negative curvatures are required in a helix. The question is whether these defects can realize the connection of two straight tubes of same diameter at a large angle of order 30°? A positive answer has recently been provided by the theoretical works of Dunlap [6] and Ihara *et al* [7]. Dunlap showed that 30° is precisely the favourable angle to connect two straight, single wall cylindrical tubes made of sp2 carbon by the introduction of just one pentagon and one heptagon at opposite sides of the tubes. Ihara et al made molecular dynamics simulations on atomic size, single wall, coiled tubules containing pairs of such 5-7 rings. They demonstrated the topological feasibility and thermodynamical stability of the structures. These were the defects also assumed by Iijima *et al* [8] to explain their observations of straight tube transformations such as U-turn reversals.

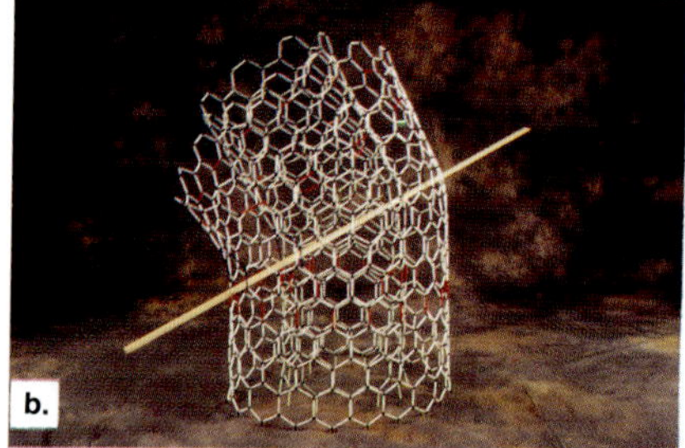

Figure 5. Molecular model of three straight cylindrical segments of a coiled single layer tube. The middle "zigzag" tube (with C-C bonds parallel to the tube axis) has 12 hexagons in its circumference and connects at 30° angle to two "armchair" tubes (with C-C bonds perpendicular to the tube axis) of same diameter 0.93 nm. The connections are realized with minimum strain by pairs of opposed pentagon-heptagon rings.

Fig.5 shows a molecular model of one possible structure. The 5-7 opposed rings are seen to produce sharp 30° bends connecting one achiral, "zigzag" (12,0) tube to two achiral, "armchair" (7,7) tubes *of same diameter* (the notations are those of ref. [6]). Chiral tubes can also be connected in a similar way [6]. Multilayer, nested tubes are obtained by aligning the 5-7 ring pairs from one layer to the next. The dimension of the tube shown in Fig.5 (about 0.9 nm diameter) can be scaled up to the sizes of our observed tubes while respecting the 0.34 nm graphite interlayer distance [5]. For instance a multilayer tube build on (12+9n,0)-(7+5n,7+5n) connections [6] with $3 \leq n \leq 15$ will resemble the tube in Fig.3b. Coiling is effected by regularly rotating the azimuth of the 5-7 connectors by a fixed angle around the cylinder axis, as shown in Fig.5. This offset angle, along with the (arbitrary) length of the straight segments determine the coil radius and pitch [5].

It remains to be seen whether further experimental evidences will confirm the proposed defect structure and whether or not other types of structural defects could accomplish the observed bending and twisting of nanotubes. Also, it will be interesting to devise kinetic [9] and dynamic mechanisms to explain the regular introduction of 5-7 point defects during the catalytic growth process.

Acknowledgments.

This work was in part supported by the Belgian Inter-University Research Program initiated by the Prime Minister Office, by the Regional Government of Wallonie and by the Belgian National Science Foundation. One of us (A.A.L.) thanks B.I. Dunlap for helpful discussions.

References.

[1] S. Iijima, Nature **354**, 56 (1991); S. Iijima and T. Ishihashi, Nature 363, 603 (1993); D.S. Bethune *et al*, Nature 363, 605 (1993).
[2] see e.g. N.M. Rodriguez, J. Mater. Res. **8**, 3233 (1993) and references therein.
[3] details of the synthesis can be found in Ivanov *et al*, Chem. Phys. Lett., in press.
[4] S. Motojima *et al*, Proc. 9th EURO-CVD (1993) and references therein.
[5] details can be found in X.B. Zhang *et al*, Europhysics Lett., in press; D. Bernaerts *et al*, Phil. Mag., submitted.
[6] B.I. Dunlap, Phys. Rev. B46, 1933 (1992); Phys. Rev. B48, 5643 (1994).
[7] S. Ihara *et al*, Phys. Rev. B48, 5643 (1993).
[8] S. Iijima *et al*, Nature 356, 776 (1992).
[9] S. Amelinckx *et al*, Science, in press.

RESONANCE RAMAN SPECTROSCOPY OF MULTISHELL CARBON NANOTUBES

J. Kastner and H.Kuzmany

Institut für Festkörperphysik, Universität Wien
Strudlhofgasse 4, A-1090 Vienna, Austria

D.N. Weldon and W. Blau

Chemistry Department and Department of Pure and Applied Physics
Trinity College, Dublin 2, Ireland

ABSTRACT

We report on the vibrational properties of multishell carbon nanotubes with an average diameter of about 40 nm studied by resonance Raman spectroscopy. Due to the rather large tube size the Raman spectra resemble strongly those of graphite crystallites. The first order spectrum exhibits two lines centered at 1582 cm^{-1} and at 1350 cm^{-1}. The former can be identified as the broadened E_{2g} line of monocrystalline graphite. The latter is disorder induced and enhanced by the curved nature of the graphitic layers in the tubes. The observed relation between the ratio of the integrated intensities of these lines and the in-plane graphite sheet size of carbon nanotubes was found to be different from the corresponding relation in poly-crystalline graphite. Position and intensity of the line around 1350 cm^{-1} strongly depend in the energy of the exciting laser. This dispersion effect is discussed in terms of a photoselective resonance process. The different resonance Raman behavior of carbon nanotubes and graphite microcrystallites is attributed to differences in the electronic structure.

1. Introduction

The unique structure of carbon nanotubes and their expected unusual properties like high tensile strength and quasi one-dimensional conductivity has led to enormous interest in this material[1,2]. Methods for large scale synthesis of multi shell tubes[3] with several tenths of nm in diameter as well as single wall tubes[4] with a diameter down to 0.7 nm have been developed. Carbon nanotubes have the physical dimensions of both fullerenes (several nm in diameter) and graphite crystallites (several hundreds of nm or μm in length), which should be reflected in the vibrational properties.

Recently Raman spectra and infrared spectra of multi shell tubes have been reported[3,5]. A strong resemblance to the vibrational properties of graphite was found. The strongest Raman line at about 1580 cm^{-1} has been derived from the E_{2g} mode from monocrystalline graphite (G line). In addition, a disorder induced line at around 1350 cm^{-1} (D line) and its overtone have been observed.

2. Experimental

For the production of carbon nanotubes a steel fullerene generator (450 Torr He atmosphere) was employed. An 8 mm ultrapure graphite rod and a plug were used as the positive and the negative electrode, respectively. After applying a DC voltage of 27 V a cylinder with a grey outer layer and a black inner core was found on the negative electrode. High resolution electron microscopy (HREM) showed that the inner core contains approximatively 70 % nanoparticles, of which nanotubes are by far the most abundant[5]. The average length of the tubes as measured by HREM is about 130 nm. Raman scattering was performed on a Dilor XY spectrometer with a liquid nitrogen cooled CCD–multichannel detector. For excitation an Ar^+, Kr^+ and Ti:sapphire laser with wavelengths ranging from 457 up to 742 nm have been used. A typical intensity for excitation was 140 W/cm^2.

3. Results and Discussion

Fig. 1 shows the Raman spectrum of carbon nanotubes. The dominating line has its maximum at 1582 cm^{-1}. It can be assigned to the above mentioned G line. The full width at half maximum (FWHM) of 20–22 cm^{-1} is only slightly higher as compared to that of highly oriented pyrolytic graphite (HOPG)(15–18 cm^{-1}), which indicates a high crystallinity and a low concentration of defects. In the first order spectrum there is an additional line at 1350 cm^{-1} and a weak feature at about 1620 cm^{-1}. These lines appear in the spectrum due to a breakdown of the wave vector selection rule resulting from finite crystal size and disorder effects[5].

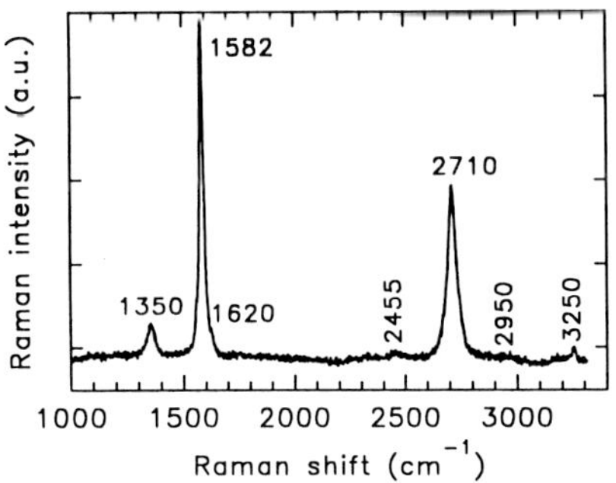

Fig. 1: Raman spectrum of carbon nanotubes in the spectral range between 1000 and 3400 cm^{-1} as excited with 514.5 nm.

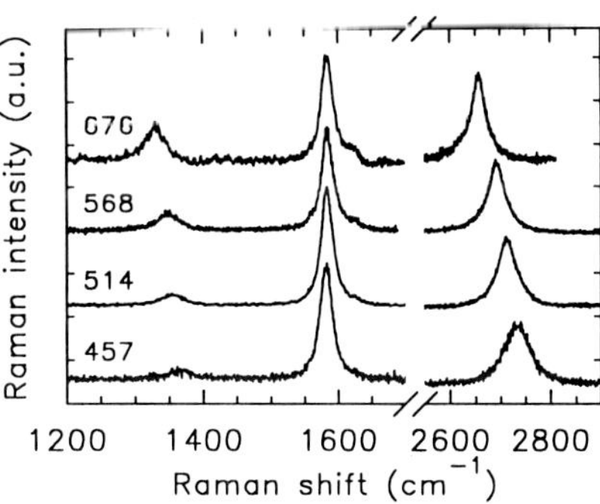

Fig. 2: Raman spectra of carbon nanotubes after excitation with different laser lines. The numbers indicate the wavelength in nm.

The relative intensity of the D line can be related to the in-plane graphite crystallite size[6]: L (in nm) = 4.4 x I_G/I_D. Since in our case the relative intensity (I_D/I_G) is

128

0.23 the theoretical nanotube length is 20 nm. This is much lower than the average obtained from HREM. We assign this discrepancy to an enhancement of the D line by the curved nature of the graphitic layers in the nanotubes.

In the second order Raman spectrum there is a dominating line at about 2710 cm^{-1} and several weaker features at 2455, 2950 and 3250 cm^{-1}. The strong line is generally regarded to be the second order of D line[5] and similarly the weaker lines can be also assigned to overtone or combination scattering (2455 $\approx$ 840 + 1620, 2950 $\approx$ 1350 + 1620, 3250 $\approx$ 2 x 1620).

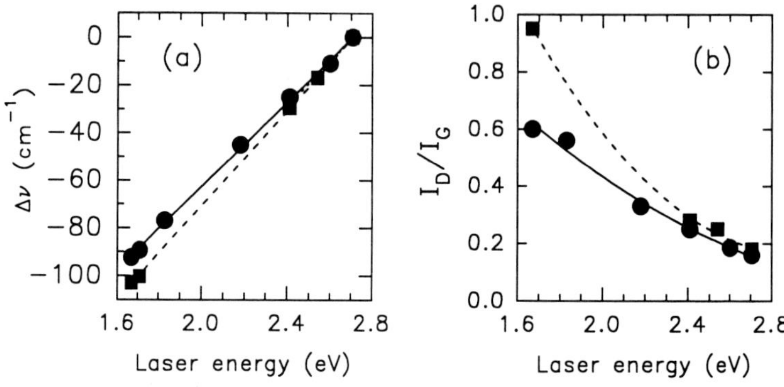

Fig. 3: Frequency shift ($\Delta\nu$) of the line around 2700 cm^{-1} (a) and integrated intensity of the D line (I_D/I_G) as compared to the G line (b) for carbon nanotubes ($\odot$) and graphite microcrystallites ($\square$) versus the energy of the exciting laser.

In Fig.2 Raman spectra of carbon nanotubes are shown as measured with different laser lines. The spectra are normalized to equal height of the G line. From the spectra it is evident that both the D and its second order shift to lower wavenumbers with increasing wavelength of the exciting laser, whereas the position of the G line remains constant. In addition, after excitation with the yellow or red laser the line at 1620 cm^{-1} appears in the spectrum more pronounced. Fig.3(a) shows explicitly the shift ($\Delta\nu$) of the second order of the D line with the energy of the exciting laser. $\Delta\nu$ for nanotubes is by about 15 % weaker as compared to graphite crystallites. The change of the relative intensity of the D line is significant as Figs. 2 and 3(b) demonstrate. In the latter figure the intensity ratio of the D line to the G line is plotted. This ratio decreases nonlinearly with increasing laser energy and behaves characteristically different for tubes and graphite crystallites.

A dependence of the Raman shift of a mode on the energy of the exciting laser is formally called dispersion effect. Two different photoselective resonance mechanisms are possible explanations for this effect:

1. Force constants, frequencies and electronic transitions might depend on the size of the nanotubes. By measuring with different laser energies one is probing selectively different parts of the sample and the overall response gives the resulting shape and

line position. This scenario is well known from conjugated polymers[7] and sp bonded linear carbon molecules[8].

2. The dispersion of the D phonon at the Brillouin zone edges at the M and K derived points might be responsible for the line shift with the exciting laser. Excitations with different laser lines result in different resonance enhancements of the contributions from regions near the K and M derived points to the D line.

In both pictures the special electronic structure of carbon nanotubes accounts for the difference in the resonance Raman behavior as compared to graphite. The change of the I_D/I_G ratio with the energy of the exciting laser can be explained similarly as in graphite[9] by the different resonance enhancements of the G line as compared to the D line. By excitation with a red laser one is far away from the π–π^* transition (for graphite in the range between 4 and 5 eV) and the G line is not any longer enhanced as compared to the D line. Applying this model to the results presented here reveals again a difference in the resonance Raman profile and thus in the electronic structure for graphite and carbon nanotubes.

Acknowledgment

This work was supported by the Osteuropaförderung des BMfWF project GZ 45.338/1-IV/6a/94.

4. References

1. S. Iijima, *Nature* **354** (1991) 56.
2. M. S. Dresselhaus, G. Dresselhaus and P. C. Eklund, *J. Mater. Res.* **8** (1993) 2054.
3. H. Hiura, T. W. Ebbesen, K. Tanigaki and H. Takahashi, *Chem. Phys. Lett.* **202** (1993) 509.
4. S. Iijima and T. Ichihashi, *Nature* **363** (1993) 603.
5. J. Kastner, T. Pichler, H. Kuzmany, S. Curran, W. Blau, D. N. Weldon, M. Delamesiere, S. Draper and H. Zandbergen, *Chem. Phys. Lett.* (1994) in print.
6. D. S. Knight and W. B. White, *J. Mater. Res.* **4** (1989) 385.
7. H. Kuzmany and J. Kastner, in *Macromolecules 1992*, ed. J. Kahovec (VSP, Zeist, 1993) p 333.
8. J. Kastner, H. Kuzmany, L. Kavan, F. Dousek and J. Kürti, *J. Chem. Phys.* (1994) submitted.
9. K. Sinha and J. Menendez, *Phys. Rev.* **B41** (1990) 10845.

ELECTRONIC STRUCTURE OF COAXIAL CARBON TUBULES

Ph. Lambin
Département de Physique, Facultés Universitaires Notre-Dame de la Paix,
61 Rue de Bruxelles, B 5000 Namur, Belgium

J.-C. Charlier and J.-P. Michenaud
Physico-Chimie et Physique des Matériaux, Université Catholique de Louvain,
1 Place Croix du Sud, B-1348 Louvain-La-Neuve, Belgium.

Abstract

We have investigated the electronic structure of multilayered carbon nano-fibers using a realistic tight-binding Hamiltonian. Metallic tubules with the so-called armchair configuration have been considered. We have analyzed the effects of interlayer interactions on the electronic structure of fibers composed of a few (2 to 5) coaxial tubules. It is shown that the interlayer couplings only weakly affect the metallicity of the tubules, except that some multilayers built around the armchair tubule with 10 atoms per ring (based on C_{60}) may become semimetallic.

1. Introduction

The electronic structure of a single-layer graphite tubule is well understood : depending on its helicity and diameter, an isolated buckytube is either a metal or a semiconductor with a small or moderate band gap.[1,2] Real carbon nano-fibers are generally composed of several coaxial layers. For the case of bi-layer fibers, recent calculations indicate that the interlayer interactions only weakly affect the electronic properties of the tubules.[3] It is our purpose to analyze these effects in more details for the case of non-helical buckytubes having their axis perpendicular to a C-C bond. Isolated tubules with this perpendicular (or armchair) configuration are predicted to be metallic.[1] The radius of such a tubule composed of n atoms arranged in n/2 pairs around a circumference scales like $\rho \sim (3d_{nn}/4\pi)n$. With the assumed fixed nearest-neighbor C-C distance $d_{nn} = 1.42$ Å, ρ increases by 3.39 Å for each addition of 10 atoms per ring. Knowing that the interlayer distance measured experimentally in coaxial tubules is 3.4 Å,[4] it appears that the perpendicular configuration is realistic for modeling multilayered tubules when it is assumed that the numbers of atoms are increasing by 10 units from one layer to the next. In Sect. 3, the band structure of bi-layer fibers consisting of n=10 and n=20 concentric tubules is investigated and is compared with that of a less commensurate 12-22 fiber. A five-layer fiber is also considered and insight into the band structure of larger systems is proposed. First, a brief description of the tight-binding Hamiltonian used for these calculations is presented in Sect. 2.

2. The Hamiltonian

The tight-binding Hamiltonian used is specialized to the sp^2 carbon network with four orbitals per atom. The 2s and 2p atomic levels and the Slater-Koster two-center parameters for intra-layer nearest-neighbor interactions have been adjusted to a recent *ab-initio* band structure of graphite.[5] The parameters obtained are $\varepsilon_s = -5.49$ eV, $\varepsilon_p = 0$ (zero of energy), $V_{ss\sigma} = -4.80$, $V_{sp\sigma} = 4.75$, $V_{pp\sigma} = 4.39$ and $V_{pp\pi} = -2.56$ eV. For tubule calculations, we have found that reducing the $pp\pi$ transfer integral lead to results in better agreement with LDA band structures. The expression used for $V_{pp\pi}$ is $-2.56/(1+1.5/\rho^2)$ (eV), where ρ (in Å) is the curvature radius of the buckytube.

Interlayer interactions are considered between π orbitals only. They are defined as $|p_\pi\rangle = l|p_x\rangle + m|p_y\rangle + n|p_z\rangle$ where l, m and n are the direction cosines of the vector normal to the carbon network. Interactions are considered between a site I on a layer and sites J of the next layer whose distances measured on that layer from the projection of site I do not exceed 1.9 Å. This value was chosen so as to have six connections per atom at most. The interlayer interactions are written as $V \cos\phi \, e^{-(d-\delta)/L}$ where ϕ is the angle between the two π orbitals, d is their separation distance, δ is the interplanar spacing in graphite (3.34 Å) and $L = 0.45$ Å.[6] The energy parameter V can take two values V_1 or V_2. All the interlayer interactions tentatively receive the parameter V_1 with the restriction that there cannot be two V_1 interactions in any "triangular path" (i.e. a closed walk through three connected sites). When this happens, the V_2 parameter is used for *both* interlayer interactions involved in the triangle (the third side of the triangle is an intra-layer bond). There is one exception to this rule : the interaction parameter keeps the V_1 value for each bond connecting two atoms whose projections on the outer layer come closer to 0.5 Å.

When this algorithm is applied to natural graphite, with its two kinds of atoms A and B, the parameter V_1 is found working to both AA and BB bonds. The corresponding interactions determined *ab-initio* are 0.36 and 0.32 eV, respectively. [5] We take $V_1 = 0.36$ eV and so correctly describe the AA interaction. The BB interaction is then reduced to 0.19 eV by the assumed exponential law, owing to its larger bond length ($d = \sqrt{\delta^2 + d_{nn}^2}$). For the second-neighbor interactions, we take $V_2 = 0.16$ eV. In natural graphite, this value leads to 0.08 eV for the AB interaction. This is about two times smaller then the *ab-initio* value (0.18 eV) [5] but better reflects the experimental data (0.02 - 0.07 eV).[7] The algorithm briefly described above has somewhat improved with respect to an earlier version that we first developed for turbostratic graphite[8] and applied to double-layer tubules.[9] There is in particular much more electron-hole symmetry in the band structures obtained by the present approach.

3. Results

We first consider a two-layer fiber. The inner and outer tubules with the armchair geometry have 10 and 20 atoms per circumference. They have symmetry D_{5h} and D_{10h} respectively, and the symmetry of the bi-layer system is at least C_5. For suitable positions of the layers, it is however possible to preserve the full symmetry of the inner tubule. One such configuration has been considered in the literature.[3] We have investigated another structure, having D_{5d} symmetry, which corresponds to a local minimum of the total energy of the system.[10] This configuration is illustrated in Fig. 1 together with the computed band structure. All the atoms of the inner tubule are equivalent and make (second-neighbor) interactions with four atoms of the external tubule. A remarkable feature of the electronic band structure of the isolated armchair tubules is the existence of two non-degenerate bands crossing at E=0 at approximately 2/3 of the first-Brillouin zone extension.[1,2] For the coupled system, there are four allowed crossings between the two pairs of non-degenerate bands derived from both tubules. This system is metallic with its Fermi level located close to E=0.

Other configurations with lower symmetries can be generated by translating and rotating the layers relative to each other. Some crossings between the four non-degenerate bands around E=0 become avoided and the system may be driven in a semimetallic state by the interlayer interactions. Fig. 2 shows a configuration of the 10-20 bi-layer having S_{10} symmetry, where the electronic structure is much affected by the coupling between layers. With the tight-binding parameters of Sect. 2, this system

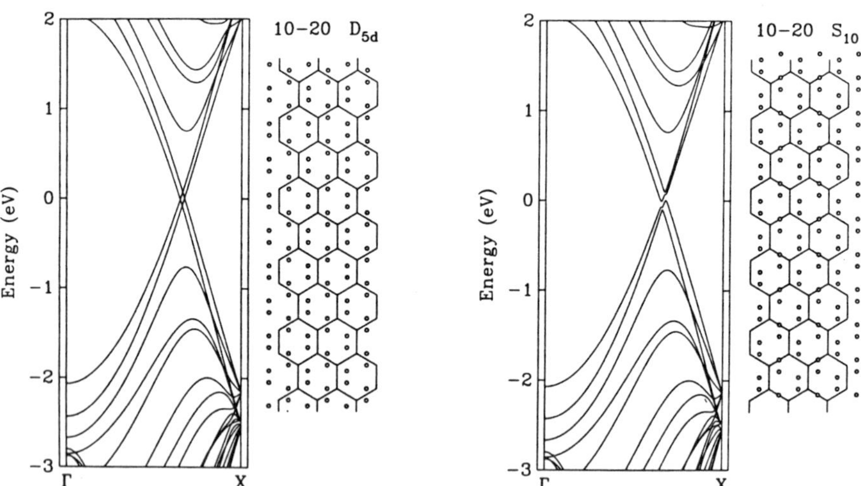

Figure 1 (left). Band structure of a fiber consisting of pair of concentric armchair tubules with 10 and 20 atoms per circumference. The configuration of the fiber projected on the inner tubule and developed on the plane is also shown on the right-hand side of the band structure. In this drawing, the cylindrical axis is along the horizontal direction. The atoms of the outer tubule have their projections indicated by circles, the atoms of the inner tubule are located at the vertices of the regular hexagonal network. The radii of the tubules are 3.41 and 6.79 Å.

Figure 2 (right). Same as in Fig. 1 for another configuration of the 10-20 bi-layer fiber having a lower symmetry.

presents disconnected electron and hole bands. The Fermi level is again found to be very close to E=0 and this leaves the fiber with a very low density of few carriers.

Fig. 3(a) shows a detailed portion of the band structure of another bi-layer where the concentric tubules have 12 and 22 atoms per ring. Their point groups D_{6h} and D_{11h} have no common elements except a possible reflection plane. Due to the low symmetry, the twofold degeneracy of the parabolic-like bands derived from each tubule is split by the interlayer coupling. In the absence of coupling, the two pairs of non-degenerate bands with linear dispersion cross each other at E=0. The pair of A_1 bands are split by the interlayer interactions whereas the A_2 bands are much less affected. There is little interactions between these two groups of bands, in contrast to the 10-20 system (Fig. 2) where there is more geometrical correlation between the layers. The band structure of a 5-layer tubule is shown in Fig. 3(b). There are 5 well-separated A_1 bands crossing a set of 5 nearly-degenerate A_2 bands near E=0. The density of states of the coaxial tubules (see Fig. 4) presents a plateau around the Fermi level ($E_F \approx 0$). The width of that low-density plateau varies linearly with inverse diameter of the fiber, and is consistent with recent measurements by scanning tunneling spectroscopy.[11]

The band structure of a N-layer armchair fiber can be sketched from first-order perturbation theory. When curvature effects are neglected, assuming tubules with not too small radii ($\rho > 10$ Å), the dispersion relation of the A_1 (or A_2) band of an isolated armchair tubule is found independent of its radius.[3] In a fiber composed of N concentric layers without interactions, the A_1 (or A_2) bands derived from the N tubules are degenerate, as a consequence. To first-order in the interaction V between adjacent layers, the energy shifts of the individual bands for a given wave vector are the eigenvalues of

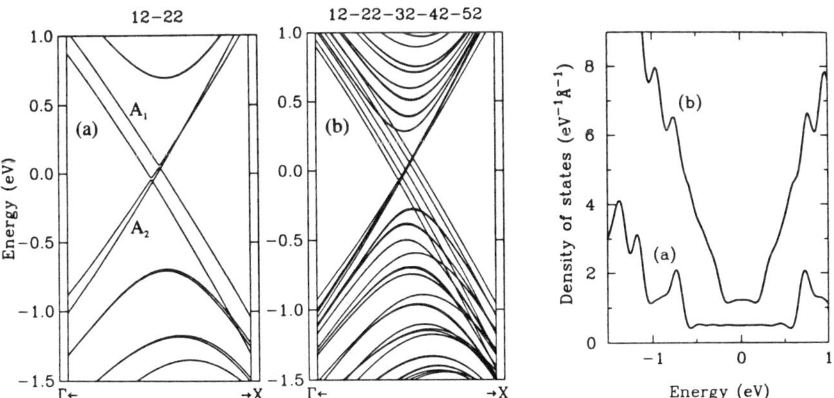

Figure 3. (left). Band structure of a fiber consisting of (a) two and (b) five concentric tubules with the perpendicular configuration. In both cases, the inner tubule has 12 atoms per circumference. This number increases by units of 10 for each additional layer. The diameter of the external tubule is 14.9 Å in case (a) and 35.3 Å in case (b). The wave vectors range from 1/2 to 5/6 of the ΓX line.

Figure 4 (right). Density of states of the two multilayered fibers of Fig. 3.

the tri-diagonal matrix

$$\begin{pmatrix} 0 & <\psi_1|V|\psi_2> & 0 & \\ <\psi_2|V|\psi_1> & 0 & <\psi_2|V|\psi_3> & \\ 0 & <\psi_3|V|\psi_2> & 0 & \\ & & & \ddots \end{pmatrix}$$

where ψ_i is the Bloch function of the A_1 (or A_2) band of the ith layer. The wavefunction with A_2 symmetry has opposite signs on two nearest-neighbor C sites around the same ring. By contrast, the wavefunctions of the A_1 bands are totally symmetric under the operations of the C_{nv} group of a general Bloch vector. This explains why the interlayer coupling affects the A_1 bands more than the A_2 bands (see Fig. 3). At the large-N limit, and assuming that the matrix elements $<\psi_i|V|\psi_{i+1}>$ tend to a constant value V, one finds that the perturbed A_1 states eventually form a continuum extending from -2V to +2V on both sides of the unperturbed branch. In planar graphite, the width 4V of the π bands is 1.4 eV at the K point of the first Brillouin zone.[5] For the five-layer tubule of Fig. 3(b), the width of the A_1 band is 0.3 eV only. This low value again traduces poor geometrical correlation between cylindrical graphene layers in nano-fibers.

Acknowledgment. This work has been supported by the Belgian Interuniversity Research Project on Sciences of Interfacial and Mesoscopic Structures (PAI/IUAP N. P3-49). Ph. L. and J.C. C. acknowledge the National Fund for Scientific Research of Belgium.

References

[1] J.W. Mintmire, B.I. Dunlap, and C.T. White, Phys. Rev. Lett. **68** (1992) 631 ; N. Hamada, S.Sawada, and A. Oshiyama, Phys. Rev. Lett. **68** (1992) 1579.

[2] R. Saito, M. Fujita, G. Dresselhaus, and M.S. Dresselhaus, Phys. Rev. B **46** (1992) 1804 ; C.T. White, D.H. Robertson, and J.W. Mintmire, Phys. Rev. B **47** (1993) 5485.

[3] R. Saito, G. Dresselhaus, and M.S. Dresselhaus, J. Appl. Phys. **73** (1993) 494.

[4] Z.G. Li, P.J. Fagan, and L. Liang, Chem. Phys. Lett. **207** (1993) 148.

[5] J.C. Charlier, X. Gonze, and J.P. Michenaud, Phys. Rev. B **43** (1991) 4579.

[6] N. Tit and V. Kumar, J. Phys.: Condens Matt. **5** (1993) 8255.

[7] E. Mendez, A. Misu, and M.S. Dresselhaus, Phys. Rev. B **21** (1980) 827.

[8] J.C. Charlier, Ph. Lambin, and J.P. Michenaud, Phys. Rev. B **46** (1992) 4540.

[9] Ph. Lambin, J.C. Charlier, and J.P. Michenaud, Comput. Mater. Sci. (1994, in press).

[10] J.C. Charlier and J.P. Michenaud, Phys. Rev. Lett. **70** (1993) 1858.

[11] C.H. Olk and J.P. Heremans, J. Mater. Res. **9** (1994) 259.

TWO NEW DISCHARGES FOR PRODUCTION OF FULLERENES AND NANOTUBES

Churilov G.N.
Institute of Physics
Akademgorodok, Krasnoyarsk, 660036, Russia

ABSTRACT

Obtained by author tow discharges of kHz frequency range have been used for the first time to produce fullerenes and nanotubes. The setup for production of fullerenes and nanotubes in spontaneously blowing out and self-focusing carbon plasma stream is presented. The setup for fullerene synthesis in magnetron discharge a.c., when the arc current and the magnetic field are in phase. UV/visible electronic spectra of fullerenes are presented.

The arc-discharge technique[1,2] yielding 50 mg/h of pure C_{60} is the most popular means of producing fullerenes. However even this most efficient method is unlikely to cover the anticipated needs of industry.

We employed two ac discharges of kilohertz frequency range to generate fullerenes. Quite a while ago we observed the effect of spontaneous ejection of a self-focusing carbon plasma jet[3]. Plasma was generated in a graphite chamber or water-cooled copper chamber with a hole (Fig.1). Power was supplied by a step-down transformer winding made as a solid metal can. An arc of 10-30 kW at $150-10^3$ A current and a frequency of 20-300 kHz was initiated by a graphite rod touching to the hole wall. The central electrode was a carbon or graphite rod, 6-12 mm in diameter. The resultant carbon deposit was found to comprise nanotubes and a small amount of fullerene (its presence was detected by A.V. Okotrub at the Institute of Inorganic Chemistry, Novosibirsk) (Fig.2).

The only plausible explanation to the ambiguous result obtained can be that the plasma jet does not mix well

Fig.1. Spontaneously ejecting, self-focusing carbon plasma jet effluent in air at atmospheric pressure. At an applied power of about 200 kW the jet length was 0.7 - 1 meter.

with air. An ordinary carbon arc spectrum exhibits bright CN lines which indicates homogeneity of the plasma-arc/air mixture. In the plasma jet spectrum these lines are not observed. The plasma jet contains insignificant amount of nitrogen and consequently oxygen and the latter does not affect fullerenes formation. We applied spontaneous carbon plasma jet in a device generating fullerenes in vacuum. A plasma jet was turned onto a water-cooled stainless steel cone placed inside a water-cooled stainless steel chamber. The internal pressure varied from 1 to 10 atm. The chamber was filled with helium. Carbon rods for spectral analysis were used as a central electrode. The external electrode was made of graphite and had a conical hole. We observed that at certain frequencies the rate of graphite rod vaporization was abnormally high. These frequencies depend on the rod diameter and we believe the effect to be associated with ultrasonic vibrations induced in the rod. Vaporization of carbon rods for many of the devices on fullerene production is known[4] not to exceed 10^{-4} g/Coul. In our design it was 10^{-3} g/Coul. The percentage of C_{60} and C_{70} fullerenes appears to be the same as when produced at 50 Hz. This is proved by optical spectra shown in Fig.3.

The other discharge applied in fullerene technology is one of a magnetron nature on a kilohertz frequency scale. A power supply diagram of the plasma generator is presented in Fig.4. The generator

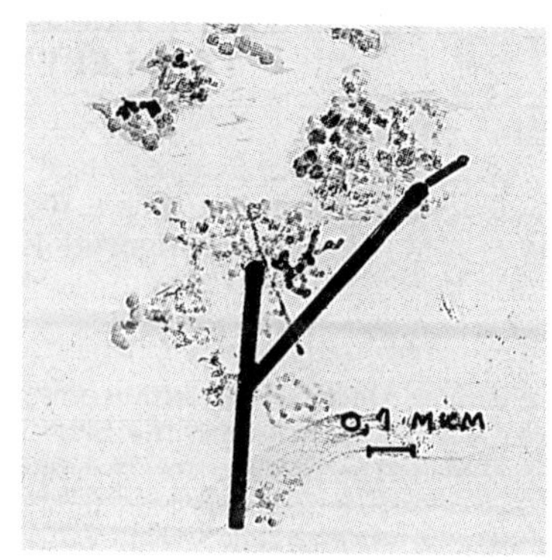

Fig.2. Electronic micrographic picture of carbon deposit from the carbon plasma jet effluent in air.

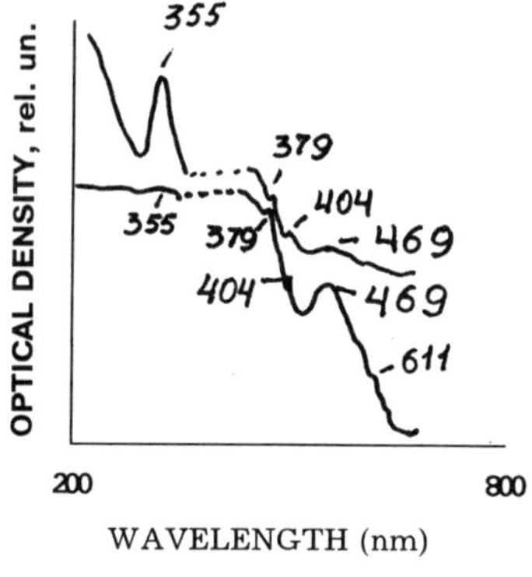

Fig.3. UV/visible electronic spectra between 200 and 800 nm in spectroscopic hexane solution of a mixture of C_{60} and C_{70}. Upper curve: reference sample, lower curve: fullerenes produced. Absorption curves was obtained with a 5 cm and 0.2 cm cells.

functions as follows. Electric generator 1 produces ac current at frequency ω which is then amplified by amplifier 2. Capacitor 3 and inductance L of the primary winding of transformer 4 form a circuit, i.e. the magnitude of capacitance C and inductance L are chosen to match the resonance frequency. Current passes through inductance 5 acting as load on transformer 4 and generates a magnetic field with axial and radial components. Zero phase shift between the arc current and the inductance magnetic field is set by phase shifter 5 to ensure

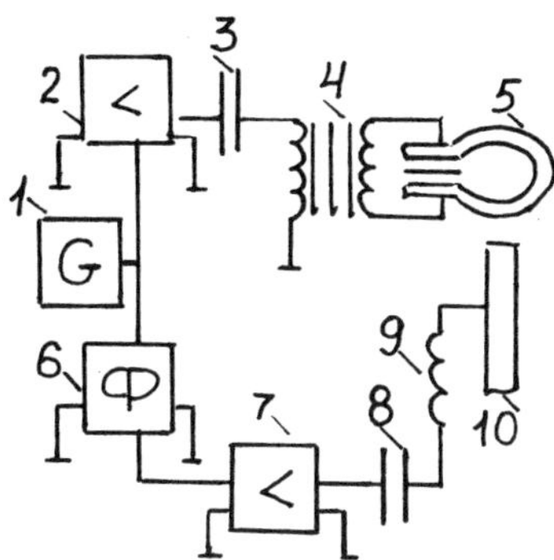

Fig. 4. Schematic diagram of connection plasma generator.

maximum Lorentz force acting on the arc. Amplifier 7 amplifies the ac power. Capacitor 8 and inductance coil 9 form a series circuit. With the central and external electrodes driven together to touch each other, the series circuit provides load on amplifier 7 and is responsible for the dropping volt/ampere characteristic of the arc and sinusoidal current behaviour. Parameters of this circuit are chosen so that frequency ω is a resonance frequency for the circuit. Due to electric field applied between electrode 10 and external electrode 5, electric current traverses the discharge gap. The discharge experiences action of the Lorentz force causing its rotation at a rate of 400 s^{-1}. The rate of rotation does not change for a wide range of current and magnetic field values. If a water-cooled chamber is placed next to a ring electrode, carbon will deposit on the chamber walls. Experiments were conducted at a discharge current of 15A and a total power supply of 5 kW. The pressure ranged from 600 to 760 Torr. At lower pressures, discharge was absolutely diffusive and carbon rod vaporization was insignificant, 10 g/Coul.

Optical spectra of fullerene solution obtained in a magnetron discharge were in agreement with the spectra shown in Fig.3. Fullerene yield in this discharge amounted to 20%.

With a slightly modified central electrode (it was shaped as a water-cooled cylinder with a hole), carbon powder was poured through the discharge and fullerene yield was much higher than that with the rod.

Conclusion

Experimental results obtained by spontaneous ejection plasma jet and their comparison with analytical tests of fullerene obtained at the Institute of Inorganic Chemistry (50 Hz, 1000 A) can be summarized as follows.

1. Qualitative analysis of the absorption spectra in the visible and UV spectral ranges revealed C and C both in the reference sample and the sample under study.

2. Chromatographic separation on silufol under laboratory condition showed high degree of homogeneity of the sample under study.

3. Fullerene generation at helium pressures ranging from 1 to 10 atmospheres.

4. The fullerene yield does not exceed 10% of the evaporated carbon weight.

5. The percentage of C and C fullerenes appears to be the same as when produced at 50 Hz.

6. The existence of resonance frequencies at which enhanced carbon rod vaporization was observed opens up new possibilities to use kilohertz discharge in the design of highly efficient fullerene generators.

7. Magnetron discharge plasma rotates in an interelectrode gap and allows to localize carbon deposit on the walls of a small water-cooled chamber. This advantage combined with the capability of the arrangement to work with carbon or graphite powders makes us optimistic about a feasibility to construct a device for continuous generation of fullerene in reasonably large quantities.

Acknowledgments

The author is grateful to Dr.A.Ya.Koretz for his assistance in obtaining optical spectra of fullerene and to Dr.A.Puzyr' for the electron micrographic pictures of carbon deposit. The author highly appreciates financial support received from Yu.I.Sheiko.

References

1. W. Kratschmer et al., *Nature* (1990) **347** p.354

2. W. Kratschmer , K. Fostiropoulos , D.R. Huffman ,*Chem. Phys. Let.* (1990) **170** p.167

3. G.F.Ignatev , G.N.Churilov, V.G.Pak, *Izvestija SO AN SSSR seriya tekhn.nauk*, **4**, p.93 (in Russian).

4. A.V.Eleckiy, B.M.Smirnov, *Uspekhi Fizicheskikh Nauk* **163**, p.33 (in Russian).

GROWTH TYPES AND STABLE CONFIGURATIONS OF SMALL CARBON CLUSTERS DETERMINED BY HIGH RESOLUTION ELECTRON SPECTROSCOPY

B. KESSLER, H. HANDSCHUH, G. GANTEFÖR, P.S. BECHTHOLD,
and W. EBERHARDT

Forschungszentrum Jülich GmbH, Institut für Festkörperforschung,
D-52425 Jülich, Germany

ABSTRACT

High resolution photoelectron spectroscopy of mass-separated negatively charged Carbon clusters reveals fine structure due to the excitation of the vibrational modes of the neutral clusters. For carefully annealed clusters, where only the most stable isomers are present in the beam, the observed vibrational modes can be assigned to four dominant growth modes and structure types: linear chains (C_5^-, C_7^-, C_9^-), monocyclic rings (C_{10}^-, C_{12}^-, C_{14}^-, C_{16}^-, C_{18}^-), bicyclic rings (C_{20}^-, C_{24}^-, C_{28}^-) and fullerenes (even $n > C_{30}^-$).

Ever since the discovery of C_{60} [1] speculations were carried out about the growth modes of small carbon clusters leading to such a unique symmetry for a 60 atom cluster, which also is different from the other known forms of solid carbon. Initial experiments showed a bimodal growth behaviour for carbon clusters C_n, where small clusters were produced by laser vaporisation under certain conditions for all n, but for n larger than about 30 only clusters with even numbers of atoms were found.[2] Recent diffusion measurements [3] however demonstrated the co-existence of different shapes, including multiple isomers of carbon clusters in this size regime: chains, monocyclic rings, polycyclic rings, fullerenes and possibly even structures with lower symmetries. Thus so far the experimental determination of the structure of carbon clusters has turned out to be difficult except for C_{60}, C_{70}, and the larger fullerenes for which macroscopic amounts of material can be generated. Accordingly the questions about the most stable configurations of small carbon clusters and the pathway to the production of the fullerenes have not been resolved yet experimentally.

Theory has also addressed the question about the nature of the growth process of fullerenes and whether ring structures may be stable precursors.[4-6] In general, in these calculations chain like structures are found to be energetically favourable for small clusters (n < 10), whereas rings start to be more stable around n = 10. This is due to the competition between the strain energy needed to bend the carbon chains into a ring and the energy gained from saturating the dangling bonds at the ends of the chain by closing the ring. This immediately implies a threshold for the existence of bicyclic structures above n = 20. Above n about 30 fullerenes are believed to be the most stable structures even though it is impossible to calculate all possible isomer structures in this size range.

Here we try to answer these questions about the most stable structures of small carbon clusters containing up to 70 atoms by performing high resolution photoelectron spectroscopy on mass selected, well annealed, negatively charged carbon cluster anions. The photoemission signal not only reflects the ground and excited state electronic structure of the corresponding neutral clusters, but in addition to that exhibits finestructure due to the excitation of the characteristic vibrational modes of the individual clusters. While there are a few photodetachment studies on carbon clusters reported in the literature, vibrational resolution so far has only been reached for the smallest carbon-cluster anions up to C_{11}^-.[7] These studies confirm the theory based expectations that the odd numbered clusters in this size range are growing as chains.

Our newly developed experimental set up produces well-annealed carbon-cluster anions C_n^- in the vibrational ground state up to n = 70. This setup consists of a laser-vaporisation source with an incorporated annealing stage in the high pressure region prior to the expansion, a time-of-flight (TOF) mass spectrometer, and a time-of-flight 'magnetic bottle' type electron spectrometer. An Excimer laser (XeCl, 4.025eV) or the higher harmonics of a Nd-YAG laser are used for electron detachment. The resolution of the 'magnetic bottle' electron spectrometer is sufficient to observe vibrational modes in the photoelectron spectra of a large number of clusters in the size range up to C_{70}.

The annealing process during the condensation phase is a crucial step in these studies. In this process the energy of the clusters that have been formed in the high pressure region is raised by an electric discharge. Afterwards the clusters are carefully cooled again using a long extender. Thus presumably the less stable isomers are transformed into the stable configurations through dissociation or direct transformation. The effect of this annealing process on the distribution of the carbon cluster anions is illustrated by the two mass spectra shown in Fig. 1. The top curve shows the spectrum produced by the source without the annealing discharge whereas the bottom curve displays the distribution of the well annealed clusters. Obviously the annealing causes drastic changes in the relative abundancies of the cluster anions.

The mass spectrum of the annealed cluster anions displayed in the bottom panel of Fig. 1 gives a first hint of the different structures of carbon clusters and the various growth regimes. Since the clusters are well annealed the relative intensity of a specific mass can be interpreted as a fingerprint of the stability of the cluster. Therefore the mass spectrum of the annealed clusters can be subdivided into four mass regimes:

i: $n < 10$,   ii: $n = 10 - 18$,   iii: $n = 20 - 28$,   iv: $n > 30$.

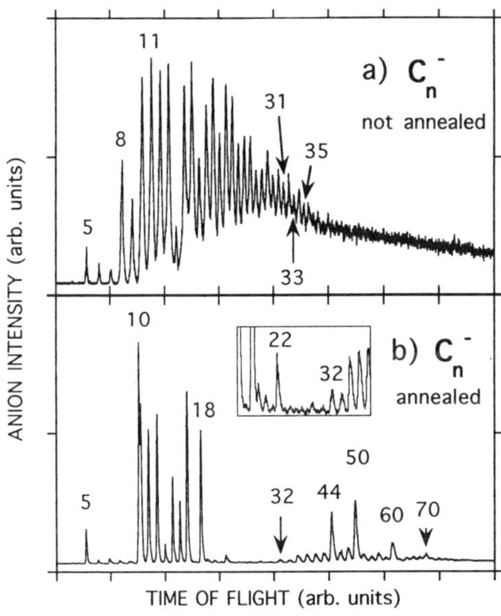

Fig. 1: Time-of-flight mass spectra of carbon cluster anions C_n^-. Panel (a) shows the output of the laser vaporisation source and panel (b) shows the resulting mass distribution when the clusters are annealed. Some cluster sizes n are indicated by numbers. The inset shows the region for $n = 16 - 40$ with an expanded y-scale (x8).

These four regimes may correspond to certain preferred geometric structures. As discussed above, from previous experimental and theoretical studies we expect that a transition from a linear chain growth to a monocyclic ring structure takes places around

142

n = 10 [3,4,8-10] and this is exactly where the first clear break in the distribution of the annealed clusters is observed.

Vibrationally resolved photoelectron spectra (PES) of these C_n^- clusters provide a direct insight into the geometric structure of each individual species. Fig. 2 displays some selected spectra representative for the four regimes. Indeed the spectra of clusters of the same regime exhibit close similarities indicating corresponding similarities of the geometric structure.[11] Without annealing the photoelectron spectra of various isomers overlap and the vibrational structures are washed out.

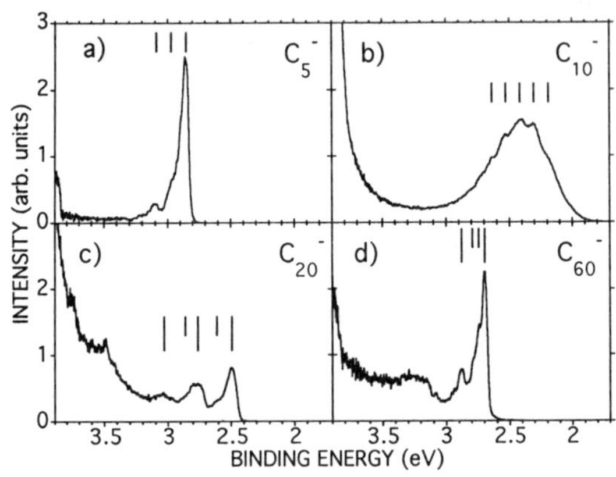

Fig. 2: Photoelectron spectra of mass selected carbon cluster anions C_n^- as a function of the binding energy (hv = 4.025eV): Representative spectra of the four different growth regions of the mass distribution are shown.

The spectra show several narrow (e.g. C_5^-) or broad (e.g. C_{10}^-) features which are assigned to electronic transitions from the ground state of the anion into the ground electronic state of the neutral cluster. For all clusters discussed here in detail, theory predicts the first excited electronic state to be located at least 0.7 eV above the ground state. This is due to the fact that we limit this discussion to electronically closed shell clusters. For a chain like structure a closed electronic shell is expected for chains containing odd numbers of carbon atoms, whereas for rings this condition is fulfilled for even numbers of atoms in the ring. Also for the fullerenes the calculated HOMO-LUMO gaps are larger than 0.7eV. Thus the fine structure of the photoemission features may be safely interpreted in terms of the excitation of vibrations of the neutral cluster. A single vibrational progression consists of several approximately equally spaced peaks (marked by bars in Fig. 2). This spacing corresponds to the vibrational frequency (typically up to 2500cm^{-1}) of a specific mode of the cluster that has been excited by the detachment process. The shape corresponds to the Franck-Condon profile of the transition between the anion and the neutral carbon cluster.

The vibrational frequencies extracted thus from the photoemission spectra are summarised in Fig. 3. This figure includes all clusters for which we actually were able to assign the substructure in the photoemission spectra to vibrational excitations. Consequently the following discussion is limited to exactly these clusters.

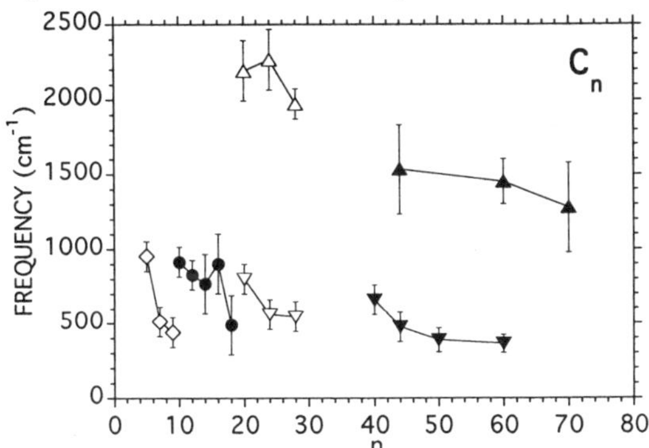

Fig. 3: Vibrational modes observed in the photoelectron spectra of C_n^-. (Lines are drawn to guide the eye.)

The spectrum shown in Fig. 2a is representative of the photoelectron spectra of C_5^-, C_7^- and C_9^-, which apart from a difference in electron affinity closely resemble each other. These data also agree with earlier results.[7] The strongest peak in the PES is assigned to the 0-0 vibrational transition from the electronic ground state of the anion in a linear configuration to the neutral ground state. The structures at the high binding energy side correspond to the excitation of symmetric stretch vibrations.[7] The frequencies displayed in Fig. 3 decrease with increasing cluster size due to the increasing mass.[7]

In region ii of the mass range for $9 < n < 20$ we only resolve vibrational substructure in the photoelectron spectra of the even numbered clusters. Again these spectra are fairly similar, apart from the change in electron affinity and in the width of the Franck-Condon profile. Fig. 2b shows the photoelectron spectrum of C_{10}^- as a representative of the spectra of these clusters. In contrast to the sharp peaks of Fig. 2a the main feature at lowest BE consists of a broad peak with vibrational fine structure. As displayed in Fig. 3,

144

the vibrations of n = 10, 12, 14, 16, and 18 do not simply prolong the series of the linear chains (n = 5, 7, and 9) but open a new branch of decreasing frequencies with increasing cluster size. Together with the overall totally different shape of the photoelectron spectra, compared to the smaller carbon clusters, we conclude that the clusters with n = 10, 12, 14, 16 and 18 are quite likely to have the structure of planar monocyclic rings.[3,4,8,9,12,13] This assignment agrees with the expectations from theory and also with the disappearance of chains in the mobility data for n > 10.[3]

In region iii of the mass spectrum of the annealed carbon clusters the intensities of the individual species are very low. Nevertheless, for C_{20}^- (Fig. 2c), C_{24}^-, and C_{28}^- (not shown) the structure of the photoelectron spectra is again completely different compared to the smaller clusters. Each spectrum is dominated by three main peaks at low BE with an energy separation of almost 0.3eV. The distances of the spectral features are assigned to two main vibrational progressions for each cluster size. The frequencies are displayed in Fig. 3 and demonstrate the similarity of the three clusters. The low frequency mode does not simply continue the frequency branch of the monocyclic rings. This and the extraordinarily high frequency modes around 2000cm^{-1} are striking differences to all other clusters discussed in this paper.

Different predictions on the geometry of C_{20} have been made, ranging from linear structures, monocyclic rings, bicyclic structures, cuplike structures to fullerenes.[3,4] The dramatic change in intensity of the annealed clusters between C_{18}^- and C_{20}^- can be explained by assuming that C_{20} is the smallest cluster with a bicyclic structure with a low threshold for fragmentation into two separate rings. As explained above, for energetic reasons these rings should contain at least ten atoms each. Mobility measurements on unannealed carbon clusters have revealed the first bicyclic structures starting at n = 20, whereby the relative intensity of these isomers has been found to be quite low also.[3]

Two structure models for bicyclic rings of C_{20} have been discussed[3] and are displayed in Fig. 4. In model (a) the carbon rings are expected to have alternating single and triple bonds ('acetylenic' with typical frequencies of 900cm^{-1} and 2050cm^{-1} respectively[14]). In the case of model (b) the carbon rings are expected to have double bonds ('cumulenic': 1650cm^{-1} [14]). The high frequency of the triple bonds (model (a)) fits the data slightly better than model (b). Model (a) is restricted to n = 20, 24 and 28; 22 and 26 cannot be constructed as long as both rings are assumed to have the same size. The photoelectron spectra for C_{22}^- and C_{26}^- (not shown) are strikingly different indeed, which also supports model (a).

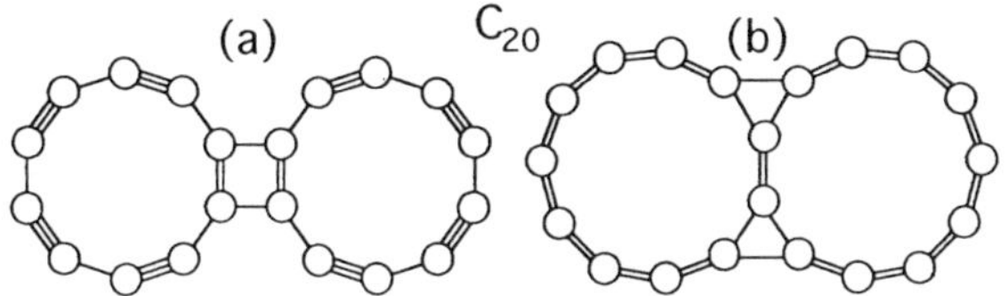

Fig. 4: Structure models for C_{20}.[3]

Smaller clusters than C_{20} are not likely to follow the construction model for bicyclic rings because the strain makes smaller rings than C_{10} very unstable. A construction of larger bicyclic structures than C_{28} is not very likely because the fullerene structure becomes more stable with increasing cluster size. An interpretation that considers C_{20}^-, C_{24}^- and C_{28}^- to be fullerenes is very unlikely since the frequency around 1500 cm^{-1} is not observed and the photoelectron spectra exhibit a different pattern. Moreover, calculations for C_{20} and C_{24} by density functional techniques including gradient corrections show that the fullerene is not energetically favoured even when compared to monocyclic rings or bowl-shaped structures.[4] Calculations of the total energy of bicyclic structures with the geometry of model (a) are not yet available.

C_n^- clusters with even numbers of n > 30 are expected to be fullerenes. This is confirmed by the mass spectrum (Fig. 1) due to the lack of odd numbered clusters and the high intensity of the magic numbers C_{44}^-, C_{50}^-, C_{60}^- and C_{70}^-. The photoelectron spectra of the most stable fullerenes C_{44}^-, C_{60}^- and C_{70}^- on the other hand give considerably more information. Low and high lying vibrational frequencies have been assigned tentatively and are plotted in Fig. 3. The higher frequencies around 1500cm^{-1} belong to stretching vibrations within the surface of the fullerenes (tangential or angular motion) where the strong σ-bonds are located. The low frequency vibrations of the fullerenes around 500cm^{-1} correspond to breathing type vibrations (radial motion) which are directed along the softer π-bonds. These frequencies are decreasing with increasing cluster size. This may be due to the fact that the π-bonds are less distorted for larger fullerenes. A tight binding calculation for C_{50}^-, C_{60}^-, C_{70}^- and C_{80}^- predicts a similar behaviour.[15]

A more careful analysis of the vibrational properties of the fullerenes has to take into account the symmetry of the individual clusters and goes beyond the scope of this paper. Since the symmetry of C_{60}^- is well known the spectrum of this cluster can be understood and will be given in more detail elsewhere.[16] Furthermore the electron-phonon coupling constant for electrons in the LUMO, which is one of the parameters used to describe the superconducting behaviour of the doped fullerenes, can be determined from this spectrum.

Concluding, stable carbon cluster anions C_n^- in the vibrational ground state are produced by laser vaporisation with subsequent annealing. Spectroscopic evidence for different topological classes of ring structures is presented: photoelectron spectra of mass selected carbon-cluster anions are recorded that exhibit a variety of vibrational frequencies. By comparison with the relative abundance in the mass spectra an interpretation in terms of linear chains for n = 5, 7, 9, monocyclic rings for even numbers of n between 10 and 18, bicyclic rings for n = 20, 24, 28 and fullerenes for n > 30 is given.

References:

1. H.W.Kroto, J.R.Heath, S.C.O'Brien, R.F.Curl, R.E.Smalley, *Nature* **318** (1985) 162.
2. E.A.Rohlfing, D.M.Cox, A.Kaldor, *J.Chem.Phys.* **81** (1984) 3322.
3. G.v.Helden, M.T.Hsu, N.Gotts, M.T.Bowers, *J.Phys.Chem.* **97** (1993) 8182, G.v.Helden,P.R.Kemper, N.Gotts, M.T.Bowers, *Science* **259** (1993) 1300, G.v.Helden, N.G.Gotts, M.T.Bowers, *Nature* **363** (1993) 60, G.v.Helden, M.T.Hsu, N.G.Gotts, P.R.Kemper, M.T.Bowers, *Chem.Phys.Lett.* **204** (1993) 15.
4. K.Raghavachari, D.L.Strout, G.K.Odom, G.E.Scuseria, J.A.Pople, B.G.Johnson, P.M.W.Gill, *Chem.Phys.Lett.* **214** (1993) 357, K. Raghavachari, J.S.Blinkley, *J.Chem.Phys.* **87** (1987) 2192, K.Raghavachari, B. Zhang, J.A.Pople, B.G.Johnson, P.M.W.Gill, *Chem.Phys.Lett.* **220** (1994) 384.
5. C.J.Brabec, E.B.Anderson, B.N.Davidson, S.A.Kajihara, Q.-M.Zhang, J.Bernholc, *Phys.Rev.* **B46** (1992) 7326.
6. J.R.Chelikowsky, *Phys.Rev.Lett.* **67** (1991) 2970.
7. D.W.Arnold, S.E.Bradforth, T.N.Kitsopoulos, D.M.Neumark, *J.Chem.Phys.* **95** (1991) 8753.
8. Shihe Yang, K.J.Taylor, M.J.Craycraft, J.Conceicao, C.L.Pettiette, O.Cheshnovsky, R.E.Smalley, *Chem.Phys.Lett.* **144** (1988) 431.
9. C.Liang, H.F.SchaeferIII, *J.Chem.Phys.* **93** (1990) 8844.
10. V.Parasuk, J. Almlöf, *Theor.Chim.Acta* **83** (1992) 227.
11. H.Handschuh, G.Ganteför, B.Kessler, P.S.Bechthold, W.Eberhardt, submitted to *Phys.Rev.Lett.*
12. J.Hunter, J.Fye, M.F.Jarrold, *Science* **260** (1993) 784.
13. K.S.Pitzer, E.Clementi, *J.Amer.Chem.Soc.* **81** (1959) 4477.
14. G.Herzberg, *Molecular Spectra and Molecular Structure II*, (Van Nostrand Reinhold Company, New York, 1945), p.195.
15. S.J.Woo, E.Kim, Y.H.Lee, *Phys.Rev.* **B47** (1993) 6721.
16. O.Gunnarsson, H.Handschuh, P.S.Bechthold, G.Ganteför, B.Kessler, W.Eberhardt, (to be published).

Part Three

STRUCTURE AND STRUCTURAL PHASE TRANSITION

ORDER AND DISORDER IN THE FULLERENES

A. B. Harris
Department of Physics
University of Pennsylvania
Philadelphia, PA 19104-6396
e-mail: harris@mohlsun.physics.upenn.edu

ABSTRACT

Orientational ordering in various fullerenes is reviewed. The restrictions on ordering given cubic symmetry are discussed. I emphasize the simple consequences of the Landau expansion describing the ordering transitions in C_{60} and C_{70}. Some results for the mean-field dynamics of C_{70} are summarized.

Here I review the basic phenomena concerning the structure, phase transitions, and dynamics of orientational ordering in various fullerene systems. Concerning the structure I discuss how the cubic symmetry implied by diffraction data restricts possible orientationallly ordered structures. Then a brief discussion is given of simple consequences which follow from the symmetry of the Landau expansion for the orientational free energy in terms of appropriate order parameters. Finally, I give a brief summary of some results of a mean field theory of the dynamics for C_{70}.

I. Orientationally Ordered Structures

Here we discuss orientationally ordered phases involving C_{60} molecules. These molecules form truncated icosahedra and are thus replicas on an atomic scale of a football (i. e. US soccer ball), as shown in the famous picture (see Fig. 1) taken from Ref. [1]. In analyzing powder diffraction data of fullerene solids a first step is to see if the spectrum can be indexed according to one of the highest symmetry (i. e. cubic) Bravais lattices. This information only depends on the angles at which diffraction is observed. In many cases the crystal structure is indeed cubic. Then from a fit to the intensities of the diffraction spectrum conclusions concerning the ordering may be drawn. It may be argued that a discussion of this procedure is superfluous since structure determinations can be done using the well-known Rietveld analysis for which elaborate computer programs are available. However, to avoid error (e. g. due to getting misled by local minima in the R-factor which characterizes the goodness of fit) and to highlight the physics involved we present here a simple discussion which is given in more detail in Refs. 2 and 3.

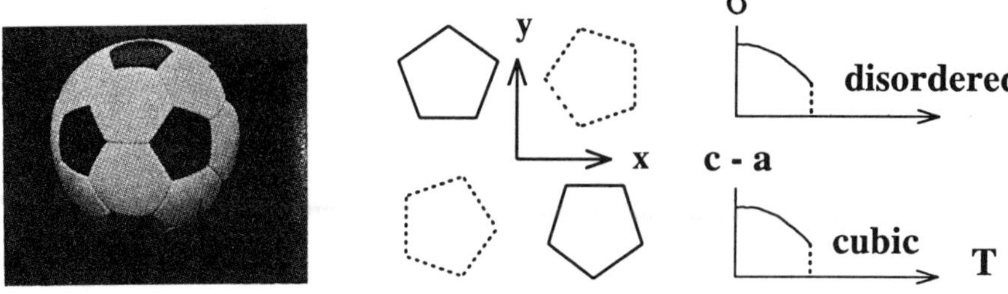

Fig. 1. Left: Soccer ball. Center: Molecules in the fcc unit cell. Molecules at $z = 0$ are solid, those at $z = 1/2$ are dotted. Right: Order parameter and $c - a$ versus temperature.

To see that there is a constraint on orientationally ordering imagine placing C_{60} molecules with their centers on an fcc lattice oriented so that they each have one of their pentagonal faces perpendicular to the z-axis, as represented in Fig. 1. In this hypothetical orientationally ordered structure the x and y directions are equivalent, but obviously the z direction is crystallographically inequivalent to x and y. Thus, if one has an ordering transition described by an order parameter σ (defined, say, to be proportional to the square root of the intensity of the orientational superlattice Bragg reflections), as shown in Fig. 1, one would inevitably have a tetragonal distortion $(c - a)$, as also shown in Fig. 1. So this hypothetical structure is obviously not cubic.

Since the disordered phase of C_{60} is fcc and the ordered phase is sc (with four molecules per unit cell) (for a review see Ref. 4), we look for orientationally ordered structures assuming 1) the molecular centers form an fcc lattice, 2) the unit cell is sc containing four molecules which are not all orientationally equivalent, and the molecules are initially perfect icosahedra, but may distort in accordance with the local site symmetry. Under these assumptions there are only three allowed space groups[5, 2] Pa$\bar{3}$, Pn$\bar{3}$, and Pm$\bar{3}$, described in Table I.

Table I. Allowed Cubic Space Groups for Oriented Icosahedra

Site	Pa$\bar{3}$		Pn$\bar{3}$	Pm$\bar{3}$
	Local Threefold Axes			Orientation
0,0,0	111	111	111	B
1/2, 1/2, 0	1$\bar{1}\bar{1}$	$\bar{1}1\bar{1}$	$\bar{1}\bar{1}1$	A
1/2, 0, 1/2	$\bar{1}\bar{1}1$	1$\bar{1}\bar{1}$	$\bar{1}1\bar{1}$	A
0, 1/2, 1/2	$\bar{1}1\bar{1}$	$\bar{1}\bar{1}1$	1$\bar{1}\bar{1}$	A

In Pm$\bar{3}$, the molecules are in their "standard" orientations shown in the left panel of Fig. 2, whereas in the other space groups, the molecules are rotated (starting from standard orientation A) through a setting angle ϕ about the local threefold axis, as indicated in the center panel. In the right panel we reproduce the figure of Ref. 5 showing how interhexagonal (double) bond in one molecule faces the electron-poor region of the center of the pentagonal facet of an adjacent molecule. This argument agrees with the quantum calculations of the charge density of a C_{60} molecule.[7]

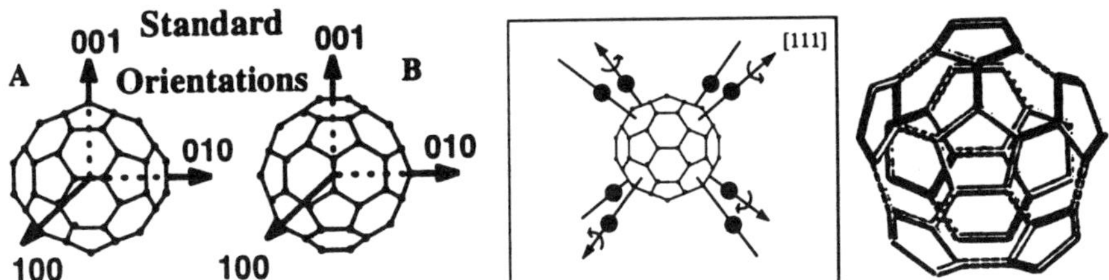

Fig. 2. Orientations of C_{60} molecules.

In Fig. 3 we show the comparison between the diffraction spectrum calculated for these space groups and the experimental data for C_{60}. It is seen that the correct structure[5] is indeed $Pa\bar{3}$. The setting angle is determined[5, 6] to be $\phi \approx 22°$.

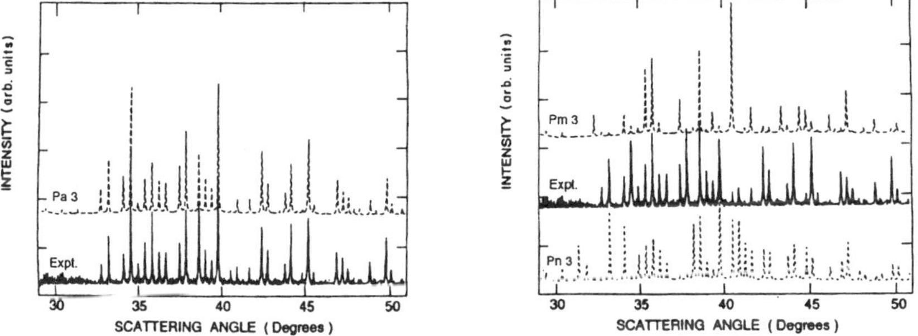

Fig. 3. Comparison[2] of experimental and calculated powder x-ray diffraction spectra.

One can carry out the same type of analysis for the system Na_xC_{60}. In this case,[8] due to the possible partial occupancy of the tetrahedral sites, one may lose the inversion symmetry of the C_{60} sites. (In the previous undoped case, breaking of inversion symmetry was not considered because this would require an accidental degeneracy leading to a simultaneous orientational ordering and inversion symmetry breaking.) For Na_xC_{60} allowing loss of inversion symmetry did not lead to improved fits to the diffraction spectrum. For $1 < x < 3$ solid solution behavior was deduced[8] in which the structure is $Pa\bar{3}$, as for the undoped system.

One can perform a similar analysis for the proposed[9] "2-a" structures having an fcc unit cell containing eight molecules which has a lattice constant twice that

152

an fcc unit cell containing eight molecules which has a lattice constant twice that of the disordered phase. This structure was proposed on the basis of several experiments which seem to be sensitive to deviations from $Pa\bar{3}$ symmetry. There was an implication that such deviations occurred for when the temperature was reduced well below the higher temperature (250K)[4] at which $Pa\bar{3}$ ordering occurs. Again we assume the molecules have their centers on an fcc lattice and that the system has cubic symmetry. Then the conclusions are[10, 3] that the axes must be arranged as in $Pn\bar{3}$ and the allowed space groups are $Fm\bar{3}$, $Fd\bar{3}$, and $Fd\bar{3}c$. These structures are thus quite different from $Pa\bar{3}$ and their diffraction spectra are not compatible with experimental results, as shown in Fig. 4.[10] Other experiments are not consistent with the proposed "2a" structure.[11]

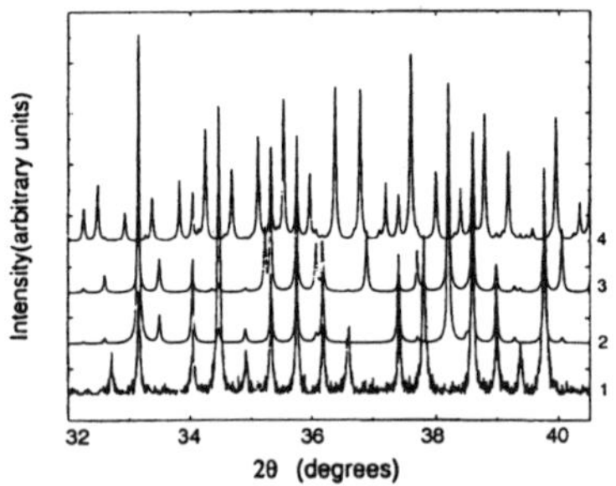

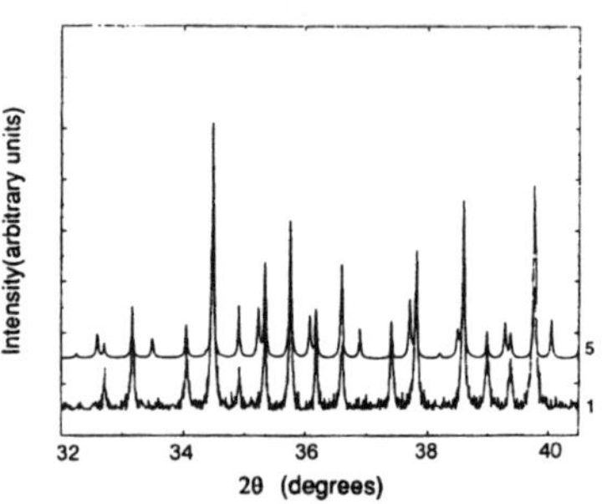

Fig. 4. X-ray diffraction intensity versus scattering angle.[3] Experimental=1. Calculated: 2=$Fd\bar{3}c$, 3=$Fd\bar{3}$, 4=$Fm\bar{3}$, and 5=trigonal, i. e. sublattices similar to $Pa\bar{3}$ with setting angles $\phi_1 = 24°$ and $\phi_2 = 84°$.

More plausibly, one might obtain such a "2-a" structure by creating two sublattices in the $Pa\bar{3}$ structure. Such a structure is NOT cubic, but rather is trigonal[3]. If the two sublattices have setting angles which are quite different, the calculated diffraction spectrum again disagrees with experiment, as shown in Fig. 4, right. Of course, if the two sublattices have setting angles $\phi_0 \pm \delta$, where ϕ_0 is that previously taken for C_{60}, and if δ is small enough, (less than, say, 0.3°), then there is no inconsistency with the observed diffraction spectrum. For more details, see Ref. 3.

II. Mean Field Theory

Here I review the main features of mean field theory as applied to the orientational ordering phase transition. For simplicity, I describe first the ordering transition[4] in C_{70} from an orientationally disordered fcc phase into a trigonal phase in which the molecules are spinning about their long axes, which are aligned parallel to a formerly cubic [1,1,1] direction. The C_{70} molecule is shown in Fig. 5. When the molecule spins about its long axis, it has the same symmetry as a liquid crystal molecule and we therefore describe its orientation by the traceless moment of inertia tensor, $\mathbf{Q}$:

$$Q_{\alpha\beta} \equiv \left\langle \sum_{i=1}^{70} r_{i,\alpha} r_{i,\beta} - \frac{1}{3} r_i^2 \delta_{\alpha,\beta} \right\rangle_T \propto S \left[\hat{n}_\alpha \hat{n}_\beta - \frac{1}{3} \delta_{\alpha,\beta} \right] , \qquad (1)$$

where $< >_T$ indicates an ensemble average at temperature T, α and β label components, $\mathbf{R}_i$ is the location of the ith atom relative to the center of mass, δ is the Kronecker delta, $\hat{n}$ is the director, and the normalized order parameter, S, indicates the degree of ordering. Values of $\mathbf{Q}$ for various situations are shown in Fig. 5.

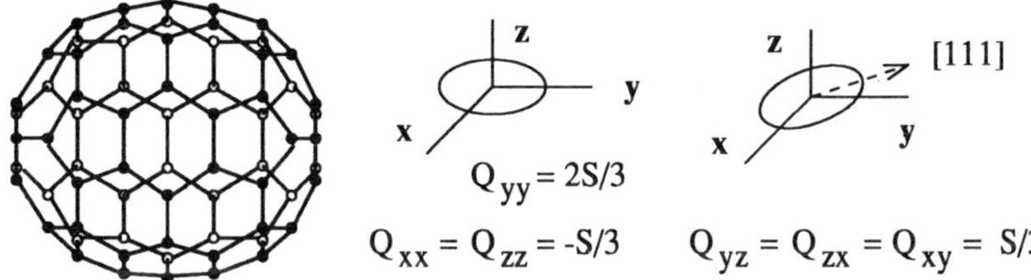

Fig. 5. Left: The C_{70} molecule. Center and right: orientations with corresponding nonzero order parameters.

The free energy per unit volume can be expanded in powers of $\mathbf{Q}$. The result must be invariant under the symmetry of the disordered fcc lattice. (Basically, this means that x, y, and z must enter symmetrically and an even number of times in each term.) Based on only symmetry arguments, this expansion of the form[12]

$$\begin{aligned} F_{\mathbf{Q}} &= A_0(T - T_0)(Q_{xy}^2 + Q_{yz}^2 + Q_{zx}^2) - B(T)Q_{xy}Q_{yz}Q_{zx} \\ &\quad + C(T)\left[Q_{xy}^2 + Q_{yz}^2 + Q_{zx}^2\right]^2 + D(T)\left[Q_{xy}^4 + Q_{yz}^4 + Q_{zx}^4\right] . \end{aligned} \qquad (2)$$

In writing this result we used the experimental fact that ordering takes place with molecules aligned along [1,1,1] directions, so that the critical order parameter components are $Q_{\alpha\beta}$ with $\alpha \neq \beta$. Since the free energy contains a term third order in $\mathbf{Q}$, the ordering transition (at T_c) must be a discontinuous one: for $T > T_c$, $S = 0$, whereas for T just below the transition temperature S suddenly assumes a nonzero value. When ordering does occur (i. e. when $\mathbf{Q}$ becomes nonzero), the form of the cubic term favors having all Q's equal in magnitude and, since $B(T) > 0$, their product is positive, as happens if any of the four [1,1,1] directions is selected for ordering. When a [1,1,1] direction is so selected, the crystal loses its cubic symmetry and must distort, of course. To include the possibility of a lattice distortion we allow coupling, $F_{\mathbf{Q}-\mathrm{el}}$ between $\mathbf{Q}$ and the elastic degrees of freedom, which involve the strains $\epsilon_{\alpha\beta} \equiv \partial u_\alpha / \partial r_\beta + \partial u_\beta / \partial r_\alpha$. Symmetry dictates that to leading order

$$F_{\mathbf{Q}-\mathrm{el}} = -E\left[\epsilon_{xy}Q_{xy} + \epsilon_{yz}Q_{yz} + \epsilon_{zx}Q_{zx}\right] + H\left[\epsilon_{xx} + \epsilon_{yy} + \epsilon_{zz}\right]\left[Q_{xy}^2 + Q_{yz}^2 + Q_{zx}^2\right] . \qquad (3)$$

154

We also include the standard form of the purely elastic free energy:

$$F_{el} = \frac{1}{2}c_{11}\left[\epsilon_{xx}^2 + \epsilon_{yy}^2 + \epsilon_{zz}^2\right] + c_{12}\left[\epsilon_{xx}\epsilon_{yy} + \epsilon_{yy}\epsilon_{zz} + \epsilon_{zz}\epsilon_{xx}\right] + \frac{1}{2}c_{44}\left[\epsilon_{xy}^2 + \epsilon_{yz}^2 + \epsilon_{zx}^2\right] . \quad (4)$$

Minimizing $F_{Q-el} + F_{el}$ with respect to the strains, we see that[12]

$$\frac{\epsilon_{xy}}{Q_{xy}} = \frac{\epsilon_{yz}}{Q_{yz}} = \frac{\epsilon_{zx}}{Q_{zx}} \sim \frac{E}{c_{44}} , \qquad \frac{\Delta V}{V} = \epsilon_{xx} + \epsilon_{yy} + \epsilon_{zz} \sim Q^2 , \quad (5)$$

where ΔV is the discontinuity in the volume V. This result suggests the following experiment: Suppose the first order transition is monitored as a function of, say, pressure, p. At the transition the jump in the order parameter, $\Delta S(p)$ will be a function of pressure. The accompanying discontinuities in the strains $\Delta \epsilon$ will also depend on p. The Landau expansion, *into which we have only incorporated symmetry,* indicates that $\epsilon_{\alpha\beta}(p) \sim \Delta S(p)$ (for $\alpha \neq \beta$) and $\Delta V(p) \sim [\Delta S(p)]^2$.

For C_{60} mean field theory is much more complicated. The Landau expansion was first given in Ref. 2, to which we refer the reader for more details. (More recently more elaborate theories have been given.[13]) Due to the higher symmetry of the C_{60} molecule, its orientational ordering is not described by a second rank tensor as in Eq. (1) above. Instead, one introduces order parameter components such as

$$\langle Y_L^M \rangle \equiv \sum_{i=1}^{60} r_i^L Y_L^M(\hat{r}_i) , \quad (6)$$

where Y_L^M is a spherical harmonic. For an undistorted C_{60} molecule $\langle Y_L^M \rangle$ vanishes for $L < 6$. Also, since ordering for C_{60} involves sublattice formation, we must keep track of spatial Fourier components at nonzero wave vector. So one introduces order parameters X, Y, and Z to describe Fourier compolnents of $\langle Y_L^M \rangle$ at wave vectors $\mathbf{q}_x = (2\pi/a)(1,0,0)$, $\mathbf{q}_y = (2\pi/a)(0,1,0)$, and $\mathbf{q}_z = (2\pi)/a(0,0,1)$, respectively. In addition, we need to introduce similar order parameters, $\bar{X}, \bar{Y}, \bar{Z}$ for the corresponding quantities for the second setting of $Pa\bar{3}$. (The two settings correspond to the two columns in Table I which pertain to $Pa\bar{3}$,) Thus the Landau expansion for the orientational ordering transition in C_{60} (up to third order terms) is

$$F = \frac{1}{2}a(T - T_0)\left[X^2 + Y^2 + Z^2 + \bar{X}^2 + \bar{Y}^2 + \bar{Z}^2\right] - w\left[XYZ + \bar{X}\bar{Y}\bar{Z}\right] . \quad (7)$$

This form of free energy also describes $Pa\bar{3}$ ordering in solid hydrogen.[14] As before, the occurrence of a third order term indicates that the orientational transition in C_{60} must be a discontinuous one. At the time of submission of Ref. 2 the data was not clear as to whether or not the transition was discontinuous. Since then, however, the experimental situation has become completely unambiguous. In Fig. 6 we show data of David et al[15] for the lattice constant and that of Heiney et al[4] for the intensity of an orientational Bragg reflection, which both indicate a discontinuous transition.

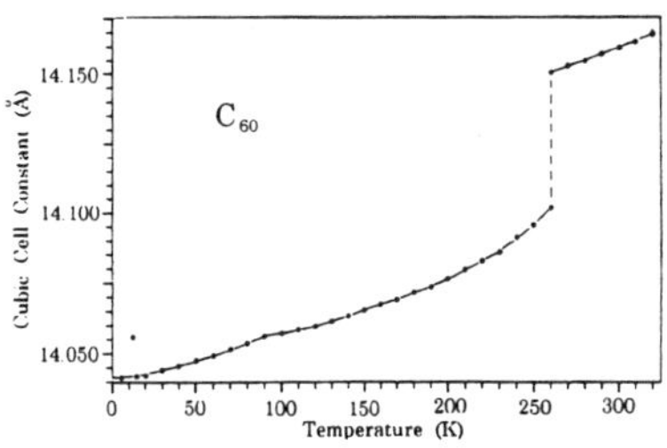
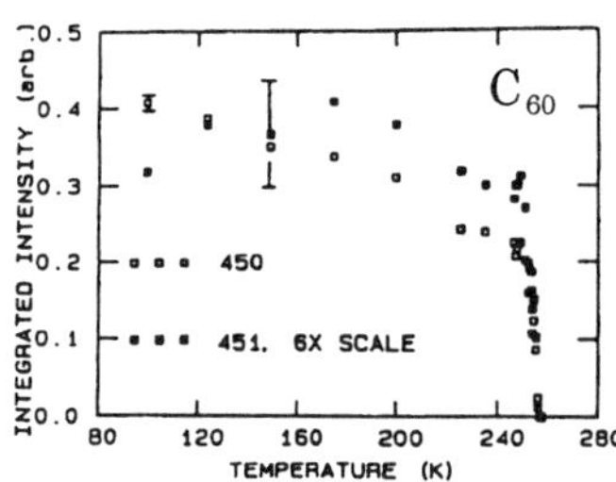

Fig. 6. Lattice parameter[6] and orientational Bragg Intensities[4] for C_{60} vs temperature.

III. Dynamics of C_{70}

Here I give an extremely brief summary of the results of a recent treatment of the dynamics near the fcc-trigonal transition in C_{70}. This theory[16] is phrased in terms of an effective Hamiltonian, $\mathcal{H}_{\text{eff}}$, for the dynamics of long wavelength fluctuations,

$$\mathcal{H}_{\text{eff}} = \int \left\{ n F_Q + \frac{\mathbf{L}^2}{2nI} - \frac{1}{2} n D_Q Q \nabla^2 Q \right\} d\mathbf{r} , \tag{8}$$

where $\mathbf{L}$ is the angular momentum density, n is the number of molecules per unit volume, I is the moment of inertia for rotation about an axis perpendicular to the the long axis of the molecule, $Q \nabla^2 Q$ indicates (schematically) the leading term in a gradient expansion of the effective Hamiltonian, and F_Q is the free energy in terms of $\mathbf{Q}$ given in Eq. (2), above. Note that $\mathbf{Q}$ is not a conserved quantity and therefore obeys a phenomenological equation of the form

$$\frac{dQ_{\alpha\beta}}{dt} = [Q_{\alpha\beta}, I_{,\gamma}] \frac{\delta \mathcal{H}_{\text{eff}}}{\delta L_\gamma} - \Gamma_Q \frac{\delta \mathcal{H}_{\text{eff}}}{\delta Q_{\alpha\beta}} , \tag{9}$$

where [ ] indicates a Poisson bracket. On the other hand, $\mathbf{L}$ is a conserved quantity and therefore obeys an equation of the form

$$\frac{dL_{\alpha\beta}}{dt} = [L_\gamma, Q_{\alpha\beta}] \frac{\delta \mathcal{H}_{\text{eff}}}{\delta Q_{\alpha\beta}} + \gamma_{ij} \nabla_i \nabla_j \frac{\delta \mathcal{H}_{\text{eff}}}{\delta L_\gamma} . \tag{10}$$

Here damping is introduced via phenomenological constants Γ_Q and γ_{ij}. These equations are reminiscent of $du/dt = p/m$ and $dp/dt = -dV(x)/dx - \gamma p$, for damped motion in a potential $V(x)$. The solution to Eqs. (9) and (10) interpolates between an overdamped regime in the orientationally disordered phase to an underdamped regime at low temperatures where long range ($\mathbf{Q}$) order is well developed. For parameters estimated for C_{70} we show (in Fig. 7) the inelastic neutron scattering cross section for temperatures near T_c (whose value we take to be 300 K).

This work was supported in part by the National Science Foundation.

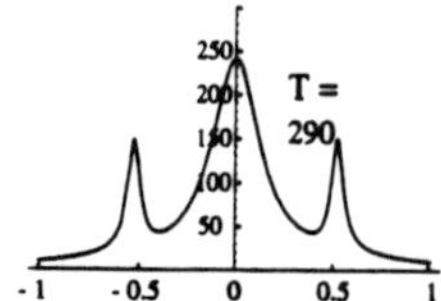

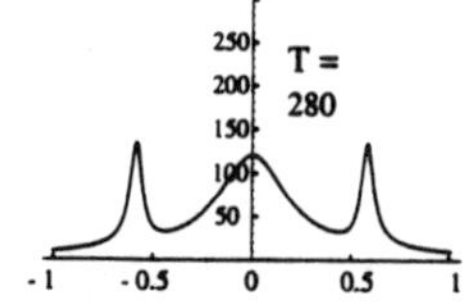

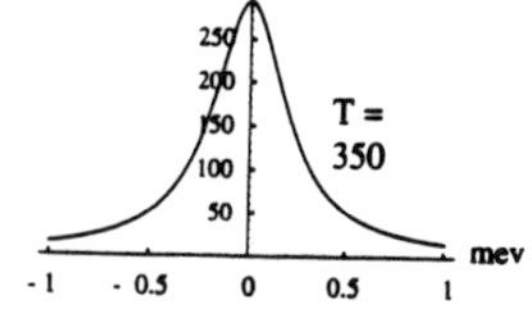

Fig. 7. Inelastic neutron scattering cross section for C_{70} for wave vector $K = 0.1 \mathring{A}^{-1}$ at various temperatures, T (in Kelvin), near $T_c = 300$ K.

References

[1] H. W. Kroto et al, *Nature* **318** (1985) 162.

[2] A. B. Harris and R. Sachidanandam, *Phys. Rev.* **B46** (1992) 4944.

[3] A. B. Harris, R. Sachidanandam, and T. Yildirim, *Phys. Rev.* in press.

[4] P. A. Heiney, *J. Phys. Chem Solids*, **53** (1992) 1333.

[5] R. Sachidanandam and A. B. Harris, *Phys. Rev. Lett.* **67** (1991) 1467.

[6] W. I. F. David et al, *Nature* **353** (1991) 147.

[7] T. Yildirim et al, *Phys. Rev.* **B48** (1993) 1888.

[8] T. Yildirim et al, *Phys. Rev. Lett.* **71** (1994), 1383.

[9] G. Van Tendeloo et al, *Phys. Rev. Lett.* **69** (1992) 1005. E. J. J. Groenen et al, *Chem. Phys. Lett.* **197** (1992) 314. P. H. M. van Loosdrecht et al, *Chem. Phys. Lett.* **198** (1992) 587. Note that R. Moret et al, *J. Phys. I*, **2** (1992) 1699 was later withdrawn: R. Moret et al, *ibid.* **3** (1993) 1085.

[10] A. B. Harris and R. Sachidanandam, *Phys. Rev. Lett.* **70** (1993) 102.

[11] J. E. Fischer et al, *Phys. Rev.* **B47** (1993) 14614.

[12] R. Sachidanandam and A. B. Harris, *Phys. Rev.* **B46** (1994) 2878.

[13] K. H. Michel et al, *Phys. Rev. Lett.* **68** (1992) 2929. P. C. Chow et al, *Phys. Rev. Lett.* **69** (1992) 2943, 3591(E).

[14] J. R. Cullen et al, *Solid State Commun.* **10** (1972 195.

[15] W. I. F. David, *Europhys. Lett.* **18** (1992) 219.

[16] T. C. Lubensky, R. Sachidanandam, and A. B. Harris, to be published.

Rotational Thermodynamics and Isotope Effect in C_{60}

G. Dresselhaus

Francis Bitter National Magnet Laboratory

Massachusetts Institute of Technology, Cambridge, Massachusetts, 02139, USA

R. Saito

Department of Electronics Eng., Univ. of Electro-Communications, Tokyo 182, Japan

M.S. Dresselhaus

Department of Electrical Engineering and Computer Science and Department of Physics,
Massachusetts Institute of Technology, Cambridge, Massachusetts 02139, USA

Abstract

A thermodynamic model of the rotational phase transition in solid C_{60} is presented. The phase transition which appears at 261 K is a first-order phase transition in which a competition occurs between entropy gain by rotation and energy gain by intermolecular attraction. The anomaly in the specific heat, and the entropy and enthalpy changes observed at the critical temperature are well reproduced in the present numerical results. Differences in the symmetry of the ^{12}C and ^{13}C nuclei in a C_{60} molecule are expected to be relevant to the rotational motion of the molecule. Symmetry selection rules sensitively affect the calculated temperature dependence of the specific heat in the temperature region below $\sim$1 K for isotopically pure $^{12}C_{60}$ and $^{13}C_{60}$.

1 Introduction

It has been shown experimentally [1-5] that below a characteristic temperature of $T_{01} = 261$ K, the C_{60} molecules in solid C_{60} lose two of their three degrees of rotational freedom. In the low temperature phase below T_{01}, the residual rotational motion occurs along the four $\langle 111 \rangle$ axes and is a hindered rotation. Whereas the structure of solid C_{60} above T_{01} is the fcc structure $Fm\overline{3}m$, the structure below $\sim$261 K is a simple cubic structure (space group T_h^6 or $Pa\overline{3}$) with a lattice constant $a_0 = 14.17$Å and four C_{60} molecules per unit cell, since the four molecules within the fcc structure become inequivalent below T_{01} [2, 3]. From a physical standpoint, the lowering of the symmetry as T is reduced below T_{01} is caused by the assignment of a specific $\langle 111 \rangle$ direction (vector) to each of the 4 molecules within a unit cell. In this paper a model for the first-order phase transition at $T_{01} = 261$ K is presented. Good agreement is obtained between the predicted changes in enthalpy and entropy at T_{01} and the experimental values for these quantities. Further, the effect of the isotopic content of the C_{60} molecule on rotational-vibrational states is discussed. Particular emphasis is given to the differences in the low temperature specific heat between $^{12}C_{60}$, $^{13}C_{60}$ and conventional C_{60} having the naturally occurring abundance of ^{13}C.

2 Model for Phase Transitions in C_{60}

Using an intermolecular model potential for C_{60}, several theoretical calculations of the molecular dynamics of the ratchet motion have been performed for the molecules in solid C_{60} [6, 7], as well as calculations of the librational and inter-molecular vibrational dispersion relations [8, 9, 10]. A Landau theory for the phase transition in solid C_{60} at $T_{01} = 261$ K has also been reported [11, 12]. Model calculations for the intermolecular C_{60}-C_{60} interactions yield a first-order transition at T_{01}, and a low temperature freezing transition broadly occurring over a wide temperature range between 90 K and 165 K range [8, 13, 14].

One model that has had some success in explaining most of the characteristics identified with the first-order phase transition in solid C_{60} at T_{01} is described below [12]. This model considers three states: (a) the freezing of all ratchet motion, (b) ratchet motion about a three-fold (C_3) axis with 3 positions per molecule for rotation, and (c) ratchet motion not only about the three-fold axis but also about a five-fold (C_5) axis, thus yielding a total of 90 rotation positions. In a C_{60} molecule, there are 10 C_3 axes, and 6 C_5 axes. If a molecule is restricted to rotate about only one $\langle 111 \rangle$ direction in the low temperature simple cubic phase, then the only possible ratchet motion is about one of the C_3 axis, ratcheting between three equivalent potential minima, each separated from one another by a potential barrier. For each C_3 axis, we in fact have 3 absolute potential minima and 6 local minima [9]. At high temperature we consider 30 symmetry operations about 6 C_5 axes in addition to the C_3 axes. Since all C_3 axes permute with each other, we thus have 90 potential minima.

From the above argument it follows that in states (b) and (c), there are 3 and 90 equivalent sites, respectively, which correspond to the potential minima for the ratchet motion. It is assumed that at a given temperature, N_1, N_2 and N_3 molecules of a total of N molecules are in (a), (b) and (c) states, respectively. The total number of states, W, is then given by,

$$W = \frac{N!}{N_1! N_2! N_3!} 3^{N_2} 90^{N_3}. \tag{1}$$

Using the Stirling formula $\log N! \sim N \log N - N$, the entropy, S, is then given by

$$\begin{aligned} S &= k_{\mathrm{B}} \log W \\ &= -N k_{\mathrm{B}}(X_1 \log X_1 + X_2 \log X_2 + X_3 \log X_3) \\ &\quad + N k_{\mathrm{B}}(X_2 \log 3 + X_3 \log 90), \end{aligned} \tag{2}$$

where,

$$N = \sum_{i=1}^{3} N_i, \quad X_i = N_i/N, \quad \text{and} \quad \sum_{i=1}^{3} X_i = 1. \tag{3}$$

The internal energy E is written as a sum of two terms

$$E = -N J_a X_1 - \frac{N z J_b}{2}(X_1 + X_2)^2, \tag{4}$$

in which J_a (> 0) stands for the energy gained from freezing the motion for the (a) states, J_b (> 0) is the inter-molecular attractive interaction, and $z = 12$ is the number of nearest-neighbor molecules. It is assumed that the attractive interaction acts only when the two molecules in the (a) or (b) states are nearest neighbors.

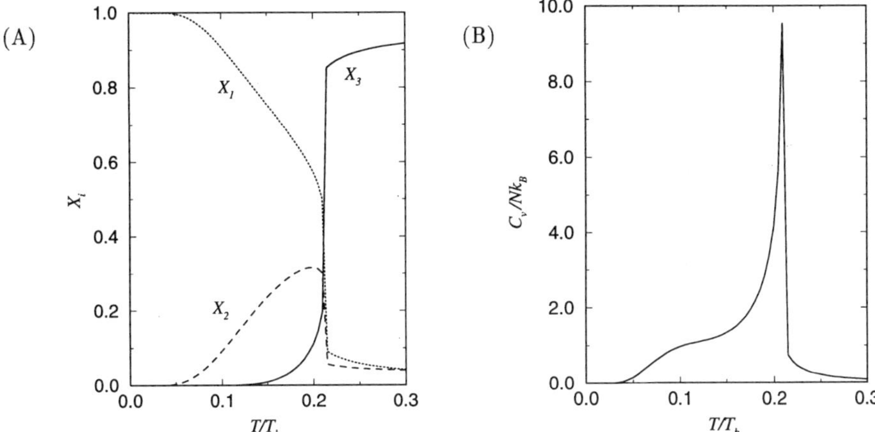

Figure 1: (A) Temperature dependence of the fractional occupation of states (a), (b), and (c) denoted by X_1 (dotted line), X_2 (dashed line) and X_3 (solid line), respectively, plotted as a function of T/T_b (see text). The large changes at $T/T_b = 0.21$ are identified with a first-order phase transition at T_{01} and the onset temperature at $T/T_b \sim 0.05$ is identified with T_{02} at lower T [13]. (B) Calculated temperature dependence of the specific heat C_v/Nk_B [12].

In thermal equilibrium, the free energy, $F = E - TS$, is minimized with respect to X_1 and X_2 while keeping the temperature fixed and imposing the condition: $X_1 + X_2 + X_3 = 1$. This energy minimization results in three simultaneous self-consistent equations which yield the temperature dependence of N_i ($i = 1, 2, 3$), the number of molecules in states (a), (b) and (c), respectively. The temperature dependences of X_1, X_2 and X_3 thus obtained are plotted in Fig. 1 (A) as dotted, dashed, and solid lines, respectively, as a function of T/T_b with a fixed ratio of $T_a/T_b = 1/3$, where $T_a = J_a/k_B$ and $T_b = zJ_b/k_B$, k_B being the Boltzmann constant. From Fig. 1 (A), it follows firstly that X_1, the probability that a molecule is frozen, decreases monotonically with increasing T in the range $0.05 < T/T_b < 0.21$, and secondly a discontinuous change is found in many properties at $T/T_b = 0.21$, characteristic of a first-order transition. The characteristic temperature at $T/T_b = 0.21$ is identified with the phase transition temperature $T_{01} = 261$ K, taking $T_b = J_b/k_B = 1243$ K. The observed first-order transition results from the competition between the entropy gain by rotation and the energy gain by intermolecular attraction. The onset value T_{02} for the lower temperature phase transition is operationally defined by the temperature where X_1 becomes 0.99. Figure 1 (A) shows that the onset value, thus defined, occurs at $T/T_b = 0.05$. Using this criterion, it is found that T_{02} is proportional to T_a. The experiments also do not specify a unique value for T_{02} which is observed to depend on many parameters, including the frequency of the measurement probe.

The temperature dependence of the order parameter X_1 may be compared with that of x-ray diffraction spots such as (4,5,1) [15] and (4,5,0) [16] which are forbidden in the fcc structure. The (4,5,1) (or(4,5,0)) peak intensity decreases from its value at very low temperature, and X_1 experiences a discontinuous jump of about half of its intensity at $T = 261$ K [15]. The T dependence of this x-ray peak is similar to that for X_1 shown in Fig. 1(A). A discontinuity in the order parameter is observed, too, in the low-energy region of the Raman spectrum ($\sim$30 cm^{-1}) [17]. It is also noted that the observed temperature dependences of the libron intensity and the uncorrelated Lorentzian scattering intensity are similar to the temperature dependence of X_2 and X_3, respectively. The model calculation described above has been applied to yield the temperature dependence of the heat capacity [12] shown in Fig. 1(B). The model predicts a first-order phase transition at T_{01} with agreement obtained between theory and experiment regarding the latent heat and entropy changes at T_{01}.

3 Isotope Effect

The rotational spectra associated with the free C_{60} molecule are exceptional and represent a classic example of the effect of isotopes on the molecular vibrational-rotational levels of highly symmetric molecules [18]. To treat these isotope effects, we consider the total wavefunction of the 60 carbon atoms (including their nuclei) Ψ which can be expressed by the product wave function

$$\Psi = \Psi_{el}\Psi_{vib}\Psi_{rot}\Psi_{ns}, \tag{5}$$

where Ψ_{el}, Ψ_{vib}, Ψ_{rot} and Ψ_{ns} refer, respectively, to the electronic, vibrational, rotational and nuclear spin factors.

In the case of $^{12}C_{60}$, the total wavefunction, Ψ, should be totally symmetric for any permutations of the $^{12}C_{60}$ nuclei, because each ^{12}C nucleus is a boson ($I = 0$). In contrast, the $^{13}C_{60}$ nuclei have totally antisymmetric states for an odd number of nuclear exchanges, since each ^{13}C nucleus is a fermion ($I = 1/2$). For a C_{60} molecule which contains both ^{12}C and ^{13}C isotopes, the proper statistics should be applied to permutations among the ^{12}C atoms and among the ^{13}C atoms. Since $I = 0$ for ^{12}C, the nuclear spin factor Ψ_{ns} for $^{12}C_{60}$ transforms as the irreducible representation A_g, and since Ψ_{el}, Ψ_{vib}, Ψ_{ns} all transform as A_g at low temperature (< 1 K), the rotational states, $E_J(rot) = \hbar^2 J(J+1)/2I$, also require that Ψ_{rot} have A_g symmetry. Thus the rotational motion is restricted to J values which contain the irreducible representation A_g of the point group I_h which restricts the J values to be $J = 0, 6, 10, 12, 16, \ldots$. Thus the lowest Pauli-allowed rotationally excited state for $^{12}C_{60}$ must have $J = 6$ which corresponds to $\sim$1.6 K. It is interesting that all rotations from $J = 1$ to $J = 5$ are not allowed by the Pauli principle. In $^{13}C_{60}$, the ^{13}C nucleus has a nuclear spin $I = 1/2$ and there are thus 2^{60} nuclear spin states with I_{tot} values ranging from $I_{tot} = 0$ to $I_{tot} = 30$. The decomposition of 2^{60} nuclear spin states into irreducible representations of $I_{tot} = I$ has been examined by Harter and his coworkers [19], noting that the totally anti-symmetric states of $^{13}C_{60}$ belong to the A_g irreducible representation of I_h. The allowed J values for the rotational spectrum for the $^{13}C_{60}$ molecules also depend on the nuclear spin states. In the case of carbon, the spin-orbit interaction or the interaction with other molecules are relatively small, and therefore the nuclear spin states have relatively long life times. Thus a large difference is expected between the rotational states of the $^{12}C_{60}$ and $^{13}C_{60}$ molecules at low temperature. Such differences in the allowed rotational states and the corresponding differences in the vibrational states lead to significant differences in the low temperature heat capacity, are shown in Fig. 2. Above about 4 K the rotations are in the classical limit and no experimental difference would be expected between the various isotopic species. Large differences occur in the quantum limit below $\sim$1 K (see Fig. 2) where $^{12}C_{60}$ (solid line) shows a large Schottky type anomaly because of the scarcity of rotational levels, whereas $^{13}C_{60}$ (dashed line) has a more continuous density of rotational states. A similar effect contributes to a Schottky anomaly at $\sim$0.03 K for the $J = 1$ rotational states, as shown in the inset to Fig. 2.

Acknowledgments: The MIT authors gratefully acknowledge NSF Grant #DMR 92-01878. One of authors (RS) acknowledges the Casio Foundation for supporting part of joint research with MIT. Part of the work by RS was supported by a Grant-in-Aid for Scientific Research in Priority Area "Carbon Cluster" (Area No. 234/05233214) from the Ministry of Education, Science and Culture, Japan.

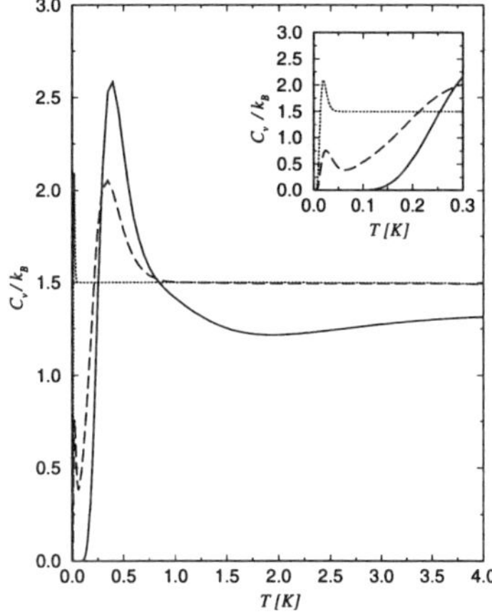

Figure: 2 Specific heat of a C_{60} molecule. $^{12}C_{60}$ (solid line), $^{13}C_{60}$ (dashed line) and C_{60} without symmetry (dotted line) plotted as a function of T. Inset shows behavior in the very low T region [20].

References

[1] J. E. Fischer, Materials Science and Engineering **B19**, 90–99 (1993).

[2] P. A. Heiney *et al.*, Phys. Rev. B **45**, 4544–4547 (1992).

[3] P. A. Heiney *et al.*, Phys. Rev. Lett. **67**, 1468 (1991).

[4] W. I. F. David *et al.*, Europhys. Lett. **18**, 219 (1992).

[5] W. I. F. David, *et al.*, Europhys. Lett. **18**, 735 (1992).

[6] A. Cheng and M. L. Klein, Phys. Rev. B **45**, 1889 (1992).

[7] M. Sprik, A. Cheng, and M. L. Klein, J. Phys. Chem. **96**, 2027 (1992).

[8] X. P. Li, J. P. Lu, and R. M. Martin, Phys. Rev. B **46**, 4301 (1992).

[9] T. Yildirim and A. B. Harris, Phys. Rev. B **46**, 7878 (1992).

[10] W. Que and M. B. Walker, Phys. Rev. B **48**, 13104–13110 (1993).

[11] A. B. Harris and R. Sachidanandam, Phys. Rev. B **46**, 4944 (1992).

[12] R. Saito *et al.*, Phys. Rev. B **49**, 2143 (1994).

[13] W. Schranz *et al.*, Phys. Rev. Lett. **71**, 1561 (1993).

[14] J. P. Lu, X. P. Li, and R. M. Martins, Phys. Rev. Lett. **68**, 1551 (1992).

[15] P. A. Heiney *et al.*, Phys. Rev. Lett. **66**, 2911 (1991). See also Comment by R. Sachidanandam and A. B. Harris, Phys. Rev. Lett. **67**, 1467 (1991).

[16] P. A. Heiney, J. Phys. Chem. Solids **53**, 1333–1352 (1992).

[17] P. J. Horoyski and M. L. Thewalt, Phys. Rev. B **48**, 11446 (1993).

[18] F. Negri *et al.*, Chem. Phys. Lett. **211**, 353 (1993).

[19] W. G. Harter and T. C. Reimer, Chem. Phys. Lett. **194**, 230 (1992).

[20] R. Saito, G. Dresselhaus, and M. S. Dresselhaus, unpublished.

INTERMOLECULAR POTENTIALS AND THEORY FOR THE OPTICAL OBSERVATION OF THE LATTICE MODES IN SOLID C_{60}

Taner Yildirim

Department of Physics, University of Pennsylvania, Philadelphia, PA 19104

ABSTRACT

We review the existing intermolecular potentials and then demonstrate the origin of the orientationally ordered $Pa\bar{3}$ structure of solid C_{60} by a simple LDA calculation using additive charge density approximation. It is found that Coulomb interaction between C_{60} molecules is maximized at the $Pa\bar{3}$ ordering, a strong contradiction of the common belief in the literature, and the short range repulsive interaction is the one which stabilizes the observed $Pa\bar{3}$ structure. The nature of the elementary excitations out of $Pa\bar{3}$ ground state (librons and phonons) are then discussed with particular emphasis on the infrared absorption and Raman intensities of these modes. Even though librons are Raman active, Raman intensities should be practically zero due to high symmetry of C_{60}, and thus the features observed and assigned to librons in many papers in the literature must be due to impurities, such as O_2, as shown by B. Burger and H.Kuzmany.

1. Introduction

The orientational properties of solid crystalline C_{60} are unique and quite interesting. This solid undergoes a phase transition at $T_m=260$ K from orientationally disordered ($Fm\bar{3}m$) to an orientationally ordered phase whose structure is that of space group $Pa\bar{3}$.[1] In this structure, each of the four molecules in the unit cell is rotated about its local [111] direction, starting from the standard orientation, through a setting angle ϕ, whose value is about 22°.

Here I first discuss the origin of the $Pa\bar{3}$ structure by a simple LDA calculation using additive charge density approximation. Then I discuss the mechanism for the infrared absorption and Raman intensities of the lattice modes in this structure.

2. Intermolecular Potentials

Clearly, to understand many of the properties of C_{60}, including $Pa\bar{3}$ ordering, it is essential to have a good intermolecular potential. Although it is customary to use Lennard-Jones atom-atom potentials for molecular solids, it was soon recognized that such a potential for solid C_{60} leads to an instability in the cubic structure[2], in contradiction to the experimental result. To remedy this defect, two improved potential models have been proposed.[3,4] In both cases the Lennard-Jones potential is supplemented by Coulomb interactions between effective charges introduced to describe the electronic charge density of a molecule. The observed Pa$\bar{3}$ structure is stabilized by adjusting these point charges. Although these models gives reasonable frequencies for librons and phonons,[5] two major unsatisfactory features were soon recognized. First, in order to stabilize Pa$\bar{3}$ structure and give a reasonable transition temperature, they require unexpectedly large effective charges in the double and single bonds. Second, as shown in Fig. 1 (a) they predict a minimum and a maximum for two nearly symmetry equivalent orientations ($\phi \approx 22°$ and $\phi \approx 80°$), a situation which does not seem physical. To study this situation, in Ref.[6], Yildirim *et al.* calculated multipoles of a C_{60} molecule from LDA charge density and attempted to obtain the intermolecular potential by a multipole expansion. We found that, although the interaction between a C_{60} and alkali ions at tetrahedral sites is well described by the multipole expansion,[6,7] C_{60}–C_{60} interaction is not, essentially due to the small charge overlaps between the molecules. This situation is very similar to the case of graphite, where the interaction between the graphite sheets are dominated by the term arising from the small charge overlap, rather than quadrupole-quadrupole interactions. Hence, to get the right potential for solid C_{60}, one has to take into account the charge overlap, which was completely neglected in the multipole expansion of Ref. [6].

Since a full LDA calculation is not possible for Pa$\bar{3}$ ordering, we will do a Gordon–Kim type of calculation,[8] in which the total energy of the system is given by

$$E[\rho] = V_C[\rho] + \frac{\pi^{4/3}3^{5/3}}{5} \int_{\Omega_o} \rho^{5/3}(r)d^3r + \frac{1}{36} \int_{\Omega_o} \frac{|\Delta\rho(r)|^2}{\rho(r)}d^3r + V_{XC}[\rho] \qquad (1)$$

where, $\rho(\mathbf{r}) = \sum_{\mathbf{R}} \rho(\mathbf{r} - \mathbf{R})$ is the total charge density which is assumed to be a superposition of the charge densities of the isolated C_{60} molecules. The first term, V_C, is the Coulomb energy of the electron density $\rho(r)$ and the core charges Z_i;

$$V_C[\rho] = \sum_{i,j,\mathbf{R}} \frac{Z_iZ_j}{|\mathbf{r}_i - \mathbf{r}_j - \mathbf{R}|} + \int_{\Omega_o} \sum_{i,\mathbf{R}} \frac{Z_i\rho(r)}{|\mathbf{r}_i - \mathbf{r} - \mathbf{R}|}d^3r + \int_{\Omega_o} \int_{\Omega} \frac{\rho(r)\rho(r')}{|\mathbf{r} - \mathbf{r}'|}d^3rd^3r'. \qquad (2)$$

The second and third terms in $E[\rho]$ are the kinetic energy and the gradient correction. The last term is the exchange and correlation energy, which is very important as in the case of graphite. A detailed report of this study will be published elsewhere. Here I will only point out two important findings of this calculation.

Fig. 1(b) and (c) shows the Coulomb and Kinetic energy respectively. As expected from symmetry, we have either maximum or minimum at both setting angles, $\phi \approx 22°, 80°$, where pentagonal and hexagonal faces are against to the double bonds of the neighboring molecules respectively. Coulomb interaction is maximum at these setting angles, a strong contradiction of the widely accepted statement; *Pa3 is stabilized by the Coulomb interaction which is minimized when an electron reach and electron poor parts of the neighboring molecules face eac other.* This statement is obviously not right because one has to consider the interactions not only between electron clouds but also between electron clouds and the carbon-core charges.(See Eq.(2)) When all these terms are added together, the resulting Coulomb energy is maximum at these setting angles as shown in Fig. 1 (b).

Fig. 1 (c) and (d) show that Kinetic energy and 12-6 potential are very similar, indicating that 12-6 potential is actually doing its job very well. Both terms have two deep minima at the setting angles $\phi = 22^o$ and 80°. This is because of the fact that these terms arise from the charge overlaps of neighboring molecules and at these particular setting angles the charge overlap is minimum.

In conclusion, we have shown that, while the Coulomb energy is important, most of the orientational ordering is dominated by the short range repulsive interactions which arise from small charge overlaps of the neighboring molecules. This also explains the high sensitivity of the ordering transition temperature to the applied pressure.

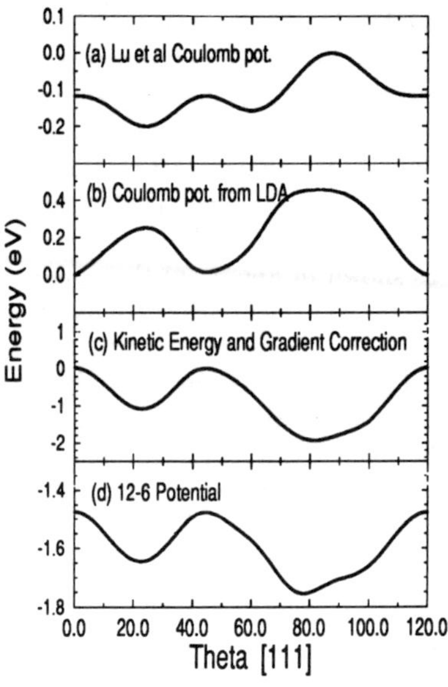

Fig. 1 Various potentials in solid C_{60} as the molecules rotate about [111] directions according to $Pa\bar{3}$ symmetry. In Fig. 1 (b) and (c) zero of energy is arbitrary.

3. Optical Observation of Lattice Modes

As some of the low-frequency lattice modes are amenable to study by infra-red and Raman spectroscopy, they are a potential source of information about intermolecular interactions. The aim of this section is to consider the possible sources of intensity of the infrared absorption and Raman scattering in orientationally ordered $Pa\bar{3}$ structure of C_{60}. When doing this, I will many times refer to analogous molecular solids of N_2 and H_2 which are useful to keep in mind in this context.

Since in $Pa\bar{3}$ structure there are four molecules in the unit cell, there will be twelve translational modes (phonons) and twelve rotational modes (librons). Here we are primarily concerned with those lattice modes which interact with long wavelenght radiation in infra-red and Raman spectroscopy. These correspond to small values of q. At $\mathbf{q} = 0$, the phonons are odd under inversion and the group representation is $\Gamma_T = A_u + E_u + 3T_u$.[5] The librations are even and given by $\Gamma_R = A_g + E_g + 3T_g$.[5] Thus at $\mathbf{q} = 0$ we have five pure translational and five pure librational modes with one onefold, one twofold, and three threefold degeneracies. Below I discuss the observation of these modes by infrared absorption and Raman scattering.

3.1. Infrared Absorption

As is well known, infrared absorption may be observed for optical modes at zero wave vector with T_u symmetry. Therefore in solid C_{60} there are two T_u translational modes which are infrared active. (The third T_u is the acoustic mode which corresponds to the translation of the crystal as a whole.) Of course, being infrared active does not imply a non-zero absorption intensity! Absorption occurs if the transition moment $< \Psi_{m'}|\mathbf{M}|\Psi_m >$ does not vanish. Expanding the electric moment, $\mathbf{M}$, of the system in terms of normal phonon coordinates, the absorption intensity can be shown to be proportional to $I \propto |\frac{\partial \mathbf{M}}{\partial Q_k}|^2$. Therefore the values $|\frac{\partial \mathbf{M}}{\partial Q_k}|$ can be regarded as the *"observables"* of the system. Now the natural question is *"where does $\mathbf{M}$ come from for solid C_{60}?"*. Due to inversion symmetry, C_{60} molecules do not have have a dipole moment. This situation is very similar to solid H_2 which has also $Pa\bar{3}$ symmetry. In solid H_2, $\mathbf{M}$ is an induced moment by the quadrupole moments of the linear molecules. Using this analogy, for solid C_{60} the first non-zero moment is the $l = 6$ *moment* and thus one possible mechanism for infrared absorption can be $l = 6$ *moment–induced dipole moment* which is supposed to be very small. The other possibility is the crystal field effect. Since the point group of $Pa\bar{3}$ is T_h, each C_{60} can have an effective quadrupole moment, like H_2 molecule, and hence the absorption intensity is proportional to the square of this *crystal field induced quadrupole moment*. It is of interest to calculate the absorption intensities from both mechanisms and compare them.

Experimentally it has been reported that C_{60} single crystal reveals two transitions[9], corresponding to two T_u optical modes in $Pa\bar{3}$ structure. The intensities of these absorptions should vanish above the orientational transition temperature since we do not have an optical phonon in the disordered phase ($Fm\bar{3}m$). Therefore, the measurement of the temperature dependence of these absorptions provides an alternative experimental way to access the temperature dependence of the order parameter.

3.2. Raman Scattering

With regard to Raman scattering, symmetry indicates that the libron modes ($A_g + E_g + 3T_g$) are Raman active. However this again does not mean that Raman intensities are definitely non-zero. Raman scattering occurs if the matrix element $< \Psi_{m'}|\mu|\Psi_m >$ of the electric field induced moment μ ($\mu = \alpha\mathbf{E}$) is not zero. After expanding the polarizability tensor α in terms of libron amplitudes, Raman intensity is proportional to $I \propto |\frac{\partial \alpha}{\partial Q_k}|^2$. Hence in the Raman scattering, the observable of the system is the change in the polarizability tensor of the system as the molecules librate. In the case of solid N_2 or solid H_2, the molecular polarizability is anisotropic,

$$\alpha = \begin{pmatrix} \alpha_{\parallel} & 0 & 0 \\ 0 & \alpha_{\perp} & 0 \\ 0 & 0 & \alpha_{\perp} \end{pmatrix} \tag{3}$$

and thus depends on the orientation of the diatomic molecules. However in the case of solid C_{60}, the molecular polarizability is isotropic; i.e. at least if distortion of the molecule away from icosahedral symmetry are neglected. Thus, in independent molecule approximation, the polarizability is independent of libron amplitudes, and hence gives rise to no Raman scattering due to single librons, in contrast to many published experimental papers in the literature.[9]

Assuming crystal field with T_h symmetry is not negligible, the polarizability tensor of C_{60} in the solid will have the form given in Eq. (3) for solid H_2 and hence the Raman intensities of the librons will be proportional to

$$I(A_g) = 0, \quad I(E_g, T_g) = (\alpha_\parallel - \alpha_\perp)^2 \tag{4}$$

where $\alpha_\parallel$ and $\alpha_\perp$ are the polarizability of the C_{60} molecule parallel and perpendicular to local $(1,1,1)$ direction respectively. Since there is no experimental evidence that such a distortion of C_{60} away from I_h symmetry is large, Raman intensities of librons given above should be practically zero, as we first reported in Ref.[5] two years ago. A recent experiment by B.Burger and H. Kuzmany[10] seems to support this conclusion. They looked at the the Raman spectrum of C_{60} before and after the crystal was exposed to air. Not surprisingly, they found almost nothing in the Raman spectrum of the pure crystal, while nice features appeared after exposing the crystal to air. Their spectrum is very similar to those published in earlier papers, such as Ref. [9], indicating that the features observed so far are due to impurities, such as O_2, H_2O.

4. Acknowledgments

I would like to acknowledge many helpful conversations with A.B. Harris, E. J. Mele, and H. Kuzmany.

5. References

1. P. A. Heiney *et al.*, Phys. Rev. Lett. **66**, 2911 (1991).
2. Y. Guo *et al.*, Nature **351**, 464 (1991).
3. M. Sprik *et al.*, J. Phys. Chem. **96**, 2027 (1992).
4. J. P. Lu *et al.*, Phys. Rev. Lett. **68**, 1551 (1992).
5. T. Yildirim and A. B. Harris, Phys. Rev. B **46**, 7878 (1992).
6. T. Yildirim *et al.*, Phys. Rev. B **48**, 1888 (1993).
7. T. Yildirim *et al.*, Phys. Rev. Lett. **71**, 1383 (1993).
8. D. P. Divincenzo and E. J. Mele, Phys. Rev. B **25**, 7822 (1982).
9. P.J. Horoyski and M. L. W. Thewalt, Phys. Rev. B **48**, 2862 (1993).
10. B. Burger and H. Kuzmany, preprint, to appear in this issue.

ORDER AND DISORDER FROM HIGH RESOLUTION RAMAN SPECTRA OF C_{60} SINGLE CRYSTALS

B. Burger and H. Kuzmany
Institut f. Festkörperphysik, Universität Wien
Strudlhofgasse 4, A-1090 Wien, Austria

ABSTRACT

We report high resolution Raman spectra of C_{60} single crystals at low temperature. The splitting of the H_g modes has been analysed by means of polarized measurements, thus assigning them to their symmetry species. Furthermore a study of the external modes of C_{60} with respect to the influence of oxygene is presented.

1. Introduction

Since the discovery of an efficient production method for fullerenes by Krätschmer and Huffman[1] in 1990, our knowledge about these materials, especially C_{60} has grown very fast. In examining Fullerites, Raman spectroscopy has been proved to be a very sensitive tool for investigating changes in lattice and electronic structure. The question we address in this contribution is, whether there is a splitting of the lines from a crystal or correlation field and whether the splitting follows a simple group theoretical analysis. We present high resolution Raman spectra of the H_g modes and show their intrinsic character. The second part of this contribution covers measurements of the external modes with emphasis on the influence of oxygene contamination on the observed Raman intensities.

2. Vibrational Analysis

From a vibrational analysis of the C_{60} molecule only 10 modes are found to be Raman active, assuming I_h symmetry. These modes correspond to the non degenerate A_g and the fivefold degenerate H_g species. In the crystal the symmetry of the molecule is reduced to T_h, which leads to a partial lowering of degeneracies, resulting in the splitting of H_g modes into E_g and F_g. The A_g modes remain unchanged. Below the ordering transition at 255 K a simple cubic lattice with a base of four, site inequivalent molecules is formed. This process does not further remove degeneracies, but yields a Davidov splitting according to the following scheme.

$$A_g \longrightarrow A_g + F_g$$
$$H_g \longrightarrow A_g + 2\,E_g + 5\,F_g$$

Since it was not only interesting to check on the amount of the line splitting but also on the species of the splitted modes, a symmetry analysis of the components is required. The intensity of the Raman signal is given by

$$I \propto |\vec{n_i} \overleftrightarrow{\alpha} \vec{n_s}|^2$$

where $\vec{n_i}$ and $\vec{n_s}$ are unit vectors in the direction of the polarization of the incident and the scattered light respectively, and $\overleftrightarrow{\alpha}$ stands for the Raman tensor. Solving this equation for the Raman tensor of T_h yields the polarization dependent scattering intesities for the various geometries. Since crystals of C_{60} have only well developed faces in { 100 } and { 111} direction, selective geometries are obtained for the setting angle ϕ as compiled in Tab. 1. The symbols $\parallel$ and $\perp$ refer to the relative orientation between the polarization of incidenting and scattered light.

ϕ		{100}	{111}
0°	$\parallel$	$A_g + E_g$	$A_g + E_g + F_g$
	$\perp$	F_g	$E_g + F_g$
45°	$\parallel$	$A_g + E_g + F_g$	$A_g + E_g + F_g$
	$\perp$	E_g	$E_g + F_g$

Tab. 1 : Modes to be seen at various scattering geometries. ϕ refers to the angle between the [100] direction and the polarization of the incident laser beam.

3. Experimental

Raman spectra were taken from single crystals grown by a double temperature gradient technique[2]. The samples were typically 2×2 mm in size with very nice, smooth faces. They were mounted in a closed cycle cryostat, with a [100] face normal to the exciting Ar^+ and a $Ti\text{-}Al_2O_3$ laser. Sample temperatures given are nominel as measured on the cold tip of the cryostat. Extinction of the total symmetric $A_g(2)$ pinch mode in crossed polarizer geometry was better than 1:100. The spectra were taken with a Dilor XY spectrometer with a liquid nitrogene cooled CCD detector.

4. Results and Discussion

4.1 Internal Modes

As expected, the H_g modes split in the crystal, even at temperatures above the ordering transition. Figure 1 presents spectra of the $H_g(1)$ mode at various temperatures. In the uppermost spectrum, taken at 415 K, one can observe a clear line with a shoulder on the low frequency side. This result remains unchanged, except for a hardening of the modes for the second spectrum, taken at room temperature. A spectrum taken with a higher resolution, using 774 nm excitation, shows two distinct lines at 268 cm^{-1} and 272 cm^{-1}. The last spectrum, excited again with 514 nm, shows the $H_g(1)$ mode at 10 K with high resolution. The spectrum could be well fitted with 5 modes in the region between 264 cm^{-1} and 274 cm^{-1}. To assign these modes, polarization dependent measurements were performed, as shown in Figure 2. The uppermost spectrum was obtained for a geometry which reveals only F_g modes. As expected, five modes are observed at 262 cm^{-1}, 264 cm^{-1}, 268 cm^{-1}, 272 cm^{-1} and 274 cm^{-1}. The second

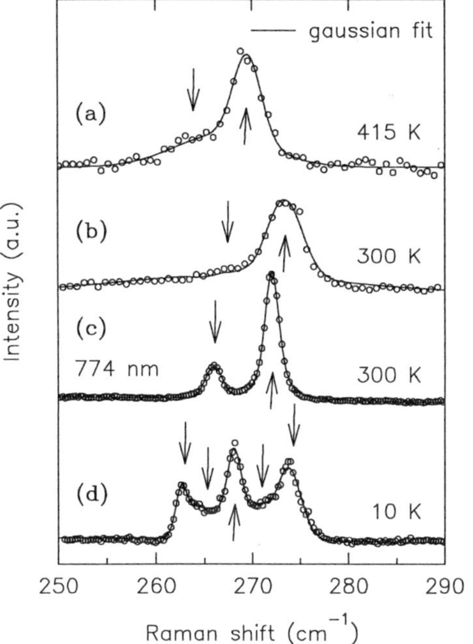

Fig. 1: Raman Spectra of the $H_g(1)$ mode at various temperatures, excited at 514 nm (100 W/cm^2). The resolution was 3.5 cm^{-1} (a,b), 1 cm^{-1} (d), and 1.1 cm^{-1} on excitation with Ti-Al$_2$O$_3$-laser at 774 nm. Peaks of the Gaussian oscillators used to fit the spectra are indicated by arrows.

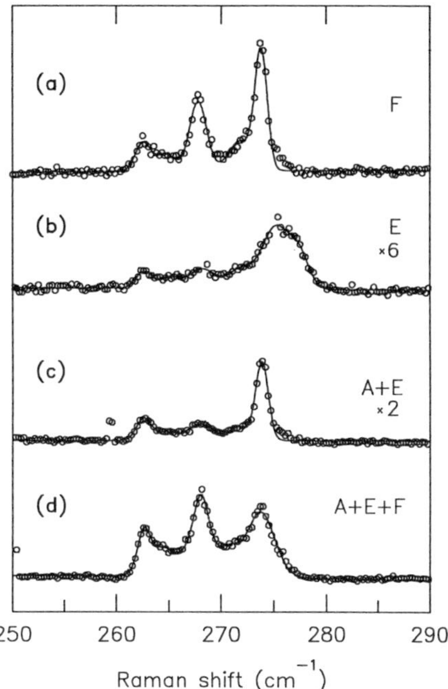

Fig. 2: Raman Spectra of the $H_g(1)$ mode at various scattering geometries, excited at 514 nm (100 W/cm^2). The resolution was 1 cm^{-1}. The symmetry species observed for each spectrum are indicated.

spectrum is supposed to show only E_g modes, which are found at 275 cm^{-1} and 276 cm^{-1}. It is upscaled by a factor 6, indicating the good suppression of other species. The third spectrum, where A_g and E_g modes are allowed, gives an assignment for the A_g species at 274 cm^{-1}. The last spectrum shows the superposition of all species, like in Fig. 1. In this way we obtained an assingment for several of the splitted H_g-modes. A table of the obtained results for all H_g modes will be published elsewhere. It should be mentioned that not all of the H_g lines show a splitting as well defined as for the $H_g(1)$ mode. The $H_g(8)$ mode for example shows only a very weak splitting and no response to a change of scattering geometry. The reason for this is unclear up to now and will be a subject of further investigations.

The spectra described above were taken at a flux of 100 W/cm^2 from a sample,

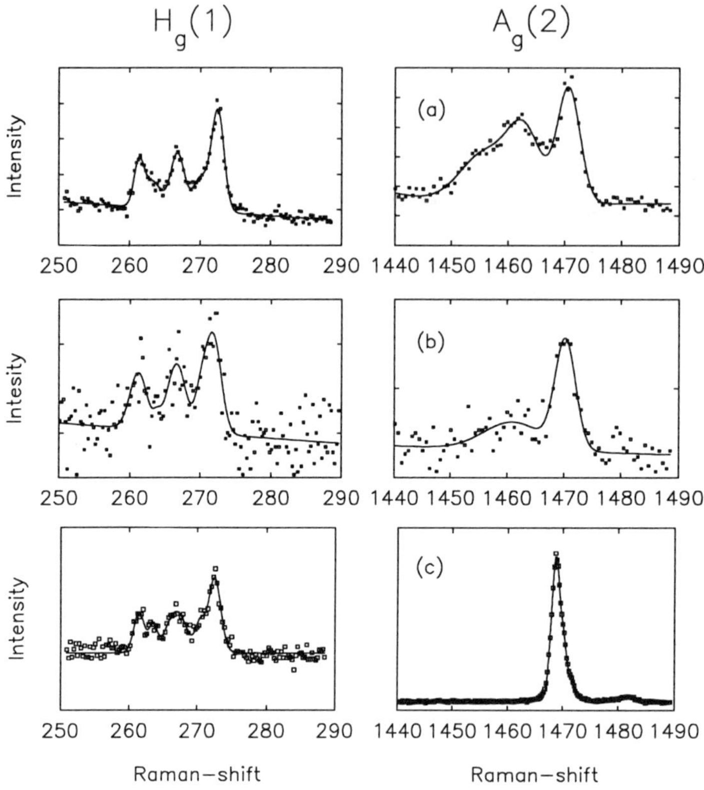

Fig. 3 : Spectra of H$_g$(1) and A$_g$(2) mode for an oxygen free sample at a flux of 100 W/cm^2 (a), 5 W/cm^2 (b) and of an oxygenated sample 4 days on air at 100 W/cm^2 (c).

which was mounted to the cryostat on air. It was important to check, to what extent the splitting may originate from a phototransformation[3] or from excited state scattering[4]. We can exclude scattering from a photopolymerized sample due to the low temperatures, where photopolymerization does not occur. To check for excited state scattering, we mounted a crystal under argon atmosphere. Figure 3 shows the results of this experiment. As can be seen, the response of the A$_g$(2) mode changes dramatically with the sample quality and light flux, whereas the H$_g$(1) mode remains unchanged. At 100 W/cm^2 strong satellite lines can be observed for the clean crystal, which may originate from an excited states scattering[4]. These lines nearly vanish at 5 W/cm^2. Exposingg the crystal to air quenches the excited state and even at 100 W/cm^2 no satellite lines are observed any more. The observed splitting as shown in

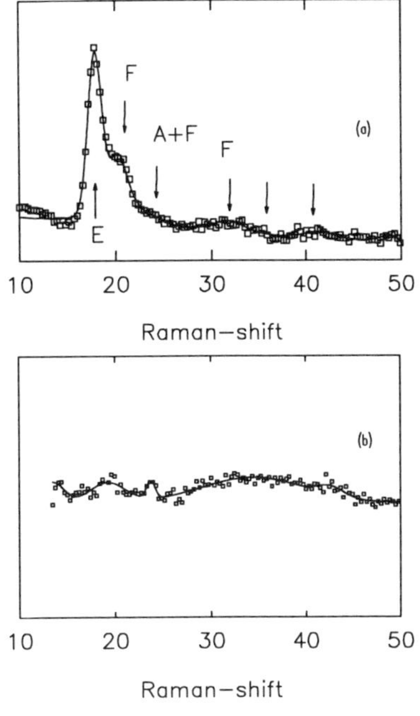

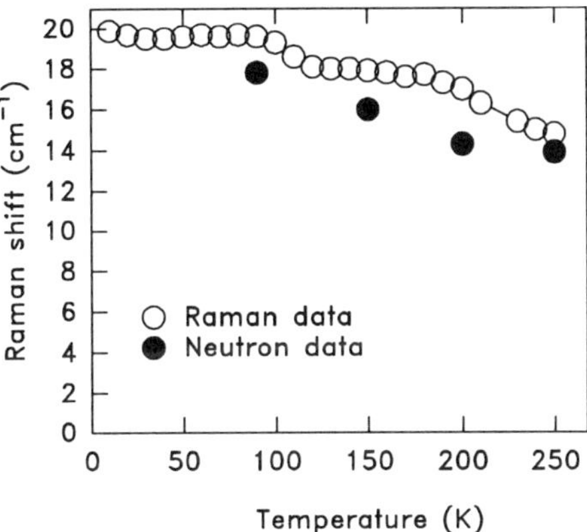

Fig. 4: Raman Spectra of the external modes taken at 793 nm from an oxygenated (a) and an oxygene free sample (b).

Fig. 5: Raman shift and FWHM of the strongest external mode between 10 and 260 K. Full drawn circles are shift data obtained from neutron scattering by Pintschovius et al.[8]

Fig.1 and 2 is consistent with predictions from group theory for the high and low temperature phase of the crystal.

4.2 External Modes

The C_{60} single crystals also show external modes at low temperatures. Only the 5 librations are Raman active. Librations are hidered rotations of the buckyballs and should therefore vanish at the ordering transition, where the molecules begin to rotate freely. We measured the external modes as a function of temperature and oxygene contamination. The upper part of Fig. 4 shows a typical spectrum from an air exposed crystal. Peaks are observed at 18 cm^{-1}, 20 cm^{-1}, 24 cm^{-1}, 31 cm^{-1}, 37 cm^{-1} and 41 cm^{-1} in reasonably good agreement with results from neutron scattering. The good agreement with neutron data and the complete and sudden quench of these Raman lines at the ordering transition proved, that they originate from librons. The extra line at 41 cm^{-1} in Fig.4a may be adscribed to one of the allowed IR active translational modes, which was Raman activated by disorder. The original tentative assignment shown in Fig.4a could not be confirmed by studying the spectra with selective geometries. No response to the selection rules could be detected. Since this effect was considered to result from disorder induced among others by oxygene, the experiment was repeated with completely oxygene free samples. The result is shown in Fig.4b. No sizeable scattering can be detected. Exposing the crystal to air for 4 days recovers the spectrum of the external modes with a slight shift. The lack of Raman scattering in oxygene free and highly ordered crystals is in contrast to previous results[5,6] but not unexpected. Yildirim et al.[7] have shown in a previous theoretical work, that the librons, though Raman active, will have a very small scattering tensor due to nearly spherical symmetry of the C_{60} molecules. They were claimed to appear in the spectra only by a symmetry breaking disorder.

5. Conclusion

We have shown, that the splitting of the H_g modes in C_{60}-Fullerite follows predictions from grou theory with respect to crystal field and correlation interaction. It has been proved, that this splittig is intrinsic and not a result of excited state scattering or photopolymerization.
Raman scattering from external modes was confirmed, but oscillator strength for the scattering was demonstrated to be disorder (oxygene) induced.

6. Acknowledgments

We thank M. Haluška for the preparation of the crystals and the Hoechst AG for the proliferation of the raw material. This work was supported by the Fond zur Förderung der wissenschaftlichen Forschung, project P09741-TEC.

7. References

1. W. Krätschmer, L.D. Lamb, K.Fostiropoulos,D.R. Huffman, *Nature*, **347**, 354 (1990)
2. M. Haluška, H. Kuzmany, M. Vybornov, P.Rogl, P.Fejdi, *Appl. Phys. A*, **56**, 161 (1993)
3. A.M. Rao, Ping Zhou, Kai-An Wang, G.T. Hager, J.M. Holden, Ying Wang, W.-T. Lee. Xiang-Xin Bi, P.C. Eklund, D.S. Cronett, M.A. Duncan, I.J.Amster, *Science*, **259**, 955 (1993]
4. P.H.M. van Loosdrecht, P.J.M. van Bentum, M.A. Verheijen, G. Meijer, *Chem. Phys. Lett.*, **198**, 587 (1992)
5. P.J. Horoyski, M.L.W. Thewalt, *Phys. Rev. B*, **48**, 2862 (1993)
6. H. Kuzmany, M. Matus, T. Pichler, and J.Winter, to be published in *Proc. Nato ARW on Physics and Chemistry of Fullerenes*, **NATO ASI -C series**
7. T. Yildirim and A.B. Harris, *Phys. Rev. B*, **46**, 7878 (1992)
8. L.Pintschovius,, B. Renker, F. Gompf, R.Heid, S.L. Chaplot, M. Haluška, H. Kuzmany, *Phys. Rev. Lett.*, **69**, 2662 (1992)

PULSE METHOD MEASUREMENTS OF THERMAL PROPERTIES OF C_{60}

V. Skákalová, P. Fedorko,
*Dept. of Chem. Physics, Fac. of Chem. Technology, Slovak
Technical University, 812 37 Bratislava, Slovakia*

L. Kubičár, V. Boháč
*Slovak Academy of Sciences, Inst. of Physics,
800 00 Bratislava, Slovakia*

and

M. Haluška
*Inst.f. Festkorperphysik, Universität Wien, Strudlhofgasse 4,
A-1090 Vienna, Austria*

ABSTRACT

Thermal properties (specific heat, thermal diffusivity and thermal conductivity) of C_{60} compressed powder were studied by pulse method. A structural phase transition and a glass transition were found at 240 K and 160 K, respectively. An influence of heat treatment on the thermal properties was observed. The temperature dependence of the phonon mean free path is discussed.

1. Introduction

Knowledge of thermal properties of a material provides informations about its structural transitions, relaxations, phonon transport and enthalpy of thermal effects. A number of papers dealing with the thermal conductivity K[1÷3], specific heat c[3÷7], sound velocity[8] and complex elastic constant[9] of single crystals as well as of compressed powders of C_{60} was published.

Most of the experiments confirmed a structural phase transition in the temperature region 235-260 K. Differences between various results were interpreted by the effect of the solvent remaining in the sample[6] or by the existence of several different crystallographic structures of C_{60} at room temperature[7]. The transition from the fcc to the sc structure at 260 K were identified as a very sharp first-order transition[1,4,6÷9]. The transition from the orthorhombic[7] and monoclinic structure were observed at lower temperatures and were less pronounced.

A glass transition of C_{60} was observed at different temperatures in the region 85-160 K[1,8,9]. According to papers[8,9] the transiton temperature may depend on the measuring technique and on the temperature variation rate.

Our work deals with the thermal properties of the C_{60} compressed powder studied by the pulse method. In addition to the c(T) dependence, this method gives informations about the basic phonon transport property of the material - its thermal diffusivity a(T), on the basis of which conclusions on the phonon mean free path can be done.

2. Experimental
2.1. Sample preparation

The C_{60} powder was dried in vacuum at 250 C during 24 hours and then compressed under N_2 atmosphere by a pressure 200 MPa. Compared with a single crystals, the porosity of the pellets was 10 %.

2.2. Principle of the pulse method

The pulse method[10] was used for measuring thermophysical parameters. A heat pulse generates a dynamic temperature field in the sample. From parameters of the temperature response to the heat pulse (usually the time and the magnitude of the temperature response maximum) specific heat and thermal diffusivity are calculated. Then, thermal conductivity is determined as $K = ac\rho$, where ρ is the density of the sample.

3. Results and discussion

Fig.1a,b,c and Fig.2a,b,c represent temperature dependences of the specific heat c(T), thermal diffusivity a(T) and thermal conductivity K(T) in various temperature variation regimes. The results can be commented from several points of view:
1. How do the phase transitions influence the measured dependences ?
2. What is the influence of heat treatment on c(T), a(T) and K(T) ?
3. What is the temperature dependence of the phonon mean free path ?

The pronounced peak at 241 K in Fig.1a,2a without any hysteresis could characterize the transition from the orthorhombic or monoclinic structural phase[7], in contrast to the sharp peak observed at 260 °C for fcc phase[5,6,7]. As far as we know there are no published data of a(T) for C_{60}. In our a(T) dependence (Fig.1b,2b) we found a minimum at 237 K. The calculated K(T) dependence has a sharp maximum at 245 K (Fig.1c, 2c). Published results for K(T) are not very consistent: a fcc single crystal showed a discontinuous jump at 260 K[1], while no anomaly of K(T) was observed for the C_{60}/C_{70} compressed powder in the critical temperature region[3].

The change of the character of the dependences at about 160 K can be attributed to the glass transition. At this temperature c(T) has a smooth shift (Fig.2a) while in a(T) a slight break (Fig.2b) is observed. Anomalies of c(T) and a(T) lead to a break or even a minimum in K(T) at 160 K (Fig.2c).

Dependences c(T), a(T) are very sensitive to the heat treatment or to the rate of heating and cooling determining the thermodynamic state of the structure. Due to rapid cooling from room temperature to 80 K (Fig.1), c(T) and a(T) below 207 K have, respectively, higher or lower values in comparison to the values obtained after slow cooling. A pronounced anomaly appears at 207 K for the samples cooled by high cooling rate and measured by low heating rate. At this temperature both c(T) and a(T) get values

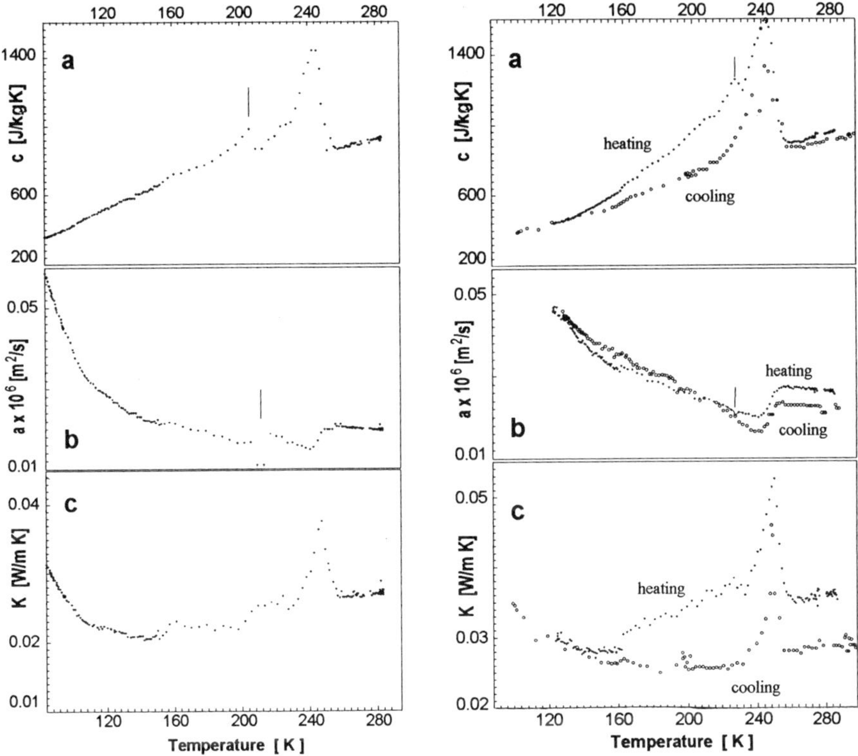

Fig.1. Temperature dependence after rapid cooling from room temperature of a) specific heat c(T), b) thermal diffusivity a(T) and c) thermal conductivity K(T)

Fig.2. Temperature dependence after slow cooling down to 230 K followed by rapid cooling to temperatures below 160 K of a) specific heat c(T), b) thermal diffusivity a(T) and c) thermal conductivity K(T).

corresponding to those obtained after slow cooling. This can be explained by a relaxation of the frozen room temperature structure, which, at low temperature, is far from thermodynamic equilibrium. At 207 K this structure relaxes to the stable state. Another anomaly in c(T) and a(T) appears at 223 K : due to slow cooling from the room temperature to 240 K followed by rapid cooling to temperature below 160 K (Fig.2), c(T) exhibits by heating a pronounced maximum at 223 K accompanied by a minimum in a(T). This anomaly can be attributed to a relaxation of stresses created during the rapid cooling of the structure corresponding to the temperatures below 230 K. A similar "double peak" anomaly in c(T) was reported without interpretation in the paper[6].

From the definition of the thermal diffusivity $a(T) = 1/3\ l(T)\ v(T)$ the dependence of the phonon mean free path l(T) can be determined if the phonon velocity v(T) is known. Taking data of $\Delta v/v$ for a monoclinic structure[8] and our data of $\Delta a/a$, and assuming that the v(T) for monocrystals and polycrystals is qualitatively the same, we obtain the temperature dependence of $\Delta l/l$ (Fig.3) with almost the same character as the a(T) dependence. It is quite the opposite to the presumption[1] that a sharp decrease of the phonon mean free path should be observed at the transition to the orientationally disordered state. This result has to be verified by a direct measurement of the sound velocity of our samples.

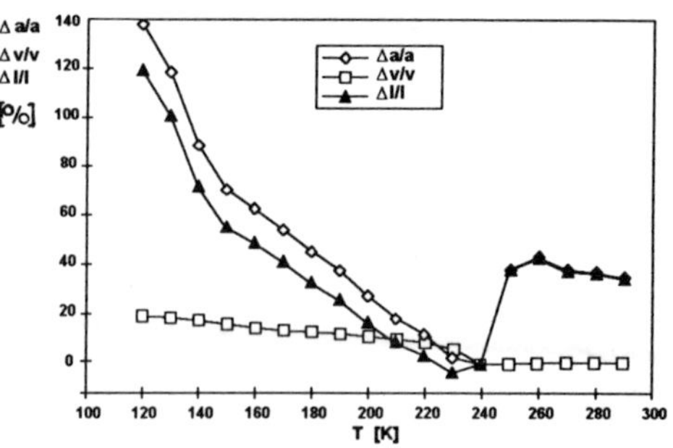

Fig.3. Temperature dependence of the relative change of thermal diffusivity $\Delta a/a$, relative change of sound velocity $\Delta v/v$ and calculated relative change of phonon mean free path $\Delta l/l$.

4. Acknowledgment

The work was supported by the Action Austria-Slovak Republic, Science and Education project No 4SR7.

5. References

1. R. C. Yu, N. H. Tea, M. B. Salamon, D. Lorents and R. Malhorta, *Phys. Rev. Lett.* **68** (1992) 2050.
2. B.Sundquist, *Phys. Rev. B* **48** (1993) 14712.
3. J. R. Olson, K. A. Topp and R. O. Pohl, *Science* **259** (1993) 1145.
4. T. Matsuo, H. Suga, W. I. F. David, R. M. Ibberson, P. Bernier, A. Zahab, C. Fabre, A, Rassat and A. Dworkin, *Solid State Commun.* **83** (1992) 711.
5. W. P. Beyermann, M. F. Hundley, J. D. Thomson, F. N. Diederich and G. Gruner, *Phys. Rev. Lett.* **68** (1992) 2046.
6. T. Atake, T. Tanaka, H. Kawaji, K. Kikuchi, K. Saito, S.Suzuki, J. Achiba and I. Ikemoto, Chem. *Phys. Lett.* **196** (1992) 321.
7. M. Chung, Yiqin Wang, J. W. Brill, X. D. Xiang, R. Mostovoy, J. G. Hou and A. Zettl, *Phys. Rev. B* **45** (1992) 13831.
8. X. D. Shi, A. R. Kortan, J. M. Williams, A. M. Kini, B. M. Savall and P. M. Chaikin, *Phys.Rev Lett.* **68** (1992) 827.
9. W. Schranz, A. Fuith, P. Dolinar, H. Warhanek, M. Haluska and H. Kuzmany, in *Electronic Properties of Fullerenes*, eds. H. kuzmany, J. Fink, M. Mehring and S. Roth (Springer-Verlag, Berlin, Heidelberg, 1993) 117.
10. L. Kubičár, *Pulse Method of Measuring Basic Thermophysical Parameters* (VEDA in coedition with ELSEVIER, Bratislava 1990).

Structure and surface studies of C_{60} crystals

M.A. Verheijen, W.J.P. van Enckevort, G. Meijer,

J.L. de Boer[†], S. van Smaalen[†], V. Petricek[‡] and M. Dusek[‡]

Research Institute for Materials, Faculty of Science, University of Nijmegen
Toernooiveld, 6525 ED Nijmegen, The Netherlands
[†] Laboratory of Chemical Physics, University of Groningen, Nijenborgh 4
9747 AG Groningen, The Netherlands
[‡] Institute of Physics, Academy of Sciences of the Czech Republic
Na Slovance 2, 180 40 Praha 8, Czech Republic

Abstract

Single crystal X-ray diffraction studies of hexagonal C_{60} crystals have shown that these crystals can be regarded as a closest packing of spheres in a h.c.p. lattice. Surface studies of cubic crystals by phase sensitive optical microscopy show slip lines, stacking fault outcrops, cleavage patterns and steps. The latter two are discussed in some detail.

1. Introduction

Over the last few years a lot of research has been carried out to elucidate the crystal structure and the phase transitions in solid C_{60}. It has been well established that above 260 K C_{60} forms an f.c.c. lattice in which the molecules have almost complete rotational freedom. Below 260 K the unit cell becomes simple cubic having 4 molecules per unit cell [1-3].

Recently, we have observed that a minority of the C_{60} crystals grown from the vapour phase show facets corresponding to a hexagonal symmetry. Here we will briefly present the room temperature structure of these hexagonal crystals.

Furthermore, we will present phase sensitive optical microscopy studies of growth phenomena on the {111} and {100} faces of cubic C_{60} crystals. Special attention will be paid to the origins of growth steps and the formation of cracks on these surfaces.

2. Crystal growth and structure

The C_{60} crystals have been grown from the vapour phase by a multiple sublimation technique, which has been described elsewhere[4,5]. Many batches of crystals have been prepared. Most of the crystals display a f.c.c. morphology having {111} and {100} faces. Three crystals, however, clearly show faces characteristic for a hexagonal symmetry. The morphology of these crystals consists of at least two types of facets, the {101} and the {100} faces.

178

Figure 1. Cleavage pattern on a {111} face introduced by cracking of the surface. The cleavage plane is {111} (magnification 980 ×).

One of these crystals has been analyzed using single crystal X-ray diffraction[6]. The diffraction experiment was performed at room temperature. Prolonged random search for reflections only gave reflections which could be indexed on a hexagonal unit cell. Averaging in Laue symmetry 6/mmm reduced 5923 measured reflections to 629 unique ones, with internal consistency $R_I = 0.045$. Extinction conditions were found to be compatible with the space group $P6_3/mmc$. The lattice parameters were determined to be $a = 10.009$ Å and $c = 16.338$ Å . The unit cell contains two molecules, having their centers at the positions (1/3, 2/3, 1/4) and (2/3, 1/3, 3/4). At room temperature these molecules have almost complete rotational freedom. Thus the crystal structure can be regarded as an ideal hexagonal closest packing of spheres.

3. Surface studies

The crystal surfaces are examined by optical Phase Contrast Microscopy (PCM) and Differential Interference Contrast Microscopy (DICM) in combination with normal and high contrast photography. These methods are capable of detecting height differences down to several Å , which means that monomolecular steps can be identified. The lateral resolution is limited to about half a micrometer. To minimize contamination and oxidation effects the quartz tubes containing the crystals are opened just prior to the optical studies. A detailed discussion of growth phenomena observed on the {100} and {111} faces is given in ref. 5. We will here focus on the origin of growth steps and the formation of cracks.

3.1. Fracture

As was already reported by Haluska et al.[7] both the {111} and the {100} faces display cracks upon illumination. Also on our crystals crack lines can be found. This fracture can be obtained in air as well as in vacuum, where the irradiance on the sample surface is more than one order of magnitude lower then the level at which polymerization occurs[8]. Therefore, the cracking must be due to an accumulation of heat in the near surface volume of the crystal, resulting in an expansion of this part of the crystal. The strains induced in this way thus cause fracture. In microscopy cracking of the surfaces can be prevented by using low level light intensities and an infrared blocking filter.

Cracking occurs along those {110} cleavage planes that are oriented perpendicularly to the crystal faces. The fact that cleavage only occurs along the perpendicular {110} planes indicates that the thermally induced tensile or shear stresses act parallel to the surface. The crack lines do not proceed through the entire crystal; for the {111} faces, at a depth of 1 - 5 μm below the surface fracture occurs parallel to the surface. As a result, the outer layer of a cracked surface consists of flakes with a thickness of several microns. When these flakes are removed a beautiful cleavage pattern can be seen (see figure 1). The step edges on the {111} cleavage plane are oriented parallel to the /1$\bar{1}$0/ directions and thus reflect the threefold symmetry of this plane. Interesting is the fact that cracking is not accompanied by the formation of slip lines or slip bands. This indicates that at room temperature dislocations are relatively immobile and not easily formed. Therefore, the crystals do not easily deform plastically and cracking occurs upon application of stresses.

3.2. Growth steps

The different growth runs are carried out at different supersaturations, which results in different step shapes. At low supersaturations low (one to a few times the monolayer thickness), curved steps are present on both types of faces. At higher supersaturations the steps on te {111} faces are faceted and many monolayers high. Here the step edges run parallel to the Periodic Bond Chains (PBC's) in the /110/ directions and are relatively high[4]. In addition a lot of stepbunching, i.e. an accumulation of lower steps, occurs. Very low, unfaceted steps have nevertheless also been observed.

As was mentioned above several step sources have been found. On many crystals the steps originate from one or maximal two distinct points near or at the edge of the crystal faces. The occurrence of these sources near or at the edge of the face can be the result of either contact nucleation or volume diffusion[5]. Another step source which has been observed frequently is a defect or a group of defects, e.g. low angle grain boundaries, planar faults and twin boundaries[5]. On a {100} face we once observed a macroscopic growth spiral (figure 2a). The origin of this spiral is a screw dislocation which is situated at the end of a low angle twist boundary. After arrest of growth the crystal was shortly etched by heating the crystal. By preferential etching the tilt boundary has been made clearly visible. Note that the steps on the {100} face are very high. We suppose this is due to the large burgers vector of the dislocation(s) at the centre of the spiral.

A last mechanism for step generation which has been observed on the {111} surfaces of C_{60} crystals is two-dimensional nucleation. In figure 2b circular steps can

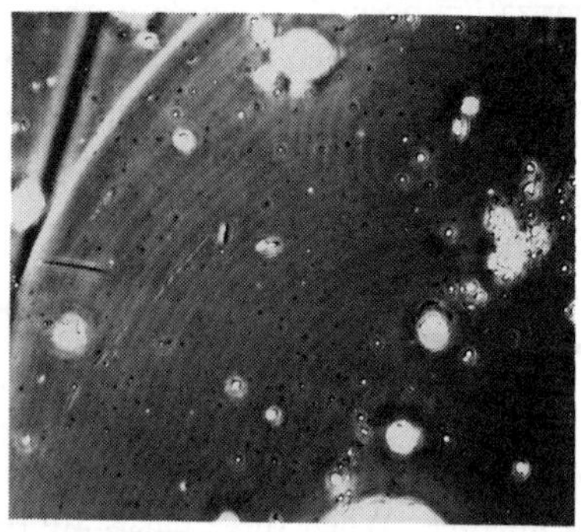

Figure 2. a) Growth spiral on a {100} face having its origin at the end of a low angle twist boundary (magnification 260 ×). b) A step source emitting very low, presumably monomolecular growth steps on a {111} face (magnification 980 ×).

be seen which are generated from a few distinct points. We therefore suppose that on this surface steps can also be created by two-dimensional nucleation at preferred sites. In contrast to the birth and spread model here the nuclei are repeatedly formed at the same point. This suggests nucleation at a preferential site, for instance at an edge dislocation or at a very small particle. As the frequency of step nucleation is low the separation between adjacent steps is large. This implies that the two-dimensional nucleation centres are easily overgrown by step trains emitted from screw dislocations, stacking faults and twins. Furthermore the small particles attached to the surface that act as a step source have a limited height. Therefore their resulting growth hillock will only become as high as the nucleus itself. Thus these particles will incidently be overgrown unless a non-vanishing defect is involved.

1. P.A. Heiney *et al.*, *Phys. Rev. Lett.* **66** (1991) 2911
2. R. Sachidanandam and A.B. Harris, *Phys. Rev. Lett.* **67** (1991) 1467
3. W.I.F. David, R.M. Ibberson, T.J.S. Dennis, J.P. Hare and K. Prassides, *Europhys. Lett.* **18** (1992) 219
4. M.A. Verheijen, H. Meekes, G. Meijer, E. Raas and P. Bennema, *Chem. Phys. Lett.* **191** (1992) 339
5. M.A. Verheijen, W.J.P. van Enckevort and G. Meijer, *Chem. Phys. Lett.* **216** (1993) 72
6. J.L. de Boer, S. van Smaalen, V. Petricek, M. Dusek, M.A. Verheijen and G. Meijer, *Chem. Phys. Lett.* **219** (1994) 469
7. M. Haluška, H. Kuzmany, M. Vybornov, P. Rogl and P. Fejdi, *Appl. Phys. A* **56** (1993) 161
8. Y. Wang, J.M. Holden, Z.-H. Dong, X.-X. Bi and P.C. Eklund, *Chem. Phys. Lett.* **211** (1993) 341

STRUCTURE OF Sb FULLERIDE SINGLE CRYSTALS

F. VALACH, M. HULMAN, A. FOREJTOVÁ
*Department of Chemical Physics, Slovak Technical University,
Faculty of Chemical Technology, 812 37Bratislava, Slovakia*

and

M. HALUŠKA, H. KUZMANY
*Institute of Solid State Physics, University of Vienna,
A-1090 Vienna, Austria*

ABSTRACT

Single crystals of Sb doped buckminsterfullerene C_{60} were prepared by the double temperature gradient technique. The crystal structure has trigonal symmetry, space group P3. The hexagonal unit cell with parameters a = 17.2(1) and c = 24.20(5)Å contains six species $Sb_{2.5} C_{60}$. One part of the Sb atoms occupies the positions on the 3-fold axis out of C_{60} icosahedra. The C_{60} units occupy the assymetrical positions of the unit cell. The shortestest Sb-C_{60} distances are 2.65 and 2.70Å. The shortest C_{60}-C_{60} separation is 9.9 Å.

1. Introduction

Metal doped buckminsterfullerene C_{60} in monocrystal state was rarely investigated by X-rays. The position of the metal atom relative to the C_{60} units in the crystal structure determines the charge transfer properties of the doped fullerene. In order to estimate the metall-fullerene bonding and to predict the possibilities of doping by other metals into the host structure, the crystal structure of Sb doped monocrystals was studied.

2. Experimental

Sb fulleride crystals were prepared from vapor phase by using a temperature profile with two oppositely oriented gradients in the reactor tube[1]. The pure C_{60} was deposited on one end of the tube and Sb on the other end before the tube was sealed under high vacuum. The advantage of this method is the possibility to change 'independently' both gradients with respect to the different vapor pressure of components used. Fulleride single crystals with smooth and shiny faces up to 1.5 x 1.0 x 0.5 mm were grown in the center of the tube. The density of the crystals was measured by the flotation method in $CHBr_3$ /CH_3OH mixture, D_m = 1.70 gcm^{-3}. Intensities of diffractions were taken using a SYNTEX P2$_1$ diffractometer, with MoKα radiation (λ = 0.71069 Å) and graphite monochromator, by the $\theta/2\theta$ scan technique ($2\theta_{max}$ = 55°). A total of 3195 reflections were measured (index range $0 \leq h \leq 6$, $0 \leq k \leq 11$, $-31 \leq l \leq 31$) of which 2026 are unique (R_{int} = 0.043 after averaging) and 514 are holding the condition $F > 1.96\sigma(F)$. Absorption correction (Ψ -scan) was applied, maximum and minimum transmission factors were 0.994 and 0.891. Intensity data were reduced using the XP21 program[2]. The structure was solved by the integrated Patterson and direct method using the program PATSEE[3] .

182

3. Results and discussion

The crystals were found to have the trigonal space group $\bar{P}3$ (No. 147)[4] . Hexagonal unit cell parameters are a = 17.2(1) and c = 24.20(5) Å. The unit cell contains Z = 6 species $Sb_{2.5}C_{60}$. The packing of the C_{60} units and a part of the Sb atoms are shown in Fig.1. Fractional coordinates of the atoms are collected in Tab. 1. Sb atoms occupy positions on the 3-fold axes and C_{60} units occupy the asymmetric positions. The remaining two atoms Sb may be disordered in asymmetric positions and on 3-fold invers axes and on the symmetry centers.

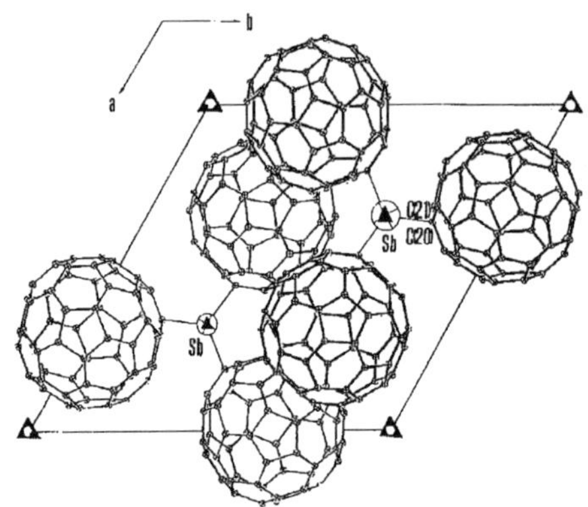

Figure 1.
Packing of the C_{60} units and positions of ordered Sb atoms.
Only two adjacent layers (001) are shown for clarity.

Within one layer each atom Sb is surrounded by three C_{60} units [bond distances Sb-C(20) and Sb-C(21) are 2.65 and 2.70 Å; the bond angle C(20)-Sb-C(21) is 28.6°]. Details of these interatomic contacts are shown in Fig. 2. The diameter of the C_{60} units is 7.0(1) Å. The mean C-C bond length is 1.41(6) Å. Mean C-C-C bond angle in hexa- and pentacycles is 120(1) and 108(1)° . Minimal C_{60} -C_{60} separations of molecular centroids are 9.9 Å (in layer) and 17.6 Å (between the layers).

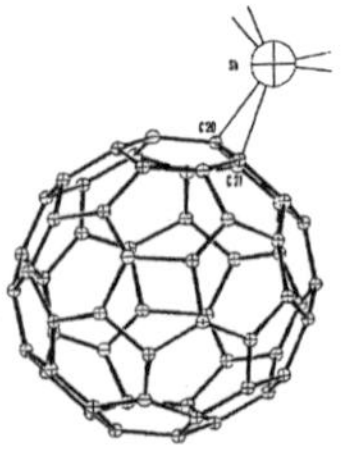

Figure 2.
Shortest Sb-C contacts in the layer of crystal structure $Sb_{2.5} C_{60}$.

Table 1.
Atomic coordinates for $Sb_{2.5}C_{60}$.

	x/a	y/b	z/c		x/a	y/b	z/c
Sb	.6666	.3333	.1666				
C(1)	1.1002	.2189	.2351	C(31)	1.0579	.1732	.0880
C(2)	1.0248	.2038	.2669	C(32)	1.1013	.2493	.0532
C(3)	.9420	.1331	.2373	C(33)	1.1849	.3183	.0829
C(4)	.9705	.1100	.1879	C(34)	1.1876	.2821	.1352
C(5)	1.0678	.1610	.1862	C(35)	1.1107	.1921	.1388
C(6)	.9429	.2831	.2930	C(36)	1.0777	.3711	.0171
C(7)	1.0224	.2733	.2933	C(37)	1.0528	.2768	.0219
C(8)	1.1002	.3654	.2886	C(38)	.9534	.2257	.0246
C(9)	1.0654	.4251	.2856	C(39)	.9235	.2893	.0216
C(10)	.9668	.3764	.2868	C(40)	.9991	.3818	.0182
C(11)	.8694	.2185	.2673	C(41)	1.1533	.4315	.0430
C(12)	.8148	.2440	.2334	C(42)	1.1547	.5073	.0719
C(13)	.7830	.1876	.1846	C(43)	1.2061	.5263	.1225
C(14)	.8158	.1258	.1895	C(44)	1.2386	.4639	.1238
C(15)	.8695	.1433	.2389	C(45)	1.2068	.4046	.0765
C(16)	.8369	.3305	.2260	C(46)	1.0818	.5176	.0745
C(17)	.9165	.4000	.2546	C(47)	1.0003	.4518	.0457
C(18)	.9653	.4787	.2211	C(48)	.9209	.4313	.0764
C(19)	.9118	.4582	.1716	C(49)	.9547	.4893	.1241
C(20)	.8349	.3683	.1751	C(50)	1.0520	.5404	.1224
C(21)	.8055	.3145	.1318	C(51)	1.1008	.5571	.1676
C(22)	1.0539	.5232	.2216	C(52)	.8524	.3484	.0778
C(23)	1.1093	.4979	.2525	C(53)	.8505	.2721	.0519
C(24)	1.1850	.5136	.2202	C(54)	.8065	.1959	.0872
C(25)	1.1804	.5497	.1671	C(55)	.7788	.2213	.1372
C(26)	1.2437	.4288	.1732	C(56)	.8424	.1005	.1431
C(27)	1.2160	.4545	.2231	C(57)	.8375	.1368	.0901
C(28)	1.1708	.3781	.2592	C(58)	.9139	.1537	.0569
C(29)	1.1700	.3018	.2325	C(59)	.9687	.1272	.0888
C(30)	.9217	.0932	.1427	C(60)	1.2170	.3358	.1785

4. Acknowledgements

This work was supported by the Osteuropafoerderung des BMfWF project GZ 45.338/1-IV/ 6a/94, Slovak Ministry of Education and Science project N0. 1/710/93 and Aktion Oesterreich - Slowakische Republik, Wissenschafts- und Erziehungskooperation project 4SR7.

5. References

1. M. Haluška, H. Kuzmany, M. Vybornov, P. Rogl and P. Fejdi, *Appl. Phys.* **A56** (1982) 2015.
2. F. Pavelčík, *XP21 Computer program for SYNTEX P21 data reduction.* Faculty of Pharmacy, J. A. Komensky Univ., 83232 Bratislava, Czechoslovakia 1978.
3. E. Egert, *Acta Cryst.* **A41** (1985) 262.
4. *International Tables for Crystallography*, Vol. A (D. Reidel, Dordrecht, Boston, 1983) p.488.

ORIENTATIONAL DISORDER OF C_{60} POWDER
BELOW THE 260 K TRANSITION

R.Glas, O.Blaschko, G.Krexner
Institut für Experimentalphysik, Universität Wien
Strudlhofgasse 4, A-1090 Vienna, Austria

and

M.Haluška and H.Kuzmany
Institut für Festkörperphysik, Universität Wien
Strudlhofgasse 4, A-1090 Vienna, Austria

Abstract

Diffuse neutron scattering measurements on C_{60} powder were performed covering the temperature range from slightly above the 260 K transition down to 20 K. In the low-temperature phase a considerable amount of elastic diffuse scattering intensity is observed beside simple cubic Bragg reflections. The results provide evidence that the occurrence of static orientational disorder is an essential characteristic of the structure of C_{60} below T_c.

1. Introduction

As is well known C_{60} undergoes a phase transition from a fcc structure at high temperatures to a sc lattice below 260 K [1,2,3]. However, the sc ordering at low temperatures is far from complete[4] and the origin and nature of the remaining disorder is still controversial .

The question is closely related to the rotational dynamics of the C_{60} molecules. Above 260 K experimental data are compatible with a fairly free rotation of the molecules[5] though even at higher temperatures their orientation is not completely random. Short-range correlations between neighboring molecules give rise to a preference for certain orientations[6].

Below 260 K the most direct evidence for orientational disorder is provided by measurements of diffuse neutron scattering. Various concepts have been discussed in order to explain the unusually high intensity of the diffuse scattering, e.g. a largely free rotation of the molecules similar to the high temperature phase, jump-diffusion between well defined orientations, and large-amplitude librational modes.

Neutron experiments investigating diffuse scattering which have been performed so far[7,8] have concentrated on the static structure factor, i.e. they integrate over the entire range of energy transfer and do not distinguish between elastic and inelastic contributions to the diffuse scattering intensity. Consequently, these measurements cannot tell whether the disorder in question is static or dynamic.

On the other hand, this distinction is essential in order to account for the peculiar behavior of C_{60} at low temperatures. The present investigation of diffuse scattering, therefore, has been done on a neutron triple-axis spectrometer.

2. Experimental

The measurements were done on the triple axis spectrometer VALSE located at a cold neutron guide position of the Laboratoire Léon Brillouin in Saclay (France). 3.55 Thz were selected as incident neutron energy and in order to eliminate higher order contaminations a pyrolytic graphite filter was put into the incident beam.

The spectrometer was operated in two basically different modes:

a) two-axes mode: in this configuration the scattered neutrons are counted without analyzing their energy thus measuring the intensity of the *total diffuse scattering* including both the *elastic* and *inelastic* fractions of the scattering intensity.

b) three-axes mode: in this configuration the energy of the scattered neutrons is determined by use of a pyrolytic graphite analyzer. If the analyzer is set to the same energy as the monochromator the inelastically scattered neutrons are discriminated and one obtains the *purely elastic diffuse scattering.*

A comparison of the data obtained in the two-axes mode and the three-axes mode of the instrument thus permits to separate clearly the elastic and inelastic contributions to the diffuse scattering and consequently to distinguish unambiguously between static and dynamic disorder of the C_{60} molecules.

The sample consisted of 1 g of C_{60} powder which was heat treated under high vacuum conditions for a few minutes at 300 °C in order to remove residual organic solvents. The powder was enclosed in an aluminium container with a wall thickness of 0.5 mm and mounted into a closed cycle cryostat.

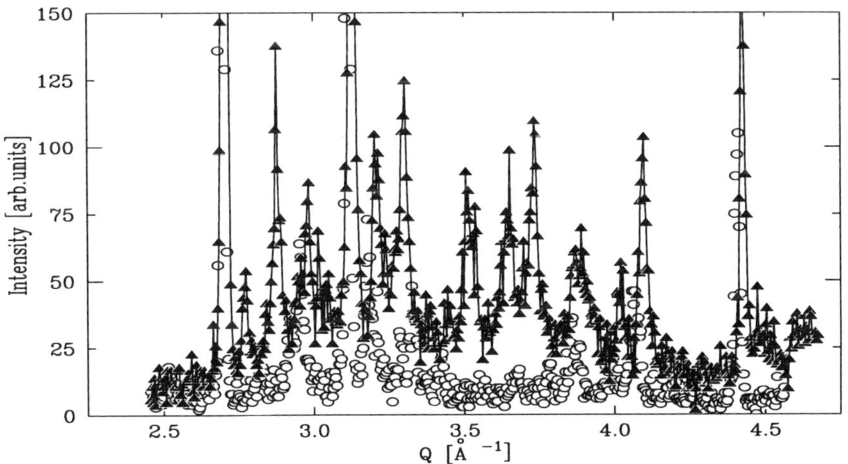

Fig. 1: High resolution measurement of the elastic diffuse scattering performed in the three-axes mode: (O) 300 K and (▲) 20 K.

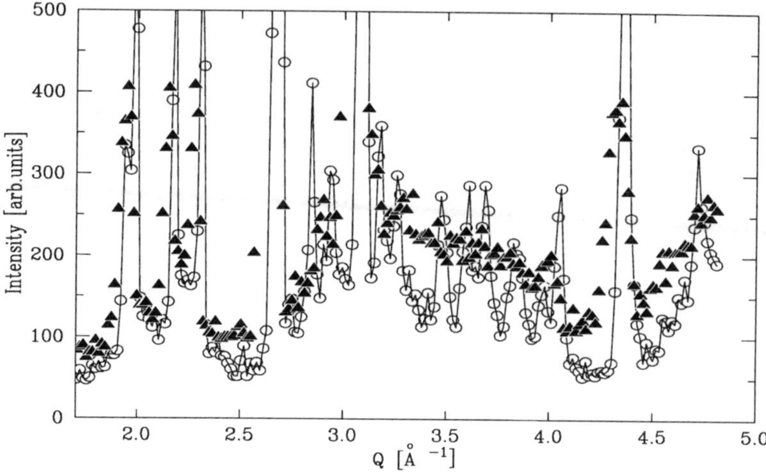

Fig. 2: Comparison of the diffuse intensity observed at 300 K (above T_c) in the two-axes mode
(▲) and at 150 K (well below T_c) in the three-axes mode (—O—).

3. Results and Discussion

Measurements of both the total and the purely elastic diffuse scattering were
performed at various temperatures above and below T_c. Immediately below T_c an
increase of the elastic diffuse intensity is observed in addition to the simple cubic
Bragg peaks of the low-temperature phase which becomes even more pronounced upon
further cooling. As a typical example Fig.1 compares the elastic diffuse scattering
measured in the high-temperature phase at 300 K and in the low-temperature phase
at 20 K (the lowest temperature measured). The increase of the elastic diffuse
scattering intensity at low temperatures is clearly visible. (The strong peaks near
2.7, 3.1, and 4.4 Å^{-1} observed in all spectra are aluminium Debye-Scherrer lines due
to the sample container and the cryostat.)

Fig.2 opposes the total diffuse scattering as determined at 300 K and the purely
elastic diffuse scattering measured at 150 K. The data obtained at 300 K in the
two-axes mode (Fig.2) are clearly different from those obtained in the three-axes
mode (Fig.1) thus showing that diffuse scattering at 300 K is mainly quasielastic.
On the other hand, the purely elastic diffuse scattering at 150 K (Fig.2) exhibits
a modulation resembling the quasielastic diffuse scattering in the high temperature
phase. This suggests that a substantial part of the quasielastic scattering observed
above 260 K condenses into elastic diffuse scattering at the phase transition and that
the static disorder found below 260 K is due to freezing of part of the C_{60} molecules
in their respective dynamically disordered orientation. Though a majority of the

C_{60} molecules becomes orientationally long range ordered during the phase transition a considerable fraction of them will not take part in the ordering but remain in a configuration representing a state of static disorder.

One might conceive of an alternative (trivial) explanation for the increase of the diffuse scattering in the low-temperature phase by referring to disturbing contributions of simple cubic Bragg intensities below T_c. However, we emphasize that the diffuse elastic scattering is observed independent of changes of the resolution of our instrument during the experiment and in, in particular, that it has been observed also in a single crystal[9] where influences of spurious Bragg tail scattering can be excluded with certainty.

The results of this work can be summarized as follows:

– In the high temperature phase the diffuse scattering is mainly quasielastic and the observed Q-modulation is compatible with a fairly free rotation of the C_{60} molecules in agreement with earlier work.

– In the low temperature phase, however, the observed diffuse scattering is purely elastic implying that it takes its origin from a state of static disorder contrary to the view usually expressed in the current literature .

Additonal data and a more detailed account of the work discussed in this communication will be presented in a separate paper[10].

Acknowledgement. This work was supported in part by the Fonds zur Förderung der wissenschaftlichen Forschung in Austria.

References

[1] P.A.Heiney, J.E.Fischer, A.R.McGhie, W.J.Romanow, A.M.Denenstein, J.P.McCauley,Jr., A.B.Smith III, and D.E.Cox, *Phys.Rev.Lett.* **66**,2911(1991).

[2] W.I.F.David, R.M.Ibberson, J.C.Matthewman, K.Prassides, T.J.S.Dennis, J.P.Hare, H.W.Kroto, R.Taylor, and D.R.M.Walton, *Nature* **353**,147(1991).

[3] R.Sachidanandam and A.B.Harris, *Phys.Rev.Lett.* **67**,1467(1991).

[4] W.I.F.David, R.M.Ibberson, T.J.S.Dennis, J.P.Hare, and K.Prassides, *Europhys.Lett.***18**,219(1992).

[5] J.R.D.Copley, D.A.Neumann, R.L.Cappelletti, and W.A.Kamitakahara, *J.Phys.Chem.Solids* **53**,1353(1992).

[6] P.C.Chow, X.Jiang, G.Reiter, P.Wochner, S.C.Moss, J.D.Axe, J.C.Hanson, R.K.McMullan, R.L.Meng, and C.W.Chu, *Phys.Rev.Lett.* **69**,2943(1992).

[7] J.R.D.Copley, D.A.Neumann, R.L.Cappelletti, W.A.Kamitakahara, E.Prince, N.Coustel, J.P.McCauley,Jr., N.C.Maliszewskyj, J.E.Fischer, A.B.Smith III, K.M.Creegan, and D.M.Cox, *Physica B* **180-181**,706(1992).

[8] F.Leclercq, P.Damay, M.Foukani, P.Chieux, M.C.Bellissent-Funel, A.Rassat, and C.Fabre, *Phys.Rev.B***48**,2748(1993).

[9] M.Haluška, H.Kuzmany, M.Vybornov, P.Rogl, and P.Fejdi, *Appl.Phys.A* **56**,161(1993).

[10] R.Glas, O.Blaschko, G.Krexner, M.Haluska, and H. Kuzmany, *Phys. Rev.B*, in print.

PHASE TRANSITIONS IN FULLERENE C_{60}

WOJCIECH KEMPIŃSKI, JAN STANKOWSKI, ZBIGNIEW TRYBUŁA,
STANISŁAW HOFFMANN, WOJCIECH HILCZER, BORYSŁAW CZYŻAK,
MARCIN KRUPSKI, SZYMON ŁOŚ, BARTŁOMIEJ ANDRZEJEWSKI
Institute of Molecular Physics, Polish Academy of Sciences,
Smoluchowskiego 17/19, Poznań, Poland

and

WOLFGANG KRÄTSCHMER
Max Planck Institute for Nuclear Physics, PoBox 103980,
W-6900, Heidelberg, Germany

and

PRZEMYSŁAW BYSZEWSKI
Institute of Physics, Polish Academy of Sciences,
Al.Lotników 32/46, Warsaw, Poland

ABSTRACT

High (close to 260K) and low temperature (in the range of 60 - 90 K) phase transitions of C_{60} samples is studied with various methods. The microwave X-band dielectrometry shows discontinuities in the resonance frequency of the TM_{012} resonator at the transition from fcc to sc phases near 260K. The same phase transition are examined with the high-pressure EPR. The linewidth of EPR signal at the transition increase and g-factor reaches minimum. The value of the pressure coefficient has been established $dT_c/dp=(10\pm1)K/kbar$. Standard EPR method are used to study the low temperature phase transition. In the vicinity of 90K changes are observed in the three basic EPR parameters. The freezing of the orientational motion is well seen as an increase of the peak-to-peak linewidth ΔB_{pp} versus temperature. The redistribution of charges from the holes to trapped electrons on the fullerene balls is seen in the dependence of the amplitude of the EPR signal versus temperature and in significant strong decrease of the g-factor on cooling. Additional phase transition appears close to 60K but only when the sample is heated. We have observed this transition with both radio frequency and the microwave dielectrometry methods. The same transition we are able to observe with the pulsed EPR for the relaxation time T_1.

1. Introduction

Different measuring techniques have been employed to examine the main phase transitions of the C_{60} fullerene. From our earlier studies of alkali metals (K or Rb) doped fullerene C_{60} [1-3] we know that even small amount of the alkali adducts makes impossible the observations of any structural or orientational phase transitions. These transitions were observed with other techniques [4-9] but only for carefully purified samples. The high temperature phase of C_{60} has face centered cubic (fcc) structure of $Fm\bar{3}m$ symmetry[10,11]

With decreasing temperature the system gets transformed first, close to 260K, to the simple cubic (sc) phase of Pa$\bar{3}$ symmetry [12,13] and then close to 90K reaches the orientationaly ordered phase of the glassy nature [7-9,14]. Using five different measuring techniques we confirmed the existence of two basic phase transitions at temperatures near 260K and 90K. Relatively strong EPR signal for the fullerene sample made by Dynamic Enterprise Ltd. enabled us to observe changes in the character of the paramagnetic center from $\dot{C}_{60}^{+}$ anion-radical above 90K to $\dot{C}_{60}^{-}$ radical with trapped electron below. Radio and microwave dielectric measurements reveal additional transition around 60K but only during the heating process.

2. Apparatus

To investigate high temperature phase transition we used two measuring techniques: microwave X-band dielectrometry[15] and high-pressure EPR[16].
The low temperature transition (near 90K) we have studied with the standard EPR equipped with Oxford cryostat. Phase transition around 60K is observed using radio and microwave dielectric measurements with Gen Rad 1689M RCL Digibridge (11Hz-100kHz), Multi Frequency Hewlett Packard 4275 LCR Meter (10kHz-10MHz) and microwave X-band dielectrometer at frequencies close to 9GHz. Confirmation of this phase transition is obtained with pulsed EPR - Bruker ESP 380E FT/CW. Samples are contributed from different sources: W.Krätschmer, P.Byszewski and Dynamic Enterprise Ltd. (with the strongest EPR signal). In order to remove the residual solvent the polycrystalline powder or pellets were treated at 480K under dynamic vacuum ($\sim$10^{-2} Torr) for seven days. The sample for microwave measurements at 260K was annealed for one hour in the conditions of helium flow at 720K measured in the oven. Only careful purification of the sample enables one to observe expected transitions. During the purification process EPR signal increases both in amplitude and width which confirms earlier Bartl's observations[17].

3. Experimental

High temperature phase transition detected with the microwave X-band dielectrometer is presented in Fig.1. The transition is accompanied by emission of the latent heat which permits the definition of the extended temperature region 256.2 - 257.3K. Arrows on Fig.1 mark successive stages of this transition. The studied system, from fcc phase first jumps to the sc phase and gets back again to fcc and then reaches permanently the low temperature sc phase. Subsequent stages of this transition are observed due to gradual heat transfer from the hindered rotators (C_{60} molecules) to the lattice and then through the flowing helium gas medium to the thermometer and the cryostat. The heat exchange between the lattice and the environment is slow enough to cause overheating of the lattice through the hindering of the rotational motion of the C_{60} molecules[18]. Discussed above transition we have monitored using the high-pressure EPR method too.

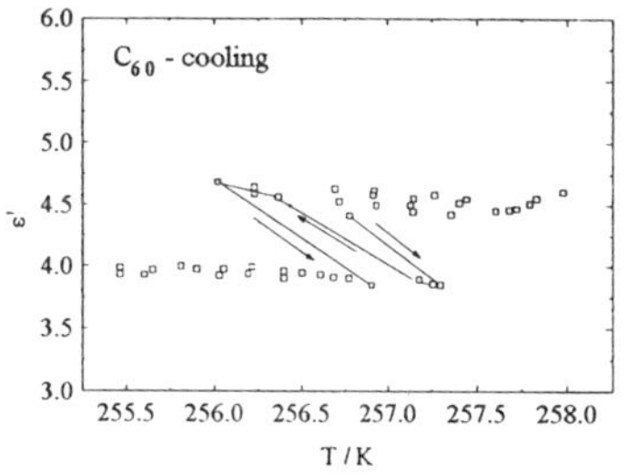

Fig. 1 Permittivity versus temperature close to structural phase transition with arrows marked subsequent stages of the transition

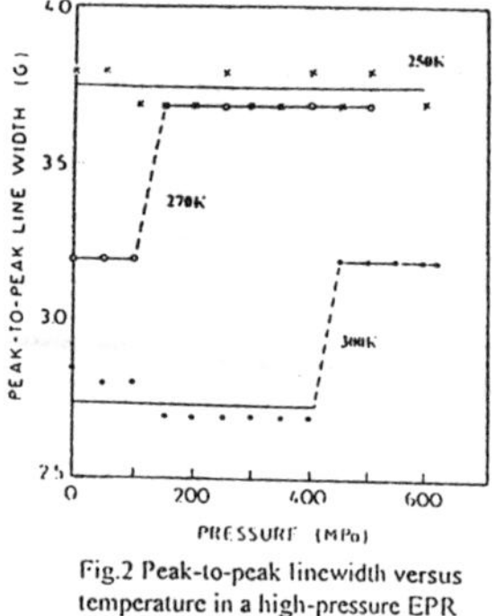

Fig.2 Peak-to-peak linewidth versus temperature in a high-pressure EPR experiment.

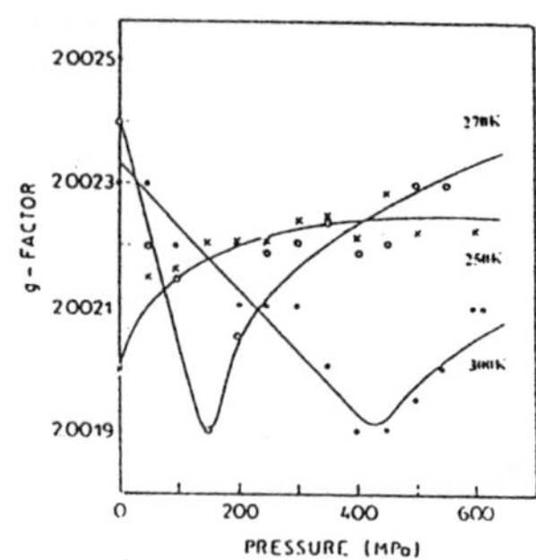

Fig.3 g-factor versus temperature in a high-pressure EPR experiment.

Fig.2 shows that EPR peak-to-peak linewidth increase - because of reduction of molecular motion at the transition. In Fig.3 g-factor versus temperature reaches minimum when the system transforms from fcc to sc phase. Minima observed at different pressures for different temperatures occur because of librational motion of the C_{60} molecules. These librations existing below the transition destroy the regular equatorial path round the C_{60} ball occupied by the paramagnetic center observed in our experiment[16]. The pressure coefficient $dT_c/dp=(10\pm1)$K/kbar is in a good agreement with those given in [19].

Low temperature phase transition at 90K was examined with the standard EPR technique. Three basic EPR parameters: the amplitude of the line (A), its peak-to-peak width (ΔB_{pp}) and the spectroscopic splitting factor (g) change drastically with the lowering of the temperature. These changes are associated with two processes - the first one is the freezing of the orientational motion of the C_{60} molecules, well seen in strong increase of ΔB_{pp} versus temperature (Fig.4) and the second one is

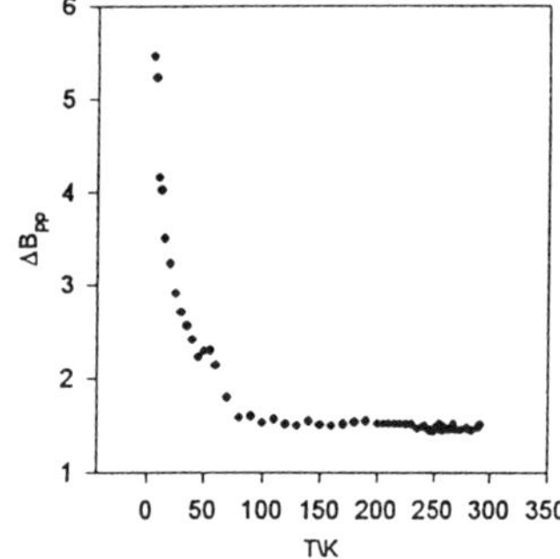

Fig.4 Peak-to-peak linewidth versus temperature

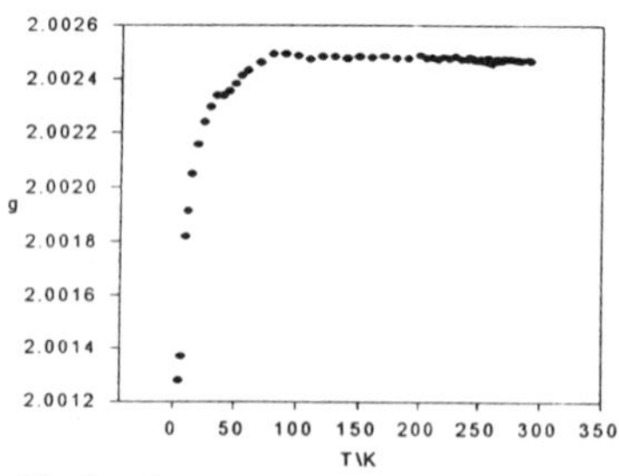

Fig.5 g-factor versus temperature

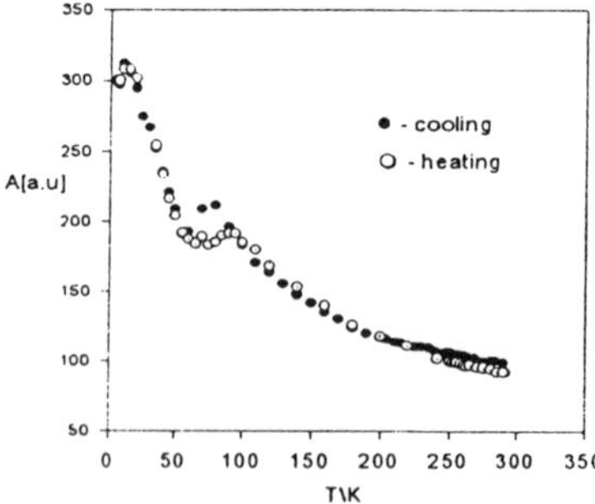

Fig.6 Amplitude of EPR signal versus temperature

connected with the redistribution of the charges from holes to trapped electrons, which means that above the transition the observed paramagnetic centers are $\dot{C}_{60}^{+}$ radicals

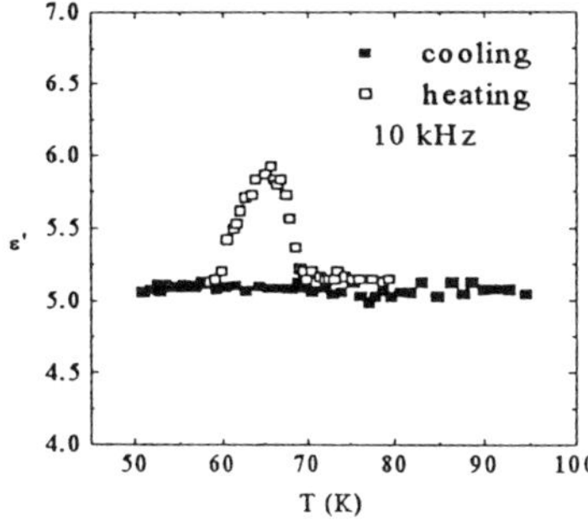

Fig. 7 Permittivity ε' vs. temperature close to 60K phase transition

whereas below the transition, at the lowest temperature region, $\dot{C}_{60}^{-}$ centers originate the EPR signal. These redistribution of charges is well seen in Fig.5 where g-factor changes from value of 2.0025 characteristic for $\dot{C}_{60}^{+}$ to g = 2.0012(6) which is characteristic for the $\dot{C}_{60}^{-}$ [2,20]. The disappearance of the $\dot{C}_{60}^{+}$ features on cooling is well seen as the maximum in the amplitude of EPR signal in Fig.6. Redistribution process extends from about 60 to 90K in both cooling and heating processes with the hysteresis, therefore this transition does not have the glass character and can be attributed to the additional phase transition which we are able to observe using dielectric method. Fig.7 shows the maximum in the permittivity ε' in the temperature range from 60 to 70K for Byszewski's sample (60-90K for sample from DEL with the strongest EPR signal). This transition does not exist on cooling and occurs only when the sample is heated. Polarization of the system could be stronger when we start to heat our sample from the lowest temperatures (below charge redistribution region) where nearly all electrons are localized on the fullerene balls. There is no frequency shift at this transition. We test it with the low frequency measurements (1,2,10 and 100kHz) for the pellets and for powder sample placed inside the microwave cavity (~9GHz). No frequency shift of T_c means no glass character of this transition. The existence of this transition is supported by the relaxation time T_1 measurements performed in the pulsed EPR experiments - Fig.8.

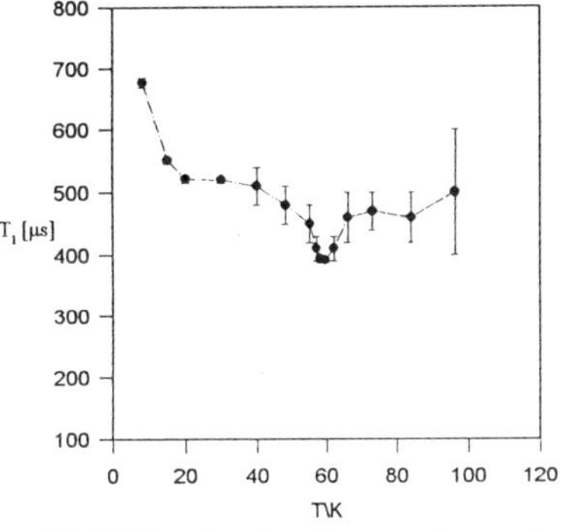

Fig.8 Relaxation time T_1 versus temperature in pulsed EPR experiment

4. Acknowledgment

This work was partly supported by the KBN Grant 2 P302006 05

5. References

1. P. Byszewski, J. Stankowski, Z. Trybuła, W. Kempiński and T. Żuk, *Journal of Mol. Structure,* **269** (1992) 175
2. J. Stankowski, P.Byszewski, W.Kempiński, Z.Trybuła and T.Żuk, *phys. stat. solidi* **178** (1993) 221
3. J. Stankowski, W. Kempiński, P. Byszewski and Z. Trybuła *Acta Phys. Polon.* **84** (1993) 1117

4. C.S.Yannoni, R.D.Johnson, G.Meijer, D.S.Bethune and J.R.Salem, *J. Phys. Chem.* **95** (1991) 9

5. R.D.Johnson, C.S.Yannoni, H.C.Dorn, J.R.Salem and D.S.Bethune, *Science* **255** (1992) 1235.

6. R.T.Tycko, R.C.Haddon, G.Dabbagh, S.H.Glarum, D.C.Douglass and A.M.Mujsce, *J. Phys. Chem.* **95** (1991) 518.

7. K.Prassides, H.W.Kroto, R.Taylor, D.R.M.Walton, W,J.F.David, J.Tomkinson, R.C.Haddon, M.J.Rosseinsky, D.W.Murphy and *Carbon* **30** (1992) 1277

8. W.I.F.David, R.M.Ibberson, T.J.S.Dennis, J.P.Hare and K.Prassides, *Europhys. Lett.* **18** (1992) 219

9. R.C.Yu, N.Tea, M.B.Salamon, D.Lorents and R.Malhotra, *Phys. Rev. Lett.* **68** (1992) 2050

10. R.M.Fleming, B.Hessen, T.Siegrist, A.R.Kortan, P.Marsh, R.Tycko, G.Dabbagh and R.C.Haddon in "Fullerenes: Synthesis, Properties and Chemistry of Large Carbon Clusters", G.S Hammond and V.J.Kuck, edittors, *American Chemical Society Symposium Series* **481** (1991) p.25.

11. R.M.Fleming, T.Siegrist, P.M.March, B.Hessen, A.R.Kortan, D.W.Murphy, R.C.Haddon, R.Tycko, G.Dabbagh, A.M.Mujsce, M.L.Kaplan and S.M.Zahurak *"Clusters and Cluster-Assembled Materials"*, **206** (Materials Research Society, Pittsburgh, 1991)

12. S.Liu, Y.Lu, M.M.Kappes and J.A.Ibers *Science* **254** (1991) 408

13. R.Sachidanandam and A.B.Harris, *Phys.Rev.Lett.* **67** (1991) 1467

14. F.Gugenberg, R Heid, C.Meingast, P.Adelmann, M. Braun and H. Wühl, *Phys. Rev. Lett* **69** (1992) 3774

15. W.Kempiński, Z. Trybuła, J.Stankowski, Sz.Łoś and W.Krätsccchmer, *Molec. Phys. Rep.* **5** (1994) 214

16. J.Stankowski, B.Czyżak, M. Krupski and B. Andrzejewski, *to be published*

17. A Bartl, B.Schandert, U. Kirbach and L.Dunsch, *to be published*

18 J.Stankowski et.al *to be published*

19. G.A. Samara, J.E. Schirber, B.Morosin, L.V. Hansen, D. Loy and A.P.Sylvester, *Phys. Rev. Lett.* **67** (1991) 3136

20. J. Stankowski, W. Kempiński, A. Koper and J. Martinek, *Appl. Magn. Reson.* **6** (1994) 145

TEMPERATURE DEPENDENCE OF THE STRUCTURE OF C_{60}- THIN FILMS ON MICA(001)

S. Henke, K. H. Thürer, J. K. N. Lindner, B. Rauschenbach, and B. Stritzker
Institut für Physik, Universität Augsburg, D - 86135 Augsburg, Germany

ABSTRACT

Thin C_{60}-films have been deposited on mica(001) substrates by thermal evaporation at substrate temperatures between room temperature and 200°C and at a constant deposition rate. The influence of the substrate temperature on the growth of C_{60}-thin films has been systematically investigated by X-ray diffraction (XRD). θ-2θ-measurements of the (111)- peaks show a decrease of the FWHM with increasing substrate temperature, leading to a minimum FWHM of 0.15° for a substrate temperature of 200°C. Oriented films with an out-of-plane mosaic spread of $\Delta\omega$= 0.2° could be grown at a substrate temperature of 150°±25°C. It can be shown that the in-plane epitaxial arrangement $C_{60}(111)$ ‖ mica(001) is determined by the seeding conditions and is independent of the substrate temperature. An increasing substrate temperature enhances the epitaxial alignment of the C_{60}-crystals oriented with a {111}-face parallel to surface and also the azimuthal alignment of the twins which are rotated by 60° about the surface normal.

1. Introduction

Numerous investigations of thin C_{60}-films deposited on different substrates have been generated by both, the fundamental scientific interest in and the possible technical applications of this new material[1]. By X-ray diffraction measurements the growth of C_{60}-thin films on sapphire substrates has been observed to be polycrystalline and a crystal grain coherence length of 60Å has been estimated[2]. Previous studies have then presented mica[3] and other substrates[4-9] as appropiate candidates for the aligned growth of C_{60}-thin films. (001)-oriented mica is well suited as substrate due to the relatively small layer-substrate misfit (0.034), available substrate sizes and the lift-off surface pretreatment. It has been shown by transmission electron microscopy that thin 50nm C_{60} films with a nearly perfect crystallite morphology can be grown[10].

The purpose of the present paper is to point out the importance of the substrate temperature for the growth of oriented C_{60}-films and the implications of defects on the structure and morphology. Therefore, systematic studies of the dependence of structure and orientation on the substrate temperature were carried out using X-ray diffraction (XRD).

2. Experimental

Thin C_{60} films were prepared by thermal evaporation on mica(001) at a base pressure in the range of 10^{-6} Torr and at a constant deposititon rate of 0.5 Å/s. The thickness of the films was kept constant at 250nm. (001)mica (muscovite) substrates 1 cm² in size were cleaved on air prior to loading into the deposition chamber. To investigate the temperature dependence of the growth process the temperature of the substrates was varied between room temperature (RT) and 200°C. The characterization of the thin films was performed

194

with a four-circle X-ray diffraction goniometer (for details see Ref.[11]). The angular re-solution was determined to be smaller than 1° in polar and azimuthal directions.

3. Results and Discussion

Fig. 1 shows the dependence of the X-ray intensity of the C_{60}-Bragg peaks (111), (311) and (222) on the substrate temperature between RT and 200°C. The formation of aligned C_{60}-films is already observed at RT, leading to detectable strong (111)- and (311)-reflections. The FWHM of both peaks are narrower than 0.8°. It can be seen that the intensity ratio $I_{(111)}/I_{(311)}$ increases significantly with increasing substrate temperature. The θ-2θ diffractogram for films grown at 100°C shows the (111)-peak and small traces of (311)- and (222)-reflections. The films prepared at 150°C reveal purly (hhh)-oriented growth and the absence of the (311)-peak. Further increase of the substrate temperature to 200°C leads to diffractograms characteristic of nearly (111)-oriented films, but with a small fraction of (311)-oriented crystallites. With increasing substrate temperature from RT to 125°C the FWHM-

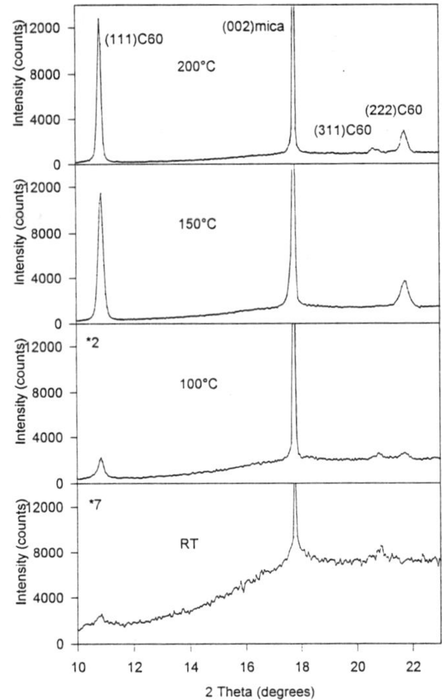

Fig. 1. X-ray diffraction spectra of C_{60}- thin films on mica(001) evaporated at a substrate temperature of a) RT (intensity *7), b) 100°C (intensity *2), c) 150°C and d) 200°C.

values decrease significantly to nearly 0.2°. At higher temperatures the FWHM is first of all constant and finally reaches a minimum of 0.15° at 200°C corresponding to a average crystal size of about 550Å calculated by the well-known Scherrer equation. The calculated average sizes of the crystallites are of the same order of magnitude as measured by atomic force microscopy and by transmission electron microscopy (for a detailed analysis see Ref.[11]). To investigate the mosaic spread of the (111)C_{60}-crystals, two kinds of rocking curves have been recorded, first the so-called ω-scan (Fig. 2) probing the (111)-planes parallel to substrate surface and second the ϕ-scan (Fig. 3) along one of the other three {111}-planes which are inclined by 71° to the (111)-plane. Fig. 2a displays ω-scans through the (111)-planes of C_{60}-films grown at different temperatures. Up to substrate temperatures of 100°C the ω-scans show only a weak maximum on a strong background. The FWHM may be determined only to 0.4°. This means a rather broad orientational distribution of (111)-planes, reflecting an aligned, but principally polycrystalline growth of the film. For comparison, a ω-scan of the (002)mica reflection leads to a FWHM of 0.13°. In contrast, the rocking curve of C_{60}-films deposited at 150°C is very narrow (FWHM is about 0.2°) and typical for well aligned films. A further increase of the substrate temperature to 200°C (Fig. 2a) leads to an increase of the FWHM-value to 0.4°. The variation of the FWHM of the ω-scans through the (111)-planes of the C_{60}-film is summarized in Fig. 2b. We observe a minimum at a temperature of 150°C ± 25°C in good agreement with the purely (111)-oriented growth demonstrated in Fig. 1. The good alignment of the (111)-planes parallel to the surface at T=150°C is obviously correlated with the suppression of the (311)-orientation at this temperature. Therefore at deposition rates of 0.5Å/s a temperature range around T_s=150°C should be best suited to for the growth of oriented thin films with (111)-planes parallel to the surface. The in-plane epitaxial arrangements were examined by monitoring the C_{60}-{111}-reflections within the ϕ-scan (Fig. 3). A comparison of the spectra shows, that with increasing temperature an increase of the peak intensities and a decrease of the background occurs, which can be ascribed to a decreasing portion of C_{60}-crystals with a [111]-surface normal, but random azimuthal orientation. It was expected that a high portion of the C_{60}-crystals would have a {111}-face oriented parallel to the surface,

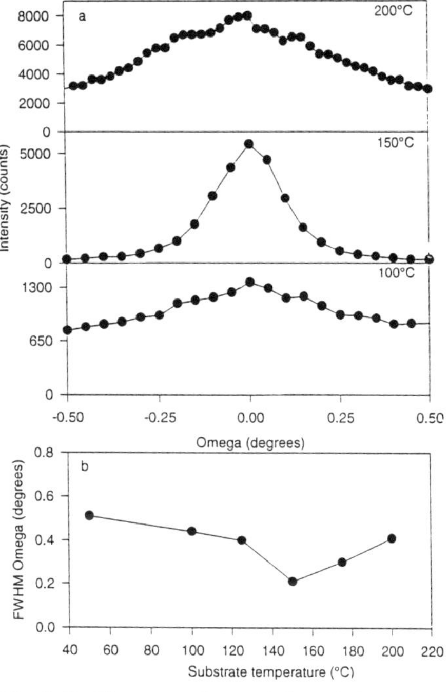

Fig. 2. Rocking curves (ω-scans) through the C_{60}-(111)-planes(a) and FWHM(b) versus substrate temperatures between 50°C and 200°C.

which means, we should observe peaks with an angular distance of 120° between two [111]-directions. But it is found that a comparable portion of crystallites (twins) are rotated by 60° about the surface normal (here [111]). The threefold symmetry in the spectra (ϕ-scans) is clearly visible. Due to the threefold symmetry there are three equivalent orientations which can be generated by rotation of 120° about the [111]-axis. Corresponding X-ray {111}-pole figures have been measured[12]. The constant ratio between intensities of the main peaks and the additional twin orientations seems independent on the deposition temperature. Therefore, it can be concluded that the starting conditions determine the growth process. Additionally one can clearly see that the azimuthal orientation of the C_{60}-grains is enhanced at higher temperatures. This shows that the higher substrate temperature significantly supports the growth of (111)-crystallites, especially their epitaxial orientation with respect to the azimuthal angle ϕ.

4. Conclusion

In summary, the influence of the substrate temperature on the growth and structural defects of oriented C_{60}-thin films on (001)mica is studied systematically by XRD-methods.

Well oriented films with an out-of-plane mosaic spread of $\Delta\omega = 0.2°$ could be grown at a substrate temperature of $150°\pm25°C$. With increasing substrate temperature from RT to 200°C a linear increase of the average grain size is observed. At this temperature only (hhh)-Bragg reflection peaks are visible and the competing (311)-grain orientation vanishes. The in-plane epitaxial arrangement $C_{60}(111)\parallel mica(001)$ is determined by the starting conditions and independent of the substrate temperature. An increasing substrate temperature enhances the epitaxial orientation of the C_{60}-crystals with a {111}-face oriented parallel to the surface and also the azimuthal orientations of the twins which are rotated by 60° about the surface normal.

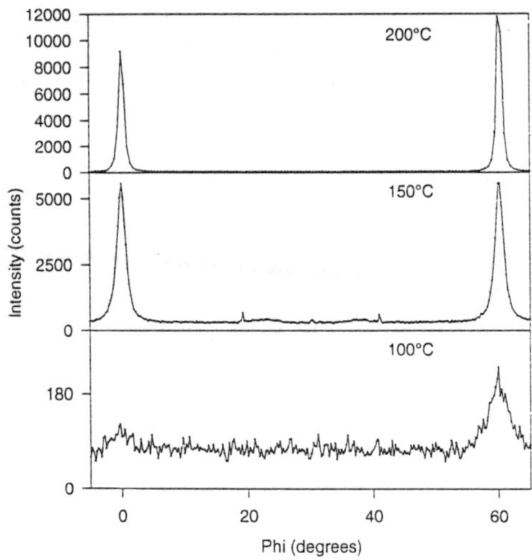

Fig. 3. ϕ- scans of the C_{60}-films with respect to the (111)-planes versus the substrate temperature between 100°C and 200°C.

5. Acknowledgements

The authors are indebted to the Hoechst AG company, Frankfurt, for providing C_{60}-material and S. Geier for assistance.

6. References

1. M. S. Dresselhaus, G. Dresselhaus, P. C. Eklund, *J. Mater. Res.*, Vol. **8**, 2054 (1993)
2. A. F. Hebard, R. C. Haddon, R. M. Fleming, and A. R. Kortan, *Appl. Phys. Lett.* **59**, 2109 (1991)
3. D. Schmicker, S. Schmidt, J. G. Skofronick, J. P. Toennies and R. Vollmer, *Phys. Rev. B* **44**, 10995 (1991)
4. M. Sakurai, H. Tada, K. Saiki, and A. Koma, *Jpn. J. Appl. Phys.* **30**, L1892 (1991)
5. G. Gensterblum, L. M. Yu, J. J. Pireaux, P. A. Thiry, R. Caudano, J. M. Themlin, S. Bouzidi, F. Coletti, J. M. Debever, *Appl. Phys. A* **56**, 175 (1993)
6. H. Xu, D. M. Chen, and W. N. Creager, *Phys. Rev. Lett.* **70**, 1850 (1993)
7. S. Fölsch, T. Maruno, A. Yamashita, and T. Hayashi, *Appl. Phys. Lett.* **62**, 2643 (1993), 8. J. A. Dura, P. M. Pippenger, N. J. Halas, X. Z. Xiong, P. C. Chow, S. C. Moss, *Appl.Phys. Lett.* **63**, 3443 (1993)
9. J. E. Fischer, E. Werwa, P. A. Heiney, *Appl. Phys. A* **56**, 193 (1993)
10. W. Krakow, N. M. Rivera, R. A. Roy, R. S. Ruoff, J. J. Cuomo, *Appl. Phys. A* **56**, 185 (1993)
11. S. Henke, K. H. Thürer, J. K. N. Lindner, B. Rauschenbach and B. Stritzker, submitted to *Applied Physics Letters*
12. S. Henke, K. H. Thürer, S. Geier, B. Rauschenbach and B. Stritzker, in preparation

PHASE TRANSITIONS IN C_{60}–CLATHRATES

G. FAIGEL, G. BORTEL, G. OSZLÁNYI, S. PEKKER, M. TEGZE

Research Institute for Solid State Physics,
H-1525 Budapest, POB. 49, Hungary

P.W. STEPHENS

Department of Physics, State University of New York,
Stony Brook, NY 11794-3800, USA

ABSTRACT

Phase transition in dimethoxymethane–C_{60} and 1bromobutane–C_{60} clathrates were studied by x–ray powder diffraction. Both clathrates show phase transitions corresponding to orientational ordering. Although the two high temperature structures have very similar orthorhombic lattice the phase transitions take place in different ways. In the dimethoxymethane–C_{60} sample there is no symmetry change while in the 1bromobutane–C_{60} an orthorhombic to monoclinic transformation of the lattice can be observed.

1. Introduction

The existence of structural phase transitions in pure C_{60} crystals attracted considerable attention. Indications for phase transitions in C_{60} compounds were also reported. However, no systematic studies exists for these compounds in contrast to the case of pure C_{60} . In this work we present a study of the phase transitions found in the C_{60} clathrates using temperature dependent x–ray powder diffraction technique.

2. Experimental

C_{60} powder was prepared and purified by the conventional method. The C_{60} clathrate crystals (C_{60}–n–pentane, C_{60}–1bromobutane, C_{60}–dimethoxymetane) were grown from saturated toluene solution of C_{60} due to the slow diffusion of the appropriate precipitant[1]. The crystallites were slightly smashed for powder diffraction[2]. Parallel beam powder diffraction measurements were performed at the X3B1 beamline of Brookhaven NSLS. For low temperature measurements a Displex closed cycle refrigerator was used.

3. Results and Discussion

The room temperature structure of the C_{60}–n–pentane has been determined in earlier single crystal and powder diffraction studies[1,3]. A one face centered orthorhombic structure was found with lattice spacing a = 10.101 Å, b = 10.163 Å, c = 31.706 Å. This structure can be built from slightly distorted square planar layers

198

of C_{60} molecules. The packing of layers is alternating between close and loose packing and generates channels for the guest molecules (Fig. 1). X–ray diffraction measurements show that the C_{60}–1bromobutane and the C_{60}–dimethoxymetane have the same structure as C_{60}–n–pentane with very similar lattice constants. Therefore one would expect a common temperature dependent behavior reflecting the same physical process in these materials. In the case of C_{60}–n–pentane a phase transition occurs at about 190 K[4]. Besides the abrupt variation of the lattice constants at this temperature there is a symmetry change from orthorhombic to monoclinic going from high to low temperature. The lowering of the symmetry involves the distortion of the square planar layers of the C_{60} molecules resulting in layers which have not only slightly different spacing between molecules in the two crystallographic directions but the angle between these directions will be different from 90 degree.

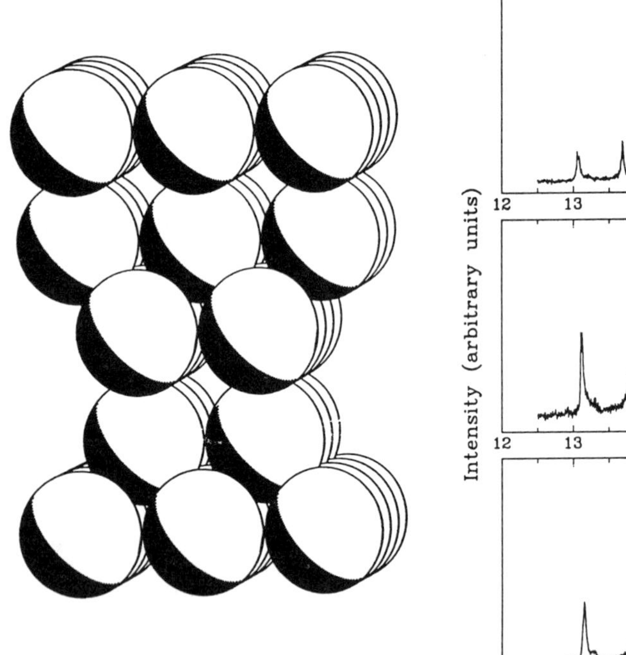

Fig. 1. The sceleton of the orthorombic structure showing only the C_{60} molecules.

Fig. 2. X–ray powder diffractograms of the C_{60}–1bromobutane sample at different temperatures.

X–ray powder diffraction measurements yield similar results for the C_{60}–1bromobutane sample (Fig. 2 and Fig. 3). However, in the case of C_{60}–dimethoxymetane we find a different behavior. Although there is an abrupt

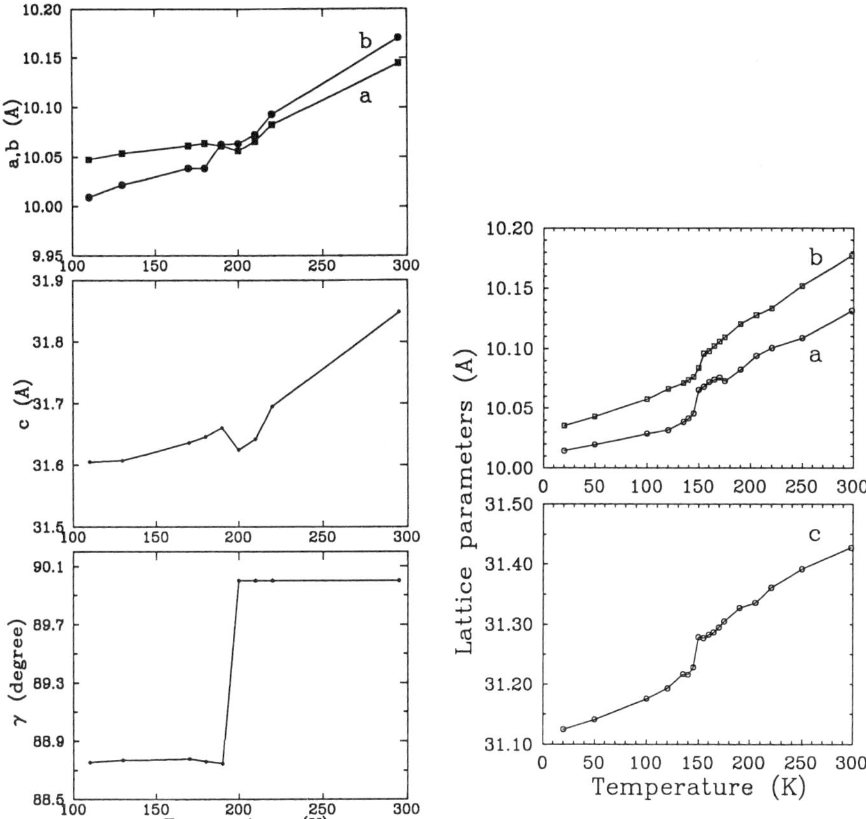

Fig. 3. Temperature dipendence of the lattice parameters of the C_{60}–1bromobutane sample.

Fig. 4. Temperature dependence of the lattice parameters of the C_{60}–dimethoxymetane sample.

variation of the lattice constants at about 150 K, this is not accompanied by a symmetry change (Fig. 4). This can be explained by the different molecular dynamics as compared to the other two clathrates. It was shown in our earlier studies that in the C_{60}–n–pentane the orthorhombic structure has a dynamic origin[3], and the n–pentane molecules are doing a flip–flop motion between two distinct orientations. Cooling the system results in the freezing in of this flip–flop motion and the pentane molecules stay in a preferred orientation causing the symmetry change. Starting from this picture the behavior of the C_{60}–dimethoxymetane sample can be explained by supposing that the dimethoxymetane molecules do not have two preferred orientations but they take every orientations with equal probability (within the space limitation given by the lattice). The abrupt change in the lattice constants are caused by the freezing of the C_{60} rotation only and this is not accompanied with the freezing of the motion of the guest molecules.

4. Acknowledgments

This work was supported in part by the Hungarian National Foudation for Scientific Research (OTKA) under contract No. 2943 and T4222. PWS was supported by the US NSF under grant DMR-92-02528. The SUNY beamline at the NSLS is supported by the US DOE grant DEFG0286-ER-45231.

5. References

1. S. Pekker et al., *Solid State Commun.* **83** (1992) 423.
2. G. Oszlányi et al., *Phys. Rev.* **B48** (1993) 7682.
3. G. Oszlányi et al., *Solid State Commun.* **89** (1994) 417.
4. G. Faigel et al., *Phys. Rev.* **B** (in press).

STRUCTURAL PHASE TRANSITIONS IN SINGLE-CRYSTAL C$_{70}$

P. Dolinar *, W. Schranz, A. Fuith, H. Warhanek

Institut für Experimentalphysik der Universität Wien, A-1090 Wien, Austria
**perm. address: J. Stefan Institute, Univ. of Ljubljana, 61111 Ljubljana, Slovenia*

C. Meingast, F.Gugenberger, G.Roth,

*Kernforschungszentrum Karlsruhe, Institut für Nukleare Festkörperphysik,
PO Box 3640, Karlsruhe, Germany*

M. Haluška, and H. Kuzmany

Institut für Festkörperphysik der Universität Wien, A-1090 Wien, Austria

Abstract

The linear thermal expansion and low frequency elastic constants of sublimation grown C$_{70}$ single-crystals have been studied using dynamical mechanical analysis in the temperature range 95-770 K and high-resolution capacitance dilatometry from 5-380 K. X-ray diffraction of our samples shows that their structure is predominantly hcp. The measurements, performed both along and perpendicular to the hexagonal c-axis, show three phase transitions. A 5-6% change in the length L occurs at 355 K on heating and at 305 K on cooling. Two other transitions were observed at 302 K and at ≈280 K. We associate the transitions with a consecutive freezing of the orientational disorder of ellipsoidal C$_{70}$ molecules. Uniaxial pressure, applied along the c-axis, shifts the 360 K transition to lower temperatures, which is in agreement with the model of orientational ordering.

1 Introduction

The ellipsoidally-shaped C$_{70}$ molecules crystallize at high temperatures into an orientationaly disordered molecular solid. Most reported samples, powdered as well as single-crystal, are a mixture of face centered-cubic (fcc) and hexagonal close-packed (hcp) structures [1, 2, 3], presumably due to a small energy difference between these two phases. The fraction of fcc phase in powdered samples can be increased by high-temperature annealing, which was taken as evidence that the equilibrium phase is fcc. However, a sizeable fraction of hcp phase always remains present and the hcp single crystals have a much lower density of stacking faults than fcc ones [2, 4].

Starting with a fcc stacking, molecular-dynamics calculations predict two phase transitions (fcc ↔ rhombohedral ↔ monoclinic) due to the consecutive ordering of the C$_{70}$ molecules about their short and long axes, respectively [5]. Since hcp and fcc crystals differ only with respect to their stacking sequence [4, 6], one expects little difference between the dynamics for these two modifications, at least at the molecular level. Indeed, x-ray diffraction experiments on both fcc and hcp material show two major phase transitions [6, 7, 4] in qualitative agreement with the simulations.

The details of these phase transitions are however still unclear. They have been studied with a variety of techniques and most of the experiments show two transitions near 280 K and 340 K, respectively. Recent modulated DSC studies [3] exhibit two strong (at 280 and 360 K) and two weak (at 300 and 350 K) transitions in sublimed C_{70} upon heating, which the authors associated with the orientational ordering transitions in the majority fcc and minority hcp phases, respectively.

In this paper we present high-resolution thermal expansion and x-ray diffraction measurements as well as pressure-dependent thermal expansion and low frequency elastic constant studies of predominantly hcp C_{70} single-crystals from the same source. We observe three phase transitions: at $T_{III} \approx 280$ K , $T_{II} \approx 300$ K and $T_I \approx 355$ K. The transition at T_I exhibits a broad thermal hysteresis of 50 K and is shifted to lower temperatures when applying a uniaxial stress along c-axis. The transitions at T_I and T_{III} are identified with the ordering of the long molecular axis along the hexagonal c-axis and the freezing of the spinning motion, respectively. Arguments are presented that the transition at T_{II} is an intrinsic transition, i.e. not due to a minority phase. This work has been reported in more detail previously [8, 9].

2 Experimental

The C_{70} single-crystals were grown by sublimation, using a double temperature gradient technique at $T_1 = 843$ K and $T_2 = 888$ K [10]. The purity of the source material was better than 99% [11]. Most of our samples have an hcp morphology, while the others adopt a shape typical for the fcc stucture. Typical dimensions of the crystals were 0.2-1 mm along the c-axis and 0.3×0.4 mm^2 perpendicular to it. The thermal expansion was measured with a high-resolution capacitance dilatometer [8, 12] in the temperature region 5-380 K and with a commercial Dynamical Mechanical Analyser (DMA 7) from Perkin-Elmer between 95 and 770 K [9]. The elastic constant C_{eff} was measured in the dynamic mode of the DMA 7 as a function of temperature at frequencies 1 Hz and 50 Hz. The geometry of our samples allowed only measurements along the hexagonal c-axis, which yields an effective elastic constant $C_{eff} = S_{33}^{-1}$.

3 Results and Discussion

X-ray analysis was performed after the thermal expansion measurements on a crystal, which was carefully divided into many small grains. The results show that the crystals are predominantly hcp, although a small fraction of fcc stacking cannot be ruled out [8].

The thermal expansion measurements of several crystals, using both above mentioned techniques, all showed very similar results. The expansion was first measured between 5 K and 350 K both parallel and perpendicular to the hexagonal c-axis, since cycling through the transition at T_I damages the crystal and smears out the transition at T_{II}. The capacitance dilatometry results upon heating are shown in Fig. 1 in terms of the relative length changes $\Delta L(T)/L_{5K}$ (Fig. 1a) and the linear thermal expansivity $\alpha(T) = 1/L \times \partial L/\partial T$ (Fig. 1b). The most distinct feature is the phase transition at $T_{III} = 285$ K, where L_a increases by ≈ 0.8 % and L_c decreases

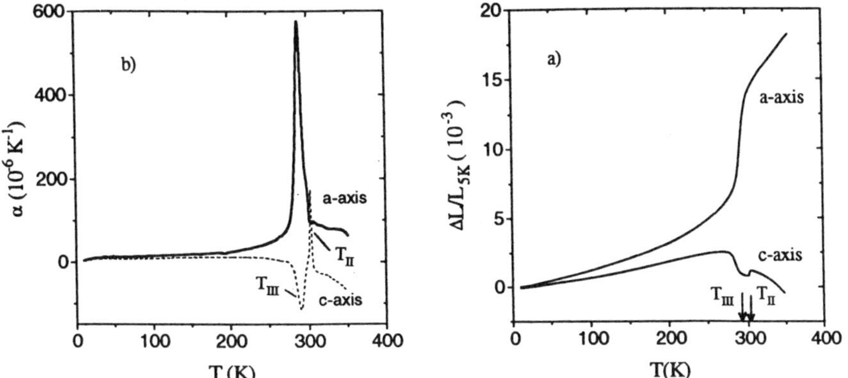

Figure 1) Temperature dependence of a) relative length change $\Delta L/L$ and b) the corresponding expansivity α measured upon heating by capacitance dilatometry.

by ≈ 0.2 %. This transition is believed to correspond to the onset of rotations about the long-axis [2, 6], which would cause an expansion perpendicular to the rotation axis and, due to the larger gap between molecules in the packing plane, a smaller separation between packing planes.

The 'c-axis' measurements exhibit a second phase transition at $T_{II} = 302$ K, where L_c increases by $\approx 0.05\%$ upon heating. It was not possible to see any effect in L_a at T_{II}, but the expansivity curve clearly shows a change in the slope of α_a at $T \approx 300$ K. Above T_{II}, the expansivity is very anisotropic; α_a is large and positive and α_c is negative. This behavior seems to be triggered by the transition at T_{II}. The dynamic elastic constant measurements (Fig. 2) also show a pronounced anomaly at T_{II}, which is reproducible and appears on both heating and cooling. Its shape

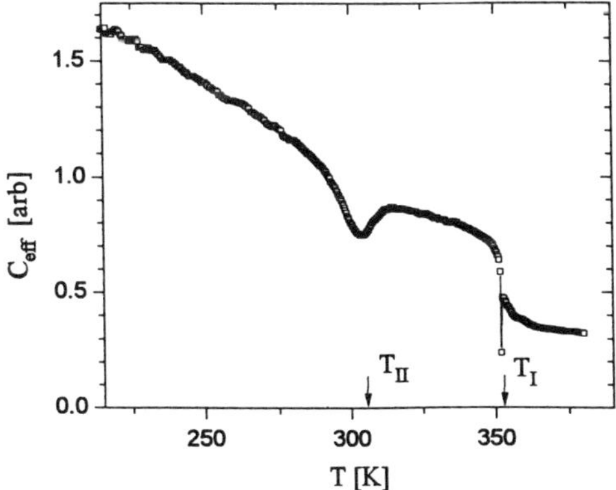

Figure 2) Temperature dependence of the real part of the dynamic elastic constant C_{eff}, measured at 1 Hz.

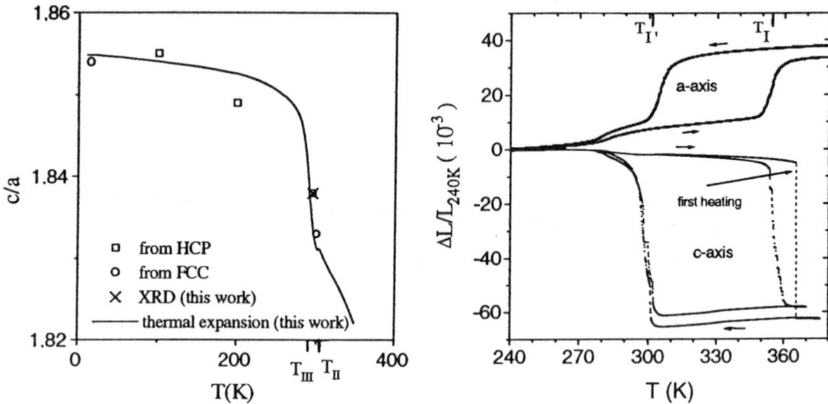

Figure 3) c/a ratio derived from our data in comparison to XRD data for both hcp- and fcc-derived structures.

Figure 4) Relative length change of the crystal between 240 and 380 K (capacitance dilatometry).

and T_{II} do not depend on the measuring frequency. On the other hand, no anomaly was observed in C_{eff} at $T_{III} = 280$ K. Together with the fact that the changes in L_c at T_{II} and T_{III} are of opposite signs, this suggests that the transitions at T_{II} and T_{III} are fundamentally different. Because of the above we do not belive that one of these transitions is due to a small fraction of minority phase (fcc in our crystals). The large and anisotropic expansivity above T_{II} must be related to a temperature dependent dynamics of the C_{70} molecules (e.g. precessional motion) and since this behavior appears to start at T_{II}, we have previously attributed this transition to the freezing of this dynamic motion [8, 9].

Fig. 3 shows the temperature dependence of the c/a ratio (solid line), which was calculated from the measured length changes and with $a_{300K} = 10.15$ Å, $c_{300K} = 18.58$ Å. The result agrees very well with c/a values [15] for both hcp (squares) [4] and fcc (circles) [6] derived structures and provides strong evidence that our sample is indeed a 'single crystal', because it shows the expected anisotropy. Fig. 2 further shows that both fcc anf hcp stacking sequences behave very similar.

Figures 4 and 5 show the expansion behavior to higher temperatures and the transition at $T_I = 355$ K upon heating and at $T_{I'} = 305$ K upon cooling becomes very apparent by the large contraction and expansion af the c-axis and a-axis, respectively. These large changes are close to what is expected when the long axes of the molecules freeze along the hexagonal c-axis [4].

The shape of the thermal expansion curve around T_I changes when uniaxial stress along the c-axis σ_c is applied. The transition becomes smeared out and is shifted to lower temperatures (Fig. 5) with a shift rate $dT_I/d\sigma_c \approx 1700$ K/GPa. No influence of uniaxial stress on the transition temperature T_{III} was noticed.

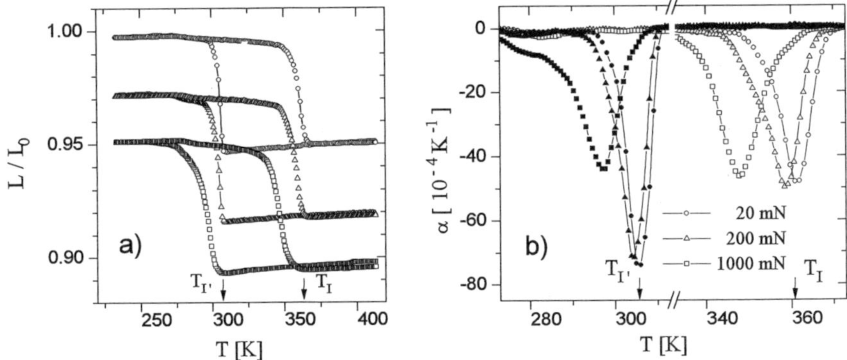

Figure 5) Temperature dependence of a) L and b) α, measured in hexagonal c-direction with various applied uniaxial forces. The open symbols in b) represent data upon heating and the full symbols upon cooling. (DMA measurements)

Hydrostatic pressure induced transitions were observed in x-ray analysis [13] and in Raman scattering studies [14], and authors associate them with the gradual freezing of the orientational disorder in a similar way as with respect to the temperature. This implies that with hydrostatic pressure P the phase transitions in C_{70} are shifted to higher temperatures. The change of the volume at T_I is positive upon heating [7], therefore a hydrostatic pressure stimulates the ordering of C_{70} molecules and shifts the transition to higher temperatures. A uniaxial stress along the c-axis, on the other hand, hinders the alignement of molecules and therefore shifts T_I to lower temperatures.

The large expansion of the c-axis upon cooling at $T_{I'}$, even when a uniaxial pressure is applied along this direction, provides an additional proof that our crystals are indeed hcp. This is because in an ideal fcc crystal there are 4 equivalent [111] directions and a uniaxial pressure along one of them should favor the alignment of molecules along the other three directions.

4 Conclusions

The results of our thermal expansion and low frequency elastic constants measurements in 'hcp' C_{70} single crystals point to a complex orientational ordering scenario. At temperatures above 360 K the C_{70} molecules are orientationally completely disordered. The first ordering transition appears at 305 K upon cooling and at 355 K upon heating. The large expansion along the c-axis and the contraction perpendicular to it arise from the alignement of the molecules along the hexagonal c-direction. In the intermediate deformed hcp phase the expansivities are highly anisotropic, presumably due to a temperature dependent precessional motion of the molecules. Freezing of this precessional motion may yield the 302 K transition, which shows up via small contraction along the c-axis, a change in the bulk expansivity and an anomaly in the elastic constant. Finally, the rotations about the long-axis freeze near 280 K. Here, the dominant effect is a contraction along the a-axis direction upon cooling.

Acknowledgements. Part of this work was supported by the Austrian "Fonds zur Förderung der wissenschaftl. Forschung" under the project number P 8285 and by the BuMiWuF projects GZ 45.212/2-27b/91 and GZ 45.223/2-27b/91. We wish to thank to Prof. Y. Achiba for supporting us with C_{70} powder material.

References

[1] G.B.M. Vaughan, P.A. Heiney, J.E. Fischer, D.E. Luzzi, D.A. Ricketts-Foot, A.R. McGhie, Y.W. Hui, A.L. Smith, D. Cox, W.J. Romanow, Jr., B.H. Allen, N. Coustel, J.P. Jr. McCauley, and A.B. Smith. *Science* **254** (1991) 1350.

[2] G.V. van Tendeloo, S. Amelinckx, J.L. de Boer, and S.V. Smaalen. *Europhys. Lett.* **21** (1993) 329.

[3] A.R. McGhie, J.E. Fischer, P.W. Stephens, R.L. Cappelletti, D.A. Neumann, W.H. Mueller, H. Mohn, and H.-U. ter Meer. *submitted to Phys. Rew.* (1993).

[4] M.A. Verheijen, H. Meekes, G. Meijer, P. Bennema, J.L. de Boer, S. van Smaalen, G.V. van Tendeloo, S. Amelinckx, S. Muto, and J. van Landuyt. *Chem. Phys.* **166** (1992) 287.

[5] M. Sprik, A.L. Cheng, and M.L. Klein. *Phys. Rev. Lett.* **69** (1993) 1660.

[6] G.B.M. Vaughan, P.A. Heiney, D. Cox, J.E. Fischer, A.R. McGhie, A.L. Smith, R.M. Strongin, M.A. Cichy, and A.B. Smith. *Chem. Phys.* **178** (1993) 599.

[7] C. Christides, I.M. Thomas, T.J.S. Dennis, and K. Prassides. *Europhys. Lett.* **22** (1993) 611.

[8] C. Meingast, F. Gugenberger, G. Roth, M. Haluška, and H. Kuzmany. *Z. Phys. (in press)* (1994).

[9] P. Dolinar, W. Schranz, A. Fuith, H. Warhanek, M. Haluška, and H. Kuzmany. *submitted to Solid State Commun.* (1994).

[10] M. Haluška, H. Kuzmany, M. Vybornov, P. Rogl, and P. Fejdi. *Appl. Phys. A* **56** (1993) 161.

[11] Y. Achiba, private communication.

[12] C. Meingast, et al. *Phys. Rev. B* **41** (1990) 3774.

[13] H. Kawamura, M. Kobayashi, Y. Akahama, H. Shinohara, and et al. *Solid State Commun.* **83** (1992) 563.

[14] A.A. Maksimov, K.P. Meletov, Yu.A. Osip'yan, I.I. Tartarovskii, Yu.V. Artemov, and M.A. Nudel'man. *JETP Lett.* **70** (1993) 816.

[15] In the monoclinic phase c/a was averaged by using an averaged $'a' = (a + b)/2$ value

APPLICATION OF μSR TO THE STUDY OF THE DYNAMICS OF C_{70}

U. BINNINGER[1], E, RODUNER[2], I.D. REID[3], C. BERNHARD[1], A. HOFER[1],
E. RECKNAGEL[1], J. ERXMEYER[4] and CH. NIEDERMAYER[1]

[1] Universität Konstanz, Fakultät für Physik, D - 78434 Konstanz, Germany

[2] Physikalisches-Chemisches Institut, Universität Zürich, CH-8057 Zürich, Switzerland

[3] Paul Scherrer Institut, CH - 5232 Villigen PSI, Switzerland

[4] Hahn-Meitner Institut GmbH, D - 14109 Berlin, Germany

ABSTRACT

Solid C_{70} was studied by the muon-spin-rotation (μSR) technique. Positive muons implanted into solid C_{70} form endohedral muonium $Mu@C_{70}$ and muonated free radicals MuC_{70}. The anisotropic hyperfine interaction of these species offers the possibility of studying the reorientational dynamics of solid C_{70}. From our measurements we can conclude that the C_{70} molecules are static below about 150 K on a time scale in the order of 30 ns. Above this temperature averaging processes due to molecular rotations are observed.

1. Introduction

Muonium, consisting of a positive muon and an electron ($Mu=\mu^+e^-$) can be considered as a light isotope of hydrogen. The muon is a short-lived particle ($\tau_\mu = 2.2$ μs), decaying into two neutrinos and a positron, the latter being emitted preferentially along the muon spin direction. Given a beam of spin-polarized muons this allows the time evolution of the spin polarization in internal or applied magnetic fields to be monitored by counting the decay positrons in a fixed direction. After the muon is implanted into a C_{70} target it rapidly thermalizes via ionization and excitation of the sample[1]. As it reaches thermal energies it can end up in one of three different states: as a diamagnetic μ^+, as a vacuum-like muonium state, encapsulated inside the molecular cage (endohedral, $Mu@C_{70}$), and a muonated free radical, MuC_{70}, a hydrogen-like Mu addition to unsaturated bonds on the carbon rings.

2. Results

2.1 Transverse field experiments

In transverse field (TF) experiments a field is applied perpendicular to the incoming muon spin direction. In a field of 3 kG we observe four of the five possible Mu adducts. The Mu addition to the carbon atom at the equator (E) (Notation: A to E going from pole to equator) is not formed. This is in agreement with recent semi-empirical molecular orbital calculations which showed that the equatorial position is relatively inert towards radical attack[2]. At room temperature the four visible radicals have the following hyperfine couplings: 275.97(6) MHz, 341.4(2) MHz, 354.1(1) MHz and

208

360.3(1) MHz, respectively. They are attributed to muon adducts to positions D, C, A and B, respectively.

The temperature dependence of the line width of radical D and C is displayed in fig. 1a. At low temperatures the line width for all radicals originating from the powder averaging over the anisotropic part of the hyperfine interaction is about 17 MHz. Above 150 K we observe motional narrowing. Due to the different angles of the dominant hyperfine interaction component (D_{zz}) with respect to the rotation axis, the narrowing of the line width parameter is different for the different radicals. The line width for the D-radical, which has a large angle between D_{zz} and the rotation axis shows a strong decrease of the line width above 150 K, while the narrowing for the C-carbon adduct is weaker. The room temperature line width for the C adduct is twice as large as for the D adduct due to this different angles. This different behavior can be explained assuming long axis spinning above 150 K possibly accompanied by a tumbling or precession of the long axis around a crystallographic direction.

At low fields (100 G) two frequencies are observed: one at 135.06(8) MHz and the other at 144.5(1) MHz. These frequencies are characteristic of free muonium and yield a hyperfine coupling constant of 4282(21) MHz, which is similar to that in C_{60} ($A_{hf}(C_{60})$ = 4341(24) MHz) and only slightly reduced from the vacuum value (A_{vacuum} = 4463 MHz). This large hyperfine coupling constant requires that the electron density in the surrounding of the muonium atom is very low and therefore these lines are attributed to muonium inside the C_{70} molecule.

Due to the elongated form of the C_{70} molecule the endohedral muonium state shows axial symmetry. This is reflected in the line width parameter which at low temperatures is about three times larger than in C_{60}[3]. At 280 K a step is clearly visible (see fig. 1b) indicating that at the structural phase transition the anisotropy of the hyperfine interaction is lowered due to additional movements of the C_{70} molecules. However, even at 350 K the line width remains at 1 MHz, indicating that at this high temperature a small anisotropy still exists. This can also be seen in avoided level crossing (ALC) experiments[4,5].

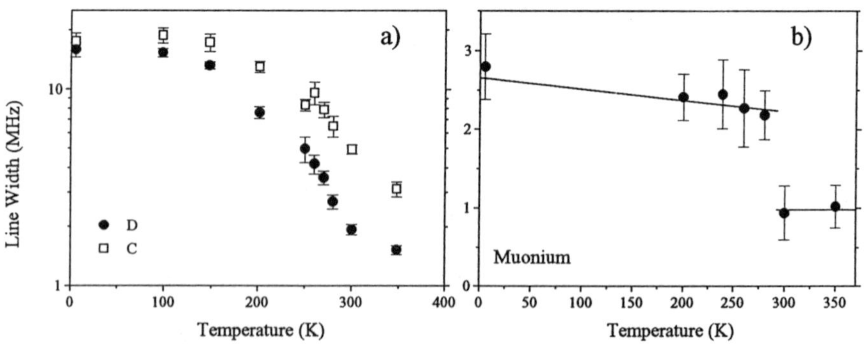

Fig 1:　a) Line width for the radical D and C. b) Line width of the endohedral muonium Mu@C$_{70}$. A step at 270 K is clearly visible.

2.2 Zero field measurements

An earlier zero field (ZF) study[6] of the endohedral muonium Mu@C_{70} revealed the appearance of an oscillation of $\nu=0.70(2)$ MHz below 270 K, signaling the presence of an axially-symmetric hyperfine interaction that lifts the triplet degeneracy. The disappearance of this frequency at 270 K is in accordance with the sudden decrease of the line width of Mu@C_{70} observed in our TF measurements.

In the present study we report similar results for the *radicals* in C_{70}. In fig. 2 a time spectrum and the corresponding Fourier transform are shown for 98K. Three zero field oscillations (1.3(1), 7.3(2) and 8.6(1) MHz) are observed which we attribute to the intra-triplet transitions of the triaxial muon-electron hyperfine interaction of the C_{70} radicals. The lines are broadened due to the slightly different anisotropies of the four radicals. The observation of a *non*-axially symmetric hyperfine interaction clearly shows that the molecules do not rotate at 98 K on a time scale of the about 30 ns (inverse of the anisotropic part of the hyperfine energy).

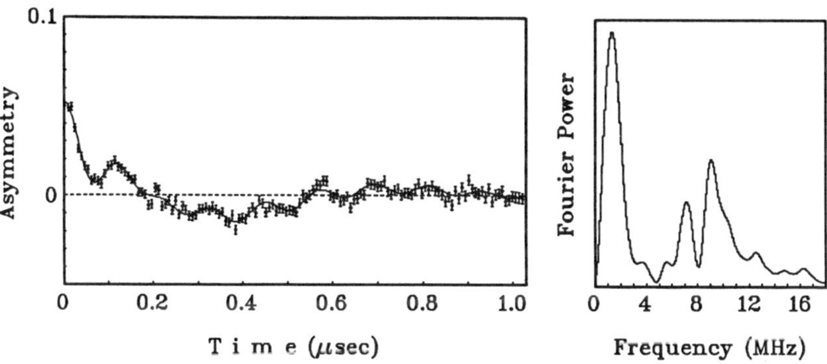

Fig 2:   Time spectrum and Fourier transform of a zero field µSR measurement in C_{70} at 98 K. The three zero field oscillations are clearly visible in the Fourier transform.

At 350 K, the low frequency lines which correspond to the intra triplet splitting due to the *anisotropic* hyperfine interaction are not observed anymore. The reason is that fast rotations average this anisotropy to zero and the hyperfine interaction appears as quasi - isotropic. However, at 350 K we observe high frequency µSR lines at 276 MHz, 340 MHz and 358 MHz which correspond to the singlet-triplet transitions of the radicals D, C and A & B. The singlet-triplet splitting is due to the *isotropic* hyperfine interaction. At lower temperature these lines are too broad and can not be seen experimentally.

3. Discussion

Our analysis of the dynamical behavior of solid C_{70} can be summarized as follows:

T≤150 K: At temperatures below about 150 K the C_{70} is *static* on the time scale of the μSR technique (in the order of about 30 ns). This is most clearly seen from the ZF measurements which show the existence of a non axialsymmetric hyperfine interaction of the MuC_{70}. The same conclusion is drawn from the constance of the line widths up to 150 K in fig. 1.

150 K ≤ T ≤ 280 K: In the temperature region between 150 K and 280 K the C_{70} molecules rotate about their long axis. These rotations can explain the averaging of the anisotropy of MuC_{70}, whereas the axially symmetric hyperfine interaction of $Mu@C_{70}$ is not affected by this kind of molecular motion. This view is supported by earlier measurements by Prassides *et al.*[6].

T≥280K: Above about 280 K the rotation around the C_{70} long axis is accompanied by some kind of tumbling or precessional motion of the C_{70} molecules around the crystal axis. The precise onset temperature of the tumbling or precessional motion of the long C_{70} axis can not be derived from our measurements, but it may well be below 280 K. Our data show that at 350 K the system is still not completely isotropic.

This results are inconsistent with high-resolution capacitance dilatometry measurements from Meingast *et al.*[7]. In their measurements they see a phase transition at 285 K which they attribute to the onset of long axis spinning and a second phase transition at 302 K attributed to long axis precession rotations. As pointed out we clearly see the onset of the uniaxial rotations at 150 K in agreement with NMR results from Arcon *et al.*[8] where the rotations appear above 50 K. The different onset temperatures are due to the different time windows of NMR and μSR.

4. Acknowledgment

This work was supported by the Bundesminister für Forschung und Technologie. We also acknowledge the help of the staff of the Paul Scherrer Institut for technical support. We want to thank A. Weidinger and K. Prassides for helpful discussions.

5. References

1)   Ch. Niedermayer, *et al. Phys.Rev. B* **47** (1993) 10923; B. Addison-Jones, *Hyp.Int.* in press
2)   E. Roduner and I.D. Reid, in preparation; I.D. Reid and E. Roduner, *Hyp. Int.* in press
3)   R.F. Kiefl, *et al. Phys. Rev. Lett* **68** (1992) 1347 (orginal), 2708 (corrected)
4)   T.J.S. Dennis, *et al. J.Phys.Chem* **97** (1993) 8553
5)   E. Roduner, *et al.* in preparation
6)   K. Prassides *et al. J. Phys. Chem* **96** (1992) 10600
7)   C. Meingast *et al.* this proceedings
8)   D. Arcon, R. Blinc, J. Dolinsek, J. Seliger, and F. Milia, this proceedings

Alkali-Intercalated Structures of C_{60}

R. M. Fleming, O. Zhou, D. W. Murphy T. Siegrist
AT&T Bell Laboratories, Murray Hill, NJ 07974, USA

ABSTRACT

Intercalated compounds of C_{60} can be formed by placing atoms in interstitial sites between the molecules. We summarize the various structures obtained for alkali intercalants, which are usually ionic salts. For alkali intercalation, only A_3C_{60} is metallic and superconducting. The neutral molecule NH_3 can be used as a to expand the lattice.

Pristine C_{60} crystallizes in a face-centered cubic (fcc) lattice with a unit cell of 14.17Å . For modeling purposes, the C_{60} molecule can be thought of as a hard sphere 7.1Å in diameter with a nearest-neighbor separation of 10 Å. The fcc unit cell contains two types of interstitial sites large enough for alkali intercalation, one octahedral site per C_{60} with a radius of 2.06 Å and two tetrahedral sites per C_{60} with a radius of 1.12 Å. Each site occupied by an alkali atom results in the transfer of one electron to the C_{60} conduction band. Two low-lying bands are available, each capable of accommodating six electrons. Singly ionized alkali atoms have radii of 0.97, 1.33, 1.47 and 1.67 Å for Na, K, Rb and Cs respectively. Full occupation of these fcc sites gives a stoichiometry of A_3C_{60} and a half-filled conduction band. A characteristic feature of the C_{60} cell geometry is the large size of the octahedral site, which is larger than any alkali atom. This feature is dramatically illustrated in Fig. 1 where the we show the experimental electron density of C_{60} in the (100) plane, obtained from a conventional Fourier synthesis of single-crystal x-ray data. One can directly observe the hollow, spherical shape of the molecule and the nearly uniform electron density that results from the rapid rotation of C_{60} at room temperature. The large size of the octahedral site is clearly evident, seen in Fig. 1 at the mid-point of the edges of the unit cell.   The smaller tetrahedral sites, located at (¼ ¼ ¼) are not visible in the figure.

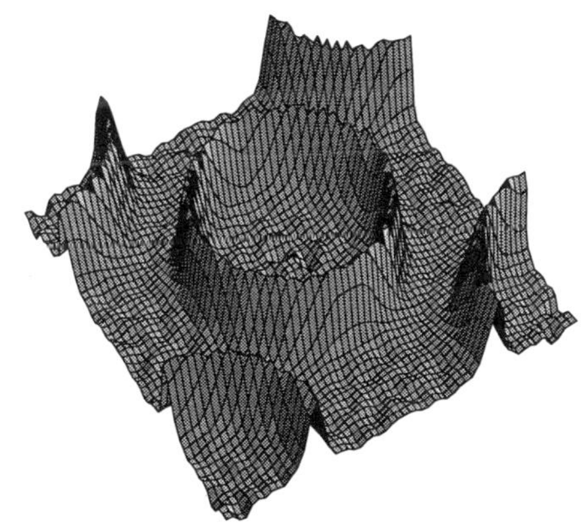

Fig. 1: Electron density in (100) plane of C_{60} obtained by Fourier analysis of single-crystal x-ray data.

A number of intercalated phases can be obtained by "decorating" the fcc structure

212

of pristine C_{60} with alkali atoms. Other structures have a different C_{60} sub-lattice cell and a different number of sites available for intercalation. A summary of the structures of many of the alkali-C_{60} and alkaline earth-C_{60}[1] compounds are shown in Fig. 2. The cubic A_1C_{60} structure shown in Fig. 2 is only formed for A = K, Rb, or Cs at temperatures above about 150 °C.[2] At lower temperatures Rb_1C_{60} forms a lower symmetry structure while K_1C_{60} phase separates[3] into C_{60} plus K_3C_{60}. Thus far, A_2C_{60} has only been synthesized for Na_2C_{60}.[4] For Na intercalation, partial occupancy of the available sites results in a solid solution Na_xC_{60} for $1 \leq x \leq 3$.[4,5] Filling all of the available fcc sites with one alkali atom per site results in the superconducting phase A_3C_{60}.[6,7] This phase exists for A = Na (T > 250K),[4] K, Rb and mixtures of Na, K, Rb, and Cs.[8] Cubic symmetry can sometimes be preserved when multiple atoms occupy the octahedral site as was first shown with the alkaline-earth salt Ca_5C_{60},[9] and later for Na_6C_{60}[4] and $Na_{11}C_{60}$.[10] Compounds with multiple metal atoms on the octahedral site may not have complete charge transfer to the conduction band due to covalancy in the metal atom clusters. The saturated alkali phase[11] for larger atoms such as A = K, Rb and Cs (as well as Ba)[12] has a body-centered cubic (bcc) C_{60} sub-lattice with a stoichiometry of A_6C_{60}. Except for the alkaline earth compound Ba_6C_{60}, which is a superconductor, all of the A_6C_{60} compounds are a filled-band insulators. The simple cubic A-15 structure (found only for the alkaline-earth, Ba)[13] has the same C_{60} sub-lattice as A_6C_{60}, but Ba atoms break the bcc symmetry by occupying half of the sites on the faces of the unit cell.

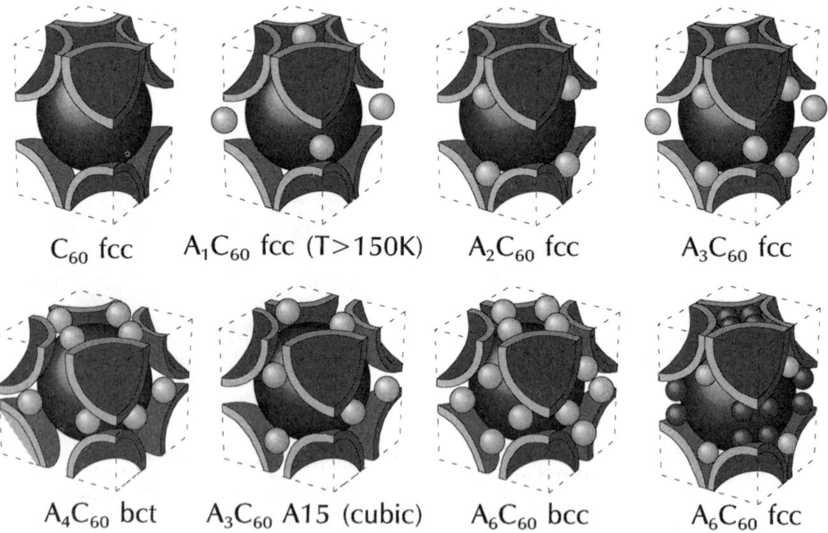

Fig. 2: Summary of structure types of alkali-intercalated C_{60}. (The A-15 structure has only been observed in the alkaline earth phase, Ba_3C_{60}.)

Finally, the C_{60} sub-lattice of the body-centered tetragonal A_4C_{60} structure[14] can be visualized as having a sublattice intermediate between fcc and bcc. Although simple band structure calculations[15] predict a partially filled band for all A_xC_{60} compounds in the range $0 < x < 6$, only alkali compounds with $x = 3$ are metallic. It is not clear whether the lack of metallic character to these other compounds is due to an unobserved gap, which should presumably cause a small structural distortion,[16] or due to the effects of electron correlations as proposed for K_4C_{60}.[17] The role of the small distortion of alkali sites of A_3C_{60} seen in NMR measurements[18] is presently not understood.

In addition to ordering of intercalant atoms on octahedral or tetrahedral sites, the C_{60} molecules themselves have different ordered arrangements. In pristine C_{60} at room temperature the molecules are rapidly rotating as first shown by NMR.[19] Fig. 3 shows a plot of experimental x-ray structure factors obtained from single-crystal diffraction intensities as a function the x-ray momentum transfer, $\mathbf{q}$. The solid line is $J_o(qR) = sin(qR)/qR$, the zero'th order Bessel function expected if the molecular rotations were

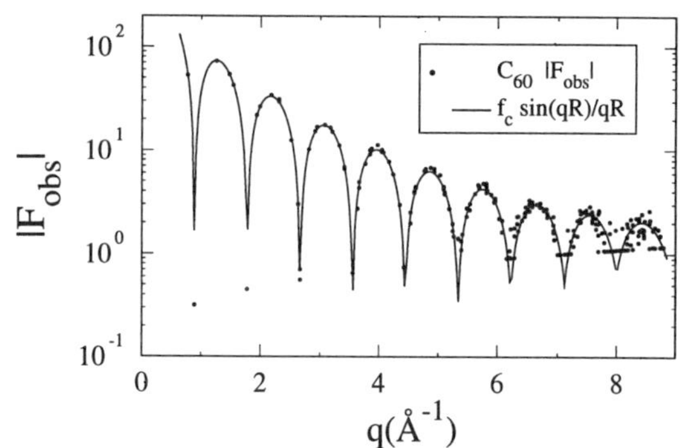

Fig. 3: Absolute value of the observed structure factor as a function of $\mathbf{q}$ for a single crystal of C_{60}. The solid line is the expected structure factor for a spherical shell.

isotropic producing a spherical shell of diameter R. The fit of the observed structure factors to J_o, independent of the direction of $\mathbf{q}$, of illustrates the nearly isotropic character of the molecular rotations. Using adapted spherical harmonics, Chow, et al.[20] have shown that there are small but significant higher-order corrections to $J_o(q\,R)$ at room temperature as a result of an activation energy associated with molecular reorientation and the icosahedral symmetry of the fullerene molecule. Below about 260 K pristine C_{60} orders[21] in $Pa\overline{3}$ symmetry which is approximately described as having the double carbon bonds on one molecule (the bonds between two pentagons) face the hexagonal face of a neighboring molecule.[22] $Pa\overline{3}$ ordering of C_{60} molecules is also observed in Na_xC_{60}, $1 < x < 3$,[4,5] as well as Na_2CsC_{60}[23] and Na_2RbC_{60}.[24,25] Other C_{60} intercalated compounds order with the C_{60} molecules having their two-fold axes (the direction normal to the C-C double bonds) parallel with principal directions of the unit cell. Examples of this type of ordering give $Im\overline{3}$ symmetry for A_6C_{60} and $Pm\overline{3}n$ symmetry for the A-15 structure. In some structures, notably most A_3C_{60} compounds,[7] a merohedral disorder of the C_{60} molecule is present. This type of disorder results because the cubic symmetry of the

214

lattice allows for a four-fold rotation axes, but the C_{60} molecule has only two-fold axes (in addition to three-fold and five-fold). Merohedral disorder results when random molecules are rotated $\pi/2$ about their two-fold axes. The effect raises the apparent symmetry of the structure from $Fm\bar{3}$ to $Fm\bar{3}m$.

Most studies of alkali intercalated C_{60} have concentrated on the superconducting phase A_3C_{60}. This series of compounds exhibits a striking correlation between the size of the unit cell and the superconducting transition temperature, T_c.[8,26] The correlation is thought to result from a scaling of the density of states with the wave function overlap from neighboring C_{60} molecules. As the unit cell size increases in size, the wave function overlap decreases causing a more narrow conduction band and a larger density of states. Because the phonons which couple to superconductivity are confined to intra-ball modes,[27,28] the T_c is insensitive to alkali phonon modes. The remarkable separation between effects of the density of states and effects of phonons is one of the distinctive features of superconductivity in these compounds. In this paper we will concentrate on the behavior of the A_3C_{60} compounds at two extremes of the T_c versus lattice curve, the small-cell compounds and the large cell compounds.

The fcc octahedral site, which is larger than the hard-sphere radius of any metal ion, can be thought of as "lattice contracting." As successively smaller alkali atoms are placed on the octahedral site,[8] the unit cell decreases to values even smaller than the 14.17 Å cell of pristine C_{60}. The smallest cell observed thus far is Li_2RbC_{60} with a lattice parameter of 13.9 Å.[29] Na_3C_{60} is anomalous in that it has a larger unit cell than would be expected from an extrapolation of the cation size. This is likely related to the fact that Na_xC_{60} exists for a range of stoichiometries, $1 \leq x \leq 3$ and also to the existence of the fcc $x = 6$ phase. Na_3C_{60} contains four Na atoms per octahedral site, disordered over the corners of a cube 2.0 Å on a side. It has also been shown that a five-atom cluster of Na is stable on the octahedral site of $Na_{11}C_{60}$.[10] Diffraction of Na_3C_{60} below 250K is qualitatively similar to that expected from a two-phase sample containing both Na_2C_{60} and Na_6C_{60} suggesting that Na diffusion from one octahedral site to another occurs as the temperature is lowered. Therefore, the lack of superconductivity in Na_3C_{60} is likely the result of phase separation and the existence of significant volumes of the sample with $x \neq 3$.

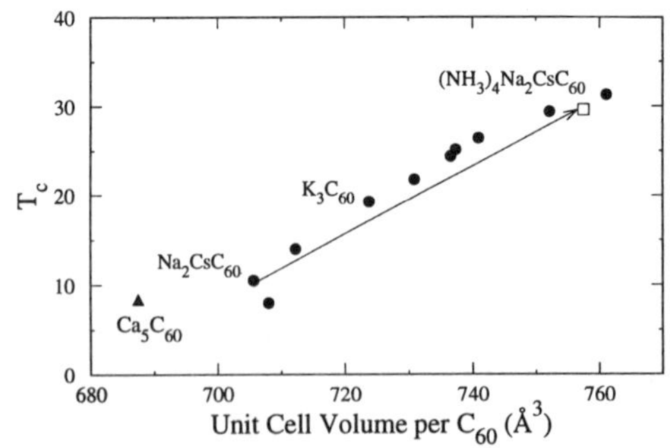

Fig. 4: T_c scales with the volume of the unit cell of fcc intercalated-C_{60} superconductors. NH_3 can be used to expand the cell and enhance T_c.

An intriguing question concerns the behavior of A_3C_{60} in the large unit cell limit. As the unit cell size increases, one expects that the superconducting transition temperature will rise, but may ultimately be limited by a metal-insulator transition when the bandwidth narrows to the Mott limit. We find that the largest unit cell for a single phase material is obtained for Rb_2CsC_{60} however, slightly higher transition temperatures have been reported for Cs_2RbC_{60}.[30] Since Cs^{+1} has the largest ionic radius of any atom and Cs_3C_{60} does not form, it is unlikely that a larger unit cell with only one atom per interstitial site will ever be synthesized. However, several features of alkali-intercalated C_{60} indicate that one can circumvent this limit and still have metallic conductivity. We have seen that Ca and Na compounds form with more than one atom in the octahedral site. This indicates that multiple-atom occupancy of the octahedral site is possible, however these samples probably have some covalent bonding between atoms in the cluster and less than one electron donation per alkali atom. Enhanced superconducting transition temperatures should still be possible if the number of electrons donated to the conduction band is preserved, i.e. if the object in the octahedral site has a valence of +1.

We have demonstrated that one can use ammonia, to achieve lattice expansion without compromising the electron count or lowering the symmetry. This class of compounds could potentially increase the size of A_3C_{60} salts beyond the limit imposed by ionic radii of single atoms. An example of this type of structure is the compound $(NH_3)_4Na_2CsC_{60}$ where ammonia intercalation increases the lattice parameter from 14.13 to 14.47 Å and T_c from 10.5 to 29.6 K.[31] Synthesis is achieved by a room temperature reaction of NH_3 gas with pre-reacted Na_2CsC_{60} followed by annealing at 100 °C. This compound maintains fcc symmetry after the addition of ammonia, but the Cs atom, which is originally on the octahedral site, exchanges with a Na atom on the tetrahedral site. The octahedral site then contains one Na atom at the center, (½ ½ ½), and four NH_3 molecules in a tetrahedron around (½ ½ ½), similar to the case of Na atoms in Na_6C_{60}. The function of NH_3 is simply that of a "lattice-expander."

The application of this technique to other A_3C_{60} structures with larger alkali atoms does not always result in a simple expansion of the cubic lattice. For example, a reaction of NH_3 with K_3C_{60} under the same conditions as described above results in $(NH_3)K_3C_{60}$.[32] The single ammonia molecule occupies the octahedral site but it displaces the K atom from the center of the site. The result is a lower symmetry unit cell (face-centered orthorhombic). This compound has a room temperature resistivity about five times higher than that of K_3C_{60}, and no superconductivity is observed. It is interesting to speculate on whether or not this compound has already undergone a Mott localization given that the compound retains three electrons in the conduction band. Reactions with NH_3 at higher pressures of ammonia result in other NH_3-K_3C_{60} phases with non-cubic symmetries.[33]

In conclusion, we have outlined the rich variety found in alkali-C_{60} intercalation compounds. Thus far, metallic character for alkali intercalated compounds has only been conclusively demonstrated for A_3C_{60}. Superconductivity is thus far limited to cubic salts with single atoms or symmetric ammonia-alkali clusters at the center of the octahedral site.   Ammonia intercalation into A_3C_{60} can be used to expand the lattice while maintaining the electron count in the conduction band, however some ammonia-A_3C_{60} structures have lower symmetry. The amount of ammonia incorporated into the structure can be varied by changing the synthesis temperature and/or pressure.

References

1. A. R. Kortan, these proceedings.
2. Q. Zhu, et al., *Phys. Rev. B.* **47** (1993) 13948.
3. R. Tycko, et al., *Science* **253** (1991) 884.
4. M. J. Rosseinsky, et al., *Nature* **356** (1992) 416.
5. T. Yildirim, et al., *Phys. Rev. Lett.* **71** (1993) 1383.
6. A. F. Hebard, et al., *Nature* **350** (1991) 600.
7. P. W. Stephens, et al. *Nature* **351** (1991) 632.
8. R. M. Fleming, et al., *Nature* **352** (1991) 787.
9. A. R. Kortan, et al., *Nature* **355** (1992) 529.
10. T. Yildirim, et al., *Nature* **360** (1992) 568.
11. O. Zhou, et al., *Nature* **351** (991) 462.
12. A. R. Kortan, et al., *Nature* **360** (1993) 566.
13. A. R. Kortan, et al., *Phys. Rev B* **47** (1993) 13070.
14. R. M. Fleming, et al., *Nature* **352** (1991) 701.
15. R. C. Haddon, et al., *Adv. Mater.* **6** (1994) 316.
16. S. C. Erwin and C. Bruder, *Physica B* in press.
17. P. J. Benning, et al., *Phys. Rev. B* **47** (1993) 13843.
18. R. E. Walstedt, et al., *Nature,* **362** (1993) 611.
19. C. S. Yannoni, et al., *J. Phys. Chem.* **95** (1991) 9; R. Tycko, et al., *ibid.,* p. 95; R. Taylor, et al., *J. Chem. Soc. Chem. Comm.* **1990** (1990) 1423.
20. P. C. Chow, et al., *Phys. Rev. Letters* **69** (1992) 2943.
21. Paul A. Heiney, et al., *Phys. Rev. Letters* **66** (1991) 2911.
22. William I. F. David, et al. *Nature,* **353** (1993) 147.
23. Kosmas Prassides, et al., *Science* **263** (1994) 950.
24. I. Hirosawa, et al., *Solid State Commun.* **87** (1993) 945.
25. K. Kniaz, et al., *Solid State Commun.* **88** (1993) 47.
26. O. Zhou, et al., *Science* **255** (1992) 833.
27. C. M. Varma, et al. *Science* **254** (1991) 989.
28. M. Schluter, et al., *Phys. Rev. Letters* **68** (1992) 526.
29. K. Tanigaki, et al., *Europhysics Letters* **23** (1993) 57.
30. K. Tanigaki, et al., *Nature* **352** (1991) 222.
31. O. Zhou, et al., *Nature* **362** (1993) 433.
32. M. Rosseinsky, et al., *Nature* **364** (1993) 425.
33. O. Zhou, et al., *to be published.*

Thermal Expansion of K_xC_{60} and Rb_xC_{60} ($x \approx 3$)

G. Burkhart, C. Meingast, B. Renker
Kernforschungszentrum Karlsruhe, Postfach 3640, 76021 Karlsruhe
Institut für Nukleare Festkörperphysik

Abstract. The thermal expansion of polycrystalline K_xC_{60} and Rb_xC_{60} ($x \approx 3$) pellets has been investigated in the temperature range 4 K - 300 K using high-resolution capacitance dilatometry. Due to the superconducting phase transition a jump in the thermal expansivity of K_3C_{60} at $T_c \approx 19$ K and Rb_3C_{60} at $T_c \approx 29$ K is observed. A lower limit of the specific heat jump at T_c is calculated using the Ehrenfest relation. There is also evidence for a broad phase transition near 100 K and 80 K for K_3C_{60} and Rb_3C_{60}, respectively.

1.Introduction

The discovery of superconductivity in alkaline doped fullerenes [1] and also in alkaline earth doped fullerenes [2] gives rise to investigate the superconducting properties in order to understand the mechanism of superconductivity in these organic compounds. As a general thermodynamical behaviour of superconductors the Ehrenfest relation

$$\frac{dT_c}{dp} = 3V_{mol}T_c\frac{\Delta\alpha}{\Delta c_p}$$

predicts a discontinuity of the linear thermal expansion $\alpha = 1/L \cdot dL/dT$ and the specific heat c_p at the superconducting transition temperature. Especially for the fulleride compounds this effect on the thermal expansion will be pronounced because of the large pressure effect [3] on the superconducting transition temperature. Furthermore, dilatometry is a sensitive method to examine phase transitions as has been demonstrated for the predominately van der Waals solid C_{60} [4], which undergoes both an orientational ordering transition at $T \approx 260$ K and an orientational glass transition near $T \approx 90$ K. In K_3C_{60} and Rb_3C_{60} the C_{60} molecules are believed to be ordered even at room temperature [5,6], which is most likely due to the stronger ionic binding. Thus no equivalent phase transition according to the temperature dependent dynamics of the C_{60} molecules are expected. Nevertheless, there are indications by ESR and NMR measurements for a phase transition in K_3C_{60} in the temperature range 160 K - 200 K [7,8] and Rb_3C_{60} at $T \approx 95$ K [9]. To identify possible phase transitions in K_3C_{60} and Rb_3C_{60} the measurement of the thermal expansion is a powerful method.

2.Experimental

K_3C_{60} and Rb_3C_{60} powder were prepared by solid-phase reaction of C_{60} powder with alkali vapor using a standard technique [10] and subsequently pressed pellets were made with 4mm diameter and height between 3mm and 6mm. Due to the extreme sensitivity of these compounds to air all work must be done in a glove box. The dilatometer measurements were made on a suitable air-tight encapsulation for which an aluminium cup is covered with a thin flexible aluminium foil. The foil, which is stretched over the sample, still allows one to measure the contraction of the sealed sample. Measurements on aluminium as well as C_{60} samples with and without the encapsulation showed that it has a negliable effect on the results. The linear thermal expansion was investigated using a high-resolution dilatometer with an absolute resolution of $\Delta L \approx 0.1 \text{Å}$. The temperature of the superconducting phase transition was determined by a SQUID magnetometer, which also allows an estimation of the superconducting fraction in the sample.

3.Results

3.1.Thermal expansion in the range 4 K - 300 K

Fig.1 shows the thermal expansivity of two different K_3C_{60} samples. The overall expansion is similar for both samples exept for the peak near 250 K in sample II, which is attributed to residual α-C_{60} phase. Slowly heating sample II to 380 K produced a large irreversible length change and subsequent measurements showed an increase of the α-C_{60} peak, which is most likely caused by the thermal decomposition of the metastable K_1C_{60} phase into α-C_{60} and K_3C_{60}. A rough estimate of the composition yields 22% α-C_{60} for sample II (after heating) and no α-C_{60} in sample I assuming that α-C_{60} and C_{60} behave similar. This correlates well with the much larger magnetic shielding fraction of sample I (84%) in comparison to sample II (19%). In difference to K_3C_{60} no α-C_{60} peak can be obtained in any Rb_3C_{60} samples in agreement with the phase diagram in which Rb_1C_{60} is stable to low temperatures [Poirier, this proceedings].
The existence of a phase transitions in K_3C_{60} [7,8] and Rb_3C_{60} [9] has been suggested by several groups using ESR and NMR techniques. Preliminary results of dilatometer measurements show a broad phase transition in the temperature range between 80 K and 130 K (both samples in Fig. 2), which also exists in Rb_3C_{60} but in a slightly lower temperature range (between 60 K and 110 K).

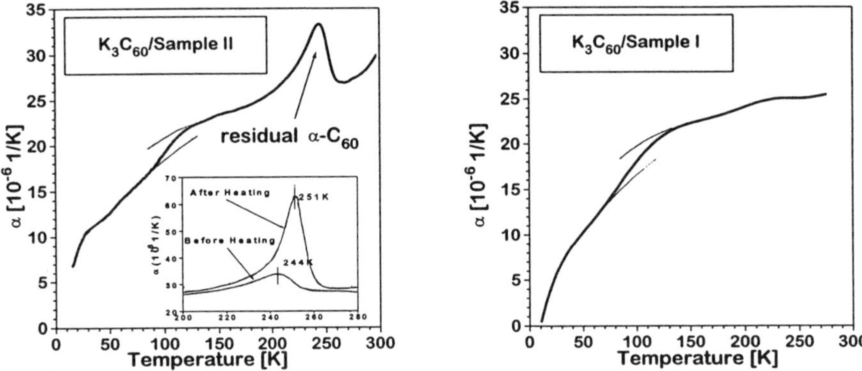

Fig. 1:*Thermal expansion of two differently prepared K_3C_{60} samples. Sample I shows 84% shielding fraction and no residual α-C_{60} whereas sample II exhibits a shielding fraction of 19% and a significant amount of α-C_{60} as indicated by the peak near 250 K. After heating sample II the α-C_{60} amount increases from an estimated 4% to 22%, which can be explained in terms of thermal decomposition of the metastable K_1C_{60}. In both samples there is evidence for a broad phase transition between 80 K and 130 K.*

3.2. Thermal expansion at T_c

As predicted by the Ehrenfest relation the linear thermal expansivity shows a jump at the superconducting phase transition in both materials. The results are illustrated in Fig. 2 for K_3C_{60} and Fig. 3 for Rb_3C_{60}:

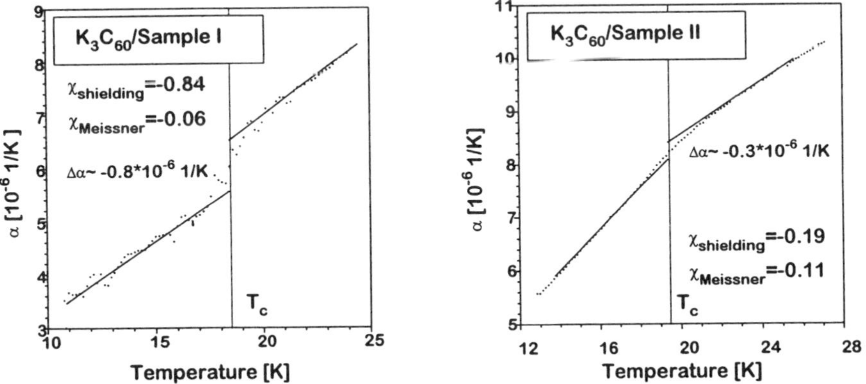

Fig. 2:*The phase transition to superconductivity in K_3C_{60} at $T_c \approx 19K$ yields a jump in the thermal expansivity. Sample I and sample II show different values of $\Delta\alpha$ which are correlated to the variable shielding fractions measured by a SQUID magnetometer.*

The jump height in Fig. 2, which should be proportional to the superconducting fraction in the samples, correlates nicely with the measured magnetic shielding signal. It is difficult to accurately determine the superconducting fraction from the magnetic measurements due to the granular nature of the samples. **Lower limits** of the expected expansion and specific heat anomalies at T_c can be estimated by assuming the super-conducting fraction is 100 percent:

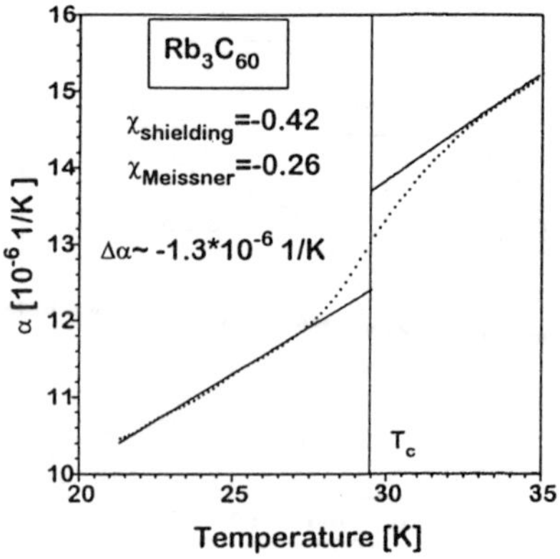

Fig.3:*The phase transition to super-conductivity in Rb_3C_{60} at $T_c \approx 29$ K.*

$$-\Delta\alpha(K_3C_{60}) \geq 0.8 \cdot 10^{-6} [\frac{1}{K}]$$

$$-\Delta\alpha(Rb_3C_{60}) \geq 1.3 \cdot 10^{-6} [\frac{1}{K}]$$

Knowing the jump in the linear thermal expansivity $\Delta\alpha$ and the pressure dependence of T_c [3] a calculation of the jump in the specific heat Δc_p is possible according to the Ehrenfest relation. The expected (minimal) values for $\Delta c_p/T_c$ are:

$$\frac{\Delta c_p(K_3C_{60})}{T_c} \geq 0.13 \ [\frac{J}{molK^2}]$$

$$\frac{\Delta c_p(Rb_3C_{60})}{T_c} \geq 0.17 \ [\frac{J}{molK^2}]$$

For Rb_3C_{60} no experimental results of the specific heat jump at T_c are available. Focusing on K_3C_{60} the predicted $\Delta c_p/T_c$ anomaly in the present work is about twice as large as has been measured (0.068 J/molK^2) [11]. Probably this is a consequence of the different sample characterization, especially concerning the fraction of the super-conducting phase. In order to avoid some of the problems with differently prepared samples the thermal expansivity and the specific heat will be measured on the same sample in our experimental division.

4.Summary

The first measurements of the thermal expansivity of K_3C_{60} and Rb_3C_{60} fullerides identify two phase transitions. In both materials the superconducting phase transition leads to a jump in the thermal expansivity at T_c in agreement with the Ehrenfest relation. The jump height depends on the superconducting fraction emphasizing a careful estimation of the sample composition. There is evidence for a broad phase transition in K_3C_{60} in the temperature range between 80 K and 130 K as well as in Rb_3C_{60} in the temperature range from 60 K to 110 K, the nature of which is still not known. In order to explain this phenomena it must be emphasized that the ionic molecular interaction in these compounds will yield completely different features in comparison to the well established molecular dynamics in (dominately) van der Waals interacting solid C_{60}.

References

[1]   K.Tanigaki et al.; Nature **356**, 419 (1992)
[2]   A.R.Kortan et al.; Nature **360**, 566 (1992)
[3]   G.Sparn et al.; Science **252**, 1829 (1991)
[4]   F.Gugenberger et al.; Phys. Rev. Lett. **69**, 3374 (1992)
[5]   P.W.Stephens et al.; Nature **351**, 632 (1991)
[6]   P.W.Stephens et al.; Phys. Rev. B **45**, 543 (1992)
[7]   Y.Yoshinari et al.; Phys. Rev. Lett. **71**, 2413 (1993)
[8]   S.E.Barett et al.; Phys. Rev. Lett. **69**, 3754 (1992)
[9]   G.Zimmer et al.; Europhys. Lett. **24**, 59 (1993)
[10]   K.Holczer et al.; Science **252**, 1159 (1991)
[11]   A.P.Ramirez et al.; Phys. Rev. Lett. **69**, 1687 (1992)

ANHARMONICITY OF LOW-FREQUENCY PHONONS IN FULLERITE AND FULLERIDES

S.-L. Drechsler

Institut für Festkörperforschung am IFW Dresden e.V., D-01171 Dresden, Germany

T. Galbaatar and N.M. Plakida

Joint Institute for Nuclear Research, Dubna, Russia

and

R. Gutierrez

Institut für Theoretische Physik, Technische Universität Dresden, Germany

ABSTRACT

Inelastic neutron scattering (INS) and specific heat data of C_{60} and related compounds are analyzed in various anharmonic models. A consistent description of these data leads to the picture of strongly anharmonic octahedral potassium vibrations in K_3C_{60} and weak local anharmonicities of librons in fullerene and fullerides. In particular, the anomaly near 3meV seen at low temperatures in K_3C_{60} can be explained by a purely lattice model and it might be *not* related to the occurence of superconductivity. Within the proposed model the temperature (T) dependence of the force constants of C_{60} is an order of magnitude smaller compared with those of the commonly used pseudoharmonic approximation. The T-dependence of the linewidths of libronic modes and alkaline isotope effects are discussed, too.

1. Multi-well model for octahedral alkaline ions

In recent INS studies on superconducting fcc K_3C_{60} [1] below 25 K an unexpected increase of spectral weight below 3 meV with decreasing T has been observed. It was ascribed to a change of the acoustic phonons caused by the occurence of superconductivity for $T < T_c$. Keeping in mind the large volume available in octahedral (O) sites and the large thermal factor [2] $\mathcal{A}(300K)=12\pi^2 < u_x^2 + u_y^2 + u_z^2 >= 16\text{Å}^2$, we suppose that K(2)-ions at O-sites see a strong anharmonic potential of multi-well character. By symmetry, 6, 8 or 12 minimum positions around the central maximum are expected. Tunneling in such a potential leads to a separate group of low-lying K(2) states. The decreasing difference of occupational numbers of those states causes the decreasing oscillator strength for the fundamental transition with increasing T (see Fig. 1) . In the case of predominant tunneling across the central barrier, we arrive at a threefold degenerate double-well problem. The potential was modelled by

$$V(r) = -\frac{V_2}{2}r^2 + \frac{V_4}{4}r^4, \qquad r = x, y, z \quad . \tag{1}$$

With V_2=2.95 eV/Å^2 and V_4=190eV/Å^4 one obtains $E_1 - E_0 = 3.45$ meV from the solution of the Schrödinger problem. The experimental value of $\mathcal{A}(300K)$ [2] was used

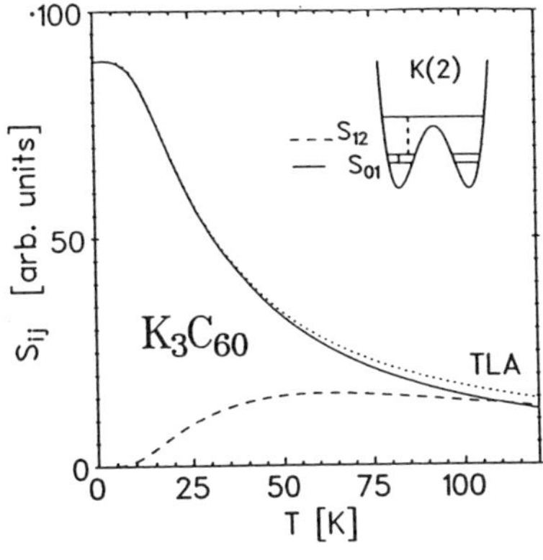

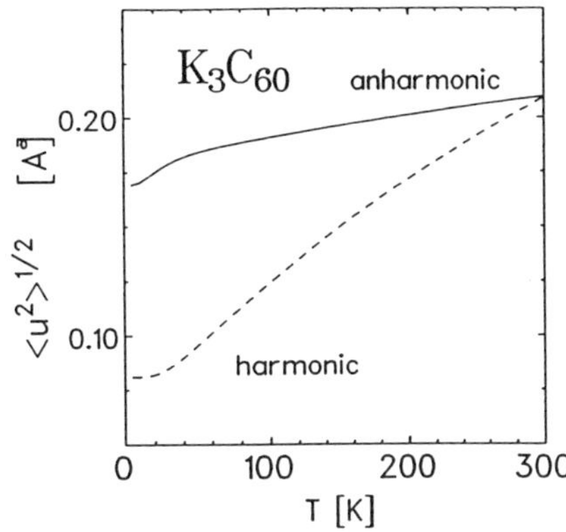

Fig. 1. Intensities vs. T of the 1st and 2nd transitions in the double-well potential.

Fig. 2. K(2) root-mean-square displacement $\propto \sqrt{\mathcal{A}}$ vs. temperature T.

to fix the parameters of the anharmonic potential of Eq. 1. With these parameters we obtained $2d_0 = 0.25$Å for the double-well separation, 9.8 meV for the barrier height and $E_2 - E_1 = 11.5$ meV for the second transition dominant at room temperature. The weak T-dependence of the anharmonic K(2)-rmsd $\sqrt{< u^2(T) >}$ and the corresponding harmonic one are shown in Fig. 2. The oscillator strength entering also the neutron cross section in the two-level approximation (TLA) reads $S_{01} \propto \tanh(\hbar\omega_{01}/2T)$ and the suppression of this transition for $T > T_c$ is accidental. Quantitatively, there is also some influence of the T-dependent interaction between K(2)-derived modes and acoustic phonons which can be described in the RPA-pseudospin-phonon picture.

2. Pseudoharmonicity and local libronic anharmonicity

The libronic INS-data are usually interpreted within the pseudoharmonic picture. However, in order to explain the T-dependences of the libron frequency and intensity for C_{60} between 0 and 265 K, the pseudoharmonic force constant should be softened e.g. due to thermal expansion by a factor of three or four, which seems to us unphysical. In order to avoid such artificial behaviour, we adopt a local effective anharmonic potential with only weakly T-dependent harmonic and anharmonic force constants

$$V(\varphi) = A(T)\varphi^2 - B(T)\varphi^4 + C\varphi^6 - D\varphi^8 + ... \quad . \tag{2}$$

The constants $A(T)$, $B(T)$, C, D = const are chosen to reproduce the $\omega_{lib}(T)$ and the rms angle derived from the INS-data [3,4]. In determining the latter one we considered also corrections (higher order cumulants) to the Debye-Waller factor. The effective cosine-like libronic potential $V(\varphi)$ obtained in such a way and the weak T-dependences of the quadratic and the quartic force constants are shown in Fig. 3

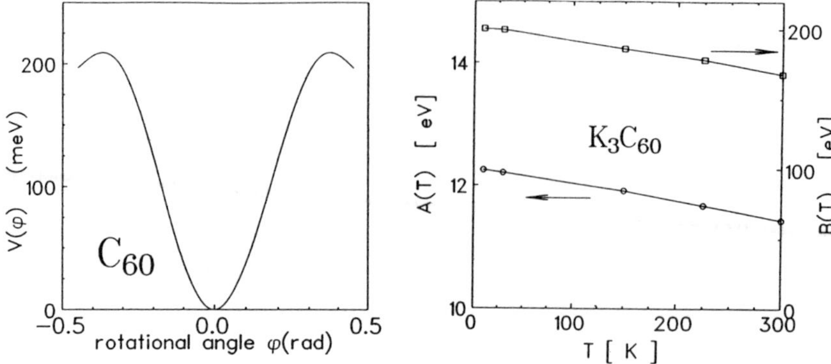

Fig. 3. Effective libronic potential.

Fig. 4. Libronic harmonic and quartic force constants vs. temperature T.

and 4, respectively. Similar results have been found for C_{60} although the anharmonicity required to fit the rms angle (RMSA) and the INS intensity [4] comes out somewhat stronger compared with K_3C_{60}. Using the results of Sec. 1 and harmonic tetrahedral K(1)-vibrations near 13 meV, the calculated specific heat is in accord with [5]. In particular, the discrepancy between the harmonic interpretation of the data of [3] and [5] about the value of $\omega_{lib}(0)$ could be resolved for weak libronic anharmonicity. The strong anharmonicity of the K(2)-ions manifests itself in a sizeable down-shift ~ 0.5 K of the onset of the C/T-curve after K isotope substitutions and in a large thermal factor $\mathcal{A}(T \to 0)$.

3. The libronic linewidth

Although the approach of Sec. 2 allows to fit $\omega_{lib}(T)$ and the RMSA, the steep increase of the libronic linewidth above $T_{on} \sim 100 \div 150$ K observed for fullerenes with Pa$\bar{3}$ or *bcc* structures requires an explicit many-body description. Since T_{on} exceeds considerably all external frequencies, a sizeable contribution from the interaction with high-energy onball-modes is suggested. We consider a *large* set of oscillators coupled anharmonically to 3 damped low-energy ones, standing for the C_{60}-librations in an orientationally disordered environment, generalizing the two-mode model [6]. Including cubic interactions with acoustic phonons, too, our Hamiltonian reads

$$H = H_{harm} + \frac{1}{2} \sum_{i=1}^{174} g_i^{ac} \varphi^2 (u_i^{opt})^2 + \sum_{j=1}^{3} g_j \varphi (u_j^{ac})^2, \tag{3}$$

where u_i denotes the displacement of the ith oscillator. Microscopically, Eq. 3 models a special type of interball-interaction. Using the equation of motion method for Green's fuctions, we evaluated the quasiharmonic frequencies Ω_i and the self-energies $S_i(\omega, \varepsilon_i)$ within 2nd-order perturbation theory, where ε_i is a disorder parameter.

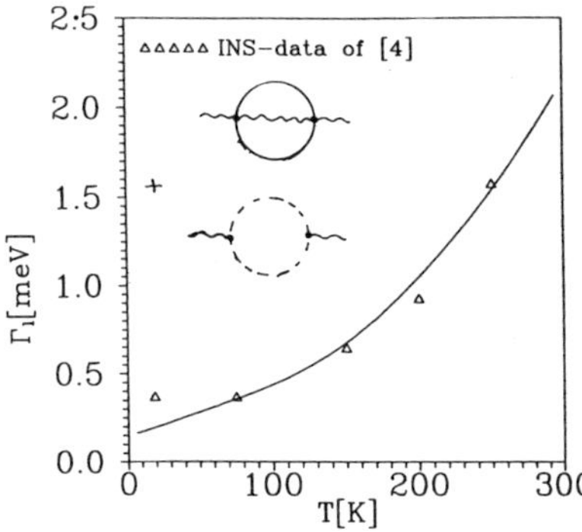

Fig. 5. Libronic linewidth Γ_l vs. T.

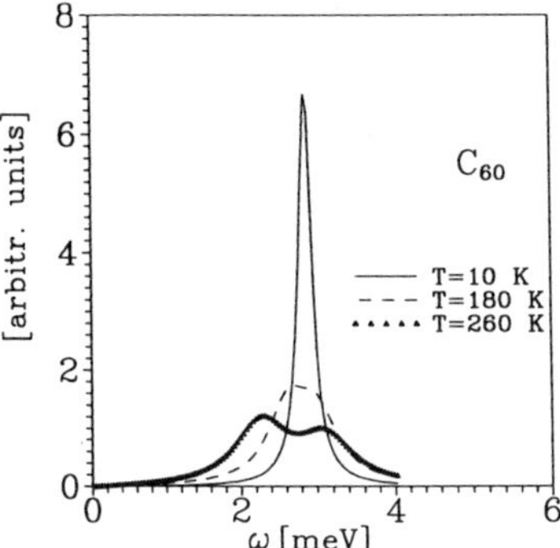

Fig. 6. Libronic spectral density vs. energy.

The corresponding diagrams are shown in the insert of Fig. 5. The acoustic phonon contribution (lower graph) was handled within the standard Klemens approximation yielding the nearly linear T-dependence of the libronic linewidth $\Gamma_l(T)$ below 100 K. The disorder parameter ε_l was taken to get $\Gamma_l(0) \sim 0.2$ meV and $\Gamma_i(0) = 2\varepsilon_i \sim 0.05$ meV otherwise, according to optical data [7]. For the sake of simplicity we described the 174 internal modes by four groups centered at 185, 155, 90, and 68 meV. Then, each group contains 44 or 43 degenerate modes. The C$_{60}$ INS-data of [4] can be reproduced quite well, adopting $g_i/M_i\Omega_i^2$=1.8,2.7,-1.9, and -4.5 . The libronic spectral density is shown in Fig. 6 for various T. Its width is given by $\Gamma_l = -\hbar \mathrm{Im} S_l(\omega_l)/I\omega_l$, where I is the C$_{60}$-momentum of inertia. Notice the double-peak structure at $T > 200$K similar to [1]. Reducing the g_i by a factor of 10 one gets a nearly constant linewidth reminiscent of K(Rb)$_3$C$_{60}$, where in the $F3m3$ structure the interball-interaction is less important compared with the case of C$_{60}$ and RbNa$_2$C$_{60}$ with $Pa\bar{3}$ space group [8]. Since the onball-modes interact with few (3) degrees of freedom, they are only weakly affected by the interaction of Eq. 3 yielding a nearly linear increase of their linewidth reaching ~ 0.5 meV at $T = 260$ K, similar to [6,7].

References

1. B. Renker et. al., Zs. Phys. B, **92**, 451 (1993); ibid. **90**, 325 (1993).
2. P.W. Stephens et al., Phys. Rev. B, **45**, 534 (1992).
3. C. Christides et al., Phys. Rev. Lett. **69**, 3104 (1992).
4. J.P. Copley et al., J. Phys. Chem. Solids **53**, 1353 (1992).
5. A.P. Ramirez et al., Phys. Rev. Lett. **69**, 1687 (1992).
6. S. Tolbert et al., Chem. Phys. Lett. **188**, 163 (1992).
7. B. Burger et al., these Proceedings (1994).
8. C. Christides et al., Europhys. Lett. **24**, 755 (1993).

DISORDER AND INTERACTIONS
IN THE DOPED FULLERIDES

E.J. Mele, S.C. Erwin, M.S. Deshpande
Department of Physics
Laboratory for Research on the Structure of Matter
University of Pennsylvania
Philadelphia, PA 19104

M.J. Rice
Xerox Webster Research Center
Webster, NY 14580

and

H-Y Choi
Department of Physics
Sung Kyun Kwan University
Suwon 440 746 Korea

ABSTRACT

A theory for the coupling of the density fluctuations of the doped conduction charge in orientationally disordered M_3C_{60} to the A_g and H_g intramolecular vibrational modes of the fullerene molecule is reveiwed. In the disordered phase, the even parity modes of the isolated molecule are experimentally accessible through dipole active external probes. The effect is used to probe of the local effects of the orientational disorder in the doped phase, and to critically examine several models for the electron phonon coupling in this system.

1. Introduction

Molecular solids derived from the C_{60} molecule have attracted the attention of physicists, chemists and materials scientists over the last several years.[1] It has been demonstrated that the undoped parent cubic phase of undoped C_{60} can be doped through an interesting series of insulating and conducting phases by the intercalation of metal species into the interstitial volume of the solid. The primary role of the metal species is to donate charge into the t_{1u} LUMO complex of the undoped molecule. In this process, the added negative charge on the C_{60}, compensated by the surrounding cation shell in the interstitial volume, significantly modifies the orientation dependence of the intermolecular potential. Thus the orientational phase of the doped solids is generally observed to be quite different from that of the undoped host. For the M_3C_{60} phases, where M is an alkali metal which we will study in this work, the C_{60} molecules appear to be quenched into an orientationally disordered state.[2] The M_3C_{60} phases alloyed with the smaller alkali Na, appear to retain the Pa3 structure, a situation which apparently competes with the superconductivity in these phases.[3] For the K_4C_{60} phase, which occurs in the body centered tetragonal structure, the simplest interpretation of the four fold symmetry of the observed diffraction pattern is that the molecules are again quenched into an orientationally disordered phase.[4] The structure recently reported for the Rb_1C_{60} orthorhombic phase, also requires a rotational relaxation of the C_{60} away from the preferred setting in the undoped cubic structure.[5]

At the outset it is clear that the degree and type of orientational order or disorder in these phases can have a very important effect on the electronic properties of any of these doped phases. Although the overall valence structure of these compounds is largely controlled by the intramolecular electronic structure, the transport properties are governed by low energy excitations of the doped carriers. For these added carriers the entire width of the effective conduction band is then due to the relatively weak amplitudes for a doped electron to hop from site to site in the t_{1u} derived conduction band. These in turn depend very sensitively on the orientations of the two C_{60} molecules connected in the hop. This sensitivity derived from both the vector character of the molecular t_{1u} state from which the conduction band is derived, and from the antibonding character of this state. The relatively high residual resistivity of the doped phase has been attributed to the scattering of the conduction electron motion by the static orientational disorder in the doped phase.[6,7] However, since these doped phases are narrow band systems in which electron electron interactions are anticipated to be quite significant, it has proven extremely difficult to definitively establish the connection between the orientational structure and the observed electronic behavior.

In this paper we briefly review a theoretical study which addresses this problem, for the orientationally disordered M_3C_{60} phases, by examining the frequency dependence of the midinfrared conductivity. The theoretical work reported here is motivated by recent experiments which reveal that in addition to a broad midinfrared peak in the conductivity, the doped phases show quite strong modulations in Re $\sigma(\omega)$ which can be correlated in frequency with the H_g intramolecular modes of the undoped molecule.[8] The appearance of even parity modes, measured through a dipole active probe, is a clear signature of the local breakdown of inversion symmetry in the doped phase. We will consider here a model in which we explicitly treat the coupled electron phonon system in an orientationally disordered phase, while ignoring the effects of direct electron electron interactions. We find that many of the experimental observations can be qualitatively and in some cases semiquantitatively explained within such a model, but that several important quantitative discrepancies remain. Refinements of the theory to deal with the effects of direct electron electron interations will therefore likely be the key to remedying these remaining discrepancies. An extended discussion of the calculations reported here has recently been submitted for publication.[9] Here we will briefly outline the theory, and summarize the essential conclusions from the study.

2. Orientational Structure of the M_3C_{60} Phase

The highest symmetry setting for a C_{60} molecule in a cubic crystal field is one which orients three mutually perpendicular two fold symmetry axes of the molecule along the Cartesian x- y- and z- directions. There are two inequivalent ways of achieving this orientation in which the molecule locally maintains the high symmetry cubic setting of the C_{60} molecule. However, these two orientations, labelled the A setting and B setting, are inequivalent, being related by a 90° rotation about any two fold axis which is not a symmetry operation of the molecule. The intermolecular contact between two fullerenes along any (110) nearest bond direction then depends on the assigments of the terminal sites to the A or B setting. This has been discussed in some more detail in reference 10.

The experimentally prepared samples are believed to be described within the Stephen's model for merohedral disorder in which each molecule adopts either the A or B orientation, but in which the assignment is essentially random from site to site in the solid.[2] There is some very weak experimental evidence for short range correlations in the orientations in which, for example, an orientation A on a reference site favors the B orientation on neighboring sites.[11] One may study the electronic structure in the conduction band of such a disordered phase, by a direct calculation on a supercell containing this form of quenched orientational disorder. Alternatively, we have developed a more analytical approach to the electronic structure of these phases in which one uses a Bethe lattice to incorporate into the theory the retraceable Feynman paths which control the asymptotic behavior of the one electron Green's function in the strongly disordered system.[7,12] The two methods yield very similar results for all the properties which we have calculated to date, and there are certain conceptual advantages provided by either of these two approaches.

3. Lattice Dynamics in the Doped Phases

The propagating conduction electrons in the doped phases have important effects on the measured spectra of intramolecular phonons in these systems. This is at first surprising, since one might have supposed that these spectra would be dominated by the strong intramolecular interactions in the molecule. However, the density fluctuations of the conduction charge do contribute an important renormalization to these phonons, and in fact the solid state effects on the symmetry, frequencies and linewidths of these modes, are governed by the coupling to the intermolecular charge fluctuations in the t_{1u} derived conduction band.

To describe the intramolecular dynamics we adopt a simple local model for the intramolecular vibrations, described by a lattice of Einstein oscillators:

$$\ddot{u}_\alpha(r,t) = -\omega_\alpha^2 u_\alpha(r,t) \tag{1}$$

The α's run over the two A_g and eight H_g on ball modes which can be linearly coupled to the density fluctuations of the t_{1u} electrons. The bare frequencies ω_α can be chosen from first principles theory, or by a direct fit to experimental data. We have chosen the former strategy and use the frequencies tabulated in the theoretical work of Faulhaber et. al. on the C_{60} trianion.[13] In the presence of doping, there is an additional contribution to the dynamics which is nonlocal in space, time and the phonon band index:

$$\ddot{u}_\alpha(r,t) = -\omega_\alpha^2 u_\alpha(r,t) + \int \frac{dr'\,dt'}{\Omega} \; K_{\alpha\beta}(r,r',t-t')\, u_\beta(r',t') \tag{2}$$

The kernel $K_{\alpha\beta}(r,r')$ entering this expression can be readily computed from the one loop density fluctuations of the conduction charge, linearly coupled to the intramolecular modes α and β at sites r and r'. Inspection of equations (1) and (2) reveals that any of the solid state information in the phonon spectra, and in particular the effects of the orientational disorder, enter the discussion through the detailed spatial dependence of the effective interaction $K(r,r')$.

The renormalized phonons of equation (2) can couple to current fluctuations in the t_{1u} manifold. The elementary process which couples these two is explicitly:

$$(3)$$

$$V(i\lambda;\omega+i\delta) = \frac{1}{N}\sum_{k}\sum_{n,n'} \frac{j_{n,n'}(k)\,\Gamma^{i\lambda}_{n'n}(k)}{E_{nk}-E_{n'k}-\omega-i\delta}[f_{nk}-f_{n'k}]$$

where $j_{nn'}$ is the matrix element of the current operator between Bloch states n and n', $\Gamma_{n'n}$ is the matrix element of the linearized deformation potential for the λ-th mode on site i, and the f's are the Fermi factors. The composite matrix element V in equation (3) is the phonon charge, which then contributes to the conductivity:

$$(4)$$

$$\mathrm{Re}\,\sigma_{ph}(\omega) = \frac{1}{\omega}\sum_{i\lambda,i'\lambda'} V(i\lambda;\omega+i\delta)\, D(i\lambda,i'\lambda';\omega+i\delta)\, V(i\lambda;\omega-i\delta)$$

where D is the renormalized phonon Green's function. It is important to notice that the matrix element displayed in equation (3) vanishes identically for the orientationally ordered structure, or indeed for any inversion symmetric structure, in which the phonon coupling involves an even parity intramolecular mode. Furthermore, a simple disorder average of (3) also yields zero, since after performing an average over the disorder, the one particle states see an effective inversion symmetric medium. However, the disorder average of the conductivity in equation (4) is *nonvanishing*, since it is seen to involve a disorder average of the square of the phonon charge. Thus the relevant microscopic scattering processes which lead to the dipole activity through the even parity intramolecular modes, are precisely those scattering processes which correlate the current fluctuations on the left and right sides of the coupled mode described in equation (4).

4. Results

We have carried out the computation outlined above, choosing for the deformation potentials Γ in equation (3) several of the popular models which have been proposed to describe the coupled electron phonon system for the doped fullerides. Our results for two of the models are compared withthe available experimental data in Figure 1.

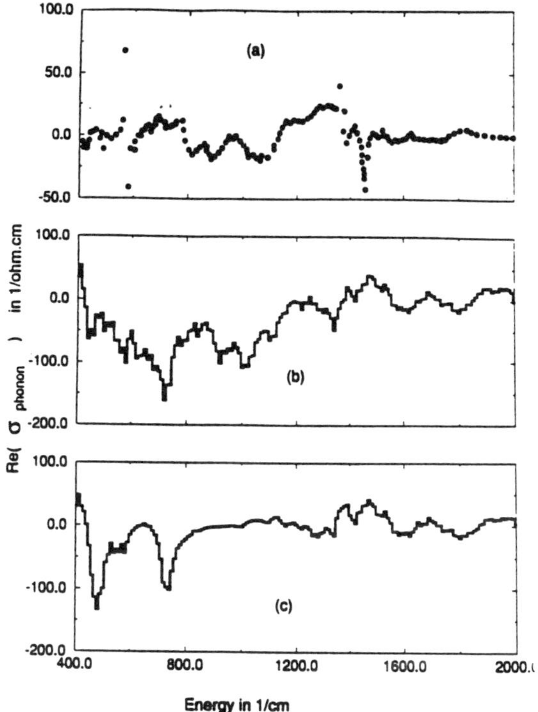

Figure 1: A comparison of the observed phonon induced modulations of the midinfrared conductivity reported from reference 8(a) with the theoretical predictions calculated from equation (4) using the two proposed electron phonon coupling models of refs. 14 (b) and 15 (c).

The experimental data obtained in this Figure for K_3C_{60}, are obtained from a Kramers Kronig analysis of the measured frequency dependent reflectivity.   To expose the modulated structure in the conductivity, we have subtracted from the measured conductivity a smoothly varying background, which contains the primary component of the midinfrared absorption. The fluctuating residual part of the conductivity has an amplitude approximately of 50 $(\Omega\text{-cm})^{-1}$, which is superposed on a broad electronic background of 500 $(\Omega\text{-cm})^{-1}$ (removed from Figure 1).

The lower two panels give the results of the computation of the phonon contribution to the midinfrared conductivity (equation 4) for two different electron phonon coupling models. The first of these models, due to Varma, Zaanen and Ragavachari,[14] posits that the important coupling between the conduction electrons and the intramolecular vibrations is dominated by the highest two modes of H_g symmetry (the H_g (7) and H_g(8) modes) which contain significant double bond stretching character. The second of these models, due to Lannoo and coworkers[15], concludes that there is substantial coupling not only through these high frequency bond stretching vibrations, but also through the midfrequency H_g modes which contain more radial and bond bending character. There have been a variety of other coupling models proposed; one finds that these can be broadly assigned to either the first (type I) or second (type II) models.

Our calculations are carried out on a series of supercells containing 32 molecular sites with quenched disorder, and some representative results are summarized in Figure 1. One finds that the two models mentioned above yield different predictions for the phonon induced modulations of the midinfrared conductivity. For the type I models, one indeed finds a very signficant modulation of the calculated specta. These modulations extend over a rather broad frequency range (400^{-1} - 1600 cm^{-1}) and the structure in these spectra can be very approximately correlated with the frequencies of the bare Hg modes which are coupled to the conduction charge. For the type II models, one finds, in addition, a rather sharp structure in the range (400^{-1}- 600 cm^{-1}). This arises since in these models one has a non-negligible coupling through the mid-frequency Raman active modes, and these occur in an interesting frequency range in which the imaginary part of the phonon self energy is relatively small in the disordered system . As a result, one obtains a rather sharp modulation of the electronic response in this region, with a more conventional Fano-type lineshape. These sharp features are of course absent in the experimental data of panel (a), leading us to conclude that the type I models provide a better approximation to the physical situation in the doped phase than the type II models.

There are some noteworthy discrepancies between the theory and the experiment. First in either model, the scale of the modulations is significantly overestimated, by approximately a factor of two in either case. In fact one observes a similar discrepancy in the overall value of the background conductivity on which these modulations are to be superposed. The merohedrally disordered tight binding theory predicts a low frequency conductivity in the range 2000-3000 $(\Omega\text{-cm})^{-1}$, while the low temperature experiments suggest a somewhat smaller value of order 1500 $(\Omega\text{-cm})^{-1}$. The theoretical prediction at low frequency is significant since it does not renormalize if one simply renormalizes the energy scale of the effective conduction bandwidth (thus this value represents an intrinsic limit to the scattering rate from static orientational fluctuations alone). Interestingly, the experimental value is very strongly temperature dependent, and decreases to approximately 500 $(\Omega\text{-cm})^{-1}$ at room temperature where the experimental data of Figure 1 were obtained.

A rather more striking discrepancy involves the coupling to the even parity fully symmetry $A_g(2)$ mode. This mode is described as a pentagonal pinch, and involves a fully symmetric tangential breathing oscillation of the pentagonal rings on the surface of the C_{60} molecule. In the fully orientationally ordered phase, this mode is left unrenormalized by the electron phonon interaction. This is easily understood, since in the $q \to 0$ limit this mode simply provides a constant shift to the average potential seen by a propagating conduction electron, and therefore it does not modulate the charge density for the doped carriers. The appearance of this mode as a sharp feature in the observed Raman (and infrared) data has oftened rationalized using this idea. However, we find that this simple argument is strongly violated in the disordered phase. In fact, in the disordered cell the $A_g(2)$ mode can and do couple strongly to density fluctuation of the conduction charge. The calculated $A_g(2)$ feature is strongly damped, with a lifetime nearly as short as that predicted for the high lying H_g modes which are experimentally found to be strongly coupled. This is in striking contrast to the experiments which do indeed show a rather narrow linewidth for the $A_g(2)$ mode. The most natural resolution to this difficulty invokes the screening of the electron phonon vertex in the presence of an additional on site direct repulsive interaction between the electrons, absent from our treatment. This screening effect should of course be expected to be strongest for the A_g modes which are locally coupled to the total charge on the fullerene molecule. A more quantitative examination of this mechanism requires an explicit treatment of the interactions in the presence of the orientational disorder of the doped phase, since as we noted above, the coupling actually vanishes by symmetry in the orientationally ordered structure. Our preliminary results indicate that the effect of the interactions is to suppress coupling to the $A_g(2)$ mode, while weakly renormalizing the coupling strength to the Hg degrees of freedom.

The effects of temperature on the theory discussed above have not yet been examined. The observed temperature dependence of these midinfrared features may indeed challenge the basic assumptions of the theory developed here. The spectral weight in the midinfrared peak decreases dramatically with increasing temperature over a very wide frequency range. Estimates of an effective mean free path, extracted from high temperature transport data, yield a value which is shorter than the radius of the C_{60} molecule, a state of affairs which appears difficult to reconcile with the model developed above. Thus while we believe that the phonon structure in the midinfrared spectra can be adequately interpreted by the theory presented here, there remain a number of issues in the experimentally measured low frequency spectra which remain to be resolved.

Acknowledgements

This work was supported by the Department of Energy under Grant DE 91 ER 45118, and by the National Science Foundation under Grant 91 20668. EJM thanks the organizers for additional support which made it possible to attend the winterschool in Kirchberg.

References

1. A.F. Hebard, Physics Today 11, 26 (1992) and references therein.
2. P. Stephens et. al., Nature 351, 632 (1991)
3. T. Yildirim et. al. , Physical Review Letters xx, yyy (1993)
4. R.M. Fleming et. al, Nature 352, 701 (1991)
5. T. Yildirim et. al., Physical Review Letters 71, 1383 (1993)
6. M.P. Gelfand and J.P.Lu, Physical Review Letters 68, 1050 (1992)
7. M.S. Deshpande et. al. , Physical Review Letters 71, 2619 (1993)
8. M. Matus, T. Pichler and H. Kuzmany, Solid State Communications 86, 221 (1993)
9. M.S. Deshpande et. al., Physical Review B (in press, 1994)
10. T. Yildirim et. al. Physical Review B 48, 12262 (1993)
11. T. Egami et. al. (unpublished)
12. E.J. Mele and S.C. Erwin, Physical Review B (in press, 1994)
13. J.C.R. Faulhaber, D. Ko, and P.R. Briddon, Physical Review B 48, 661 (1993)
14. C.M. Varma, J.Zaanen and K. Raghavachari, Science 254, 989 (1991)
15. M. Schluter, M. Lannoo, M. Needels, G. Baraff, Phys. Rev. Lett. 68, 526 (1992)

ORIENTATIONAL ORDER/DISORDER IN Na_xC_{60}

T. Yildirim[1,3], J. E. Fischer [2,3], P. W. Stephens[4], and A. R. McGhie[3]

[1]*Department of Physics,* [2] *Materials Science Department and*
[3] *Laboratory for Research on the Structure of Matter*
University of Pennsylvania, Philadelphia, PA 19104
[4]*Department of Physics, State University of New York, Stony Brook, NY 11794*

ABSTRACT

Na_xC_{60} and Na_2MC_{60} have many unique and interesting properties compared to heavy alkali compounds, as we show by presenting differential scanning calorimetry (DSC) and high-resolution powder X-ray diffraction (XRD) results. For $x < 1$, both XRD and DSC show phase separation into essentially pure C_{60} and Na_1C_{60} at 300 K while for $1 < x < 3$ we find solid solution behavior and orientational ordering transitions at T_m's above that of pure C_{60}. The ordered phases are the same as in pure C_{60} but with more restricted orientations in the disordered phase. We predict similar behavior in the Na_2MC_{60} ternaries with M = Rb and Cs, which has been recently observed. The effect of the orientational ordering on superconductivity in these compounds is also discussed.

1. Introduction

The properties of Na_xC_{60} intercalation compounds differ considerably from their heavy alkali counterparts[1-3]. The saturated phase remains fcc[2] with $x = 10 - 11$, and superconductivity has not been reported at any x so far[1,4,2]. Most surprising of all, tetrahedral Na stabilizes an ordered $Pa\bar{3}$ structure[3] to rather high T_m in contrast to heavy alkalis (K, Rb) which lock the C_{60}'s into one of two standard orientations with no long range order[5]. Understanding these unique and interesting features of Na_xC_{60} is important in relation to other M_xC_{60} phases and especially for the mechanism of superconductivity since it is absent in Na_xC_{60}. Here we report DSC and XRD studies of Na_xC_{60} with $x < 3$, and discuss the effect of the orientational ordering on superconductivity by studying T_c as a function of x in the $Pa\bar{3}$–ordered quarternaries $Na_2Rb_xCs_{1-x}C_{60}$.

2. Differential Scanning Calorimetry

DSC samples (5-10 mg) were heated in hermetically sealed aluminum pans under an argon atmosphere at 5 K/min from -90° C to 100° C. Cooling was accomplished using liquid nitrogen. The results are shown in Fig. 1. The top curve from a sample with $x < 0.2$ shows a large transition at -14°C characteristic of pure C_{60} plus a second very small one at 57°C., indicating phase separation already at this small x. The sample with $x = 0.5$ exhibits two clear endothermic peaks on heating. The heat flow and transition temperature of the first peak, when compared with pure C_{60}, strongly suggest that about 50 % of the sample is essentially pure C_{60}. Since x is nominally 0.5, the other half of the sample should be Na_1C_{60}, to which we attribute the second transition with a heat flow $\Delta H(x = 1) = 4.4$ J/g. Thus we conclude that for $x < 1$ the system phase separates into essentially pure C_{60} (or α–C_{60}) and Na_1C_{60}. This is supported by high resolution XRD discussed below. For $x > 1$ the transition corresponding to $\alpha - C_{60}$ is no longer observed; instead we have a single transition which becomes weaker and shifts to lower temperature as x approaches 3, eventually disappearing at $x = 2.8$ as shown in the bottom curve of Fig. 1. Single transitions with x-dependent T_m's indicate solid solution behavior for $1 < x < 2.8$.

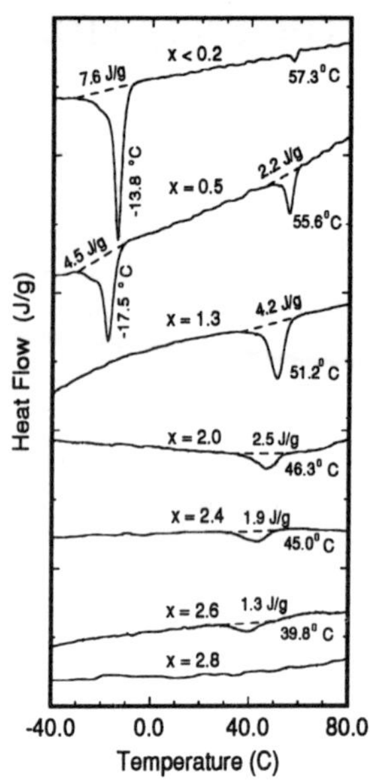

Fig. 1 DSC heating curves of Na_xC_{60} for different x.

3. X-Ray Diffraction

We now discuss X-ray data and Rietveld refinement on the same samples, measured in evacuated quartz capillaries at the NSLS beam line X3B1 (wavelength 1.15 Å). First we studied $Na_{0.5}C_{60}$ vs. temperature up to 500° C (Fig. 2). At 300 K (top panel) XRD clearly reveals two phases, consistent with DSC. The lattice constant and relative integrated intensities of peaks denoted by '*' are ascribed to $\alpha - C_{60}$, while all the remaining reflections correspond to $Na_{1.3}C_{60}$ as reported in Ref. 3. We conclude that $Na_{0.5}C_{60}$ is a two-phase mixture of essentially pure C_{60} and Na_1C_{60} at 300 K, in agreement with DSC. The middle panel shows that the two phases are better resolved at 200° C, no doubt due to the different thermal expansions. At 450° C these dissolve into a single phase (bottom panel), indicating solid solution behavior.

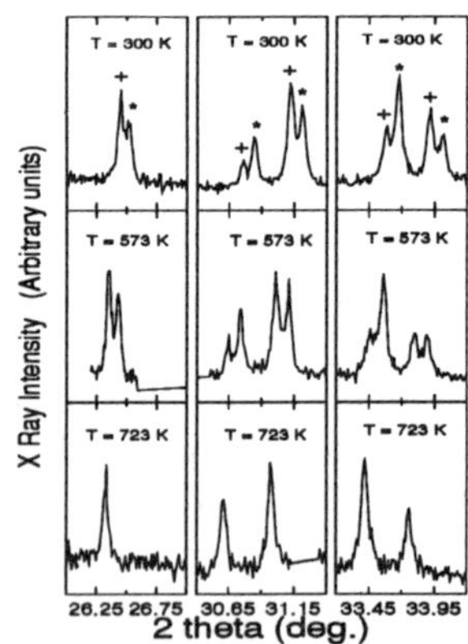

Fig. 2 T-dependent X-ray data for $Na_{0.5}C_{60}$. Peaks denoted by '+' and '*' correspond to Na_1C_{60} and pure C_{60} respectively.

Figure 3 shows partial 300 K profiles of Na_xC_{60} with $1 < x < 3$; these can be indexed as simple cubic (sc) for $x < 2.8$ and face-centered cubic (fcc) for $x > 2.8$. Peaks denoted by "+" are the four strongest sc reflections. It is important to note that these sc intensities become weaker as x approaches 3, eventually vanishing at $x = 2.8$. Comparison of the DSC transitions and the sc integrated intensities as a function of x strongly indicates that we are dealing with a phase transition from orientationally disordered fcc to orientationally ordered sc phase, as in the case of pure C_{60}. Rietveld refinements[3,6] indicate that $Pa\bar{3}$ is the sc space group which gives the best fit to the data. With increasing x, first the tetrahedral sites are preferentially occupied by Na, then for $x > 2$ the octahedral sites become singly occupied but Na_{oct} moves off the high symmetry point $(1/2,0,0)$ along random (111) directions by a specific distance (y,y,y), placing it closer to some C_{60} molecules than others. This gives rise to orientational frustration, which becomes stronger as the octahedral occupancy increases (i.e. as x approaches 3) thus weakening the sc reflections.

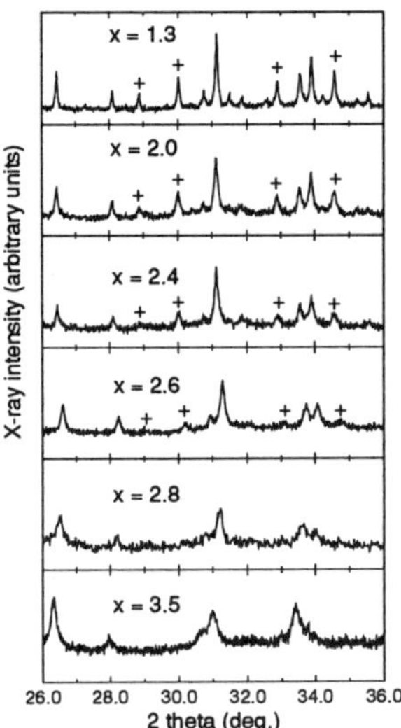

Fig. 3 Partial 300 K profiles of Na_xC_{60} for different x. The 4 strongest $Pa\bar{3}$ peaks are denoted by '+'.

At $x = 2.8$, or $f(Na_{oct}) = 0.8$, the frustration is strong enough to destroy the $Pa\bar{3}$ long range order, raising the symmetry to $Fm\bar{3}m$ in which the molecules are well-described by spherical shells.

The reason that Na stabilizes the $Pa\bar{3}$ structure while K and Rb do not can be understood as follows. Large tetrahedral ions (such as K or Rb) lock the C_{60}'s *at random* into one of two standard orientations due to repulsive core overlap. For smaller tetrahedral ions the Coulomb attraction dominates; in the present case, both the Na–C_{60} and the C_{60}–C_{60} interactions are optimized for a rotation angle of $\approx 22°$ about $< 111 >$ away from the standard orientations, so there is no frustration between these interactions and T_m is slightly increased relative to C_{60} by the Na–C_{60} contribution.

4. Effect of Orientational Order on Superconductivity

The $M_xM'_{3-x}C_{60}$ (M,M'=K,Rb,Cs) superconductors adopt the $Fm\bar{3}m$ structure with merohedral disorder[5], and T_c is controlled by $N(E_F)$, the density of states at the Fermi level within the framework of BSC theory, so that T_c increases with a according to a universal curve as shown in Fig. 4. Recently it has been found that, even though T_c of Na_2CsC_{60} lies on this universal curve, T_c of Na_2RbC_{60} is much lower than expected. We now address the reason for this anomaly.

We have shown that tetrahedral Na produces an ordered Pa$\bar{3}$ ground state, with T_m greater than that of C_{60}. This is due to the attractive Na–C_{60} Coulomb interaction. Since the Na in Na$_2$MC$_{60}$ is essentially confined to tetrahedral sites, we thus expect these ternary compounds to also adopt the Pa$\bar{3}$ structure (assuming that the octahedral M, larger than Na, remains at the high symmetry site thus making a negligible contribution to the Coulomb interaction). Indeed it has recently been shown that both Na$_2$RbC$_{60}$[7] and Na$_2$CsC$_{60}$ have the Pa$\bar{3}$ structure. Since $N(E_F)$ is sensitive to orientational order, one expects different "universal curves" for merohedral disorder (Fm$\bar{3}$m) and Pa$\bar{3}$ ordering. To support this idea, T_c versus a for Pa$\bar{3}$–ordered superconductors can be probed by hydrostatic pressure[9].

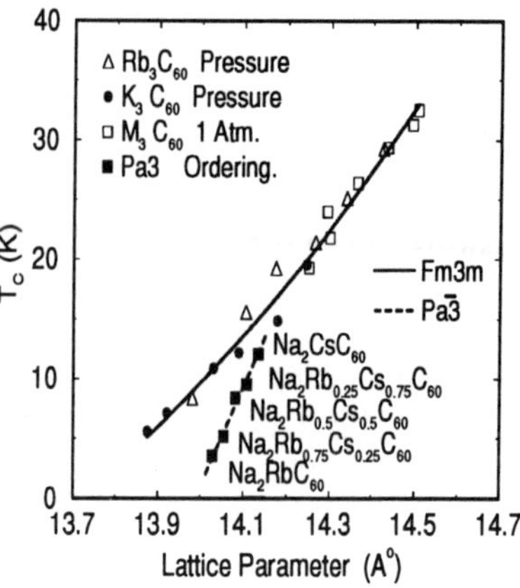

Fig. 4. T_c versus lattice constant for merohedrally-disordered (Fm$\bar{3}$m) and Pa$\bar{3}$ ordered superconductors.

Another approach which we have begun taking is to measure T_c and a as a function of x in quarternaries Na$_2$Rb$_x$Cs$_{1-x}$C$_{60}$.[6] As shown in Fig. 4, we indeed find a new curve with larger slope for Pa$\bar{3}$-ordered superconductors,[6] indicating that T_c, as controlled by $N(E_F)$, is a function of both orientational order/disorder and lattice constant (there being no reason to suppose that the coupling constant varies with x in this series of compounds). This not only explains the anomalously low T_c of Na$_2$RbC$_{60}$ but also encourages us to search for new compounds with Pa$\bar{3}$ symmetry which have larger lattice constant and consequently a higher T_c.

This work was supported by the National Science Foundation MRL Program under Grant No. DMR91-20668.

5. References

1. M. J. Rosseinsky *et al.*, Nature **356** 416 (1992).
2. T. Yildirim *et al.*, Nature **360**, 568 (1992).
3. T. Yildirim *et al.*, Phys. Rev. Lett. **71**, 1383 (1993).
4. K. Tanigaki *et al.*, Nature **356**, 419 (1991).
5. P. W. Stephens *et al.*, Nature **351**, 632 (1991).
6. T. Yildirim *et al.*, Science, submitted (1994).
7. K. Kniaz *et al.*, Solid State Comm. **88**, 47 (1993).
8. K. Prassides *et al.*, Science **263**, 950 (1994).
9. J. Mizuki *et al.*, preprint.

ALKALINE-EARTH DOPED FULLERITES

A. R. Kortan, N. Kopylov, E. Özdas,
A. P. Ramirez, R. M. Fleming, R.C. Haddon
AT&T Bell Laboratories, Murray Hill, NJ 07974, USA

and

K. M. Rabe
Yale Univ. New Haven, CT 06520, USA

ABSTRACT

Alkaline-earth (AE) metals calcium, strontium, and barium are intercalated in high purity C_{60} fullerite, to form $(AE)_x C_{60}$ compounds. AE-cation size is found to play a major role in stabilizing a particular structure, and the divalent character of the dopant AE atoms manifests itself in a smaller intermolecular separation and a complex, hybrid electronic band structure. Calcium intercalation leads to a solid solution fcc phase, where the calcium cations populate interstitial sites of the fcc fullerite. Here, the octahedral site becomes multiply occupied, and calcium atoms shift to (.41,.41,.41) diagonally opposite symmetry equivalent positions. Near the nominal composition x=5, fcc system phase transforms to a simple cubic phase and becomes a superconductor with T_c = 8.4K. Strontium intercalation leads to competing fcc and bcc A15 phases near x=3, and a superconducting (T_c = 4K) bcc phase near x=6. Largest size cation, barium forms only bcc phase compounds with fullerite, including an A15 phase, where the neighbouring molecules aligned orthogonal to each other such that interstitial sites are selectively surrounded by electron deficient five-membered rings of the molecules. The x=6 bcc phase of the barium is a superconductor with a T_c=7K. These results, combined with other experimental and theoretical studies suggest that, energy states including those contributing to the superconductivity in these novel materials are hybrid and complex in nature.

1. Introduction

Intercalating the C_{60}-fullerite with electron donor atoms leads to novel fulleride compounds with interesting electronic and structural properties. In particular, alkali-metals (A) Li, Na, K, Rb, Cs, and their alloys have been successfully intercalated[1] in fullerite, and their properties have been extensively studied. Most remarkably, the fcc $A_3 C_{60}$ compound with A=K, Rb, Cs and some of their alloys exhibit superconductivity at record high temperatures for organic compounds. A simple charge transfer picture from the cations to the molecular orbital derived band structure have been successful in qualitatively explaining the electronic properties.

Here, we describe the synthesis, structures and some properties of the next-column elements, alkaline-earth metals calcium, strontium and barium doped fullerites. Although $(AE)_x C_{60}$ compounds show some similarities to alkali-metal fulleride compounds, in general, they are more difficult to synthesize, reveal new structures, and exhibit more complex electronic properties.

240

2. Experimental

All samples are prepared in a controlled atmosphere glove box where the oxygen and water vapor levels are maintained below few parts per million. Starting materials, C_{60} was chromatographically purified, calcium metal was high purity (99.99%) and dendritically solidified, strontium metal was 99.9% pure (Johnson Matthey), and barium metal was also of high purity.

Alkaline-earth metals have relatively high melting temperatures, calcium 838 °C, strontium 769 ° and for barium 714 °C. At reaction temperatures, typically in the range 400-740 °C, C_{60} fullerite has significantly higher vapor pressure, and compound formation occurs through C_{60} gas - solid metal reaction. It is therefore important to have a fine grain size for starting metals, for compositional uniformity. Metal powders are produced by using custom made tantalum or tungsten files, and mixed with C_{60} powder in preweighted proportions. Iron containing files are found to severely contaminate the metal powders. Powder mixtures are then compacted by pressing into pellets in tantalum or tungsten cells. For calcium, best samples are prepared in tungsten cells. Samples are loaded into quartz tubes without removing them from the compaction cells, and sealed under a vacuum of 10^{-6} Torr. A total of 44 calcium, 40 barium and 30 strontium samples are prepared. Other samples were prepared by loading the quartz tubes with unpressed powder mixtures.

Metals are intercalated at temperatures 400 - 740°C, and for periods ranging from minutes to weeks. In calcium and strontium samples, C_{60} is found to decompose rapidly, at temperatures above 600 °C. Curiously, little decomposition is observed in barium samples even at temperatures as high as 800°C. An insitu x-ray intercalation furnace is also used to determine the right reaction conditions. Best reaction conditions are determined as, several hours at 550°C for calcium, few days at 550°C for strontium, and few days at 720-740°C for barium samples. Following this reaction step, powders are removed from the compaction or quartz cells, loaded into quartz x-ray capillaries, and again sealed under high vacuum.

Structures of various phase formations are monitored by powder x-ray diffraction measurements, with Mo K_{α} or Cu K_{α} radiation on a 12kW rotating anode generator equipped with a triple axis goniometer. Flat ZYA graphite crystals were used to monochromatize and analyze the x-ray beam with a longitudinal resolution of 10^{-2}Å^{-1}. Samples are rocked 6 degrees at each two-theta data point to obtain a statistically averaged powder pattern. Structural analysis of the x-ray powder intensities were carried out using the DBWS, and GSAS Rietveld refinement packages.

3. Results
3.1 Calcium Fulleride

Calcium intercalation[2] leads to the appearance of new fcc peaks in the x-ray diffraction pattern, indicating that calcium start diffusing into the interstitial sites of the fcc fullerite. Simultaneous with the appearance of these peaks, diffraction peaks shift towards larger angles continuously with increasing calcium incorporation into the lattice. This change in the lattice constant with intercalation proceeds rapidly and near $Ca_{1.5}C_{60}$ reaches a minumum at $a_0 = 14.02$ A. Further increase in calcium concentration upto Ca_6C_{60} does not cause significant changes in the lattice constant but somewhat effects the diffraction intensities. These indicate that the calcium-C_{60} system behaves like a solid-solution system, where some interstitial sites are randomly but uniformly occupied starting from very dilute limit. Considering the large amounts of calcium incorporated into the lattice, the two tetrahedral and one octahedral site per C_{60} occupancies need to be reexamined. X-ray diffraction pattern analysis reveals that the octahedral site is always occupied off-centered close to the (.41,.41,.41) position. This allows for a multiple occupancy of the octahedral site, since there are eight such corner sites for each octahedral site. Near Ca_5C_{60}, new diffraction peaks that belong to the primitive group appears and the system becomes a superconductor below 8.4 K. These peaks appear at small angles and therefore indicate that they may be due to a cation ordering rather than C_{60} ordering which alters the diffraction pattern mostly at large angles. Superconductivity is verified by magnetic field dependent microwave absorption and magnetic susceptibility measurements. A diamagnetic shielding fraction of 45% in 5 gauss, and a Meissner fraction fraction of 10% are measured at 4.5 K.

3.2 Strontium Fullerides

Strontium intercalation[3] leads to the appearance of fcc and bcc phases simultaneously, starting from dilute cation concentrations. This is due to the presence of a new, stable bcc-A15 phase which competes with the fcc phase in the same composition range. Both fcc and A15 phases have an equilibrium stochometry of Sr_3C_{60}. Earlier it was suggested[4,5] that, for x < 3 an fcc phase, and for x > 3, a bcc based A15 phase would be preferentially stabilized. We were not able to obtain single phase fcc samples by preparing samples with nominal compositions slightly lower than the nominal composition x=3, and the relative fraction of these two phases have essentially remained the same. A small energy difference between the two phases would explain the difficulties in separating these phases. Refinements were, therefore, carried out using coexisting two phases, and yielded reasonable results. For the nominal composition Sr_3C_{60} sample, we find that the fcc ($Fm\overline{3}$) phase has a lattice constant of 14.144 Å, and an off centered octahedral occupancy, similar to the calcium case. The octahedral site (4b) at 1/2, 1/2, 1/2 becomes displaced to 0.386, 0.386, 0.386 along the {111} axes at 32f sites. This and other refinements carried out on samples with different compositions suggest that, the system phase transforms from fcc to bcc-A15 at a dopant concentration that is less than three atoms per molecule, or before all fcc tetrahedral interstitial sites become fully occupied.

For the bcc-A15 (space group $Pm\overline{3}$) phase, we find a lattice constant of 11.140 Å, and a unity occupation for all sites. We have also investigated relative orientation of the molecules in separate refinements. In a true A15 structure (space group $Pm\overline{3}n$), the molecule at the body center position is rotated 90° with respect to its neighbours. Refinements could not make a clear distinction between different orientations, because of the disorder caused scattering in the experimental data. In contrast, a well ordered A15 phase with 90° rotationally ordered neighbouring molecules is observed in Ba_3C_{60}.

Increasing the strontium composition above x=3, we find a bcc $Im\bar{3}$ phase. At the nominal composition of Sr_6C_{60}, Rietveld refinements find a simple $Im\bar{3}$ phase, identical to the one found in alkali-metal fullerides. Refinements yield a lattice constant of, $a_0 = 10.975$, and a unity occupancy for all site. These samples also exhibited superconductivity below 4K, as measured by a SQUID magnetometer. The superconducting volume fraction reaches a maximum value of 6 %, at 2.2 K, and the Meissner fraction is measured as few percent for our best sample. This measured value for the shielding fraction should be taken as a lower bound, due to finite size effects in granular superconductors.

3.3 Barium Fullerides

In constrast to calcium fullerides which are fcc at all compositions, barium fullerides only exist in bcc-based structures[6,7]. These phases are an A15 (space group $Pm\bar{3}n$) for Ba_3C_{60}, and a bcc (space group $Im\bar{3}$) for Ba_6C_{60}. For the A15 phase, structure refinements[6] find a lattice constant of 11.33Å and unit occupancy for all sites. Additionally, the good quality of the data allows us to investigate the orientational ordering of the molecules. Here, the relative orientation of the near-neighbour molecules defines different environments for the cations in the interstitial sites. Best fits to the data are obtained when the near-neighbour molecules are rotated $90°$ with respect to each other, and the cations occupy only those sites which are surrounded by five-membered rings of the neighbouring molecules.

For the Ba_6C_{60} $Im\bar{3}$ phase, refinements[7] yield a lattice constant of 11.182Å, and cation positions of (0.22, 0.5, 0) and symmetrically equivalent positions. The size of a doubly ionized barium, 1.33 Å placed at (0.22, 0.5, 0) requires a lattice constant of 11.754 Å assuming spherically symmetric molecules. The relatively small lattice constant measured indicates a strong orbital overlap between the neighbouring molecules. Superconductivity below 7 K is observed in samples within a narrow range of compositions about x=6. The superconducting volume fraction is found as 12%, and a Meissner fraction of 2% for our best sample.

4. Discussion and Conclusion

Alkaline-earth and alkali metal intercalated fullerides form similar structures. The relative stability of a particular phase can qualitatively be understood by the relative size of the cations in the interstitial sites. The size of the dopant cation in the tetrahedral interstitial site (1.12Å for fullerite) is particularly very important. Small cations like Ca^{++} (0.99Å) and Na^+ (0.97Å) form only fcc based structures. Increasing the composition beyond three cations per molecule (two in the tetrahedral and one in the octahedral), these systems prefer multiply occupying the octahedral site rather than phase transforming to the bcc structure which can accommodate upto six cations per molecule at its interstitial sites. For intermediate size cations like Sr^{++} (1.14Å), K^+ (1.33Å), and Rb^+ (1.47Å) both fcc and bcc structures are stabilized at compositions below and above three cations per molecule respectively. Here, energetically it becomes more favorable to expand the lattice to a bcc lattice rather than forcing a multiple occupancy into the octahedral site. For large size cations like Ba^{++} (1.33Å) and Cs^+ (1.67Å), the system can only accommodate these cations in bcc structures, where the interstitial sites are 31% larger than the tetrahedral sites of the fcc structure. The ionic state of the cations also play an important role in determining the stability of different phases. Although Ba^{++} and K^+ are the same size, the lattice constant of Ba_6C_{60} is $0.21\overset{\circ}{A}$ smaller than K_6C_{60} one. A decrease in the inter-ball distance with increasing charge transfer was predicted earlier[8,9]. A comprehensive structural energy consideration[4,5,10] should include Madelung energy differences, orientation dependence of the C_{60}-cation interaction and relaxational energy gains induced by the distortion.

Alkaline-earth fullerides are a new and interesting family in fulleride compounds, with remarkably rich structures and properties. They form a unique A15 phase in barium fulleride, where molecules reorient themselves to surround the cations with their electron-deficient five-membered rings. Their octahedral site of the fcc phase in calcium and strontium fullerides is off-centered, and multiply occupied. They exhibit superconductivity at high doping levels, and in new symmetries. Their electronic properties[11-15] show a departure from the simple band-filling picture of the alkali-fullerides, and a strong hybridization takes place between the dopant and the fulleride bands.

5. References

1. For a recent review see, "Fullerenes", edited by H. Ehrenreich and F. Spaepen, Solid State Physics Vol. 48, Academic Press, Boston, 1994.

2. A.R. Kortan, N. Kopylov, S. Glarum, E.M. Gyorgy, A.P. Ramirez, R. M. Fleming, F.A. Thiel, R.C. Haddon, Nature, 355,529 (1992).

3. A.R. Kortan, N. Kopylov, E. Ozdas, A.P. Ramirez, R. M. Fleming, and R.C. Haddon, (preprint).

4. K. M. Rabe, J. C. Phillips, and J. M. Vandenberg, Phys. Rev. B, 47 (1993) 13067.

5. D. W. Murphy, J. Phys. Chem. Solids, 53 (1992) 1321.

6. A.R. Kortan, N. Kopylov, R.M. Fleming, O. Zhou, F.A. Thiel, R.C. Haddon, K.M. Rabe, Phys. Rev. B. 47, 13070 (1992).

7. A.R. Kortan, N. Kopylov, S. Glarum, E.M. Gyorgy, A.P. Ramirez, R. M. Fleming, O. Zhou, F.A. Thiel, P.L. Trevor, R.C. Haddon, Nature, 360,566 (1992).

8. S. Saito and A. Oshiyama, Phys. Rev. Lett. 66 (1991) 2637.

9. S. C. Erwin, M. R. Pederson, Phys. Rev. Lett. 67 (1991) 1610.

10. W. Andreoni, F. Gygi, and M. Parrinello, Phys. Rev. Lett. 68, 823(1992).

11. S. Saito and A. Oshiyama, Solid State Comm., 83, 107 (1992).

12. S. Saito and A. Oshiyama, Phys. Rev. Lett. 71 (1993) 121.

13. Y. Chen, D.M. Poirier, M.B. Jost, C. Gu, T.R. Ohno, J.L. Martins, J.H. Weaver, L.P.F. Chibante, R.E. Smalley, Phys. Rev. B15, Sept.(1992).

14. R.C. Haddon, G.P. Kochanski, A.F. Hebard, A.T. Fiory, R.C. Morris, Science 258, 1636 (1992).

15. R.C. Haddon, G.P. Kochanski, A.F. Hebard, A.T. Fiory, R.C. Morris, A.S. Perel, Chem. Phys. Lett. 203, 433 (1993).

Part Four

SINGLE CHARGED FULLERENES

ALKALI-METAL-FULLERIDE PHASE DIAGRAM DETERMINATION WITH X-RAY PHOTOEMISSION SPECTROSCOPY

D.M. POIRIER and J.H. WEAVER
Department of Materials Science and Chemical Engineering
University of Minnesota, Minneapolis, MN 55455 USA

ABSTRACT

The binary phase diagrams of the alkali-metal-fullerides, A_xC_{60}, are discussed in light of recent experimental results concerning the A_1C_{60} phases and their transformations. We proposed a eutectoid model for K_1C_{60} since this phase transforms into α-C_{60} + K_3C_{60} below 425 K. The A_1C_{60} phases of Rb- and Cs-fullerides transform from NaCl to non-cubic structures upon cooling and their phase diagrams require more involved solutions. The transformations in all these systems have been characterized with x-ray photoemission spectroscopy and phase diagrams are proposed.

1. Introduction

Equilibrium phase diagrams are invaluable to materials scientists developing processes for isolating phases or controlling microstructures. Phase diagrams are developed mainly empirically and are thus only possible to define after a considerable amount of experimental effort has been put forth, generally involving a variety of techniques. The fullerides of K and Rb are the most thoroughly studied systems in the fulleride family. Although it is a nontraditional technique for phase diagram study, x-ray photoemission spectroscopy (XPS) has provided valuable insight in the case of the fullerides. XPS was used to show that samples of K_xC_{60}, with compositions $0<x<3$ were composed of phase-separated K_3C_{60} and α-C_{60},[1] a conclusion subsequently confirmed by numerous spectroscopic studies.[2-5] This extremely important feature of the phase diagram proved difficult to observe using the more traditional technique of x-ray diffraction.[6,7]

In this report we present XPS data for the Rb_xC_{60} system in the region $0<x<3$, paying particular attention to the Rb_1C_{60} phase which exhibits a high temperature NaCl structure and a room temperature structure that is non-cubic.[6-8] We also discuss briefly the transformations observed upon cooling for Cs_1C_{60} and K_1C_{60}, reiterating that K_1C_{60} is not a thermodynamically stable phase at room temperature. This discussion culminates in a schematic phase diagram for the $x\sim1$ region.

2. Experimental Details

The XPS measurements were performed in ultra-high vacuum using monochromatized Al K_α radiation ($h\nu$=1486.6 eV). C_{60} was evaporated from Ta boats and condensed onto heated GaAs(110) substrates that were cleaved *in situ*. Film thicknesses of 400 to 1000 Å were used. During C_{60} deposition, the substrates were held at 450 K to produce highly crystalline samples.[9] A tungsten filament was used to heat the samples, and the temperature was measured with a chromel-alumel thermocouple attached to the sample holder. Fulleride formation was accomplished by exposing the C_{60} films, held at 450 K, to the flux from degassed SAES alkali dispensers at chamber pressures in the low 10^{-10} Torr range. After alkali vapor exposure, film compositions were measured as a function of time at 450 K. In-diffusion was signaled by decreasing relative intensity of the

248

alkali XPS features. K incorporation produced films that were unchanging at 450 K after less than one hour of annealing. Rb-containing films required longer to equilibrate and Cs-containing films showed the most sluggish in-diffusion, often requiring >12 hours of annealing to equilibrate. Some Rb_xC_{60} films with x<1 showed occupation of tetrahedral sites which is not expected until x>1. Annealing at temperatures >500 K caused depopulation of tetrahedral sites, but this was accompanied by fullerene desorption in a process that we call distillation.[10]

3. XPS as a Probe of Phase Composition

The utility of XPS for phase study in the fullerides is due to the ionic nature of these materials. In an XPS experiment the measured quantity is electron binding energy, which for ionic materials can be related to the electrostatic or Madelung potential at the ion site in the crystal.[11] In Fig. 1(a) we show the terms contributing to the binding energy of a core electron photoejected from an ionic solid referenced to the vacuum level, E_{VAC}. E_{FA} is the binding energy of a core electron in the free atom, which is determined by the nuclear charge and the other electrons in the atom. Ionizing the atom will shift the binding energy to that for the free ion, E_{FI}. The electron binding energy is increased in the case of a positive ion (as shown), or decreased in the case of a negative ion. Putting the ion in a crystal introduces a Madelung term, Φ, which shifts the binding energy to its value in the crystal E_C. (We have neglected a small term due to electron cloud repulsion for neighboring ions.) Since positive ions are surrounded by negative ions in the crystal, the effect of the Madelung term is generally to shift the binding energy from the free ion value toward the free atom value thus reducing the effect of ionization. These terms define the initial state of the system. The process of photoejection is accompanied by relaxation of the surrounding electrons in the solid toward the photohole, resulting in the final state shift E_R. The sum of these terms gives the binding energy measured by photoemission, E_{PE}.

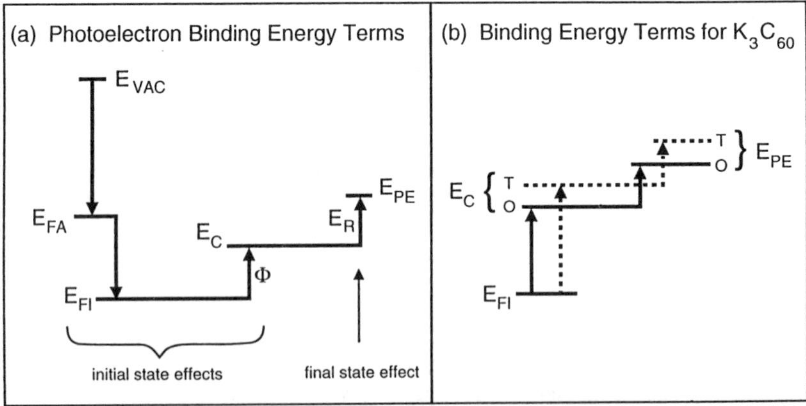

Fig. 1 Schematic representation of terms contributing to the binding energy of a core electron in an ionic solid for (a) the general case and (b) for K_3C_{60}. See the text for definitions of terms.

Different Madelung potentials lead to different binding energies for electrons originating from tetrahedrally- or octahedrally-coordinated K ions in K_3C_{60}. This is illustrated in Fig. 1(b). Calculations indicate a potential difference of 1.16 eV for ions in the two types of sites.[12] This energy difference will dictate the initial state binding energies and, given approximately equal final state shifts, the XPS spectrum is then expected to show the tetrahedral and octahedral site features separated by this amount. The experimentally-observed separation is 1.17 eV. If only one type of interstitial site is occupied in a given phase, only one photoemission feature should be expected. This is anticipated and found for the A_1C_{60}, A_4C_{60}, and A_6C_{60} compounds. However, since the crystal structures and their screening properties are different for these phases, the resulting Φ and E_R terms give different core electron binding energies, and thus unique signatures for each phase.

4. K_xC_{60}, $0<x<3$

We comment only briefly here on the K-C_{60} phase diagram since it has been studied with many techniques and discussed at length elsewhere. It has been demonstrated with XPS,[1,13,14] Raman spectroscopy,[4,15] NMR spectroscopy,[2,16] x-ray diffraction,[7] and recently with IR transmission spectroscopy[17] that there is phase separation between α-C_{60} and K_3C_{60} at room temperature. This changes above ~425 K when K_1C_{60} in the NaCl structure becomes stable.[6,7,13,15-17] A lower temperature was reported in Raman work with thin films,[15] but Winter and Kuzmany have found that the eutectoid temperature is 410-430 K for K incorporation in a C_{60} single crystal.[18] X-ray diffraction results have confirmed that the high temperature K_1C_{60} phase is in the NaCl structure.[6] The transformation of this phase into α-C_{60} + K_3C_{60} upon cooling through 425 K was examined extensively with XPS, and it was shown that the transformation is eutectoid in nature.[13] That work also indicated that it was possible to quench, to some extent, a K_1C_{60} phase at room temperature. While x-ray diffraction results suggested that a K_xC_{60} phase existed at room temperature over a large range of stoichiometries,[6] it has been determined that this structure is non-equilibrium.[7] The phase diagram for K_xC_{60} shown in Fig. 2, with the x<3 region based on spectroscopic results, remains consistent with experimental observations.

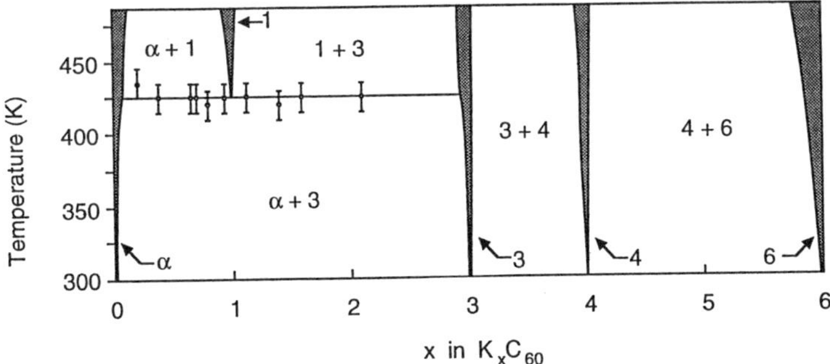

Fig.   2   Phase diagram of K_xC_{60} illustrating the eutectoid transformation of K_1C_{60} to α-C_{60} + K_3C_{60} below 425 K. Data points are measured transformation temperatures as described in Ref. 13. The compositional boundaries of the phases are schematic only.

5. Rb_xC_{60}, $0<x<3$

X-ray diffraction studies suggested that Rb_xC_{60} in the range $0<x<3$ was based on an fcc arrangement of C_{60} molecules.[19] The remaining question was how the Rb ions were distributed within this lattice. Here again, XPS was of value, showing that only one type of interstitial site was occupied for $x<1$.[14] This was interpreted in terms of Rb_1C_{60} in the NaCl structure. For $x>1$ the signature of Rb_3C_{60} began to emerge. We thus proposed phase separation between α-C_{60}, Rb_1C_{60}, and Rb_3C_{60} depending on the overall stoichiometry. More recent studies have shown that the Rb_1C_{60} structure is indeed the NaCl structure at high temperatures, but that the lattice is non-cubic at room temperature.[6-8] The crystal has variously been described as a rhombohedral distortion of the NaCl structure [6] or an orthorhombic structure.[8]

Motivated by the success of the technique for describing the K_xC_{60} system, we studied Rb_1C_{60} in detail with high resolution XPS. Examination of these data shows a subtle difference between spectra obtained for the two Rb_1C_{60} phases. Figure 3 shows the Rb 3d core level spectra measured at 480 and 300 K for a pure phase Rb_1C_{60} sample prepared by distillation.[10] The single spin-orbit-split doublet present at both temperatures indicates a single site for Rb ions in each structure. Significantly, there is a binding energy shift between spectra taken at the two temperatures. Such a shift could be attributed to lattice expansion that would occur continuously. However, as shown in Fig. 3(b), binding energy measurements at closely spaced temperature intervals show that the shift occurs abruptly and with hysteresis upon cooling or warming. This behavior indicates a phase transformation, as had been observed previously in other studies.[6,20-22]

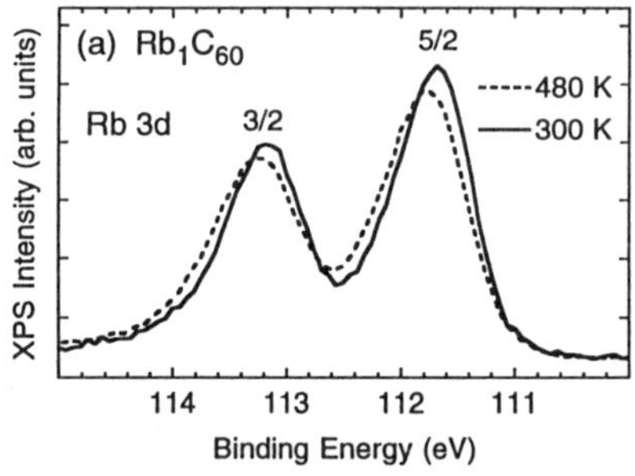

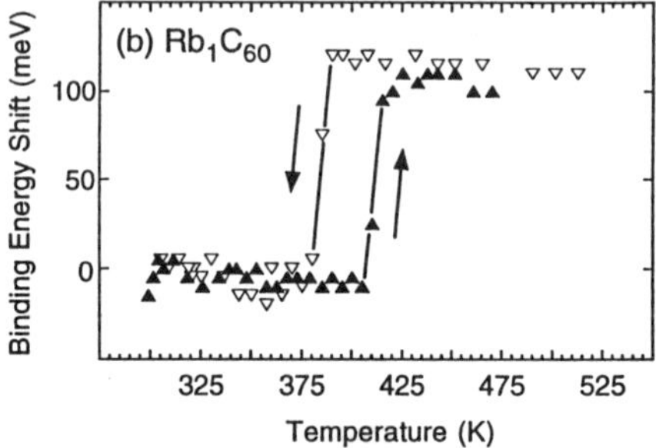

Fig. 3 (a) Rb 3d XPS spectrum of Rb_1C_{60} and (b) the binding energy shift of this feature as a function of temperature. The abrupt and hysteretic change in binding energy indicates a phase transformation. Lines through the data points are a guide to the eye.

Given that the transformation can be observed with XPS, we set out to track any shifts in the transformation temperature with composition. Such a shift was notably absent in K_xC_{60}, leading us to propose a eutectoid transformation for K_1C_{60}.[13] Results show that the behavior of Rb_1C_{60} is quite different. Figures 3-6 show spectroscopic results and measurements of the transformation for a variety of samples. There is some ambiguity in defining the transformation temperature, since it occurs over a finite temperature range.

Here, we will take the onset of the binding energy shift measured upon heating as the transformation temperature, and thus define a temperature of 405 K for Rb_1C_{60} from the results of Fig. 3(b).

In Fig. 4(a) we show the C 1s spectrum for a sample with composition $Rb_{0.5}C_{60}$. Such a sample is composed of a mixture of α-C_{60} and Rb_1C_{60}. The data points show the measured $Rb_{0.5}C_{60}$ spectrum, dashed lines are scaled spectra from α-C_{60} and Rb_1C_{60}, and the solid line is the sum of these two contributions. Taking the integrated areas under the two contributions and assuming compositions of x=0.0 for α-C_{60} and x=1.0 for Rb_1C_{60} gives a composition of x=0.6. This is within the experimental uncertainty of the composition x=0.5 determined from the Rb to C photoemission intensity ratio. The Rb 3d spectrum contains only a Rb_1C_{60} contribution since the α-C_{60} contribution is negligible. The Rb 3d core level binding energy shift is illustrated in Fig. 4(b). The same measurements were performed for a sample with composition $Rb_{0.2}C_{60}$ and the transformation shown in Fig. 4(c) was measured. Note that the transformation temperature is identical for these two compositions, with an onset at 390 K, even though the overall Rb concentration has decreased by a factor of 2.5 from (b) to (c). This provides additional evidence for the phase separation suggested by the C 1s decomposition since the transformation temperature of Rb_1C_{60} in equilibrium with α-C_{60} should be constant regardless of the relative amounts of the two species. In contrast, this observation does not bode well for the suggestion[6,19] that a solid solution exists for 0<x<1 since the transformation temperature should be affected by compositional variation.

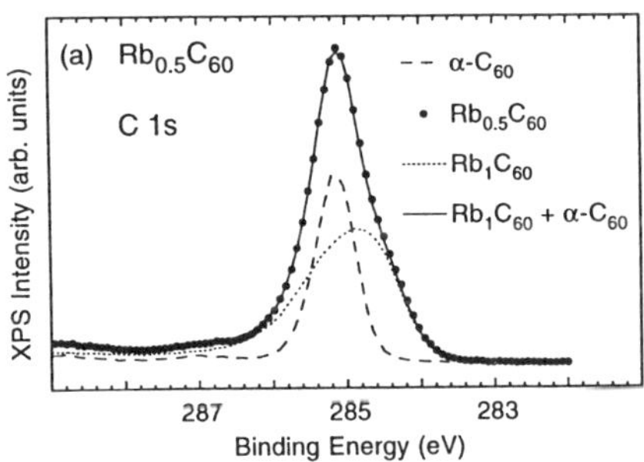
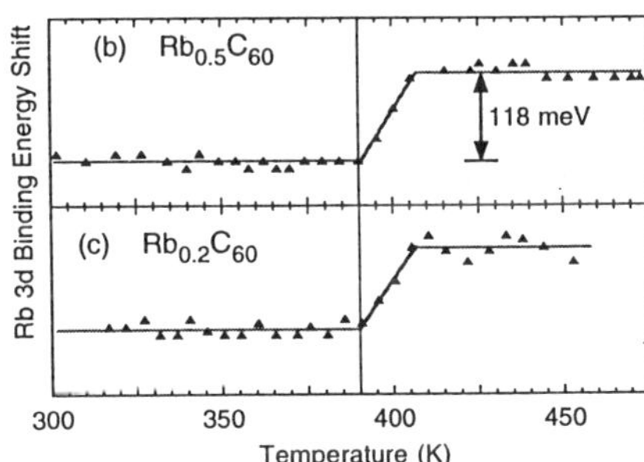

Fig. 4 (a) C 1s spectrum for $Rb_{0.5}C_{60}$ decomposed into its α-C_{60} and Rb_1C_{60} components. (b) The binding energy shift of the Rb 3d XPS feature for $Rb_{0.5}C_{60}$ measured upon heating. (c) As in (b) for $Rb_{0.2}C_{60}$. Identical transformation temperatures are found for the two compositions.

At higher compositions, 1<x<3, Rb_1C_{60} will be in equilibrium with Rb_3C_{60} and will have a different chemical potential than when in equilibrium with α-C_{60}. This requires a range of stoichiometries for the Rb_1C_{60} phase that may be reflected in the transformation temperature. Indeed, evidence for this is already apparent from Figs. 3 and 4 since the transformation temperature for x=1 was 15 K higher than for x=0.5 or 0.2. A sample with composition x=1.5 was examined to further explore this issue. In this case the Rb 3d spectrum contains contributions from Rb_3C_{60} as well as Rb_1C_{60}. The spectrum is decomposed in Fig. 5(a) into its Rb_1C_{60} and Rb_3C_{60} components. Fitting this spectrum with three peaks A, B, and C, we see the anticipated binding energy shifts [Fig. 5(b)] only for components A and B which have a Rb_1C_{60} contribution. Peak C arises entirely from the Rb_3C_{60} contribution and shows no abrupt shift. The curve superimposed on the data points for peak C shows the expected continuous shift due to lattice expansion. This curve was derived using the Madelung energies calculated by Zhang *et al.*,[12] the thermal expansion data of Zhou *et al.*,[23] the polarizability for C_{60}^{3-} determined by Quong and Pederson,[24] and the model for the screening term suggested by Lof *et al.*[25] for pure C_{60} which includes only nearest neighbor terms. The shape of the curve is defined entirely by these quantities while an adjustable offset defines the room temperature binding energy. The agreement between this curve and the data indicates that the Rb_3C_{60} portion of the sample behaves as a single phase, expanding and contracting with temperature changes.

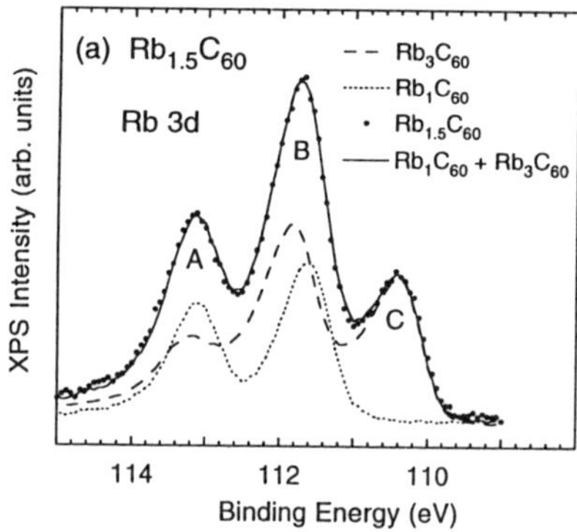

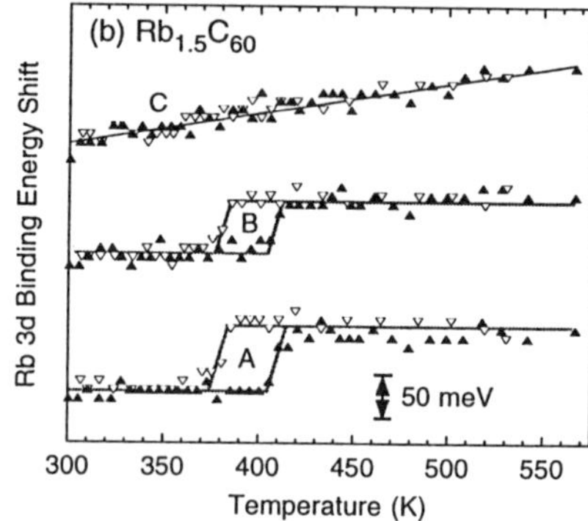

Fig. 5 (a) Rb 3d spectrum for $Rb_{1.5}C_{60}$ decomposed into its Rb_1C_{60} and Rb_3C_{60} components. The spectrum exhibits three peaks A, B, and C. In (b) the binding energies of these components are plotted versus temperature. Peaks A and B contain Rb_1C_{60} contributions and exhibit abrupt shifts (lines are a guide to the eye). Peak C exhibits a continuous shift at a rate consistent with calculated thermal expansion (curve shows calculation).

In Fig. 6, we show the Rb 3d spectrum for a Rb_3C_{60} sample together with binding energies of the three components measured as a function of temperature. At this composition there are no abrupt binding energy shifts, consistent with a negligible amount of Rb_1C_{60}. Lines through the data are calculated lattice expansion shifts, as in Fig. 5(b). We note also that while the phase transformation in Rb_1C_{60} exhibits very subtle effects in the XPS spectra, any transformation in Rb_3C_{60} is apparently even more subtle. We are thus unable to detect the inequivalence of tetrahedrally coordinated Rb ions reported in NMR studies of Rb_3C_{60} below ~370 K.[26]

The x=1 region of the equilibrium Rb_xC_{60} phase diagram can be mapped out using the data presented in Figs. 3-5. We use the abbreviations α, 1, 1′, and 3 to denote α-C_{60}, Rb_1C_{60} in the NaCl structure, Rb_1C_{60} in its equilibrium low temperature structure, and Rb_3C_{60}. The boundary conditions imposed by the XPS data are: (1) coexistence of 1 with α for x<1 and T>390 K, (2) coexistence of 1′ with α for x<1 and T<390 K, (3) coexistence of 1 with 3 for x>1 and T>405 K, and (4) coexistence of 1′ with 3 for x>1 and

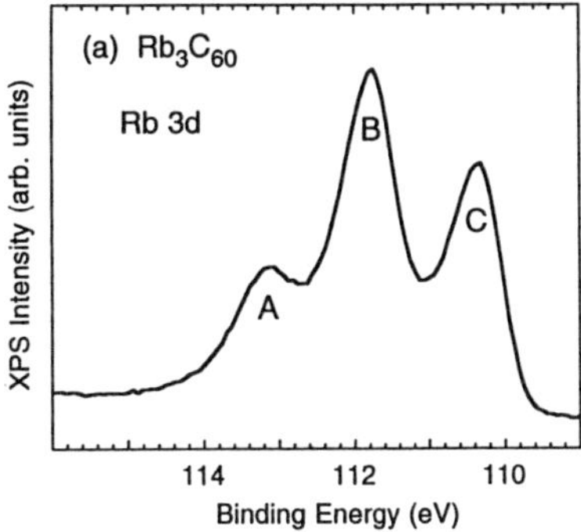

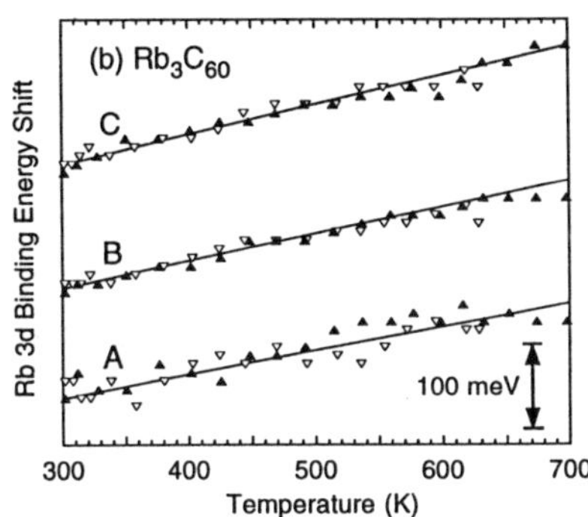

Fig. 6 (a) Rb 3d spectrum of Rb_3C_{60} and (b) the binding energy shifts of the three peaks as a function of temperature. Curves through the data points are calculated shifts due to lattice expansion.

T<405 K. We can satisfy these boundary conditions with the phase diagram illustrated in Fig. 7. This construction is the simplest topology that provides the observed two-phase regions and is consistent with the thermodynamic underpinnings of phase equilibria. The Rb_1C_{60} phase transformation is revealed to consist of two phase transformations. A eutectoid defines the low temperature limit of stability for the NaCl structure, while a peritectoid defines the high temperature limit of the non-cubic structure.

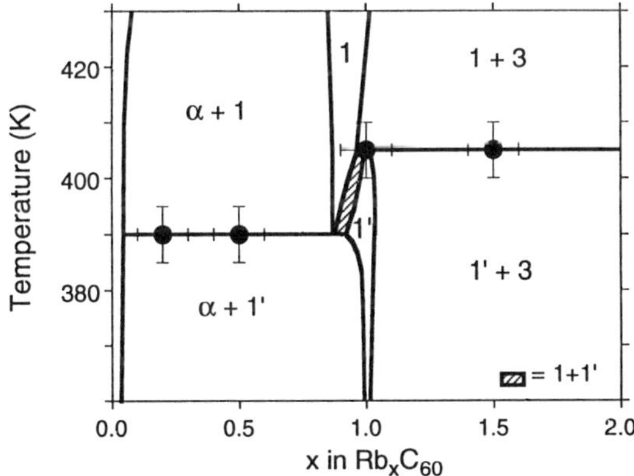

Fig. 7 Proposed phase diagram for Rb_xC_{60}. 1 and 1′ are the high temperature and low temperature phases, respectively. The shaded region corresponds to coexistence of 1 and 1′. The 1 phase terminates, upon cooling, with a eutectoid temperature. The 1′ phase terminates, upon heating, with a peritectoid transformation. The invariant temperatures are 390 K and 405 K.

254

6. Cs_xC_{60}, $0<x<4$

Like Rb_1C_{60}, Cs_1C_{60} was shown to be stable at room temperature,[14] and was shown to transform from the NaCl structure to a non-cubic structure upon cooling to room temperature.[6] The corresponding 1′-1 transformation is also observable with XPS, and studies analogous to those discussed for Rb_xC_{60} have been carried out for Cs_xC_{60}. The energy shifts of the Cs $4d_{5/2}$ core level for three different compositions are shown in Fig. 8(a). These were measured for the heating half of the temperature cycle. At the compositions x=0.8, 1.0, and 1.3, the transformation onsets were 345, 390 and 410 K. This suggests that the Cs_1C_{60} portion of the Cs-C_{60} phase diagram possesses the same topology as that for Rb-C_{60} as drawn in Fig. 8(b). A notable exception must be that the Cs_4C_{60} phase would be in equilibrium with the 1 and 1′ phases of Cs-C_{60} for x>1, as shown, since there is no stable Cs_3C_{60} phase. Another obvious difference is that the two invariant temperatures are separated by 65 K for Cs_1C_{60} compared to 15 K for Rb_1C_{60}. This difference is probably correlated to the fact that a Rb_3C_{60} phase is stable while Cs_3C_{60} is not, and it may imply that the compositional width of Cs_1C_{60} is significantly larger than Rb_1C_{60}. Experimentally, we see that the eutectoid temperature is suppressed moving from K_1- to Rb_1- to Cs_1C_{60}. This would appear to reflect an increasingly favorable match of the cation size to the octahedral interstitial site, thus stabilizing the cubic NaCl structure.

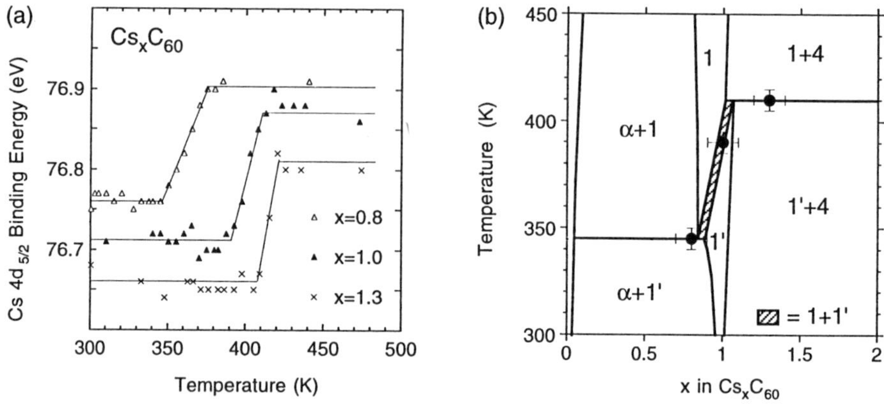

Fig. 8 (a) Binding energies measured for the Cs $4d_{5/2}$ photoemission feature of Cs_xC_{60} and (b) the proposed phase diagram for Cs_xC_{60} in the region of x=1. The separation between eutectoid and peritectoid temperatures is considerably larger than that found for Rb_xC_{60}.

We note that rather broad transitions have been reported for the 1 to 1′ phase transition as measured by differential scanning calorimetry[6] or NMR spectroscopy.[20] The transformations measured with XPS are comparatively sharp, but the transformation onsets for different x-values occur over a range of temperatures that is at least 65 K. We therefore suggest that the powder samples employed for the other measurements may be inhomogeneous and exhibit a range of local x-values and transformation temperatures.

This would not be surprising given the difficulty associated with distributing Cs uniformly in film samples, a difficulty that will be compounded for large grained powders or single crystals.

7. Conclusions

The A_xC_{60} binary phase diagrams exhibit considerably more structure than originally anticipated and spectroscopic work, XPS in particular, has provided key information. The results described here for Rb_xC_{60} and Cs_xC_{60} have allowed the construction of the x~1 portion of the phase diagrams. This picture is considerably different than that for K_1C_{60} which has no stable structure at room temperature. Taken together, the results for the various fullerides point to the fine energy balance that exists upon filling the octahedral site of the C_{60} lattice with ions of increasing size. The NaCl structure distorts at lower temperatures for Cs_1C_{60} than for Rb_1C_{60}, presumably because Cs is a better match to the size of the octahedral site. The K ion has the poorest size match and even the distorted K_1C_{60} phase is not stable at low temperatures.

8. Acknowledgements

This work was supported by the National Science Foundation. D.M.P. acknowledges partial support by a University of Minnesota Graduate School Fellowship. Purified C_{60} was generously provided by R.E. Smalley and L.P.F. Chibante.

9. References

1. D.M. Poirier, T.R. Ohno, G.H. Kroll, Y. Chen, P.J. Benning, J.H. Weaver, L.P.F. Chibante, and R.E. Smalley, Science **253**, 429 (1991).
2. R. Tycko, G. Dabbagh, M.J. Rosseinsky, D.W. Murphy, R.M. Fleming, A.P. Ramirez, and J.C. Tully, Science **253**, 884 (1991).
3. C.T. Chen *et al.*, Nature **352**, 603 (1991).
4. T. Pichler, M. Matus, J. Kürti, and H. Kuzmany, Phys. Rev. B **45**, 13841 (1992).
5. W.L. Wilson, A.F Hebard, L.R. Narasimhan, and R.C. Haddon, Phys. Rev. B **48**, 2738 (1993).
6. Q. Zhu, O. Zhou, J.E. Fischer, A.R. McGhie, W.J. Romanow, R.M. Strongin, M.A. Cichy, and A.B. Smith III, Phys. Rev. B **47**, 13948 (1993).
7. J.E. Fischer and P.A. Heiney, J. Phys. Chem. Solids **54**, 1725 (1993).
8. O. Chauvet, G. Oszlányi, L. Forro, P.W. Stephens, M. Tegze, G. Faigel, and A. Jánossy, Phys. Rev. Lett. **72**, 2721 (1994).
9. Y.Z. Li, M. Chander, J.C. Patrin, J.H. Weaver, L.P.F. Chibante, and R.E. Smalley, Science **253**, 429 (1991); Y.B. Zhao, D.M. Poirier, and J.H. Weaver, J. Phys. Chem. Solids **54**, 1685 (1993).
10. D.M. Poirier, Appl. Phys. Lett. **64**, 1356 (1994); M. Knupfer, D.M. Poirier, and J.H. Weaver, Phys. Rev. B **49**, 8464 (1994); D.M. Poirier, D.W. Owens, and J.H. Weaver, Phys. Rev. B (submitted).
11. P.H. Citrin and T.D Thomas, J. Chem. Phys. **57**, 4446 (1972).
12. W. Zhang, H. Zheng, and K.H. Bennemann, Solid State Commun. **82**, 679 (1992).
13. D.M. Poirier and J.H. Weaver, Phys. Rev. B **47**, 10959 (1993).

14. D.M. Poirier, T.R. Ohno, G.H. Kroll, P.J. Benning, F. Stepniak, J.H. Weaver, L.P.F. Chibante, and R.E. Smalley, Phys. Rev. B **47**, 9870 (1993).
15. J. Winter and H. Kuzmany, Solid State Commun. **53**, 1321 (1992).
16. R. Tycko, J. Phys. Chem. Solids **54**, 1713 (1993).
17. D. Koller, M.C. Martin, and L.Mihaly, preprint.
18. J. Winter and H. Kuzmany, private communication.
19. Q. Zhu, O. Zhou, N. Coustel, G.B.M. Vaughan, J.P. McCauley, W.J. Romanow, J.E. Fischer, and A.B. Smith III, Science **254**, 545 (1991).
20. R. Tycko, G. Dabbagh, D.W. Murphy, Q. Zhu, and J.E. Fischer, Phys. Rev. B **48**, 9097 (1993).
21. A. Jánossy, O. Chauvet, S. Pekker, J.R. Cooper, and L. Forró, Phys. Rev. Lett. **71**, 1091 (1993).
22. M.C. Martin, D. Koller, X. Du, P.W. Stephens, and L. Mihaly, Phys. Rev. B **49**, 10818 (1994).
23. O. Zhou, Q. Zhu, G.B.M. Vaughan, J.E. Fischer, P.A. Heiney, N. Coustel, J. McCauley Jr., A.B. Smith III, and D.E. Cox, Mat. Res. Soc. Symp. Proc. Vol. **270**, 191 (1992).
24. A.A. Quong and M.R. Pederson, Mat. Res. Soc. Symp. Proc. Vol. **270**, 209 (1992).
25. R.W. Lof, M.A. van Veenendaal, B. Koopmans, H.T. Jonkman, and G.A. Sawatsky, Phys. Rev. Lett. **68**, 3924 (1992).
26. R.E. Walstedt, D.W. Murphy, and M. Rosseinsky, Nature **362**, 611 (1993).

STRUCTURAL STUDY OF Rb_1C_{60} AND K_1C_{60} FULLERIDES

M. TEGZE, G. BORTEL, G. FAIGEL
Research Institute for Solid State Physics,
H-1525 Budapest, POB. 49, Hungary

L. FORRÓ, G. OSZLÁNYI[†]
Laboratorie de Physique des Solides Semicrystallins, IGA,
Department de Physique,
Ecole Polytechnique Federale de Lausanne,
CH-1015 Lausanne, Switzerland

A. JÁNOSSY
Technical University of Budapest, H-1521 Budapest, Hungary and
Research Institute for Solid State Physics,
H-1525 Budapest, POB. 49, Hungary

P. W. STEPHENS
Department of Physics, State University of New York,
Stony Brook, NY 11794-3800, USA

ABSTRACT

The room temperature equilibrium structures of the Rb_1C_{60} and K_1C_{60} have been studied by x-ray diffraction. It is shown that both alkali fullerides posess a pseudo body centered orthorhombic lattice. The $C_{60}-C_{60}$ intermolecular separation is the shortest among all the known alkali-fullerides. The phase transition of Rb_1C_{60} from the high temperature fcc to the room temperature orthorhombic phase is followed by x-ray diffraction and it shows the characteristics of a typical first order phase transition.

1. Introduction

Among the alkaline metal doped fullerenes many phases appear with different compositions. Most of the efforts were concentrated on the A_3C_{60} type compounds since they show superconductivity. However, recently the interest turned to the A_1C_{60} compounds, since they have unusual phase transitions accompanied with drastic changes of the transport properties[1-5].

Zhu et al.[2] have prepared first single-phase samples of Rb_1C_{60} as well as K_1C_{60} and Cs_1C_{60} . They find that these have an fcc (rock salt) structure at elevated temperatures, and suggested a rhombohedrally distorted fcc lattice for both Rb_1C_{60} and K_1C_{60} at room temperature, but structural details were not given.

[†] on leave from the Research Institute for Solid State Physics, H-1525 Budapest, POB. 49, Hungary

In this work we present detailed structural studies of Rb_1C_{60} and K_1C_{60}. We show that the room temperature lattice of Rb_1C_{60} and K_1C_{60} is orthorhombic with an unusually short C–C intermolecular separation (9.1 Å).

2. Experiments

Samples were prepared by a solid state reaction of high purity C_{60} powder and alkali metal in sealed quartz capillaries at 260 °C for 80 days. The samples used in our studies had nominal compositions: $Rb_{0.5}C_{60}$, Rb_1C_{60} and K_1C_{60}. Parallel beam x-ray powder diffraction measurements were performed at the X3B1 beamline of the Brookhaven National Synchrotron Light Source.

3. Results

The high temperature (465 K) diffraction pattern of the Rb_1C_{60} sample is shown in Fig. 1. Three phases can be clearly distinguished: pure C_{60}, Rb_1C_{60} and Rb_3C_{60}. All three have fcc structures with lattice parameters

$$a(C_{60}) = 14.172(5) \text{ Å},$$
$$a(Rb_1C_{60}) = 14.077(5) \text{ Å},$$
$$a(Rb_3C_{60}) = 14.466(5) \text{ Å}.$$

The room temperature diffraction data of the same sample leads to surprising new results. Fig. 2 shows the powder pattern of the sample after slow cooling to room temperature. Three phases can be identified again: pure C_{60}, a new orthorhombic phase of Rb_1C_{60} and Rb_3C_{60}. The refined lattice

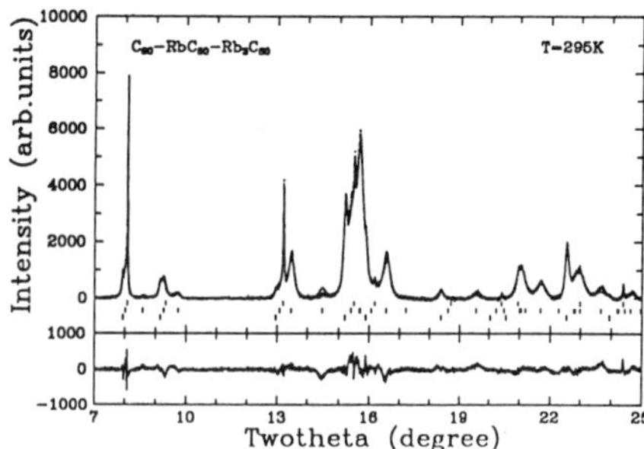

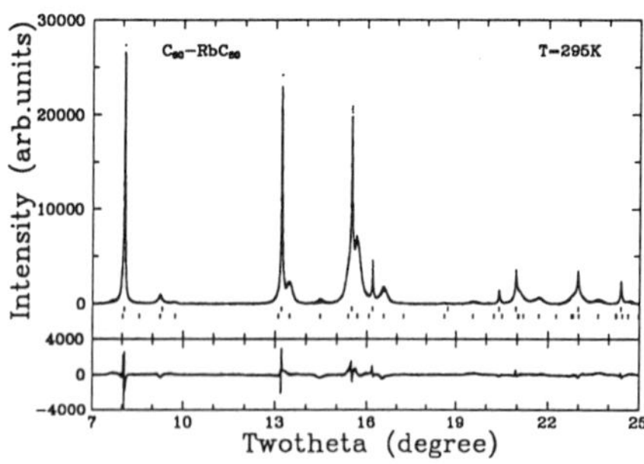

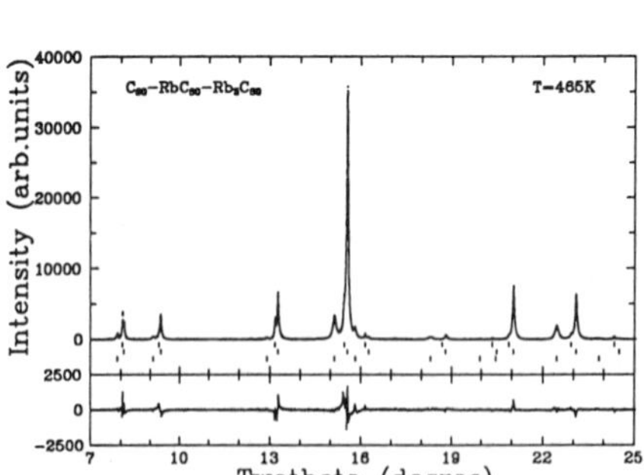

Fig. 1. The diffraction pattern and the Rietveld fit of the Rb_1C_{60} sample at 465 K.

Fig. 2. The diffraction pattern and the Rietveld fit of the Rb_1C_{60} (upper panel) and the $Rb_{0.5}C_{60}$ (lower panel) samples at room temperature.

parameters are: $a(C_{60}) = 14.152(5)$ Å, $a(Rb_1C_{60}) = 9.134(10)$ Å, $b(Rb_1C_{60}) = 10.099(10)$ Å, $c(Rb_1C_{60}) = 14.234(10)$ Å and $a(Rb_3C_{60}) = 14.417(5)$ Å. In the orthorhombic Rb_1C_{60} structure (see Fig. 3) the 9.13 Å interfullerene separation along a is unusually short. The most likely space groups are Immm or Pnnm, depending on the orientation of the fullerene molecules. The short intermolecular separation suggests chemical bonding (formation of polymer chains along the (100) direction[6]) which would be compatibile with space group Immm.

In the case of K_1C_{60} a result similar to Rb_1C_{60} was obtained (Fig. 4). The room temperature equilibrium phase of K_1C_{60} is pseudo body centered orthorhombic with lattice parameters: a = 9.093(10) Å, b = 9.994(10) Å and c = 14.425(10) Å.

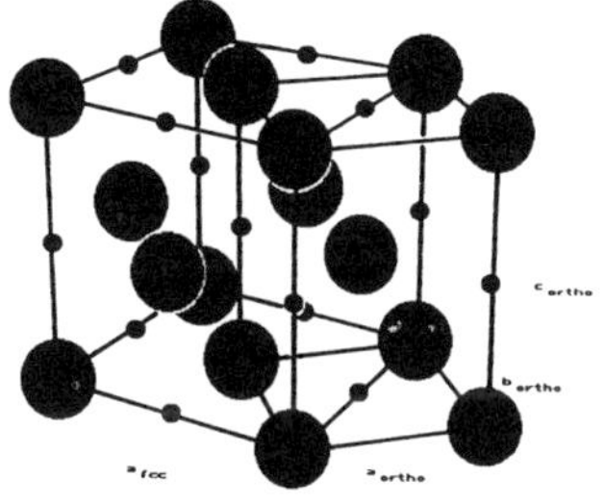

Fig. 3. The connection between the high temperature fcc and the room temperature orthorhombic cell.

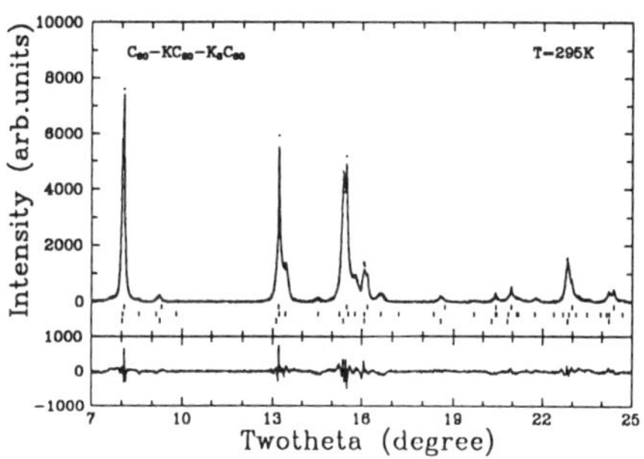

Fig. 4. The diffraction pattern and the Rietveld fit of the K_1C_{60} sample at room temperature.

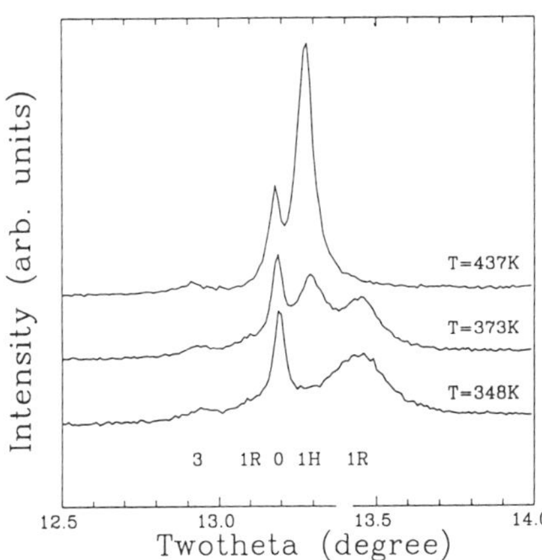

Fig. 5. A characteristic 2Θ range of the powder pattern of the Rb_1C_{60} sample above, at, and below the phase transition temperature.

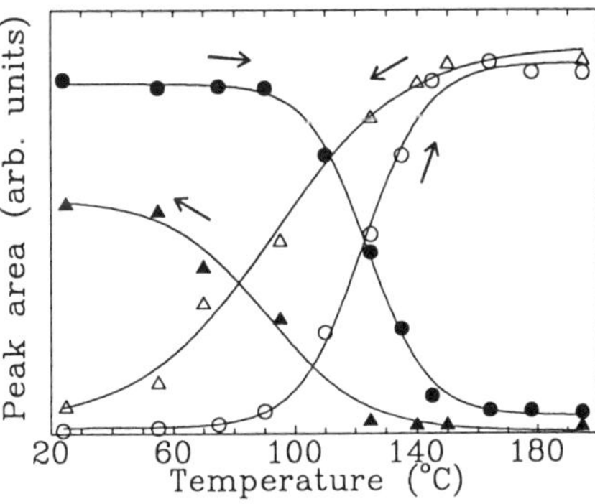

Fig. 6. The peak areas of the fcc Rb_1C_{60} on heating (empty circles), on cooling (empty triangles) and the orthorhombic Rb_1C_{60} on heating (full circles), on cooling (full triangles) versus the temperature.

In order to determine how the phase transition takes place, we carried out a series of short x-ray diffraction scans as a function of the temperature on the Rb_1C_{60} sample. Three typical spectra are shown in Fig. 5. The phase transition takes place at 380 K. The evolution of the phase transition can be followed in Fig. 6 where the integrated intensity of the peaks corresponding to the high temperature fcc and room temperature orthorhombic phases are shown as a function of the temperature. The hysteretic difference between the data collected on heating and cooling is characteristic to a first order phase transition.

4. Summary

In this paper we showed that the room temperature equilibrium structure of both Rb_1C_{60} and K_1C_{60} is orthorhombic. This result is based on detailed Rietveld refinement studies. In these structures the intermolecular separation is the shortest among all the known C_{60} compounds. The phase transition from the high temperature fcc to the room temperature orthorhombic structure is shown to be of first order.

5. Acknowledgements

This work has been supported by OTKA grants under contract numbers 2943, 2932, 2979 and E–012313. The SUNY X3 beam line at NSLS is supported by the DOE under grant No. DEFG-0286-ER-45231. Part of this research was carried out at the National Synchrotron Light Source which is supported by U.S. DOE, Division of Material Sciences and Division of Chemical Sciences. Work at SUNY was supported by the National Science Foundation Division of Materials Research under grant No. DMR 92-02528.

6. References

1. J. Winter and H. Kuzmany, *Solid State Commun.* **84** (1992) 935.

2. Q. Zhu, O. Zhou, J. E. Fischer, A. R. McGhie, W. J. Romanow, R. M. Strongin, M. A. Chicy and A. B. Smith III, *Phys. Rev.* **B47** (1993) 13948.

3. D. M. Poirier and J. H. Weaver, *Phys. Rev.* **B47** (1993) 10959.

4. A. Jánossy, O. Chauvet, S. Pekker, J. R. Cooper and L. Forró, *Phys. Rev. Lett.* **71** (1993) 1091.

5. R. Tycko, G. Dabbagh, D. W. Murphy, Q. Zhu and J. E. Fischer, *Phys. Rev.* **B48** (1993) 9097.

6. S. Pekker, L. Forró, L. Mihály and A. Jánossy, *Solid State Commun.* **90** (1994) 349.

PHASE TRANSITION IN A Rb_1C_{60} SINGLE CRYSTAL FROM IR REFLECTIVITY

T. Pichler, R. Winkler and H. Kuzmany

Institut für Festkörperphysik, Universität Wien
Strudlhofgasse 4, A-1090 Vienna, Austria

ABSTRACT

In situ IR reflectivity was measured for the (111) surface of a Rb doped C_{60} single crystal. By iterative doping and annealing processes we obtained a nearly pure phase of Rb_1C_{60} with a thickness of about 4 μm. The change of the four F_{1u} modes was analysed as a function of temperature. A splitting was observed within a temperature range of 2 K at 404/415 K for a cooling and heating cycle, respectively. A qualitative analysis of the reflectivity spectra revealed the line positions for the F_{1u} modes. From a quantitative analysis for the $F_{1u}(2)$ and the $F_{1u}(4)$ mode a good agreement with the charged phonon model with respect to line shift and plasma frequency and with respect to previous measurements on thin films is found.

1. Introduction

Doping C_{60} with alkali metals leads to an accommodation of the latter in octahedral and tetrahedral lattice sites. For A_3C_{60} a metallic phase with fcc crystal structure[1] and for further doping a semiconducting phase A_4C_{60} with a bct and A_6C_{60} with a bcc lattice[2] is formed. An A_1C_{60} phase with fcc rocksalt structure was found by Raman, X-ray and PES and IR experiments[3-6]. Whereas for K_1C_{60} on cooling a phase seperation into C_{60} and K_3C_{60} was observed[3], very recent X-ray experiments on Rb_1C_{60} revealed a first order phase transition at 350 K to an orthorombic phase Rb_1C_{60} on slowly cooling the sample. Quenching experiments revealed a metastable low temperature fcc phase[7,8].

We performed detailed temperature dependent measurements of the IR reflectivity in the range between 400 and 5000 cm^{-1} for a doped Rb_1C_{60} single crystal to analyse this first order phase transition by IR spectroscopy.

2. Experimental

The single crystal we used was vacuum grown as described previously[9]. It had the shape of a parallel plate with an (111) face of about 4×5 mm^2 and a thickness of 0.9 mm. The sample was fixed with silver paste to the copper block of a IR cell which was inserted in a BRUKER 66V spectrometer and allowed temperature dependent measurements between 80 and 460 K. Doping was performed from Rb vapor generated in the cell by heating purified Rb to 390 K. Two electrical contacts were provided on the crystal for a measurement of the resistance of the sample during the doping process. The doping and equilibration process was done in the same way as already reported[10]. The reflectivity of the C_{60} single crystal was measured with a resolution of 0.5 cm^{-1} in the MIR region between 5000 cm^{-1} and 400 cm^{-1}. All temperature dependent measurements were done by slowly cooling (heating) the sample.

3. Results and Discussion

Fig. 1 shows the reflectivity spectra of Rb_1C_{60} for three different temperatures. The line positions of the F_{1u} modes in the high temperature Rb_1C_{60} rocksalt structure were identified from a comparison to thin film measurements[10]. At temperatures below the fcc-orthorombic phase transition dramatic changes in the spectra are found. A splitting of the modes into three components and some new second order structures indicated by the asterics occur. The lineshape of the $F_{1u}(3)$ mode at $1182\,cm^{-1}$ at low temperatures is not fully understood at the moment. The difference to the other F_{1u} modes may be due to the very small oscillator strength of this mode. Therefore this mode is only visible in reflectivity due to interference effects.

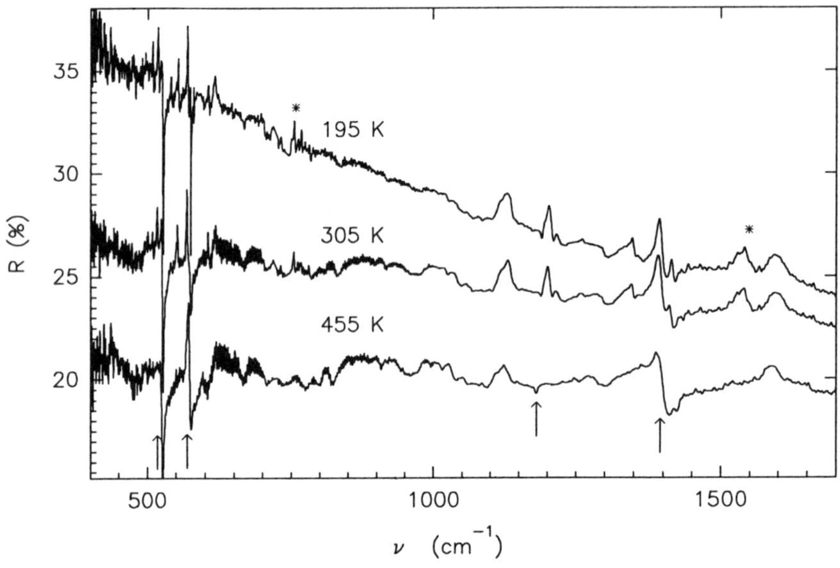

Fig. 1: Reflectivity spectra of Rb_1C_{60} in the range of the 4 F_{1u} modes for three different temperatures. The arrows indicate the positions of the 4 F_{1u} modes.

To analyse the transition in greater detail we performed temperature dependent measurements with high resolution. The temperature dependence of the resistance of the single crystal is shown in Fig. 2. The resistance shows a weakly activated temperature dependence with a jump at the first order phase transition at 404 K by slowly cooling and at 415 K by slowly heating the sample. At this transition the dramatic changes in the reflectivity spectra occur within only two degrees. This is shown in Fig. 3(a,b) for the $F_{1u}(1,2)$ and the $F_{1u}(4)$ modes. In addition to the

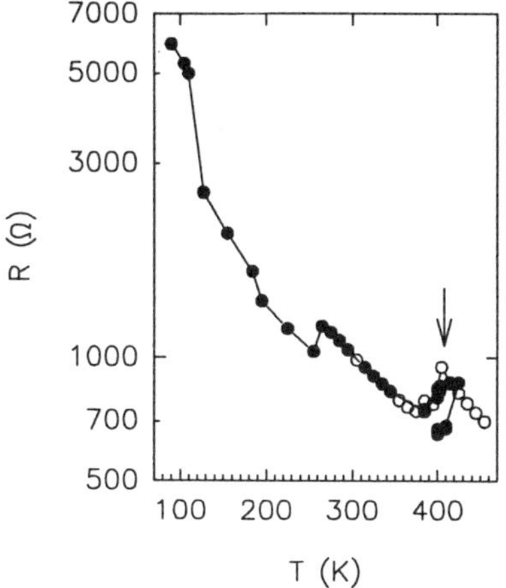

Fig. 2: Temperature dependence of the resistance of Rb_1C_{60} single crystal. The arrow indicates the phase transition. The first cooling process is shown by the open circles. The additional heating and second cooling process is indicated by the full circles.

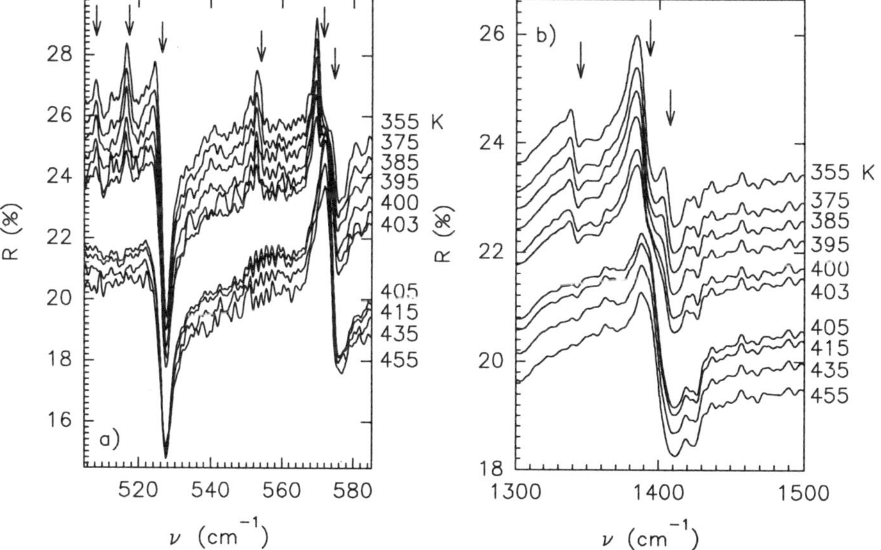

Fig. 3: Reflectivity spectra of a Rb_1C_{60} in the extended range of the $F_{1u}(1,2)$ (a) and of the $F_{1u}(4)$ (b) modes for temperatures indicated by the numbers on the right side of the figures. The arrows indicate the position of the splitted F_{1u} modes.

splitting of these modes (arrows) some new structures as mentioned above and an overall increase of the background reflectivity at the phase transition are found.

At high temperatures in the fcc rocksalt structure the lineshift and the enhancement of the $F_{1u}(2,4)$ modes in the Rb_1C_{60} single crystal can be described by a charged phonon effect due to the coupling of the F_{1u} phonons to the doping induced electronic $t_{1u} \rightarrow t_{1g}$ transition[11]. At 445 K the parameters of the $F_{1u}(2)$ mode are $\omega_p = 54\,cm^{-1}$, $\gamma = 3.55\,cm^{-1}$ and $\omega_T = 573.7\,cm^{-1}$. For the $F_{1u}(4)$ mode $\omega_p = 181\,cm^{-1}$, $\gamma = 20.35\,cm^{-1}$ and $\omega_T = 1398\,cm^{-1}$ with $\epsilon_\infty = 7.03$ are found. This is in good agreement with the values found for thin films in[10]. At 350 K (below the phase transition) the values of the three components of the strongly splitted $F_{1u}(4)$ mode are: $\omega_{p1} = 53\,cm^{-1}$, $\gamma_1 = 6.21\,cm^{-1}$, $\omega_{T1} = 1341\,cm^{-1}$, $\omega_{p2} = 132\,cm^{-1}$, $\gamma_2 = 9.58\,cm^{-1}$, $\omega_{T2} = 1390.1\,cm^{-1}$, $\omega_{p3} = 102.6\,cm^{-1}$, $\gamma_3 = 11\,cm^{-1}$and $\omega_{T3} = 1405.9\,cm^{-1}$. The total oscillator strength of all three componets of this splitted line is the same as for the degenerated F_{1u} mode above the phase transition.

In conclusion we observed dramatic changes in the IR reflectivity at the first order phase transition in Rb_1C_{60} with a transition width of less than 2 K at 404/415 K. Below that phase transition the F_{1u} modes are splitted into three components with the same total oscillator strength.

Acknowledgement. Valuable discussions with L. Mihaly, M.C. Martin, M.J. Rice and H. Y. Choi are greatly acknowledged. We thank the Hoechst AG for supplying C_{60} raw material and M. Haluska for preparation of the crystals. This work was supported by the Österreichische Nationalbank, project P4507.

4. References

1. P.W. Stephens, L. Mihaly, P.L. Lee, R.L. Whetten, S.M. Huang, R. Kaner, F. Deiderich, and K. Holczer, *Nature* **351**, (1991) 632.
2. O. Zhou, J.E. Fisher, N. Coustel, S. Kycia, Q. Zhu, A.R. McGhie,W.J. Romanow, J.P. McCauley Jr,A.B. Smith III and D.E. Cox, *Nature* **351**, (1991) 462.
3. J. Winter and H. Kuzmany, *Solid State Commun.* **84**, (1993) 935.
4. Q. Zhu, O. Zhou, J.E. Fisher, A.R. McGhie, W.J. Romanow, R.M. Strongin, M.A. Cichy and A.B. Smith III, *Phys. Rev.* **B 47**, (1993) 13948.
5. D.M. Poirier and J.H. Weaver, *Phys. Rev.* **B 47**, (1993) 10959.
6. T. Pichler and H. Kuzmany, Proceedings of IWEP-NM93, *Springer Series in Solid State Sciences* **117**, (1993) 281.
7. O. Chauvet, G. Oszlanyi, L. Forro, P.W. Stephens, M. Tegze, G. Faigel and A. Janossy preprint.
8. M.C. Martin, D. Koller, X. Du, P.W. Stephens and L. Mihaly preprint.
9. M. Haluska, H. Kuzmany, M. Vybornov, P. Rogl, P. Fejdi, *Appl. Phys.* **A 56**, (1993) 161.
10. T. Pichler, R. Winkler, and H. Kuzmany, to be published in *Phys. Rev.* **B**.
11. M.J. Rice and H.Y. Choi, *Phys. Rev.* **B45**, (1992) 10173.

OPTICAL TRANSMISSION STUDIES OF Rb_1C_{60}

LASZLO MIHALY, DANIEL KOLLER AND MICHAEL C. MARTIN
Department of Physics, State University of New York at Stony Brook
Stony Brook, NY 11794–3800, USA

Abstract

Infrared spectroscopy was performed on Rb_1C_{60} compounds at temperatures from 10K to 500K, with heat treatments including slow cooling and quenching from high temperature. The main resonant features in the high temperature state are derived from the four IR active molecular modes. Upon cooling several new vibrational lines appear. The resonances and the overall transmissions of the quenched and slow cooled samples are distinctly different, in accordance with the linear chain structure of the slow cooled state, and suggesting and even stronger symmetry breaking in the quenched state.

The rich phase diagram of the A_xC_{60} system has attracted a considerable interest over the last year [1, 2, 3]. The recently discovered orthorhombic, linear chain structure in the slow cooled samples [4, 5, 6] in the neighborhood of $x \sim 1$, and the near-metallic or metallic electrical conductivity of these compounds [7] motivates further research on the A_1C_{60} material. In the present review we will summarize the most recent infrared transmission results obtained on thin films of approximately $x = 1$ composition, with a particular emphasis on the new vibrational modes appearing in the low temperature state of these compounds. We will also discuss the temperature and thermal history dependence in the overall transmission background, indicative of the properties of the conduction electrons in the system.

The experiments were performed in a sample chamber [8] equipped with silicon windows for the study of the IR transmission properties of the alkali fullerides (Fig 1.). The first versions of the chamber were made with "Torr-Seal" epoxy [8]. More recently, we eliminated the use of the epoxy, opening the way for higher temperature heat treatments. The cell safely survives rapid cooling from high temperatures using liquid nitrogen. Throughout this work the transmission is defined as the point by point ratio of the spectrum of the chamber with the sample to the spectrum of the empty chamber.

The details of the sample preparation are described elsewhere [8]. The key element is the in-situ monitoring of the F_{1u} vibrational lines of the C_{60} molecules during the doping process, while separately controlling the temperature of the sample and the temperature of the alkali metal film deposited on the spherical appendix to the sample chamber (Fig. 1). The sample was kept at 225C; the temperature of the alkali metal was varied so that the doping proceeded in a controlled fashion where the position and the strength of the $F_{1u}(4)$ mode was used to characterize the composition of the sample [9, 10]. Our goal was to prepare films where the $F_{1u}(4)$ line at $1395cm^{-1}$, corresponding to the x=1 composition, is the strongest. Weaker resonances at $1428cm^{-1}$

266

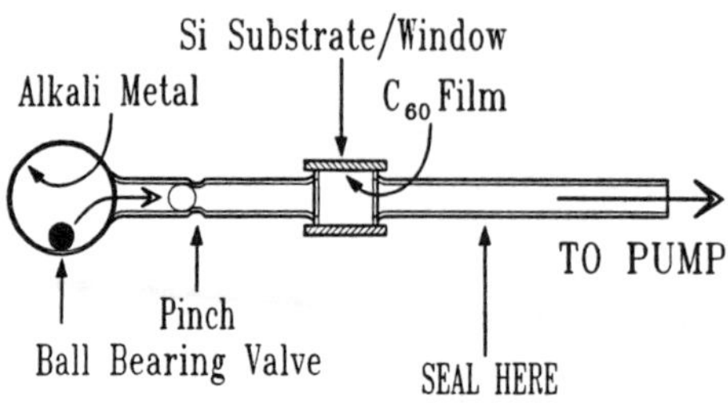

Figure 1. Drawing of the sample chamber. Area of Si window is $\sim 1\text{cm}^2$.

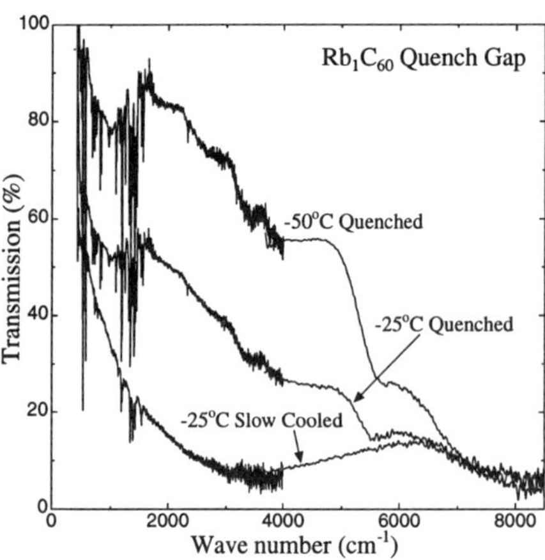

Figure 2. Representative IR spectra recorded on a thin film sample of RbC$_{60}$.

$(x = 0)$, 1363cm^{-1} $(x = 3$ or $4)$ and 1340cm^{-1} $(x = 6)$ indicated the presence of small amounts of other phases (frequencies quoted are for T=225C).

Figure 2 shows a a few characteristic spectra obtained in various states of the RbC$_{60}$ compound. The overall frequency dependence of the high temperature and slow-cooled state is quite similar. Model calculations on a system consisting of Drude electrons and mid infrared oscillators suggest that the dc conductivity of the system would be in the range of 10mΩcm. The cut-off in transmission at frequencies above 8000 wavenumber is due to the t_{1u}-t_{1g} interband transition of the electrons, also seen in other alkali metal fullerides [11]. The gradual drop and the smooth upturn of the transmission in the mid IR range could be either due to the Drude contribution or due to a broad distribution of mid-IR oscillators, or the combination of these. The granular nature of the sample may also lead to a transmission increase at low frequencies. Measurements at lower frequencies could help to sort out these explanations, but the relatively low transmission is clearly indicative of the presence of mobile electrons in the material. The slow cooled sample has a linear chain structure for the C$_{60}$ molecules [5, 6], and it is expected to have an anisotropic conductivity. Since the films in this study are polycrystalline, we can only measure directionally averaged properties, and the intrinsic conductivity along the chains could be significantly higher than the average value evaluated from the transmission.

In the quenched state at room temperature the structure and the optical transmission is identical to the high temperature fcc phase [4, 7]. At lower temperatures the quenched sample has a dramatically enhanced transmission at low frequencies [7], and the x-ray diffraction pattern indicates a structure distinctly different from the high temperature structure [12]. Also, near room temperature the quenched state is unstable, and there is a relaxation towards the stable, linear chain structure [7, 13].

The enhanced transmission corresponds to a reduced conductivity, due to the removal of conduction electrons. The detailed evaluation of the high frequency part of the transmission spectrum also indicates that there is an energy gap of about $5000 cm^{-1}$ in the electronic system [14].

In the frequency range below $1600 cm^{-1}$ all spectra exhibit a number of sharp resonance lines. It is well known that the presence of conduction electrons makes the detection of the phonon resonances harder, especially for reflectivity measurements [15] or for transmission on thicker films, like the first one investigated in our laboratory [7]. However, on thinner films we were able to detect the resonance structures even in the slow cooled samples, where the conduction electron shielding is significant. This is illustrated in Fig. 3, where the spectra of the same film in the quenched and slow cooled state is plotted. The vertical scale is normalized so that the resonance line corresponding to the x=6 impurity phase (at $1342 cm^{-1}$) appears approximately equal in magnitude for the two spectra [16]. Thus, in spite of the electronic shielding effects, other lines of equal spectral strength should show up in similar magnitude. Indeed, the 1429 cm^{-1} resonance from the x=0 impurity phase and other resonances derived from the IR allowed F_{1u} molecular vibrations behave as expected. It is therefore significant that the quenched sample has many more resonances and some of the lines seen in the slow cooled state are missing in the quenched state. Furthermore, the fine structure of the $F_{1u}(4)$ derived mode is clearly different for the two specimens.

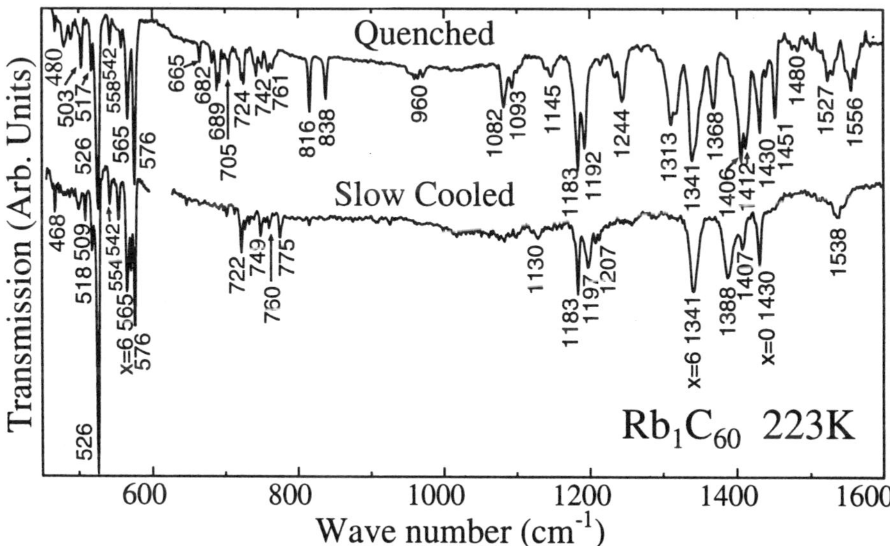

Figure 3. Comparison of resonance features in the slow cooled and quenched states of RbC$_{60}$. Rb$_6$C$_{60}$ and C$_{60}$ are labeled x=6 and x=0, respectively. The curves are scaled so that the impurity phase resonances apear approximately equal in strength.

It is instructive to compare the IR active frequencies of the doped compound to the fundamentals seen in pure C_{60} compound [17]. (Note that the IR study of the pure C_{60} was performed on thick single crystals. The weak resonances seen in that measurement are normally invisible on films comparable in thickness to the sample in the present study). It is well known that the four T_{1u} modes in the solid C_{60} are close to the F_{1u} modes of the single C_{60} molecule. Furthermore, the charge transfer to the C_{60} molecule causes a relatively small change in the resonance frequency of the T_{1u} modes [9, 10]. Encouraged by these observations, we attempt to obtain a qualitative understanding of the vibrational features of the slow cooled and quenched RbC_{60} samples by starting from the experimentally measured resonances of the pure fcc C_{60} solid [17]. Implicit to this approach is the assumption that the chemical and structural changes represent relatively small perturbation for the molecules involving a large number of C atoms.

As we argued earlier [13], the $15 \mathrm{cm}^{-1}$ splitting of the $F_{1u}(4)$ line in the slow cooled state can be understood in terms of the linear chain structure. The formation of chemical bond between the C_{60} molecules is a much larger perturbation than the Van der Waals forces acting in the fcc solid, and the linear structure of the chains breaks the cubic symmetry, leading to the removal of the threefold degeneracy. In first approximation the two directions perpendicular to the chain direction remain equivalent, thus two of the three T_{1u} derived modes may stay quasi-degenerate.

In addition to the splitting of the $T_{1u}(4)$ line, the slow cooled sample has a few new lines in the $500 \mathrm{cm}^{-1}$ and $700 \mathrm{cm}^{-1}$ range. We believe that these are derived from the *ungerade* modes of the C_{60}. Since the linear chain structure does not remove the inversion symmetry of the C_{60}, the *gerade* modes are not expected to show up. In principle, we may see the 87 symmetry allowed vibrational modes, derived from the 23 *ungerade* modes of the C_{60}; however, many of these modes may have such a small dipole activity, that they are below the noise level of the measurement.

A smaller, but clearly visible, splitting of the $F_{1u}(4)$ line appears in the quenched state, and a much larger number of new resonances appear in the spectrum. Some of these lines appear to be close to the known A_g and H_g frequencies, indicating that the inversion symmetry of the single molecule is broken. Although the final word on the structure of this phase will come from X-ray diffraction studies [12], we argued [7] that the dimerization of the C_{60} molecules provides a possible explanation for the large number of new modes in the quenched state. Intermolecular charge transfer between the members of the dimer carries a dipole moment while the vibrational frequency is close to a *gerade* mode frequency of the isolated molecule. The dimerization would also be quite consistent with the removal of the conduction electrons, as a half filled electronic band splits into a full and empty band.

A comprehensive picture of the various stable and metastable RbC_{60} phases is presented schematically in Fig 4. The state of the material is characterized by the infrared transmission; large transmission corresponds to insulator, while low transmission represents a conductor. Upon slow cooling the transmission is approximately

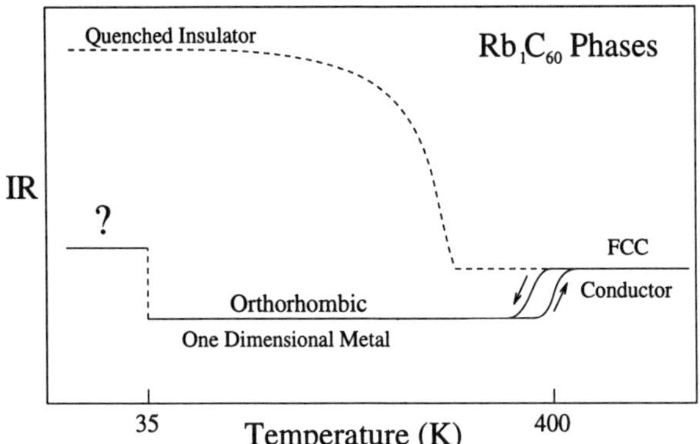

Figure 4. Schematic illustration of the various states of RbC$_{60}$ in terms of IR transmission.

independent of temperature in the fcc phase, drops at the fcc-orthorhombic phase transition [7] and stays independent of temperature down to 34K [18]. At this point there is a distinct jump in the transmission, indicating yet another transition in the electronic state of the sample. The electron spin resonance studies on RbC$_{60}$ were interpreted in terms of a magnetic transition in this temperature range [4]. It remains to be seen how the jump in the IR transmission at 34K is related to the continuous transition in the ESR signal at 50K, but we notice that the IR spectrum was measured in zero magnetic field, while the ESR was investigated in a field of 0.35T. When quenched to room temperature, the sample is still in the fcc phase, but when quenched below 300K, the material becomes an insulator. The onset of insulating state follows a continuous curve.

CsC$_{60}$ and KC$_{60}$ share many of the features of RbC$_{60}$ discussed here with differences discussed in Ref. [13]. Further studies on these compounds are clearly warrented to clarify the surprisingly complex nature of the various stable and metastable states.

Acknowledgments We are indebted to Laszlo Forro, Peter W. Stephens, András Jánossy and Sándor Pekker for valuable discussions. This work has been supported by the NSF Grant 9202528, and by a grant from the US- Hungarian Joint Fund, JF225.

References

1. Q.Zhu, O. Zhu, J.E. Fisher, A.R. McGhie, W.J. Romanow, R.M. Strongin, M.A. Cichy, A.B. Smith II, Phys. Rev. B **47**, 13948 (1993); J.E. Fischer and P.A. Heiney, J. Phys. Chem. Solids (to be published).
2. T. Pichler, R. Winkler, H. Kuzmany, Phys. Rev. B (to be published) and this conference.
3. D.M. Poirier and J. H. Weaver, Phys. Rev. B **47**, 10959 (1993); J.H. Weaver and D.M. Poirier in *Fullerene Fundamentals, Solid State Physics*, Vol 48, eds. H. Ehrenreich and F. Spaepen, Acad. Press, New York, 1994.
4. O. Chauvet, G. Oszlányi, L. Forro, P. Stephens, G. Faigel, M. Tegze, A. Jánossy, Phys.

Rev. Lett. **72**, 2721 (1994).

5. S. Pekker, L. Forro, L. Mihaly and A. Janossy, Solid State Commun, in print (1994).

6. P.W. Stephens, G. Bortel, G. Faigel, M Tegze, A. Jánossy, S. Pekker, G. Oszlány and L. Forro, to be published.

7. Michael C. Martin, Daniel Koller, X. Du, P.W. Stephens, L. Mihaly, Phys Rev. B **49**, 10818 (1994).

8. Daniel Koller, Michael C. Martin, L. Mihaly, Rev. Sci. Instrum. **65**, 760 (1994).

9. Michael C. Martin, Daniel Koller, L. Mihaly, Phys. Rev. B **47**, 14607 (1993).

10. J. Winter, H. Kuzmany, Solid State Commun., **84**, 935 (1992); H. Kuzmany, M. Matus, T. Pichler, J. Winter, Proc. Nato ARW on physics and chemistry of fullerenes, Crete, (1993).

11. Y. Iwasa *et al.*, J. Phys. Chem. Solids **54**, 1795 (1993); K.-J. Fu *et al.*, Phys. Rev. B **46**, 1937 (1992).

12. To our knowledge, to date no successful Rietweld refinements were performed on the existing X-ray spectra. See G. Faigel *et al.*, this conference, Zhu *et al.* Phys. Rev. B **47**, 13948 (1993) and Q. Zhu, private communication.

13. Daniel Koller, Michael C. Martin, L. Mihaly to appear in Mol. Cryst. Liq. Cryst.

14. Daniel Koller, Michael C. Martin and L. Mihaly, to be published.

15. This purely experimental limitation has nothing to do with the magnitude of electron phonon interaction; the conduction electron response always shields the phonon resonance.

16. Kuzmany and co-workers reported in this conference that the the x=1 sample in its slow-cooled state has an intrinsic resonance at this position. Our measurements confirmed this conclusion. However, for the film discussed here the x=6 line is much stronger than the intrinsic line, and therefore it can be safely used for normalization purposes.

17. See Michael C. Martin, Xiaoqun Du and Laszlo Mihaly, this conference and Phys. Rev. B (accepted for publication).

18. Michael C. Martin, Daniel Koller, L. Mihaly, G. Oszlányi and L. Forro to be published.

DOPING AND VIBRATIONAL ANALYSIS OF SINGLE CRYSTAL FULLERIDE PHASES

J. Winter and H. Kuzmany

Institut für Festkörperphysik, Universität Wien
Strudlhofgasse 4, A-1090 Vienna, Austria

ABSTRACT

In situ Raman measurements are reported for potassium doped C_{60} single crystals. The measurements were performed in a modified Haddon type cell which allows heating and cooling of the C_{60} crystals. The doping process was studied at high temperatures and at room temperature with additional annealing.

Up to the penetration depth of the laser homogeneous single phases of K_1C_{60} and K_3C_{60} could be obtained. The doped phases were identified by the position of the $A_g(2)$ pinch mode, which is very sensitive to the charge transfer. The high frequency H_g modes disappeared in both cases. At cooling the high temperature phase K_1C_{60} was observed to undergo a sharp transition to C_{60} and K_3C_{60} between 430 K and 410 K.

1. Introduction

Raman scattering was from begin on a key technique for probing and studying the doping process in fullerites[1-3]. The various doped phases can be identified by the position of the $A_g(2)$ pinch mode which is very sensitive to the charge transfer in the alkali metal doped systems [1].

In the case of potassium doping three different phases exist at room temperature: K_3C_{60}, K_4C_{60} and K_6C_{60} [1-6]. At temperatures above 430 K an additional phase K_xC_{60} with x = 1 is observable [7-9]. Usually the doped samples are powder samples or polycrystalline films. The following results are from *in situ* Raman measurements of doped single crystals. Pure doped phases could be obtained at least up to the penetration depth of the laser.

2. Experimental

The fullerite single crystals were grown by a sublimation-condensation method as described by Haluška *et al.*[10]. Raman spectra were taken from {111} and {100} crystal faces with a size of 2 mm x 3 mm. By using silver paint the crystals were mounted on the sample holder in a modified Haddon type cell which allowed the heating of the crystals up to a temperature of 480 K. Before starting the doping experiment the crystals were annealed for one day at 450 K to purify them from oxygen and

272

polymerized phases. The doping process was done in a vacuum better than $9 \cdot 10^{-7}$ mbar and at a sample temperature of 450 K.

Raman spectra were excited with a conventional argon laser at 514.5 nm. Typical intensities for excitation were 45 W/cm^2. The scattered light was analysed with a Dilor XY spectrometer in a backscattering geometry, and a nitrogen cooled CCD – multichannel detector.

3.   Results and Discussion

Fig. 1 shows a typical behavior of the Raman response for the pinch mode during doping the crystals with potassium at a temperature of 450 K at various times. The figure shows clearly the decrease of the pinch mode of the undoped material at 1467 cm^{-1} and the increase of a new line at 1459 cm^{-1}. After 6 hours doping single phase K_1C_{60} could be obtained up to the penetration depth of the laser. The ratio between the integrated intensities of the $A_g(2)$ pinch mode of the two different phases is about 25. The homogeneity of the K_1C_{60} phase at the surface of the crystal was proved by Raman measurements at various spots on the crystals. Cooling the sample showed the decay of the K_1C_{60} phase into the two well known phases C_{60} and K_3C_{60}. Fig. 2 illustrates this behavior for different temperatures. The Raman spectra showed a sharp phase transition temperature between 430 K and 410 K which is in good agreement with measurements on thin epitaxial films from Poirier et $al.$ [8]

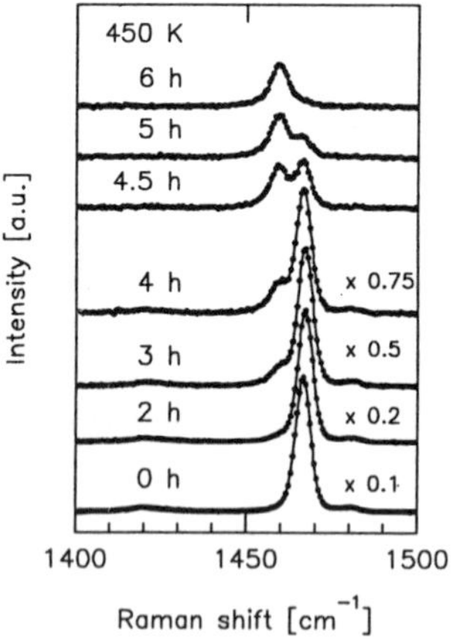

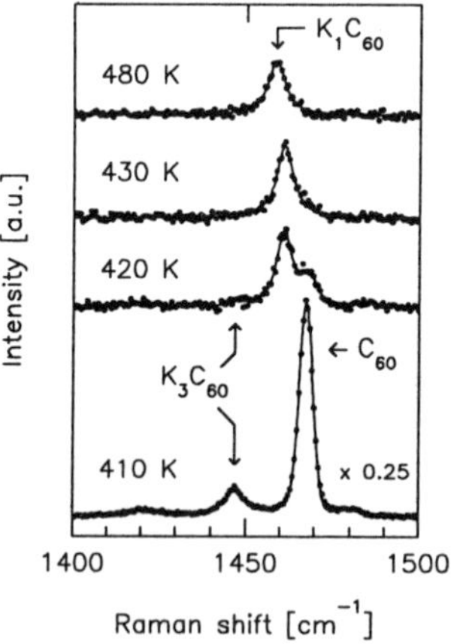

Fig. 1 : Raman response for the pinch mode of single crystal C_{60} after doping for 0 (undoped), 2, 3, 4, 4.5, 5 and 6 hours with potassium at 450 K.

Fig. 2 : Temperature dependence of the pinch-mode in K_1C_{60} on cooling from 480 K to 410 K. The small shift of the $A_g(2)$ pinch mode of the K_1C_{60} in the spectra between 480 K and 430 K is temperature induced.

but slightly higher as compared to results obtained from Raman measurements on conventional thin films [11].

Fig. 3 and Fig. 4 show the doping and equilibration process at high temperatures and at room temperature. The first spectrum of Fig. 3 is for the homogeneous K_1C_{60} phase from Fig. 1. After 12 hours equilibration at 450 K without further doping a small amount of undoped material could be observed. This behavior was interpreted as a diffusion of the potassium from the K_1C_{60} phase into the bulk by forming α-C_{60}. The original state could be reached again by additional doping. Further doping even lead to a small amount of K_3C_{60} at the surface of the crystal. After an equilibration time of another hour the $A_g(2)$ pinch mode of the K_3C_{60} phase at the surface had nearly disappeared. This was interpreted as a diffusion of the potassium into the crystal and thus forming a thicker layer of K_1C_{60}.

A similar behavior is shown in Fig. 4. After a total of about 30 hours doping a single phase Raman spectrum for the K_3C_{60} was observed. Cooling the crystal to room temperature the Raman measurements showed that some undoped material is present. This indicates a temperature dependence of the absorbtion of the scattered

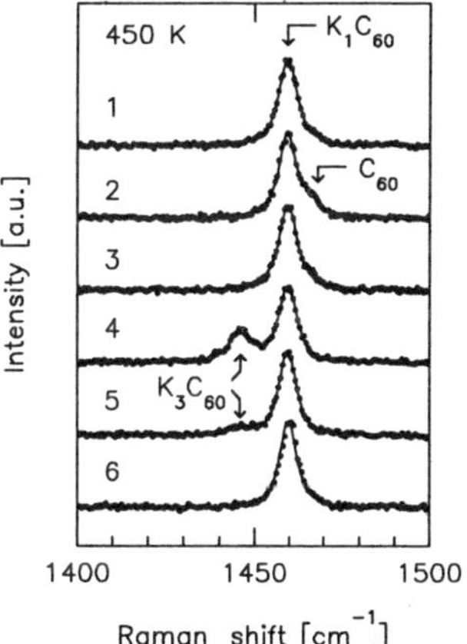

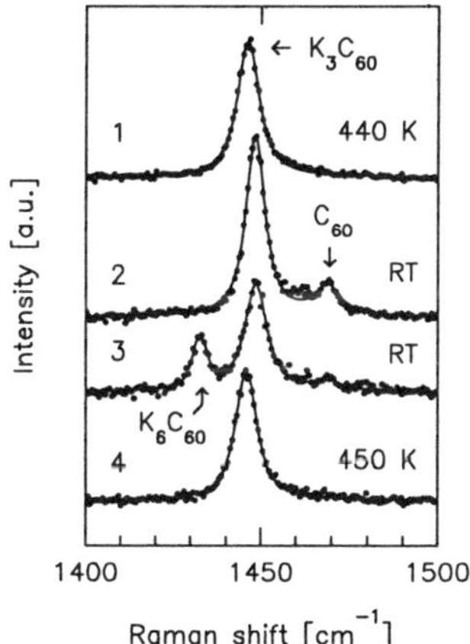

Fig. 3 : Doping and equilibration process:
(1) Single phase K_1C_{60} after doping the crystal 6 hours with potassium at 450 K.
(2) After 12 hours equilibration without further doping.
(3) After additional doping of 1 hour.
(4) After further doping K_3C_{60} could be observed at the surface of the crystal.
(5) After additional equilibration.
(6) After another hour the single phase K_1C_{60} was observed again.

Fig. 4 : Doping at room temperature and equilibration process at high temperature:
(1) Single phase K_3C_{60} at 440 K .
(2) Room temperature: K_3C_{60} and a small amount of undoped material.
(3) Doping at room temperature leads to a small amount of K_6C_{60} at the surface of the crystal.
(4) Annealing at 450 K showed that the K_6C_{60} dissolved by diffusing into the bulk and pure K_3C_{60} phase was formed.

274

light. Further doping at room temperature was carried out to check if some K_6C_{60} at the surface could be produced and dissolved at higher temperatures. The last spectrum of Fig. 4. shows the result of this experiment and indicates that even the potassium from the phase K_6C_{60} diffused into the bulk at high temperature.

Summarizing we have shown that *in situ* Raman measurements are very useful for studying the doping process of fullerites. Up to the penetration depth pure phases K_1C_{60} and K_3C_{60} could be obtained but the Raman spectra showed that the doped phases are not stable at high temperature. By diffusion of the potassium from the doped phases into the bulk more α-C_{60} is formed. Between 430 K and 410 K the Raman spectra showed a decay of the high temperature phase K_1C_{60} into C_{60} and K_3C_{60} which was very sharp compared to the result from Raman measurements on conventional films.

Acknowledgment. We are grateful to M. Haluška for the preparation of the C_{60} single crystals and to Hoechst AG for supporting us with C_{60}. Valuable discussions with D.M. Poirier and financial support by the FFWF, project P09741-TEC, is acknowledged.

References

1. R.C. Haddon, A.F. Hebard, M.J. Rosseinsky, D.W. Murphy, S.J. Duclos, K.B. Lyons, B. Miller, J.M. Rosamilia, R.M. Fleming, A.R. Kortan, S.H. Glarum, A.V. Makhija, A.J. Muller, R.H. Eick, S.M. Zahurak, R. Tycko, G. Dabbagh, and F.A. Thiel, *Nature* **350** (1991) 320.
2. P.C. Eklund, P. Zhou, K.A. Wang, G. Dresselhaus, M.S. Dresselhaus, *J. Phys. Chem. Solids* **53** (1992) 1391.
3. T. Pichler, M. Matus, J. Kürti, and H. Kuzmany, *Phys. Rev.* **B 45** (1992) 13841.
4. O. Zhou, J.E. Fischer, N. Coustel, S. Kycia, Q. Zhu, A.R. McGhie, W.J. Romanow, J.P. McCauley, A.B. Smith, and D.E. Cox, *Nature* **351** (1991) 462.
5. R.M. Fleming, M.J. Rosseinsky, A.P. Ramirez, D.W. Murphy, J.C. Tully, R.C. Haddon, T. Siegrist, R. Tycko, S.H. Glarum, P. Marsh, G. Dabbagh, S.M. Zahurak, A.V. Makhija, and C. Hampton, *Nature* **352** (1991) 701.
6. Q. Zhu, O. Zhou, N. Coustel, G.B.M. Vaughan, J.P. McCauley, W.J. Romanow, J.E. Fischer, and A.B. Smith, *Science* **254** (1991) 545.
7. J. Winter and H. Kuzmany, *Solid State Communication* **84** (1992) 935.
8. D.M. Poirier and J.H. Weaver, *Phys. Rev.* **B 47** (1993) 10959.
9. Q. Zhu, O. Zhou, N. Bykovetz, J.E. Fischer, A.R. McGhie, W.J. Romanow, C.L. Lin, R.M. Strongin, M.A. Cichy, and A.B. Smith, *Phys. Rev.* **B47** (1993) 13948.
10. M. Haluška, H. Kuzmany, M. Vybornov, P. Rogl, and P. Fejdi, *Appl. Phys.* **A** **56** (1993) 161.
11. J. Winter and H. Kuzmany, *Springer Series in Solid-State Sciences* **117** (1993) 273.

OPTICAL, RAMAN AND INFRARED STUDY OF TDAE-C_{60}

D.Mihailovič, P.Venturini, A.Hassanien, J.Gasperič, K.Lutar and S.Miličev
J.Stefan Institute, University of Ljubljana, Jamova 39, 61111 Ljubljana, Slovenia

and

V.Srdanov
*Department of Chemistry, University of California Santa Barbara,
CA-93106, USA*

ABSTRACT

The organic magnetic material TDAE-C_{60} in the form of powders and thin films has been investigated by various optical measurements. Using optical absorption on thin films we have found excited state absorption in TDAE-C_{60} which is indicative of a populated t_{1u} of C_{60}. The Raman spectra when excited with 514.5 nm laser excitation show strong sensitivity to light which is somewhat different to the behaviour observed in pure C_{60}. The Raman spectra of pentagonal pinch mode do not show the expected 6 cm^{-1} frequency shift for C_{60}^-. Instead we observe two modes, centered at approximately one half of the expected shift, which suggests the possibility of a different charge transfer mechanism to alkali metal doped C_{60}. IR transmission spectra show no Drude absorption down to 200 cm^{-1}, indicating non-metallic behaviour at 300K in spite of having a partially filled band. For the lowest three of the four T_{1u} IR vibrations we observe behaviour which is similar to K_1C_{60}, while we are not able to identify a T_{1u} (4) mode which, for C_{60}^- is expected to be strong and around 1400 cm^{-1}.

Introduction

Some optical properties of the tetrakis(dimethylamino)ethylene (TDAE) doped C_{60} fullerene are presented in view of characterizing the charge transfer properties and vibronic couplings in this material. We try and answer the following questions: 1) is there a full charge transfer to the C_{60} from TDAE, 2) is the material a metal (i.e. does it show a low-frequency Drude absorption), and 3) do the frequency shifts of IR and Raman vibrations with doping also follow the general behaviour that has been observed in K_1C_{60} or Rb_1C_{60}[1,2,3].

Optical absorption in the near infrared

Doping of C_{60} by TDAE is expected to promote an electron into the previously unoccupied LUMO level. New transitions are then expected from this t_{1g} level to the higher levels of t_{1u} symmetry, the first of which is 1.15 eV higher in energy. Since the optical transitions are parity allowed, we expect the new feature at 1.15 eV in the absorption spectrum to be strong. As it is not possible to investigate the absorption spectra in the near-infrared region of powder samples because of light scattering, thin films were produced by dipping oxygen-free C_{60} films into TDAE for various lengths of time in an oxygen-free atmosphere. The observed absorption on a doped C_{60} film approximately 1000 Å thick is shown in Figure 1. Although the feature at 1.15 eV (indicated by the arrow in Figure 1) is clearly visible, it is not nearly as strong as the other dipole allowed transitions at higher energies (left panel), presumably indicating that the thin film is not fully doped. The bandwidth of the transition is seen in Figure 1 (right panel) to be ~ 0.06 eV.

We find that prolonging the doping time does not intensify the absorption feature significantly, which suggests that once a few surface layers are doped, the diffusion of TDAE into the bulk must be very slow. Since Raman measurements probe the whole depth of the film, we can confirm this hypothesis by Raman measurements. Indeed these show spectra of pristine C_{60}, in particular, we observe no frequency shift of the 1469 cm^{-1} pinch mode. We estimate the extent of the doping to be only a few (perhaps 10) molecular layers or 10% of the film.

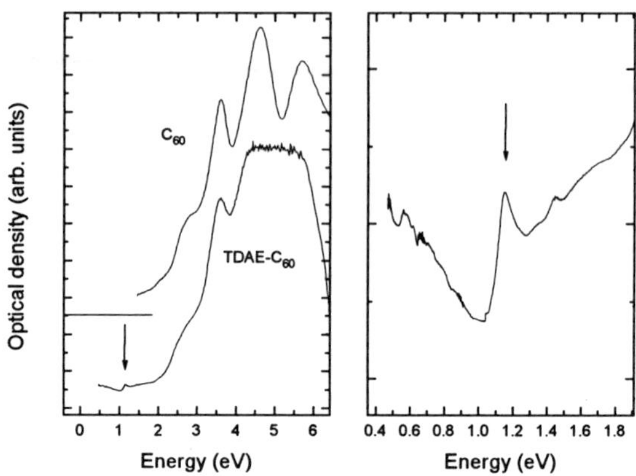

Figure 1. Optical absorption spectra of TDAE-C$_{60}$. The right panel is an enlargement of the spectrum on the left. C$_{60}$ is shown for comparison in the left panel. Extra absorption features fill in the peaks above 3 eV.

Raman spectra of TDAE-C$_{60}$.

Since thin films are not doped in depth, we have measured Raman spectra only on powder. Just as in C_{60}, with 514.5 nm laser excitation we observe a very strong photo-effect. The Raman spectra in the vicinity of the pentagonal pinch mode are plotted at different excitation power densities in Figure 2.

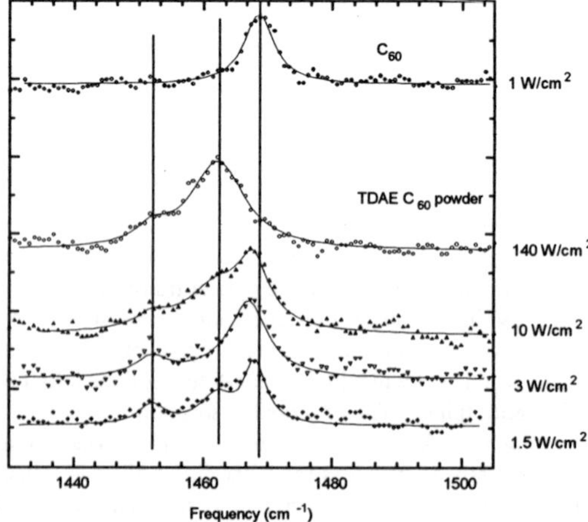

Figure 2. Raman spectra of TDAE-C$_{60}$ at different laser powers taken with a 514.5 nm laser. The spectrum taken at 140 W/cm^2 shows a significant photo-induced change. A C$_{60}$ spectrum is shown for comparison (top trace).

At very low power (below 3 W/cm^2) three modes are seen in the region 1420-1520 cm^{-1}: at 1467 $\pm$ 1 cm^{-1}, at 1463 $\pm$ 1.5 cm^{-1} and a weaker mode at 1452 $\pm$ 2 cm^{-1}. At higher power, only one peak is apparent, at 1463 cm^{-1}, which is presumed to be the signature of the photo-degraded material and not the intrinsic mode as has been suggested[4]. Since the structure of TDAE-C$_{60}$ is monoclininc C2[5] with two C$_{60}$ molecules per unit cell, we can expect the A_g pentagonal pinch Raman mode to split into two modes because of the interaction between the two molecules. We tentatively identify the two modes to be at 1467 and 1463 cm^{-1}. We would expect that the mean frequency of the two modes would correspond to the shift in frequency due to the extra charge on the molecule. The value 1465 implies a smaller shift than in K$_1$C$_{60}$ or Rb$_1$C$_{60}$, where it is 1463 cm^{-1}. The 1452 cm^{-1} peak does not show any significant laser power dependence. The photosensitive behaviour is similar to C$_{60}$, but we don't see a complete reduction in intensity of the 1467 cm^{-1} mode and we cannot identify any intensity in the 1459 cm^{-1} mode which first appears in pristine C$_{60}$ upon photoexcitation in vacuum[6].

IR transmission spectra

The IR transmission spectra are shown in Figure 3. The two low-frequency modes apparently do not show any splitting due to the monoclinic distortion, only a shift which is comparable in size to the one in alkali-doped C$_{60}$.

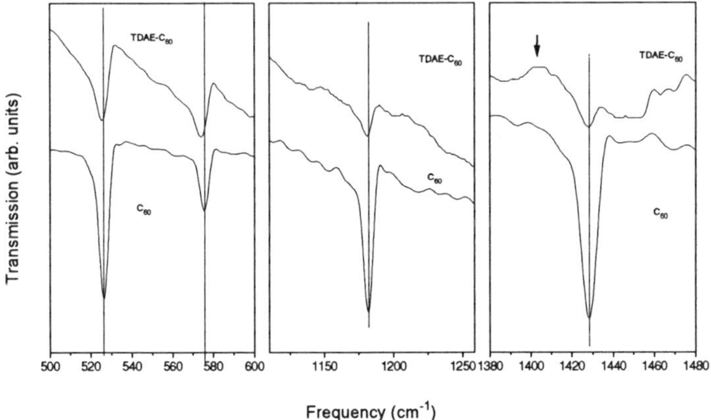

Figure 3. IR Spectra of TDAE-C$_{60}$ and pure C$_{60}$ in the region of the four C$_{60}$ modes. The arrow points to the position of the mode in K$_1$C$_{60}$.

The mode at 1182 cm^{-1} shows no shift or splitting, and although the mode overlaps with a mode from TDAE, we see no TDAE modes, so we assume it comes from C$_{60}$. So far the behaviour is almost exactly as expected for C$_{60}^-$. The odd one out is the 1428 cm^{-1} mode which is expected to broaden, become more intense and shift to around 1400 cm^{-1}. Its expected position is indicated by the arrow in Figure 3.

We observe only an unshifted mode at 1428 cm^{-1}, but unfortunately since it is possible that there is some unreacted C$_{60}$ present in our sample, it would be hard to give a definite assignment of the 1428 cm^{-1} mode *to the doped molecule*. However, in repeated measurements on different samples we have not seen any *other* mode in the vicinity which could be interpreted as the T_{1u} (4) mode.

Importantly, in the IR spectra on both thin films and powder samples we have *observed no low-frequency absorption*, which would be indicative of metallic conductivity.

278

Conclusion

The frequency shifts of IR and Raman vibrational lines in TDAE-C_{60} are somewhat different that those of the K_1C_{60}. The discrepancy is mainly in the Raman 1469 cm^{-1} pinch mode and the IR 1428 cm^{-1} mode. The behaviour in the Raman can be understood in terms of a mode splitting due to the two inequivalent ions per unit cell and a smaller shift than in K_1C_{60}[1].

The behaviour of the $T_{1u}(4)$ mode in TDAE-C_{60}, whose position is at 1428 cm^{-1} in pure C_{60} is at present not understood. It is either not shifted from 1428 cm^{-1}, or it is perhaps broad and for some reason weak and therefore difficult to detect. Given that the other three modes are well visible, its behaviour thus remains a puzzle at present, and improved sample quality (e.g. good quality thin films or crystals) may be necessary before this issue is resolved.

The absence of a Drude low-frequency absorption in this material down to 220 cm^{-1} is fairly clear evidence that the material is non-metallic. Even in the unlikely event that a significant portion of the sample were degraded (the samples were checked by AC susceptometry measurement before, but not after the IR measurement) we would still see a significant absorption at low frequencies. The optical absorption at higher frequencies suggests a populated t_{1u} LUMO state of C_{60} with a bandwidth of < 0.06 eV at 300K for the $t_{1u} \rightarrow t_{1g}$ transition..

Acknowledgment. We wish to acknowledge H.Kuzmany for a valuable discussion of the Raman data.

References.

[1] J.Winter and H.Kuzmany Sol.Stat.Comm. **84**, 935 (1992)

[2] T.Pichler and H.Kuzmany, p. 281, "Electronic Properties of Fullerenes", Eds. H.Kuzmany, J.Fink and S.Roth (Springer Verlag 1993), also T.Pichler, M.Matus and H.Kuzmany, Sol.Stat.Com.**86**, 221 (1993)

[3] M.C.Martin, D.Koller and L.Mihaly, Phys.Rev.B **47**, 14607 (1993)

[4] D.V.S.Muthu et al, Chem.Phys.Lett. **217**, 146 (1994)

[5] P.W.Stephens et al., Nature **355**, 331(1992)

[6] L.Akselrod, H.J.Byrne, C.Thomsen and S.Roth, Chem.Phys.Lett. **215**, 131 (1993)

Zero Field μSR and
NMR on magnetic TDAE - C_{60}

L. Cristofolini, M. Riccò, R. De Renzi
Dipartimento di Fisica, Università di Parma, Parma, Italy.

G.P. Ruani , S. Rossini, C. Taliani
Istituto di Spettroscopia Molecolare, CNR, Bologna, Italy.

Abstract

Recently the presence of a magnetic moment has been demonstrated in TDAE-C_{60} with a magnetic transition at T_g=16.1K [1,2,3,4]. We present a study on the magnetically ordered phase of TDAE-C_{60} by local probes sensitive to the microscopic magnetic field and to its dynamics. Both high field NMR and zero field μSR indicate the absence of long range order and suggest a spin glass picture.

Muon spin relaxation The sample was characterized by SQUID magnetometry giving results fully consistent with the literature [1,2].

In figure 1 we show the time evolution of the muon polarization at low temperatures. The amplitude is temperature independent and accounts for **all** the implanted μ^+. It corresponds to muons in a **diamagnetic** environment (no radicals are formed).

Of the three species formed in pure C_{60}, endohedral muonium Mu@C_{60}, adduct radical Mu-$C_{60}^{\bullet}$ and diamagnetic Mu-C_{60}, the latter clearly is dominating in TDAE-C_{60}. The high temperature behaviour of the depolarization rate, indicating motional narrowing above T=180K, in conjunction with the ^{13}C NMR data discussed below, indicates freezing of the C_{60} reorientations in the low temperature regime and confirms the addition of muons to the latter molecules rather than to TDAE units.

Between 180K and 20K the muon relaxation is due to the nuclear dipolar interaction with the methyl protons of TDAE and it should display the Kubo-Toyabe lineshape with the characteristic recovery of $\frac{1}{3}$ of the polarization at long times. The absence of this recovery in the data of figure 1 indicates a distribution of sites for the muon, reflecting the Mu-C_{60} bond orientational disorder in the c-centred monoclinic lattice. Two sites, with different second moment of the nuclear dipolar field distribution are sufficient in practice to fit the data.

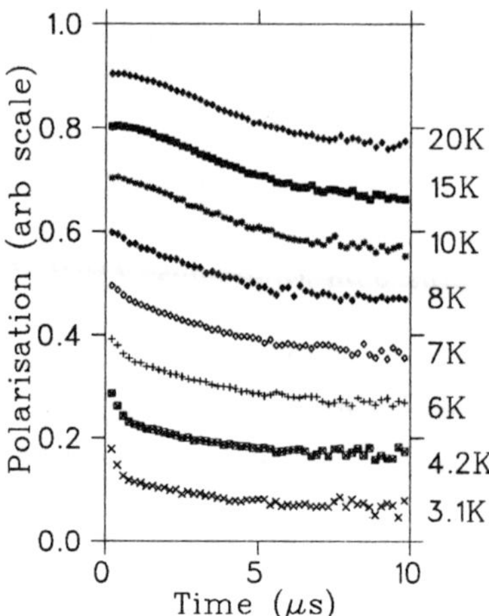

Figure 1: *μSR depolarization function at different temperatures.*

Below 20K we see a rapid increase of σ which is of magnetic origin. We do **not** observe any precession signal down to 3.1K, as shown in figure 1. This rules out any long range order for TDAE-C_{60}. However the strong dynamical relaxation at low temperatures indicates the presence of quasistatic electronic magnetic moments. Below 10K this magnetic relaxation mechanism dominates, while from 10K to 16K the deviation from a purely nuclear mechanism reduces and becomes finally negligible. This indicates relatively fast electronic fluctuation in this second range.

The best known example of muon zero field relaxation due to short range order magnetism is the spin glass.

Fitting the data to the homogeneous spin glass relaxation [6] with two different local field distributions, related to the two muon sites determined above T_g gives a non-vanishing order parameter q below 8-10K.

The relaxation in longitudinal fields up to 500G further distinguishes the behaviour at 4.2 and 8K. While at 8K the relaxation is mainly driven by the static nuclear dipolar fields (exchange narrowing of the electronic fluctuations) at 4.2K the depolarizing effect of the fluctuating electronic moment is clearly predominant.

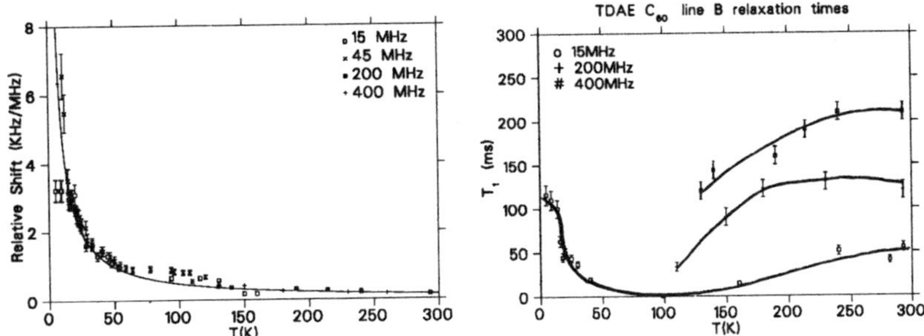

Figure 2: *Left: Relative paramagnetic shift times T as a function of T for different applied fields. Right: Longitudinal relaxation time of line B.*

Nuclear Magnetic Resonance Two NMR probes are available in this compound: ^{13}C and ^{1}H. The main contribution to the ^{13}C NMR signal comes from the C_{60} molecule, while the proton signal arises from TDAE. ^{13}C probes mainly the dynamics of the C_{60} molecule (through the anisotropy of the chemical shift tensor).

The full width at half maximum of the ^{13}C resonance (measured in a field of 7.0T) is almost temperature independent above 180K, while it increases below this temperature. This must be due to the slowing of the C_{60} molecular rotations.

The proton spectrum, taken at 0.3, 1.0, 4.7 and 9.4 T, consists of 2 lines in partial agreement with other recent NMR works [4,5,6]. Line A shows a paramagnetic shift and a Gaussian lineshape, while line B is unshifted and its roughly triangular shape may be fitted by the Fourier Transform of a stretched exponential.

Figure 2 (left) shows the temperature dependence of the A frequency, together with a fit to a Curie law which is followed for T>16K.

The ratio between the area of line A and of line B is temperature independent down to 50K.

The same Curie temperature behaviour is observed for the linewidths of both lines, which increase on lowering the temperature. However the FWHM does not scale linearly with the applied field.

At low temperatures the inhomogeneous contribution to the line broadening, separated by a Hahn echo experiment, dominates, indicating the presence of a static distribution of fields. while at room temperature the linewidth is of homogeneous nature.

The longitudinal relaxation time (T_1) of line B has been studied at different fields and temperatures. It is shown in figure 2 (right) for applied fields of 0.3T, 4.7T and 9.4T. It may be interpreted as exchange narrowing of magnetic fluctuations at high temperature, and their freeze out as the temperature is decreased. This interpretation is however not fully consistent with exchange narrowing observed down to 10K by μSR.

The recovery of the magnetization after an inverting pulse is a single exponential for high field and high temperature ($T > 150K$ for an applied field of 4.7T or 9.4T) while it is a stretched exponential at lower temperatures. At a field of 0.3T the recovery is stretched exponential at all the temperatures. This behaviour of the spin lattice relaxation is consistent with a spin glass picture.

Conclusions In ^{13}C NMR the linewidth strongly increases on cooling below 180K, showing that the C_{60} molecule rotation is frozen below that temperature. The spin glass picture is supported by the strong magnetic depolarization observed in μSR below 10 K, by the stretched exponential lineshape and relaxation function for one of the two proton lines, and by the minimum of its T_1. The paramagnetic behaviour of the other proton line abruptly stops at 10K, where the shift saturates; this indicates that the two proton lines probe different aspects of the same phenomenology.

References

[1] P. M. Allemand, F. Wudl et al., Science 254 (1991) 826.
[2] P. W. Stephens, F. Wudl et al., Nature 355 (1992)331.
[3] K. Tanaka, A.A. Zakhidov et al., Phys. Lett. A 164 (1992) 221, and Phys.Rev. B (to be published).
[4] P. Venturini et al., Int. J. of Modern Phys. B 6 23 (1992)3947, and submitted to Solid St. Comm.
[5] R. Blinc et al. submitted to Solid St. Comm.
[6] Y.J Uemura et al. Phys. Rev. B 31 (1985) 546, and H. Pinkvos, A. Kalk, Ch. Schwink, Phys. Rev. B 41 (1990) 590
[7] R.Blinc et al, present conference

PULSED ESR AND NMR IN MAGNETIC TDAE-C$_{60}$

R. Blinc, P. Cevc, D. Arčon, J. Dolinšek, D. Mihailovič, and P. Venturini

J. Stefan Institute, University of Ljubljana, Ljubljana, Slovenia

Abstract

Pulsed ESR as well as ^{13}C and ^{1}H NMR measurements show that the magnetic transition in TDAE-C$_{60}$ is spin glass like rather than ferromagnetic or antiferromagnetic. The observation of a slow component in the time decay of the remanent magnetization below T_c - observed by ESR after the magnetic field $H_0 \approx 20$ Gauss has been switched off - supports this observation.

1. Introduction

The nature of the magnetic transition[1,2] in TDAE-C$_{60}$ (TDAE = tetrakis-dimethyl-amino-ethylene) at $T_c = 16$ K is still not completely understood. The question whether the unpaired spin density is on the C$_{60}$ molecule or on the counter-ion is as well not definitely settled.

To check whether we deal with

a) a ferromagnetic transition

b) an antiferromagnetic transition, or

c) a spin glass transition

we decided to perform a pulsed ESR study and compare the obtained results with high field and low field CW ESR data. To check on the spatial distribution and location of the unpaired electron spin density, a pulsed ^{1}H and ^{13}C NMR study was also made. Finally we measured via ESR the time decay of the remanent magnetization after switching off the magnetic field to check on the possible existence of glassy dynamics below T_c.

2. Experimental Results

2.1. *High field and low field CW ESR:*

The intensity of the high field ESR line at room temperature is compatible with the presence of one unpaired spin per C$_{60}$ ion. The position of the line is the one expected

284

for the case that the major part of the spin density is located on the C_{60}^- ion. The temperature dependences of the width of the high field ($\omega_L/2\pi = 9$ GHz) and low field ($\omega_L/2\pi = 45$ MHz) ESR spectra of TDAE-C_{60} are shown in Fig.1 for the region close to T_c.

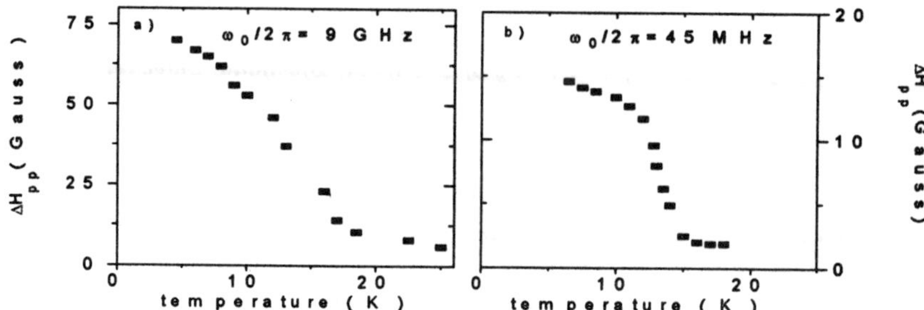

Figure 1: Temperature dependence of the width of the high field (a) and low field (b) ESR spectra of TDAE-C_{60} in the vicinity of T_c.

In both cases there is a large increase in the width of the ESR signal at T_c. There are, however, important differences. The high field transition occurs at a higher T_c value than the low field one. This confirms the previous observation[1,2] that T_c is field dependent and shifts with increasing magnetic field to higher temperatures as expected for a ferromagnetic ordering. Another important difference is that the width of the ESR line below T_c is of the order of 70 Gauss at $\omega_L/2\pi = 9$ GHz (i.e. in the X-band) whereas it is only 16 Gauss at $\omega_L/2\pi = 45$ MHz. This shows that at least part of the magnetization below T_c is induced by the magnetic field.

Another interesting point is that the shift of the center of the resonance line (Fig. 2a) below T_c does <u>not</u> scale with the magnetic field as expected if it would be due to a change in the g-factor.

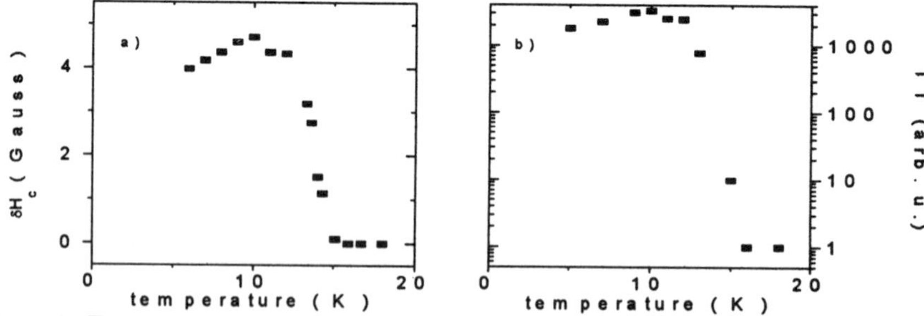

Figure 2: Temperature dependence of the position of the center of the low field ESR line (a) and the intensity (b) of this line around T_c.

It is of the order of 3.8-4 Gauss in the low field case (Fig. 2a). This is much too small for a classical ferromagnetic transition.

The transition at T_c is also connected with a tremendous increase ($\approx 10^3$) in the intensity of the low field ESR line (Fig. 2b). Since the ESR intensity is proportional to the electronic susceptibility, this means a huge increase in the local correlations leading to an increase in the susceptibilty. There is however no diverging susceptibility characteristic for a ferromagnet but rather a rounded shoulder as expected for a spin glass in a magnetic field.

2.2. Pulsed ESR

CW ESR does not discriminate between a homogeneous and inhomogeneous line broadening. This distinction is critical as an inhomogeneous line broadening would reflect a local distribution $P(h)$ which is signature of a spin glass transition. Such discrimination is possible by pulsed ESR[3].

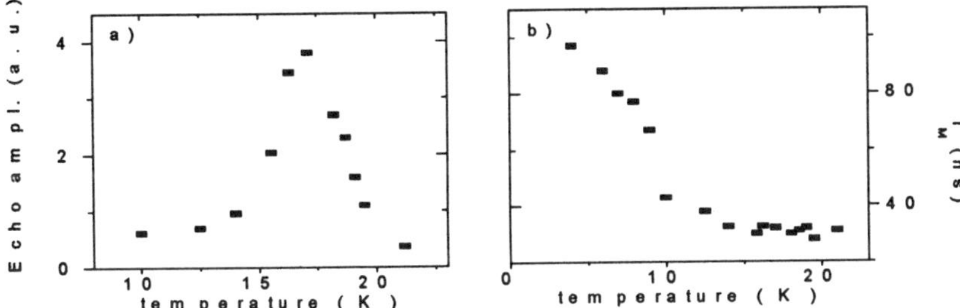

Figure 3: Temperature dependence of the electron spin echo amplitude (a) and the phase memory time (b) in TDAE-C_{60}.

The temperature dependence of the electron spin echo amplitude at the Larmor frequency $\omega_L/2\pi = 9.84$ GHz is shown in Fig. 3a. No electron spin echo was detected above 30 K. The echo became detectable around 22 K and the echo amplitude reached a maximum at $T_c \approx 16.5$ K. The presence of an echo demonstrates the existence of internal inhomogeneities in the spin system, i.e. the ESR line can be considered to be composed of homogeneous "spin packets" with different Larmor frequencies. For T$>$ 30 K the ESR line is homogeneous and no spin echo is found. The width of the homogeneous spin packets is given by the "phase memory" time T_M. The temperature dependence of T_M is shown in Fig. 3b. It should be noticed that T_M is T - independent between 22 K and 16 K and slowly increases with decreasing T between T_c and 9 K. At 9 K there seems to be a sharp increase in T_M. Below this temperature T_M continues to increase more gradually with decreasing T.

The comparison of T_M and the high field CW ESR linewidth (Fig. 1a) demonstrates that the ESR line is indeed inhomogeneously broadened around and below T_c. The inhomogeneity seems to be connected with the formation of ferromagnetically correlated clusters of different sizes and directions of magnitization. This also agrees with the S-shaped form of the magnetization versus magnetic field curve which shows that clusters formed by $10^2 - 10^3$ spins become ferromagnetically correlated below

286

T_c^4.

2.3. ^{13}C and ^1H NMR

The proton NMR spectra[5] as well as the ^{13}C NMR spectra of TDAE-C_{60} are composed of two lines (Fig. 4a). In both cases the position of one (B) of the two lines is practically temperature independent, whereas the second (A-line) undergoes a strong paramagnetic shift on approaching T_c from above (Fig. 4b).

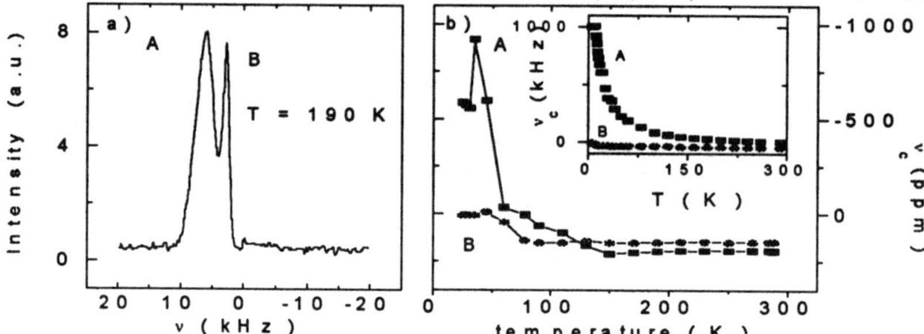

Figure 4: (a) ^{13}C NMR spectrum of TDAE-C_{60} at 190 K and $H_0 = 6.34$ T. (b) Temperature dependence of the positions of the two ^{13}C NMR lines with respect to TMS. The insert shows the T-dependence of the shifts of the two proton NMR lines.

This demonstrates that the unpaired spin density is located both on the C_{60}^- and on the TDAE$^+$ ions. The fact the maximum value of the ^{13}C paramagnetic shift is one order of magnitude smaller than maximum value of the ^1H shift could be due to the fact that the unpaired Π-electron spin density on the C_{60}^- ions would be in the first approximation zero at the carbon sites and is non-zero only due to higher order and finite curvature effects.

The T_1 for the ^{13}C A-lines is about 25 ms between 300 K and 25 K whereas the corresponding value for the proton A-line in the same temperature interval is about 1 ms.

The value of T_1 for the proton B line, on the other hand, is about 1 sec at 200 K. At this temperature the value of the T_1 of the ^{13}C B-line is also more than 1 sec. At lower temperatures the ^{13}C B-line T_1 approaches that of the A-line and becomes short (≈ 25 ms) whereas the ^1H B-line T_1 remains rather long and very different from the ^1H A-line T_1.

Both the ^1H as well as the ^{13}C NMR lines are inhomogeneously broadened as demonstrated by 2D "separation of interactions" NMR.

2.4. Magnetization Decay

The time dependences of the magnetization in a spin glass changes dramatically at T_c. Above T_c the entire magnetization responds rapidly to a change in the magnetic field, whereas below T_c the frozen out spins respond extremely slowly.

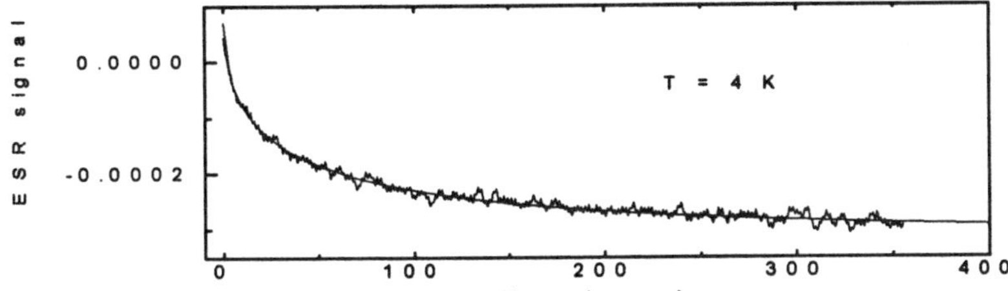

Figure 5: Time dependence of the decay of the remanent magnetization in TDAE-C$_{60}$ at T= 4 K after the field H$_0 \approx$ 20 Gauss was switched off. The solid line represents the fit to the stretched exponential form Eq. 1.

To check on that the isothermal decay of the remanent magnetization has been measured via the zero field ESR signal after the small external field (H$_0 \approx$ 20 Gauss) has been switched off. Above T$_c$ no time persistent remanent magnetization or slowly relaxing component could be seen on the time scale (1-10^3 sec) of this experiment. Below T$_c$, on the other hand, both a slowly relaxing magnetization component and a remanent magnetization could be observed (Fig. 5). The magnetization decay curves could not be described by a single exponential. They could be however well represented by a stretched exponential decay form

$$M(t) = M_0 + M^* \exp\left(-(t/\tau)^\alpha\right) \quad , \quad T < T_c \quad . \tag{1}$$

The remanent magnetization M$_0$ decreases with increasing T and was too small to be observable above 12 K. The value of α changed from 0.14 in the 4-7 K T-interval to 0.07 above 8 K, whereas τ was found to be thermally activated and reached 75 s at 4 K.

The above dynamic features are characteristic of a spin glass like freezing process[6] rather than of a ferromagnetic or antiferromagnetic type phase transition.

3. Discussion

The above results seem to show that we deal in TDAE-C$_{60}$ with a spin glass transition which can be formally described with a random bond Sherrington-Kirkpatrick Hamiltonian[6]

$$\mathcal{H} = -\sum_{i<j} J_{ij} S_i S_j - \sum_i H_i S_i \tag{2}$$

where

$$P(J_{ij}) = \frac{1}{\sqrt{2\pi J'^2}} \exp\left(-(J_{ij} - J_0')^2/2J'^2\right) \tag{3}$$

with $J' = J/\sqrt{N}$ and $J_0' = J_0/N$. The corresponding magnetic phase diagram is schematically shown in Fig. 6. It is possible that the mean

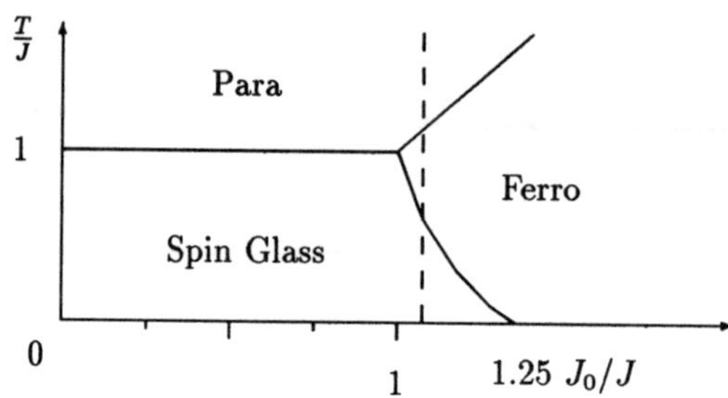

Figure 6: Schematic magnetic phase diagram[6] for Ising spins interacting via an infinite ranged gaussian distribution of exchange forces with a variance J and a mean J_0. The dotted line shows the possible phase transition sequence in TDAE-C_{60}.

J_0 of the coupling constant distribution is larger than the variance J so that the transition sequence is not simply paramagnetic - spin glass but rather paramagnetic - glassy ferromagnetic - spin glass. This could explain the apparent existence of two phase transitions, one arround 14-16 K and another arround 8-10 K.

References:

1. P.M. Allemand, K.C. Khemani, A. Koch, F. Wudl, K. Holczer, S. Donovan, G. Gruner, and J.D. Thompson, *Science* **253**, 301 (1991).

2. P.W. Stephens, D. Cox, J. W. Lauher, L. Mihaly, J.B. Wiley, P.M. Allemand, A. Hirsch, K. Holczer, Q. Li, J.D. Thompson, and F. Wudl, *Nature* **355**, 331 (1992).

3. P. Cevc, R. Blinc, D. Arčon, D. Mihailovič, P. Venturini, S.K. Hoffman, and W. Hilczer, to be published.

4. K. Tanaka, A.A. Zakhidov, K. Yoshizawa, K. Okahara, T. Yamabe, and K. Yakushi, *Phys. Rev.* **B 47**, 7554 (1993).

5. R. Blinc, J. Dolinšek, D. Arčon, D. Mihailovič, and P. Venturini, *Solid State Communications* **89**, 487 (1994).

6. K. Binder, and A.P. Young, *Rev. Mod. Phys.* **58**, 801 (1986).

SUSCEPTIBILITY ABOVE AND BELOW T_c IN TDAE-C_{60}.

D.Mihailović, V.Kraševec, D.Arčon, K.Lutar, P. Venturini and P.Cevc
J.Stefan Institute, University of Ljubljana, Slovenia

ABSTRACT

Measurements of the susceptibility χ as a function of frequency of the external field are presented which show that the material exhibits many of the characterisitics of a spin glass: frozen magnetic moments below $T_c \sim 16$ K, a lack of long-range order and somewhat unusual low frequency magnetic relaxation. In addition, we point out that above T_c the spin susceptibility measured by ESR and the macroscopic susceptibility measured by AC susceptibiity are found to display very different behaviour, the former being predominantly Curie-like, while the latter being virtually temperature-independent i.e. Pauli-like.

Introduction

Since the discovery of a magnetic state below 16 K in the doped fullerene tetrakis(dimethylamino)ethylene (TDAE-C_{60})[1], interest in this and other organic magnetic materials based on fullerenes has significantly increased. On the basis of the non-linear magnetization curve, a dramatic increase in the susceptibility below 16 K and the field-induced enhancement of T_c it was suggested[1] that the material is a soft itinerant ferromagnet. Somewhat unusual were the conspicuous absence of a remanent magnetization and the absence of any significant frequency shift of the ESR line due to an internal field, which should be present in a ferromagnet with long range magnetic order. Thus a number of groups proposed alternative suggestions to the original interpretation: Zakhidov et al[2] suggested that the material is a superparamagnet, while Blinc et. al.[3] suggested it was a spin-glass. In this paper we show using a.c. susceptibility measurements that the material shows many of the properties expected of a spin-glass, namely low-frequency dynamics and frequency-dependent susceptibility. In addition we point out some unusual features in the temperature dependence of the susceptibility above and below T_c.

Measurements

The material was prepared in the usual way, where extreme care was taken to avoid oxygen contamination. For the AC susceptomery, the samples were sealed in quartz ampoules which were designed to fit snugly into the sapphire sample holder of an ACSUS susceptometer. All the measurements were done with temperature cycling rates < 2 K/min. Additional measurements were taken with a Faraday balance type susceptometer, where the samples were placed in sealed calibrated holders. The ESR and magnetic properties of the samples did not change in the 6 months during which experiments were performed. ESR measurements were taken at 9 GHz in a Bruker 300 ESR spectrometer and an Oxford Instruments cryostat.

Low-temperature AC susceptibility

In Figure 1 we show the real and imaginary parts of the AC susceptibility of TDAE-C_{60} powder at different frequencies of AC field. We observe a very strong frequency dependence of the susceptibility. The extremely slow dinamics which imply a magnetization relaxation on the millisecond timescale could also possibly be due to powder rearrangment in an applied field. To eliminate this possibility, we have sealed the powder in a calpilary containing Ar gas. At low temperature the Ar solidifies and eliminates any possible movement of the powder particles. The frequency dependence behaviour was found to be unchanged, from which we conclude that the effects we observe are intrinsic to the material.

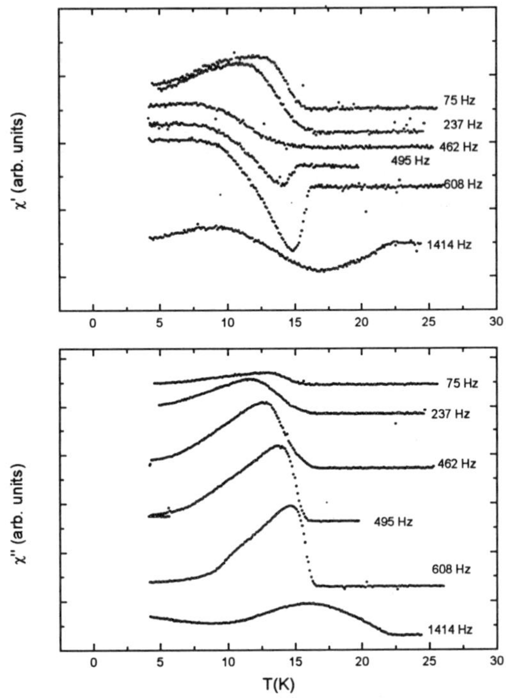

Figure 1. Real and imaginary parts of the susceptibility measured at different frequencies of applied AC field. The data were taken on cooling where the rate was the same for all curves.

The unusual feature of the different curves is the apparent negative χ' below T_c at frequencies above 500 Hz. We suggest that rather than signifying a diamagnetic susceptibility it implies a dispersive component of the relaxation with a center frequency near 500 Hz. The imaginary part correspondingly shows a peak in dissipation at the same frequency. In spin glasses, we usually observe a shift in the peak of the susceptibility with frequency, while the present data suggest the existence of a dominant relaxation process near 500 Hz. Further, the curves in Figure 1 show that the the relaxation rate is strongly dependent on temperature. We have plotted χ' and χ'' as a function of frequency at 12 K in Figure 2.

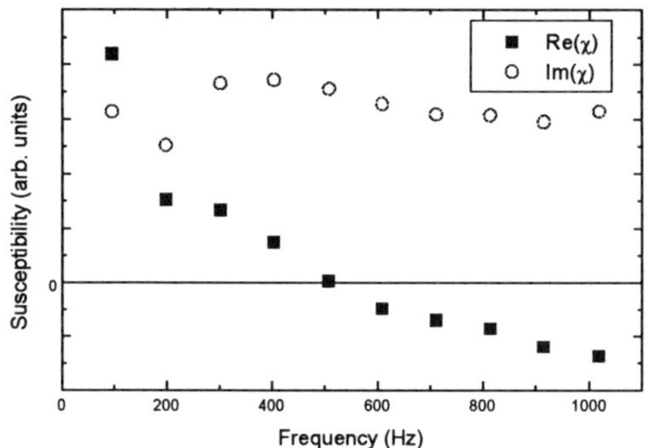

Figure 2. The frequency dependence of χ' and χ'' at 12K as a function of frequency of applied AC field. The shape of the curves is distorted by the non-linear frequency response of the susceptometer.

Unfortunately due to the limitations of our instrument, we have not been able to trace the entire curve, but the zero crossing is clearly observed. We note also an unusual feature observed in Figure 2, namely that the value of T_c - as defined by the point where the susceptibility starts to deviate from the high-temperature behaviour - is strongly dependent on the AC measuring frequency. The origin of this behaviour is possibly related to ordering which occurs on temperature cycling.

Pauli-like versus Curie-like susceptibility

All measurements of macroscopic susceptibility in TDAE-C_{60}, whether this is done by a DC SQUID, AC susceptibility or a Faraday balance technique give very similar, Pauli-like susceptibiity, which is virtually temperature independent (Figure 3 a).

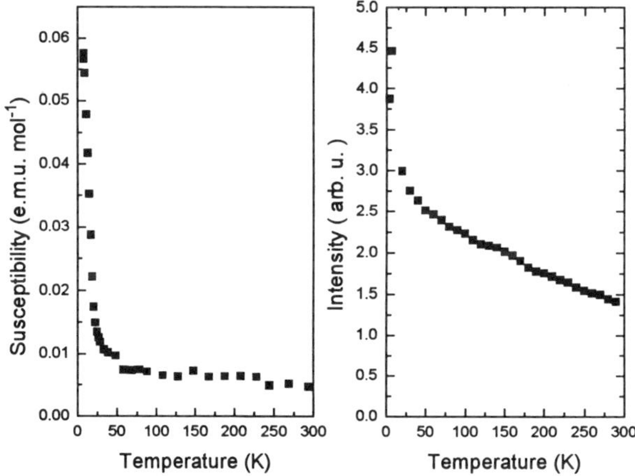

Figure 3. a) The susceptibility of TDAE-C_{60} measured macroscopically by Faraday balance and b) the spin susceptibility measured by ESR intensity.

Although in the original report[1], χ was not corrected for the sample holder, subsequent measurements have been, and we have found the original data to be reproduced by both AC susceptibility measurements and Faraday balance measurements. In contrast, the spin susceptibility, (also shown in Figure 3) as measured by the ESR intensity as a function of temperature (in agreement with previous reports[2]) shows very different behaviour, namely predominantly a Curie-like susceptibility. The detailed shape of this curve we have found to be somewhat dependent on thermal history of the sample between room temperature and T_c, which suggests that the orientational disorder of the C_{60} molecules, (which from NMR we know orders at around 170K) plays a role in the spin ordering.

Discussion

Usually there are three defining properties for a spin glass[4]: (i) frozen-in magnetic moments below a certain freezing temperature T_f accompanied by a peak in the susceptibility, (ii) lack of periodic long-range magnetic order and (iii) remanence and magnetic relaxation on macroscopic time scales below T_f when there are changes in magnetic field.
Of these, we clearly observe (i) a peak in the a.c. susceptibility which unambiguously comes from frozen magnetic moments, we can deduce a lack of magnetic long range order (ii) from the absence of a large internal field in zero field ESR measurements (only a small g-value shift is observed below the transition, which is inconsistent with a large internal field), and finally (iii) magnetic relaxation which is reflected in the strong frequency dependence of the real and imaginary parts of the susceptibility.
Given that the glass-like behaviour is observed, we suggest a possible microsopic origin to be due to a Fermi-glass-like distribution of localized state energies arising from the disorder in the orientation of C_{60} molecules. At low energies, some of these states are singly occupied, which leads to a Sherrington-Kirkpatrick[5] distribution of exchange interactions between spins on neighbouring buckyballs and a glassy spin ordering.

Conclusion

A time-dependent study of the macroscopic susceptibility reveals characteristic spin-glass behaviour for the organic magentic material TDAE-C_{60}. The origin of the glassy behaviour of the spins is suggested to be in the orientational disorder of the C_{60} molecules.
We also wish to point out the apparent discrepancy between the Curie-like spin susceptibility measured by ESR and Pauli-like behaviour in the macroscopic SQUID and AC susceptibility which shows that a significant component of susceptibility seems still unaccounted for.

References

[1] P-M.Allemand et al, Science, **253**, 301 (1991)

[2] K.Tanaka et al, Physics Letters A **164**, 221 (1992)

[3] P.Venturini et al, Int.J.Mod.Phys. B **6**, 3947 (1992)

[4] K.Binder and A.P.Young, Rev.Mod.Phys. **58**, 801 (1986)

[5] D.Sherrington and S.Kirkpatrick, Phys.Rev.Lett. **35**, 1972 (1975)

DOPING OF C_{60} WITH TERTIARY AMINES: A COMPARATIVE STUDY - EVIDENCE FOR ITINERANT META-MAGNETISM?

H. KLOS, I. RYSTAU, A. SCHILDER, W. SCHÜTZ AND B. GOTSCHY
Physikalisches Institut der Universität Bayreuth, GERMANY

and

A. SKIEBE AND A. HIRSCH
Institut für organische Chemie II der Universität Tübingen, GERMANY

ABSTRACT

The fullerene radical anion salt (TDAE)C_{60} exhibits a transition into a ferromagnetic ground state below 16 K. The interesting properties of (TDAE)C_{60} stimulated us to search for other amines which form together with C_{60} magnetic fullerites. Three tertiary amines where investigated. The doping process was studied by *in situ* EPR and absorption spectroscopy (UV/VIS/NIR). It is a complex composition of several subsequent reaction steps.
The solid state properties were investigated by EPR and AC-susceptibility, the electrical properties were investigated contactless by microwave conductivity.

1. Motivation

The high electron affinity of C_{60} implies a facile access to salts containing fullerenid anions. One method for the preparation of organic charge transfer complexes of fullerene-C_{60} is the chemical reduction of C_{60} with strong organic donors like tetrakis(dimethylamino)-ethylene (TDAE) [1].
The radical anion salt (TDAE)$^{+\cdot}C_{60}^{-\cdot}$ was the first soft ferromagnetic C_{60}-system discovered [2]. Its Curie-temperature of about 16 K is the highest transition temperature of a pure organic material reported. The magnetic properties of (TDAE)C_{60} were interpreted on the basis of itincrant ferromagnetism [3,4], which is unusual for organic ferromagnets. The success of C_{60}-doping with the tertiary amine TDAE stimulated us, to investigate the amine (A) route in more details. We will present magnetic resonance data and studies of the reaction kinetics of new charge transfer salts prepared by doping of C_{60} with the tertiary amines 1,8-diazabicyclo(5.4.0)undec-7-ene (DBU) [5, 6], 1,5-diazabicyclo(5.3.0)non-5-ene (DBN) [7] and for comparison TDAE.

2. Results

The UV/Vis/NIR-spectra of a benzonitrile solution of all three system show an absorption in the NIR with a maximum at about 1078 nm. The intensity of this diagnostic peak, which is attributed to the C_{60} mono anion [8], versus time monitors the formation and in the case of (DBx)C_{60} the annihilation of the charge transfer salt (A)$^{+\cdot}C_{60}^{-\cdot}$. Figure 1, right shows the time dependence of the 1078 nm absorption in (DBU)C_{60}. The reaction can be separated into two subsequent processes, a fast charge transfer from the amine to C_{60} (characterised by k_1) and a slower radical recombination process (characterised by k_1^*). More precisely the C_{60} doping with amines can be best described

294

with a sequence of two irreversible processes of first order. The radical anion salt $(A)^{+\cdot}$ $C_{60}^{-\cdot}$, which is responsible for the magnetism, is only an intermediate product of the reaction. An analytical solution for the C_{60}^- concentration can be found for an excess of A.

$$\left[C_{60}^{-\cdot}\right] = \frac{k_1 \cdot [C_{60}]}{k_1^* - k_1} \cdot \left(e^{-k_1 t} - e^{-k_1^* t}\right) \tag{1}$$

The quantity, which is experimentally accessible, is the extinction E which is proportional to the concentration. The solid line in figure 1, right is a fit of the measured absorption data at 1078 nm in $(DBU)C_{60}$ in benzonitril to equation 1. The validity of these interpretations is further confirmed by EPR measurements *in situ* of the C_{60}-amine-doping in benzonitril which reflect closely the time dependence of the 1078 nm absorption. Values of k_1 and k_1^* for DBU, DBN, TDAE are compiled in table 1.

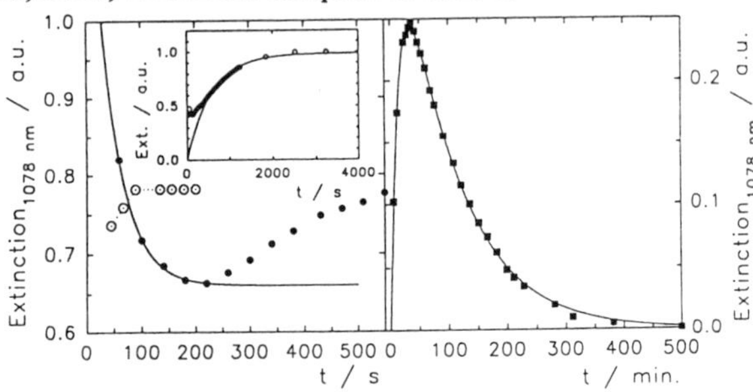

Figure 1, left: Extinction of $(TDAE)C_{60}$ vs. time in benzonitril at 1078 nm (closed circles), open circles: C_{60}^{2-} extinction; right: Extinction of $(DBU)C_{60}$ vs. time in benzonitril at 1078 nm; .

Table 1

	$(TDAE)C_{60}$	$(DBN)C_{60}$	$(DBU)C_{60}$
k_1/s^{-1}	>0.1	$1.9 \cdot 10^{-2}$	$1.1 \cdot 10^{-3}$
k_1^*/s^{-1}	0	$1.0 \cdot 10^{-3}$	$1.8 \cdot 10^{-2}$

As can be seen from the insert of figure 1, left the radical recombination in $(TDAE)C_{60}$ is missing $(k_1^* \to 0)$. A nucleophilic amine addition as in the case of $(DBx)C_{60}$ is impossible. In addition we find for TDAE a peculiarity , which was not found for DBx. $(TDAE)C_{60}$ has a very high extinction at 1078 nm for very short times. The extinction nearly reaches the saturation value for long times. Thus, the time constant for the single electron transfer must be beyond the limit of the spectrometer plus the time to mix TDAE and C_{60}/benzonitril (total *deadtime* $\approx$ 10 s). The C_{60}^- absorption at 1078 nm decreases below 200 s (see figure 1, left). Obviously also more than one reaction path exists in TDAE. In this time domain we find an increased absorption at wavelengths, where the C_{60}^{2-} absorption is expected [8] (C_{60}^{2-}: open circles in figure 1, left). But no pronounced peaks can be resolved - the C_{60}^- absorption still dominates. But the dianions are only metastable and finally only the mono anions survive.

The differences in the reaction kinetics between TDAE and DBx are due to the different chemical natures of these amines and the redox potentials. These differences can also be seen in the different velocities for the formation of solid $(A)C_{60}$. The precipitation time decreases form DBU to DBN to TDAE (the CT-salt is insoluble in benzene). Therefore the determining step for the reaction is the electron transfer, which depends on the strength of the donor.

The conclusions from the NIR-absorption experiments are consistent with observations from EPR measurements in the solid state. For $(TDAE)C_{60}$ with $k_1^* \rightarrow 0$ one spin per formula unit in the solid state was already reported [2]. In contrast in the case of $(DBx)C_{60}$ we get very inhomogeneous solids with a variable amount of spins. However, we find also a clear trend. In $(DBN)C_{60}$ with the intermediate k_1^* typically 14 % spins survive, whereas in $(DBU)C_{60}$ with the largest k_1^* we find only about 4 % spins per formula unit. Probably a small amount of $(DBx)^+ \cdot C_{60}^-$ is trapped within diamagnetic clusters and the second reaction step is prevented by immobilisation.

At higher temperatures pristine polycrystalline $(TDAE)C_{60}$ exhibits a solid broad line of about 22 G [9], whereas pristine polycrystalline $(DBx)C_{60}$ exhibits two EPR-lines. One broad line of about 40 G and a narrow line of about 2 G. The broad lines exhibit a narrowing with a final line width of 5.0 G at 5 K. The susceptibility in $(TDAE)C_{60}$ is of simple Curie-type and exhibits a strong increase below 16 K (figure 2, left). For the broad line in $(DBU)C_{60}$ we find a Curie behaviour at higher temperatures. Anyhow, at about 70 K a cusp in $\chi \cdot T$ evolves (figure 2, right). This cusp is beyond the error of the measurement, it is reproducible and also seen in measurements of the static susceptibility. The magnitude of the change in χ corresponds to the pairing of two independent spins one half to a triplet ground state. For $(DBN)C_{60}$ the susceptibility of the broad line is Curie-Weiss like with a Curie-temperature of about -4 K (weak anti-ferromagnetic).

Figure 2, left: AC-susceptibility × temperature in $(TDAE)C_{60}$ vs. temperature at 10 kHz; right: EPR-susceptibility × temperature in $(DBU)C_{60}$ vs. temperature at 10 kHz.

3. Conclusions

For TDAE, DBU, DBN we found pronounced magnetism. However, the strong electronegativity of C_{60} opens a new reaction path. As a consequence of such a radical recombination, which is prevented in TDAE by steric hindering, only a few percent of the nominal spin concentration survive in $(DBU)C_{60}$ or $(DBN)C_{60}$ and only a short range magnetic order in $(DBU)C_{60}$ evolves below 70 K. One further important finding came from EPR studies in the solid state. We found a clear difference for the EPR line width at room temperature and the temperature dependence of the EPR line width between $(TDAE)C_{60}$ and $(DBx)C_{60}$. The EPR-line width ΔB decreases much slower with temperature T in $(DBx)C_{60}$ ($\Delta B \propto T$ at high temperatures) than in $(TDAE)C_{60}$ ($\Delta B \propto T^3$). Obviously ΔB is not only an intramolecular property of C_{60}^-. We suppose, that instead the electron - phonon coupling is responsible for the different EPR line widths (electron - phonon coupling is also responsible for the superconductivity in fullerene salts) and we propose to adopt a slightly different interpretation of the magnetism in $(TDAE)C_{60}$. Instead of itinerant ferromagnetism an itinerant metamagnetism [10] could prevail and the huge moment below 16 K would be due to an interplay between magnetic and elastic energy and not to any peculiarity in the band structure.

The C_{60}-C_{60} spacing in (TDAE)C_{60} is smaller than in pure C_{60} [3]. Thus, already the first interpretations of the magnetism in this system were based on itinerant moments. But so far conclusive conductivity data were missing. (TDAE)C_{60} is very air and moisture sensitive and prevails as an amorphous powder. Even contacting of pressed pellets seems to be dangerous due to the high reactivity of TDAE. So we decided to use a contact-less method - microwave conductivity (μW-σ) in X-band. This method is absolute reliable concerning the temperature dependence, but since sample geometry and cavity filling factor (we used loosely shot samples) enters in a crucial way the absolute value can be underestimated by a factor of 10. Figure 3 shows the temperature dependence of the μW-σ in (TDAE)C_{60} vs. temperature. σ shows an activated behaviour with an activation energy of $\Delta E \approx 60$ meV. Above 320 K the activation energy changes which we attribute to a decomposition of the sample. It is hard to relate such a small activation to any intrinsic energy in C_{60}. ΔE can be for example a mobility gap and metallic behaviour of (TDAE)C_{60} can not be ruled out.

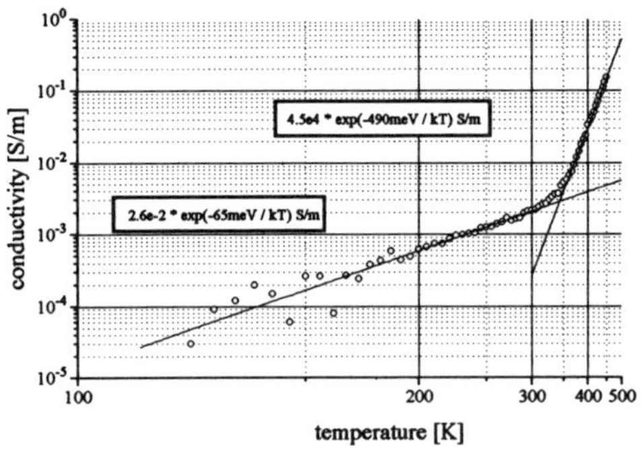

Figure 3: X-band microwave conductivity in (TDAE)C_{60} vs. temperature.

4. References

1. Haddon, A. F. Hebard, M. J. Rosseinsky, D. W. Murphy, S. H. Glarum, T. T. M. Palstra, A. P. Ramirez, S. J. Duclos, R. M. Flemming, T. Siegrist, R. Tycko, ACS Symp.Ser., **481** (1992) 71.
2. Allemand, K. C. Khemani, A. Koch, F. Wudl, K. Holczer, S. Donovan, G.Grüner,J. D. Thompson, Science, **253** (1991) 301.
3. Stephens, D. Cox, J. W. Lauher, L. Mihaly, J. B. Wiley, P. M. Allemand, A. Hirsch, K. Holczer, Q. Li, J. D. Thompson, F. Wudl, Nature **355** (1992) 331.
4. Sparn, J. D. Thompson, P. M. Allemand, Q. Li, F. Wudl, K. Holczer, P. W. Stephens, Solid State Commun. **82** (1992) 779.
5. Guggisberg, Helv. Chim. Acta **61** (1978) 1050.
6. C. Pigenet, H. Lumbroso, Bull. Soc. Chim. Fr. (1972) 3743.
7. H. Oediger et. al., Chem. Ber. **99** (1966) 2012.
8. M. Baumgarten, A. Gügel, L. Gherghel, Adv. Mat. **5** (1993) 458.
9. Tanaka, A. A. Zakhidov, K. Yoshizawa, K. Okahara, T. Yamabe, K. Yakushi, K. Kikuchi, S. Suzuki, I. Ikemoto, Y. Achiba, Physics Letts. A **164** (1992) 221.
10. Duc, D. Govord, C. Lacroix, C. Pinettes, Europhys. Letts. **20** (1992) 47.

$\left(P(C_6H_5)_4\right)^+_2 C_{60}^- X^-$ A MODEL SUBSTANCE FOR SINGLE CHARGED C_{60} MOLECULES

H. KLOS, W. BRÜTTING, A. SCHILDER, W. SCHÜTZ and B. GOTSCHY
Physikalisches Institut der Universität Bayreuth, GERMANY

G. VÖLKL
Fakultät für Physik der Universität Leipzig, GERMANY

B. PILAWA
Physikalisches Institut der Universität Karlsruhe, GERMANY

A. HIRSCH
Institut für organische Chemie II der Universität Tübingen, GERMANY

ABSTRACT

For physical investigations it is desirable to have stable single crystals. Therefore we have prepared stable C_{60}^- single crystals of the C_{60}^--Tetraphenylphosphoniumhalogenides via electro-crystallisation. The electronic properties of this salts are quite similar to the ferromagnetic $TDAE^+C_{60}^-$ above T_c. Both compounds show a strong temperature dependence of the spin dynamics. At 40 K these salts exhibit three EPR resonance lines with an anisotropic behaviour. Below 30 K magnetic fluctuations are observed.

1. Introduction

Magnetic resonance techniques have been successfully applied both to pure C_{60}-crystals and to the new class of $A_x C_{60}$ compounds like $K_x C_{60}$ or $TDAE^+ C_{60}^-$. The similar compound Tetraphenylphosphonium$^+ C_{60}^-$Iodide$^-$ excited our interest due to the physical and chemical properties of this salt. $\left(P(C_6H_5)_4\right)^+_2 C_{60}^- I^-$ is a model substance for single charged C_{60} molecules in the solid.

We performed ^{1}H-NMR, CW-EPR, pulsed EPR, static/EPR-susceptibility, DC/AC conductivity, microwave conductivity and specific heat measurements to answer fundamental questions about the molecular dynamics of the C_{60}^- in the solid state.

2. Results

$\left(P(C_6H_5)_4\right)^+_2 C_{60}^- X^-$ was prepared by electrochemical crystallisation yielding stable single crystals. The results of the elemental analysis and X-ray data can be best reconciled with $\left(P(C_6H_5)_4\right)^+_2 C_{60}^- I^-_{0.35}$ [1] and $\left(P(C_6H_5)_4\right)^+_2 C_{60}^- Cl^-$ [2]. They are truly ionic salts with each anion surrounded by cations and vice versa.

I^- are located in the centre of the cubic unit cell. $\left(P(C_6H_5)_4\right)$-cations are located on $\bar{4}$ centres at $(0,\frac{1}{2},\frac{1}{4})$. The C_{60}^- anions are centred around $4/m$ centres at $(0,0,\frac{1}{2})$. The I_h point group has no element of 4-fold symmetry. Hence, two orientations at 90° relative to each other were refined with 50% occupancy factor. C_{60}^- as well as I^- anions are surrounded by a tetragonally distorted cube of $\left(P(C_6H_5)_4\right)$-cations. In turn each cation is surrounded by two interpenetrating tetragonally distorted tetrahedra one formed by C_{60}^--anions the other by I^- anions [1].

The DC-conductivity was measured between 40 K and room temperature. Both $\left(P(C_6H_5)_4\right)^+_2 C_{60}^- Cl^-$ and $\left(P(C_6H_5)_4\right)^+_2 C_{60}^- I_{0.35}$ exhibit the same thermal activation energy of

E_a = 250 meV. The activation energy obtained from microwave conductivity is E_a = 91 meV. AC-conductivity measurements were also performed at frequencies between 100 Hz and 10 MHz at 40 K, 60 K, 70 K, 290 K. In contrast to EPR or ^{1}H-NMR measurements no significant frequency dependence and no temperature dependent dynamics of the AC-conductivity was observed.

In the specific heat no first order transition appears between 20 K and room temperature. The sample $\left(P(C_6H_5)_4\right)_2^+ C_{60}^- I_{0.35}^-$ has a specific heat of 830 J/K mol at 200 K composed of a contribution from C_{60} and $\left(P(C_6H_5)_4\right)^+ Cl^-$

Figure 1 exhibits data of pulsed EPR and CW-EPR. There are two sorts of electronic spins.

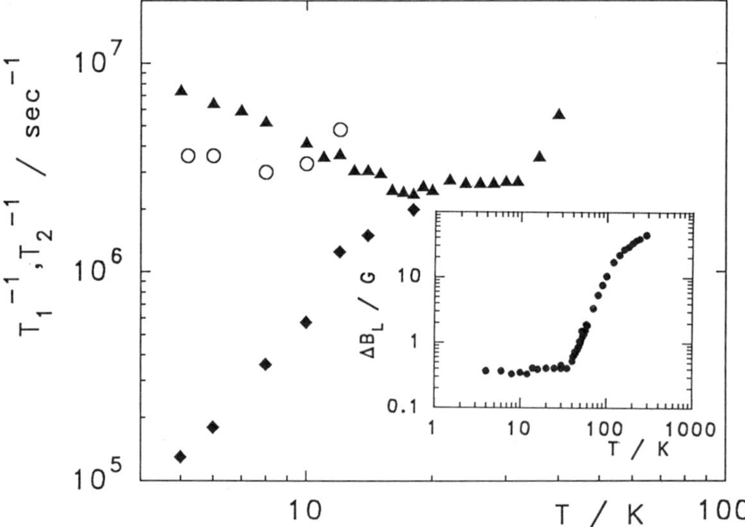

Fig. 1.: T_1^{-1} and T_2^{-1} data from pulsed EPR of C_{60}-Tetraphenylphosphoniumchloride. The insert shows CW-EPR data in a double logarithmic plot of the line width ΔB_L vs. T.

One sort which is localised $T_1^{-1} \propto T^{2.2}$. The other sort of spins is exposed to strong magnetic fluctuations characterised by a T_1^{-1}-minimum at 20 K (closed triangle). There are also two different $T_M \approx T_2$ times observed. The insert shows the line width ΔB_L of $\left(P(C_6H_5)_4\right)_2^+ C_{60}^- Cl^-$. ΔB_L is proportional to $1/T_2$. The line width is proportional to T^3 between 40 K and ~200 K. This behaviour of the line width is also observed in TDAE$^+$C$_{60}^-$ [3] and it seems to be typical for single charged fullerenes.

At low temperatures the single crystal spectra consist of three significant lines. The resonance positions of these EPR-lines are temperature independent. The three lines can be resolved between 4.2 K and 70 K for $\left(P(C_6H_5)_4\right)_2^+ C_{60}^- Cl^-$ and between 4.2 K and 90 K for $\left(P(C_6H_5)_4\right)_2^+ C_{60}^- I_{0.35}^-$. Probably the positions of the resonance lines are temperature independent also above these temperatures, because the g-factor of the EPR-line (which is the envelop of the three low temperature lines) is 1.9981 in $\left(P(C_6H_5)_4\right)_2^+ C_{60}^- Cl^-$ at room temperature in good agreement with the literature [4].

If a single crystal of $\left(P(C_6H_5)_4\right)_2^+ C_{60}^- I_{0.35}^-$ (c-axes perpendicular to B_0) is rotated in an homogeneous B_0-field the three resonance lines exhibit a characteristic rotation pattern [5]. The powder EPR-spectra of this salt reveal, that the three resonance lines belong to magnetical or spatial different species.

Figure 2 exhibits T_1^{-1} data from pulsed Overhauser shift measurements [6,7] of C_{60}^--Tetraphenylphosphoniumchloride and T_1^{-1} data from conventional solid state ^{1}H-NMR of C_{60}^--Tetraphenylphosphoniumiodide.

After a saturating rf-pulse at the Lamor frequency the pulsed Overhauser shift method probes the spin-lattice relaxation of protons which are very close to the buckyballs. The pulsed Overhauser shift is almost independent of the temperature between 4 K and 40 K. On the other hand the solid state ^{1}H-NMR measures an averaged proton magnetisation. The ^{1}H-NMR data exhibit two points at 25 K and 68 K, where a change of the temperature dependence of the spin-lattice relaxation is observed.

In a plot $1/T \cdot T_1$ vs. T of the solid state ^{1}H-NMR the data follow the electronic susceptibility.

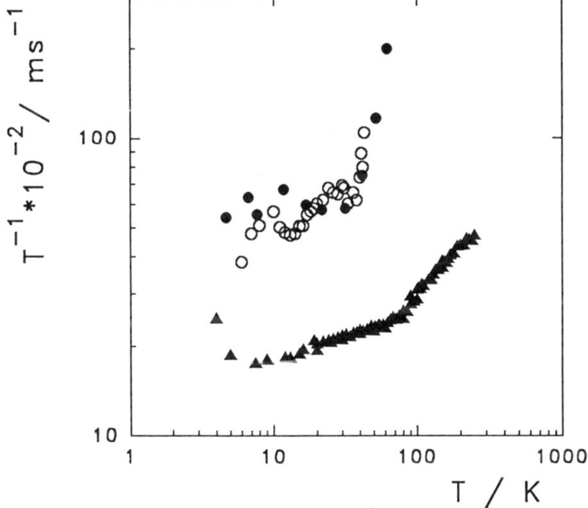

Fig. 2.: T_1^{-1} data from pulsed Overhauser shift measurements of $\left(P(C_6H_5)_4\right)_2^+ C_{60}^- Cl^-$ (open circle, 35 GHz ; closed circle 9.8 GHz) and T_1^{-1} data from conventional solid state ^{1}H-NMR (closed triangle, 200 MHz) of $\left(P(C_6H_5)_4\right)_2^+ C_{60}^- I_{0.35}^-$.

Measurements of the static susceptibility indicate, that the data do not follow a Curie-Weiss behaviour. Susceptibility data taken from EPR measurements are, within experimental accuracy, compatible with static susceptibility data. From Pascal constants we calculated for the diamagnetic contribution to the susceptibility χ_{Dia} = -1025.26 emu/mol.

The difference $\chi - \chi_{Curie} - \chi_{Dia}$ can be explained as a temperature independent Pauli like contribution consistent with the NMR-data. χ_{Pauli} corresponds to about 70 % of the total susceptibility. We find a very narrow conduction band [8] due to the large C_{60}^--C_{60}^- distance. Thus, the system is weak metallic and Mott-Hubbard localisation is prevented by non-stoichiometry. χ_{Curie} corresponds to 30 % of the total susceptibility.

300

3. Conclusions

Stable single crystals of C_{60}^--Tetraphenylphosphoniumhalogenides prepared via electro-crystallisation exhibit very interesting electronic properties. Measurements of the DC/AC- and microwave conductivity of the C_{60}^--Tetraphenylphosphoniumhalogenides and of $TDAE^+C_{60}^-$ give a thermal activated behaviour of the conductivity [9]. In the specific heat of C_{60}^--Tetraphenylphosphoniumiodide no first order transition appears between 20 K and room temperature. There are three EPR-resonance lines in the single crystal of C_{60}^--Tetraphenylphosphoniumiodide with a characteristic rotation pattern and at 70 K a dramatic change of the electronic spin dynamics is observed, where the EPR-line width increases proportional to T^3. Exactly the same behaviour is found in $TDAE^+C_{60}^-$. An explanation of this behaviour could be, that as soon as the C_{60}-balls start to rotate the Jahn-Teller-distortion of C_{60} is averaged out and we see the onset of rotational broadening or a transition from static to dynamic Jahn-Teller distortion. The beginning of the rotation is seen in the Overhauser shift measurements and in the microwave conductivity due to a strong additional microwave absorption within a few K around 70 K. Pulsed EPR reveals a coexistence of localised spins and strongly fluctuating spins. Pulsed ^{1}H-NMR-measurements of C_{60}^--Tetraphenylphosphoniumiodide reveal strong magnetic fluctuations below 30 K. Below 30 K we found a dramatic increase of the ^{1}H-NMR line width and the ^{1}H-NMR line becomes very inhomogeneous. Because of the high stability, the possibility to prepare single crystals, the similarity to other C_{60} radical anion salts (like $TDAE^+C_{60}^-$) the C_{60}^--Tetraphenylphosphoniumhalogenides seem to be model substances for single charged C_{60} molecules in the solid state.

4. References

1.   A. Pénicaud, A. Peréz-Benítez, R. Gleason, V. E. Munoz, P. R. Escudero, J. Am. Chem. Soc. **115** (1993) 10392.
2.   U. Bilow, M. Jansen, Chem. Com. in press.
3.   H. Klos, I. Rystau, A. Schilder, W. Schütz, B. Gotschy, A. Skiebe, A. Hirsch, *Proceedings of the IWEP/NM 94* submitted (1994).
4.   P. M. Allemand, G. Sardanov, A. Koch, K. Khemani, F. Wudl, J. Am. Chem. Soc. **113** (1991) 2780.
5.   M. Keil, University Bayreuth, diploma thesis 94.
6.   V. Dyakonov, G. Rösler, H. Klos, G. Denninger, Synth. Metals **55-57** (1993) 3214.
7.   H. Klos, U. Becker, V. Dyakonov, G. Rösler, B. Gotschy, G. Denninger, A.Hirsch, *Springer Series in Solid State Science* **117** (1993) 344.
8.   U. Becker, G. Denninger, V. Dyakonov, B. Gotschy, H. Klos, G. Rösler, A. Hirsch, H. Winter, Europhys. Letts. **21** (1993) 267.
9.   A. Schilder, H. Klos, I. Rystau, W. Schütz, B. Gotschy, Solid State Comun. submitted (1994)

Part Five

ELECTRONIC STRUCTURE AND ELECTRON SPECTROSCOPY

ELECTRONIC STRUCTURE OF C_{60} FULLERITES AND NANOTUBES

Steven G. Louie

Department of Physics, University of California, and
Materials Sciences Division, Lawrence Berkeley Laboratory
Berkeley, California 94720 USA

ABSTRACT

We present results on the electron excitation energies of solid C_{60} calculated using a first-principles quasiparticle theory. The effects of electron correlations and molecular orientational disorders are included. The single-particle properties of solid C_{60} are found to be well described in a quasiparticle band picture with orientation disorders. Electron-hole interactions are shown to be very strong in optical processes resulting in Frenkel excitons and sizable difference between the optical and photoemission gaps. The electronic and structural properties of carbon and other nanotubes are also investigated. Total energy results predict that the tubules are stable with respect to the formation of strips down to very small radii. Hybridization of the σ^* and π^* states is shown to be as important as band-folding effects in determining the metallicity of small radius tubules.

1. Introduction

The electronic properties of solid C_{60} have been a subject of numerous experimental and theoretical investigations[1]. Issues of importance include the size of the energy gap, the degree of band dispersion, the nature of electron correlation effects, and the effects of electron interactions and molecular orientation disorder on optical and photoemission properties. Photoemission experiments have given conflicting interpretations on the width of the electron bands, and optical and photoemission experiments yielded different band gap values indicating strong electron-hole interactions[2-8]. For the carbon nanotubes, although there are presently few direct experimental data, their electronic structures have been predicted to be highly unusual and sensitive to their geometric structures[9], varying from metals to semiconductors depending on size and chirality. Furthermore, single-wall tubules with a narrow distribution of diameters have been synthesized recently[10].

In this short review, we discuss some results of electron excitation energies in undoped solid C_{60} calculated using an *ab initio* quasiparticle approach[11]. The approach is based on an expansion of the electron self energy (the many-electron correction to an electron excitation energy) to lowest order in the dynamically screened Coulomb interaction, the so-called GW approximation. We also present results on the small-radius single-wall nanotubes. Questions of structural stability and effects of σ-π hybridization are examined.

2. Band Gaps, Photoemission, and Optical Properties of Solid C_{60}

We investigated[12] the electronic structure of solid C_{60} in three different geometric structures: a hypothetical $Fm3$ structure with all the molecules identically oriented in an fcc lattice, the experimental low temperature $Pa3$ structure with four molecules per cell, and the room temperature randomly oriented structure. The calculations of the electron excitation energies were carried out for the *Fm3* structure using the *ab initio* quasiparticle self-energy method developed in Ref. 11. A Slater-Koster Hamiltonian was obtained by fitting to the *Fm3* quasiparticle results. This Slater-Koster Hamiltonian was then used to examine the effects of molecular orientation and to compute direct and inverse angle-resolved photo-emission spectra.

Photoemission experiments[3-6] have given values of 2.3-2.6 eV for the band gap of solid C_{60}. These values are significantly different from that of a 1.83 eV gap observed in optical experiment[8] and a theoretical value of ~1 eV from local density functional (LDA) band calculations[13]. Another interesting experimental finding is the apparent lack of dispersion in the spectral peaks in angle-resolved photoemission spectra[3,7]. This observation has been used by some as evidence for nondispersive bands in this material.

Figure 1 compares the calculated quasiparticle HOMO (H_u) and the first two LUMO (T_{1u} and T_{1g}) bands in the *Fm3* structure with the LDA results. Many-electron effects give rise to two major corrections to the LDA results. The energy gap is enlarged by a factor of two from 1.04 eV to 2.15 eV. This enhancement of the gap brings the theoretical result into good agreement with the photoemission data and is consistent with trends in previous quasiparticle calculations for semiconductors and insulators. The self-energy operator seen by a quasielectron is very different from that of the quasihole. The many-body effects are not reproduced well by the simple LDA exchange-correlation potential which depends only on the local electron density. Another notable many-body correction is

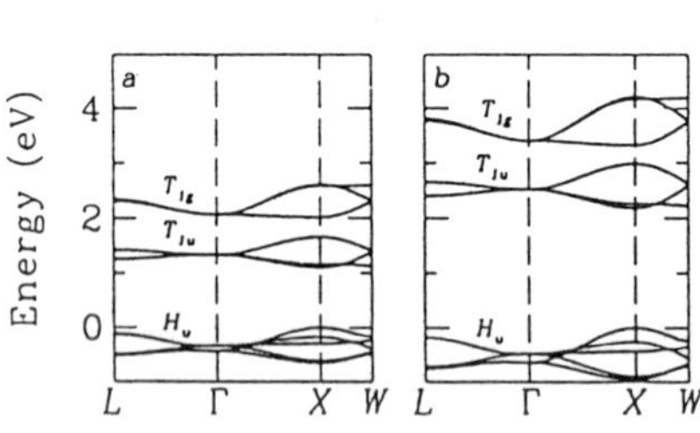

Fig. 1. *Fm3* band structure of solid C_{60} obtained in LDA (a) and in quasiparticle approach (b).

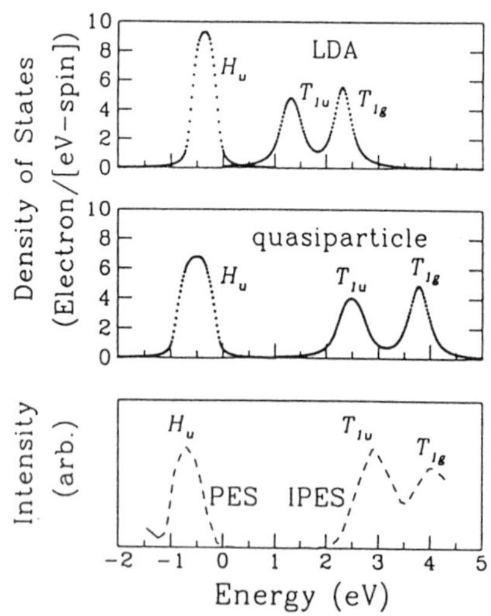

Fig. 2. DOS of solid C_{60} based on LDA results, quasiparticle results, and experimental data[5].

that the band widths of both the HOMO and LUMO complexes are increased by approximately 30%.

The quasiparticle and LDA density of states are presented in Fig. 2 along with typical room temperature photoemission (PES) and inverse photoemission (IPES) data[5]. The theoretical results are for the random structure and were simulated using an ensemble of supercells containing 512 C_{60} molecules. Orientation randomness basically removes the sharp features existed in the density of states (DOS) of the ordered $Fm3$ structure without significantly altering the band gaps and widths. We note the remarkable improvement in agreement with PES/IPES results for the quasiparticle results as compared to LDA for both the minimum band gap and the H_u - T_{1u} peak-to-peak distance. The present theory correctly calculates the excitation energies as the energies needed to add a quasiparticle or quasihole to the interacting many-electron system.

To obtain the optical gap, one must add in the interaction between the quasielectron and quasihole. We used a simple model Hamiltonian[14]:

$$H = H_e + H_h + H_{e-h} , \qquad (1)$$

where H_e and H_h are tight-binding Hamiltonians for the independent H_u quasiparticle and T_{1u} quasihole states given above. The third term is a screened electron-hole Coulomb interaction. For electron and hole at different molecules, H_{e-h} is given by the bare Coulomb interaction screened by the C_{60} lattice in a self-consistent point-dipole model. The intramolecular interactions are taken to be the Lowdin π-π Coulomb integrals in quantum chemistry, but scaled by a single-parameter to match the known on-buckyball attraction measured in the gas phase[15].

Within this model, electron-hole interactions are found to be very large, resulting in a series of Frenkel exciton bands deep in the quasiparticle gap. For the singlet states in the $Pa3$ structure, the lowest exciton level is of T_{2g} symmetry with excitation energy of near 1.6 eV and wavefunction mostly localized on a single molecule[14]. This large exciton binding energy explains the observed discrepancy between the quasiparticle gap and optical absorption threshold. Further, we find the onset of the triplet exciton series is at 0.26 eV below that of the singlet series, in good agreement with the observed singlet-triplet splitting of 0.28 eV[16].

Angle-resolved photoemission spectra were simulated by computing the k-dependent quasiparticle DOS for an ensemble of ten, randomly oriented 12x12 surface primitive cells that were 6 layer thick, oriented in the (111) direction. Such spectra may be compared to the angle-resolved inverse photoemission (ARIPES) results in Ref. 7 on epitaxially grown C_{60} films. The calculated spectra in the $\overline{\Gamma}$-$\overline{K}$-$\overline{M}$ direction of the surface Brillouin zone are given in Fig. 3a. We note that even though there are several dispersing bands in the T_{1u} and the T_{1g} complexes, the calculated spectral peaks show negligible $k_{\parallel}$ dependence. In Fig. 3b, the dispersions of the theoretical peaks are compared with those reported in Ref. 7. The absence of strong, visible dispersion effects is seen in both theory and experiment. Thus the dispersive T_{1u} and T_{1g} states can lead to apparently non-dispersive ARIPES spectra. The absence of dispersion in the spectra is caused by several

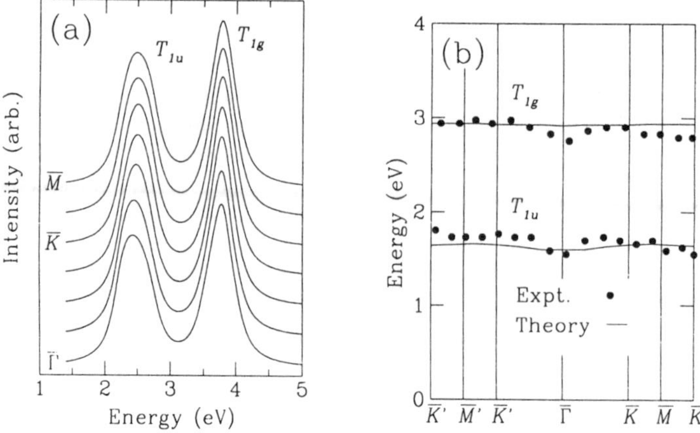

Fig. 3. Simulated angle-resolved inverse photoemission spectra for the T_{1u} and T_{1g} complexes in the $\overline{\Gamma}$-$\overline{K}$-$\overline{M}$ direction (a); peak position as function of $k_{\parallel}$ (b). Experimental data are from Ref. 7.

factors: orientational disorder, the required Brillouin-zone integration normal to the surface, finite-resolution effects, and the multi-band nature of the complexes studied. Hence, it would be incorrect to deduce that this is an extremely narrow-band system based solely on non-dispersive ARIPES data. Also, the results are in good agreement with experiment both in the peak widths and the peak-to-peak distance between the T_{1u} and T_{1g} complexes. This level of agreement is achieved only when the energy-dependent self-energy corrections are included in the quasiparticle energies. As mentioned above, the corrections lead to an ~30% increase in the band widths and in the separation between the T_{1u} and T_{1g} complexes.

For the HOMO complex, our calculated the k-dependent DOS predicted[12] that the band features should be observable in photoemission experiments with fine angular resolutions. Recent high resolution ARPES experiments[17] have indeed observed dispersive features consistent with a bandwidth given by the quasiparticle calculation.

3. Nanotubes

We focus the discussion here on small diameter single-wall nanotubes which, for carbon, have been synthesized recently using transition metal catalysts[10]. Our LDA calculations[18] revealed that carbon tubules are energetically stable with respect to strips down to quite small radii. For example, for a (6,0) tube (in the notation of Saito *et al*[9]) with diameter of d=4.78 Å, its calculated energy is lower by more than 0.2 eV/atom compared to that of the corresponding strip. The elastic energy in the tubule (which scales as d^{-2}) is still considerably lower than the dangling bond energy in forming the strips at d~4 Å, in agreement with a previous classical force-field calculation[19].

The large curvature of the small diameter tubes results in the hybridization of the σ^* and π^* states[18]. The electronic properties of small tubes are significantly altered from those obtained in previous tight-binding (TB) calculations. Strongly modified low-lying conduction band states are introduced into the band gap of insulating tubes because of strong σ^*-π^* hybridization. As a result, the LDA gaps of some tubes are lowered by more than 50%, and the (6,0) tube which previously predicted to be semiconducting is shown to be metallic. (See Table 1). The band structure and DOS for the (6,0) are shown in Fig. 4.

Table 1. Band gap (in eV) of selected carbon tubes.

Tubes	TB	LDA
(6,0)	0.05	(-0.83)
(7,0)	1.04	0.09
(8,0)	1.19	0.62
(9,0)	0.07	0.17

The state labeled (a) is especially interesting. It corresponds to a strongly hybridized σ^*-π^* state (see Fig. 5) whose energy at Γ is 0.83 eV below the doubly degenerate state that forms the top of the valence band in TB calculations.

Similar calculations[20] for nanotubes of BN and other combinations of B, N and C have been carried out. We find that, for BN, BC_3, and BC_2N, the relative energy differences between a tubule and a sheet and between a tubule and a strip are very similar to those of the carbon tubules of similar diameters, making the experimental synthesis of these tubules likely. However, in contrast to the carbon tubes, we find that all BN tubes are insulators and have a nearly constant quasiparticle band gap of ~5.5 eV independent of diameter and helicity for tubes with $d \geq 9.5$ Å. Furthermore, the bottom of the conduction band of the BN tubes is predicted to be a new type of free-electron-like tubule state with wavefunction concentrated in the interior of the tube.

4. Summary

We have performed *ab initio* calculations on the energy gap and quasiparticle ener-

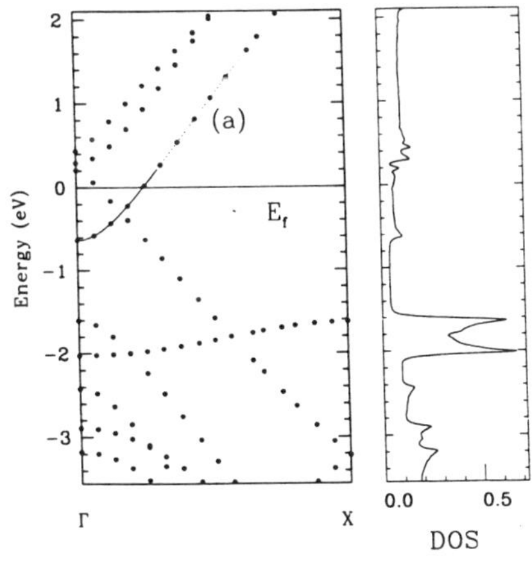

Fig. 4. Band structure and DOS for the carbon
nanotube (6,0).

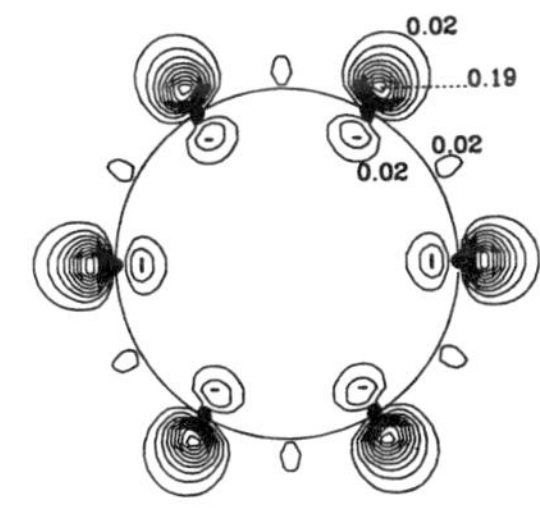

Fig. 5. Charge density of state (a) in Fig. 4.

gies in solid C_{60} showing that both many-electron and molecular orientation disorder effects are important. The single-electron excitation spectra of this material may be understood in a standard quasiparticle band picture. The description of the optical gap however requires the inclusion of electron-hole interactions (exciton effects). We have also performed studies on the electronic and structural properties of the carbon nanotubes using the LDA and uncovered a large σ-π hybridization effect on the metallicity of small diameter tubes which is as important as band-folding effects considered by earlier theories. Further our total energy calculations show that formation of other nanotubes such as those involving B, C, and N may be possible.

Acknowledgments. I would like to acknowledge collaboration with E. L. Shirley, X. Blase, L. X. Benedict, A. Rubio, Y. Miyamoto and M. L. Cohen on various parts of the work reviewed here. This work was supported by NSF Grant No. DMR91-20269 and by the U.S. DOE under Contract No. DE-AC03-76SF00098. Supercomputer time was provided by the San Diego Supercomputer Center and by the National Energy Research Supercomputer Center.

References
1. See, for examples, articles in this Proceedings and in the special issue of J. Phy. Chem. Solids, vol. **54**, no. 12 (1993).
2. J. Wu, *et al.*, Physica (Amsterdam) **197C**, 251 (1992).
3. P. J. Benning, *et al.*, Science **252**, 1417 (1991).
4. T. Takahashi, *et al.*, Phys. Rev. Lett. **68**, 1232 (1992).
5. R. W. Lof, *et al.*, Phys. Rev. Lett. **68**, 3924 (1992).
6. J. H. Weaver, J. Phys. Chem. Solids **53**, 1433 (1992).
7. J.-M. Themlin, *et al.*, Phys. Rev. **B46**, 15602 (1992).
8. G. Hartmann, *et al.*, to be published.
9. N. Hamada, S. Sawada, and A. Oshiyama, Phys. Rev. Lett. **68**, 1579 (1992); R. Saito, M. Fujita, G. Dresselhaus, and M. S. Dresselhaus, Appl. Phys. Lett. **60**, 2204 (1992); J. W. Mintmire, B. I. Dunlap, and C. T. White, Phys. Rev. Lett. **68**, 631 (1992).
10. S. Iijima and T. Ichihasi, Nature **363**, 603 (1993); D. S. Bethune *et al.*, Nature **363**, 605 (1993).
11. M. S. Hybertsen and S. G. Louie, Phys. Rev. Lett. **55**, 1418 (1985); Phys. Rev. **B34**, 5390 (1986).
12. E. L. Shirley and S. G. Louie, Phys. Rev. Lett. **71**, 133 (1993); S. G. Louie and E. L. Shirley, J. Phys. Chem. Solids **54**, 1767 (1993).
13. N. Troullier and J. L. Martins, Phys. Rev. **B46**, 1754 (1992) and references therein.
14. E. L. Shirley, L. X. Benedict and S. G. Louie, to be published.
15. R. E. Haufler, *et al.*, Chem. Phys. Lett. **179**, 449 (1991).
16. G. Gensterblum, *et al.*, Phys. Rev. Lett. **67**, 2171 (1991).
17. G. Gensterblum, *et al.*, Phys. Rev. **B48**, 14756 (1993); P. J. Benning, C. G. Olson, D. W. Lynch and J. H. Weaver, to be published.
18. X. Blase, L. X. Benedict, E. L. Shirley, and S. G. Louie, Phys. Rev. Lett. **72**, 1878 (1994).
19. S. Sawada and N Hamada, Solid State Commun. **83**, 917 (1992).
20. X. Blase, A. Rubio, Y. Miyamoto, S. G. Louie and M. L. Cohen, to be published.

Electronic structure of fullerenes from high-energy spectroscopy

M. S. Golden, M. Knupfer and J. Fink
*Institut für Festkörperforschung, IFW Dresden e.V., P. O. Box 16, D-01171 Dresden,
Federal Republic of Germany.*

J. F. Armbruster, T. R. Cummins, H. A. Romberg, M. Roth,
M. Schmidt and M. Sing
*Institut für Nukleare Festkörperphysik, Kernforschungszentrum Karlsruhe,
P. O. Box 3640, D-76021 Karlsruhe, Federal Republic of Germany.*

R. Michel, J. Rockenberger, F. Hennrich, H. Schreiber and M. M. Kappes
*Institut für Physikalische Chemie, Universität Karlsruhe, D-76128 Karlsruhe,
Federal Republic of Germany*

ABSTRACT

The electronic structure of fullerenes in the solid state has been derived from high-energy spectroscopic measurements. Photoemission and electron energy-loss spectroscopy in transmission are shown to provide a detailed picture of the occupied and unoccupied electronic states and valence band excitations in these molecular solids. In addition, angle resolved photoemission measurements of C_{60} show significant k-dependent structure, particularly in the feature derived from the HOMO-1 molecular orbitals.

1. Introduction

The electronic structure of fullerenes continues to be one of the central points in fullerene research.[1] Not only does the novel nature of these new materials make such studies of intrinsic value, but a detailed knowledge of the electronic structure also plays a vital role in the understanding of many of the fascinating and potentially useful physical properties of the fullerenes and their compounds. In this contribution we highlight some of the information about fullerene electronic structure that can be gained from high-energy spectroscopies such as photoemission (PES) and electron energy-loss spectroscopy (EELS) in transmission using data from a wide range of fullerenes (C_{60}, C_{70}, C_{76} and C_{84}) as cases in point. The topical question of the measurement of band dispersion in solid C_{60} using angle resolved photoemission (ARPES) is also discussed in the light of new data from the (111) surface of ordered C_{60} thin films.

2. Experimental

Fullerene-containing soot was produced using the Krätschmer/Huffman carbon arc method.[2] After extraction with toluene, the C_{60} and C_{70} were separated using chromatography on an alumina column with 10% toluene in hexane as eluant. The remaining higher fullerenes were purified and separated using state-of-the-art liquid chromatography.[3] The EELS experiments were performed in transmission with a primary beam energy of 170keV in a purpose-built spectrometer.[4] The samples are made by vacuum sublimation of the appropriate fullerene onto a substrate (usually an alkali halide single crystal) held at elevated temperature (~200°C). After deposition of ~1000Å of fullerene, as monitored by a quartz crystal thickness monitor, the films are floated off the substrates in distilled water, mounted on standard electron microscopy grids and transferred into the EELS spectrometer. The energy resolution of the spectrometer was set to 140meV. For the valence level excitations the momentum resolution of the instrument was set to $0.04Å^{-1}$ and for the core level excitations to $0.2Å^{-1}$. All EELS experiments were conducted at room temperature. The same samples were also characterised using TEM. For the higher fullerenes C_{76} and C_{84}, electron diffraction analysis in the TEM indicated an average face centred cubic (fcc) structure with $a_0(C_{76})=15.3\pm0.1Å$[5] and $a_0(C_{84})=15.8\pm0.1Å$.[6]

The photoemission experiments were carried out using a commercial spectrometer (VSW HA54), together with a noble gas discharge lamp providing radiation at energies of 21.22 and 16.8eV for operation with He and Ne, respectively. The electron analyser is mounted on a two axis goniometer and has a total angular acceptance of 2°. The total energy resolution was set to either 25 or 60meV. Fullerene films of ~100Å in thickness are prepared *in situ* by sublimation from a Knudsen cell onto a heated substrate. The samples fall into two main categories: those deposited onto freshly evaporated polycrystalline gold films, and those grown on the cleaved (001) surface of GeS single crystals. The latter, at least for C_{60}, form ordered single crystalline films,[7] whilst the former are polycrystalline in morphology. Ordered samples may additionally be characterised using low energy electron diffraction (LEED).

3. Results and discussion

The now familiar photoemission profile of a thin film of C_{60} deposited on gold[8] is shown in Fig. 1(a), together with the analogous spectra of C_{70} and C_{84}.[6] What is striking in these data are the relatively sharp and well-separated features in the low energy region of the spectrum of a solid. This points in the case of C_{60} to the high degeneracy of the electronic levels as expected from the high symmetry of the truncated icosahedral molecular geometry and in general to the weak interaction between the fullerene molecules in the solid state. This last point is further borne out by the similarity of the C_{60} and C_{70} spectra shown in Fig.1(a) to the PES spectrum of the respective fullerenes in the gas phase.[9,10] It can be seen from Fig. 1(a) that the spectra of these fullerenes contain three main groups of features. The first is the set of bands located closest to the Fermi energy, E_F, at ~1.5-4eV binding energy (BE). These are derived from the highest unoccupied molecular orbital (HOMO) of each fullerene, together with some of the deeper lying π^* MO's, e.g. the HOMO-1 of C_{60}. This is followed by a group located at ~6eV and a third lying deeper in energy between ~7-9eV BE. This shows that the gross features of the electronic structure are similar from one fullerene to the next, as would be expected from their common molecular architecture. However, the spectra do differ, most noticeably in the first set of bands at lowest BE which results from the greater sensitivity of the highest lying MO's to the details of the individual fullerene geometry and structure in the solid state. The lower molecular symmetry of the higher fullerenes is expressed in the reduced degeneracy of the molecular energy levels and thus an increased width of the observed bands in the PES spectra. It is possible to compare the photoemission spectra with calculations of the density of states (DOS) - and this has been done with some success for C_{60}.[8,11] However, the molecular nature of solid fullerenes ensures that the electronic states up to as much as 100eV above the Fermi level, which form the final states for valence level photoemission, remain significantly structured.[12] This leads to spectral features which are very sensitive to the photon energy used to record them, and thus which do not provide a simple measure of the occupied density of states. Consequently, it could be misleading to compare valence level PES with DOS calculations in order to attempt to distinguish between, for example, different structural isomers of a higher fullerene.

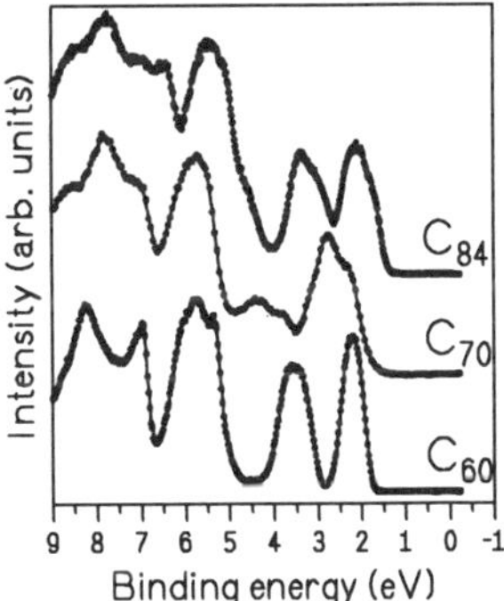

Fig. 1(a) He I PES spectra of fullerene thin films deposited on gold. $\Delta E_{1/2}$=25meV, T=300K.

Fig 1(b) C1s excitation spectra of fullerene thin films recorded using EELS in transmission. $\Delta E_{1/2}$=140meV.

The highly degenerate nature of the electronic levels of C_{60} and to a lesser extent C_{70},[13] C_{76}[5] and C_{84}[6] is also seen in the unoccupied states. Figure 1(b) shows the C1s core level excitation spectrum of these fullerenes measured using EELS in transmission. In such an experiment, electrons are excited from the C1s core level into carbon-derived unoccupied states of appropriate symmetry. In analogy with other conjugated carbon systems,[4] the structure at energies below ~290eV is due to transitions into the bands derived from the unoccupied π^* MO's of the fullerene. The data for C_{60} and C_{70} are similar to those recorded using XAS[14] and IPES.[15] The step-like structure at ~290eV corresponds to the onset of transitions into the unoccupied σ^* levels, the energy of which is insensitive to the particular fullerene in question. It can be seen from Fig. 1(b) that the energy of the onset of the $C1s{\rightarrow}\pi^*$ transitions decreases as the fullerene size increases. This has been related to the decrease in curvature and concurrent decrease in C2s character in the π electronic system as the larger fullerenes become more graphitic in nature.[5,6]

Further information about the distribution of the π and π^* bands can be obtained from the valence level excitations observed in EELS. The loss functions for low momentum transfer of C_{60},[13] C_{70},[13] C_{76}[5] and C_{84}[6] are shown in Fig. 2. The general structure is familiar from studies of graphite[16] and conjugated polymers,[4] in which transitions between the occupied and unoccupied π and σ-derived states give rise to peaks in the loss function, resulting in maxima centred at ~6 and ~25eV which are often referred to as the π and $\pi+\sigma$ plasmon, respectively. The number of $\pi{\rightarrow}\pi^*$ transitions resolved for C_{60}, and to a lesser extent for the higher fullerenes, is further evidence of the degree to which the molecular nature of the electronic structure of the fullerenes is retained in the solid state. The onset of the loss function can be seen to decrease from 1.8eV for C_{60} to 1.2eV for C_{84} indicating the smaller gap in the higher fullerenes. In each case, the gap measured using EELS in the solid state is similar to that observed in solution, and is significantly smaller than the transport gap derived from PES/IPES data. This indicates the excitonic nature of the gap transition observed in EELS experiments. From the difference between the EELS and PES/IPES gap values, the exciton binding energy can be estimated to be ~0.5-0.8eV for C_{60}, C_{70} and C_{84}.[5]

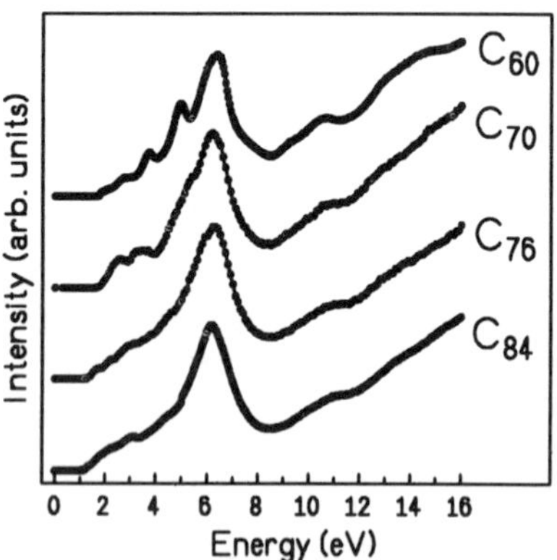

Fig. 2. Loss function of fullerene thin films recorded using EELS in transmission with q=0.15Å^{-1}.

The preceding discussion highlights the tendency of solid C_{60} and other fullerenes to behave in a localised, molecular manner. Taken together with the fullerenes' well known propensity for orientational disorder, this raises the question as to whether fullerenes in the solid state are able to support extended, band-like electronic states. One possible test of this is the observation, or otherwise, of band dispersion in C_{60} using ARPES. Bare LDA calculations predict a dispersion of ~0.5eV within the HOMO-derived bands,[17] or ~0.9eV when correction for the self-energy is included.[18] A dispersion of ~0.9eV is predicted for the (HOMO-1)-derived bands.[19] Consequently measurement of dispersion of this order by ARPES would normally be expected to be well within the grasp of the present state-of-the-art photoemission instrumentation.

Until lately, the existing experimental evidence from angle resolved studies pointed towards small or negligible dispersion of the occupied[20] and unoccupied[21] bands of C_{60}. However at last year's Winter School, Gensterblum *et al.*[22] presented ARPES data recorded with hv=8.1eV which revealed significant structure in the HOMO-derived feature. This structure was interpreted in terms of an initial state HOMO feature comprised of two main components, one of which remains relatively constant in energy and one which disperses by ~400meV along the $\overline{\Gamma} - \overline{K}$ direction of the surface Brillouin zone (SBZ).

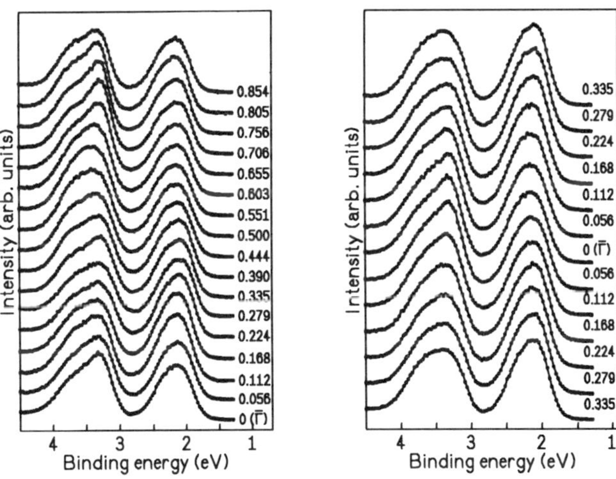

Fig. 3. EDC's of the HOMO and (HOMO-1)-derived features of $C_{60}(111)$ recorded with hv=16.8eV. The labels indicate the $k_{\parallel}$ value (in Å^{-1}) associated with photoemission from the HOMO feature. Left-hand panel: $\overline{\Gamma} - \overline{K} - \overline{M}$ direction; right-hand panel: $\overline{M} - \overline{\Gamma} - \overline{M}$ direction. $\Delta E_{\frac{1}{2}}$=60meV, T=300K.

In order to investigate this question further we have carried out ARPES measurements of ordered thin films of C_{60} on GeS(001) using Ne I radiation (hv=16.8eV). The energy distribution curves (EDC's) corresponding to emission from the

HOMO and (HOMO-1)-derived bands of C_{60} as a function of wavevector component parallel to the surface, $k_\parallel$, are shown in Fig. 3. The left- and right-hand panels show spectra recorded in the high symmetry $\overline{\Gamma}-\overline{K}-\overline{M}$ and $\overline{M}-\overline{\Gamma}-\overline{M}$ directions of the SBZ, respectively. It is clear from the figure that there are changes in the shape of the HOMO, and more pronouncedly, the (HOMO-1)-derived peaks as the value of $k_\parallel$ is altered.

The (HOMO-1)-derived peak appears to be composed of two features (most visible at the zone centre and at $\overline{M}$ in the second SBZ ($k_\parallel \approx 0.7 Å^{-1}$)). No splitting of the (HOMO-1)-derived feature was observed in our photoemission studies of polycrystalline C_{60} films (e.g. see Fig. 1(a)). The smooth nature of the changes in the HOMO-1 feature as a function of $k_\parallel$ would argue in favour of dispersion (either in the initial or final states) as the cause. It is interesting to note, however, that the EDC's recorded at the symmetry equivalent $\overline{M}$ points in the first and second SBZ are markedly different ($k_\parallel$(HOMO-1)=0.335 in the $\overline{M}-\overline{\Gamma}-\overline{M}$ series and 0.705Å^{-1} in the $\overline{\Gamma}-\overline{K}-\overline{M}$ series, respectively). These differences cannot be due to dispersion in $k_\parallel$, but signal instead the importance of k-dependent matrix element effects. While it is evident that there are changes in the spectral profiles of these features as a function of $k_\parallel$, it is far from simple to analyse such changes in terms of an $E(k)$ dispersion relation for the HOMO or HOMO-1 sub-bands of C_{60}. There are a number of reasons why this is the case, the most important being the three-dimensionality, large real-space lattice and multi-band nature of solid C_{60}. The small Brillouin zone resulting from the large real space lattice of C_{60} means that in order to work with acceptable momentum resolution, the ARPES experiments need to be carried out with low photon energies.[23]

The consequence of this is that with hv=8-17eV, the final states accessed in the experiment lie only 6-15eV above E_F. However, the data of Fig. 1(c), as well as IPES[15] and PES[12] experiments show that such electronic states remain highly structured. Consequently, the ARPES data should be interpreted in terms of transitions between structured, dispersive initial *and* final states - i.e. $k_\parallel$-dependent structure observed in the spectra need not necessarily be a product of dispersion in the initial DOS alone. In addition, only $k_\parallel$ is strictly conserved in ARPES (resulting from the loss of translational invariance at the surface, which must be crossed by the escaping photoelectron).[24] Therefore, although the value of $k_\parallel$ can be selected by choice of appropriate final state energies and emission angles, the value of $k_\perp$ is not easily determined, and as C_{60} is cubic, the dispersion in k perpendicular to the plane of the surface cannot be disregarded. Additional difficulty is conferred by the multi-band nature of the HOMO and the (HOMO-1)-derived features. For example, at the $\overline{M}$ point one would need to fit the spectrum of the HOMO-derived feature with 5 peaks - one for each of the non-degenerate sub-bands formed from the h_{1u} molecular orbitals. This problem may be exacerbated by the presence of satellites due to phonon excitations (which are clearly observed in gas phase photoemission measurements[9] and in photoemission from the t_{1u}-derived bands of A_3C_{60} (A=K, Rb)[25,26]), leading to the need to synthesise the experimental HOMO-derived feature in each EDC with a *minimum* of five components.

Thus, a number of complications conspire such that it is not possible to assign a particular feature in the EDC's to a corresponding sub-band of the HOMO or HOMO-1 manifold. Very recently, Benning *et al.*[27] arrived at a similar conclusion in a thorough ARPES study of $C_{60}(111)$ using very low energy photons (hv=8.1-10.8eV), thus conferring a $k_\parallel$ resolution more than two times better than in the present study. They clearly show the importance of a combination of final state dispersion together with favourable matrix elements in producing sharp structure of the type observed by Gensterblum *et al.*[22] with hv=8.1eV. In addition, they conclude that the observed widths of the HOMO and (HOMO-1)-derived features (both ~1eV) are a result of the underlying bandwidth and not a result of vibronic sidebands.

4. Conclusions

Photoemission and electron energy-loss spectroscopy in transmission provide a valuable source of information about the occupied and unoccupied electronic structure of fullerenes in the solid state. The spectra of different fullerenes exhibit a number of similar features as a result of their common molecular architecture. However, the lower degeneracy of the higher fullerenes in comparison to C_{60}, resulting from their lower molecular symmetry, is clearly visible both in the valence band photoemission, the C1s core level excitation spectra and the valence band excitations. The latter also clearly show the trend of reducing band gap on increasing fullerene size. Whilst providing evidence for dispersive bands underlying the (HOMO-1)-derived feature, ARPES measurements of $C_{60}(111)$ indicate the importance of k-dependent matrix elements in the determination of the overall spectral profile of the states derived from the fullerene molecular orbitals.

5. Acknowledgements

Our thanks to the authors of Refs. [19] and [27] for communication of their work prior to publication, and to Christoph Meingast for assistance with the printing of the manuscript.

6. References

[1]For good reviews of the field of solid state and electronic structure studies of fullerenes, see for example: J. H. Weaver and D. M. Poirier, in *Solid State Physics*, edited by H. Ehrenreich and F. Saepen (Academic, New York, 1994), Vol. 48, p.1, and references therein; and J. H. Weaver, *J. Phys. Chem. Solids* **53** (1992) 1433

[2]W. Krätschmer, L. D. Lamb, K. Fostiropoulos, and D. R. Huffman, *Nature (London)* **347** (1990) 354

[3]R. Michel *et al.*, unpublished

[4]J. Fink, *Adv. Electron. Electron. Phys.* **75** (1989) 121

[5]J. F. Armbruster, H. A. Romberg, P. Schweiss, P. Adelmann, M. Knupfer, J. Fink, R. H. Michel, J. Rockenberger, F. Hennrich, H. Schreiber and M. M. Kappes, Z. Phys. **B**, submitted

[6]J. F. Armbruster, M. Roth, H. A. Romberg, M. Sing, M. Schmidt, M. S. Golden, P. Adelmann, P. Schweiss, J. Fink, R. Michel, J. Rockenberger, F. Hennrich and M. M. Kappes, Phys. Rev. **B**, submitted

[7]G. Gensterblum, Y.-M. Lu, J.-J- Pireaux, P. A. Thiry, R. Caudano, J.-M. Themlin, S. Bouzidi, F. Coletti and J.-M. Debever, Appl. Phys. **A56** (1993) 175

[8]M. Merkel, M. Knupfer, M. S. Golden, J. Fink, R. Seemann and R. L. Johnson, Phys. Rev. **B47** (1993) 11470

[9]D. L. Lichtenberger, K. W. Nebesny, C. D. Ray, D. R. Huffman and L. D. Lamb, Chem. Phys. Lett. **176** (1991) 203

[10]D. L. Lichtenberger, M. E. Rempe and G. Gogosha, Chem. Phys. Lett. **198** (1992) 454

[11]J. H. Weaver, J. L. Martins, T. Komeda, Y. Chen, N. Troullier, T. R. Ohno, G. H. Kroll, R. E. Haufler and R. E. Smalley, Phys. Rev. Lett. **66** (1991) 1741

[12]P. J. Benning, D. M. Poirier, N. Troullier, J.-L. Martins, J. H. Weaver, R. E. Haufler, L. P. F. Chibante and R. E. Smalley, Phys. Rev. **B44** (1991) 1962

[13]E. Sohmen, J. Fink and W. Krätschmer, Z. Phys. **B86** (1992) 87; E. Sohmen and J. Fink, Phys. Rev. **B47** (1993) 14532

[14]L. J. Terminello et al. Chem. Phys. Lett. **182** (1991) 491; C.-T. Chen, L. H. Tjeng, P. Rudolf, G. Meigs, J. E. Rowe, J. Chen, J. P. McCauley, A. B. Smith, A. R. McGhie, W. J. Romanow and E. W. Plummer, Nature (London) **352** (1991) 603

[15]M. B. Jost, N. Troullier, D. M. Poirier, J.-L. Martens, J. H. Weaver, L. P. F. Chibante and R. E. Smalley, Phys. Rev. **B44** (1991) 1966

[16]K. Zeppenfeld, Z. Phys. **243** (1971) 229

[17]S. Saito and A. Oshima, Phys. Rev. Lett. **66** (1991) 2637

[18]E. L. Shirley and Steven G. Louie, Phys. Rev. Lett. **71** (1993) 113

[19]K.-P. Bohnen, unpublished

[20]J. Wu, Z.-X. Shen, D. S. Dessau, R. Cao, D. S. Marshall, P. Pianetta, I. Lindau, X. Yang, J. Terry, D. M. King, B. O. Wells, D. Elloway, H. R. Wendt, C. A. Brown, H. Hunziker and M. S. de Vries, Physica (Amsterdam) **C197** (1992) 251

[21]J.-M. Themlin, S. Bouzidi, F. Coletti, J.-M. Debever, G. Gensterblum, L.-M. Yu, J.-J. Pireaux and P. A. Thiry, Phys. Rev. **B46** (1992) 15602

[22]G. Gensterblum, L.-M. Yu, J.-J. Pireaux, P. A. Thiry, R. Caudano, T. Buslaps, R. L. Johnson, G. Le Lay, V. Aristov, R. Günther, A. Taleb-Ibrahimi, G. Indlekofer and Y. Petroff, Phys. Rev. **B48** (1993) 14756

[23]The value of $k_\parallel$ corresponding to the distance from the zone centre to the $\overline{M}$ and $\overline{K}$ points of the SBZ of $C_{60}(111)$ are $[\sqrt{2}(2\pi)]/[\sqrt{3}(a_0)]=0.362\text{Å}^{-1}$ and $[\sqrt{2}(4\pi)]/[3(a_0)]=0.418\text{Å}^{-1}$, respectively. Thus each measurement at normal emission with $h\nu=16.8\text{eV}$ averages over ~15% and ~13% of the SBZ in these high symmetry directions, respectively.

[24]For example: E. W. Plummer and W. Eberhardt, Adv. Chem. Phys. **49**, Eds.: I. Pirigone and S. A. Rice (Wiley, New York 1982); Studies in Surface Science and Catalysis vol. **74**, Ed.: S. D. Kevan (Elsevier, Amsterdam 1992)

[25]M. Knupfer, M. Merkel, M. S. Golden, J. Fink, V. P. Antropov and O. Gunnarsson, Phys. Rev. **B47** (1993) 14734

[26]P. J. Benning, F. Stepniak, D. M. Poirier, J. L. Martins, J. H. Weaver, L. P. F. Chibante and R. E. Smalley, Phys. Rev. **B47** (1993) 13843

[27]P. J. Benning, C. G. Olson, D. W. Lynch and J. H. Weaver, Phys. Rev. Lett., submitted

Corrections to Migdal's theorem for photoemission spectra

O. Gunnarsson

Max-Planck-Institut für Festkörperforschung, D-70506 Stuttgart,Germany

and

V. Meden and K. Schönhammer

Institut für Theoretische Physik, Universität Göttingen, D-37073 Göttingen, Germany

Abstract

The electron spectral function is calculated for a model including electron-phonon coupling to Einstein phonons, to study corrections to Migdal's theorem. A cumulant expansion is used for the time-dependent Green's function, and the second and fourth order cumulants are studied. For a one-band model it is illustrated that Migdal's theorem is valid in the large band width limit for a fixed coupling constant. In this limit, the spectrum has at most one satellite in addition to the main peak. As the band width is reduced, the spectrum calculated with the fourth order cumulant develops multiple satellites, if ε_k is close to the Fermi energy. Although the interband electron-phonon coupling is shown to be substantial, and the effective band width may therefore be argued to be large, it is shown that Migdal's theorem is nevertheless not valid for C_{60}.

1. Introduction

The electron-phonon coupling is believed to be important for many properties of doped C_{60} compouds (A_3C_{60}, A=K, Rb). The theoretical treatment of the electron-phonon coupling is greatly simplified for systems where Migdal's theorem[1] is valid. This theorem states that it is sufficient to treat the lowest order diagram in the self-energy if the electronic band width is sufficiently large compared with the largest phonon energy. For typical metals, the band width may be two to three orders of magnitude larger than the largest phonon energy, and Migdal's theorem is then expected to be well satisfied. The spectral function has been studied extensively for the case when Migdal's theorem is assumed to be valid[2]. For A_3C_{60} compounds the largest intra-molecular phonon energy is of the order 0.2 eV, while the width of the partly occupied t_{1u} sub band is about 1/2 eV. Since these two energies are of the same order, it is highly questionable if Migdal's theorem can be used for these systems.

To go beyond Migdal's theorem, one may try to calculate more complicated diagrams for the self-energy. For Einstein phonons and in the case when the dispersion of the electronic states goes to zero, e.g., for a core-level[3, 4], this approach leads to a self-energy with an incorrect analytical structure, and a qualitatively wrong spectral function[5]. We have therefore instead used a cumulant expansion[6, 7] of the time-dependent Green's, as explained in Section 2. In Section 3 we present results and in Section 4 the validity of Migdal's theorem for A_3C_{60} is discussed. This work has been presented in more detail elsewhere[8] and exact results have been presented for a one-dimensional model[9].

2. Cumulant expansion

We consider the time-ordered Green's function $G(k,t)$. Since we are interested in photoemission, we focus on times $t < 0$ and states k with a noninteracting energy ε_k below the Fermi energy E_F. From the Green's function we can obtain the spectral function. We write

$$G(k,t) = G_0(k,t)\exp[\sum_{n=1}^{\infty} \frac{g^n}{n!}C_n(k,t)] \tag{1}$$

where $G_0(k,t) = i\exp(-i\varepsilon_k t)$ is the noninteracting Green's function. We then use a diagramatic approach to calculate the Green's function to a certain order N,

$$G(k,t) = \sum_{n=0}^{N} \frac{g^n}{n!}G_n(k,t). \tag{2}$$

By expanding the exponent in Eq. (1) and by requiring that the result is identical to Eq. (2) for all terms up to order N in g, we can determine $C_n(k,t)$ for $n \leq N$. Since, however, $C_n(k,t)$ appears in the exponent in Eq. (1), this expression contains contributions of all orders in g.

For the core-level problem, already the cumulant expansion to second order gives the correct result[4]. In the opposite limit, where the band width is made so large, for a fixed coupling, that Migdal's theorem becomes valid, the second order cumulant expansion is also correct.

3. Results

We have considered a model where the phonons couple linearly to the electrons

$$H = \sum_{k} \varepsilon_k c_k^\dagger c_k + \sum_{q} \omega_0 b_q^\dagger b_q + \tilde{g} \sum_{k,q} c_{k+q}^\dagger c_k(b_{-q}^\dagger + b_q). \tag{3}$$

Here the first two terms describe the electrons and phonons, respectively, and the last term describes the electron-phonon coupling. We introduce the total coupling $g^2 = \sum_q \tilde{g}^2$.

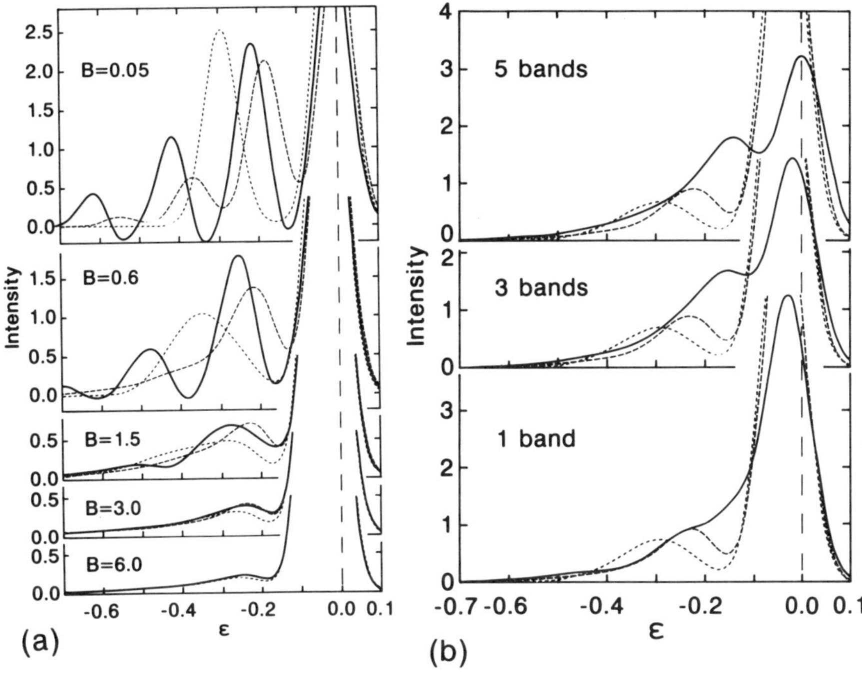

Figure 1: The spectrum for removing an electron from a state with $\varepsilon_k = 0.05$ eV. The phonon energy $\omega_0 = 0.2$ eV was used and a Gaussian broadening with the half-width 0.05 eV has been introduced. The figure shows results using the second order self-energy (dotted) and the second (dashed) and fourth (full line) cumulant expansions. The left hand figure (a) shows the spectrum as a function of the band width B for $(g/\omega_0)^2 = 1$, and the right hand figure (b) shows the spectrum for one, three or five sub bands, including spin degeneracy for $(g/\omega_0)^2 = 0.5$.

In Fig. 1a we show results for the spectral function as a function of the band width B. For $B = 6$ eV in the lower part of the left hand figure, the spectral function obtained from the second order self-energy agrees well with the results from the second and fourth order cumulants, suggesting that Migdal's theorem is well satisfied for this value of B and the parameters used here. As B is reduced, the difference between the second-order self-energy and the fourth order cumulant becomes increasingly larger. For $B = 0.6$ eV, appropriate for C_{60}, the difference is large and Migdal's theorem is apparently not valid. For small values of B, the fourth order cumulant result develops multiple satellites, of a type that is also seen for the core-level spectrum. For small values of B the spectrum is negative at some energies. This is unphysical, and indicates problems in the formalism. The reason for the negative spectral weight has been discussed elsewhere[8].

We have also performed calculations for a system with spin degeneracy. The lower part of Fig. 1b shows results for $B = 0.6$ eV. In this case the fourth order cumulant gives a substantial broadening of the main peak. Since $\varepsilon_{\mathbf{k}}$ is less than ω_0 below the Fermi energy, broadening via scattering of the hole under emission of a real phonon is energetically not allowed. However, the hole can scatter under the emission of a virtual phonon, which later decays in an electron-hole pair in such a way that energy is conserved. This is a fourth order process, which does not show up in the second order self-energy or cumulant approaches. This broadening process is very inefficient if spin degeneracy is not included, since a direct and an exchange diagram then largely cancel.

4. Interband electron-phonon coupling

It has been suggested that there might be an appreciable interband electron-phonon coupling in A_3C_{60} and that the effective band width may therefore be given by the total width of all the π-derived states, which is of the order 15 eV. Since this width is much larger than the largest phonon energy (0.2 eV), it has been argued that Migdal's theorem is valid for A_3C_{60}.

To test this, we have calculated the interband coupling using a simple model for the phonons[10], and a tight-binding description of the electrons[11]. It should be emphasized that the calculation of the electron-phonon coupling is sensitive to the details and that the treatment used here can only give the qualitative behaviour of the coupling. Using this model we calculate the change of the Hamiltonian when a given phonon mode is excited. We then calculate the expectation value of the Hamiltonian between a state in the partly occupied t_{1u} band and a state belonging to another subband ν. Squaring this expectation value, summing over all the states belong to the t_{1u} band, over all the states belonging to the band ν and over all phonon modes, we

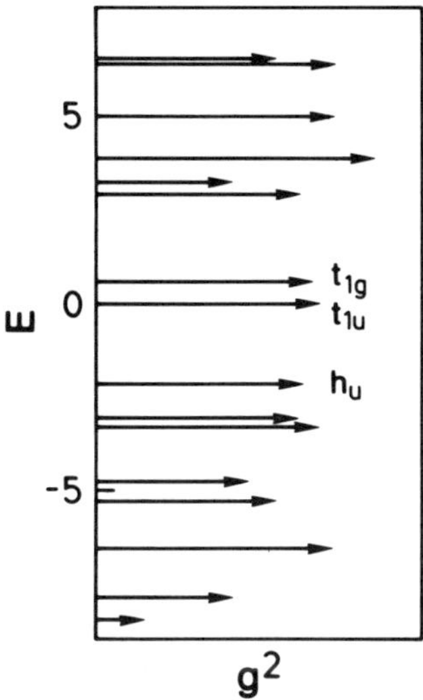

Figure 2: The square of the electron-phonon coupling strength g^2 for scattering an electron from the t_{1u} band to some other π-derived sub band under the emission of a phonon. The horizontal axis gives the coupling strength and the vertical axis the energy of the sub band in a tight-binding description

obtain the total coupling between the t_{1u} band and the band ν.

In Fig. 2 we show the calculated coupling strength. The coupling to all sub bands is comparable, and if the band gaps between the sub bands can be neglected, one may argue that the effective band width is about 15 eV.

To test the validity of Migdal's theorem for A_3C_{60} we have considered a model with several sub bands, with the fixed width 0.6 eV and the separation of the centers 1.5 eV. The results are shown in Fig. 1b. We can see that as the number of sub bands is increased, the accuracy of the second order self-energy result does not improve. This is in contrast to the results in Fig. 1a where we kept the number of sub bands fixed and equal to one, but increased the width of this sub band. In that case, the states coupling to the phonons gradually move away from the Fermi energy, and the effects of the coupling is reduced. For large B the effective coupling is then so small that it is sufficient to consider the lowest order diagram. On the other hand, if the band width and the other parameters of the one band model have values approriate for C_{60}, the coupling is too strong to allow the neglect of higher order diagrams. Adding additional bands, keeping the width of and coupling to the t_{1u} band fixed, does not reduce the effective coupling, and does not make Migdal's theorem valid.

5. Summary

We have introduced a cumulant expansion as a method for including effects beyond Migdal's theorem in the calculation of the spectral function. It was illustrated that as the band width is increased, keeping the total number of states fixed, the validity of the assumptions behind Migdal's theorem gradually improves. On the other hand, if the band width is increased by increasing the number of sub bands, keeping the width and coupling to each sub band fixed, the applicability of Migdal's theorem does not increase. Due to the large interband electron-phonon coupling for A_3C_{60}, one may argue that the effective band width is large. This does not mean, however, that Migdal's theorem is valid for A_3C_{60}.

References

[1] A.B. Migdal, Soviet Phys. JETP **7**, 996 (1958).

[2] S. Engelsberg and J.R. Schrieffer, Phys. Rev. **131**, 993 (1963).

[3] B.I. Lundqvist, Phys. Kondens. Materie **9**, 236 (1969).

[4] D. Langreth, Phys. Rev. B **1**, 471 (1970).

[5] P. Minnhagen, J. Phys. C **7**, 3013 (1974);

[6] R. Kubo, J. Phys. Soc. Japan **17**, 1100 (1962).

[7] L. Hedin, Physica Scripta **21**, 477 (1980).

[8] O. Gunnarsson, K. Schönhammer and V. Meden, (Phys. Rev. B, submitted).

[9] V. Meden, K. Schönhammer and O. Gunnarsson, (Phys. Rev. Lett., submitted)

[10] Z.C. Wu, D.A. Jelski, and T.F. George, Chem. Phys. Lett. **137**, 291 (1987); D.E. Weeks and W.G. Harter, Chem. Phys. Lett. **144**, 366 (1988).

[11] O. Gunnarsson, S. Satpathy, O. Jepsen and O.K. Andersen, Phys. Rev. Lett. **67**, 3002 (1991).

Electron Energy Loss Spectroscopy of Free Fullerenes

H. Delfs, A. W. Burose, and A. M. Ding

*Optisches Institut, Technische Universität Berlin,
Str. des 17. Juni 135, D-10623 Berlin, Germany*

Abstract. Gas phase measurements on the electronic structure of C_{60} and C_{70} have been performed using electron energy loss spectroscopy. The excitation energies are compared to results from other groups on solid fullerite and to theoretical calculations. We also developed a refined dielectric shell model based on the dielectric properties of graphite, which compares well with our results on the collective excitations (plasmons) of C_{60}.

1. Introduction

We investigate the electronic properties of fullerenes by electron energy loss spectroscopy (EELS) which allows to vary the momentum transferred to the scattering object as compared to optical spectroscopy. This makes it possible to probe not only optically allowed transitions, but also dipole forbidden transitions and to identify collective excitations by using different primary energies for the incident electrons. By applying this method to fullerenes in the gas phase we investigate purely molecular excitations.

The similarity between the configuration of the carbon atoms in the spherical C_{60} molecule and a planar sheet of graphite led us to a refinement of the dielectric shell model for fullerenes. Instead of assuming a simple homogeneous dielectric medium we incorporated radial and tangential anisotropy as known from the dielectric properties of graphite into the collective model.

2. Experimental Remarks

The experimental apparatus has been described in detail elsewhere[1,2]. In our spectrometer the monochromator and analyzer both consist of a pair of pseudo hemispherical energy analyzers. For primary electron energies between 20 and 1100 eV the analyzer had a typical angular acceptance of 0.5° and an energy resolution of 30 meV. To conduct the investigation the fullerenes in the gas phase, powder of pure C_{60} or C_{70} was vaporized in an oven at temperatures between 400 and 600°C.

3. Results

At low primary electron energy (E_P = 20 or 50 eV) the EEL spectra for both fullerenes show optical forbidden transitions at about 2 eV energy loss (peak A in Fig. 1) which vanish with higher primary energy. In contrast, at high primary energies collective electronic excitations (plasmons) at about 6 and 10 eV energy loss were observed (peak D) which show smaller intensities at low E_P. In C_{70} more transitions are observed due to the

324

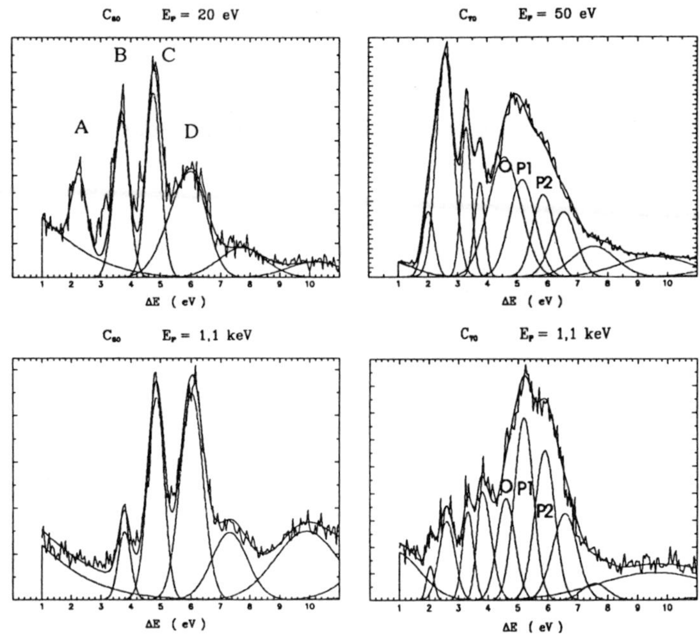

Fig. 1: EEL-Spectra of C_{60} and C_{70} taken at low and high primary electron energies together with curve fits of superpositioned Gaussian functions.

reduced symmetry of the molecule, specifically the split plasmon peak (denoted P1 and P2 in Fig. 1).

In gas phase EELS the measured signal is proportional to the differential cross section for electron scattering. In order to compare our results to those of other groups we transformed our data into generalized oscillator strength f according to the formula[3]

$$d\sigma / d\Omega|_{\theta=0} = 4e^4 E_\mathrm{P}(\Delta E)^{-3} f \ . \tag{1}$$

Table 1: Comparison of transition energies (in eV) from gas phase and solid state EELS and theoretical calculations:

Transition	our measurements		Bulliard[4]	Sohmen[5]		Lucas[6]	Bertsch[7]
	$d\sigma/d\Omega$	f	$d\sigma/d\Omega$	$Im(-1/\varepsilon)$	$\omega Im(\varepsilon)$	$Im(-1/\varepsilon)$	theory
$h_u \rightarrow t_{1u}$	2.25	2.48	2.24	2.15	2.15	2.2	2.2
$h_u \rightarrow t_{1g}$				2.7	2.7		2.8
$h_g \rightarrow t_{1u}$	3.74	3.80	3.77	3.6	3.5	3.7	3.1
$h_u \rightarrow h_g$	4.84	4.40	4.88	4.9	4.5	4.8	4.3
$g_g, h_g \rightarrow t_{2u}$		5.48		6.4	5.5		
π-plasmon	6.04	6.16	6.1	6.5		6.3	
	7.58	7.52			7.5	7.6	
	9.84	10.45				10.2	
σ-plasmon		18.5		21.6		28	

This transformation results in a slight upward shift of the peak positions, particularly at higher energies. It also brings up another strong collective excitation found as a wide band from 15 to 23 eV (Fig. 2), the so called σ-plasmon.

A comparison of our values with those obtained by other groups on solid fullerite (Table 1) shows good agreement of the transition energies. However, the origin of the remaining deviations is not clear since different response functions were measured in the experiments. We measured the HOMO-LUMO transition at 2.48 eV whereas solid state data and theoretical calculations resulted in a lower value at about 2.2 eV. An even lower excitation has been found[6] at 1.55 eV. These differences in the low energy data might be attributed to band structure effects in solid fullerite. Another significant difference remains for the high energy plasmon with values from 18.5 to 28 eV. We conclude that the close similarity between gas phase and solid state data further documents fullerites as a molecular solid.

4. The Dielectric Shell Model

The collective excitations of C_{60} can be approximated quite well by transferring the dielectric properties of graphite to the hollow spherical structure of fullerenes (warped graphitic layer). We consider a spherical shell with inside and outside radius of $R_i = 2.4$ Å and $R_o = 4.6$ Å, respectively. The dielectric continuum between these sharp boundaries is characterized by the dielectric function $\varepsilon(\omega)$. This determines the dipolar polarizability[8]

$$\alpha(\omega) = 4\pi\varepsilon_0 R_o^3 \frac{[\varepsilon(\omega)-1][2\varepsilon(\omega)+1]}{2[\varepsilon(\omega)-\varepsilon_-][\varepsilon(\omega)-\varepsilon_+]}$$

$$\text{with} \quad \varepsilon_\pm = \frac{-(5+4\rho)\pm 3\sqrt{1+8\rho}}{4(1-\rho)}, \quad \rho = \left(\frac{R_i}{R_o}\right)^3. \tag{2}$$

The molecular cross section for photoabsorption, which has to be compared to the oscillator strength extracted from our data, is then

$$\sigma_A = \frac{\omega}{\varepsilon_0 c} \text{Im}[\alpha(\omega)]. \tag{3}$$

In our model we approximate the anisotropic dielectric function of graphite for external fields parallel and perpendicular to the c-axis by a superposition of appropriate Lorentzian oscillators. A choice of oscillators at 4.5 and 11 eV for $\mathbf{E} \parallel c$ and at 4.5 and 14 eV for $\mathbf{E} \perp c$ together with a contribution of free electrons gives a good resemblance of experimental data on graphite[9]. Integrating over the varying radial and tangential components of an electric field applied to the spherical graphitic layer leads to an effective dielectric function for the shell of

$$\varepsilon_{shell} = \tfrac{2}{3}\varepsilon_{tangential} + \tfrac{1}{3}\varepsilon_{radial}. \tag{4}$$

Calculating the photoabsorption cross section in this way shows strong collective excitations from each of the single electron transitions in the graphite model (Fig. 2). The $\pi - \pi^*$ and the $\sigma - \sigma^*$ transitions in graphite at 4.5 and 14 eV lead to the so called π- and σ-plasmon peaks at 5.8 and 18.4 eV, whereas the $\pi - \sigma^*$ and $\sigma - \pi^*$ transitions, which occur in graphite at 11 eV, result in the peak at 10.7 eV. Here we accounted for less binding in the single layer for $\mathbf{E} \parallel c$ as compared to stacked layers by shifting the

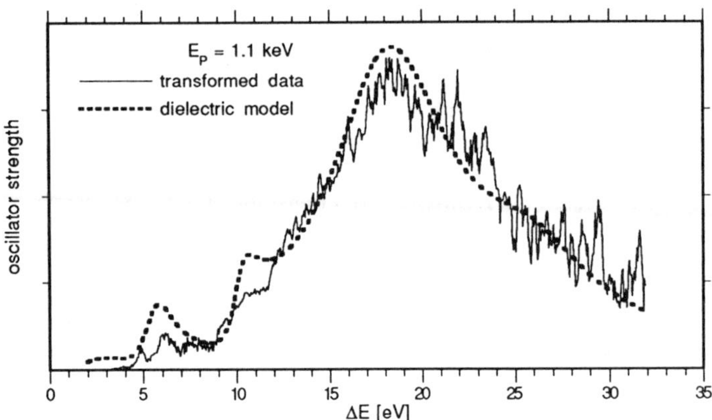

Fig. 2: Transformed experimental data proportional to the photoabsorption cross section of C_{60} and calculated curve from the dielectric shell model.

Lorentzian oscillator down to 9.5 eV. Besides this and additionally the broadening of the $\sigma - \sigma^*$ transition to 9 eV we did not change any parameter when warping graphite into C_{60}. Compared to our experimental resulted, this simple model accounts surprisingly well for the peak positions and the overall structure.

Other collective models treat the π- and σ-electrons of C_{60} either identically[8] or independently[10]. Here the double surface of the shell leads to two possible oscillatory modes which result in a double plasmon peak. However, with this straight forward incorporation of more fine structure into the dielectric model, it becomes clear that the radial or antisymmetric mode for the π- plasmon is strongly reduced. This can already be seen from a two-fluid model of the shell similar to that of Barton and Eberlein[11].

This work was possible through financial assistance from the German Research Council (DFG Sonderforschungsbereich 337).

References

[1] C. Becker, A. Burose, and A. Ding, *Z. Phys.* **D 20**, 35 (1991)

[2] A. Burose, T. Dresch, and A. M. Ding, *Z. Phys.* **D 26**, 294 (1993)

[3] M. Inokuti, *Rev. Mod. Phys.* **43**, 297 (1971)

[4] C. Bulliard, M. Allan, and S. Leach, *Chem. Phys. Lett.* **209**, 434 (1993)

[5] E. Sohmen, J. Fink, and W. Krätschmer, *Z. Phys.* **B 86**, 87 (1992)

[6] A. Lucas, G. Gensterblum, J. J. Pireaux, P. A. Thiry, R. Caudano, J. P. Vigneron, Ph. Lambin, and W. Krätschmer, *Phys. Rev.* **B 45**, 13694 (1992)

[7] G. F. Bertsch, A. Bulgac, D. Tomanek, and Y. Wang, *Phys. Rev. Lett.* **67**, 2690 (1991)

[8] Ph. Lambin, A. A. Lucas, and J.-P. Vigneron, *Phys. Rev.* **B 46**, 1794 (1992)

[9] J. Daniels, C. v. Festenberg, H. Raether, and K. Zeppenfeld, *Springer Tracts mod. Phys.* **54**, 77 (1970); H. Venghaus, *Phys. Stat. Sol. (B)* **71**, 609 (1975)

[10] D. Östling, P. Apell, and A. Rosén, *Z. Phys.* **D 26**, 282 (1993)

[11] G. Barton and C. Eberlein, *J. Chem. Phys.* **95**, 1512 (1991)

High energy spectroscopic studies of fulleride compounds

T. R. Cummins, J. F. Armbruster, H. A. Romberg, M. Roth and M. Sing
*Kernforschungszentrum Karlsruhe, Institut für Nukleare Festkörperphysik,
Postfach 3640, D-76021 Karlsruhe, Federal Republic of Germany*

M. S. Golden, M. Knupfer and J. Fink
*Institut für Festkörperforschung, IFW Dresden e.V.,
Postfach 16, D-01171 Dresden, Federal Republic of Germany*

ABSTRACT

Photoemission and electron energy-loss (EELS) investigations of the t_{1u}-derived electronic states of K_3C_{60} are reported. Earlier observation of an anomalous temperature dependence of the occupied LUMO-derived spectral weight in photoemission is confirmed, and its possible causes are discussed. EELS measurements of the charge carrier plasmon in K_3C_{60} as a function of momentum transfer reveal negligible dispersion in contrast to what is expected both for a simple metal, and from recent theoretical predictions. Preliminary EELS data of Na_xC_{60} are also reported, in which intercalation beyond x=6 is shown to result in no extra charge transfer to the fullerene-derived electronic levels.

1. Introduction

High energy spectroscopies such as photoemission (PES) and electron energy-loss spectroscopy (EELS) in transmission have played a valuable role in the elucidation of the solid state electronic structure of the fullerenes and their compounds.[1] In this contribution we present photoemission and EELS in transmission data from K_3C_{60}, focusing on the half-filled t_{1u} (LUMO)-derived conduction bands. Careful PES studies as a function of temperature confirm our previous result that there is an anomalous temperature dependence of the t_{1u}-derived spectral weight, with a continuous transfer of spectral weight to higher binding energy (BE) with increasing temperature.[2] EELS measurements reveal little or no dispersion of the ~0.6eV charge carrier plasmon excitation as a function of momentum transfer, in contrast to recent theoretical predictions. In addition we present preliminary EELS measurements of the Na_xC_{60} system, which reveal a number of important differences compared to the A_xC_{60} (A=K, Rb) fullerides.

2. Experimental

Thin films of global stoichiometry $K_{2.2}C_{60}$ of thickness ~150Å were prepared *in-situ* for photoemission studies on polycrystalline gold substrates as described previously.[3] Photoemission was carried out using He I radiation (hv=21.22eV) and a 50mm mean radius hemispherical analyser (VSW HA54), operating at a total energy resolution of 60meV. The EELS experiments were carried out in transmission with a primary beam energy of 170keV in a dedicated UHV spectrometer which is described in detail in Ref. [4]. Electron diffraction measurements were used as a tool for the *in-situ* structural characterisation of the samples. The energy resolution was set to 90meV for the plasmon dispersion studies and 140meV otherwise. The momentum resolution was chosen to be 0.04Å^{-1} for electron diffraction and valence band excitations, and to 0.2Å^{-1} for core level excitations. Free-standing films of C_{60} with a thickness of ~1000Å were prepared by sublimation in UHV and subsequently transferred into the EELS spectrometer.[5] Intercalation was performed *in-situ* by sublimation of the appropriate metal onto the fullerene film, which is held at elevated temperature, followed by a post-anneal to improve homogeneity and crystallinity.

3. Results and discussion

Figure 1 shows the photoemission profile of $K_{2.2}C_{60}$ within ~2eV of the Fermi level (E_F) as a function of temperature (15K<T<425K). The well documented phase separation in the K_xC_{60} system ensures that this region of the spectrum represents the spectral weight due to the t_{1u}-derived bands of K_3C_{60} (as the alpha phase has negligible spectral weight below ~1.5eV BE).[3,6] Care was taken to remain well below x=3 to avoid the premature formation of K_4C_{60} which has been reported.[7] The spectrum at low temperature is analogous to those reported earlier for K_3C_{60} and Rb_3C_{60}.[2,7] It can be seen from Fig. 1 that the width of the spectral weight due to the half-filled t_{1u}-derived conduction bands of K_3C_{60} is ~1.3eV. This is considerably greater than is expected from bandstructure calculations[8] which predict a width of ~0.3eV, and thus gives rise to a comparatively low density of states (DOS) at E_F. Thus in this case, the photoemission profile does not just give a simple picture of the DOS of K_3C_{60}, but rather reflects the DOS modulated by the coupling of the outgoing photoelectron both to molecular phonon modes and to the collective excitation of the charge carriers.[2]

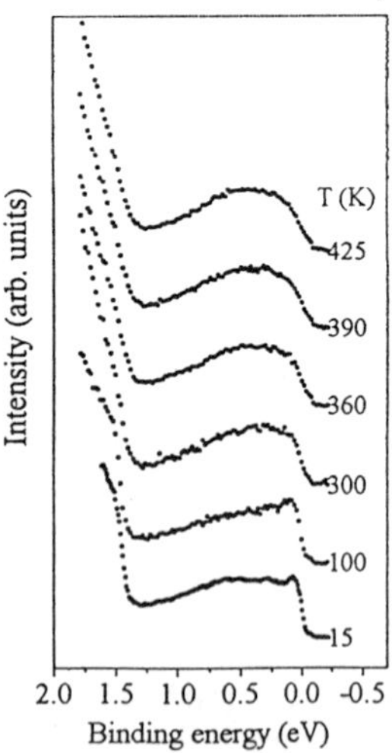

Fig. 1. PES spectra of the t_{1u}-derived bands of K_3C_{60} in the temperature range 15-425K.

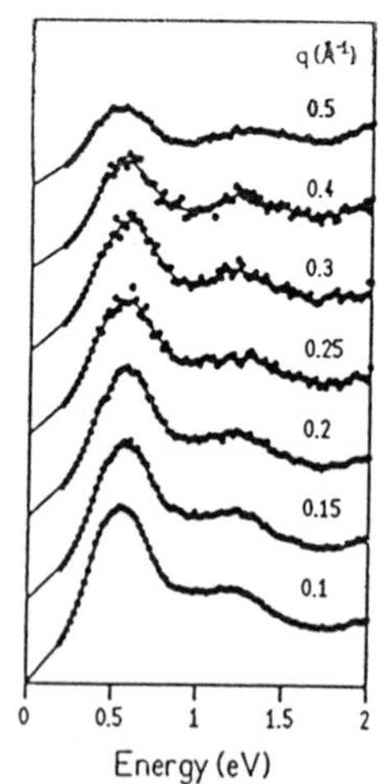

Fig. 2. q-dependent EELS measurements of the loss function of K_3C_{60}.

Following this interpretation, the features at ~0.25 and ~0.6eV BE are satellites resulting from the excitiation of the high energy A_g or H_g phonons and the charge carrier plasmon in the photoemission final state of K_3C_{60}, respectively. In a previous study[7] we noted that the photoemission spectral weight of the t_{1u}-derived bands of Rb_3C_{60} shows a marked temperature dependence. We now return to study this phenomenon in more detail for K_3C_{60}. This alkali metal fulleride has the advantage of phase separation, such that if the sample is kept below the K_1C_{60} formation temperature[9] (~150°C), the spectral weight observed within 1.5eV of E_F is due solely to K_3C_{60}.

It can be seen from Fig. 1 that with increasing temperature there is a shift in weight from the coherent, quasiparticle peak at E_F higher binding energy. Although this shift is not as large as was observed in the Rb fulleride, it is more than would be expected from the temperature dependent broadening of the Fermi cutoff alone. In addition, although there is a finite DOS at E_F in the spectra at higher temperature, the cutoff is broader than that given by the Fermi-Dirac distribution. Significantly, the changes in the spectra are reversible, and continuous over the six temperatures sampled used in the experiment. As was suggested earlier for Rb_3C_{60},[7] this argues strongly against temperature-dependent phase transitions as the cause of the spectral weight transfer. In addition, the fact that a similar temperature dependence is observed in both the K and Rb fullerides when the phase diagrams are distinctly different already points to a more subtle cause than phase transitions. Calculation of the effects of increasing temperature within a harmonic approximation (with linear coupling), has shown that changing the population of the A_g and H_g phonons and the broadening of the phonon sidebands are not adequate to account for the spectral weight transfer to higher BE.[7] It remains an open question as to the contribution from coupling to low energy anharmonic, alkali metal optical phonons or librations - whose population would be more effected by temperature changes of this order. An alternative origin of the temperature-dependent spectral weight transfer is a temperature induced metal to non-metal transition. This could arise from the formation of a pseudo-gap at E_F, driven by Anderson localisation due to increased disorder at higher temperatures. However, recent calculations predict that the Fermi surface remains intact despite the orientational disorder in A_3C_{60}, with the electronic states near E_F well described as propagating and Bloch-like with a mean free path of ~20Å.[10] A clear resolution of this question awaits detailed measurements of physical properties such as the resistivity and susceptibility of the A_3C_{60} materials up to high temperature.

Further insight into the electronic structure of the conduction band in K_3C_{60} can be gained by EELS measurements of the valence band excitations. In particular, the use of EELS in transmission uniquely allows the measurement of such excitations as a function of momentum transfer. Fig. 2 shows the loss function between 0 and 2eV as a function of momentum transfer, q. The features observed at ~0.5eV and ~1.2eV for low momentum transfer are assigned to the collective excitation of the conduction band electrons (the charge carrier plasmon) and interband transitions between the partially filled t_{1u}-derived bands and the t_{1g}-derived bands, respectively.[4] On increasing q up to 0.5Å^{-1}, the intensity of both features decreases. In the case of the plasmon, this damping is due to its decay into electron-hole excitations (interband transitions). The intensity of the $t_{1u} \rightarrow t_{1g}$ interband transitions decreases as at higher q, monopole and quadrupole transitions become increasingly favoured, at the expense of optically allowed transitions. In addition to the changes in intensity, we are able to investigate the energy position of the charge carrier plasmon as a function of q, i.e. the plasmon dispersion. It can be seen from Fig. 2 that there is no pronounced dispersion of the K_3C_{60} charge carrier plasmon which is in keeping with earlier measurements of C_{60}/C_{70} mixtures.[4] We can place an upper limit on the plasmon dispersion of ~20meV for q up to 0.5Å$_{-1}$. For $q>0.1$Å^{-1}, the uncertainties introduced due to the subtraction of the direct beam are small, and the measured q-range is greater than that required to reach the first Brillouin zone boundary, thus this lack of dispersion is surprising. For a free electron gas, the conduction band plasmon energy increases quadratically with q,[11] and would thus be expected to shift some ~340meV to higher energy on increasing q from zero to 0.4Å^{-1}. This clearly is not seen in the q-dependent loss function of K_3C_{60}, in contrast to the case of the graphite intercalation compounds, in which free-electron-like behaviour is clearly observed.[12]

Recently, Kresin and Kresin[13] have proposed that the confinement of the charge carriers to the surface of the fullerene molecules - i.e. the molecular or zero-dimensional nature of the A_3C_{60} materials - results in an unusual q-dependent behaviour of the charge carrier plasmon. In a calculation based upon the random-phase-approximation (RPA), they predict a negative plasmon dispersion law. The magnitude of the predicted negative dispersion is such that in Fig. 2 the plasmon would have shifted by up to 150meV to lower energy by a q value of ~0.5Å.[13] Thus the predicted negative plasmon dispersion appears not to occur.

Turning now from the familiar potassium/C_{60} fulleride system, we report a preliminary study of the unoccupied electronic structure of the Na_xC_{60} system using EELS in transmission. Fig. 3 shows electron diffraction profiles of a series of Na_xC_{60} samples, with $x=0$, ~3, ~5 and ~10. The intercalation level was determined from comparison of the intensity of the C1s and Na2p core level excitations (see below) and is estimated to be accurate within ±10%. Due to the small ionic radius of sodium, it is possible to exceed the usual alkali metal saturation intercalation level of six, and furthermore the Na_xC_{60} system remains fcc for both x=6 and for x up to 11.[14,15]

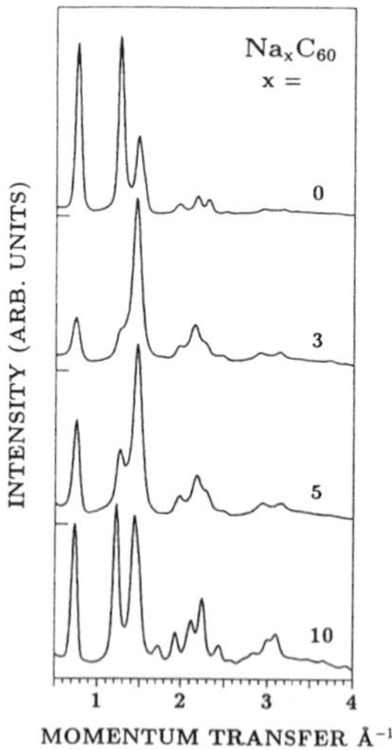

Fig. 3. Electron diffraction profiles of Na_xC_{60}

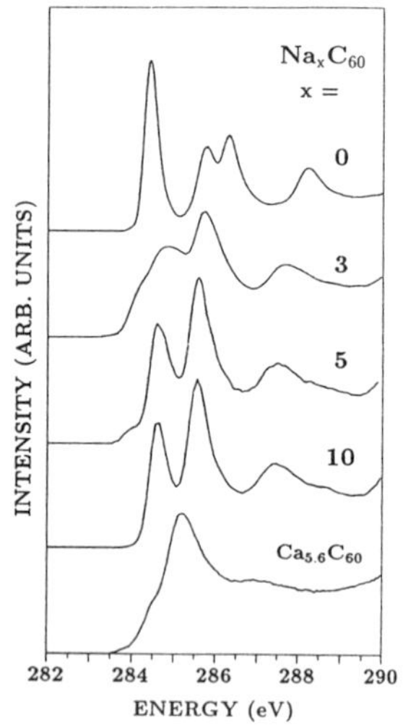

Fig. 4. C1s core level excitation spectra of Na_xC_{60} and $Ca_{5.6}C_{60}$

In the latter two cases, multiple occupancy of the fcc octahedral interstitial site has been proposed from refinements of x-ray diffraction data, giving a tetrahedral,[14] or nine-member body-centred Na cluster[15] for Na_6C_{60} or $Na_{11}C_{60}$, respectively. The electron diffraction patterns of Fig. 3 are consistent with the bulk x-ray diffraction data of Yildirim *et al.*.[15] New reflections are observed at 1.75Å[-1] and 3.05Å[-1] for the sample with highest sodium content, which correspond to the (400) and (444) reflections in Ref. [15], and appear only for $x>6$, consistent with the $x \approx 10$ stoichiometry derived from the Na2p shallow core level excitations.

Fig. 4 shows the corresponding C1s absorption edges of Na_xC_{60} together with that of $Ca_{5.6}C_{60}$.[16] These spectra are composed of excitations from the C1s core level into unoccupied C2p states and, neglecting the effect of the core hole, they represent the C2p-derived unoccupied DOS. For C_{60} the peaks are assigned to transitions to the bands derived from the π^* molecular orbitals, with those into the t_{1u}-derived bands appearing at lowest energy (284.4eV).[5] Upon Na intercalation to $x \approx 3$, and further to $x \approx 5$, all the transitions into π^*-derived bands are successively shifted to lower energy, accompanied by a decrease in the intensity of the first transition. This behaviour is familiar from the K and Rb fullerides where intercalation leads to a progressive filling of the t_{1u}-derived bands. For K and Rb this band is full at $x=6$, and the first transition channel no longer exists, thus the lowest lying feature is due to transitions into the t_{1g}-derived bands.[5] The spectrum of Na_5C_{60} shown in Fig. 4 closely resembles those of $(K/Rb)_6C_{60}$,[5] although the complete filling of the t_{1u}-derived bands has not yet been fully achieved, and so a small contribution from the transition to the t_{1u}-derived band remains as a shoulder at low energy. These data imply that Na intercalation up to $x=6$ occurs with complete charge transfer from sodium to the fullerene, in agreement with photoemission[17], Raman[18] and ^{13}C NMR[19] studies as well as recent *ab initio* calculations.[20] The core level excitation spectrum for $x=10$ is remarkably similar to that of Na_5C_{60}, and is identical to those of K_6C_{60} and Rb_6C_{60}. This suggests that additional Na intercalation beyond $x=6$ results in no measurable extra charge transfer from the Na to the fullerene electronic levels. This is consistent with the calculations of Andreoni and co-workers,[20] who predict new lowest-lying unoccupied states in Na_6C_{60} associated with the Na tetrahedron in the octahedral site (the 'O states'). In addition, reduced $Na \rightarrow C_{60}$ charge transfer for $x>6$ has been suggested from PES measurements of isolated C_{60} molecules adsorbed on a Na film held at 40K.[7] It is of interest to contrast this situation with that in another fulleride with multiple octahedral site occupancy, namely Ca_5C_{60}.[21] The C1s core level excitation spectrum of $Ca_{5.6}C_{60}$ is shown for comparison in Fig. 4. It can be seen at once that the unoccupied electronic levels of the fullerene host are quite differently effected by Ca intercalation,[16] where a decrease in the energy of the transitions into the unoccupied π^*-derived bands is accompanied by a decline in the intensity of the leading core excitation, thus giving a clear signal for the partial occupation of the t_{1g}-derived bands for $x \sim 5$, as has been directly observed in photoemission studies.[22,23]

The hybridisation observed between the fullerene and Ca electronic levels in the alkaline earth fulleride[16,22,23] raises the question whether the Na levels play an active role in the electronic structure near E_F in Na_xC_{60} for $x \geq 6$. Fig. 5 shows the Na2p shallow core level excitation spectra of Na_xC_{60} which are related to the Na(3s,3d) unoccupied DOS. It can be seen that there is no great change with increasing x, other than an increase in intensity, due to the higher Na content. There appear to be no signs of the new lowest-lying unoccupied states related to the Na tetrahedron in Na_6C_{60} predicted in Ref. [20]. We add that there are no new transitions in the valence band excitations of Na_xC_{60} at or above $x=6$ (not shown), suggesting that if there exist interband transitions between the occupied t_{1u}-derived bands and the O states, they do not have high cross-section, as was clearly stated by Andreoni *et al.*.[20]

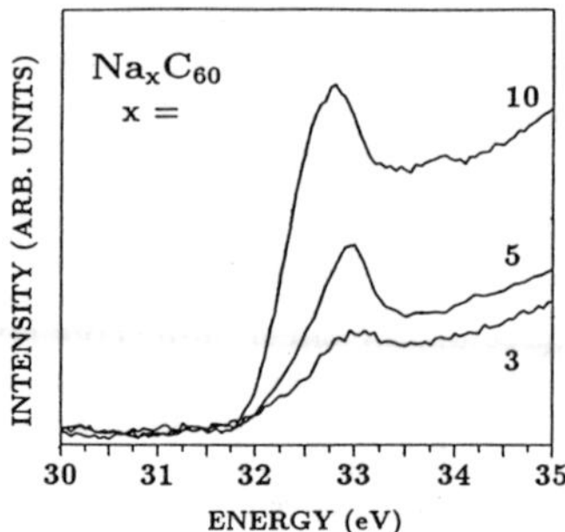

Fig. 5. Na2p core level excitation spectra of Na_xC_{60}

4. Conclusions

The t_{1u}-derived spectral weight in photoemission displays an anomalous temperature dependence with a spectral weight transfer to higher BE at high temperatures which appears unrelated to any phase transitions and may be indicative of a temperature dependent metal to insulator transition. Charge carrier plasmon dispersion in the q-dependent loss function of K_3C_{60} is less than 20meV up to $q=0.5\text{Å}^{-1}$, in contrast to the behaviour expected for conventional metals. This observation does not support recent theoretical predictions of a negative plasmon dispersion relation. EELS measurements of Na_xC_{60} indicate complete charge transfer from Na→C_{60} for compositions x≤6. For x>6 no additional charge transfer is observed, with the t_{1g}-derived fullerene bands remaining unoccupied.

5. Acknowledgements

We are grateful to P. Adelmann for the production of the C_{60} used in these studies and to the authors of Ref. [10] for communication of their results prior to publication.

6. References

[1] For a good introduction and recent review of the field see for example J. H. Weaver and D. M. Poirier, in *Solid State Physics*, ed. H. Ehrenreich and F. Saepen (Academic, New York, 1994)

[2] M. Knupfer, M. Merkel, M. S. Golden, J. Fink, O. Gunnarsson, and V. P. Antropov, *Phys. Rev.* **B47** (1993) 13944

[3] M. Merkel, M. Knupfer, M. S. Golden, J. Fink, R. Seemann and R. L. Johnson, *Phys. Rev.* **B47** (1993) 11470

[4] J. Fink, *Adv. Electron. Electron Phys.* 75 (1989) 121

[5] E. Sohmen, J. Fink and W. Krätschmer, *Europhys. Lett.* **17** (1992) 51; E. Sohmen and J. Fink, *Phys. Rev.* **B47** (1993) 14532

[6] J. H. Weaver, P. J. Benning, F. Stepniak and D. M. Poirier, *J. Phys. Chem. Solids* **53** (1992) 1707

[7] P. J. Benning, F. Stepniak and J. H. Weaver *Phys. Rev.* **B48** (1993) 9086

[8] e.g. S. Saito and A. Oshima, *Phys. Rev. Lett.* **66** (1991) 2637

[8]e.g. S. Saito and A. Oshima, *Phys. Rev. Lett.* **66** (1991) 2637

[9]D. M. Poirier and J. H. Weaver, *Phys. Rev.* **B47** (1993) 10959

[10]E. J. Mele and S. C. Erwin, unpublished

[11]see for example H. Raether, *Excitation of Plasmons and Interband Transitions by Electrons,* in *Springer Tracts in Modern Physics* **88** (Springer-Verlag, Berlin, 1980)

[12]J. J. Ritsko, *Phys. Rev.* **B25** (1982) 6452

[13]V. V. Kresin and V. Z. Kresin, *Phys. Rev.* **B49** (1994) 2715

[14]M. J. Rosseinsky, D. W. Murphy, R. M. Fleming, R. Tycko, A. P. Ramirez, T. Siegrist, G. Dabbagh and S. E. Barrett, *Nature* **356** (1992) 416

[15]T. Yildirim, O. Zhou, J. E. Fischer, N. Bykovetz, R. A. Strongin, M. A. Cichy, A. B. Smith III, C. L. Lin and R. Jelinek, *Nature* **360** (1992) 568

[16]H. Romberg, M. Roth and J. Fink, *Phys. Rev.* **B49** (1994) 1427

[17]C. Gu, F. Stepniak, D. M. Poirier, M. B. Jost, P. J. Benning, Y. Chen, T. R. Ohno, J-L. Marins and J. H. Weaver, *Phys. Rev.* **B45** (1992) 6348; D. M. Poirier, T. R. Ohno, G. H. Knoll, P. J. Benning, F. Stepniak, J. H. Weaver, L P. F. Chibante and R. E. Smalley, *Phys. Rev.* **B47** (1993) 9870

[18]S. J. Duclos, R. C. Haddon, S. Glarum, A. F. Hebbard and K. B. Lyons, *Science* **254** (1991) 1625

[19]D. W. Murphy, M. J. Rosseinsky, R. M. Fleming, R. Tycko, A. P. Ramirez, R. C. Haddon, T. Siegrist, G. Dabbagh, J. C. Tully and R. E. Walstedt, *J. Phys. Chem. Solids* **53** (1992) 1321

[20]W. Andreoni, P. Giannozzi and M. Parrinello, *Phys. Rev. Lett.* **72** (1994) 848

[21]A. R. Kortan, N. Kopylov, S. Glarum, E. M. Gyorgy, A. P. Ramirez, R. M. Fleming, F. A. Thiel and R. C. Haddon, *Nature* **355** (1992) 529

[22]Y. Chen, D. M. Poirier, M. B. Jost, C. Gu, T. R. Ohno, J.-L. Martins, J. H. Weaver, L. P. F. Chibante and R. E. Smalley, *Phys. Rev.* **B46** (1992) 7961

[23]G. K. Wertheim, D. N. E. Buchanan and J. E. Rowe, *Science* **258** (1992) 1638

TOPOLOGY AND ELECTRONIC STRUCTURE IN C_{60}

Mayumi OKADA* and Osvaldo GOSCINSKI**

Department of Quantum Chemistry, Uppsala University,
Box 518, S-751 20 Uppsala, SWEDEN

ABSTRACT

Single excitations of C_{60} are studied theoretically including some electronic correlations and effects of σ-electrons in order to investigate the singular electronic structure of C_{60}. Some dipole-allowed transitions are studied from the ground state. The excited states of some fragment molecules, radialene, pyracylene and corannulene are also studied. The electronic structures in three dimensional π-conjugated hydrocarbons are discussed. It is suggested that the charge distributions within the five membered rings are related to the appearance of the electronic excited state like an alternant π-conjugated hydrocarbon in the non-alternant ones. The singular electronic structure of C_{60} is consequence of not only the skeleton structure with the I_h symmetry of but also the ideal configuration of five membered rings all over the molecule.

Introduction

C_{60}, which consists of 12 five- and 20 six-membered rings has, a three dimensional (3D) skeleton structure with I_h symmetry. It has been attracting much interest as a novel 3D π-conjugated hydrocarbon. The net atomic charge distribution on each carbon is equal. The results at the level of Hückel[1] and Pariser-Parr-Pople[2] approximations show that the net π charge on each carbon is 1. This π charge distribution is just like those of alternant hydrocarbons to which the Coulson-Rushbrooke theorem[3] can be applied. However, C_{60} is a non-alternant hydrocarbon due to the presence of five-membered rings. The molecular orbital (MO) energies shown in Fig. 1 are not those of alternant hydrocarbons. On the other hand, it is predicted, based on the theoretical calculations for the higher fullerenes which have lower structural symmetries[4], that polarised charges, which are reminiscent of ordinary non-alternant hydrocarbons, distribute even in the neutral states. It is suggested that the differences of the charge distributions between C_{60} and the higher fullerenes might have some relation to the differences of the physical properties in the alkali-metal and TDAE doped states, namely, the electric conductivity and soft-ferromagnetism. These theoretical works[1,2,4] indicate that C_{60} has a very unusual electronic structure among the fullerene families.

In this study, in order to investigate the singularity of C_{60}, the singly excited electronic singlet state is theoretically studied including some electronic correlations

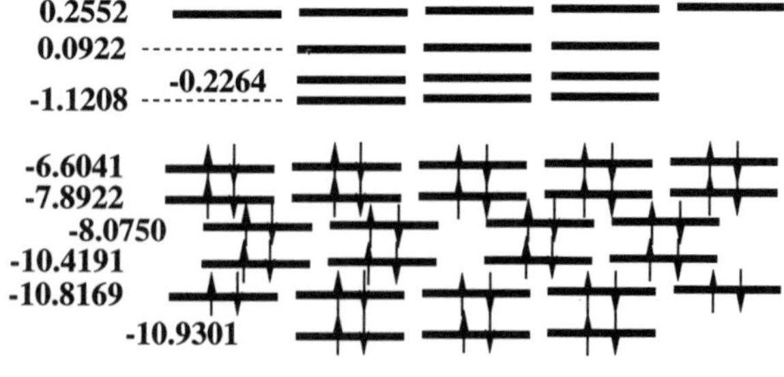

Fig. 1   MO energies near the frontier levels of C_{60}

*Also belongs to the Institute for Fundamental Chemistry, 34-4, Nishihiraki-cho, Takano, Sakyo-ku, Kyoto 606, Japan
** To whom correspondence should be addressed.

and effects of σ-electrons. The excited states of some fragment molecules, radialene, pyracylene and corannulene are also studied. The electronic structures in 3D π-conjugated hydrocarbons are discussed.

Table 1 The molecular geometries (in Å).

Computational Methods	Five memberd rings			Six memberd rings		
	C – C	C ⋯ C	C = C	C – C	C ⋯ C	C = C
Benzene					1.397	
Radialene	1.455				1.391	
Pyracylene	1.492	1.397	1.346	1.443	1.397	1.360
Corannulene	1.415			1.449		1.359
C_{60}	1.415				1.391	

ZINDO (Zerner's modified INDO) program package-1988 Version[5] was performed at the level of INDO/1-CI method including all the valence electrons with Mataga-Nishimoto approximation to obtain the wave functions of the ground and singly excited singlet states. The employed molecular geometries are tabulated in Table 1. Two dimensional (2D) molecules are lying on the xy plane. The C_5 symmetry axis in 3D molecules is taken to be parallel to the z-axis. All C-H bond lengths are fixed at 1.08Å. The excitations in configuration interaction (CI) calculation were confined near the frontier levels, since the results can describe the UV-visible absorbtion spectroscopic data of C_{60} very well[5]. Each oscillator strength (f_n) from the ground state to the n-th excited one is expressed by[6],

$$f_n = 1.033 \times 10^{11} \omega_n \times (\mu_n)^2. \tag{1}$$

In Eq. (1), ω_n is an excitation energy (in cm^{-1}) and a transition moment (μ_n),

$$(\mu_n)^2 = (\mu_n{}^x)^2 + (\mu_n{}^y)^2 + (\mu_n{}^z)^2. \tag{2}$$

is given by the x, y and z components in the cartesian coordinates. A partial dipole polarisability (α_n) is given by

$$\alpha_n = f_n / (\omega_n)^2. \tag{3}$$

Dipole polarisability (α)[7], which shows the influence of electronic correlation effects for the ground and excited states, is an important physical parameter. α is given by

$$\alpha = \sum_n f_n / (\omega_n)^2. \tag{4}$$

A large α indicates that the excited state has a large dipole polarisability. It is a "reactivity" parameter also.

Results and Discussion

The f_n and ω_n of benzene and C_{60} in this calculation agree well with the UV-spectroscopic data[5]. Since f_n is in proportion to a transition probability[6,7], an excitation with a large transition probability, $f_n \geq 0.5$, is paid attention to. The CI results are shown in Table 2. Similar results have been obtained for $f_n < 0.5$. One can also follow the dependency of f_n on the CI size[5]. Degenerate excited states have been obtained in C_{60}. The degeneracy seen in an alternant hydrocarbon such as benzene has been also obtained in radialene and corannulene. These fragment molecules as well as C_{60} are non-alternant hydrocarbons. However, the net atomic charges within the five-membered rings, although the polarised charge distribute in the whole molecules, are equal as shown in Figs. 2(a) and (c). The polarised charge distributions conventionally seen in a five membered ring vanish. These molecules no longer have the conventional electronic structures of five-membered ring.

The transition moment in Table 2 appears both in the x and y directions due to the structures with D_{5h} symmetry of radialene, while in only one direction due to the structures with the D_{6h} symmetry of benzene. μ_z is zero because of the 2D skeleton structures. Corannulene has a large transition moment in x and y directions. The transition with both μ_x and a small μ_z shows interaction between the σ- and π-orbitals due to the 3D structure. A non-degenerate excitation with a large μ_z has also been obtained, although f_n is lower than 0.5. However, μ_z of C_{60} is zero, since the transition moments in the z direction are cancelled out due to the I_h symmetry. Therefore, a non-degenerate excited state seen in corannulene would disappear. The slight differences of ω_n are reminiscent of the non-alternant MO energies shown in Fig.1. The polarised charge distribution seen in the fragment molecules also disappear. Because of an equal configuration of five membered rings obeying the isolated pentagon rule[8], C_{60} can be divided into 12 radialene or corannulene parts.

Table 2   The CI Results

	CI size	μ^n_x	μ^n_y	μ^n_z	ω_n	f_n	α_n	α
Benzene	15x15	0.0	-3.414	0.0	131025.8	0.719	2.018	
		-3.414	0.0	0.0	131025.8	0.719	2.018	
		0.0	4.048	0.0	154800.7	1.195	2.401	
		4.048	0.0	0.0	154800.7	1.195	2.401	
								8.838
Radialene	19x18	-0.003	6.087	0.0	40117.5	0.700	20.951	
		-6.087	-0.003	0.0	40117.7	0.700	20.953	
								41.904
Pyracylene	18x18	-8.898	-0.005	0.0	45897.8	1.710	39.130	
		0.023	8.078	0.0	48038.7	1.476	30.815	
		3.167	0.001	0.0	118685.2	0.561	1.917	
								71.862
Corannulene	14x14	-7.664	0.0	-0.004	40712.8	1.126	32.722	
		0.0	7.660	0.0	40716.0	1.125	32.688	
		0.0	5.502	0.0	46714.2	0.666	14.701	
		5.495	0.0	-0.001	46717.3	0.664	14.661	
								94.772
C_{60}	25x20	0.698	-6.631	0.0	39649.8	0.830	25.434	
		-6.637	-0.688	0.0	39654.4	0.832	25.471	
		-1.463	-9.156	0.0	41828.4	1.693	46.621	
		9.032	-1.585	0.0	41831.7	1.657	46.601	
		13.231	6.184	0.0	45844.6	4.605	105.541	
		6.189	-13.234	0.0	45845.3	4.608	105.609	
		-3.545	-4.597	0.0	50760.1	0.805	15.061	
		-4.612	3.558	0.0	50766.4	0.811	15.160	
								385.498

Overlap of the parts in filling C_{60} with them could give a competition between the positive and negative charges on each carbon. As a result, the competition would cause a cancellation between the charges. The charge distributions in the fragments are affected by the hydrogen atoms as shown in Figs. 2 (a) and (c). In C_{60} having no hydrogen atom, however, the cancellation would occur more completely on each carbon to finally give a neutral charge. On the other hand, the excited state of pyracylene shown in Table 2 is a non-alternant one. The polarised charge distributions from the five-memberd ring remain in Fig. 2(b). Similar results have been obtained in the deformed C_{60} and fragments with the lower symmetry.

The above results might suggest that the delocalisation of the charge distribution within a five membered rings, which is caused by the skeleton structure with C_{5v} and D_{5h} symmetries, is related to the appearance of an excited electronic state like an alternant hydrocarbon in the non-alternant ones. And furthermore, in C60, the equal configuration of the five membered rings would play an important role in the appearance of the neutral charge distributions. Therefore, the singular electronic structure of C_{60} is consequence of not only the structure with the I_h symmetry of but also the ideal configuration of five membered rings all over the molecule. The electronic structures in fullerenes might depend on the electronic structures and configuration of the five memberd rings.

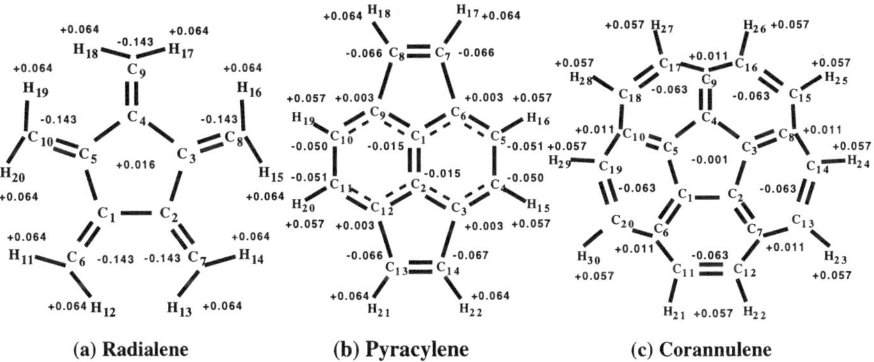

(a) Radialene   (b) Pyracylene   (c) Corannulene

Fig. 2 Net atomic charge distributions of the ground states

Acknowledgements

We are grateful to Prof. Kenichi Fukui for his encouragement and valuable suggestions. This work has been supported by the Swedish Natural Research Science Council (NFR) and the Japan Society of the Promotion of Science (JSPS). We acknowledge very valuable help from Sten Lunell and Christer Enkvist.

References

1. R. C. Haddon, L. E. Brus and K. Raghavachari, Chem. Phys. Lett., **125** (1986) 459.
2. I. László and L. Udvardi, Chem. Phys. Lett., **136** (1987) 418.
3. C. A. Coulson and G. S. Rushbrooke, Proc. Cambridge Phil. Soc., **36** (1940) 193.
4. K. Tanaka, M. Okada, K. Okahara and T. Yamabe, Chem. Phys. Lett., **202** (1993) 394. ; M. Okada, K. Okahara, K. Tanaka and T. Yamabe, Chem. Phys. Lett., **209** (1993) 91.
5. R. D. Bendale, J. D. Baker and M. C. Zerner, Intern. J. Quantum Chem., Quantum Chem. Symp., **25** (1991) 557 and references therein.
6. R. S. Mulliken and C. A. Rieke, Repts. Prog. Phys., **8** (1941) 231.
7. J. O. Hirschfelder, W. Byers Brown and S. T. Epstein, Adv. Quantum Chem., **1** (1964) 256. ; A. T. Amos and G. G. Hall, Theor. Chim. Acta., **6** (1966) 159.
8. H. W. Kroto, Nature, **329** (1987) 529.

ON THE ELECTRICAL PROPERTIES OF UNDOPED C_{60} SOLIDS

V. NÁDAŽDY, E. PINČÍK

Institute of Physics, SAS, Dúbravská cesta 9, 842 28 Bratislava

and

M. HALUŠKA, H. KUZMANY

Institute of Solid-State Physics, University of Wien, Strudlhofgasse 4, A-1090 Wien, Austria

and

D. BARANČOK

Department of Physics EF STU, Ilkovičova 3, 812 00 Bratislava, Slovakia

ABSTRACT

We studied electrical properties of undoped C_{60} solids using charge transient spectroscopy (QTS), current - voltage (I vs U) and current - temperature (I vs T) measurements. The QTS peak, which reflects a thermally activated relaxation process, is inherent to the crystal form only. On the base of performed experiments, the QTS peak can be attributed to the charge emission from bulk traps. We observed on crystals two kinds of traps. We found that the resistivity characterization of the undoped crystals is directly connected with the parameters of traps - the activation energy ΔE_T, frequency factor e_0 and the concentration N_T.

1. Principle of the QTS method

The principle of the QTS method[1] consists, in analogy to the original deep level transient spectroscopy[2], in correlating the charge transient induced by periodic rectangular voltage pulses with an optional time constant of the spectrometer while scanning the temperature (fig.1a). The transient $Q(\tau, t)$ is sampled at two times t_1 and t_2 after the pulse and the QTS signal is the difference between sampled values. There is zero difference for either very slow or very fast transients, corresponding to low or high temperatures, respectively. When the transient-time constant τ is on the order of $t_2 - t_1$, a difference signal - QTS peak - is generated. The time constant of the relaxation process at the peak maximum can be expressed through times t_1 and t_2: $\tau_m = (t_2 - t_1) \, / \, \ln(t_2/t_1)$.

2. Experiment and Results

Our measurements were carried out on single crystals grown from vapor phase[3] with different electrode materials: Au, Al, Ag-paste. Electrodes were made onto the [111] planes of the crystals in order to create $M/C_{60}/M$ capacitors. The geometrical capacitance of such a system, supposing a relative static permittivity of 4.4[4], is about 50 fF. Comparative measurements on films were performed with gold electrodes. The shape of I vs U characteristics as well as QTS responses are not influenced by the electrode material. An original finding was deduced from the QTS experiment. The representative spectra on both C_{60} film and crystal are shown in fig.1b. On films of thiknesses up to $4\mu m$ no QTS peak was detected. On the other hand we have observed two different thermally activated charge relaxations on crystals. For the first, the activation energy ΔE_T, frequency factor e_0 and the charge amplitude ΔQ were 0.28 eV, $2 \times 10^9 s^{-1}$ and 7×10^{-13} C, respectively (fig.2a).

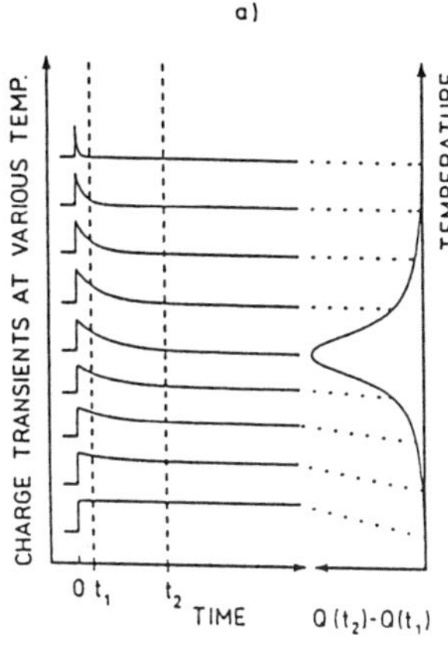

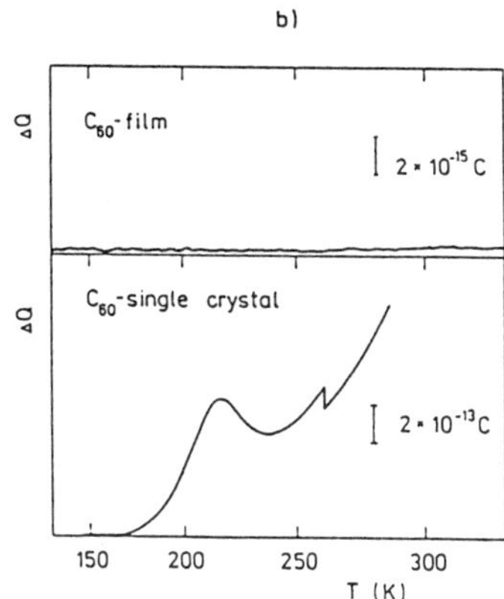

Fig. 1. a) Implementation of a DLTS rate window by means of a double-boxcar integrator with gates actuated at times t_1 and t_2. b) Comparison of QTS responses obtained on both C_{60} film and single crystal.

The second relaxation occurred with parameters $\Delta E_T = 0.42 eV$, $e_0 = 3 \times 10^{12} s^{-1}$ and $\Delta Q_{max} = 5 \times 10^{-13} C$ (fig. 2b). Note that ΔQ_{max} is proportional to the concentration N_T of relaxing centers. As shown in fig.2, activation energies ΔE_T are in good agreement with activation energies ΔE_A of the resistivity in the cases a) and b) when N_T is relatively high. In the case c) twice the value of ΔE_A gives $1.58 eV$, consistent with the electronic structure calculations that yielded C_{60} band gap $E_G = 1.5 eV$ [5].

3. Origin of the QTS peak

A thermally activated relaxation with similar parameters like ours ($\Delta E = 0.27 eV$ and $e_0 = 10^{14} s^{-1}$) was observed through the low frequency dielectric response of undoped C_{60} single crystals by Alers et al.[7]. They assume that the dielectric relaxation has its origin in the existence of a dipole moment in C_{60} induced by the high degree of frozen-in orientational disorder of the C_{60} molecules. As the QTS method directly detects displacement currents or emitted charges, our result speaks against the dielectric relaxation. The reason consists in the fact that the corresponding capacitance dispersion $\Delta C_{Tmax} = \Delta Q_{max} / \Delta U$ is on the order of the static capacitance while usually values around 10^{-2} are found for dipoles.

Other possible mechanisms are connected with a trap - limited charge emission, either from bulk traps or those located in the contact regions. The measurements with different electrode materials and complementary measurements on films suggest the origin of the QTS peak in a bulk trap level, induced by defects. This statement draws support also from complementary QTS peaks, due to emission and filling, which arise from transients after the leading edge and trailing edge of the rectangular voltage pulses, respectively. Namely, from the DLTS theory it follows that the emission peak is situated at a higher temperature than the filling one. Let the voltage polarity be related to the top electrode. When the emission peak is registered after the trailing edge of a pulse which

340

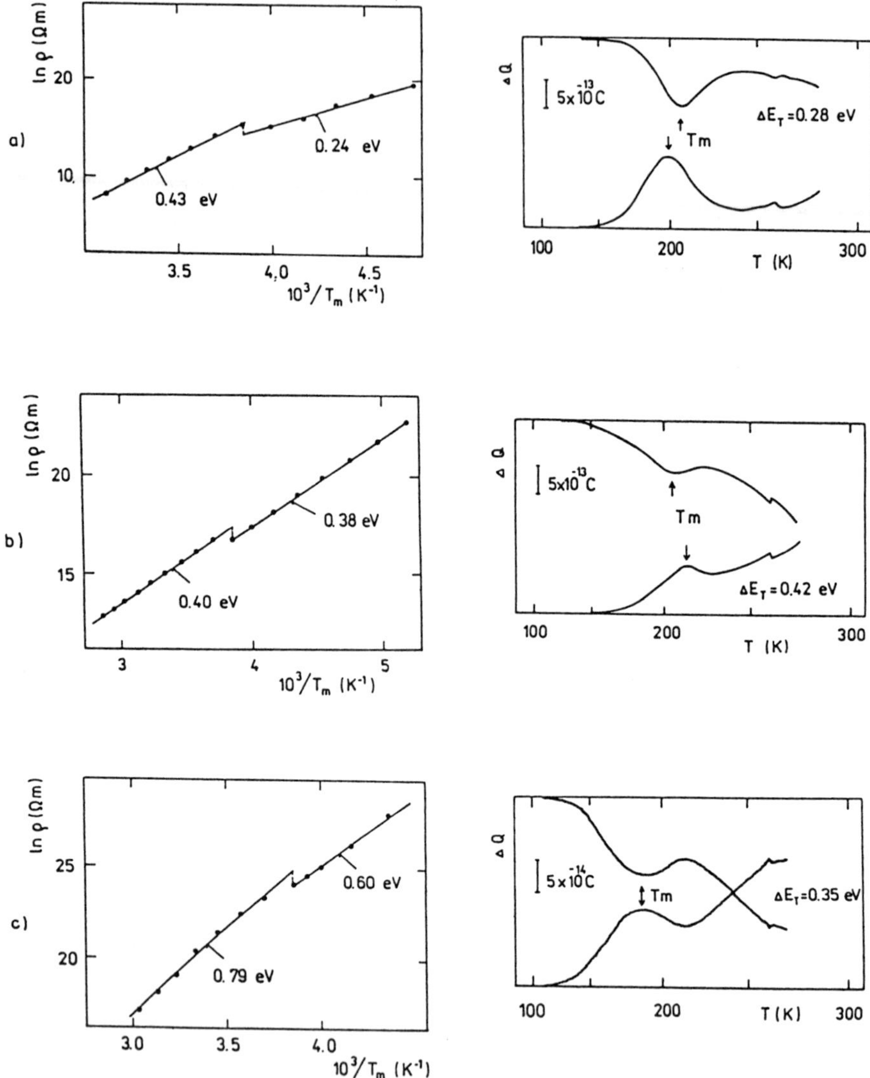

Fig. 2. The ($\ln \rho$ vs $1/T$) characteristics as well as complementary QTS spectra for three crystals: a) n-type, b) p-type, c) intrinsic, the decrease of ΔE_T from 0.42 eV to 0.35 eV is caused by the successive oxygen diffusion into the crystal [6].

drives the top electrode to a more negative potential, crystal has n - type conductivity (fig. 2a). When the emission takes place after the leading edge, crystal is p - type (fig. 2b). In the case 2c) we can speak about an intrinsic C_{60} crystal. Thus the two different relaxations can be attributed to two kinds of bulk traps.

4. Summary

The results can be summarized in following points:

- bellow the phase transition at 260 K, QTS method revealed on undoped single crystals two bulk traps: 1. $\Delta E_T = 0.28eV$, $e_0 = 2 \times 10^9 s^{-1}$

       2. $\Delta E_T = 0.42eV$, $e_0 = 3 \times 10^{12} s^{-1}$

- dependence of the conductivity type on the kind of the bulk trap and its concentration
- good agreement of the energies ΔE_T with the activation energies ΔE_A of the resistivity
  in the cases of relatively high trap concentrations.

5. Acknowledgements

This work was supported by the Osteuropa Förderung des BMfWF project GZ 45.338/1-IV/6a/94 and by the Slovak Grant Agency for Science grant No. 1/990176/93.

6. References

1. J.W. Farmer, C.D. Lamp, J.M. Meese, *Appl. Phys. Lett.* **41** (1982) 1064.

2. D.V. Lang, *J. Appl. Phys.* **45** (1974) 3014.

3. M. Haluška, H. Kuzmany, M. Vybornov, P. Rogl, P. Fejdi *Appl. Phys. A* **56** (1993) 161.

4. A.F. Hebard, R.C. Haddon, R.M. Fleming, A. Kortan, *Appl. Phys. Lett.* **59** (1991) 2109.

5. S. Saito, A.Oshiyama,*Phys. Rev. Lett.* **66** (1991) 2637.

6. D. Barančok, M. Haluška, H. Kuzmany, V. Nádaždy, *Solid State Commun.* **89** (1993) 123.

7. G.B. Alers, B. Golding, A.R. Kortan, R.C. Haddon, F.A. Theil, *Science* **257**, (1992) 511.

ELECTRONIC STRUCTURE OF TWO-DIMENSIONAL C_{60} LAYERS ADSORBED ON METAL SURFACES

Petra Rudolf[1], Michael Hunt[2], Stefano Cerasari[3] and Silvio Modesti[1,3]
1)*Laboratorio TASC-INFM, Padriciano 99, I-34012 Trieste, Italy.*
2) *Cavendish Laboratory, University of Cambridge, Cambridge CB3 OHE, U.K.*
3) *Dipartimento di Fisica, Università di Trieste, I-34127 Trieste, Italy.*

ABSTRACT

The charge transfer per molecule in a C_{60} monolayer on Au(110), polycrystalline Ag, Ni(110) and on various alkali coverages on Au(110) has been determined by studying the electronic excitations from the valence bands and the vibrational modes of the systems by electron energy loss spectroscopy. The electrons transferred are 1 ± 1 on noble metal substrates, 2 ± 1 on transition metals, and up to 6 ± 1 electrons on alkali-covered Au. C_{60} monolayers with a charge transfer of 3 ± 1 electrons/molecule exhibit metallic electronic excitation spectra.

The physical properties, such as superconductivity, of C_{60} based compounds strongly depend on the number of electrons transferred to the carbon molecule.[1] This number in turn depends on the environment of the molecule. In vacuum neutral, singly and doubly charged molecules are stable[2], in solid compounds $C_{60}{}^{3-}$ and $C_{60}{}^{6-}$ ions are stabilized by the Madelung potential[3], in liquid solutions $C_{60}{}^{n-}$ anions (n=1,2,3,4) are observed[4]. Theoretical calculations predict neutral, singly and doubly charged states for molecules adsorbed on flat metal surfaces with decreasing work function, if chemical bonding with the substrate is excluded[5]. The charge state of a C_{60} layer in contact with a metal plane is important for subjects such as the dependence of superconductivity on dimensionality and geometry, or the physical properties of superlattices made by C_{60} and metals.

We have determined the charge transfer between a C_{60} monolayer (ML) and substrates with strongly different work function and chemical reactivity such as polycrystalline Ag, Ni(110), Au(110), K/Au(110)c(2x2) and a disordered K multilayer. The charge transfer has been estimated from the spectrum of the low energy electronic excitations (0-8 eV) and from the energy shift of the dipole-active vibrational modes. We show that a ML of C_{60} on top of a metallic substrate can accept up to 6 ± 1 electrons per molecule.

99% pure C_{60} was sublimated from a crucible at 750 K onto a clean Ag, Ni(110) or Au(110)(1x2) surface. The coverage (θ) was determined from the ratio of the metal to C Auger peaks and calibrated from the ML coverage obtained by heating a C_{60} multilayer above 700K.[6] On Ni(110) dosing C_{60} at a substrate temperature (T_s) of 630K, a (5x3) LEED pattern develops for $\theta \geq 0.2$ ML.[7] Increasing θ above 0.6 ML, a quasi-hexagonal phase in which is the [1̄10] direction of the C_{60} fcc lattice is parallel to the [1̄10] direction of the substrate nucleates and finally covers the whole surface.[7] A hexagonal C_{60} ML on polycrystalline Ag was grown by dosing C_{60} at T_s=600K.[7] The same deposition conditions where also used to produce a well-ordered quasi-hexagonal C_{60} ML on Au(110).[6] Well-ordered multilayers of C_{60} with (111) orientation were obtained dosing on top of an ordered ML on Au(110) at 300K.[6] Potassium was evaporated from a well out-gassed SAES getter source. The c(2x2) overlayer, which is the saturation coverage of a single K layer on Au(110) and corresponds to 0.5 K atoms per surface Au atom,[8] was produced dosing at T_s=480K and monitoring the LEED pattern as well as the ratio of the K (251eV) to the Au (69eV) Auger peak. The ordered C_{60} ML (hexagonal LEED pattern) on top of this substrate was achieved dosing C_{60} at the same T_s. The K multilayer was grown at T_s=120K, T_s at which the multilayer is stable but disordered[8]. On this substrate the C_{60} ML was dosed at 120K and the absence of intermixing[9] was carefully checked monitoring the coverage dependence of the intensities of the Auger peaks. The base pressure during the measurements was $4x10^{-11}$ mbar and rose to $2x10^{-9}$ mbar during C_{60} sublimation and to $6x10^{-10}$ mbar during K evaporation.The vibrational and low-energy electronic excitation spectra were collected with a high resolution electron energy loss (HREEL) spectrometer (ELS 22) in the specular geometry (electron incidence angle 18°) with a primary electron energy of $\approx$2eV (10 eV) and an energy resolution of 4-7meV (15-20meV) for the vibrational (electronic) excitation spectra. The spectra relative to K multilayers were taken at 120K; all others at 300K.

The electronic excitation spectrum of an ordered 5 layer C_{60} film is shown in Fig. 1. It is semiconductor-like, showing a gap of $\approx$2eV, sharp dipole-forbidden excitonic peaks in the range 1.5-2eV and band to band transition peaks at 2.2, 2.7, 3.0, 3.7, 4.4 and 6.7eV, in agreement with previously published data[10]. On the contrary, the spectrum of a ML of C_{60} on Au(110) is metallic-like, without any excitonic structure, but with a continuum, peaked at 0.9eV, of transitions in the gap of bulk C_{60}. If a gap exists, it is masked by the vibrational excitations of C_{60} (sharp peaks below 200meV) and of CH containing contaminants (360meV). All other features in the spectrum are much broader than the respective ones of bulk C_{60} and slightly displaced in energy. The spectra of 1 ML C_{60} on Ni(110) and polycrystalline Ag (not shown) are similar to that on Au(110), but the features are even broader and the lowest energy structure higher in intensity. In a ML of C_{60} on K/Au(110) c(2x2) the lowest energy feature still grows in intensity and transforms into a smooth background centered at $\approx$0.5eV, while a new strong peak at 1.3eV develops. The latter peak energy differs less than 0.05eV from that of the second structure of the EEL spectra of K_3C_{60}[11] and Rb_3C_{60}[12]. The first feature of K_3C_{60} and $Rb_3 C_{60}$ at 0.55eV is broadened in the adsorbed layer. All the other structures in the spectrum, two weak shoulders at 2.4 and 3.4eV (see arrows) and two peaks at 4.6 and 6.3eV, have energies within 0.2eV of those of the corresponding features in K_3C_{60}.[13] In a ML of C_{60} on top of 2 disordered layers of K on Au(110) the excitations below 1eV have disappeared, and the main structures are now at 1.35, 2.6, 3.0, 4.3 and 5.8eV.The first peak is at exactly the same energy as the corresponding one measured for K_6C_{60}[11], Rb_6C_{60}[12] and Ca_3C_{60}[3]; all the other excitations differ by less than 0.1eV from those of K_6C_{60}.

In K_3C_{60} (charge state C_{60}^{3-}) the 0.55eV peak is the plasmon associated to the electrons which half fill the LUMO t_{1u} band[11], while the 1.3 eV peak is due to transitions from the LUMO t_{1u} to the t_{1g} level. In K_6C_{60}[11], Rb_6C_{60}[12] and Ca_3C_{60}[3] (charge state C_{60}^{6-}) the t_{1u} level is filled, therefore the 0.55eV plasmon vanishes and only the interband t_{1u}-t_{1g} transitions at 1.3 eV survive[3,11,12]. The presence of the t_{1u} plasmon in the C_{60} ML on Au(110), Ni(110) and

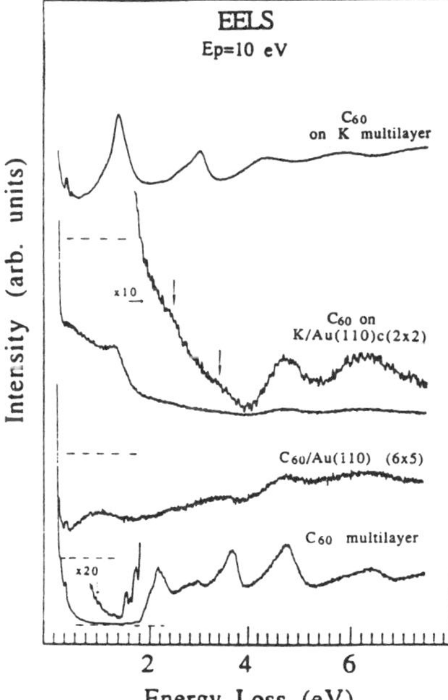

Fig. 1   EEL spectra of bulk C_{60} and of C_{60} monolayers on different substrates. The dashed line under each spectrum marks the zero level. The broad feature below 1eV in the second and the third spectrum is the t_{1u} plasmon.

Ag indicates therefore that the t_{1u} level begins to fill in these systems. The similarity between the spectra of C_{60} on K/Au(110)c(2x2) and K_3C_{60} and the intensity of the t_{1u} plasmon, comparable to that of the 1.3eV peak as in K_3C_{60}, show that the LUMO is roughly half filled in this case. The absence of the t_{1u} plasmon in C_{60} on the K multilayer and the strong similarity of the spectrum of this system with that of K_6C_{60} indicate that the sixfold degenerate LUMO is essentially filled on a K surface.

A quantitative confirmation of the CT to adsorbed C_{60} molecules comes from the large energy shift of the molecular vibrational modes. The bottom curve in Fig. 2 shows the HREEL spectrum of a 5 ML C_{60} film. The main peaks at 66, 146 and 179 meV and the shoulder at 71meV are the four dipole-active t_{1u} modes.[14,15] For all spectra the energies are obtained by least-square fittings with Gaussian peaks and the error bar is ±0.5meV.[16]

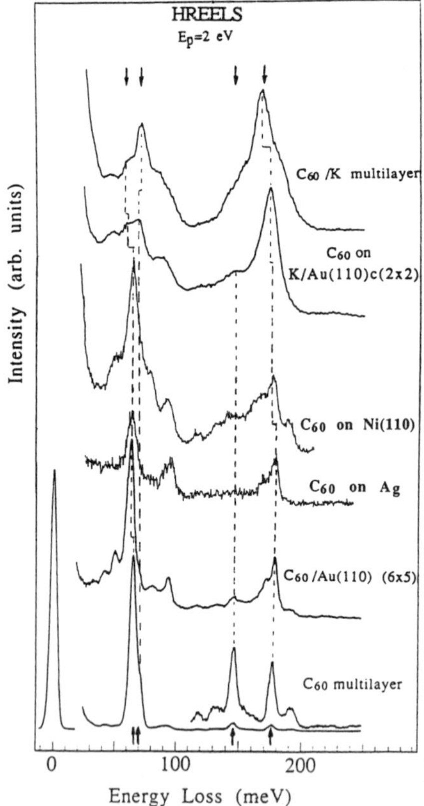

Fig. 2   HREEL vibrational spectra of bulk C_{60} (x50 and x500) and of C_{60} monolayers on Au(110) (x300), polycrystalline Ag (x380), Ni(110) (x800), K/Au(110)c(2x2) (x350), and K multilayer (x50), measured in the specular geometry. The arrows at the bottom indicate the energies of the dipole-active t_{1u} modes in bulk C_{60}, and those on top indicate the corresponding modes in bulk K_6C_{60}.

The lowest energy t_{1u} mode is strongly attenuated with respect to the others and shifts by 3meV when C_{60} is adsorbed on Au(110) or on Ag, by 4meV on Ni(110), by 6meV when it adsorbs on K/Au(110)c(2x2) and by 7meV on a K double layer. The energies of the 2nd and 3rd t_{1u} mode, at 71meV and 146meV respectively, are not appreciably affected by adsorption, but the relative intensity of the 2nd t_{1u} mode increases. On the contrary, the highest energy t_{1u} mode shows the strongest substrate dependence: it shifts from 178meV for C_{60} multilayer and 1 ML C_{60} on Au(110) and Ag, to 175meV for 1 ML C_{60} on Ni(110) and on K/Au(110)c(2x2), and to 165 meV for 1ML C_{60} on a K double layer or mul-tilayer, steadily growing in relative intensity. This behaviour is again strikingly similar to what was observed for alkali-doped C_{60}.[14] In fact, for K_6C_{60} the 1st and the 4th t_{1u} mode shift by 7 and 11meV, respectively, with re-spect to pure C_{60}, while the energies of the other two modes are scarcely affected by the CT. The simiarity also extends to the substrate dependence of the intensities of the 2nd and the 4th t_{1u} modes, which are enhanced by nearly 2 order of magnitude by alkali doping both in bulk compounds[14] and in our system. Theoretical calculation[17] shows that the coupling to virtual t_{1u}-t_{1g} electronic transitions causes a linear dependence of the energy shifts of the t_{1u} modes on the number of transferred electrons (by -1.8 meV per electron for the 178meV mode), while the intensities have a quadratic dependence. Ab-initio molecular dynamics calculations[18] predict different CT shifts for the four t_{1u} modes, in agreement with our results: Only the 1st and the 4th t_{1u} modes soften in a $Li_{12}C_{60}$ cluster where 12 extra electrons are added to C_{60}. The measured energies of the t_{1u}

modes for bulk C_{60}[14], K_3C_{60}[19] and K_6C_{60}[14] shift nearly linearly with CT: -1.25meV/ electron and -1.8 meV/electron for the 1st and the 4th mode, respectively. These shifts can be used as a calibration for an estimate of the CT in the adsorbed C_{60} layers: the number of transferred electrons calculated this way is 0 for C_{60} on Au(110) and Ag, 1.7 for C_{60} on Ni(110) and on K/Au(110)c(2x2), and 6.7 on the K multilayer, respectively, if the 4th mode is used. With the 1st mode one obtains 2.3 transferred electrons to C_{60} on Au(110) and Ag, 3.2 on Ni(110), 4.4 on K/Au(110)c(2x2) and 5.6 on the K multilayer. The error bar is ±0.7 electron in both cases. The discrepancies in the numbers obtained from the two modes are due to some nonlinearity in the shifts, probabely caused by the distortion of the molecule due to the bonding to the substrate. However, the averages of the values given by the two modes, 1.2 on Au and Ag, 2.4 on Ni, 3.1 on K/Au(110)c(2x2) and 6.2 for the K multilayer, are fully consistent with the estimates based on the electronic excitations. We can safely conclude that the CT is between 0 and 2 electrons on Au(110) and Ag, between 1 and 3 electrons on Ni(110), between 2 and 4 electrons K/Au(110)c(2x2), and between 5 and 7 electrons on K films.

The difference in CT when the C_{60} molecules are adsorbed on noble metal and transition metal surfaces suggests that the metal d-orbitals play an important role in C_{60} bonding. Considering the surface densities of K and C_{60} it appears that for the K/Au(110) c(2x2) substrate more than half of the K electronic charge available for bonding is transferred to the C_{60} overlayer. On the K multilayer the charge transferred is at least the entire amount available for bonding in the top K layer. The work functions of a ML and of a multilayer K film are the same (2.3 eV) within 0.2 eV.[16] Therefore the effect of work function change is not significant and we attribute the difference in the number of transferred electrons to the competing effect of bond formation between K and Au which sensibly decreases the electron density in the vacuum side of an alkali ML.[20] An additional contribution to CT in the multilayer could come from surface disorder[5].

References

1) See for instance, *The Fullerenes*, edited by H. W. Kroto, J. E. Fischer, and D. E. Cox (Pergamon, Oxford, 1993).

2) R. L. Hettich, R. N. Compton, and R. H. Ritchie, Phys. Rev. Lett.**67** (1991) 1242.

3) O. Zhou *et al.*, Nature **351** (1991) 462; H. Romberg, M. Roth and J. Fink (unpublished).

4) D. Dubois *et al.*, J. Am. Chem. Soc. 113 (1991) 4364.

5) E. Burstein *et al.*, Physica Scripta **T42** (1992) 207.

6) J. K. Gimzewski, S. Modesti and R. R. Schlittler, Phys. Rev. Lett. **72** (1994) 1036.

7) M. Hunt *et al.*, (unpublished)

8) D. K. Flynn-Sanders *et al.*, Surf. Sci. **253** (1991) 270.

9) D. M. Poirier *et al.*, Science **253** (1991) 646.

10) G. Gensterblum *et al.*, Phys. Rev. Lett. **67** (1991) 2171.

11) E. Sohmen, J. Fink and W. Krätschmer, Europhys. Lett. **17** (1992) 51.

12) E. Sohmen and J. Fink, Phys. Rev. **B47** (1993) 14532.

13) In Fig. 1 all the structures in the spectra correspond to excitations in the C_{60} layers because the momentum transfer is high enough to strongly depress the contributions from the substrates (see also ref. 16).

14) K.-J. Fu *et al.*, Phys. Rev. **B46** (1992) 1937 and references therein.

15) G. Gensterblum *et al.*, Appl. Phys. **A56** (1993) 175

16) P.Rudolf, S.Cerasari and S. Modesti, (unpublished)

17) M. J. Rice and Han-Yong Choi, Phys. Rev. **B45** (1992) 10173.

18) J. Kohanoff, W. Andreoni, and M. Parrinello, Chem. Phys. Lett. **198** (1992) 472.

19) T. Pichler, M. Matus and H. Kuzmany, Solid State Commun. **86** (1993) 221 and T. Pichler *et al.*, these proceedings; the energy of the first t_{1u} mode has been obtained from the maximum of the weak feature at 480 cm^{-1} in K_3C_{60}.

20) H. Ishida and K. Terakura, Phys. Rev. **B38** (1988) 5752.

Part Six

INFRARED AND RAMAN SPECTROSCOPY, LUMINESCENCE, TRANSIENT OPTICS, AND NLO

IR REFLECTION-TRANSMISSION OF VIBRATIONAL MODES IN SINGLE CRYSTAL C_{60}

R. Winkler, T. Pichler, and H. Kuzmany
Institut für Festkörperphysik, Universität Wien
Strudlhofgasse 4, A-1090 Vienna, Austria

Abstract

IR reflectivity and transmission was measured at crystallographic {111} and {100} planes of C_{60} single crystals. The applied sample geometry allowes the simultaneous investigation of the four infrared active F_{1u} fundamental modes together with some weak silent modes and the whole set of second- and third-order combination modes by a reflection-transmission technique. The orientational phase transition at 260 K is revealed by a dramatic jump in the line widths of the F_{1u} modes within a fraction of a degree. This result is explained by a broadening mechanism due to collisions with librons.

1. Introduction

From the early work of Krätschmer et al. we know that the four F_{1u} modes, the only ones which are IR allowed by the I_h symmetry of the free molecule, are known to dominate thin film spectra of solid C_{60}[1]. High resolution single crystal data, however, show a multitude of additional lines[2,3]. In our case the spectra turned out to be highly anisotropic due to differences in geometric resonances. Oscillator strength, line position, and width of the fundamental modes have been obtained from fits using a harmonic oscillator model in a temperature range from 80 K to 460 K.

2. Experimental

Crystals of reasonably large size with well expressed {111} and {100} faces were obtained by vacuum sublimation technique. Some of the crystals had two large parallel {111} faces, up to 5 mm×6 mm, and a thickness of about 1 mm. IR reflectivity was measured from both types of faces in a BRUKER 66V Fourier spectrometer between 50 and 5000 cm^{-1} with a resolution of 0.5 cm^{-1}. In this spectral range the 46 fundamentals and all combination modes of C_{60} up to third order are expected.

For the purpose of temperature dependent measurements a parallel plate shaped crystal was fixed on a heatable copper block within a vacuum chamber by means of silver paste. The sample holder was in limited thermal contact with a liquid nitrogen reservoir. This arrangement allowed to vary the temperatur continuously between 80 K and 460 K. During the measurements a dynamical vacuum of $< 10^{-4}$ Pa was maintained in the sample chamber.

3. Results

The reflection-transmission spectrum from a {111} face of a single crystal is shown in Fig. 1. The arrows indicate the position of the four F_{1u} modes, $527, 576, 1183$, and

350

1429 cm^{-1}. The full spectrum consists of two obviously different parts, 0-3000 cm^{-1}, and 3000-4600 cm^{-1}. A striking peculiarity of the spectrum are the different line shapes of the F_{1u} modes on the one hand and the remaining features on the other hand. This can be seen more clearly from Fig. 2, which compares reflectivity from {111} and {100} faces of one single crystal in a reduced wavenumber range: Whereas the F_{1u} modes appear as perfect reflection lines, all others have the shape of transmission lines. Besides this the latter differ from the F_{1u} modes by a surprising anisotropy.

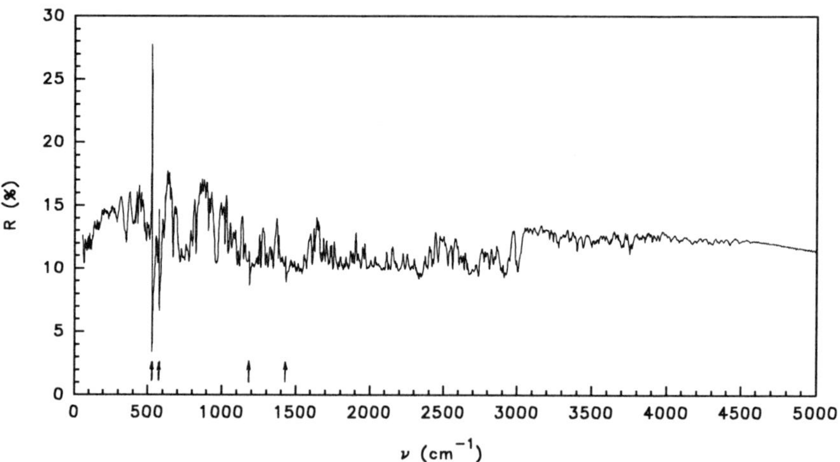

Fig. 1. Full range reflection-transmission spectrum.

To understand all these results we first discuss the origin of the extra lines. Below 3000 cm^{-1} they have already been assigned mainly to second-order combination modes and to silent fundamentals (273-1576 cm^{-1})[3,4]. The latter could become IR active by the lowering of molecular symmetry due to e. g. the natural abundance of ^{13}C isotopes or other symmetry breaking contaminations like hydrogen. Our measurements even show the third-order modes in the spectral range up to $\approx 4600\,\mathrm{cm}^{-1}$. The oscillator strengths of all these modes are a factor of more than 100 down as compared to the F_{1u} modes and therefore practically unobservable in pure reflectivity spectra. However, because of their weak absorption light sufficiently far away from the strong F_{1u} resonances penetrates the whole crystal and is reflected at its rear face. So in the case of parallel plate shaped crystals effective transmission contributes to the measured reflectivity. A detailed study of the multiple beam interferences shows, that the total reflectivity consists of a weighted sum of pure reflectivity and transmission[5]. The observed anisotropy of the transmission contribution is explained by the geometric form of the measured crystals: In $\langle 111 \rangle$ direction the crystal was parallel shaped, whereas in $\langle 100 \rangle$ direction it was much thicker with not so well expressed parallel faces. Therefore parallel plate interferences were suppressed.

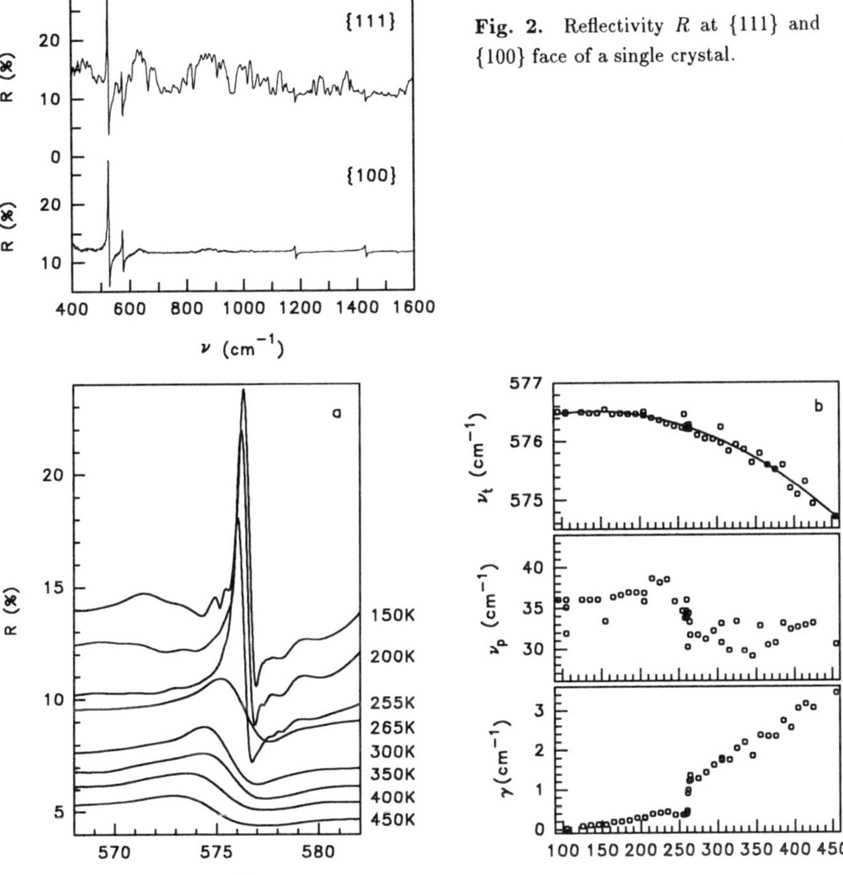

Fig. 2. Reflectivity R at {111} and {100} face of a single crystal.

Fig. 3. Reflectivity in the spectral surrounding of $F_{1u}(2)$ mode at various temperatures (**a**). Oscillator parameters of $F_{1u}(2)$ mode versus temperature T (**b**).

The high quality of the crystal faces allowed a detailed investigation of the temperature dependence of all visible modes by the reflection-transmission technique. Above the orientational phase transition all lines show a uniform up-shifting and narrowing with decreasing temperature. At the transition ($T_c = 260\,\mathrm{K}$) the lowest and the highest F_{1u} mode split into at least two components. A splitting of the $F_{1u}(3)$ mode was not observed in our case but might be possible at even lower temperatures according to a recent result[6]. The $F_{1u}(2)$ mode definitely remains unsplit but narrows drastically. A series of reflection lines at various temperatures is shown in Fig. 3 (a) for the $F_{1u}(2)$ mode. More details can be obtained from a oscillator fit. Fig.

3 (b) presents the so obtained values of (transversal) phonon frequencies ν_t, oscillator strengths ν_p^2, and widths γ. ν_t varies continuously at the transition point, whereas γ shows a dramatic jump. Away from T_c ν_p remains constant as one should expect from theory. Whether or not there is a discontinuity of physical significance remains uncertain at the moment. For a more detailed description will be given in a forthcoming paper [5].

4. Discussion

The temperature dependence of the measured line parameters are expected to give some details about the orientational phase transition. The most striking feature is the discontinuity of the line widths at T_c. This observation suggests that the damping is not determined by anharmonicities in the intramolecular potential but rater involves lattice modes. A qualitative explanation can be given as follows:

The optical phonon interacts with a libron via anharmonic coupling and thus is kicked out of the center of the Brillouin zone. In the low temperature phase libron frequencies are $> 15 \mathrm{cm}^{-1}$, much larger than the dispersion range of the internal modes. Therefore energy conservation prohibits inelastic phonon-libron collisions (libron absorption). Energy and crystal momentum conservation are maintained by elastic collisions only. These represent at least fourth-order processes (the fourth-order expansion coefficient of the interaction potential is involved). In the high temperature phase the librons become degenerated and soften extremely (paralibrons). This might provide the possibility of inelastic collisions. The latter are third-order processes and therefore probably more likely. Thus the damping increases.

The temperature behaviour of line width away from the transition point is governed by the occupation numbers of the libron states following the Bose-Einstein distribution and thus linear at temperatures $> 50\mathrm{K}$.

Acknowledgement

We thank M. Haluška for growing the crystals from C_{60} material generously provided by Hoechst AG, and A. B. Harris, and S.-L. Drechsler for enlightening discussion. This work was supported by the Österreichische Nationalbank, project P4507.

5. References

1. W. Krätschmer, L. D. Lamb, K. Fostiropoulos, and D. R. Huffman *Nature* **347**, 354 (1990).
2. P. Bowmar, M. Kurmoo, M. A. Green, F. L. Pratt, W. Hayes, P. Day, and K. Kikuchi *J. Phys.: Condens. Matter* **5**, 2739 (1993).
3. M. C. Martin et al. *this volume*.
4. K. A. Wang, A. M. Rao, P. C. Eklund, M. S. Dresselhaus, and G. Dresselhaus *Phys. Rev.* **B 48**, 11375 (1993).
5. R. G. Winkler, T. Pichler, and H. Kuzmany, in preparation.
6. C. C. Homes, P. J. Horoyski, M. L. W. Thewalt, and B. P. Clayman *Phys. Rev. B*, in press.

INFRARED STUDIES OF SINGLE CRYSTAL C_{60}: SILENT AND HIGHER ORDER VIBRATIONAL MODES

MICHAEL C. MARTIN, XIAOQUN DU AND LASZLO MIHALY
Department of Physics, State University of New York at Stony Brook
Stony Brook, NY 11794–3800, USA

Abstract

We report the measurement of infrared transmission of large C_{60} single crystals. The spectra exhibit a very rich structure with over 180 vibrational absorptions visible in the $100 - 4000 \text{cm}^{-1}$ range. Many silent modes are observed to have become weakly IR-active. We also observe a large number of higher order combination modes. The temperature $(77K-300K)$ dependence of these modes were measured and are presented. Careful analysis of the IR spectra in conjunction with Raman scattering data showing second order modes and neutron scattering data, allow the selection of the 46 vibrational modes C_{60}. We are able to fit *all* of the first and second order data seen in the present IR spectra and previously published Raman data (~ 300 lines total), using these 46 modes and their group theory allowed second order combinations.

The truncated icosohedral structure of C_{60} fullerenes belongs to the icosohedral point group, I_h, and has 4 infrared active intramolecular vibrational modes with F_{1u} symmetry. There are also 10 Raman-active vibrational modes: 2 with A_g symmetry and 8 with H_g symmetry. All 14 of these modes have been experimentally observed with great precision by many researchers [1, 2, 3, 4]. 32 additional modes that are IR and Raman forbidden (silent modes) are $1A_u$, $3F_{1g}$, $4F_{2g}$, $5F_{2u}$, $6G_g$, $6G_u$, and $7H_u$.

We present the infrared transmission data on a C_{60} single crystal at 300 and 77K in Figure 1. There are approximately 150 vibrational mode absorptions visible in the room temperature spectrum and many of the modes have a fine structure at low T. The four F_{1u} modes are seen to be so strong that they saturate with zero transmission in a range around the usual 527, 576, 1182 and 1429cm^{-1} positions. Every feature in this spectra and their temperature dependencies has been recently confirmed by two different groups measuring C_{60} crystals [5, 6]. Higher order and silent lines have also been observed in Raman [7] and IR [8] measurements on thin films.

Many of the known Raman vibrational modes can readily be identified in our infrared spectra using their well known positions: $H_g(2)$ at 431cm^{-1}, $H_g(3)$ at 709cm^{-1}, $H_g(4)$ at 775cm^{-1}, $H_g(5)$ at 1102cm^{-1}, $H_g(8)$ around 1576cm^{-1}, and $A_g(2)$ at 1470cm^{-1}. $H_g(7)$ at 1425cm^{-1} might also be active, but it is hidden under the extremely strong $F_{1u}(4)$ mode nearby. $H_g(1)$, $H_g(6)$ and $A_g(1)$ do not appear in our IR spectra.

The appearance of the Raman lines in the IR spectra strongly suggests that other

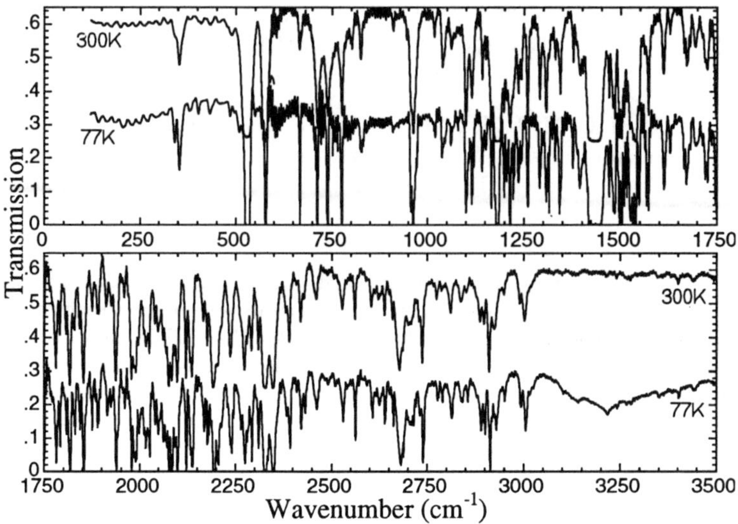

Figure 1. C_{60} single crystal infrared transmission spectra at 0.5cm^{-1} resolution obtained at 300K and 77K. The 300K spectrum is shifted up by 0.25 for clarity.

IR-inactive resonances could appear in the spectra as well. However, the total number of lines far exceeds the number of vibrational modes, and the extension of the spectral features up to ~ 3200cm^{-1} indicates that combination modes are also present. The binary combinations (involving two fundamentals) are expected to be stronger than third or higher order combinations. Indeed, the apparent absence of significant features above 3200cm^{-1} is in good agreement with the expected highest frequency of ~ 1600cm^{-1} for the fundamental modes [9]. A thicker crystal has shown a band of third order modes between 3200 and 4800cm^{-1} [6].

All the fundamental vibrational modes may become weakly IR-active due to an isotopic impurity in some C_{60} molecules [7]. From the natural abundance of ^{13}C one would expect that 34.4% of the fullerene molecules have one ^{13}C atom, (*i.e.* the molecule would be $^{13}C^{12}C_{59}$). The addition of a single ^{13}C lowers the symmetry of the molecule and is expected to make *all* of the previously silent modes active. [10]. Weak IR activity may also appear due to the molecule being located at an FCC symmetry site in the solid. The crystal fields reduce the icosohedral symmetry of the fullerenes

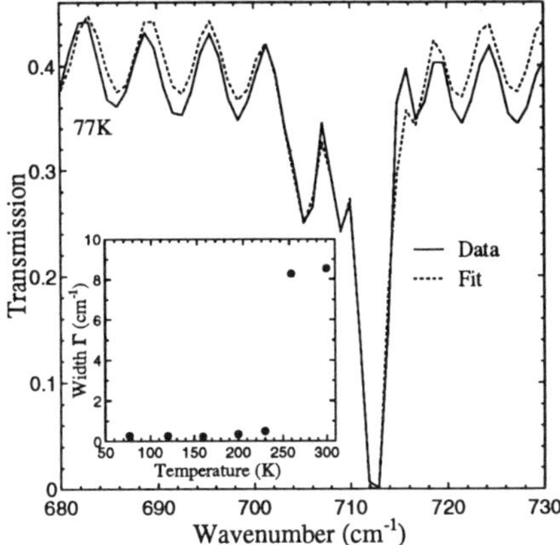

Figure 2. IR transmission spectrum (solid) and fit (dashed) of a C_{60} crystal at 77K. Inset shows the temperature dependence of the width Γ of this mode.

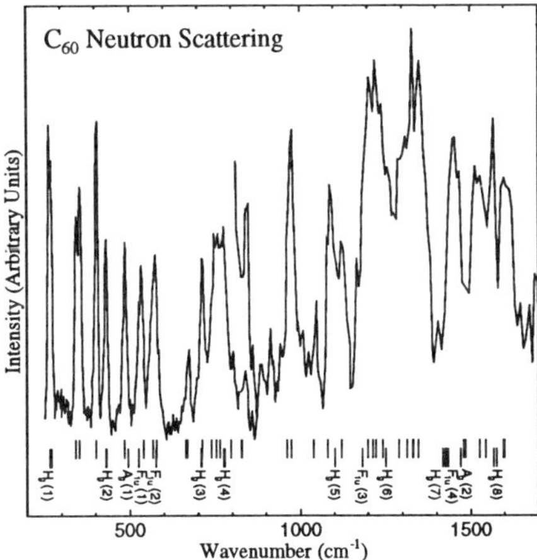

Figure 3. Comparison of the 46 modes extracted to neutron measurements (taken from Ref. [16]). The known Raman and IR modes are offset and labeled.

and activates normally silent odd-parity modes. Symmetry breaking from electric field gradients due to surface effects, impurities and dislocations, also provide ways for silent modes to appear in the IR spectrum.

The orientational ordering of the C_{60} molecules is known to influence the line shapes and intensities of the IR [11, 12] and Raman [13, 14, 12] modes. Comparison of the 300K spectrum to the 77K spectrum suggests that most of the low frequency lines exhibit dramatic narrowing at low temperature, while the width of typical high frequency lines change much less. Detailed analysis (Figure 2) shows that in most of the cases the total oscillator strength of the lines does not change, but the linewidth drops sharply at around the orientational ordering transition temperature $\sim$ 250K (see inset of Figure 2). This is in contrast to the gradual increase of peak heights reported by Kamarás *et al.* [12]. In many cases the broader high temperature line splits to several sharper low temperature resonances.

We followed the method of extracting the 46 fundamental vibrational mode frequencies described previously [15]. When *every* IR and Raman mode was fit by this procedure, 46 good candidates for the vibrational modes of C_{60} had been obtained. Finally the probable specific symmetry groups for each mode were chosen based on which other modes it successfully mixed with. The frequencies and assigned symmetries are listed in Ref. [15].

Figure 3 demonstrates how our 46 modes are in agreement with the neutron data. When comparing the IR and neutron results, one has to consider that the neutron time-of-flight spectrometry integrates over all wavenumbers, while the optical fundamentals are probed at $k=0$. Some of the fundamentals derived in our work are also in agreement with the high resolution energy electron energy loss spectroscopy (HREELS) data of Lucas *et al.* [17]. It appears that the HREELS lines at 355, 444, 686, 758, 968, 1097, 1258, and $1565cm^{-1}$ have corresponding resonances in our spectra at frequencies shifted to lower values by $2\text{-}15cm^{-1}$.

In summary, we have measured a very rich and complex IR spectrum of thick C_{60} crystals between 300 and 77K. We have assigned the 46 vibrational modes that agree with neutron measurements. We have used group theoretical predictions for second order combination modes to assign *all* the observed (IR and Raman) vibrational modes to specific symmetry fundamentals and combination modes.

The knowledge of all fundamentals is crucial for testing model calculations of fullerene vibrations. By the study of weakly active fundamentals one can also measure most of the Raman-active resonances (including the oxygen and polymerization sensitive $A_g(2)$ mode) without exposing the sample to a laser beam which could induce unwanted changes in the sample.

This work was supported by NSF grant DMR9202528. We would like to thank Philip B. Allen and K. Kamarás for valuable discussions. Gwyn P. Williams' and G. Larry Carr's assistance in running the spectrometers at the NSLS is appreciated. Thanks to Gene Dresselhaus, Peter Eklund, K. Kamarás, R. Winkler, and H. Kuzmany for providing us with their results prior to publication.

References

1. W. Kratschmer it et al., Chem. Phys. Lett. **170**, 167 (1990); W. Kratschmer *et al.*, *Dusty Objects in the Universe*, eds. E. Bussoletti, A. A. Cittone, Kluwer, Dordrecht 1991; S. Bethune *et al.*, Chem. Phys. Lett. **179**, 181 (1991)

2. W. Kratschmer, L. D. Lamb, K. Fostiropoulos, D. R. Huffman, Nature **347**, 354 (1990).

3. S. J. Duclos *et al.*, Science **254**, 1625 (1991).

4. T. Pichler, M. Matus, J. Kurti, and H. Kuzmany, Phys. Rev. B **45**, 13841 (1992).

5. K. Kamarás, K. Matsumoto, M. Wojnowski and E. Schönherr, *unpublished*.

6. R. Winkler, T. Pichler and H. Kuzmany, *unpublished*.

7. Z.-H. Dong *et al.*, Phys. Rev. B **48**, 2862 (1993)

8. K.-A. Wang *et al.*, Phys. Rev. B **48**, 11375 (1993).

9. For a review, see A. F. Hebard, Physics Today, November 1992, p.26.

10. P.H.M. van Loosdrecht, *et al.*, Chem. Phys. Lett. **198**, 587 (1992).

11. C.C. Homes *et al.*, Phys. Rev. B **49**, 7052 (1994).

12. K. Kamarás *et al.*, Chem. Phys. Lett. **214**, 338 (1993).

13. P.H.M. van Loosdrecht, *et al.*, Phys. Rev. Lett. **68**, 1176 (1992).

14. H. Kuzmany, M. Matus, T. Pichler, and J. Winter, Proc. NATO ARW, Crete 1993.

15. Michael C. Martin, Xiaoqun Du, John Kwon and Laszlo Mihaly, Phys. Rev. B, *to be published*.

16. C. Coulombeau *et al.*, J. Phys. Chem. **96**, 22 (1992).

17. G. Gensterblum *et al.*, Phys. Rev. Lett. **67**, 2171 (1991).

INFRARED SPECTRA OF C_{60} AND $(Ph_4P)_2C_{60}I$ CRYSTALS

K. KAMARÁS

Solid State Physics Laboratory, University of Groningen,
Nijenborgh 4, NL-9747 AG Groningen, The Netherlands

K. MATSUMOTO, M. WOJNOWSKI, E. SCHÖNHERR

Max-Planck-Institut für Festkörperforschung, Heisenbergstrasse 1,
D 70569 Stuttgart, Federal Republic of Germany

H. KLOS and B. GOTSCHY

Lehrstuhl für Experimentalphysik II, Universität Bayreuth,
Universitätsstrasse 30, D 95440 Bayreuth, Federal Republic of Germany

ABSTRACT

We present infrared spectra of crystalline C_{60} and $(Ph_4P)_2C_{60}I$. In thick pure C_60 crystals an extreme wealth of vibrational structure shows up in transmission. The orientational phase transition influences only a small number of vibrations, which we assign to odd-parity fundamentals. In $(Ph_4P)_2C_{60}I$, an insulating charge-transfer salt of C_{60}, we observe the F_{1u} modes of C_{60}^-, with comparable intensity to those of the counterion $(Ph_4P)^+$.

1. Introduction

Thanks to the enormous progress in the preparation of fullerenes, high quality single crystals became available for optical study during the last year. Infrared spectroscopy on single crystals provides detailed information on the phonon structure of solid C_{60}. In fulleride salts, the vibrations are highly sensitive to the oxidation state of C_{60}; by selecting adequate model compounds with well characterized chemical composition, we can obtain reference data for use in unknown materials. In the present paper, we provide an example for both of these applications on C_{60} and $(Ph_4P)_2C_{60}I$, respectively.

2. Experimental

Crystals of C_{60} were grown by vapor transport [1], while those of $(Ph_4P)_2C_{60}I$ were produced by electrochemical reduction in solution [2]. In pure C_{60}, where the principal infrared lines are well known, we were looking for more subtle effects by measuring the transmission of a 0.5 mm thick crystal as a function of temperature. In $(Ph_4P)_2C_{60}I$, on the other hand, where the size of the samples made the use of an infrared microscope necessary, we performed reflectance measurements at room temperature in order to obtain the main features undistorted. Spectra were taken by two different Bruker FTIR instruments, one equipped with an IR microscope, the other one extending in frequency to the far infrared. Temperature was varied in a flow-through liquid He cryostat.

3. Results and discussion

3.1. C_{60}

The infrared transmission spectrum of thick C_{60} crystals consists of numerous weak lines besides the four allowed F_{1u} modes [3, 4]. These lines either correspond to combination and overtone modes or to forbidden fundamentals. The former are a consequence of anharmonicity in the crystal and have been observed also by Raman spectroscopy [5]. The latter can become IR-active because of symmetry lowering of the molecule itself, due to the presence of ^{13}C atoms; such a mechanism can affect at least 85 out of the 174 molecular vibrations [6]. Another possible reason is the site symmetry in the crystal being lower than that of the molecule [7]. In this case, only odd-parity modes can be activated since the inversion center is retained. The two types of modes can be distinguished based on their behavior at the orientational phase transition. This transition has obviously no effect on the molecular symmetry but influences the site symmetry; also, because the fast rotation of the molecules stops, the effect of the crystal field is expected to be more pronounced.

By comparing the spectra just above and below the transition temperature, shown in Fig. 1., we find new lines in the Pa$\bar{3}$ phase at 340, 752, 758, 764, 1079, 1314 and 1566 cm^{-1}. Their temperature dependence suggests they are odd-parity fundamentals, although further work is needed to make a complete assignment of their symmetries. No coincidence is found when comparing the above frequencies with those of the weak lines in recently published gas-phase emission spectra [8], further corroborating the role of crystal field in activating these particular vibrations.

3.2. $(Ph_4P)_2C_{60}I$

Rice and Choi suggested that in fulleride salts electron-phonon interaction develops between the extra electrons residing on the C_{60} ball and the F_{1u} vibrations, its magnitude depending on the charge x of the ion [9]. There have been several infrared studies on the concentration dependence of the F_{1u} modes in the alkali salts [10, 11, 12], but in those systems the situation is complicated by the presence of multiple phases and by the fact that the stoichiometry cannot be controlled directly. We performed measurements on single crystals of $(Ph_4P)_2C_{60}I$ grown electrochemically [2]. In Fig. 2 we show the infrared reflectivity measured by an infrared microscope and the absorbance obtained from it by Kramers-Kronig analysis. Two of the F_{1u} modes fall into our measurement range: $\nu_3=1179$ cm^{-1} showing only a minimal frequency

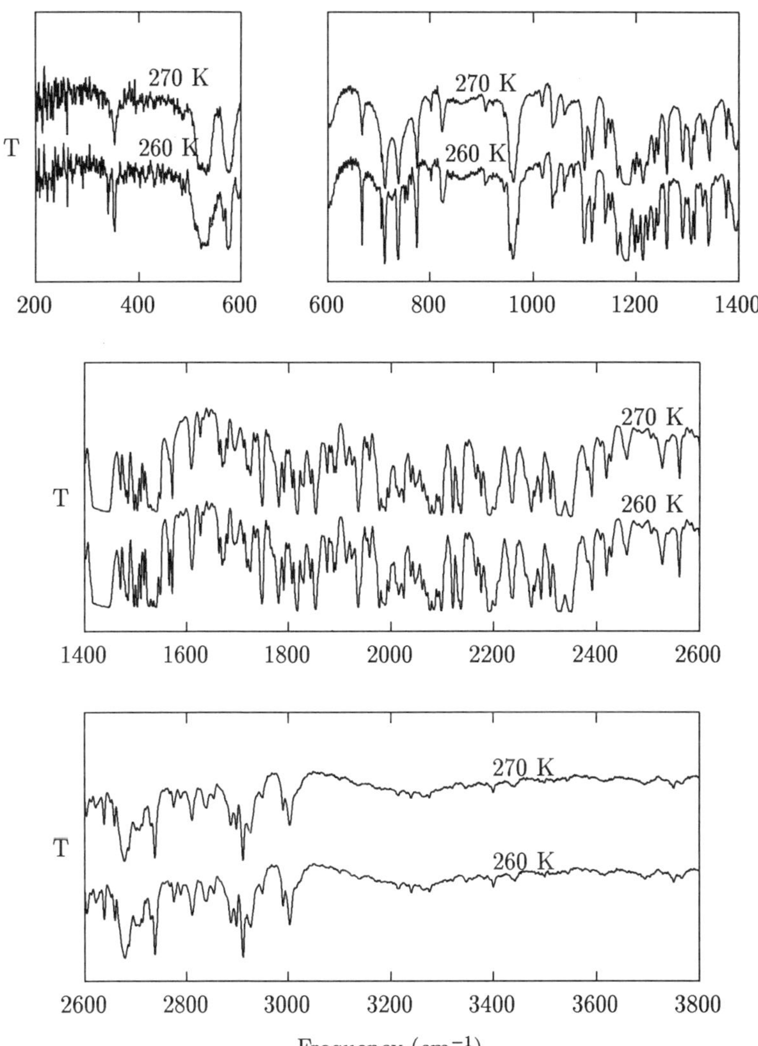

Figure 1: Transmission of a C_{60} crystal above (270 K) and below (260 K) the orientational phase transition.

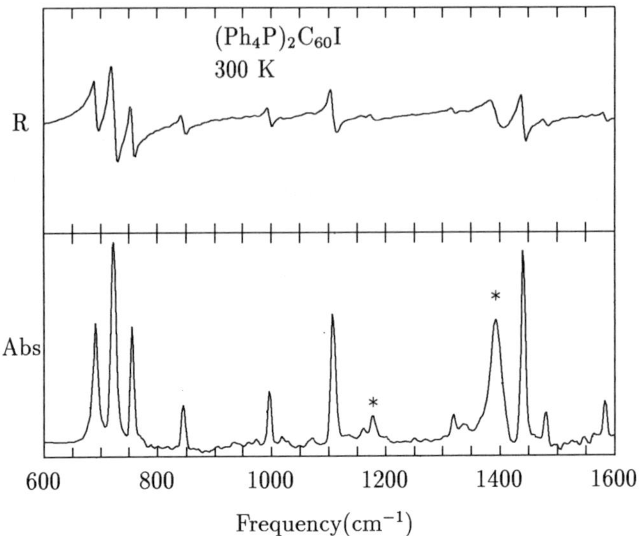

Figure 2: Reflectance and absorption of $(Ph_4P)_2C_{60}I$. Absorbance peaks assigned to C_{60}^- are denoted by asterisks.

shift relative to neutral C_{60}, and $\nu_4=1394$ cm^{-1}, found before in RbC_{60} [10]. Since the oxidation state in our compound is determined by the chemical composition, we can unambigously assign this vibration to the C_{60}^- monoanion. The intensity of the line is comparable to the modes of the counterion $(Ph_4P)^+$, unlike in one-dimensional organic salts of similar cations with TCNQ [13]. This finding is in accordance with the theoretical estimation of the monoanion having the smallest coupling constant among the reduced C_{60} species [9].

4. Conclusions

We provided two examples for the use of infrared spectroscopy in fullerene research. We could identify crystal-field induced fundamental absorption lines in pure C_{60}, and two principal IR-allowed modes of the monoanion in $(Ph_4P)_2C_{60}I$. The full complexity of the spectra is, however, far from resolved. Both the vibrational structure of C_{60} itself and the interaction of the phonons with electrons in the fulleride salts are areas where a combined effort of theory and different experimental methods (including Raman and neutron scattering as well as high resolution electron-energy loss spectroscopy) is required.

5. Acknowledgements

We thank A. Breitschwerdt for expert technical help with the infrared microscope. This work was supported by the Bundesminister für Forschung und Technologie (Bonn, Federal Republic of Germany), by the European Community EE-EC grant No. ERB 3510 PL 922716, and by the Hungarian National Research Foundation under contract numbers 2950 and T4222.

6. References

[1] K. Matsumoto, E. Schönherr and M. Wojnowski, *J. Crystal Growth* **135** (1994) 154.

[2] M. Keil, H. Klos, A. Schilder, W. Schütz and B. Gotschy, this volume.

[3] M. C. Martin, X. Du, J. Kwon and L. Mihaly, *Phys. Rev. B*, to be published.

[4] R. Winkler, T. Pichler and H. Kuzmany, this volume.

[5] Z.-H. Dong, P. Zhou, J. M. Holden, P. C. Eklund, M. S. Dresselhaus and G. Dresselhaus, *Phys. Rev. B* **48** (1993) 2862.

[6] D. E. Weeks, *J. Chem. Phys.* **96** (1992) 7380.

[7] G. Dresselhaus, M. S. Dresselhaus and P. C. Eklund, *Phys. Rev. B* **45** (1992) 6923.

[8] L. Ncmcs, R. S. Ram, P. F. Bernath, F. A. Tinker, M. C. Zumwalt, L. D. Lamb and D. R. Huffman, *Chem. Phys. Lett.* **218** (1994) 295.

[9] M. J. Rice and H.-Y. Choi, *Phys. Rev. B* **45** (1992) 10173.

[10] T. Pichler and H. Kuzmany, *Springer Series in Solid-State Sciences* **117** (1993) 281.

[11] T. Pichler, M. Matus and H. Kuzmany, *Solid State Commun.* **86** (1993) 221.

[12] M. C. Martin, D. Koller and L. Mihaly, *Phys. Rev. B* **47** (1993) 14607.

[13] J. Petzelt, K. Král, N. Ryšavá, L. Dobiašová, J. Kroupa, A. Oswald, A. Szyszkowski and A. Graja, *Solid State Commun.* **32** (1979) 1315.

CALCULATION OF THE INFRARED AND RAMAN SPECTRA OF SHORT LINEAR CARBON CHAINS

Jenö Kürti, Csaba Magyar, András Balázs and Péter Rajczy
Department of Atomic Physics, R.Eötvös University,
H-1088 Budapest, Puskin u. 5-7, Hungary

ABSTRACT

Beside fullerenes, linear carbon clusters also attracted much interest in the last years, both experimentally and theoretically. In a recent work Raman spectroscopy of electrochemically prepared carbyne was reported. The Raman spectra can be explained — as usual for conjugated polyenes — by the effective conjugation length model, taking into account the distribution of the conjugation lengths. We carried out vibrational analysis for short linear chains: $H\text{-}C_{2n}\text{-}H$ ($n = 1, 2, 3, 4, 5$) and $H_2\text{-}C_{2n}\text{-}H_2$ ($n = 1, 2, 3, 4, 5$), modelling the alternating carbyne, and the cumulene type carbon chains, respectively. Geometries were optimized by using *ab initio* Hartree-Fock theory with Dunning's double-ζ basis set. IR- and Raman-intensities were obtained by calculating dipole- and polarizability derivatives, respectively. No resonance Raman effect was taken into account.

1. Introduction

The discovery of fullerenes gave new impulse to the theoretical and experimental investigation of other carbon modifications, too. It is known that in the gas phase small carbon clusters form linear chains (odd number of carbon atoms) or rings (even number of carbon atoms),[1,2] whereas larger clusters consist of multiple rings and the most stable form of clusters with $n \geq 33$ carbon atoms are cagelike molecules.[3] However, in the solid phase the linear "carbyne" modification can be stabilized for short and long chains as well.[4] The infinite linear carbon chain attracted considerable theoretical interest because it is the simplest possible 1D conjugated polymer.[5-7]

One possible method to prepare linear carbon chains in the solid phase is the dehydrochloridation of either polyvinylchloride[8] or chlorinated polyacetylene[9] using organic strong base. The existence of pure linear carbon chains in these samples was proved by Raman and IR spectroscopy. The most characteristic feature of both spectra was the appearance of a strong band at around 2100 cm^{-1}. However, the spectra consisted of many other bands corresponding to e.g. polyene parts of the sample. The relative intensity of the band at around 2100 cm^{-1} was rather low.

Another way for preparation of linear carbon chains is the electrochemical dehalogenation of poly(tetrafluoroethylene) (PTFE-teflon) by alkali metal amalgams.[10,11]

Dehalogenation of the PTFE chains results in the appearance of bare carbon chains separated by alkali fluorides. The latter do not completely hinder crosslinking between the carbon chains, hence the carbyne – alkali fluoride films always contain some amount of amorphous carbon. However, there are no other organic byproducts like polyacetylene in the sample. It was shown recently[12] that the relative intensity of the Raman band at around 2000 cm^{-1}, corresponding to the stretching mode of the linear carbon chain, is very large compered with the intensity of the Raman band characteristic for amorphous carbon between 1000-1600 cm^{-1}. This means that the carbyne content in this material is unusually high.

The position and the intensity of the Raman band at around 2000 cm^{-1} strongly depends on the quantum energy of the laser excitation.[12] This is similar to the well known "dispersion effect" in polyenes, which can be interpreted as a photoselective resonance enhancement for various segments of the polymer. The evaluation of the spectra can be carried out by the effective conjugation length model, taking into account the distribution of the conjugation lengths.[13,14] For quantitative results one needs the chain length dependence of both the electronic and the vibronic properties. They were calculated in Ref.12. by a specific Hückel-type Hamiltonian with geometry relaxation. Combining these calculations with the experimental Raman spectra resulted in an average conjugation length of 8-14 carbon atoms. Our aim was to obtain the vibrational frequencies for various short carbon chains more accurately using the more sophisticated *ab initio* method.

2. Results and Discussion

The calculation of pure carbon molecules at the *ab initio* level is very complicated and time consuming, because one inevitably has to take into account correlation effects in order to obtain reliable results. Even the ground state structure of small carbon clusters with even number of C atoms is a matter of controversy, see e.g. Refs 2,15. On the other hand, because of the preparation conditions, it can be assumed with good reason that the electrochemical treatment of the PTFE results in formation of linear carbon chains. In contrast to the gas phase, the interaction between the carbon clusters and their environments is not negligible and can stabilize the linear form of the clusters. Effects caused by end groups or crosslinks are also absent in the gas phase. Therefore, it has to be emphasized that we choosed model molecules to describe linear carbon chain segments of various length in the solid phase.

Because of the controversion in the literature whether the chains in carbyne are cumulenic or alternating, both possibilities have been taken into account by using appropriate end groups. Calculations were carried out for linear chains: $H\text{-}C_{2n}\text{-}H$ ($n = 1, 2, 3, 4, 5$) and $H_2\text{-}C_{2n}\text{-}H_2$ ($n = 1, 2, 3, 4, 5$) using *ab initio* Hartree-Fock theory with Dunning's double-ζ basis set.

Geometry optimization resulted in alternating and cumulenic structure for the first and the second series, respectively. The C-C bond lengths in Angström e.g.

for H-C$_{10}$-H and H$_2$-C$_{10}$-H$_2$ are R$_1$=1.1934,   R$_2$=1.3897,   R$_3$=1.1958,   R$_4$=1.3842, R$_5$=1.1966; and R$_1$=1.3089,   R$_2$=1.2715,   R$_3$=1.2799,   R$_4$=1.2715,   R$_5$=1.2780, respectively. The numbering begins at the chain ends in both cases.

Vibrational frequencies as well as Raman and infrared intensities were calculated for all of the above mentioned model molecules. From the full vibrational analysis, those normal modes which affect mainly the carbon skeleton, can be determined. For the carbon backbone with $2n$ atoms there are $2n$-1 stretching modes (n Raman active and n-1 IR active) and $4(n$-1) bending modes ($2(n$-1) Raman active and $2(n$-1) IR active), for both kind of model molecules with alternating and cumulenic structure.

There are various segment lengths in a real polymer the spectrum of which can be accounted for by adding up the spectra of the individual molecules. Fig. 1 shows the equally weighted sum of the calculated Raman and IR stick spectra over all the model molecules. The summation is restricted to the carbon skeleton vibrations. It is well known that *ab initio* results always overestimate the experimental observable frequencies. Thus, the calculated stretching and bending frequencies were scaled by the usual factors 0.9 and 0.8, respectively in Fig.1. Higher frequencies correspond to stretching modes, lower frequencies correspond to bending modes.

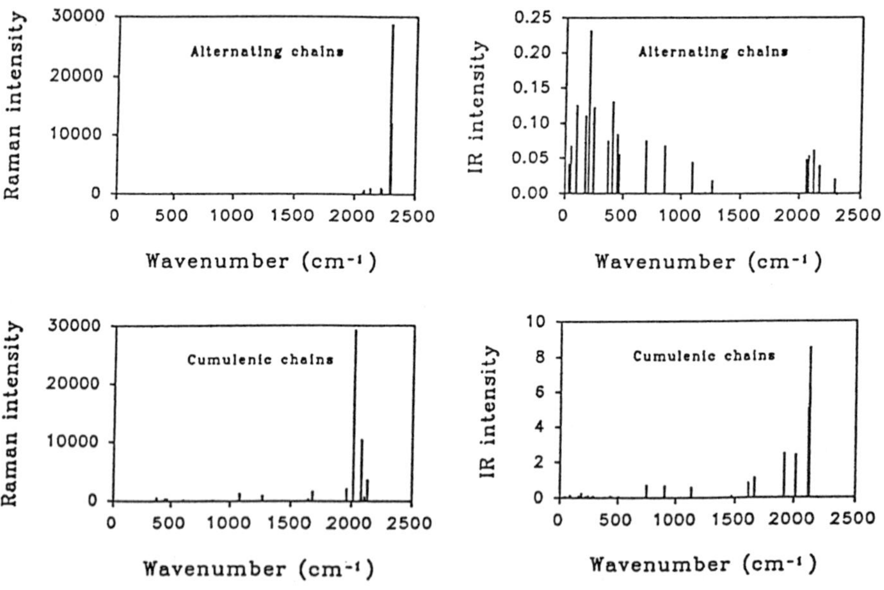

Fig.1 The sum of the calculated Raman and IR spectra for alternating and cumulenic chains with 2, 4, 6, 8, 10 carbon atoms. Only the stretching and bending modes of the carbon backbone are taken into account.

It is clearly seen that the experimentally observed Raman band at around 2000 cm^{-1} is really characteristic for linear carbon chains. There are some other weak bands for cumulenic chains but practically nothing for the alternating chains. However, there is hardly any chance to distinguish between alternating and cumulenic structures based on Raman spectrum, because the intensity of the band at around 2000 cm^{-1} — even without taking into account resonance enhancement — is much higher than the intensity of all other bands. The small difference in the absolute position of the strongest Raman band for the two case is not decisive because the frequencies can be slightly modified by interactions in the solid phase.

On the other hand, the differences between the two cases in the infrared spectra are obvious. Therefore, to be able to distinguish between the alternating and cumulenic structure, it would be worth while to investigate thoroughly also the IR spectrum of the electrochemically prepared carbyne.

3. Acknowledgements

This work was supported by the grant T 014405 (OTKA). Valuable discussions with L. Kavan, J. Kastner and G. Pongor are gratefully acknowledged.

4. References

1. W. Weltner, Jr, and R. J. van Zee, *Chem. Rev.* **89** (1989) 1713.
2. J. Hutter, H. P. Lüthi and F. Diederich, *J. Am. Chem. Soc.* **116** (1994) 750.
3. W. Eberhardt, B. Kessler, H. Handschuh, G. Ganteför, P. S. Bechtold, M. Biermann and S. Krummacher, this volume.
4. P. B. Heimann, J. Kleiman and N. M. Salansky, *Carbon* **22** (1984) 147.
5. M. Kertesz, J. Koller and A. Azman, *J. Chem. Phys.* **68** (1978) 2779.
6. A. Karpfen, *J. Phys. C: Solid State Phys.* **12** (1979) 3227.
7. M. J. Rice, S. R. Phillpot, A. R. Bishop and D. K. Campbell, *Phys. Rev.* **B 34** (1986) 4139.
8. V. V. Berdyugin, Yu. P. Kudryavtsev, S. E. Evsyukov, Yu. V. Korshak, P. P. Shorygin, and V. V. Korshak, *Dokl. Akad. Nauk SSSR* **305** (1989) 362.
9. K. Akagi, Y. Furukawa, I. Harada, M. Nishiguchi, and H. Shirakawa, *Synth. Met.* **17** (1987) 557.
10. J. Jansta and F. P. Dousek, *Carbon* **18** (1980) 433.
11. L. Kavan, Electrochemical Carbonization of Fluoropolymers, in (P. A. Thrower Ed.) *Chemistry and Physics of Carbon* Vol. 23, (M. Dekker, New York, 1991), p. 69, and references therein.
12. J. Kastner, H. Kuzmany, L. Kavan, F. P. Dousek and J. Kürti, to be published
13. H. Kuzmany, *Pure and Appl. Chem.* **57** (1985) 235.
14. J. Kürti and H. Kuzmany, *Phys. Rev.* **B44** (1991) 597, and references therein.
15. J. M. L. Martin, J. P. Francois and R. Gijbels, *J. Chem. Phys.* **94** (1991) 3753.

OPTICAL STUDIES OF PHOTOPOLYMERIZED SOLID C_{60} AND C_{70} FILMS

P.C. Eklund, A. M. Rao, Ying Wang, Ping Zhou, M.J. Holden and K.A. Wang

Department of Physics and Astronomy and Center for Applied Energy Research, University of Kentucky, Lexington, KY 40506

ABSTRACT: Irradiation of oxygen-free C_{60} films or C_{70} films with visible or ultraviolet light phototransforms the solid fullerenes to a new polymeric phase. We present experimental evidence which suggests the formation of covalent C=C bonds between the fullerene molecules in the polymeric phase. In the case of photopolymerized C_{60} films, the rate for thermal dissociation of these intermolecular bonds is found to follow an Arrhenius' type behavior with an activation energy $E_a = 1.25$ eV.

In this paper, we review some of our recent work on the photoinduced polymerization in solid C_{60} and C_{70} films. First, we discuss the results of the laser desorption mass spectroscopy (LDMS), infrared (IR) spectroscopy and Raman spectroscopy for solid C_{60} and photopolymerized C_{60} films. Next, we present experimental evidence which suggests that the cross-section for the photoinduced polymerization in C_{70} is considerably smaller than observed for C_{60}.

I. Photoinduced polymerization of solid C_{60}

As is now well known, pristine solid C_{60} is a van der Waals-bonded molecular solid whose electronic and vibrational properties are strongly connected to the properties of the molecule itself, which exhibits icosahedral (I_h) symmetry[1]. Several experiments on pristine C_{60} indicate that below $T_o \sim 260$ K, the C_{60} molecules in the solid undergo an orientational ordering[1]. The data have been interpreted to indicate that (1) for $T > T_0$, the C_{60} molecules spin freely about randomly oriented axes on fcc lattice positions, and (2) for $T < T_0$, two of the three rotational degrees of freedom in the high T phase are lost, and the molecules undergo a "ratchet-like" orientational hopping about four specific (111) directions, as determined by x-ray and neutron studies[1]. At T=90K, a second transition to a merohedrally disordered glass phase (MDGP) occurs[1] leading to an estimate of the number of molecules oriented such that an electron-rich double bond on one molecule faces the electron-deficient opening in a pentagon (~83%) or hexagon (~17%) of the adjacent molecule.

For an isolated C_{60} molecule, out of the 180-6 = 174 intramolecular vibrational modes, only 46 distinct mode frequencies are possible[1]. The IR spectrum obtained at room temperature on thin oxygen-free solid films of C_{60}, vacuum deposited onto a KBr substrate shows four prominent narrow lines at 526, 576, 1182 and 1428 cm^{-1} that are associated with the $4F_{1u}$ intramolecular modes[2]. The Raman spectrum obtained at room temperature of a C_{60} film exhibits twelve features, ten of which are identified with the Raman allowed $2A_g$ (493 and 1469 cm^{-1}) and $8H_g$ vibrations[2]. Thicker C_{60} films allow the detection of numerous higher order infrared-active[3] and Raman-active modes[4] which have been reported recently.

Previously, on the basis of Raman scattering studies, we reported the observation that Ar ion laser radiation at 514.5 and 488.0 nm exceeding ~50 mW/mm^2 appeared to initiate a photoinduced transformation of an oxygen-free, fcc C_{60} to a different solid phase with a richer Raman spectrum[5, 6]. In particular, the high frequency A_g-symmetry "pentagonal-pinch" (PP) mode was observed to broaden and downshift at room temperature from 1469 cm^{-1} to ~1460 cm^{-1}, while the polarization ratio for this mode deteriorated from 100% in pristine C_{60} to ~80% in the phototransformed phase[5]. Similar downshifting of the PP mode was also observed when oxygen-free C_{60} films were irradiated with UV-visible light using a

300 W Hg arc source[6]. In contrast to vibrational spectra of the pristine C_{60} fcc phase, the Raman and IR spectra of the phototransformed phase exhibited many more lines, indicating that the icosahedral symmetry of the C_{60} molecule has been lowered, consistent with the proposed photopolymerization process[6] discussed below. In the polymerized C_{60} films, a new Raman-active mode is observed at ~118 cm^{-1} which is identified with the stretching of these cross-linking bonds[6,7] and molecular dynamics calculations for the C_{60} dimer report this interball mode at 104 cm^{-1}[8]. Furthermore, unlike pristine C_{60}, the phototransformed C_{60} films were found to be insoluble in toluene, suggesting that the weak van der Waals bonding between the neighboring C_{60} molecules has been altered by the UV-visible irradiation[6]. Broadening of the photoluminescence and UV-visible absorption bands upon phototransformation have also been reported[9].

Laser desorption mass spectroscopy (LDMS) spectrum collected on a *phototransformed* C_{60} film (d= 2000 Å , 12 hr exposure to radiation from a 300 W Hg arc lamp) showed a succession of 20 clear peaks which were identified with clusters of cross-linked fullerene molecules $(C_{60})_N$ [6, 10]. The LDMS spectrum of a *pristine* film taken under similar condition was observed to exhibit a series of mass peaks out to N=5[6, 10]. This suggests that at high laser power, the N_2 desorption laser itself is capable of producing polymerized C_{60}. For comparison to this latter result, the pristine film under the reduced (laser power) desorption conditions exhibited only the N=1 peak in its LDMS spectrum[10]. Other evidence subsequently found in support of a polymeric form of C_{60} include photoelectron spectroscopy of C_{60} films that were irradiated with UV light[11]. Using IR spectroscopy as a probe, Yamawaki *et al.*[12] observed additional weak lines when a C_{60} film was subjected to high pressure (7 GPa). The origin of these weak additional lines was attributed to the pressure-induced formation of the C_{60} polymer[12] as the IR-spectrum was found similar to the C_{60} polymer formed by phototransformation[6].

The "2+2 cycloaddition" mechanism is a well known photochemical reaction resulting in the covalent attachment of adjacent molecules[13]. This mechanism is active in molecular solids when carbon double bonds on adjacent molecules are oriented parallel to one another and separated by less than ~4.2 Å [12]. By photochemical assistance, an excited molecular state is formed, and both these double bonds are broken and reform as a four sided ring. Two C_{60} molecules might dimerize by the "2+2 cycloaddition" mechanism to form the C_{60} dimer as shown in Fig. 1. Since a C_{60} molecule contains 30 reactive double bonds tangential to the ball surface, and these double bonds on adjacent molecules can be separated by as little as 3.5 Å in pristine solid C_{60}[1], then solid C_{60} can be seen to satisfy the general topochemical requirement for "2+2 cycloaddition" in a constrained medium [13], *but only for* $T > T_0$. At low T, the double bonds on adjacent molecules avoid each other[1], and according to the topochemical requirement, the reaction should be suppressed[14]. For $T > T_0$, however, the freely spinning molecules enjoy 30(30) = 900 favorable orientations to promote the "2+2 cycloaddition" reaction[14].

Furthermore, the photophysics of C_{60} also seems to favor the proposed "2+2 cycloaddition" reaction mechanism. First, strong, dipole-allowed, singlet-singlet absorption is observed above ~2.3 eV[1]; second, a ~100% efficient intersystem crossing[15,16] is needed to populate significantly the first excited triplet state (T_1); and third, a sufficiently long T_1 lifetime (~40 μsec[15]) is needed to maintain a significant number of molecules in the reactive triplet state. In kinetics studies which we discuss below, we find that the rate limiting step in the photodimerization process is the photoexcitation of the monomer to excited singlet state S_1[17], i.e., one dimer is found for very 488 nm photon absorbed. This result follows directly from the ~100% efficient intersystem crossing.

Another experimental result consistent with the importance of the T_1 state in the dimerization pathway is the observation that intercalated dioxygen "hardens" C_{60} against phototransformation, i.e., high laser flux was not observed to produce phototransformation when dioxygen was present in the lattice [2,5]. The stabilizing influence of dioxygen is undoubtedly connected with the reported quenching of the $C_{60}(T_1)$ state via an interaction with the (O_2) $^3\Sigma$ state [16], thereby suppressing photopolymerization.

Finally, we discuss the kinetics of the photoinduced polymerization of C_{60} films deduced from the time evolution of Raman spectrum in the vicinity of 1469 cm^{-1} mode. The C_{60} film was exposed at T = 300 K to photon flux $\Phi_0 \sim 8.3$ mW/mm^2 at 488 nm from an Argon ion laser. This irradiation level was chosen to both slowly phototransform the film, as well as to stimulate the Raman spectrum. A least-squared Lorentzian lineshape analysis of the Raman spectra was used to determine the relative concentration of the C_{60} monomers M(t) (integrated area under the 1469 cm^{-1} line) and dimers D(t) (integrated area under the $\sim$1459 cm^{-1} line) as a function of time t. Details regarding the lineshape analysis and the experiment are reported in Ref. [17]. Initially, a single line is observed at 1469 cm^{-1} which is 100% polarized[5] and identified with the "pentagonal pinch" mode (A_g) of pristine C_{60}. Early in this phototransformation process it is presumed that dimers, and later, trimers, etc., form, eventually leading to the polymeric phase $(C_{60})_N$. The progression from the pristine film to one containing dimers can be monitored by the shift in spectral weight from the 1469 cm^{-1} line to the $\sim$1459 cm^{-1} line. After $\sim$600 minutes, the phototransformation to the polymer is complete, and the 1469 cm^{-1} line is absent from the Raman spectrum[17] indicating the formation of cross linking C=C bonds between C_{60} molecules via a "2+2" cycloaddition mechanism discussed above[6,14].

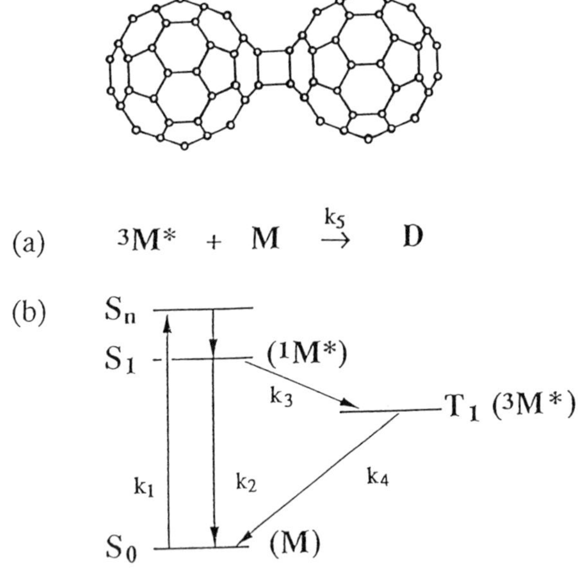

Fig. 1: A C_{60} dimer produced by a photoinduced "2 + 2 cycloaddition" reaction. (a) Bimolecular photochemical reaction in which a C_{60} monomer in the excited triplet state ($^3M^*$) reacts with a monomer in the ground state (M) to form a dimer (D). (b) Schematic energy level diagram for the C_{60} monomer, showing the principal levels involved. The k_i are the respective rate constants, and the other symbols are described in the text.

Using the molecular energy level diagram presented in Fig. 1, the kinetics of the photoinduced polymerization in solid C_{60} can be understood from a conventional triplet state photochemical mechanism, where a monomer in the excited triplet state ($^3M^*$) reacts with a monomer in the ground state (M) to yield the dimer (D), i.e., $^3M^* + M \rightarrow D$; this reaction is assumed to proceed as a normal bimolecular reaction with rate constant k_5 (Fig. 1a). Rate constants k_i ($i = 1$-4) are required to describe the rate at which $^3M^*$ molecules can be produced per incident photon (Fig. 1b). First, the system must be excited by optically pumping the monomer to excited singlet states S_j, where the pumping rate constant k_1 depends on the incident light flux Φ_0. The S_j states then usually decay very rapidly to S_1[17], so we ignore the internal conversion rate constants for $S_j \rightarrow S_1$. Once the system is in S_1 (or $^1M^*$), it can return to the ground state S_0 (or M) by radiative recombination (k_2). However, the singlet exciton recombination has a low quantum yield (~ 0.07% [17]), therefore, the system decays to T_1 (or $^3M^*$) by a nearly 100 % efficient intersystem crossing ($k_3 \sim 3 \times 10^{10}$ s^{-1} [17]). Finally, $^3M^*$ can return either to the monomeric ground state M via k_4, or the $^3M^*$ state can undergo a bimolecular reaction with M (k_5) to form the dimer D. The coupled set of equations which describe the time evolution of this system are[17] :

$$M(t) \approx (M_0)\exp(-2k_1 t) \quad \text{and} \quad D(t) \approx (1/2)(M_0)(1-\exp(-2k_1 t)), \tag{1}$$

where M_0 is the initial density of monomers. From the form of these solutions it can be seen that the optical pumping (rate k_1) must be the rate limiting step, and one absorbed photon produces one dimer.

In-situ temperature dependence of the Raman spectrum for an initially photopolymerized C_{60} film in the vicinity of the symmetric "pentagonal pinch mode" showed that the C_{60} polymer can be driven back towards pristine, van der Waals bonded C_{60} solid[7]. Details regarding the thermal decomposition of the C_{60} polymer are available elsewhere[7]. A least squares Lorentzian lineshape analysis was carried out to determine values for the T-dependence of the integrated intensity (I), frequency (ω) and linewidth (Γ) for the 1469 cm^{-1} (pristine C_{60}) and 1459 cm^{-1} (C_{60}-polymer) lines.

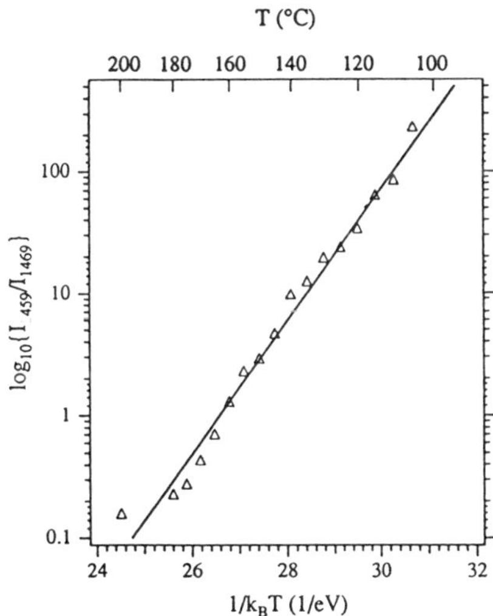

Fig. 2: Intensity ratio $R = \log_{10}\{I_{1459}/I_{1469}\}$ versus $1 / k_B T$. The triangles represent the data, and the solid curve is the result of a linear least-squares fit from which activation energy $E_a = 1.25$ eV is obtained.

Below ~100 °C, the linear increase of Γ_{1459} was assigned to thermal broadening[7]. However, above ~100 °C, Γ_{1459} exhibited an anomalous decrease with increasing temperature, which is interpreted as evidence that the polymer has dissociated to a solid containing primarily monomers and dimers (i.e., for T > 100 °C, the contribution to Γ from inhomogeneous broadening decreases). Assuming a fixed number

370

density of C_{60} shells (M_0), and under steady state conditions, when the system is dominated by monomers and dimers, the intensity ratio I_{1459}/I_{1469} was shown to be proportional to the inverse thermal decomposition rate $(k_T)^{-1}$ (Fig. 2). Anticipating an exponential temperature dependence (i.e. Arrhenius' behavior) for k_T (i.e., $k_T = A \exp(-E_a/k_B T)$, where k_B is Boltzmann's constant and A is a pre-exponential constant with units 1/time), a plot of $\log_{10}(I_{1459}/I_{1469})$ vs. $1/k_B T$, showed a nearly linear relationship over the entire temperature range where reliable intensity information can be obtained [7]. A linear fit to the data in Fig. 2 results in the activation energy $E_a = 1.25$ eV[7]. We suggest that the observed linear behavior of the ratio I_{1459}/I_{1469}, over the larger T-range as evidence that the thermal energy required to dissociate a monomer from all other oligomers is approximately the same as that needed to dissociate a dimer into two monomers. Therefore, we interpret the 1.25 eV activation energy obtained from the fit to the data as the thermal energy required to dissociate a monomer from any oligomer in C_{60}-polymer.

II. Photoinduced polymerization of solid C_{70}

Laser desorption mass spectroscopy (LDMS) results on C_{70} films (d ~2000 Å) deposited on stainless steel substrates and irradiated in vacuum using a Hg arc lamp is shown in Fig. 3. A pulsed N_2 laser (337 nm, 8 ns, 10 mJ per pulse) with parameter values similar to those used for the LDMS of photopolymerized C_{60}[6,10] was used as the desorption laser. The LDMS data exhibits a strong mass peak at m/z = (70 x 12) 840 amu, along with a series of weaker peaks at m/z = N x 840, where N=2, 3 and 4. The N= 2-4 peaks are identified with the C_{70} dimers, trimers, etc. The presence of the weak peak at m/z = 720 amu is, of course, associated with the presence of C_{60} in the sample. For comparison, it should be recalled that in similar studies on photopolymerized solid C_{60}[6,10], a series of mass peaks N x 720 amu, for N as high as 21, were observed.

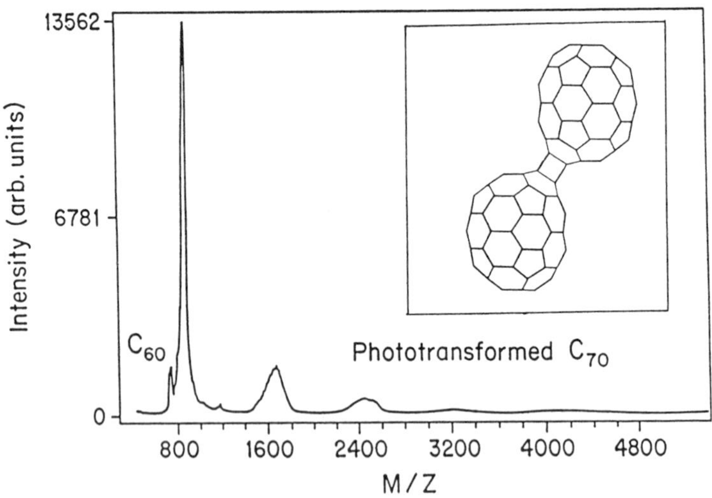

Fig. 3: LDMS spectrum of a C_{70} film (d ~ 2000 Å) phototransformed with UV-visible radiation from a 300 W Hg arc lamp for 10 days. The mass peaks following the strong C_{70} peak at m/z = 840 amu, are identified with 2-4 crosslinked C_{70} molecules. Inset: A C_{70} dimer produced by a photoinduced "2 + 2 cycloaddition" reaction.

The mass spectrum for a pristine (non-irradiated) C_{70} film also showed a series of peaks spaced by

840 amu but the intensity of the $N > 1$ peaks was substantially reduced to those seen in Fig. 3[18]. *A comparison of the LDMS results indicate that the phototransformation of solid C_{70} has a much lower cross-section than observed for solid C_{60}*. In contrast to our studies on phototransformed C_{60} [6,10], we can find no clear signature for the phototransformation of solid C_{70} in either the Raman spectrum or in the UV-visible electronic absorption spectrum. However, the Fourier transform infrared (FTIR) spectrum of irradiated solid C_{70} exhibits a new, broad band (possibly a vibrational mode continuum) in the mid-IR at ~ 1100 cm^{-1} [18].

Similar to that observed for solid C_{60} films, the irradiation of C_{70} films (in the absence of oxygen) with 488 nm Argon ion laser radiation (500 mW/mm^2, 6 h) was noticed to significantly reduce the solubility of solid C_{70} in toluene (24 °C). The unirradiated area of the C_{70} film dissolved immediately, revealing the substrate surface below the film. Furthermore, similar to photopolymerized C_{60}, the toluene-insoluble photopolymerized C_{70} films when heated under vacuum to 250 °C for 3 hours were then found to be highly soluble in toluene again. This behavior is consistent with the thermal scissoring of the covalent bonds between fullerene monomers, as observed in our recent Raman scattering experiments on photopolymerized C_{60}[7].

Based on the orientational order of the C_{70} molecules in the solid at room temperature, we proposed that the photoinduced crosslinking of the monomers in solid C_{70} also occurs via the four membered ring, as shown in the inset to Fig. 3[18]. Additionally, as discussed in Ref. 18, in solid C_{70} only 5 C=C double bonds in a polar cap can effectively participate in the formation covalent bonds between C_{70} monomers. As noted above, there are 30 such reactive C=C double bonds in a C_{60} molecule. We therefore propose that one reason to expect more difficulty in photopolymerizing solid C_{70} is simply the reduced number of reactive double bonds. Finally, it should be mentioned that the tightbinding molecular dynamics calculations by Menon and Subbaswamy[19] for the C_{70} dimer find the four membered ring, and the interball C=C bond length of 1.51 Å, both identical to those obtained earlier for the lowest energy configuration for the C_{60} dimer[8].

Acknowledgements : The research at UK was supported in part by the UK Center for Applied Energy Research, US DOE #DE-FC22-90PC90029 and by NSF Grant No. EHR-91-08764.

References
1.   M.S. Dresselhaus,G. Dresselhaus, and P.C. Eklund, J. Mat. Res. 8, 2054 (1993) and references therein.
2.   P. C. Eklund *et al.*, J. Phys. Chem. Solids 53, 1391 (1992) and references therein.
3.   K. -A Wang *et al.*, Phys. Rev. B 48, 11375 (1993); Martin *et al.* Phys. Rev. B (in press).
4.   Z.H. Dong *et al.*, Phys. Rev. B 48, 2862 (1993).
5.   P. Zhou *et al.* Appl. Phys. Lett 60, 2871 (1992).
6.   A.M. Rao *et al.*, Science 259, 955 (1993).
7.   Y. Wang *et al.*, Chem. Phys. Lett. 217, 413 (1993) and references therein.
8.   M. Menon, K. R. Subbaswamy, and M. Sawtarie, Phys. Rev. B (in press).
9.   Y. Wang *et al.*, Phys. Rev. B (submitted).
10.   D.S. Cornett *et al.*, J. Phys. Chem. 97, 5036 (1993).
11.   A. Ito, T. Morikawa, and T. Takahashi, Chem. Phys. Lett. 211, 333 (1993) .
12.   H. Yamawaki *et al.*, J. Phys. Chem. 97, 11161 (1993).
13.   K. Venkatesan & V. Ramamurthy, Photochemistry in organized and Constrained media p.133 (1991).
14.   P. Zhou, Z.H. Dong, A. M. Rao, and P.C. Eklund, Chem. Phys. Lett. 211, 337 (1993).
15.   J.W. Arbogast *et al.*, J. Phys. Chem. 95, 11-12 (1991).
16.   R.R. Hung, and J.J. Grabowski, J. Physical Chem. 95, 6073-6074 (1991).
17.   Y. Wang *et al.*, Chem. Phys. Lett. 211, 341 (1993) and references therein.
18.   A. M. Rao *et al.* , Chem. Phys. Lett. (submitted) and references therein.
19.   M. Menon *et al.*, Phys. Rev. Lett. (submitted).

THE RAMAN SPECTROSCOPY OF C_{60} SOLID FILMS

KELLY L. AKERS, and MARTIN MOSKOVITS
Department of Chemistry and the Ontario Laser and Lightwave
Research Centre, University of Toronto,Toronto, M5S 1A1, Canada

ABSTRACT

Two distinct Raman spectra have been reported for solid C_{60}. They differ in the exact position and relative intensities of the spectral lines. Specifically, the frequency of the pentagonal pinch mode occurs at 1469 cm^{-1} and 1459 cm^{-1} in the two spectra. Several explanations have been offered for the existence of a second spectrum including: the influence of interstitial oxygen, a change in phase, the coexistence of singlet C_{60} with laser pumped triplet C_{60} and photopolymerization. We show that there are two Raman spectra for solid films of C_{60} with varying relative intensities according to the sample temperature even in the absence of oxygen. The reversibility of the two Raman spectra together with the observation of the 1459 cm^{-1} spectrum at low temperature in regions of the samples which had never been exposed to laser radiation are inconsistent with the photopolymer model.

1. Introduction

The unambiguous identification of the Raman spectrum of solid C_{60} has been complicated by the following influences: oxygen exposure, sample temperature and laser irradiance. To date there have been several studies on the Raman spectroscopy of C_{60}[1-5]. In spite of the simplicity of the Raman spectrum there remains controversy concerning the exact frequencies of some vibrational modes of the molecule and, in particular, the pentagonal pinch mode. Bethune et al.[1,2] reported the first Raman spectrum of a C_{60} film; it featured an intense peak at 1469 cm^{-1} which was assigned to the totally symmetric pentagonal pinch mode. Subsequently, other researchers reported a Raman spectrum of a C_{60} film sublimed and maintained in a vacuum environment[3] with the pinch mode at 1459 cm^{-1}. Photo-polymerization of solid C_{60} films has been suggested[4] to explain the spectral changes observed upon exposure of C_{60} films and crystals to moderate irradiation with 514.5 and 488 nm light. We have reported the temperature dependance of the Raman spectrum of solid films of C_{60} in UHV demonstrating the coexistence of two distinct phases over a wide temperature range.[5] Such behaviour is indicative of an order-disorder phase transition. In this paper we report new results of an UHV Raman study of thin solid films of C_{60} as a function of temperature and laser irradiance that supports the order-disorder phase transition model.

2. Experimental

All experiments were performed in an UHV chamber ion-pumped to a base pressure of 8 x 10^{-11} torr. High purity C_{60} (99.9% pure) was used. C_{60} films were evaporated at temperatures within the range 400 - 450°C. The substrate temperature was maintained at 298 or 57 K. Raman spectra were excited with an argon ion laser (488 and 514.5 nm laser lines). The spectrometer scan speed was 1 cm^{-1}/second setting the range of exposure times to between 3 and 30 minutes for an average scan between 1400 - 1600 cm^{-1} or 100 - 2100 cm^{-1}.

3. Results and Discussion

In our previous high irradiance Raman spectra[5], a laser spot size of 150 μm was used. Although care had been taken to ensure that the same spot was being probed at room and low temperatures we were concerned that thermal expansion might have resulted in different positions of the sample being probed in the low and room temperature experiments. To eliminate this possibility an unfocused laser beam (1.5 mm diameter) was used to record the room temperature Raman spectrum. With low laser irradiance (30 W/cm^2) the pentagonal pinch mode appeared at 1459 cm^{-1} (Figure 1A) exactly as in the high irradiance spectra previously reported. Upon cooling the sample to 50 K in the absence of laser irradiation the spot size was reduced by a factor of 10 by means of an aperture. The

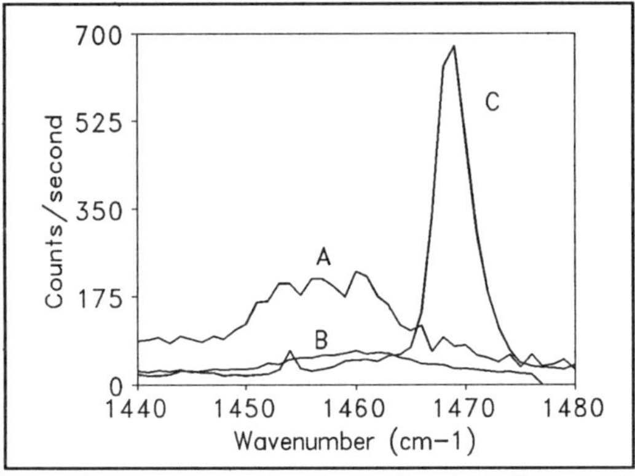

Figure 1: Raman spectrum of a C$_{60}$ film at 298 K (trace A) and at 57 K (trace B). Irradiance = 28 W/cm^2. Trace C is at 57 K after 30 minutes of exposure to radiation with irradiance = 2800 W/cm^2.

irradiance was maintained at 28 W/cm^2, however, the area of the film probed was significantly reduced to ensure that the area sampled at low temperature fell within the region irradiated and sampled at room temperature. The low irradiance (28 W/cm^2) Raman spectrum was identical to the one obtained at high temperature (Figure 1B). The film was then exposed to 30 minutes of high laser irradiance (2800 W/cm^2) and the spectrum recorded at this higher irradiance. The Raman spectrum of the low temperature form (1469 cm^{-1} pinch mode) was obtained (Figure 1C) and the experiment was repeated with a low irradiance. As before, the initial low temperature Raman spectrum was identical to the room temperature spectrum. However, the spectrum transformed in time, under continued irradiation to the 1469 cm^{-1} form. The initial rate of transformation was found to be approximately proportional to the laser irradiance. A sample kinetic curve is shown in Figure 2. This was obtained by following the change in the Raman intensity at 1469 cm^{-1} as a function of time. With an irradiance of 65 W/cm^2 the rate of the transformation was very slow. The Raman spectrum recorded after 7 hours irradiation with the sample temperature maintained at 50 K throughout was identical to Figure 1C (1469 cm^{-1} form). It is important to contrast this behaviour with what is observed when a sample which had previously been transformed to the 1469 cm^{-1} form is heated. In that case the change to the 1459 cm^{-1} spectrum occurs immediately. It is clear that the 1469 cm^{-1} spectrum is carried by the thermodynamically stable phase at low temperatures and the transition can be quickly attained at high laser irradiance. At room temperature the Raman spectrum of the C_{60} films exhibits the pinch mode at 1459 cm^{-1} at irradiances as low as 30 W/cm^2. The spectrum was found to be stable even after being irradiated with up to 5000 W/cm^2. At low temperature (50 K) two distinct Raman spectra were found with the 1469 cm^{-1} being the more stable. The demonstrated reversibility of the two spectra as a function of temperature is not consistent with the polymer model. It is unlikely that a depolymerization reaction would occur only at low temperatures.

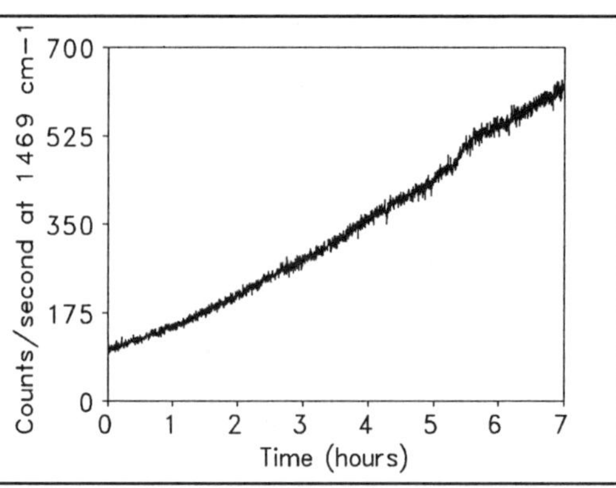

Figure 2: The time dependence of the intensity at 1469 cm^{-1} at low temperature. Irradiance = 65 W/cm^2.

More compelling evidence against the identification of the 1459 cm^{-1} spectrum with polymeric C_{60} was found by repeating the experiment at low temperatures on a spot that had not been exposed previously to laser radiation. It has been reported that the photo-polymerization of C_{60} can only occur at temperatures at which the molecules are rotating, since at low temperature the simple cubic structure prevents the 2+2 cycloaddition of parallel double bonds necessary for the proposed reaction[25]. If the 1469 cm^{-1} spectrum is the Raman spectrum of the "unpolymerized" C_{60} and polymerization occurs only above 250 K then it should be impossible to obtain the 1460 cm^{-1} spectrum at low temperature on a film which had not been previously exposed to laser radiation at room temperature. The low temperature

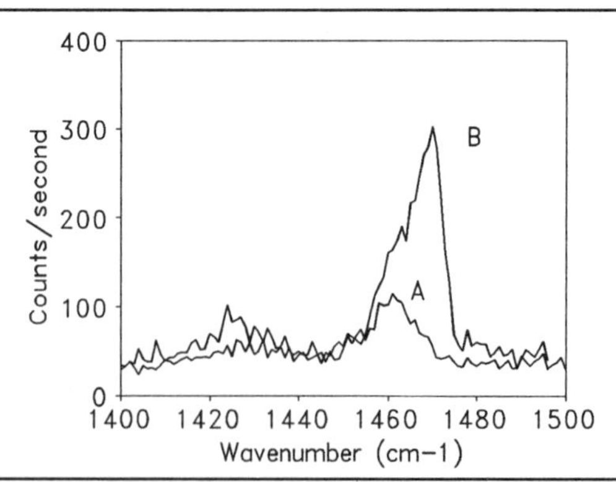

Figure 3: Raman spectrum taken at 57 K of a spot which had never been exposed to laser radiation at room temperature: before (trace A) and after (trace B) 6 hours exposure to laser radiation. Irradiance = 28 W/cm^2.

behaviour was identical to that found for samples exposed to laser radiation at room temperature (Figure 3). The observation of the 1459 cm^{-1} peak, at low temperature, on a sample that had no prior exposure to radiation argues against the interpretation of the 1459 cm^{-1} spectrum as that of a polymer. We maintain our original interpretation that the two Raman spectra shown in Figure 1 reflect an ordered phase of C_{60} molecules which are engaged in cooperative Raman scattering (low temperature) and a disordered phase (high temperature). However, it is obvious that there is a kinetic barrier to the transformation of the high temperature form to the low temperature form. This barrier is surmounted by photoactivation, likely due to the electronic excitation of the fullerene. The reverse process, in which the low temperature form is converted to the high temperature form appears to proceed with little or no barrier.

4. Acknowledgements

The authors wish to thank NSERC and the Centres of Excellence in Molecular and Interfacial Dynamics for financial support.

5. References

1. D.S. Bethune, G. Meijer, W.C. Tang, H.J. Rosen, *Chem. Phys. Lett.* 174 (1990) 219.
2. D.S. Bethune, G. Meijer, W.C. Tang, H.J. Rosen, W.G. Golden, H. Seki, C.A. Brown, M.S. de Vries, *Chem. Phys. Lett.* **179** (1991) 181.
3. S.J Duclos, R.C. Haddon, S.H. Glarum, A.F. Hebard, K.B. Lyons, *Solid State Communications*, **80** (1991) 481.
4. A.M. Rao, P. Zhou, K. Wang, G.T. Hager, J.M. Holden, Y. Wang, W.T. Lee, X. Bi, P.C. Eklund, D.S. Cornett, M.A. Duncan, I.J Amster, *Science* **259** (1993) 955.
5. K. Akers, K. Fu, P. Zhang, M. Moskovits, *Science* **259** (1993) 1152.

OPTICAL INVESTIGATION OF CARBON ONIONS

P. Fedorko, V. Skákalová
Faculty of Chemical Technology, Slovak Technical University
81237 Bratislava, Slovakia

O. Foltin
Faculty of Electrical Engineering, Slovak Technical
University, 81219 Bratislava, Slovakia

F. Kolenič
Welding Research Institute, 83259 Bratislava, Slovakia

and

V. Šmatko
Institute of Electrical Engineering, Slovak Academy of
Sciences, 84239 Bratislava, Slovakia

ABSTRACT

UV-VIS absorption and reflection spectra of materials prepared by heat treatment and intensive electron beam irradiation of carbon soot are presented. A characteristic peak at 3.7 μm^{-1} previously attributed to fullerene-like nanometric carbon particles (carbon onions) is observed in all the spectra. Irradiation has no substantial effect on the peak. No red shift of the peak, expected for individual onions, is observed in a suspension in water. Effects of clustering and of the concentration of carbon onions on the spectra are discussed.

1. Introduction

It was shown recently[1] that spherical, concentric-shelled, onion-like carbon particles consisting from up to 70 graphitic shells can be formed in carbon soot under intensive electron bombardment in a transmission electron microscope. There has been interest in the optical properties of the carbon onions, mainly because they were proposed to be possible candidates responsible for the observed interstellar dust absorption. Theoretical investigation of the optical absorption of the carbon onions[2,3] does not exclude this possibility, but there are very few experimental data on their absorption spectra, due mainly to the difficulty of preparing well structured carbon onions in sufficient quantity. In the only paper reporting measurements of the absorption spectra of carbon onions[4], the onions were prepared by heat treatment of carbon soot at temperatures over 2000 °C. Onions produced by this method differ from the ones created by electron beam irradiation: they are hollow, faceted and consist of at most 4-8 shells.

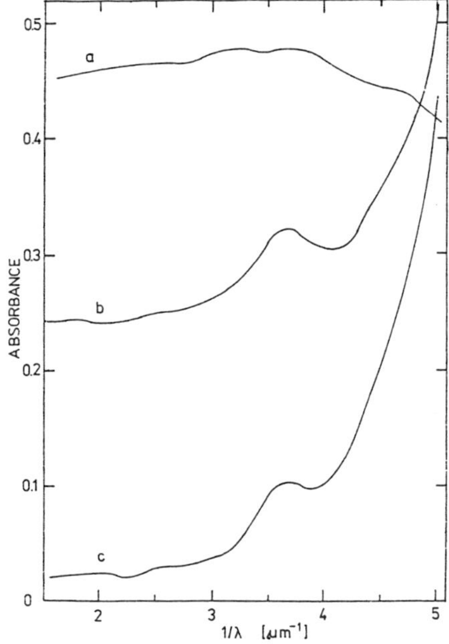

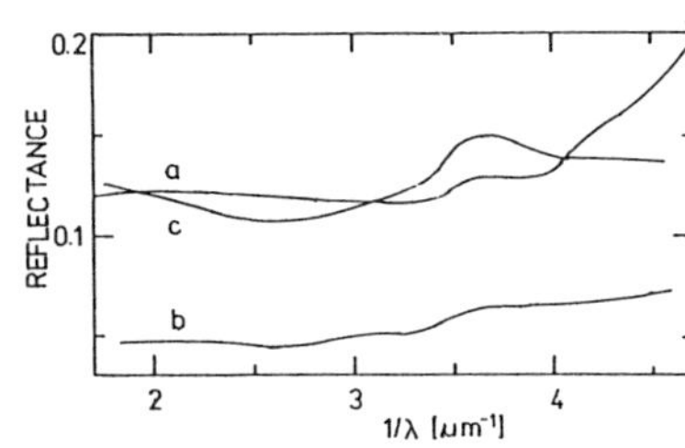

Fig.1. Absorption and reflection spectra:
a) row soot,
b) soot annealed at 1000 °C (1 hour),
c) electron beam irradiated soot at 1000 °C
   (1 hour, current density 4 mA/cm^2).

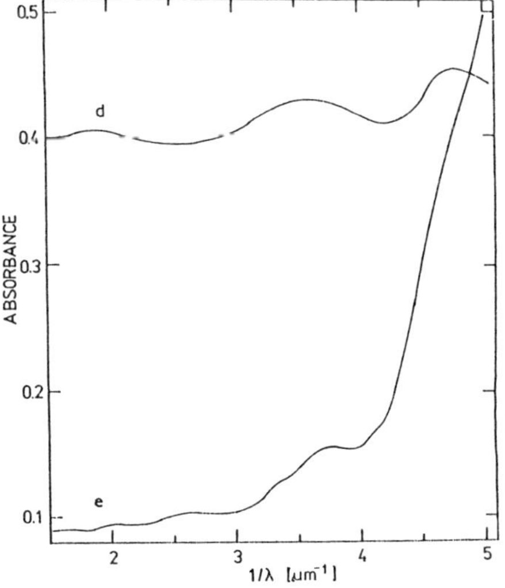

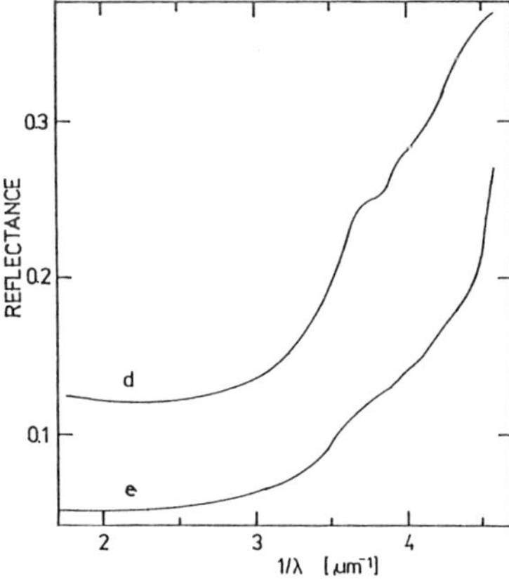

Fig.2. Absorption and reflection spectra:
d) soot annealed at 2100 °C (1/2 hour),
e) electron beam irradiated soot at 2100 °C
   (1/2 hour, current density 25 mA/cm^2).

378

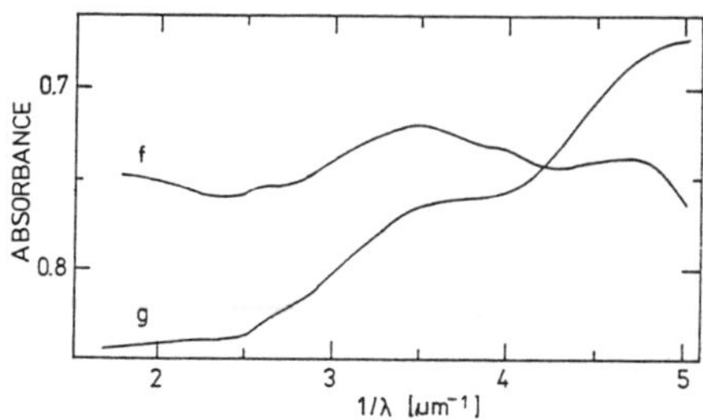

Fig.3. Absorption of a dried soot suspension
(a thin film on quartz): a) annealed at 2100 °C,
b) electron beam irradiated at 2100 °C.

In the absorption spectra of the soot treated in this way, an absorption peak centered at
3.78 μm^{-1} was observed.

The aim of this work was to investigate by the optical UV-VIS absorption and
reflexion spectroscopy the following problems: 1. Is it possible to produce, by electron
beam irradiation, carbon onions in spectroscopically observable quantities? 2. Are the
spectra of electron beam irradiated and of simply annealed carbon soot, treated at the
same temperature, similar or not, i.e. to what extent do structural differencies in the
onions produced by the two methods influence their optical properties? 3. Can the
absorption peak observed in the paper[4] be unambiguously related to the absorption of
carbon onions?

2. Experimental

Carbon soot, produced in arc discharge as a 1 μm thick film on a graphite
substrate, was irradiated (by a direct irradiation) and heated (by an indirect irradiation) in
an electron beam welding equipment working with an accelerating voltage 60 kV in a
scanning regime. Temperature of the sample was measured pyrometrically. Diffused
reflectance of the surface of the sample was measured in the configuration 45/0 with
$BaSO_4$ as a referencee sample. For the optical absorption measurement, material was
crushed and dispersed ultrasonically in water, and the absorption of the suspension was
compared with the absorption of pure water. Suspension was then deposited on a quartz
window, water was evaporated and the spectra were measured again.

3. Results and Discussion

As can be seen from Fig.1, in comparison with the untreated soot, the
characteristic peak discussed in the literature appears at approximately 3.7 μm^{-1} already at
1000 °C, after irradiation as well as after simple heating of the sample, in both "wet"
absorption (suspension in water) and "dry" reflection spectra of the soot. Annealing at
higher temperature 2100 °C (Fig.2) has some influence on the spectra in the region above
4 μm^{-1}, but the feature at 3.7 μm^{-1} appears in all the spectra. In the absorption spectra of
the dried suspension (Fig.3), a feature is visible at 3.5 μm^{-1}, i.e. at slightly longer

wavelength, but certainly no expected[3,4] blue shift of the peak due to the absence of water is observed.

Several conclusions can be done from the fact that the characteristic feature at 3.7 μm^{-1} appears in all the spectra. No substantial difference between the spectra of the annealed and of the irradiated soot could be explained by a relatively low total current density of our electron beam in comparison with the one used in an electron microscope (100-200 A/cm^2) where creation of the onions was observed. The insensibility of the position of the peak to the presence or absence of the solvent means that the absorption peak is not due to individual onions dispersed in water, where a red shift of the peak should appear [3,4]. This result supports the idea[3] that the greatest part of the onions remains clustered even after long ultrasonic dispersion in water. The most striking, however, seems to be the result that the peak appears already at a relatively low temperature 1000 °C , when, according to the microscopic investigation[4], no carbon onions should be formed. For a better understanding of this problem, as well as for having a better idea on the concentration of the onions in the soot, further microscopic study is needed. It is quite possible that due to their small concentration, the onions do not influence the spectra and that the peak at 3.7 μm^{-1} reflects some other structures, for instance graphitization of the soot under annealing, as graphite is known to have a peak in the reflection spectra at this wavelength[5].

4. Acknowledgement

The work was supported by the Action Austria-Slovak Republic, Science and Education, project n° 4SR7.

5. References

1. D. Ugarte, *Nature* **359** (1992) 707; *Europhys. Lett.* **22** (1993) 45; *Chem. Phys. Lett.* **208** (1993) 99.
2. L. Henrard, A. A. Lucas and Ph. Lambin, *Astrophys. J.* **406** (1993) 92.
3. A. A. Lucas, L. Henrard and Ph. Lambin, *Phys. Rev.* **B** (in press).
4. W. A. de Heer and D. Ugarte, *Chem. Phys. Lett.* **207** (1993) 480.
5. D. L. Greenaway, G. Harbeke, F. Bassani and E. Tossatti, *Phys. Rev.* **178** (1969) 1341.

POLARIZABILITY OF ONION-LIKE FULLERENE

L. Henrard, P. Senet[*], Ph. Lambin, A.A. Lucas

Institute for Study of Interface and Surface Sciences, Rue de Bruxelles 61, 5000 Namur,
Belgium
* Max-Planck-Institut für Strömungsforchung, Bunsenstrasse 10, 37073 Göttingen,
Germany

ABSTRACT

A discrete dipole approximation is developed for computing the polarizability of the
onion-like hyperfullerenes. A tensorial polarizability is assigned to each carbon site.
Assuming the dominant sp^2 character of the carbon-carbon bonds in the fullerene
compounds, we transfer the uniaxial 'atomic' polarizability characteristic of graphite
anisotropy in a locally uniaxial tensor in spherical coordinates. Plasmon excitations
of the fullerene are represented by a polarization wave of such an ensemble of
fluctuating dipoles. The eigenvectors of different modes are analysed.

The serendipitous discovery of a new form of carbon [1,2] led to an intensive study of
the fullerene compounds. The original quest of the composition of the interstellar dust is
still unresolved since C_{60} molecule does not provide final breakthrough. Early in the
fullerene history, Kroto[3] proposed the onion-like particle as a possible candidate for the
UV absorption band at 5.7 eV. Although the formation of this cluster is still unclear, its
high stability and its formation under intense electron irradiation[4] allow to speculate its
possible presence in interstellar dust grains[5].

The object of this paper is the understanding of the U.V. response properties of this
class of 'spherical' anisotropic material. The dynamical polarizability due to π-plasmons
oscillations allow to interpret the optical data and the electron energy loss spectroscopy
spectra.

We describe the electronic collective oscillations by assigning a local dipolar
polarizability $\alpha_k(\omega)$ at each carbon site k. The dipolar moment at site k is

$$p_k(\omega) = \alpha_k(\omega)\, E_k^l(\omega) \tag{1}$$

where $E_k^l(\omega)$ is the local field at site k, i.e. the sum of external field (E^e) and the electric
field induced by the other fluctuating dipoles.

$$E_k^l(\omega) = E^e(\omega) + \sum_{k' \neq k}^{N} T(r_k - r_{k'})\, p_{k'}(\omega) \tag{2}$$

where $T(r)$ is the usual dipolar tensor and N is the number of carbon atoms.

The total dipolar moment of the molecule is the direct sum of the atomic dipolar moments, which for site k write (using Eq. 1 and Eq. 2)

$$p_k = \left\{ \sum_{k \neq k'}[1-\alpha_k(\omega)T(r_k-r_{k'})]^{-1}\alpha_{k'}(\omega) \right\} E_e(\omega) \qquad (3)$$

The atomic polarizability tensor $\alpha_k(\omega)$ is deduced from the dielectric properties of graphite by the using a Clausius-Mossotti relation adapted to the hexagonal lattice of graphite[6]. We take the Cartesian uniaxial tensor of graphite as a spherical local uniaxial tensor for the fullerene molecule, i.e. $\alpha_k(\omega) = \alpha_{//}(\omega)\, z\, z\ + \alpha_{\perp}(\omega)\,(x\,x + y\,y)$ in graphite becomes $\alpha_k(\omega) = \alpha_{//}(\omega)\, r\, r\ + \alpha_{\perp}(\omega)\,(\theta\,\theta + \varphi\,\varphi)$ for the fullerenes. It appears that the tensor character of $\alpha_k(\omega)$ is crucial for matching experimental data for the single wall C_{60}[6]. It is obvious that this importance is enhanced for multilayered fullerenes. A more detail description of this formalism is given by P. Senet et al [6].

The experimental dielectric function of graphite is modeled by two lorentzian resonances, at 4.4 eV and 14.3 eV, for the in-plane component and, at 5.0 eV and 11.0 eV, for the out of plane component. A damping parameter of 0.01 eV has been taken for each resonance. This small value leads to very sharp peaks in the molecular polarizability (fig 1) and allows us to separate each mode from the others.

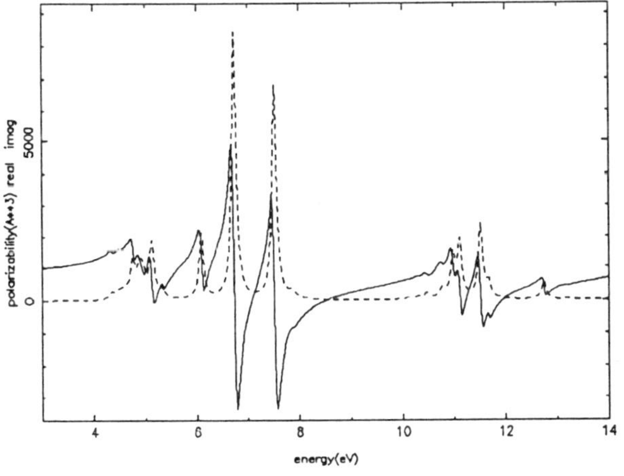

Fig1. Real part (full curve) and imaginary part (dashed curve) of the polarizability of three-layer onion-like molecule versus the external field frequency. See text for details

Fig.1 shows the dipolar polarizability of a three layers onion ($C_{60}@C_{240}@C_{540}$) between 3 and 14 eV. C_{240} and C_{540} are assumed to be perfect truncated icosahedral molecules and alternance of short bonds (hexagon-hexagon fusions) and long bonds (pentagon-hexagon fusions) is introduced for C_{60} only. From fig.1, one can clearly distinguish three energy ranges : from 4.3 eV to 5.3 eV, from 6 eV to 8 eV and around 11 eV. The first and the last regions are associated with the two resonances in the radial atomic polarizability, the second with the resonance in the tangential polarizability. The dynamical polarizability of single shell fullerenes ($C_{60}, C_{70}, C_{240}, ..$) can also be grouped in

382

these classes of resonances. C_{60} for instance shows only three peaks at 5.05 eV and 11.32 eV (radial modes) and 7.21 eV (tangential mode). The radial modes of C_{240} are split due to the deviation from sphericity. In the multi-shell fullerenes, the number of distinct modes increases as the number of layers increases.

Three dominant tangential modes are seen for the three-layer onion. From the dipolar map (fig.2a), one can see that the lower energy mode is mainly due to the outer shell of carbon (C_{540}). The higher energy mode (fig.2b) is attributed to the C_{60} (inner shell) polarization. We can thus expect as many modes as the number of shells (even for a small onion) with decreasing oscillator strengths due to the radial oscillation of the polarizability of different shells : For 7.45 eV mode, all dipolar moments are in phase. For the 6.05 eV mode, the C_{240} shell is anti-polarised. The C_{60} shell is polarised like the external shell. This alternance can explain the decreasing oscillator strength.

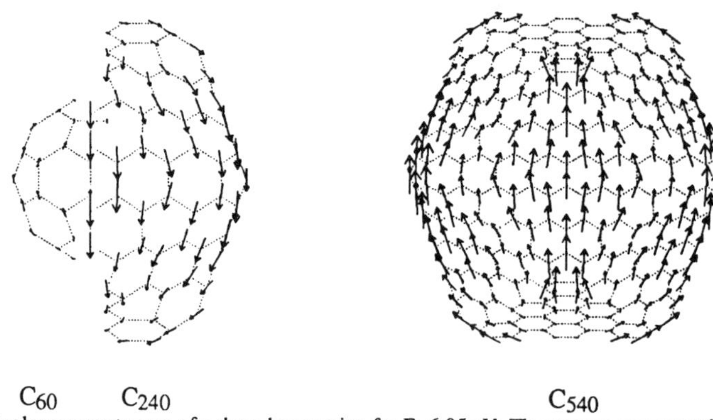

Fig 2a. Dipolar moments map of a three-layer onion for E=6.05 eV. The vectors represent the intensity and the direction of the dipolar moments on individual carbons. The field is along the vertical direction (z-direction). The map is a projection in the y=0 plane (drawing plane).

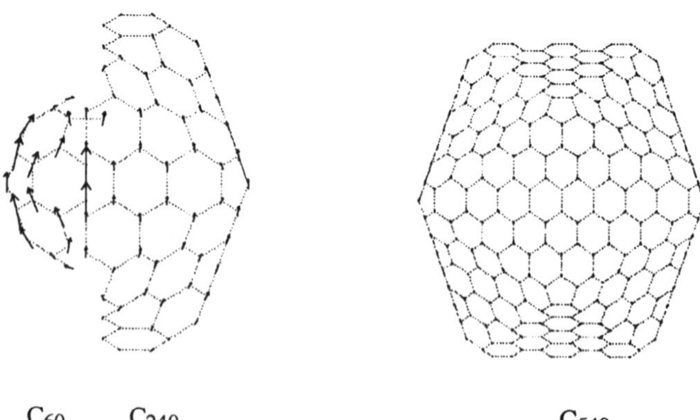

Fig 2b. Same as in fig 2a for E=7.45 eV

Using the macroscopic dielectric model [7] with the same dielectric function gives the same trends. As a discrete number of layer can not be described within a continuum model, the macroscopic polarizability presents more modes that the microscopic one. Tangential modes with decreasing oscillator strengths can also be separated.

Apell et al.[8] have performed an hydrodynamical model to analyse collective excitations in multilayer fullerenes. In their description, the carbon layers were respresented by isotropic electronic shells of thickness adjusted to 1.4Å which were separated from each other by 1.95Å. Apell at al. found two classes of modes (around 6 eV and around 11 eV) which are akin to the slow and fast mode of a continuum dielectric shell[9]. The 6 eV modes of the hydrodynamical model are analoguous to our tangential modes. The 11eV modes are associated with radial oscillations of charges in the continuum model whereas the peaks around 11 eV in fig.1 are due to a resonance in the radial component of the carbon polarizability. The characteristic spherical anisotropy of onion-like compounds is of pure geometrical origin in ref [8] whereas it is described by a microscopic polarizabiliy derived from graphite in our model.

In conclusion, we presented a discrete dipole approximation (DDA) of the polarizability of onion-like fullerenes. The crucial improvement with respect to Wright's DDA[10] is the presence of the hollow central cavity. The hollow character has proved to be essential in predicting uhe polarizability of hyperfullerenes[5,7] Moreover, The spherical symmetry was not assumed in the present formalism and we were thus able to investigate irregular shaped single- or multi-layer fullerenes.

Acknowledgement

This work was funded by the national program on Inter-University Research Project (PAI) on Interface Sciences initiated by the Belgian State Prime Minister Office Science Policy Programming) and has benefited from a grant from the Walloon Region. The authors acknowledge the use of the Namur Scientific Computing Facility. Ph.L. acknowledges the National Fund for Scientific Research of Belgium.

References

1. H.W. Kroto et al., Nature 318 (1988) 162.
2. W. Krätschmer et al., Nature 347 (1990) 354
3. H.W. Kroto, K. Mc Kay, Nature 331 (1988) 328
4. D. Ugarte, Nature 359 (1992) 707
5. L. Henrard, A.A. Lucas, Ph. Lambin, Astrophysical J. 406 (1993) 92
6. P. Senet , L. Henrard, Ph. Lambin, A.A. Lucas. in these proceedings
7. A.A. Lucas, L.Henrard, Ph. Lambin, Phys. Rev. B. In Press
8. P. Apell, D. Östling, G. Mukhopadhyay, Solid State Comm. 87 (1993) 219
9. Ph. Lambin, A.A. Lucas, J.P. Vigneron, Phys. Rev. B 46 (1992) 1794
10. E.L. Wright, Nature 336 (1988) 227

REDOX STATES OF C_{60} AND C_{70} MEASURED BY EPR AND OPTICAL ABSORPTION SPECTROSCOPY

M. Baumgarten*, L. Gherghel

Max-Planck-Institut für Polymerforschung,
AckermannWeg 10, D-55 128 Mainz, FRG

Abstract

We have measured EPR and visible-NIR spectra during reduction and oxidation of C_{60} and C_{70} in solution. Characteristic VIS-NIR absorption bands can be used for the clear assignment of individual redox states and their mixtures in solution. These optical transitions reflect the energy separation of the HOMO-LUMO orbitals which partially loose their degeneracy upon charging. The assignment of one sharp EPR signal in solution to C_{60}^- (ΔH = 0.025-0.05 mT) is supported by the same findings for a monosubstituted derivative, where a small broadening of this signal is found from additional ^1H hyperfine interaction (ΔH = 0.1 mT). Furthermore the g-values of the radical anions of C_{60} increase with the amount of charges, indicating largest contributions from spin-orbit-coupling for the monoanion.

1. Introduction:

Since the discovery and large scale production of C_{60} and higher fullerenes [1,2] an increasing number of studies has dealt with the characterization and derivatization of these molecules [3-5]. Redox chemical studies have shown that C_{60} and C_{70} readily accept 1-6 electrons each to form fulleride anions [6,7].

For a better understanding of the redox states of C_{60} and C_{70}, we have chosen the parallel application of EPR and optical absorption spectroscopy [8]. Separation of C_{60} and C_{70} was achieved using chromatography over polystyrene gel [9]. The anions were prepared under high vacuum in tetrahydrofuran (THF) solution by contact on a potassium mirror in sealed quartz tubes. Despite the low solubility of C_{60} and C_{70} in THF [8,10], the solvent is stable at high reduction potentials even in contact with potassium and the fulleride ions are better soluble than the neutral molecules up to the trianion.

2. Results and discussion:

The VIS-NIR spectra can easily be used to identify the reduction state of a solution of fulleride ions from C_{60}^- to C_{60}^{6-} as we have shown recently [8]. This is demonstrated here for the mono-, di -, and trianion and their intermediate redox states (Fig.1A-E). C_{60}^- is characterized by λ_{max} = 1072 nm together with fine structure from vibronic couplings. Similar spectra in different media have been reported quite frequently [11,12], suggesting negligible counter ion effects. An increase in absorption intensity is also found in the range around 600 nm, where the neutral compound exhibits only some very weak forbidden transitions. Dianionic contributions are indicated as soon as the band at 934 nm gains intensity, becoming gradually more intense than each of the peaks of the monoanion, accompanied by raise of a very weak band at 1305 nm (Fig1B). An isosbestic point (IP) for C_{60}^-/C_{60}^{2-} has been found at 980 nm. Upon trianion formation new absorption bands appear at 772 and 1336 nm (Fig.1D,E), while the dominating band in between decreases and is gradually shifted bathochromically to 967 nm (IP: C_{60}^{2-}/C_{60}^{3-} at 800 nm).

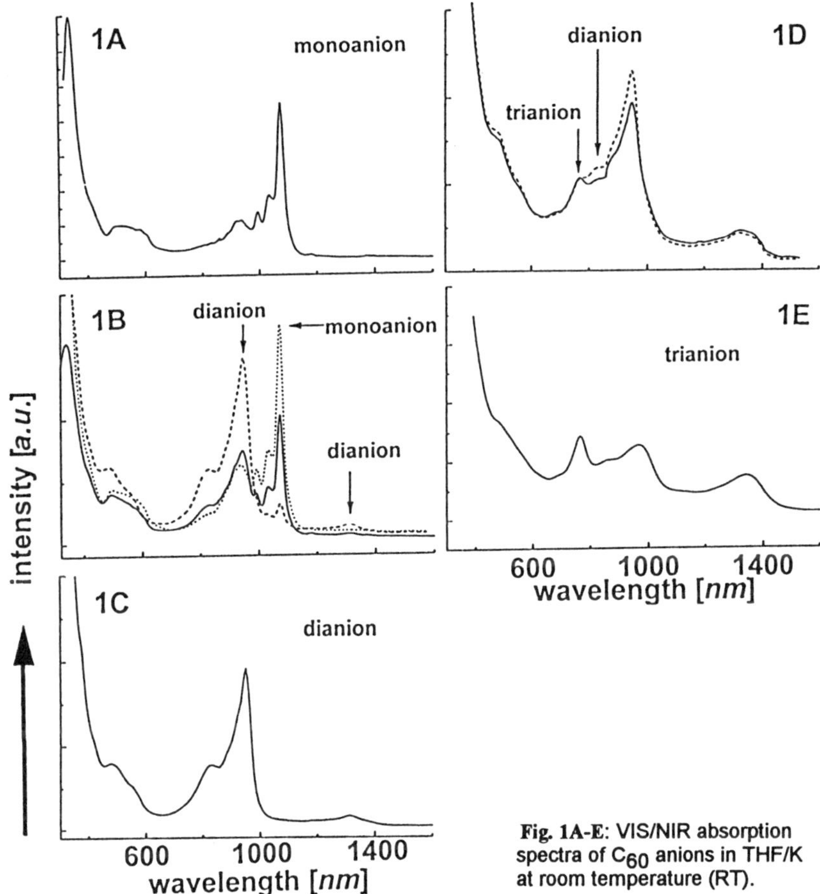

Fig. 1A-E: VIS/NIR absorption spectra of C_{60} anions in THF/K at room temperature (RT).

The EPR spectrum of the monoanion exhibits a single sharp line with g = 2.0001±0.0002, and linewidth of ΔH = 0.025-0.05 mT at RT where no further underlying broad signal can be detected. Upon freezing the linewidth increases by an order of magnitude ΔH = 0.25-0.5 mT (150K) and an additional broad underlying feature is found.

Dianion formation is accompanied by a new EPR signal with g = 2.0006, which only exists in the mixture of $C_{60}^{1-/2-}$ while it dissappears upon complete reduction to

C_{60}^{2-} where the EPR spectrum consists of a single line at g = 2.0009±0.0002 and ΔH = 0.03 mT. In frozen solution (T = 130 K) the biradical formation is evidenced by additional zero field splitting (zfs) components with one new large splitting of 5.4 mT, together with those of 2.8 mT and 1.3 mT reported earlier. This has recently been confirmed by Reed et al.[13] who also found that the zfs-components reach a maximum at T = 85 K, establishing a thermal activated triplet. Whether the central line stems from another biradical species with higher symmetry (for cubic symmetry D~0) or from monoradical contributions is not yet clear. Further temperature dependent measurements of the corresponding absorption spectra may help in clarifying this ambiguity.

Slightly further reduction leads to a loss of all zfs-components and the EPR spectrum at RT exhibits a signal at g = 2.0012 (C_{60}^{3-}); thus the g-values increase with number of charges ($g(C_{60}^-) < g(C_{60}^{2-}) < g(C_{60}^{3-})$). While the tetra- and hexaanion are diamagnetic, preliminary studies on higher reduction states have demonstrated that parallel to pentaanion formation sometimes protonation from solvent occurs, leading to dihydro-fullerene, as evidenced by a three line hyperfine pattern with a coupling of $a_H \sim 1.5$ mT.

As in nonalternant hydrocarbons the cations of C_{60} should differ essentially from the anions already suggested from a simple Hückel-MO picture [14], with five fold degeneracy of the HOMO (h_u) and threefold degeneracy of the LUMO (t_{1u}). The cations have been generated in methylenechloride upon addition of $SbCl_5$. C_{60}^+ exhibits an optical absorption at 983 nm and the corresponding EPR spectrum yields a single line at g = 2.0030 with a linewidth of ΔH = 0.153 mT at RT. This g-value is much closer to the g-value of the free electron - as usual for aromatic π-systems - and contrasts dramatically with the anion characteristics. While there is one report on a dication biradical[15], cyclic voltammetry has shown irreversible cation formation[16]. Therefore larger amounts of $C60^+$ are necessery to elucidate its composition.

In order to find out the specific influences of organic substitutents on the electronic characteristics of C_{60} we have chosen the Diels-Alder monoadduct of 1,2-Dimethoxy-o-xylylene with C_{60} 1 [17]. The reduced monoadducts 1^- and 1^{2-} show similar VIS-NIR

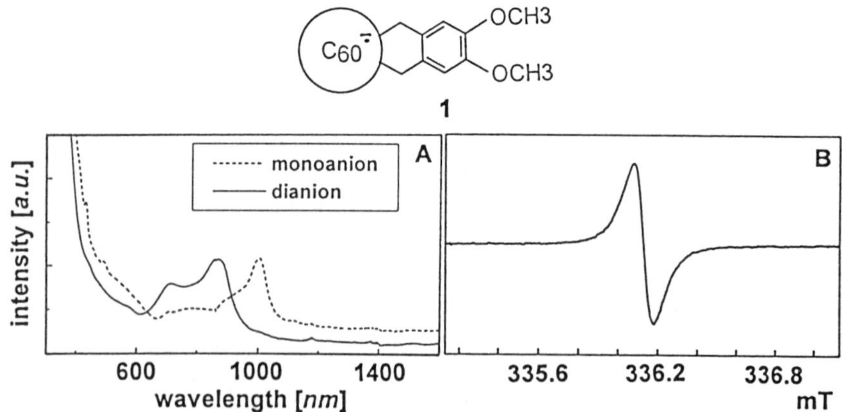

Fig.2: A) VIS-NIR spectra of 1^- and 1^{2-}, **B)** EPR spectrum of 1^- in THF/K.

absorption bands as found for $C_{60}^{-/2-}$, but they are generally shifted hypsochromically. The anions exhibit absorptions at λ_{max} = 1006 nm for 1^- and λ_{max} = 858 nm with a shoulder at 718 nm for 1^{2-} (Fig. 2A), indicative for a slightly enhanced HOMO-LUMO energy separation (t_{1u}-t_{1g}).

The EPR spectrum of 1^- in solution gives rise to a single sharp line with g = 1.9999±0.0002; ΔH = 0.10 mT at RT and again no underlying broad feature can be detected (Fig. 2b). The slight broadening of the EPR signal compared to C_{60}^- is easily explained by some unresolved proton couplings from the four methylene protons in β-positions. This allows unambiguous assignment of the narrow EPR signal to the C_{60} monoanion in dilute solution[18]. Further reduction towards C_{60}^{2-} is accompanied by the same features as observed for C_{60}, with new signals in the EPR spectrum at g = 2.0006 and finally g = 2.0009.

Neutral C_{70} exhibits a dominant visible band at λ_{max} = 469 nm, while upon reduction one new NIR absorption is found at λ_{max}= 1369 nm. Even further reduction diminishes the intensity of this band, and a new transition at 1165 nm gains intensity (Fig. 3). Finally the sample turns yellow and an even broader absorption with λ_{max} = 1070 nm is obtained, which presumably belongs to the trianion. This differs from the results by Lawson et al. [19] who report constant spectral features for the di- to tetraanion of C_{70} reduced electrochemically (λ_{max} = 670 and 1170 nm for each species in benzonitrile), which fact may find an explanation by the lower solubility of the higher charged species.

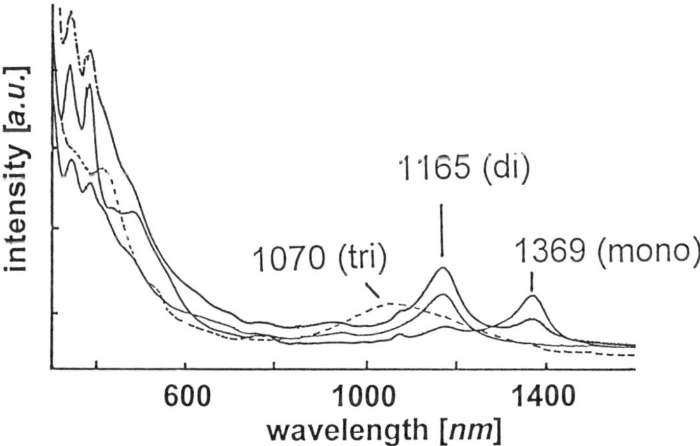

Fig. 3: VIS-NIR bands of C_{70}^-, C_{70}^{2-}, and C_{70}^{3-} in THF/K.

The EPR spectrum of C_{70}^- is characterized by a dominating line at g = 2.0023 and ΔH = 0.016±0.002 mT (C_{70}, Hoechst gold label, 99.4%) while additional signals sometimes observed are most probably due to contaminations from C_{60} and higher fullerenes rather than to anisotropic components. The paramagnetic state of the dianion with g = 2.004 but missing zfs-components in the frozen state deserves further attention.

3. Conclusion:

The VIS-NIR bands of reduced C_{60} and C_{70} enable a clear assignment of the redox states of the charged species in solution, where the transition energies reflect the splittings of the partially degenerate orbitals upon charging[11,12,20]. Counter ion effects on the optical absorption spectra are negligible small, since they have been found under various conditions, but they are certainly important in understanding the EPR features. The assignment of a narrow EPR signal to C_{60}^- in solution finds further support from comparison with a mono-substituted derivative **1**[-], where this signal is slightly broadened due to proton hyperfine interaction. Further reduced states exhibit signals with larger g-values, and the triplet state dianions of C_{60} and C_{70} seem to be thermally activated.

As in non alternant aromatic π-systems, the cations behave very different compared to the anions with an EPR signal at $g > g_e$, but further structural proof is necessary.

Acknowledgement: We wish to thank Dipl.-Chem. P. Belik for a well characterized sample of the 4,5-dimethoxy-o-xylylene adduct **1**[17].

4. References:

1. W. Krätschmer, L. D. Lamb, K. Fostiropoulos, D. R. Huffman, *Nature* **347** (1990)   354.
2. R. Taylor, J. P. Hare, A. K. Adult-Sad, H. W. Kroto, *J. Chem. Soc.   Chem. Commun.* **(1990)** 1423.
3. F. Diederich, Y. Rubin, *Angew. Chem.* **104** (1992) 123.
4. F. Wudl, *Acc.Chem. Res.* **25** (1992) 157.
5. A. Hirsch, *Angew. Chem.* **105** (1993) 1189.
6. Q. Xie, E. Pérez-Cordero, L. Echegoyen, *J. Am. Chem. Soc.* **114** (1992) 3978.
7. F. Zhou, C. Jehoulet, A. J. Bard, *J. Am. Chem. Soc.* **114** (1992) 11004.
8. M. Baumgarten, A. Gügel, L. Gherghel, *Adv. Mater.* **6** (1993) 458.
9. A. Gügel, M. Becker, D. Hammel, L. Mindach, J. Räder, T. Simon, M. Wagner, K. Müllen, *Angew. Chem.* **104** (1992) 666.
10. R. S. Ruoff, D. S. Tse, R. Malhotra, D. C. Lorents, *J. Phys.Chem.* **97** (1993) 3379.
11. M. A. Greaney, S. M. Gorun, *J. Phys. Chem.* **95** (1991) 7142.
12. G. A. Heath, J. E. McGrady, R. L. Martin; *J. Chem. Soc. Chem. Commun.* **(1992)** 1272.
13. P. Bhyrappa, P. Paul, J. Stinchcombe, P. D. W. Boyd, C. A. Reed, *J. Am. Chem. Soc.* **115** (1993) 11004.
14. R. C. Haddon, *Acc. Chem. Res.* **1992,** *25*, 127.
15. H. Thomann, M. Bernardo, G. P. Miller, *J. Am. Chem. Soc.* **114** (1992) 6593.
16. T. Suzuki, Y. Maruyama, T. Akasaka, W. Ando, K. Kobayashi, S. Nagase, *J. Am Chem. Soc.* **116** (1994) 1359.
17. P. Belik, A. Gügel, J. Spickermann, K. Müllen, *Angew. Chem.* **105** (1993) 95.
18. A line broadening of this signal also occurs in ^{13}C enriched samples: P. Dinse, private communication; see also discussion by H. Moriyama, H. Kobayashi, A. Kobayachi, T. Watanabe, *J. Am. Chem. Soc.* **115** (1993) 1185.
19. E. R. Lawson, D. L. Feldheim, C. A. Foss, P. K. Dorhout, C. M. Elliot, C. R. Martin, B. Parkinson, *J. Phys. Chem.* **96** (1992) 7175.
20. K. Harigaya, *Phys. Rev. B* **45** (1992) 13676; K. Harigaya, Prog. Theoret. Phys. Suppl. **113** (1993) 229

CALCULATION OF THE UV-VIS SPECTRA OF
$C_{60}H_{2n}$ $(n = 1, 2, 3)$

László Udvardi[1], Péter R. Surján[1,2], Jenő Kürti[3], Sándor Pekker[4]

[1] Quantum Theory Group, Institute of Physics, TU Budapest
H–1111, Budafoki út 8, Budapest, Hungary
[2] Laboratory for Theoretical Chemistry, Eötvös University
H–1518 Budapest 112, POB 32, Hungary
[3] Department of Atomic Physics, Eötvös University
H–1088 Budapest Puskin u. 5–7, Hungary
[4] Research Institute for Solid State Physics
H–1525 Budapest P.O.B. 49

Abstract

The UV-Vis spectra of Diels-Alder adducts of C_{60} show characteristic differences with respect to isolated C_{60} spectra. New bands appear between 400 and 500 nm, as well as at around 700 nm. In order to explain the origin of these bands we performed quantum chemical calculations for model molecules. The first band is a singlet–singlet transition while the lowest energy band can originate only from a triplet level and spin-orbit coupling should be responsible for this excitation. Symmetry arguments indicate that such transitions remain forbidden for pure C_{60}.

1. Introduction

Structure and properties of various Diels-Alder adducts of C_{60} has been studied recently by several authors.[1–3] We present here the experimental UV-Vis spectrum of the $C_{60} - (Anthracene)_3$ adduct (Fig. 1.). The low energy region of this spectrum is very similar to those reported in the literature for other Diels-Alder adducts (see e.g. Ref.1.). Compared to the UV-Vis spectrum of pure C_{60}, we observe two extra bands at 700 nm (1.77 eV)and 433 nm (2.85 eV). Appearance of the weak 700 nm band is especially interesting as no allowed transition is expected below 2 eV for C_{60}.[4,5]

The universality of these features of the UV-Vis spectra permitted us to substitute the complicated $C_{60} - (Anthracene)$ adduct with the title compounds in the course of quantum chemical calculations we have performed. To modellize C_{60} and n anthracene, we considered C_{60} and n H_2 with symmetries C_{2v}, D_{2h} and C_{3v} for $n = 1, 2$ and 3, respectively. The geometries of the model compounds have been optimized by the semiempirical all-valence electron AM1 method.[6] The UV-Vis spectra are evaluated within the CNDO/S-CI model[7] including 2100 singly excited configurations.

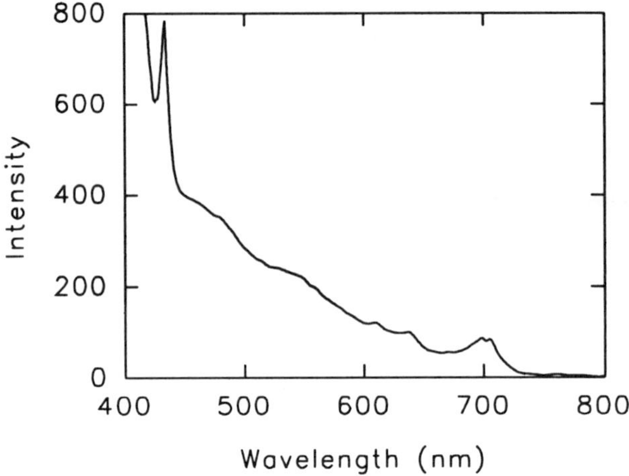

Fig. 1. Absorption spectrum of $C_{60} - (Anthracene)_3$.

2. Results

The calculated absorption curves due to singlet transitions can be seen in Fig. 2. The absorption peak at 2.85 eV shown by the experiment can be assigned to singlet–singlet transitions between 2 eV and 3 eV in the theoretical spectra. The difference between the positions of experimental and theoretical peaks can be the consequence of substituting the original compound by model molecules. In the case of the pure C_{60} this relatively strong band is not present, only weak Herzberg–Teller transitions can be found in this region of the spectrum.[4]

Both for the model molecules and for the C_{60} there is no singlet–singlet transition predicted below 2 eV. We have also determined the triplet states, the results are displayed in Table 1. They suggest that the low lying band belongs to a singlet–triplet transition. The interaction making this transition allowed is spin–orbit coupling. The corresponding dipole matrix element has the following form (neglecting the second and higher order terms):

$$\langle \psi_{S0} | \vec{R} | \psi_{T0} \rangle = \sum_{S} \langle \phi_{S0} | \vec{R} | \phi_S \rangle \frac{\langle \phi_S | V_{SO} | \phi_{T0} \rangle}{E_S - E_{T0}} + \sum_{T} \frac{\langle \phi_{S0} | V_{SO} | \phi_T \rangle}{E_T - E_{S0}} \langle \phi_T | \vec{R} | \phi_{T0} \rangle \qquad (1)$$

where ψ_{S0} is the ground state of the molecule, ψ_{T0} is the final triplet state, ϕ_S and ϕ_T are the virtual unperturbed singlet and triplet states, E_S and E_T are the corresponding energies.

The triplet states of the C_{60} molecule below 2 eV are all gerade and, since the spin–orbit coupling does not change the behavior of the wavefunctions against inversion, all terms in Eq.(1) remain zero. That is, even if spin–orbit coupling is considered, the singlet–triplet transitions remain forbidden for pure C_{60}. This explains the lack

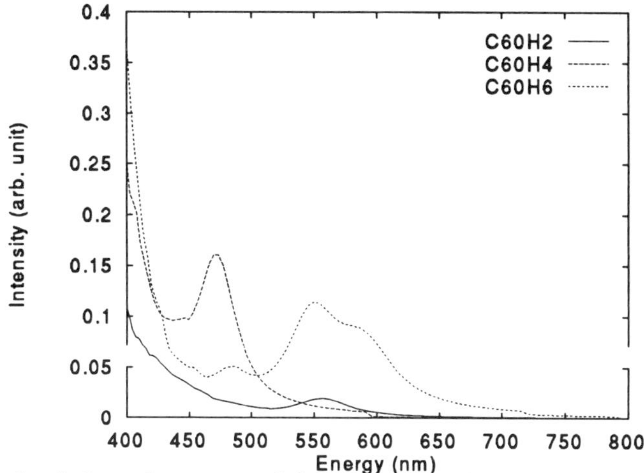

Fig. 2. Calculated absorption spectra of the title compounds with one, two and three H_2. Bands are plotted by using Lorentzian distributions with 0.1 eV half-widths.

of the absorption bands below 2 eV.

However, the situation is very different for the Diels–Alder adducts studied which either do not have inversion symmetry or show spatially allowed g–u transition in the investigated energy range (Table 1). At about 1.6 eV (770 nm) all the Diels–Alder compounds have a state with considerable dipole transition matrix element. Consequently evaluating spin–orbit coupling by Eq.(1), one can get nonzero contributions to singlet–triplet intensities.

C_{60}		$C_{60}H_2$		$C_{60}H_4$		$C_{60}H_6$	
E_{T0} (eV)	Dipol.sq.	E_{T0} (eV)	Dipol.sq.	E_{T0} (eV)	Dipol.sq.	E_{T0} (eV)	Dipol.sq.
1.179	0.0	1.198	0.001	1.173	0.000	1.081	0.000
1.607	0.0	1.297	0.003	1.297	0.000	1.287	0.120
1.805	0.0	1.358	1.936	1.377	0.000	1.289	0.096
1.945	0.0	1.531	0.000	1.381	0.000	1.484	1.164
		1.611	0.025	1.601	0.056	1.487	1.128
		1.614	0.200	1.609	0.858	1.555	0.002
		1.628	2.615	1.623	11.548	1.578	11.863
		1.953	0.003	1.894	0.000		
		1.996	0.083	1.978	0.001		

Table 1. Triplet energies and dipole squares (without spin integration) of the investigated molecules.

Our conclusion is that the upper band between 400 nm and 500 nm can be assigned to singlet–singlet transitions and the low lying band at about 700 nm belongs to a singlet–triplet transition which become allowed by spin–orbit coupling. The existence and the position of the two bands do not depend strongly on the type of adducts. They originate from the change of the bonding structure of the C_{60} molecule on one hand, and from the lowering of the icosahedral symmetry on other hand.

3. Acknowledgement

We are indebted to K. Németh for his help in the AM1 calculations. This work was partly supported by the grants OTKA 517/1990, T012820 and T014405.

4. References

1. Y.Rubin, S.Khan, D.I.Freedberg and C.Yeretzian, J.Am.Chem.Soc. **115** (1993) 344.
2. S.Pekker et al, to be published.
3. J.Averdung, J.Mattay, H.Moon, V.H.Müller, H-U.Meer, this Volume.
4. S.Leach, M.Vervolet, A.Despres, E.Breheret, J.P.Hare, T.J.Dennis, H.W.Kroto, R.Taylor, D.R.M.Walton, Chemical Physics **160** (1992) 451.
5. J.P. Hare, T.J.Dennis, H.W. Kroto, R. Taylor, D.R.M. Walton, Chem. Phys. Letters **183** 283 (1991).
6. J.P.Stewart, QCPE (1983) 455.
7. J.Del Bene, H.H.Jaffé, J. Chem. Phys. **48**. 1807 (1968).

A ONE PARAMETER MODEL OF THE UV SPECTRA OF CARBON

P. SENET*, L. HENRARD, Ph. LAMBIN and A.A. LUCAS

Institute for Study of Interface Science, Rue de Bruxelles 61, 5000 Namur, Belgium

Max-Planck-Institut für Strömungsforschung, Bunsenstrasse 10, 37073 Göttingen, Germany

ABSTRACT

The deformation of the valence electronic charge density of a carbon lattice induced by an external electric field is modeled by a polarization wave of discrete dipoles centered at the atomic sites. This approach requires only the knowledge of the frequency-dependent dipolar polarizability of the carbon atom. We deduce this atomic parameter for carbon sp^2 from the measured dielectric tensor of graphite by means of an original Clausius-Mossotti relation. We show that this graphitic (anisotropic) atomic polarizability is transferable to fullerenes. The bulk loss function of the C_{60} solid measured by EELS is well reproduced by our model. Although the C_{60} molecule is isotropic, it can be viewed as made up of carbon atoms having different polarizabilities along the radial and tangential local axes which are related to distinct electronic excitations observed experimentally.

1. Introduction

In the last few years, much work has been devoted to the measurement of the UV dielectric function ϵ of fullerites because of its close relation to their electronic properties[1-4]. In particular, Sohmen et al. have measured the loss function of C_{60}, $\mathrm{Im}(-1/\epsilon)$, by electron-energy-loss-spectroscopy in transmission (EELS)[2]. Their spectrum reproduced in Fig.2 shows many features unresolved in earlier studies. To our knowledge, a theoritical interpretation of *all* these features has not been given yet despite of many calculations using *ab initio* techniques[5,6] or dielectric continuum models[7,8]. The purpose of the present paper is to find their origin by using a simple microscopic model. The basic idea is that the dielectric properties of isolated fullerenes are expected to be quite similar to those of a closed curved graphitic sheet. The dielectric properties of graphite can be properly transfered to fullerene molecules by providing a transferable anisotropic atomic polarizability. An application of this model is presented for C_{60}.

2. Anisotropic Discrete Dipole Model

The polarization of a fullerene C_N induced by an external electric field $E^e(r,t)$ is approximated as a polarization wave of N discrete dipoles $p_\kappa(t)$ centered at the atomic equilibrium positions r_κ. For weak external electric fields a linear response is assumed:

$$p_\kappa(\omega) = \sum_{\kappa'}^{N} \alpha_{\kappa\kappa'}(\omega) E^e_{\kappa'}(\omega) \tag{1}$$

$E^e_{\kappa'}(\omega)$ is the Fourier component with frequency ω of the external field evaluated at the equilibrium atomic position $r_{\kappa'}$. Equation (1) defines a set of N non-local dipole-dipole electronic polarizabilities $\alpha_{\kappa\kappa'}(\omega)$ that are (3x3) complex tensors. The dipolar molecular polarizability $\alpha(\omega)$ is immediately deduced from (1) by assuming an uniform electric field:

$$\alpha(\omega) = \sum_{\kappa}^{N} \sum_{\kappa'}^{N} \alpha_{\kappa\kappa'}(\omega) \tag{2}$$

At this stage there are, N (3x3) matrices $\alpha_{\kappa\kappa'}(\omega)$ as parameters. This problem is considerably reduced by assuming a local relation between the atomic momentum $p_\kappa(\omega)$ and the local electric field $E^l_\kappa(\omega)$:

$$p_\kappa(\omega) = \alpha_\kappa(\omega) E^l_\kappa(\omega) \tag{3}$$

$E^l_\kappa(\omega)$ is by definition the sum of the external electric field $E^e_\kappa(\omega)$ and the field generated by the induced dipoles $p_{\kappa \neq \kappa'}(\omega)$. From the identification of equations (3) and (1), we find:

$$\alpha_{\kappa\kappa'}(\omega) = [\delta_{\kappa\kappa'}\alpha_\kappa(\omega)^{-1} + T_{\kappa\kappa'}]^{-1} \tag{4}$$

T is equal to zero for $\kappa = \kappa'$ and to the usual dipolar tensor otherwise. In this way, the local atomic polarizability $\alpha_\kappa(\omega)$ (3) is the unique parameter that we have to evaluate for each frequency ω.

3. Results

The cartesian components $\alpha_{uu}(\omega)$ of a carbon atom in a planar sp^2 lattice are deduced from the measured uniaxial dielectric tensor $\epsilon_{uu}(\omega)$ of graphite[9] by using the following Clausius-Mossotti relation:

$$4\pi \frac{N}{V} \alpha_{uu}(\omega) = \frac{(\epsilon_{uu}(\omega) - 1)}{(1 + B_u(\epsilon_{uu}(\omega) - 1))} \tag{5}$$

N/V is the atomic density. We found numerically $B_z = -0.606$ and $B_x = B_y = 0.803$ for the depolarization factors of the hexagonal lattice. We transfer the values of these graphitic polarizabilities (5) to the C_{60} molecule by transforming $\alpha_{uu}(\omega)$ in an uniaxial spherical polarizability tensor i.e we take $\alpha_{rr}(\omega) = \alpha_{zz}(\omega)$ and $\alpha_{\theta\theta}(\omega) = \alpha_{\phi\phi}(\omega) = \alpha_{xx}(\omega)$. The dynamical polarizability of C_{60} (2) computed with these data is shown in Fig.1.

We restrict the discussion to $\text{Im}(\alpha(\omega))$. The electronic transitions giving rise to the peaks at about 6.4 and 20.8 eV correspond to a mostly tangential polarization of the molecule. We assign these excitations to $\pi \rightarrow \pi^*$ and $\sigma \rightarrow \sigma^*$ transitions, respectively, since for a tangential excitation only these transitions are allowed in the dipolar approximation. For comparison, the corresponding interband transitions in graphite are responsible for two peaks in the imaginary part of ϵ_{xx}, at 4.2 and 14.8 eV, respectively[9]. The structures at about 10.9 and 17.0 eV in Fig.1 arise from two resonances in the radial graphitic atomic polarizability. Therefore, by analogy with graphite, we assign these peaks to $\sigma \rightarrow \pi^*$ and $\pi \rightarrow \sigma^*$ transitions. In graphite these transitions give two peaks in the imaginary part of ϵ_{zz} at 10.5 and 15.1 eV[9]. The peak at 10.9 eV is well resolved and should be clearly observed in gas phase. We emphasize that it is not due to the hollow character of C_{60} as proposed recently[8] but arises from the local anisotropy of the molecule.

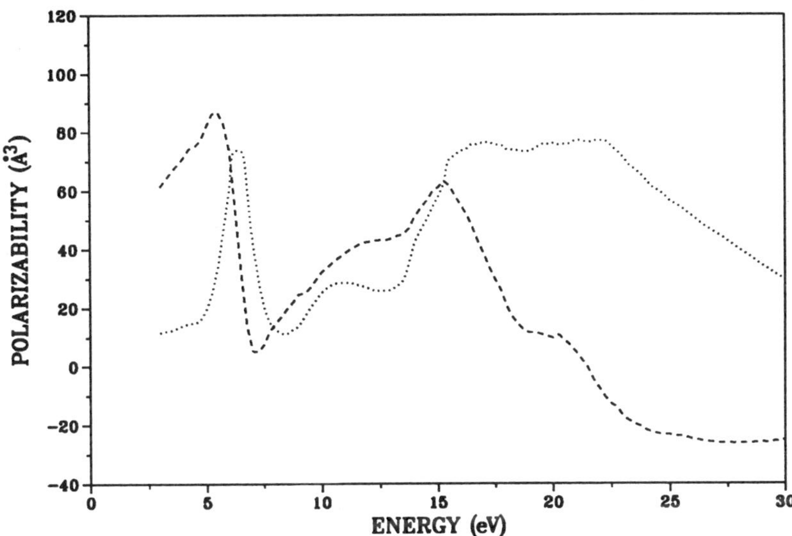

Fig.1 Real part (dashed line) and imaginary part (dotted curve) of the computed C_{60} polarizability.

The dielectric constant of the C_{60} cubic crystal has been estimated from the results of Fig.1 by using the relation (5) in which $\alpha_{uu}(\omega)$ is now the polarizability of the molecule and $B_x = B_y = B_z = 1/3$. The loss function obtained in this way is compared to the measured one[2] in Fig.2. There is a nearly one per one correspondance between the observed features and our theoritical predictions except in the optical region $(\omega < 6eV)$ of the spectrum for which the graphite dielectric data are not appropriate. We attribute the remaining disagreements in the UV part of the spectra to the fact that the calculated loss function represent only the dipolar part of the total loss function of the solid. We assign the loss peaks in Fig.2 as follows.

The peaks at c.a. 6.5 and 26 eV arise from the so-called π and $\sigma + \pi$ tangential plasmons. These excitations indeed possess a collective character in the C_{60} crystal since for these frequencies $\epsilon(\omega)$ is minimum in our model. To the contrary, the structures at about 10.4, 14.5 and 18-19 eV are associated to peaks in the imaginary part of the calculated dielectric constant at about 11.2, 15.5 and 19.8 eV. They are directly related to the corresponding electronic excitations of an isolated molecule discussed above. Their origin as being due to the anisotropy of the sp^2 carbon polarizability is made obvious from comparison with the dotted curve in which the atomic polarizability has been taken isotropic and equal to the tangential component of the anisotropic graphitic polarizability.

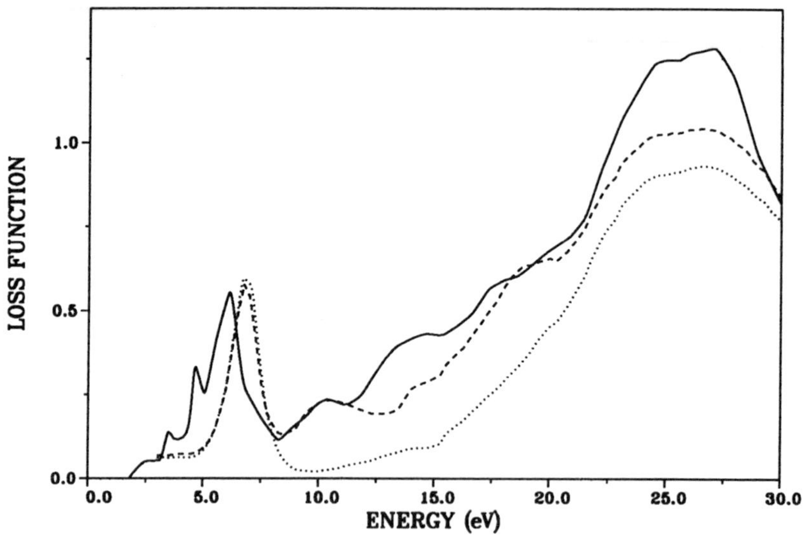

Fig.2 Experimental[2] (full line) and computed loss function of the C_{60} solid (see text)

In conclusion, we have shown by using a discrete dipole model that the local anisotropy of the sp^2 carbon atom is one of the key for the understanding of the UV spectra of C_{60}. Work is in progress to extend this conclusion to other fullerenes[10].

References

1. G. Gensterblum et al., *Phys. Rev. Lett.* **67** (1991) 2171
2. E. Sohmen, J. Fink and W. Krätschmer, *Z. Phys.* B **86** (1992) 87
3. E. Sohmen and J. Fink, *Phys. Rev.* B **47** (1993) 14532
4. M.K.Kelly et al., *Phys. Rev.* B **46** (1992) 4963
5. Y.-N. Xu et al., *Phys. Rev.* B **44** (1991) 13171
6. N. Ju et al., *Phys. Rev.* B **48** (1993) 9071
7. Ph. Lambin, A.A. Lucas and J.P. Vigneron, *Phys. Rev.* B **46** (1992) 1794
8. D. Östling, P. Apell and A. Rosen, *Europhys. Lett.* **21** (1993) 539
9. R. Klucker, M. Skibowski and W. Steinmann, *Phys. Stat. Sol.* B **65** (1974) 703
10. P. Senet et al., to be published; L. Henrard et al., in these proceedings

TRIPLET EXCITONS IN MOLECULAR CRYSTALS: THE CASE OF CRYSTALLINE C_{60}

E.J.J.GROENEN AND J.SCHMIDT
Huygens Laboratory
University of Leiden
P.O.Box 9504
2300 RA Leiden, The Netherlands

ABSTRACT

The theory previously developed for excitons in molecular crystals is reviewed. The case of the mini-excitons in isotropically mixed naphtalene crystals is described and the analogy with the triplet excitons in crystalline C_{60} is indicated.

INTRODUCTION

The first EPR spectrum of C_{60} molecules photo-excited to their lowest triplet state was reported by Wasielewski et al.[1]. In their sample, the C_{60} molecules were present at low concentration in a glassy solution of toluene. The observation of a zero-field splitting of the sublevels demonstrated that the C_{60} molecule in its lowest triplet state is subject to a Jahn-Teller distortion, as predicted by theory.

More recently it was shown that triplet states can be observed in a single crystal of C_{60} using pulsed laser excitation and pulsed EPR detection at 95 GHz[2]. The spectra can be interpreted by assuming again that C_{60} in the metastable triplet state undergoes a lowering of its symmetry (from I_h to D_{2h}). Surprisingly, the excitation appeared to be delocalized over a pair and possibly over a chain of C_{60} molecules i.e. the triplet excitation has the characteristics of a Frenkel exciton[3]. Unfortunately the extent of the delocalization and the size of the exciton bandwidth could not be determined as yet. In the theoretical models developed so far it is assumed that the bandwidth originates in electron transfer amongst the C_{60} molecules in the crystal[4,5]. In analogy with the description for molecular crystals of aromatic molecules one might expect that the bandwidth arises from the exchange interaction between adjacent C_{60} molecules[6]. The attraction of the latter approach is that it might explain the one-dimensional character of the excitons. A particularly interesting question is whether the observed Jahn-Teller distortion plays a part in the formation of the collective triplet excitation.

In this contribution we take the point of view that Frenkel excitons in the C_{60} single crystal can be explained on the basis of the theoretical model developed for describing singlet and triplet excitons in molecular crystals. The attraction of this description is that it provides a simple physical insight into the interactions responsible for the band structure. Then we discuss the special case of the "mini-exciton" i.e. the case of an electronic excitation distributed over two adjacent, translationally inequivalent molecules. We will argue that the triplet excitons in C_{60} single crystals may be described in a similar way.

Frenkel Excitons in Molecular Crystals

The concept of excitons was first introduced by Frenkel to describe the collective electronic excitations in pure molecular crystals. This description is based on the idea that the interaction between atoms on the same molecule is orders of magnitude larger than that between atoms on adjacent molecules. This means that the ground state and the electronically excited states in the molecular crystal can be described as linear combinations of electronic excitations localized on the individual molecules. These linear combinations are characterized by the wave vector $\mathbf{k}$ defined in the reciprocal lattice. Frenkel called these states "excitons". Owing to the fact that the "electrons" and "holes", which in a simple MO-picture make up the excited states, are localized on one molecule, the Frenkel exciton is also designated as the "tight-binding" exciton.

Let us consider a large crystal of N molecules with one molecule in the unit cell. The Hamiltonian relevant for our problem can be described as

$$H = \sum_n H_n + \tfrac{1}{2} \sum'_{nm} V_{nm}. \tag{1}$$

Here H_n is the Hamiltonian for a molecule at site n and V_{nm} is the interaction energy operator of two molecules at positions n and m. The prime in the second summation means that terms with $n = m$ are excluded from the sum. If, for a moment, we neglect V_{nm} we find for the total wave function describing the ground state of the crystal

$$\psi^0 = \Pi_m \, \phi^0_m, \tag{2}$$

with ϕ^0_m the wave function of the isolated molecule. The excited states corresponding to the f-th excitation of the molecule at position n are given by

$$\psi^f_n = \phi^f_n \, \Pi_{m \neq n} \phi^0_m \tag{3}$$

The states ψ^f_n are N-fold degenerate because any of the N molecules can be in the f-th excited state. The translational symmetry allows us to introduce N orthonormalized Bloch functions,

$$\psi^f(\mathbf{k}) = N^{-\frac{1}{2}} \sum_n \psi^f_n \, e^{i\mathbf{k}\cdot\mathbf{n}}. \tag{4}$$

The N-fold degeneracy of the states $\psi^f(\mathbf{k})$ is removed under the influence of the intermolecular interaction V_{nm}. Since in molecular crystals V_{nm} is small compared to the intramolecular interactions, we may use the wave functions ψ^0 and ψ^f_n to calculate the energy of the ground state and the excited states of the crystal. If we then take the difference between the energy of the ground state and the excited states we obtain the excitation energy

$$E_f(\mathbf{k}) = \Delta\epsilon_f + D_f + L_f(\mathbf{k}) \ \ . \tag{5}$$

Here $\Delta\epsilon_f$ is the excitation energy of a free molecule, D_f is the change in the interaction energy of one molecule with all the surrounding molecules in its transition to the f-th excited state, and $L_f(\mathbf{k})$ is an addition to the excitation energy which is a function of the wave vector $\mathbf{k}$.

$$L_f(\mathbf{k}) = \Sigma_m \, \beta^f_{nm} \, e^{i\mathbf{k}.(n-m)} \qquad , \qquad (6)$$

where

$$\beta^f_{nm} = \langle \phi^0_n \, \phi^f_m \, | \, V_{nm} \, | \phi^0_m \, \phi^f_n \rangle \qquad . \qquad (7)$$

In large crystals adjacent values of $\mathbf{k}$ differ very little and the N values of the excitation energy $E_f(\mathbf{k})$ form a quasi-continuous band of excited states. Each of the excited states, labeled with $\mathbf{k}$, represents a collective excited state of the entire crystal. These elementary excitations, or excitons, are characterized by a quasi-momentum $\hbar\mathbf{k}$ and energy $E_f(\mathbf{k})$ and are usually referred to as quasi-particles. They are distributed throughout the entire crystal and are not concentrated on one molecule. The states associated with excitations of a limited region of the crystal are represented by wave packets, i.e. linear combinations of $\psi^f(\mathbf{k})$. For instance,

$$\psi^f = \Sigma_k \, W(\mathbf{k}) \, \psi^f(\mathbf{k}) . \qquad (8)$$

Here $W(\mathbf{k})$ differs from zero only in a small region of $\mathbf{k}$-space. The smaller the region of excitation (Δr), the greater the uncertainty in the wave vector (Δk),

$$\Delta r . \Delta k \approx 1 \qquad (9)$$

When the crystal is irradiated with visible light the wavelength is considerably larger than the lattice constant $\mathbf{a}$. Consequently, the wave vector of the light $\mathbf{Q}$ is small ($\mathbf{Q}.\mathbf{a} < 1$) and we see that only excitons with $\mathbf{k} \approx 0$ can be created by light.

In order to find the bandwidth one has to calculate the matrix element β^f_{nm} for excitation transfer between molecules n and m. It involves the intersite interaction operator V_{nm} which describes the Coulomb interaction of electrons and nuclei of both molecules. For molecular crystals the molecules are neutral, and have no permanent electric dipole moment in the ground or excited states. In a multipole expansion, the first non-zero terms then are the electric dipole-dipole interactions of the dipole moments related to the transition from the ground state to the f-th excited state. Clearly the transition dipole moments for the allowed singlet-singlet transitions are much larger than those for the spin-forbidden triplet-singlet transitions. For aromatic hydrocarbon crystals the calculations typically give singlet exciton bandwidths of several hundred cm^{-1} in good agreement with experimental values[6]. For triplet excitons however bandwidths not greater than 10^{-3} cm^{-1} are predicted, whereas in practice values of about 10 cm^{-1} are found.

To explain the bandwidth of triplet excitons in aromatic, molecular crystals one usually considers the intermolecular exchange interaction, which depends on the overlap of the wave functions of adjacent molecules. It can be taken into account by appropriate antisymmetrization of the wave functions (2) and (3)[7]. It cannot be excluded however that the mixing of the low-lying excited states with high-lying charge-transfer states contributes to the excitation transfer interaction in the case of triplet excitons[8,9]. Calculations along either of these lines yield bandwidths of triplet excitons in aromatic, molecular crystals of the order of tens of wavenumbers.

For more than one molecule in the unit cell the description becomes slightly more complex. We then start by taking Bloch functions $\psi^f_\alpha(\mathbf{k})$ similar to (4) for every molecule α in the unit cell. If there is also an interaction between the different molecules α it is necessary to construct linear combinations of $\psi^f_\alpha(\mathbf{k})$, each of which describes a different branch of the exciton band. The number of these branches is equal to the number of inequivalent molecules in the unit cell, and the energy splitting between the $\mathbf{k} = 0$ states of these branches is usually called the Davydov splitting[6]. The absolute value of this splitting is proportional to the interaction between the molecules within the unit cell. One can easily distinguish between the different branches if the interaction between inequivalent molecules dominates that between translationally equivalent molecules.

The above considerations apply to an ideal crystal without vibrations and without any lattice defect. When the translational symmetry is destroyed, for instance by a lattice defect or by lattice vibrations, the coupling of the exciton states to the phonon bath leads to a scattering of the $\mathbf{k}$-states, the rate of which increases with increasing temperature. As a result of this scattering not only the mean $\mathbf{k}$-value of a wave packet changes in time but also its width $\Delta\mathbf{k}$. As a consequence of the uncertainty relation (9), the degree of localization increases with increasing width of the exciton wave packet. When the localization becomes of the order of the lattice constant, the excitation becomes localized on one molecule and its motion can be described as a random walk on the lattice. One usually refers to this situation as "incoherent" excitons in contrast to the "coherent" excitons described above.

Mini-Excitons in Isotopically Mixed Naphthalene Crystals

The "mini-excitons" in isotopically mixed naphthalene bear a close resemblance to the excitons observed in the C_{60} single crystal and for this reason it is interesting to discuss them in some detail. When doping a crystal of naphthalene-d_8 (N-d_8) with a small amount of naphthalene-h_8 (N-h_8) the guest molecules, in their photo-excited triplet state, form traps which are located about 100 cm^{-1} below the triplet exciton band of the host crystal[10]. If the concentration is increased to about 1 % one also observes the triplet excitation of pairs of naphthalene-h_8 guest molecules which occupy the two adjacent, inequivalent sites in the unit cell[11,12]. Such a pair can be excited in two triplet states which correspond to the symmetric and antisymmetric combination of the triplet excitations of the individual molecules,

$$\Psi_\pm = 2^{-\frac{1}{2}}\left\{ \phi^T_1 \phi^0_2 \pm \phi^0_1\phi^T_2 \right\} . \qquad (10)$$

The energies of these two states, which are separated by 2.5 cm^{-1}, are also about 100 cm^{-1} below the triplet exciton band of the host. Thus at liquid helium temperatures the excitation is localized at the pair of N-h_8 molecules. The separation of the two states was measured by phosphorescence excitation spectroscopy from which it was established that the symmetric component is the lowest in energy. The splitting of the two states is $2\beta_{12}$. The value of the excitation transfer integral $\beta_{12} = -1.25$ cm^{-1} is typical for triplet excitons in aromatic molecular crystals.

The fact that the energy levels correspond to mini-excitons, i.e. to coherent states of the type described by (10), was further supported by EPR and ODMR[13] studies at 1.2 K where only the lowest, symmetric component Ψ_+ has an appreciable population. From the EPR experiments it was established that the principal axes of the magnetic dipole-dipole interaction, responsible for the zero-field splitting, correspond to crystal axes which are the averages of the principal axes of the individual molecules. Moreover the zero-field splitting parameters turn out to be compatible with the average of the zero-field splitting parameters of the individual molecules. Subsequent, temperature-dependent ODMR and ESE experiments showed that phonons perturb the coherent superposition. At 1.2 K this perturbation is very small but at temperatures of about 10 K phonons detrap the excitation to the exciton band of the host and the mini-exciton is lost.

Triplet Excitons in Crystalline C_{60}

The C_{60} crystal undergoes a phase transition at about 260 K to a simple-cubic structure, space group $Pa\bar{3}$. There are four interpenetrating simple-cubic sublattices that are occupied by one of four types of differently oriented molecules A, B, C and D.

The first question concerning the triplet excitons observed at helium temperatures in the single crystals of C_{60} is whether the excitation is delocalized over two of these translationally inequivalent molecules (e.g. AB) or over a chain (ABAB...). We have demonstrated that the direction of the principal axes of the dipole-dipole interaction of the triplet spins and the values of the zero-field splitting can be explained as resulting from an averaging process involving the inequivalent molecules A and B (or A and C etc)[2]. Unfortunately this result does not give information about the number of molecules participating in the averaging process. At present we prefer to think in terms of a pair in much the same way as the mini-exciton in the isotopically mixed crystal N-h_8 in N-d_8. The only difference is that we believe that in the case of C_{60} the trap depth is much larger. This is supported by our observation that the EPR and ODMR spectra of the triplet state are visible up to about 60 K. An observation which is difficult to reconcile with a triplet excitation delocalized throughout the crystal.

The next problem is that of the exciton bandwidth. In the case of the dimer this is equivalent to the splitting $2\beta_{12}$ of the symmetric and antisymmetric states. So far we have not observed any variation, neither in the zero-field splitting nor in the spin-lattice relaxation rates, with temperature which would be indicative of a thermally excited process between the two Davydov components of the dimer. This means that so far we have not been able to gain insight in the size (or the sign) of the excitation transfer integral β_{12}.

An interesting experiment which would give direct information about the extent of the delocalization of the excitation is a measurement of the hyperfine interaction with the $I=\frac{1}{2}$ nuclear spin of the ^{13}C isotope. As can be seen from equation(4) the density of the wave function at a particular molecule is proportional to N^{-1}, where N is the number of molecules in the chain. This experiment should be performed on a ^{13}C-enriched single crystal of C_{60}, because the natural abundance of ^{13}C of only 1.1 % leads to hyperfine lines in the EPR spectrum which are to weak to be observable.

Conclusion

The EPR experiments on C_{60} single crystals, photo-excited to their lowest triplet state have revealed the presence of triplet Frenkel excitons. At present we believe that this exciton is trapped and delocalized over a pair of inequivalent C_{60} molecules. The value and sign of the excitation transfer integral have remained unclear as yet. The one-dimensionality of the exciton lends support to the idea that the overlap of the wave functions of neighbouring molecules provides the dominating contribution to the excitation transfer integral and that the effect of charge transfer is small. An intriguing question is whether the observed Jahn-Teller deformation in the triplet state plays a part in the directionality and the size of the excitation transfer integral.

Acknowledgement

The authors gratefully acknowledge the stimulating cooperation with Dr. G. Meijer in the C_{60} project.

This work was performed as part of the research program of the "Stichting voor Fundamenteel Onderzoek der Materie" (FOM) with financial support from the "Nederlandse Organisatie voor Wetenschappelijk Onderzoek" (NWO).

References

1. M.R. Wasielewski, M.P. O'Neil, K.R. Lykke, M.J. Pellin and D.M. Gruen, J. Am. Chem. Soc. **113** (1991) 2774.
2. E.J.J. Groenen, O.G. Poluektov, M. Matsushita, J. Schmidt, J.H. van der Waals and G. Meijer, Chem. Phys. Lett. **197** (1992) 314.
3. J. Frenkel, Phys. Rev. 37 (1931) 1276.
4. R.W. Lof, A. van Veendendaal, B. Koopmans, H.T. Jonkman and G.A. Sawatzky, Phys. Rev. Lett. **68** (1992) 3924.
5. E.L. Shirley and S.G. Louie, Phys. Rev. Lett. **71** (1993) 133.
6. A.S. Davydov, in Theory of Molecular Excitons, (Plenum Press, New York, 1971).
7. D.P. Craig and S.H. Walmsley, in Excitons in Molecular Crystals, (W.A. Benjamin Inc., New York, 1968).
8. S.I. Choi, J. Jortner, S.A. Rice and R.J. Silbey, J. Chem. Phys. **41** (1964) 3294.
9. J. Jortner, S. Rice, J.L. Katz and S.I. Choi, J. Chem. Phys. **42** (1965) 309.
10. M. Schwoerer and H.C. Wolf, Mol. Cryst. **3** (1967) 177.
11. D.M. Hanson, J. Chem. Phys. **52** (1970) 3409.
12. C.L. Braun and H.C. Wolf, Chem. Phys. Lett. **9** (1971) 260.
13. B.J. Botter, C.J. Nonhof, J. Schmidt and J.H. van der Waals, Chem. Phys. Lett. **43** (1976) 210.

PHOTOEXCITED TRIPLET STATE OF DIMETHOXYBENZENE-C$_{60}$: A PULSED EPR INVESTIGATION

M. BENNATI, A. GRUPP AND M. MEHRING

2.Physikalisches Institut, Universität Stuttgart
Pfaffenwaldring 57, D-70550 Stuttgart, Germany

and

P. BELIK, A. GÜGEL AND K. MÜLLEN

MPI für Polymerforschung, Mainz, Germany

ABSTRACT

Photoexcitation of the donor-acceptor molecule dimethoxybenzene (DMB) /C$_{60}$ in toluene was investigated by pulsed EPR technique. We observe a highly polarized triplet state at temperatures below the matrix glass point (T = 10 - 180 K). From the analysis of the EPR-lineshape we are able to assign the signal to a localized triplet state on the C$_{60}$ subunit with no evidence for charge separation on the time scale of our experiment. Triplet finestructure parameters D and E were determined from spectra simulations. We observe that the donor side-group linked to C$_{60}$ has a strong influence on the E value and the dynamics of the molecule. The time evolution of the spectra at T = 10 K shows strong anisotropic spin-lattice relaxation. Furthermore there is no significant change of the lineshape in the temperature range T = 10 - 180 K in contrast to the behaviour of pure ^{3}C$_{60}$. This indicates that no fast molecular rotation or pseudorotation due to the Jahn-Teller effect takes place.

1. Introduction

Photoexcitation of pure C$_{60}$ is known to result in the triplet state with a quantum yield of nearly one[1]. A highly polarized triplet EPR spectrum was observed whose EPR-lineshape seems to be determined by characteristic molecular dynamics such as rotation and pseudorotation[2,3,4]. The EPR investigation on the photoexcited state of a new chemically modified C$_{60}$ molecule is an attractive approach for understanding the electronic properties and the dynamics of C$_{60}$ as well as of new C$_{60}$ derivative compounds.

In (DMB)-C$_{60}$ the electron donor 4,5-dimethoxybenzene is covalently linked to C$_{60}$ via a rigid dimethane bridge[5]. Donor and acceptor are spatially separated and do not show any ground state intramolecular electronic interactions (charge-transfer)[6]. The influence of the rigid donor side-group on the C$_{60}$ excited states can be studied by a line shape analysis of the photoexcited EPR spectra.

2. Experimental

Degassed solutions of DMB-C$_{60}$ in toluene and dichloromethane were irradiated at λ = 532 nm and 355 nm by a pulsed Nd:YAG laser with an energy of $\simeq$ 2 mJ per pulse at a pulse repetition frequency of 30 Hz.

The EPR measurements were carried out using a home built electron spin echo spectrometer with a slotted tube resonator and a helium flow cryostat. Triplet EPR powder spectra were recorded with a Hahn-echo microwave pulse sequence by integrating the echo signal while sweeping the magnetic field. The lineshape of the spectra corresponds directly to the EPR line. The time evolution of the EPR-line was obtained by varying the delay time between laser and first microwave pulse.

3. Results

We observe a highly polarized EPR spectrum after laser excitation of the DMB-C_{60} samples (figure 1 and 2). The lineshape neither depends on the solvent used (toluene or CH_2Cl_2) nor on the excitation wavelength at all temperatures and time delays. The EPR-echo signal disappears as soon as the matrix becomes liquid (T $\geq$ 180K) and no FID can be detected in the liquid phase in contrast to pure C_{60}-triplet[7,8].

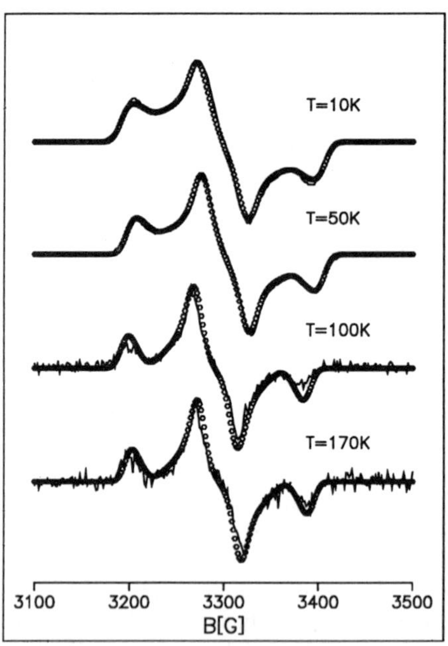

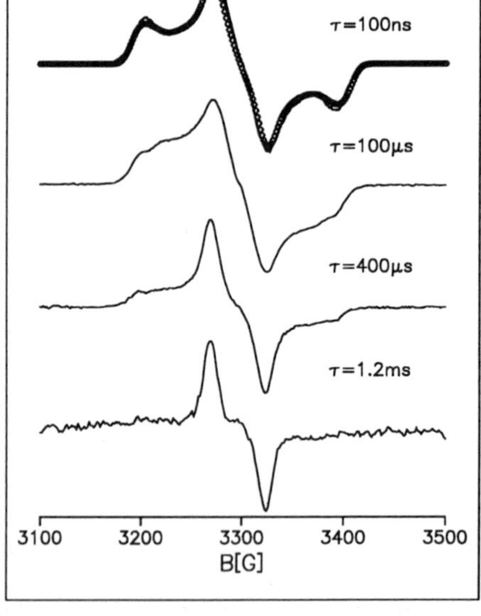

Fig. 1: Temperature dependence of the EPR powder spectra at $\tau \simeq$ 700 ns (solid lines) and simulations (circles)

Fig. 2: EPR powder spectra at T = 10 K at different delays after the laser pulse (solid lines) and simulation (circles)

The EPR-spectrum shows a polarization pattern with absorptive peaks at the low field side and emissive peaks at the high field side. From the spectral width (edge to edge) at T = 10 K (static case) we can estimate a D-value of $\simeq$ 110 G. These features were found to be characteristic for $^3C_{60}$ [2,3,4] although the lineshapes of $^3C_{60}$ and $^3(DMB-C_{60})$ are not identical and in this case the spectral position of the main peaks indicates a considerable asymmetry parameter E.

The assignment of this signal to either a localized molecular triplet or to a triplet radical pair $DMB^{+\cdot}$ - $C_{60}^{-\cdot}$ can be decided considering the signal kinetics (figure 3). The time evolution of the EPR line recorded on a specific spectral position (B $\parallel$ Z) can be described by a biexponential function with $\tau_1 = 0.020$ ms $\tau_2 = 0.34$ ms. The signal growth can be only due to selective decay of the triplet sublevel population. The monoexponential signal decay with a time constant of $\tau = 0.34$ ms seems to be too long for a radical pair with a separation of $\simeq$ 7 Å. Therefore, considering the signal intensity, the polarization pattern and the D-value, the EPR spectrum can be attributed to a molecular triplet state localized mostly on the C_{60} molecule.

We simulated the triplet EPR powder spectra by solving the eigenvalues of the Hamilton-operator:

$$\hat{H} = g\beta SB - (XS_x^2 + YS_y^2 + ZS_z^2) \tag{1}$$

The polarization and intensity for each transition were calculated from the population of the zero field energy levels and the transition probabilities for a microwave field perpendicular to the static magnetic field.

In addition T_1 relaxation was taken into account. The spectra at T $\geq$ 50 K can be simulated considering relaxation due to isotropic molecular motion. In this case the angular dependence of the transition probabilities w and W ($\Delta m = 1$ and $\Delta m = 2$ transitions) can be obtained from the matrix elements of the Hamilton-operator in the laboratory frame which are proportional to the $S_\pm$ and $S_\pm^2$ operators respectively. For a highly polarized triplet with D $\gg$ E at $\tau \approx 700$ ns after the laser pulse T_1 can be approximated by :

$$1/T_1 = 3w \qquad or \qquad 1/T_1 = (1/T_{1iso})\, cos^2\theta sin^2\theta \tag{2}$$

where θ is the azimut angle between laboratory and molecular frame, $1/T_{1iso}$ is the isotropic part of the relaxation rate. The spectra simulations are represented in figure 1 by the circles. The obtained finestructure values and isotropic spin-lattice relaxation times T_{1iso} are listed in table 1.

The triplet EPR spectrum and its temperature dependence at $\tau \simeq 700$ ns after the laser pulse (figure 1) can be simulated very well considering a single triplet state. The D-value decreases slightly with increasing temperature but the lineshape does not change significantly. This is in contrast to the behaviour of pure C_{60} where we observe a thermally activated second triplet state with increasing temperature[3,7]. The results support the assignment of the second triplet in pure C_{60} to an exchange process due to thermally activated rotational jumps or pseudorotation which is obviously hindered in DMB-C_{60}.

Figure 2 represents the time evolution of the EPR-spectrum at T = 10 K after the laser pulse. The lineshape changes significantly and it is not possible to simulate the

Table 1: Simulation parameters for ^{3}(DMB-C_{60}) and $^3C_{60}$ (..)

T [K]	$\mid D \mid$	$\mid E \mid$	T_{1iso}
	$[cm^{-1}]$	$[cm^{-1}]$	$[\mu s]$
10	0.00989 (0.0117)	0.00168 (0.00049)	-
50	0.00971 (0.0117)	0.00168	15
100	0.00933 (0.0105)	0.00168	2.5
170	0.00915 (0.0089)	0.00168	1.5

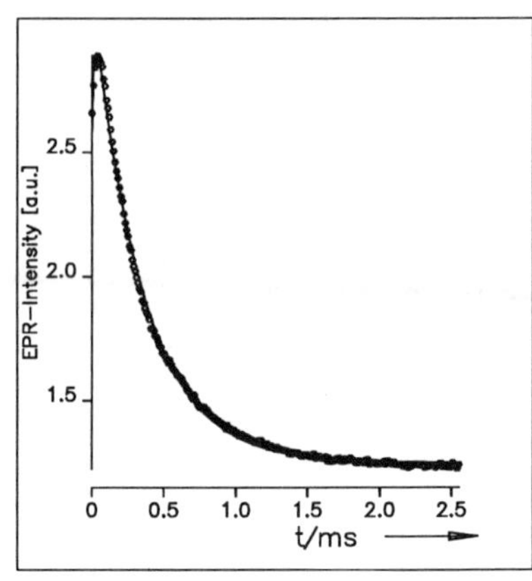

Fig. 3: Time evolution of the EPR signal

spectra only by considering T_1 relaxation due to isotropic motion. Spectra calculation considering anisotropic molecular dynamics, which is expected for this particular molecular structure, are in progress.

Acknowledgements

Financial support by the Deutsche Forschungsgemeinschaft (SFB 329) and the Fonds der Chemischen Industrie is gratefully acknowledged.

References

1. J.W. Arbogast, A.P. Darmanyan, C.S.Foote, Y.Rubin, F.N. Diederich, M.M. Alvarez, S.J. Anz, R.L. Whetten, *J. Phys. Chem.* (1991), **95**, 11
2. G.L. Closs, P. Gautam, D. Zhang, P.J. Krusic, S.A. Hill, E. Wasserman, *J. Phys. Chem.* (1992), **96**, 5228
3. M. Bennati, A. Grupp, M. Mehring, K.P. Dinse, J. Fink, *Chem. Phys. Lett.* (1992), **200**, 440
4. A. Regev, D. Gamliel, V. Meiklyar, S. Michaeli, H. Levanon, *J. Chem. Phys.* (1993), **97**, 3671
5. P. Belik, A. Gügel, A. Kraus, J. Spickermann, V. Enkelmann, G. Frank, K. Müllen, *Adv. Mat.* (1993), **5**, 854
6. P. Belik, private communication
7. A. Grupp, M. Bennati, M. Mehring, in *Electronic Properties of Fullerenes*, p.367, Springer Verlag 1993
8. M. Bennati, A. Grupp, P. Bäuerle, M. Mehring, submitted to Chem. Phys.

LOCATION OF THE LOWEST FORBIDDEN EXCITON IN SOLID C_{60} BY NON LINEAR SPECTROSCOPY

C. TALIANI[†], M. MUCCINI[†], R. ZAMBONI[†] AND F. KAJZAR[*]

[†]*Istituto di Spettroscopia Molecolare del C.N.R., via P. Gobetti, 101 - 40129 Bologna, Italy.* [*]*Commissariat à l'Energie Atomique, Direction des Technologies Avancées LETI, DEIN/LPEM, CEN Saclay, 91191 Gif sur Yvette Cedex, France.*

ABSTRACT The location of the lowest excited singlet state is performed by two non-linear spectroscopic techniques. First we report on the third harmonic generation spectrum of C_{60} thin film. Fitting of the experimental points with a three-level model suggests the presence of a low lying forbidden state at 1.87 eV. Preliminary data of two-photon excitation spectroscopy of C_{60} single crystal confirm this finding and allow to locate the level at 1.846 eV which is assigned to a T_{1g} excited state.

INTRODUCTION

The similarity between the spectra of isolated C_{60} molecules and the solid implicates that the intermolecular interactions are weak, due to Van der Waals forces, and therefore the solid is a typical molecular crystal. Neutral electronic excitations of the molecule give rise therefore to tight-binding Frenkel excitons in the solid[1]. The spectrum of a thin film was reported for the first time by Krätschmer et al.[2]. The absorption spectrum consists of a very weak and broad system extending from 1.9 to 2.7 eV (650-450 nm) and of a series of sharp bands at 3.65 eV (339 nm), 4.70 eV (264 nm) and 5.74 eV (216 nm). The solution spectrum[3] is very similar to the solid spectrum[2,4,5] and has been analyzed in terms of Herzberg-Teller vibronic coupling[6]. The forbidden nature of the low-lying excitations in C_{60} makes it impossible to locate directly the electronic origin by linear spectroscopy. Here we report on the identification of the low energy excitations in solid C_{60} by means of non-linear spectroscopies. It has been recognized long ago[7] that in π conjugated molecules like C_{60} inclusion of e-e correlation is essential to reproduce the low energy excitations. In the multi-electron approach, the electronic states are classified according to the molecular point group which, in the case of the hycosahedral C_{60} molecule, is I_h. Semiempirical Hartree-Fock quantum chemical calculations parametrized for treatment of π electrons[5] gives the following sequence of states in the range of 2.29-5 eV: $1T_{2g}$, $1T_{1g}$, $1G_g$, $1H_g$, $1T_{2u}$, $1H_u$, $1G_u$, $1T_{1u}$, $2G_u$, $2H_u$, $2T_{1u}$, $3T_{1u}$, $4T_{1u}$, $5T_{1u}$. There are four gerade forbidden states at low energy which were tentatively located by many investigators[3,4,5,6,8] in the region at about 2eV. The mechanism that is active in making these transitions observable in absorption, is the Herberg-Teller (H-T) vibronic coupling[9]. This mechanism implies that transitions between two states may occour with a finite probability at nuclear coordinates away from the equilibrium geometry for particular normal coordinates of appropriate symmetry. The vibrations which are active in the H-T coupling for a transition to a T_{1g} level belong to the symmetry representations a_u, t_{1u} and h_u. For instance the transition to the vibronic level $T_{1g}t_{1u}$, is H-T allowed and is called a "false origin". The intensities of H-T active transitions to vibronic levels may be calculated by means of a perturbative approach[6]. In the case of $1^1T_{1g} \leftarrow 1^1A_g$ transition the dominant vibronic false origins are t_{1u} at 1437 cm^{-1} (relative intensity 75) and h_u (1646 cm^{-1}, rel. int. 19). The photoluminescence (PL) spectrum of C_{60} single crystal has been indeed analyzed in this framework[10]. This experiment sets indirectly the phonon-less electronic origin at 15089 cm^{-1} (i.e. 1.871eV). A direct determination of the origin can be made by means of non-linear spectroscopic techniques.

RESULTS AND DISCUSSION

Third Harmonic Generation Spectroscopy (THG)

By illuminating a thin film of C_{60} with a tunable laser beam and investigating the dependence of the third harmonic intensity upon the tilting angle of the film one may derive the third order electric susceptibility $\chi^{(3)}(-3\omega;\omega,\omega,\omega)$ which, from time dependent perturbation approach[11,12], is given by the following expression:

$$\chi^{(3)}(-3\omega;\omega,\omega,\omega) \ \alpha \ e^4 \sum_{n \neq g} \sum_{m \neq g} \sum_{l \neq g} f_{gn}f_{nm}f_{ml}f_{lg} \ x$$

$$\left[\frac{1}{(E_{ng}-3\omega)(E_{mg}-2\omega)(E_{lg}-\omega)} + \frac{1}{(E_{ng}+\omega)(E_{mg}-2\omega)(E_{lg}-\omega)} + \right.$$

$$\left. + \frac{1}{(E_{ng}+3\omega)(E_{mg}+2\omega)(E_{lg}\pm\omega)} + \frac{1}{(E_{ng}+\omega)(E_{mg}+2\omega)(E_{lg}-\omega)} \right] \qquad (1)$$

where f_{ij} are the transition matrix elements between states i and j, and E_{ij} are the energies of these transitions and ω is the photon energy of the incoming beam. It is evident that $\chi^{(3)}$ depends critically on the energy denominators. When the energy of the incoming photon is three or two times lower than the energy of the allowed transition from the ground state to an excited state, one expects a peak in the THG spectrum which is due to a three-photon or a two-photon resonance respectively. THG could therefore be used as a tool to search for two-photon states. Experimental details are reported in previous papers[13,14].

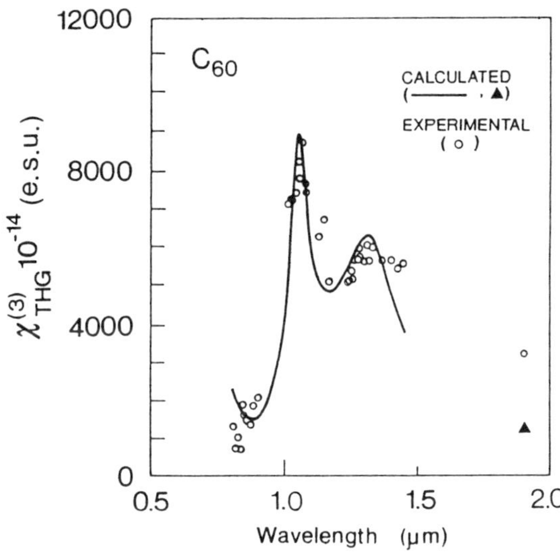

Fig. 1. Third harmonic generation spectrum of C$_{60}$ thin film (dotted line). The full line represents a fitting by using a three-level model.

The measured cubic susceptibility $\chi^{(3)}$ ($-3\omega;\omega,\omega,\omega$) is plotted in Fig. 1 as function of fundamental laser wavelength. The solid line indicates the calculated $\chi^{(3)}$ dispersion taking into account a three-level model with a *gerade* state at 1.87 eV and an *ungerade* state at 3.5 eV. The agreement with the experiment is quite good. The resonance at 1.064 μm (1.165 eV) is clearly a three-photon resonance, the triple of the fundamental photon energy being equal to the T_{1u} one-photon allowed transition at 3.5 eV. The second broad resonance at around 1.3 μm falls in the range of a significantly weaker one-photon absorption with a characteristic appearance of forbidden transitions made partially allowed by H-T coupling. We therefore assigned the maximum of the THG spectrum at 441 nm harmonic wavelength, or 1.323 μm fundamental wavelength to a two-photon resonance with the electronic origin of the T_{1g} forbidden state located in the solid at 1.87 eV[15].

410

Two-Photon Excitation Spectroscopy (TPE)

The direct location of the lowest singlet exciton band is carried out by two-photon excitation (TPE) spectroscopy on the C_{60} single crystal at 4K.

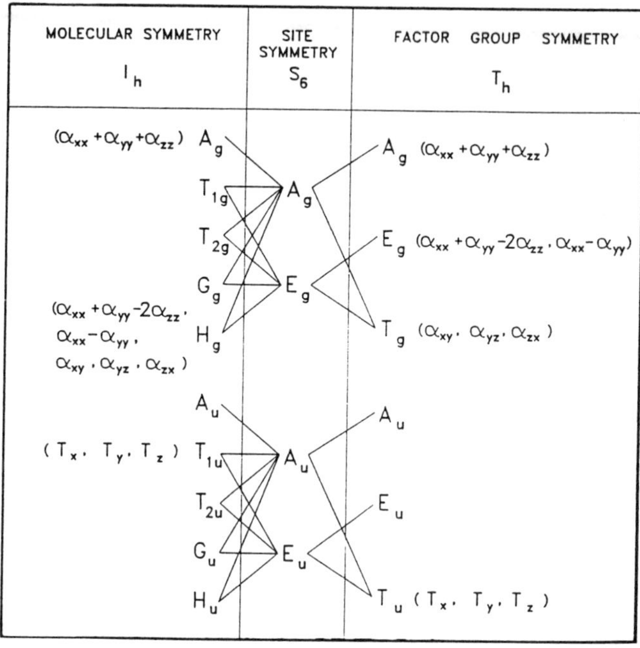

Fig. 2. Correlation diagram of the I_h Point group of the isolated C_{60} molecule with the Factor group for solid C_{60} via the S_6 Site symmetry.

The correlation among the symmetry of the free molecular *wfs* and the symmetry of the crystal *wfs* is shown in Fig. 2. Since the site symmetry is S_6 each molecular irreducible representation on the left side of Fig. 2 correlates via the Site group symmetry (center) to the irreducible representation of the crystal in a way which preserves the parity character. The correlation is indicated by the lines. The T_{1g} (molecular) for instance, correlates with A_g, E_g and T_g (Factor). Since in the T_h point group these representations transform like polarizability tensor components then the T_{1g} level, which is two-photon forbidden in the free molecule, become partially allowed in the crystal due to this correlation effect which is also known as *crystal field effect*. Preliminary results of two-photon excitation of C_{60} single crystal at 4.2K in the range of 0.85-1 eV indicate that there is a two-photon absorption activity starting at approximately 1.85 eV. This band is tentatively assigned to a pure electronic transition (zero-phonon transition of the k=0 Frenkel singlet exciton). By comparing these results with those obtained by photoluminescence[10] and THG spectroscopy[15] it is possible to unambiguosly assign the lowest energy Frenkel exciton at 1.85 eV to the T_{1g} state.

REFERENCES

1. a) J. Frenkel, Phys. Rev. <u>37</u>, 17, 1276 (1931); b) A.S. Davydov, *Theory of molecular excitons* (Plenum, New York, 1971); c) D.P. Craig and S.H. Walmsley, *Excitons in molecular crystals* (W.A. Benjamin, Inc. New York, 1968).

2. W. Krätschmer, L.D. Lamb, K. Fostiropoulos and D. R. Huffman, Nature <u>347</u>, 354 (1990).

3. S. Leach, M. Vervloet, A. Desprès, E. Breheret, J.P. Hare, T.J. Dennis, H.W. Kroto, R. Taylor and D.R.M. Walton, Chem. Phys. <u>160</u>, 451 (1992).

4. A. Skumanich, Chem. Phys. Lett. <u>182</u>, 486 (1991).

5. C. Reber, L. Yee, J. MacKiernan, J.I. Zink, R.S. Williams, W. M. Tong, D. A. A. Ohlberg, R.L. Whetten and F. Diederich, J. Phys. Chem. <u>95</u>, 2127 (1991).

6. F. Negri, G. Orlandi and F. Zerbetto, J. Chem. Phys. <u>97</u>, 6496 (1992).

7. P.O. Löwdin, Phys. Rev. <u>97</u>, 1474 (1955).

8. K. Yaban and G.F. Bertsch, Chem. Phys. Lett. <u>197</u>, 32 (1992).

9. G. Herzberg and E. Teller, Z. Phys. Chem. <u>B 21</u>, 410 (1933).

10. W. Guss, J. Feldmann, E.O. Goebel, C. Taliani, H. Mohn, P. Haeusslerand and H-U.ter Meer, Phys. Rev. Lett. (in press), (1994).

11. B. J. Orr and J. F. Ward, Mol. Phys. <u>20</u>, 513 (1971).

12. P.N. Butcher and D. Cotter, *The Elements of Nonlinear Optics*, in Cambridge studies in Modern Optics 9, Cambridge University, Cambridge (1990).

13. F. Kajzar, C. Taliani, R. Zamboni, S, Rossini and R. Danieli, in: Fullerenes: Status and Perspectives, eds. C. Taliani, G. Ruani and R. Zamboni (World Scientific, Singapore, 1992) p. 75.

14. F. Kajzar, C. Taliani, R. Zamboni, S, Rossini and R. Danieli, Synth. Metals <u>54</u>, 21 (1993).

15. a) F. Kajzar, C. Taliani, R. Zamboni, S, Rossini and R. Danieli, SPIE Proc., 2025 (1993); b) F. Kajzar, C. Taliani, R. Danieli, S. Rossini and R. Zamboni, Chem. Phys. Lett. <u>217</u>, 418 (1994).

ELECTRONICALLY EXCITED STATES OF THE C_{60} VAN DER WAALS DIMER: LOCALIZED *versus* CHARGE TRANSFER EXCITATIONS

Péter R. Surján[1,2], László Udvardi[2] and Károly Németh[1]

1: Dept. Theor. Chem., Eötvös University H–1518 Budapest 112, POB 32, Hungary
2: Quantum Theory Group, Institute of Physics, Technical University, Budapest

Abstract

We analyze the nature of electronic excitations in the $C_{60}...C_{60}$ dimer. The low-lying states are localized and correspond to Frenkel excitons. Above 3.5 eV charge transfer (Wannier-type) excitons also appear. The Davydov splitting is in the order of 10 meV.

1. Introduction

Electronic excitations in C_{60} crystals can be classified according to their localized characters: the excited electron may remain on the same C_{60} molecule, or it may jump to another one. Rigorous treatment of this effect is quite difficult as it needs an accurate description of the excited states. To approach the problem, we have studied the system of two C_{60} molecule at van der Waals distance (3.4 Å) from each other. We selected a random relative orientation: the second molecule was rotated by 30 degrees around two perpendicular axes so the dimer had no symmetry. Quantum chemical calculations at the semiempirical CNDO/S-CI level[1] were performed to determine various excited states of the dimer.

2. Method

The excited state wave functions for the dimer are obtained by a configuration interaction (CI) procedure involving singly excited configurations (Tamm-Dankoff Approximation):

$$|K> = \frac{1}{\sqrt{2}} \sum_{i,k^*} C_{i,k^*}^K \left(a_{k^*\alpha}^+ a_{i\alpha} + a_{k^*\beta}^+ a_{i\beta} \right) |0> \tag{1}$$

where $|0>$ is the Hartree-Fock (HF) ground state. In order to analyze the nature of these excited states, we have expressed them in terms of excitations between localized molecular orbitals (LMOs)[2]. To determine the LMOs for the dimer, we performed first two separate HF - SCF calculations for the monomers. Exact localized dimer orbitals are obtained by projecting them from the monomer orbitals; i.e. $\psi_{iA} = \hat{P}\phi_{iA}$ $(i = 1, 2 \dots N_A)$ and $\psi_{iB} = \hat{P}\phi_{iB}$ $(i = 1, 2, \dots N_B)$ for occupied MOs, where $\hat{P} = \sum_{i=1}^{N} |i><i|$ is the projector to the dimer occupied subspace. Virtual LMOs are projected by $(1-\hat{P})$. These set of LMOs is not orthogonal, but it was orthogonalized subsequently by Löwdin's procedure[3].

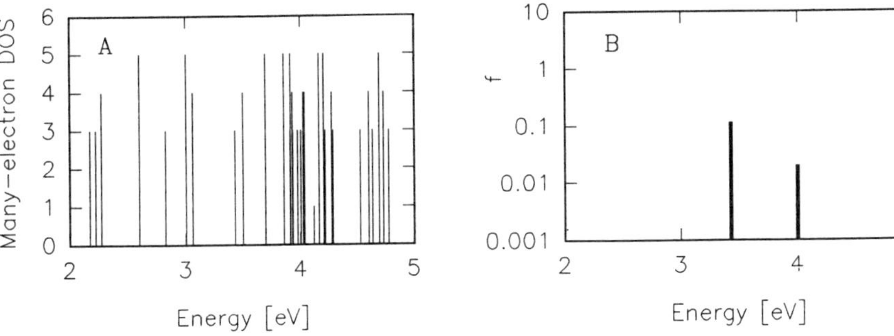

Figure 1. Density of many-electron states (A) and oscillatory strengths (B) vs energy in a single C_{60}

The connection between the canonical and localized dimer orbitals is given by the unitary matrix **L**, by which the excited states in terms of localized MOs are written as:

$$|^{1,3}K> = \frac{1}{\sqrt{2}} \sum_{p,q^*} D_{p,q^*}^K \, [\psi_{q^*\alpha}^+\psi_{p^*\alpha}^- + \psi_{q^*\beta}^+\psi_{p^*\beta}^-]|0> \qquad (2)$$

where $D_{p,q^*}^K = \sum_{i,k^*} C_{i,k^*}^K L_{kq^*} L_{ip}$ is the coefficient of state K in terms of the transformed excitations $p \rightarrow q^*$. These transitions can be classified as A $\rightarrow$ A, A $\rightarrow$ B, etc. type, where A and B stand for the two monomers. To characterize the excited many-electron states, we define the weights of the corresponding excitations: $\eta_{A\rightarrow A}^K = \sum_{i,k^*}^{A\rightarrow A}(D_{i,k^*}^K)^2$ and $\eta_{A\rightarrow B}^K = \sum_{i,k^*}^{A\rightarrow B}(D_{i,k^*}^K)^2$ and similarly for $\eta_{B\rightarrow A}^K$ and $\eta_{B\rightarrow B}^K$. Further, $\eta_{LOC}^K = \eta_{A\rightarrow A}^K + \eta_{B\rightarrow B}^K$

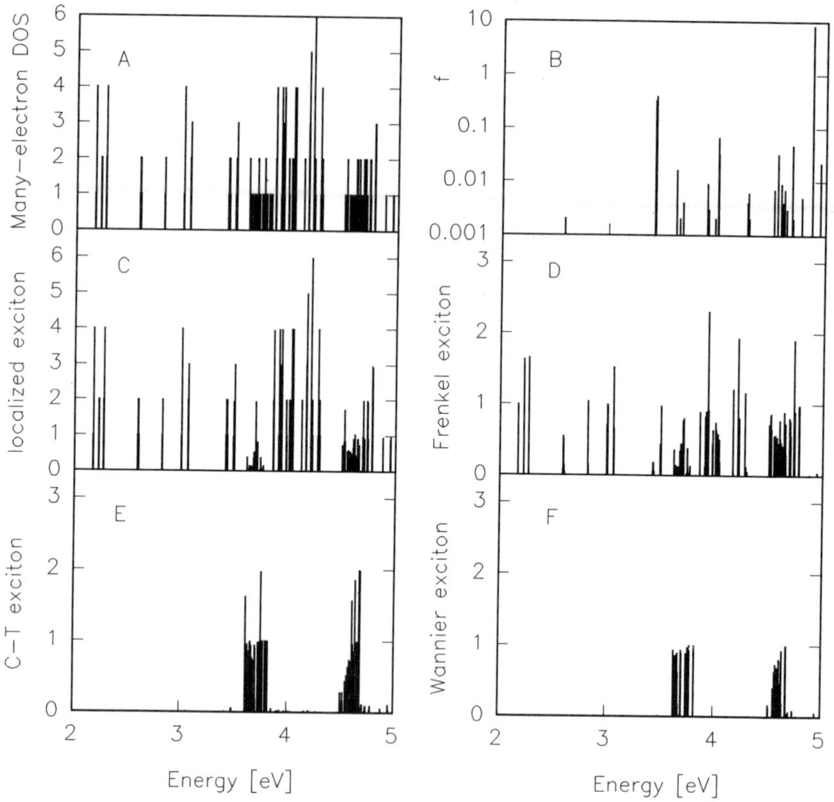

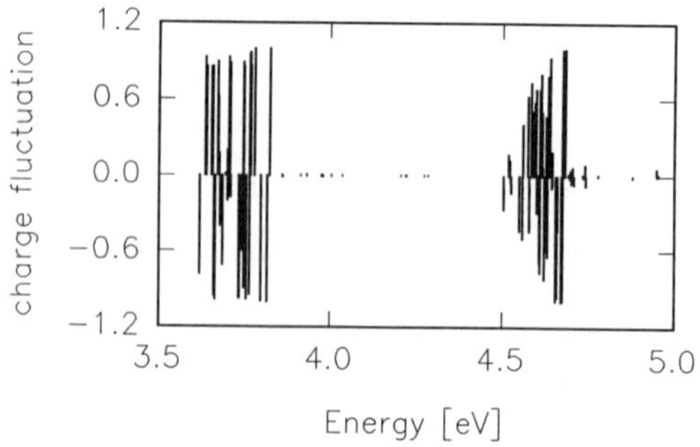

Figure 2. Density of many electron states (A), oscillatory strength (B), localized excitons (η_{LOC}) (C), Frenkel excitons(η_F) (D), charge-transfer excitons (η_{CT}) (E) and Wannier-excitons (η_W) (F) in the dimer

Figure 3. Charge of a monomer (Q_A) in the dimer in various excited states

and $\eta_{CT}^K = \eta_{A \to B}^K + \eta_{B \to A}^K$ as defined as the weights of localized and charge transfer characters of state K. The net charge on momomer A writes: $Q_A^K = \eta_{B \to A}^K - \eta_{A \to B}^K$. Finally, $\eta_F^K = | \eta_{A \to A}^K - \eta_{B \to B}^K |$ is the rate of Frenkel-excitons.

3. Results

The monomer states are given in Fig.1. Allowed transitons are seen from the oscillatory strengths, while we show all excitations in histogram form in Fig.1.A (many- electron density of states). The results agree well with those of Negri et al.[4].

The dimer states are analyzed in Fig. 2. They appear at around the same energies as the monomer transitions in slightly split degenerate pairs. Davydov splittings amount to cca. 10 meV. The oscillatory strength, however, change more drastically; it is noteworthy e.g., that, as compared to the monomer states, a new weakly allowed transition appears at 2.6 eV. The many-electron states are partitioned to Fig.2 C and E: it seems that the lower transitions are all localized while the CT levels are accumulated between 6.6 and 6.8 eV and 4.5 and 4.7 eV. Some of the localized transitions contain only pure A-A or B-B excitations (Fig.2.D): they are precursors of Frenkel-excitons. Unsymmetric A-B or B-A transitions are precursors of Wannier excitons (Fig. 2.F). These states produce net charges on the monomers. For symmetrical configurations they appear in $Q_A = \pm 1$ degenerate pairs, but the unsymmetric configuration used in the present calculations produced a significant charge fluctuation as a function of energy, see Fig. 3.

Acknowledgement. This work was supported by the grants OTKA 517/1990 and T012820.

4. References

[1] J.Del Bene and H.H.Jaffe, J.Chem.Phys **48** 1807, 4050 (1968)

[2] János Ángyán and Péter R. Surján, Theoretica Chimica Acta **63** 43 (1983)

[3] P.O.Löwdin, Phys.Rev. **97** 1474 (1955)

[4] F.Negri, G.Orlandi, F.Zerbetto, J.Chem.Phys. **97** 6496 (1992)

PHOTOLUMINESCENCE STUDIES OF SOLID C_{60}

J. Feldmann, W. Guß, E.O. Göbel
Fachbereich Physik and Zentrum für Materialwissenschaften,
Philipps-Universität, Renthof 5, 35032 Marburg, Germany

C. Taliani
Istituto di Spettroscopia Molecolare, CNR, 40129 Bologna, Italy

H. Mohn, W. Müller, H.-U. ter Meer
Hoechst AG, Angewandte Physik, 65926 Frankfurt, Germany

ABSTRACT

We have studied the luminescence properties of solid C_{60} for different morphologies such as C_{60} films, polycrystalline C_{60} powder, and C_{60} single crystals. The inhomogeneous linewidths of the fluorescence lines are drastically reduced in case of the single crystals. The richly structured fluorescence spectrum of C_{60} single crystals is composed of several sub-spectra each belonging to a particular X-trap. The fluorescence lines belonging to each X-trap show the characteristic vibronic structure of molecular C_{60} emission.

1. Introduction

Several groups have studied the photoluminescent properties of C_{60} molecules in solution[1,2] and of solid C_{60} in different morphologies such as films, polycrystalline powder,[3-12] and single crystals.[13] However, different optical spectra have been reported even for nominally equal morphologies. Thus, the microscopic interpretation of the fluorescence spectra, in particular for the solid phase, has been controversial up to now. In this paper, we show that the differing optical spectra are a direct consequence of X-trap fluorescence from solid C_{60}.

Negri et al.[16] showed that the fluorescence spectrum of molecular C_{60} can be explained in a Herzberg-Teller scheme, i.e., the lowest singlet excited state T_{1g} acquires ungerade character from energetically higher T_{1u} states by adiabatic vibronic coupling and thus the optical $S_1 - S_0$-transition becomes partially dipole-allowed.[16] As a consequence, so-called false origins determine the peak positions in the fluorescence and absorption spectra of C_{60}. However, only two false origins (209 cm^{-1} separated from each other) originating from t_{1u} and h_u vibronic couplings have appreciable oscillator strengths and are expected to dominate the optical spectra.

We show in this paper that each C_{60} sub-spectrum belonging to a specific X-trap is indeed mainly composed of two spectrally adjacent fluorescence lines and can be understood in terms of the Herzberg-Teller coupling scheme.

2. Experimental

C_{60} was produced by the carbon-arc method of Krätschmer and Huffman.[17] The fullerene containing soot was extracted and chromatographically purified by a proprietary process. The C_{60} used is 99.4% pure. This material was sublimed in an inert gas atmosphere to give crystals of $1-2$ mm diameter. The thin films were sublimated from C_{60} in ultra-high vacuum onto quartz substrates. To avoid oxygen contamination all C_{60} samples were kept under Argon atmosphere before mounting them in an evacuated helium flow-cryostat to perform low-temperature luminescence experiments. The excitation photon energy is $2eV$ and we use low excitation intensities of 1 mW focused to a spot size of approximately 100 μm in diameter.

3. Results

The influence of the morphology of solid C_{60} on the emission spectrum is clearly seen in Fig.1a-c, where normalized fluorescence spectra taken at $T=10K$ are shown for a C_{60} film (a), polycrystalline C_{60} powder (b), and for a C_{60} single crystal (c). The spectrum of the C_{60} film exhibits only two strongly broadened emission bands at

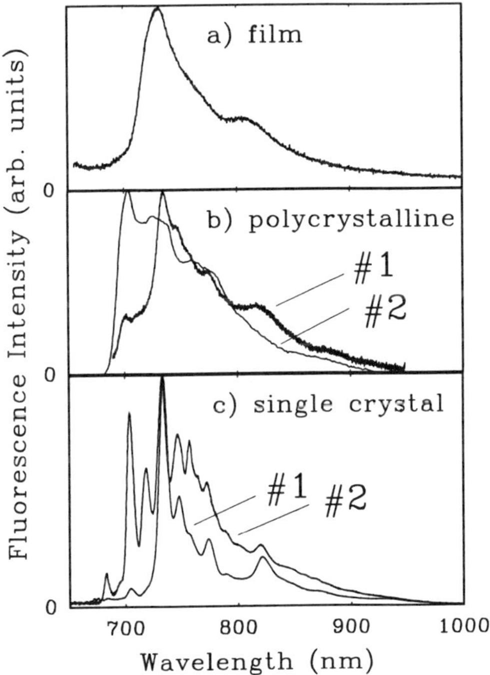

Fig.1: Photoluminescence spectra taken at T=10K for different morphologies of solid C_{60}: (a) C_{60} film, (b) polycrystalline C_{60}, and (c) C_{60} single crystal. Spectrum #1 of Fig.1b is taken from Ref. [9]. Spectra #1 and #2 of Fig.1c are taken for different positions of the excitation light on the same single crystal.

approximately 730 nm and 810 nm. It is reasonable in the case of the film to assume that fluorescing C_{60} molecules are located in statistically varying environments leading to pronounced inhomogeneous broadening effects in the optical spectra.

The number of resolvable fluorescence peaks drastically increases in the case of the polycrystalline powder and, in particular, for the single crystal. The series of well-resolved luminescence peaks for the single crystal is a consequence of reduced inhomogeneous broadening of the individual optical transitions.

Spectrum #1 of Fig.1b for polycrystalline C_{60} powder is taken from Ref. [9]. The spectrum peaks at approximately 730 nm and shows a similar but more structured shape than the fluorescence spectrum of the C_{60} film shown in Fig.1a. The fluorescence spectrum of the polycrystalline C_{60} material synthesized as described above is labelled #2 and peaks at approximately 703 nm, i.e., at shorter wavelengths than spectrum #1. In the case of the single crystal we also find differing spectra. The difference between spectrum #1 and spectrum #2 shown in Fig.1c is only the spatial position of the laser excitation spot on the single crystal. This spatially inhomogeneous behavior of the fluorescence spectrum definitely shows that inhomogeneously distributed crystal imperfections such as chemical impurities or crystal defects influence the luminescence process. In a spatial scan over the crystal mostly spectra similar to the one labelled #1 in Fig.1c are observed. In addition, such a spatial scan shows that a change in intensity for a particular fluorescence line is connected with a simultaneous change in intensity of a spectrally adjacent emission line. In other words, the spectrum seems to be composed of several pairs of emission lines spectrally separated by about 33 meV.

4. Discussion

We state that each pair of fluorescence lines originates from C_{60} molecules in a particular crystal environment. The main line at approximately 732 nm is interpreted as fluorescence from C_{60} molecules located in a perfect C_{60} environment, i.e., it stems from bulk C_{60}. Other pairs of peaks are assigned to C_{60} molecules adjacent to chemical impurities, to crystal defects, or to crystal surfaces. The imperfect crystal environment induces an energetic shift of the electronic states and thus of the fluorescence of the particular C_{60} molecule. However, the defect does not change the characteristic vibronic structure of the fluorescence of the adjacent C_{60} host molecule. Such *defect related luminescence from host molecules* are known for anthracene single crystals[18−20] and are called X-traps.

Negri et al.[16] showed that a pair of emission lines corresponding to 1437 cm^{-1} t_{1u}- and 1646 cm^{-1} h_u-related false origins are indeed expected to dominate the fluorescence spectrum of molecular C_{60}. The calculated spectral spacing between these two false origins is 26 meV,[16] whereas the measured fluorescence spectrum of molecular C_{60} in solution provides a spacing of 31 meV[2,16] in good agreement with the observed spacings for the C_{60} single crystal in Fig.1c. Accordingly, the observed pairs of emission lines are assigned to the t_{1u}- and h_u-related false origins for C_{60} in the respective crystal environment. In Table 1, the experimental values for the false origins (t_{1u} and h_u) are listed in the second and third column. In addition, the spectral spacing Δ between the two experimentally observed t_{1u} and h_u false origins is given in the last column of Table 1. Knowing the spectral positions of the t_{1u} and h_u false origins, the respective true 0-0 origins can be calculated.[16] We obtain 1.871 eV for the 0-0 origin of C_{60} in a perfect crystal environment as listed in the first column and row C of Table 1.

Three distinct pairs of emission lines corresponding to three distinct C_{60} X-traps are clearly observed when scanning over the single crystal. These X-traps are labelled X_1, X_2, and X_5 in Table 1. The Δ-values for these X-traps are comparable to the one found for bulk C_{60} emission (compare with row C). There seem to be at least two more emission centres in the spectral range around 750 nm. However, their identification is more difficult since the corresponding false origin emission lines overlap. As can be seen in Table 1 we assume two further X-traps X_3 and X_4 to account for the intensity variations around 750 nm. Due to the spectral overlap we cannot directly identify the h_u-related fluorescence lines of the X_3- and X_4-traps from the spectra shown in Fig.1c.

Table 1: List of emission centres (C: bulk C_{60}, X_i: X-traps). The values for the two false origins, t_{1u} and h_u, are taken from the experimental fluorescence spectra, whereas the true origin 0-0 is calculated using the t_{1u}-frequency from Ref. [16]. Δ is the spectral spacing between the two experimentally determined false origins.

	0 - 0	t_{1u}		h_u		Δ
	eV	nm	eV	nm	eV	eV
X_1	1.993	683.2	1.815	695.1	1.784	0.031
X_2	1.940	703.7	1.762	718.4	1.726	0.036
C	1.871	732.4	1.693	747.4	1.659	0.034
X_3	1.839	746.5	1.661	-	-	-
X_4	1.816	757.0	1.638	-	-	-
X_5	1.781	773.5	1.603	789.8	1.570	0.033

Altogether, we find two X-traps (X_1 and X_2) at higher photon energies and three X-traps (X_3-X_5) at lower photon energies as compared to the energetic position of the bulk C_{60} emission (C). The fact that fluorescence from higher energy X-traps (X_1 and X_2) can be observed means that excitation transfer from these X-traps to, e.g., low-energetic bulk C_{60} molecules is hampered at low temperature. We attribute these X-traps at higher energy to surface-related exciton states as is the case for crystalline anthracene.[21] The low-energetic X-traps are most likely due to C_{60} molecules around

420

specific crystal defects.
We note that Matsushita et al.[22] also interpret their C_{60} luminescence results in terms of X-traps. Our interpretation that the observed pairs of fluorescence lines are due to false origins mediated by the Herzberg-Teller coupling has been confirmed recently by two-photon absorption experiments performed by Taliani et al.[23]

5. Acknowledgements

We thank F. Zerbetto and U. Lemmer for helpful discussions and M. Preis for expert technical assistance. The work at Marburg University is financially supported by the Deutsche Forschungsgemeinschaft through the Leibniz Förderpreis.

6. References

1. M.R. Wasielewski et al., *J. Am. Chem. Soc.* **113**, 2774 (1991).

2. Y. Wang, *J. Phys. Chem.* **96**, 764 (1992).

3. C. Reber et al., *J. Phys. Chem.* **95**, 2127 (1991).

4. K. Pichler et al., *J. Phys. Condens. Matter* **3**, 9259 (1991).

5. P.A. Lane et al., *Phys. Rev. Lett.* **68**, 887 (1992).

6. M. Matus, H. Kuzmany, E. Sohmen, *Phys. Rev. Lett.* **68**, 2822 (1992).

7. S.P. Sibley, S.M. Argentine, A.H. Francis, *Chem. Phys. Lett.* **188**, 187 (1992).

8. T. Zhao, J. Liu, Y. Li, and D. Zhu, *Appl. Phys. Lett.* **61**, 1028 (1992).

9. J. Feldmann et al., *Europhys. Lett.* **20**, 553 (1992).

10. H.J. Byrne et al., *Appl. Phys.* **A56**, 235 (1993).

11. M. Diehl, J. Degen, H. Schmidtke, *Ber. Bunsenges. Phys. Chem.* **97**, 908 (1993).

12. E. Shin et al., *Chem. Phys. Lett.* **209**, 427 (1993).

13. Y. Iwasa, T. Koda, S. Koda, Synth. Met. **55**, 3033 (1993).

14. T.W. Ebbesen, K. Tanigaki, S. Kuroshima, *Chem. Phys. Lett.* **181**, 501 (1991).

15. M. Lee et al., *Chem. Phys. Lett.* **196**, 325 (1992).

16. F. Negri, G. Orlandi, F. Zerbetto, *J. Chem. Phys.* **97**, 6496 (1992).

17. W. Krätschmer et al., *Nature* **347**, 354 (1990).

18. A. Brillante et al., *Chem. Phys. Lett.* **31**, 215 (1975).

19. V.A. Lisovenko, M.T. Shpak, V.G. Antoniuk, *Chem. Phys. Lett.* **42**, 339 (1976).

20. D.P. Craig and J. Rajikan, *J. Chem. Soc.* **74**, Faraday Transactions II, 292 (1978).

21. M.S. Brodin, M.A. Dudinskii, S.V. Marisova, *Opt. Spectrosc.* **34**, 651 (1973).

22. M. Matsushita et al., *Optically detected magnetic resonance and the luminescent properties of crystalline* C_{60}, this volume.

23. C. Taliani et al., *Identification of the lowest forbidden exciton in* C_{60} *single crystal*, this volume.

LIGHT-INDUCED INSULATOR TO METAL TRANSITION IN FULLERENES

W.K. Maser*, H.J. Byrne, M. Kaiser, W.W. Rühle, L. Akselrod, A.T. Werner, J. Anders, X.-Q. Zhou, G. Mahler[+], T. Kuhn[+], A. Mittelbach and S. Roth

Max-Planck-Institut für Festkörperforschung,
Heisenbergstraße 1, 70569 Stuttgart, FRG

[+]Institut für Theoretische Physik und Synergetik, Universität Stuttgart,
Pfaffenwaldring 57, 70569 Stuttgart, FRG

*present Address: School of Chemistry and Molecular Sciences, University of Sussex,
Falmer, Brighton BN1 9QJ

ABSTRACT

Photoluminescence and photoconductivity measurements are performed at high excitation densities. A highly nonlinear dependence of the luminescence emission efficiency is observed above a threshold input intensity. This nonlinear emission is dependent on the cube of the input intensity, redshifted and has a long, intensity dependent lifetime. Furthermore, with the onset of the nonlinear emission, the photoconductive response increases with the cube of the input intensity and the photocurrent behaves largely independent of temperature. This behaviour is suggestive of an optically driven insulator to metal transition. A phenomenological model points to the important role of exchange and correlation energies in fullerenes at high excitation densities.

1. Introduction

Optical absorption, photoluminescence, Raman spectroscopy as well as transport measurements show strong indications that the solid state packing of C_{60} exerts a minimal perturbation on the properties of the molecule[1-3]. In this paper, the response of solid state fullerenes to high excitation densities is reported. The properties are seen to undergo dramatic, nonlinear changes which are indicative of a transition to a highly optically and electronically active state. The observed behaviour is described and discussed in terms of collective excited state phenomena.

2. Results and Discussion

Details of sample preparation and experimental technique are given elsewhere[3]. Under low illumination conditions a weak luminescence output which may be associated with an intramolecular process is observable from fullerene crystals. The temporal decay of the luminescence fits well to a single exponential of lifetime 1.2 nsec[3], consistent with the intersystem crossing time in solution[4] and no long lived component is observable. On increasing the input intensity, a dramatic change in the emission spectrum is observable, which is also reported by[5]. Figure 1 shows the evolution of the spectrum over the range 0.1 mW to 3.0 mW. The spectral evolution is accompanied by a broadening of the spectrum and a dramatic increase in the luminescence output. Investigations of the temporal decay of the emission indicate that the spectral changes may be associated with the emergence of a long lived emission component. The magnitude of this component increases with the cube of the average input power, as shown in figure 2. Furthermore, within the time window of the measurement (0-10 nsec), the decay time of the luminescence, extracted from a fit to a single exponential decay, is seen to increase with intensity, with a dependence of at least squared[3]. The intensity onset of this behaviour appears to be a threshold value and varies with position on the sample by a factor of the order of two.

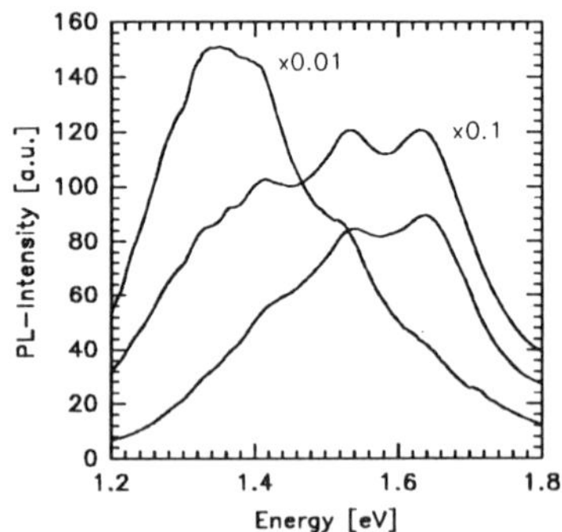

Figure 1: Spectra of C_{60} crystal illuminated with 0.1, 0.5 and 3.0 mW at room temperature.

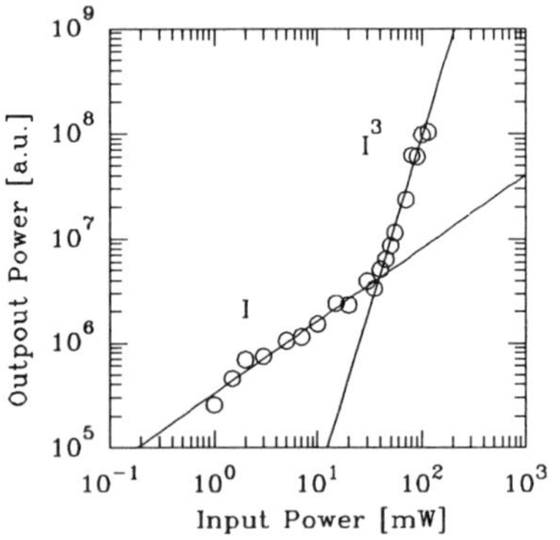

Figure 2: Dependence of the long lived luminescence component on intensity.

Under low intensity illumination, the temperature dependence of the photoconductive response of fullerene crystals may be seen as thermally activated, typically of a molecular insulator or semiconductor. The activation energy varies from sample to sample, and, for the example shown in figure 3(a), has a value ~110 meV. The observation of this molecular insulator-like behaviour is consistent with the assignment of the luminescence emission to intramolecular transitions. At elevated intensities, the intensity dependence of the photoconductive response initially reduces to a square root dependence due to the contribution of bimolecular recombination processes[6]. As is shown in figure 4, however, this square root dependence is transformed to a cubic dependence upon further increase in intensity. Furthermore, as in the photoluminescence, with increasing the intensity a second, longer lived component in the transient photoconductive response emerges. This component also increases with intensity[7]. The onset of the nonlinear dependences of the photoconductive response coincides with the visible onset of the cubic photoemission process described above. The nonlinear behaviour of the transport properties may, therefore be associated with the nonlinearity in the optical properties.

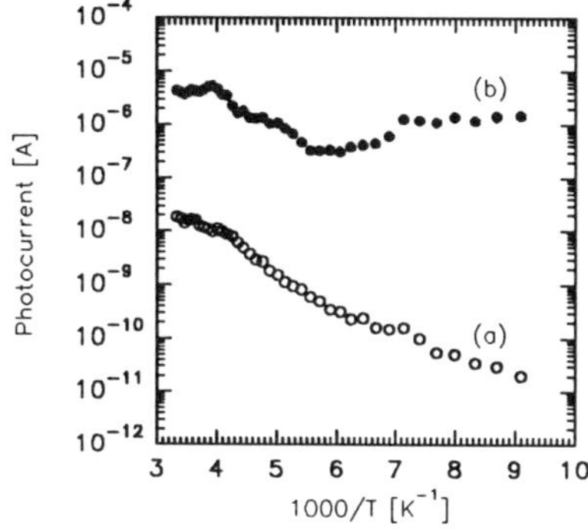

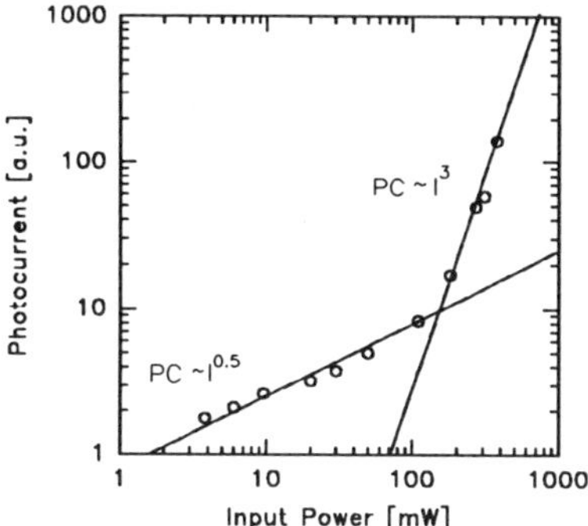

Figure 3: the photoconductive response. (a) low and (b) high illumination conditions.

Figure 4: Intensity dependence of the photocurrent in fullerene crystal.

The increase of the photoconductive response over two orders of magnitude is indicative of a dramatic change in the nature of the photoconductive state. In figure 3(b), the temperature dependence of a crystal illuminated with 300 mW average intensity, is compared to that of the same crystal illuminated with 3 mW (figure 3(a)). The measurement was performed by sweeping the temperature and monitoring the high and low intensity response at each temperature. At each point, adjustment of the laser spot on the sample was performed to optimise the response. Whereas the low intensity photoconductive response decreases by three orders of magnitude over the temperature range measured, the high intensity response remains constant within an order of magnitude. Such a temperature independence of the conductivity may be characteristic of a metallic-like behaviour and a Mott-like insulator to metal transition is inferred.

Those effects exhibited by fullerenes are a strong departure from the molecular like behaviour of the weakly excited state and point towards the influence of the interaction of molecules in the highly excited state. Such many body processes have been extensively investigated in indirect inorganic semiconductors[8]. Especially the similarities of the photoluminescence and the photoconductive behaviour of Germanium and Silicon under high intensity illumination to the phenomena described above are remarkable. The photoluminescence emission is red shifted[9] and its decay rate was shown to be highly nonexponential, increasing with time[10]. Furthermore, above a threshold, both the photoluminescence emission and the photoconductive response show a cubic dependence on the excitation intensity[11,12].

The nonlinear behaviour in indirect semiconductors has been explained by the interactions of excitons[13], which here are the fundamental optical excitations. In the model proposed by Mott[14], as the density of excitons is increased, the electron-hole interaction is considered screened by plasma environment until, at a critical density ($n_c{\sim}0.001/a_o{}^3$), they cease to exist. Indirect band gap semiconductors are particularly amenable to the production of the high exciton densities. In the supra-Mott regime, exchange/correlation energies, E_{xc}, become important and at high excitation densities, E_{xc} can dominate over

424

the exciton binding energy, resulting in the formation of an electron-hole plasma or liquid, which is responsible for the nonlinear properties in those materials.

The similarities of the properties of the highly excited state to those of Ge and Si at low temperatures prompt an evaluation of the contribution of correlation and exchange energies in fullerenes. The "exciton" binding energy in fullerenes is ~0.7 eV[15]. Band calculations of fullerenes in the solid state yield effective masses of the electron and hole to be 1.41 and 1.1, respectively[16], corresponding to a reduced mass $\mu = 0.65$. The exciton Bohr radius is given by $a_0 = h^2\varepsilon/\mu e^2$, where ε is the dielectric constant. Taking $\varepsilon = 4.3$[17], the resultant value of $a_0 = 0.35$ nm is consistent with the dimensions of a molecularly localized excited state, analogous to a Frenkel exciton in inorganic semiconductors. The exciton binding energy, given by $E_x = e^2/8\pi\varepsilon_0\varepsilon a_0$ may be calculated to be ~0.5 eV and is consistent with the separation of the HOMO-LUMO transition and the first solid state feature, or bandgap, in fullerenes[15].

In such a system, the critical, or so-called Mott density, at which the exciton binding energy becomes screened is given by[14] $n_c a_0^3 \sim (0.25)^3 \sim 10^{-2}$ and may be calculated to be ~1.4×10^{20} cm^{-3}. The molecular density is ~3.6×10^{20}cm^{-3}, and so this corresponds to an excitation density of ~40%. Both the exchange and correlation energies are strongly dependent on band structure and anisotropy, but it has been shown that their sum is largely independent, enabling a universal description of their contribution in the form[18], $E_{xc}(r_s) = (a + br_s)/(c + dr_s + r_s^2)$ where $a = 4.8316$, $b = 5.0879$, $c = 0.0152$, $d = 3.0426$ and r_s is the mean distance between excitations. Using the parameters above, the exchange/correlation energy may be calculated as a function of excitation density, and is plotted in figure 5. At a density of 1.4×10^{20} cm^{-3}, E_{xc} increases above that of the exciton binding energy and, at a density of 4×10^{20} cm^{-3}, the excess energy has a value of 150 meV.

Comparison of the exchange and correlation energies at high excitation densities to the low intensity molecularly localized energies therefore supports the existence of a Mott-like transition at a critical density of ~1×10^{20} cm^{-3} and the increasing prevalence of E_{xc} can result in an energy shift of 150 meV, consistent with that observed in the photoluminescence of fullerenes at high intensities. Such a calculation combined with a comparison of the cubic dependence of the photo-luminescence output and the photocurrent, in fullerenes and both Ge and Si, strongly supports the proposition that the nonlinearities observed are governed by similar interactions.

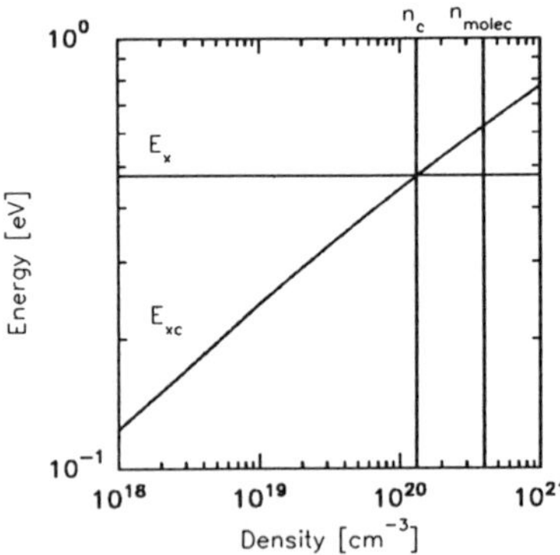

Figure 5: Plot of the sum of exchange and correlation energies as a function of excitation density.

3. Conclusions and Outlook

Under high intensity illumination, both the optical and transport properties of C_{60} undergo dramatic changes. These changes are associated with an optically induced insulator to metal transition in the highly excited state of fullerenes. The consideration of a phenomenological model supports the association of the observed behaviour with the onset of intermolecular interactions at high excitations densities. Furthermore, the nonlinearities of the highly excited state of fullerenes prompts an investigation of its applications potential, in particular in the form of an electroluminescent device[19].

4. References

1. P.A. Lane, L.S. Swanson, Q.-X. Ni, J. Shinar, J.P. Engel, T.J. Barton and L. Jones, Phys. Rev. Lett. **68** (1992) 887.
2. C. Reber, L. Yee, J. McKiernan, J.I. Zink, R.S. Williams, W.M. Tong, D.A.A. Ohlberg, R.L. Whetten and F. Diederich, J. Phys. Chem. **95** (1991) 2127.
3. H.J. Byrne, W.K. Maser, W.W. Rühle, A. Mittelbach and S. Roth, Appl. Phys. A **56** (1993) 235.
4. T.W. Ebbesen, K. Tanigaki and S. Kuroshima, Chem. Phys. Lett. **181** (1991) 501.
5. J. Feldmann, R. Fischer, E.O. Göbel, S. Schmitt-Rink and W. Krätschmer, Europhys. Letters **20** (1992) 553.
6. M. Kaiser, W.K. Maser, H.J. Byrne, A. Mittelbach and S. Roth, Solid State Commun. **87** (1993) 281.
7. H.J. Byrne et al., these proceedings.
8. See for example "Physics of Highly Excited States in Solids", M. Ueta and Y. Nishina eds., Lecture Notes in Physics vol. **57**, Springer Verlag Hdlbg. (1976).
9. R.B. Hammond, T.C. McGill and J.W. Mayer, Phys. Rev. **B13** (1976) 3566.
10. R.M. Westervelt, T.K. Lo, J.L. Staehli and C.D. Jeffries., Phys. Rev. Lett. **32**, (1974) 1051.
11. Y.A. Pokrovskii, Phys. stat. sol. (a) **11** (1972) 385.
12. M. Glicksman, M.N. Gurnee and J.R. Meyer, in [8], pp 219.
13. L. Keldysh, Proc. IX. Internat. Conf. Phys. Semicond., Moscow 1968, pp 1307.
14. N.F. Mott, "Metal-Insulator Transitions", Taylor and Francis Ltd. (1974).
15. R.W. Lof, M.A. van Veenendaal, B. Koopmans, H.T. Jonkman and G. Sawatsky, Phys.Rev. Lett. **68**, (1992) 3924.
16. W.Y, Ching, M.-Z. Huang, Y.-N. Xu, W.G. Harter and F.T. Chan, Phys. Rev. Lett. **67** (1992) 2045.
17. M.K. Kelly, P. Etchegoin, D. Fuchs, W. Krätschmer and K. Fostiropoulus, Phys. Rev. **B46** (1992) 4963.
18. P. Vashista and R.K. Kalia, Phys. Rev. **B25** (1982) 6492.
19. A.T. Werner et al., these proceedings.

NONLINEAR TRANSIENT PHOTOCONDUCTIVE RESPONSE OF FULLERENE CRYSTALS

H.J. Byrne, A.T. Werner, D. O'Brien* and S. Roth.
Max-Planck-Institut für Festkörperforschung, Heisenbergstrasse 1,
70569-Stuttgart, Germany.

and

*School of Science of Materials, Trinity College, Dublin 2, Ireland.

ABSTRACT

The intensity dependence of the transient photoconductive response of fullerene crystals is reported to show similar behaviour to their photoluminescence and steady state photoconductivity at high excitation densities. At low intensities, the response decays exponentially within a time of 15nsec. At increased intensities, a second, delayed component emerges and evolves nonlinearly into a long lived component of lifetime 100-200nsec. The system is fitted with a simple three level model in accordance with the Mott-like transition invoked to explain the previously reported nonlinear optical and transport properties.

1. Introduction

At high excitation densities, the optical and electronic properties of fullerenes have been shown to undergo dramatic changes in their photoluminescence output, and steady state photoconductive response[1]. The effects exhibited point towards a departure from the molecular like behaviour of the weakly excited state through the influence of the interaction of molecules in the excited state. Such many body processes have been extensively investigated in inorganic semiconductors, and often give rise to Mott-like insulator to metal transitions[2]. Comparison of the governing criteria in such semiconductors and those of fullerenes at high excitation densities supports the existence of a Mott-like transition at a critical density of $\sim 1 \times 10^{20} cm^{-3}$.

In order to further investigate the nature of the electronic properties of fullerenes at high excitation densities, the transient photoconductive response has been investigated as a function of intensity. In particular it is hoped to gain further insight into the dynamics of the formation and decay of the highly excited state.

1. Experimental

C_{60} crystals were generated by vacuum sublimation of fullerene powder as described previously[3]. The sample employed was of dimensions $\sim 200 \times 200 \times 200 \mu m^3$ and was mounted using silver paste on a substrate with two gold strip lines in an

Auston switch geometry of gap width of 100µm. All processing was performed under inert atmosphere, and the sample was mounted in a cryostat which was cooled to liquid nitrogen temperatures.

As excitation source, a cw modelocked Nd^{3+}:YAG laser was used to seed a regenerative amplifier operating at 30Hz. The frequency doubled output, of wavelength 532nm, had pulse width ~70psec and pulse energies of up to 200mJ. The pulses were focused to a spot size on the sample of ~200µm in diameter. The sample was connected to a variable voltage source and the transient response was monitored using a Hewlett-Packard 54111D digitising oscilloscope. For the measurements reported here a voltage of 100V was employed.

3. Results and Discussion

The transient photoconductive response was measured over a range of intensities. At low intensities, a weak, rapidly decaying signal was observable as shown in Figure 1 for the example of 1µJ incident pulse energy. The response fits well to a single exponential decay of lifetime 15nsec, the temporal limit of the detection. The response remains single exponential to excitation energies of ~2µJ, whereupon a second, longer lived component emerges which is delayed with respect to the excitation. Figure 2 shows the temporal response of the photoconductive signal at component increase with intensity. Furthermore, upon the emergence of the longlived component, the dependence of the integrated photocurrent on intensity becomes superlinear as shown in figure 3. The solid line shows a slope of 3. The progression is cyclable upon decreasing and increasing the intensity.

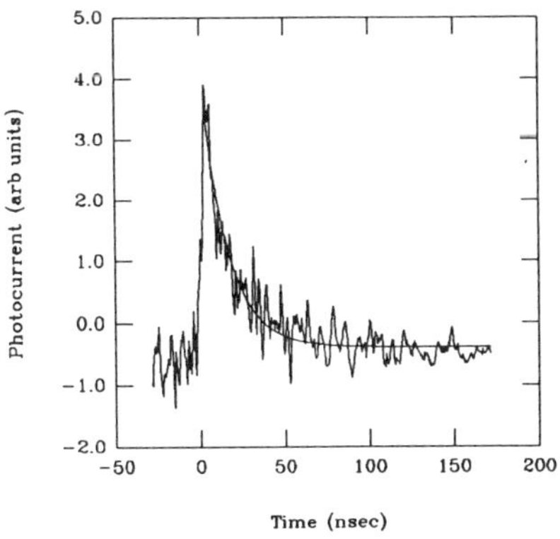

Fig.1; Transient Photocurrent at 1µJ

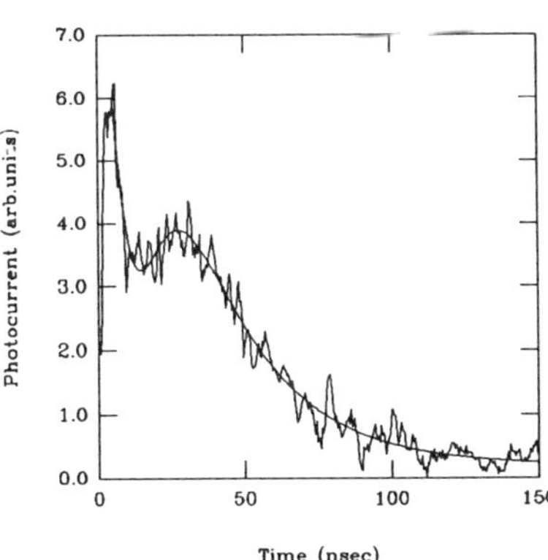

Fig .2; Transient Photocurrent at 3.5µJ

In discussing the nature of the intensity dependence of the photoconductive response described above, it should be noted that a remarkably similar behaviour was

428

observed in $La_2CuO_{4+\delta}$, where δ was near that required for the insulator-metal transition[4]. The process was discussed in terms of an optically induced insulator-metal transition resulting from the shifting of the Fermi level from below a mobility edge into the region of extended states by optical pumping.

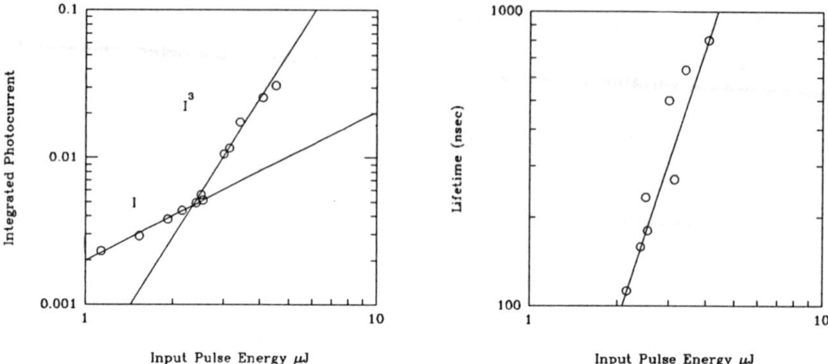

Fig. 3; Intensity dependence of integrated photocurrent   Fig. 4; Intensity dependence of long lifetime.

In the case of fullerenes, excitation at 532nm is intermediate between the transition to the LUMO, a state (S_1) which is molecularly localised[5] and the transition at 440nm to a state (S_2) which has been interpreted as extended or band-like[6]. The state addressed contributes to the photoconductive response until it decays rapidly to the LUMO. Although longlived, any contribution to the photocurrent from this state relies on hopping or tunnelling[7]. Thus, in terms of the photoconductive response, a two level excited state can be constructed. At low intensities, the photocurrent I_{pc} decays at the depopulation rate of S_2 (k_{21}), and thus,

$$I_{pc}(t) \propto S_2(t) = S_2(0)\exp(-k_{21}t) \tag{1}$$

Assuming the absorption to be linear, I_{pc} is linear in intensity. The nonlinear character of the photocurrent indicates the contribution of a state whose conductivity increases superlinearly with its population. Associating this nonlinearity with that of the photoluminescence[1], a nonlinear dependence of I_{pc} on the population of S_1 may be inferred. Thus I_{pc} takes the form

$$I_{pc}(t) = AS_2(t) + BS_1(t)^n \tag{2}$$

$$S_1(t) = (k_{21}/k_{10}-k_{21})S_2(0)(\exp(-k_{21}t)-\exp(-k_{10}t)) \tag{3}$$

and A and B are constants. The solid lines of figures 1 and 2 show fits of such a nonlinear temporal dependence of the photocurrent to the experimental data. Fits to the responses over the range of intensities were performed with all parameters free. Over the range, n varied between 3 and 5. k_{21} remained constant in the range 10-20nsec. Notably, k_{10}, the decay time of the highly excited state increased significantly with intensity. Figure 4 shows the dependence of the fit parameter on intensity. The solid line shows a slope of 2.5, a nonlinearity which is in close agreement with the nonlinearity of the photoluminescence emission lifetime from fullerenes at high excitation densities[1].

4. Conclusions

The model employed to describe the nonlinear intensity dependence of the photoconductive response is primitive and preliminary. k_{21}, as measured here, is limited by the detection and so is not in reality a material parameter. It is assumed to be intensity independent and no explicit nonlinearity of k_{10} is included in the model. Nevertheless, the analysis shows the behaviour to be in good agreement with that observed in the previously described experiments. The interpretation, based on a transition between molecularly localised "excitonic-like" states at low excitation densities, to delocalised states at high intensities is upheld by the model, indicating the importance of intermolecular exchange and correlation energies at high energy densities.

References

1. W.K. Maser, H.J. Byrne, M. Kaiser, W.W. Rühle, L. Akselrod, A.T. Werner, J. Anders, X.-Q. Zhou, G. Mahler, T. Kuhn, A. Mittelbach and S. Roth, *these proceedings and references therein.*
2. See for example *Physics of Highly Excited States in Solids*, eds. M. Ueta and Y. Nishina, Lecture Notes in Physics vol. 57, (Springer Verlag,Heidelberg 1976).
3. A.T. Werner, J. Anders, H.J. Byrne, W.K. Maser, M. Kaiser, A. Mittelbach and S. Roth, *Appl.Phys.* **A57** (1993) 81
4. G. Yu, C.H. Lee, A.J. Heeger, S.-W. Cheong and Z. Fisk, *Physica* **C190** (1992) 563
5. P.A. Lane, L.S. Swanson, Q.-X. Ni, J. Shinar, J.P. Engel, T.J. Barton and L. Jones, *Phys. Rev. Lett.* **68** (1992) 887.
6. R.W. Lof, M.A. van Veenendaal, B. Koopmans, H.T. Jonkman, G.A. Sawatzky, *Phys. Rev. Lett.* **68**, (1992) 3924
7. M. Kaiser, J. Reichenbach, H.J. Byrne, J. Anders, W. Maser, S. Roth, A. Zahab and P. Bernier, *Synth. Metals* **51** (1992) 251

REVERSIBLE, METASTABLE, ULTRAFAST PHOTOINDUCED ELECTRON TRANSFER IN CONJUGATED POLYMER AND FULLERENE COMPOSITES AND HETEROJUNCTIONS: A COMPARATIVE STUDY

N. Serdar SARICIFTCI and Alan J. HEEGER
Institute for Polymers & Organic Solids, University of California,
Santa Barbara, California 93106, USA

ABSTRACT

Experimental results on the metastable, reversible, ultrafast photoinduced electron transfer between conjugated polymers and buckminsterfullerene are reviewed. Comparative studies with different semiconducting polymers as donors demonstrate that in the degenerate ground state polymers soliton excitations and in polydiacetylenes the excitons form before the electron transfer can occur; thereby inhibiting charge separation. In non-degenerate ground state systems, photoinduced electron transfer occurs in less than 10^{-12} s, quenching the photoluminescence as well as the intersystem crossing. Utilizing thin films of the semiconducting polymer (donor) and buckminsterfullerene (acceptor) to form a heterojunction interface, we fabricated diode bilayers which functioned as photodiodes and as photovoltaic cells.

1. Introduction

The parallel interests of understanding natural photosynthesis in biological systems and of efficiently harvesting solar energy as an alternative to fossil fuels have led to a substantial, multidisciplinary effort in the field of photoinduced electron transfer phenomena in physics, chemistry, biology and in their regimes of overlap [1]. Light harvesting polymer systems utilizing photoinduced energy and/or electron transfer mechanisms are attracting more and more attention within the scientific community [2]. Polymers are particularly interesting in this field of photophysics/photochemistry because new synthetic methods (as developed through polymer chemistry and material science) make it possible to modify, functionalize and derivatize donor and acceptor units.

Recently, we reported the evidence for a photoinduced electron transfer and metastable charge separation in semiconducting polymer/C_{60} composites [3-10]. This forward electron transfer from the semiconducting polymer (as donor) onto C_{60} (as acceptor) occurs in less than 1 picosecond, thereby decreasing the luminescence by three orders of magnitude and eliminating intersystem crossing to triplet state. Using this inter-molecular photoinduced charge transfer from donor to acceptor at the interface between a donor-acceptor bilayer consisting of films of the semiconducting polymer, poly(2-methoxy,5-(2'-ethyl-hexoxy)-p-phenylene) vinylene, (hereafter referred to as MEH-PPV) and C_{60}, we were able to fabricate diodes with rectification ratios of approximately 10^4 which operated both as photodiodes and as photovoltaic cells [4].

We review the results on semiconducting polymer - C_{60} composites using a series of semiconducting polymers as the donor. In each case, we have investigated the semiconducting polymer alone and in composites with C_{60}.

2. Experimental

All soluble derivatives of poly (*para*-phenylene)vinylene (PPV) are purchased from UNIAX Corp., Santa Barbara and used without further purification. Synthesis and preparation of highly uniform poly (3-octylthiophenes) (P3OT) are described in early reports [5]. Soluble derivatives of polyacetylene (poly (1,6-heptadiyne), PHDK) are synthesized as described in [11]. C_{60} powder was purchased in high purity (99.99%) from Polygon Enterprises, Texas and later from MER Corp., Tucson, Arizona. Detailed experimental procedures of all the studies are summarized in [10].

3. Results and Discussion

The sub-picosecond photoinduced absorption spectra for P3OT as well as soluble PPV's in their pristine state as well as in composite with C_{60} are displayed in Fig. 1. The effect of C_{60} is clearly observable in the early time spectra in a sub-picosecond time scale. This direct observation clearly demonstrates that the photoinduced electron transfer is indeed ultrafast ($\tau_{ET} < 10^{-12}$ s) in these materials confirming the estimation from the luminescence quenching in our initial reports [3,5]. Furthermore, the femtosecond photoinduced absorption studies show an increase of the lifetimes of the observed charge transferred state with increasing C_{60} concentration in the polymeric host [6]. The effect of the photoinduced electron transfer on the millisecond time domain clearly demonstrates the quenching of the triplet-triplet absorption at 1.05 eV creating charged polarons in P3OT [5]. Since the absorption spectrum persists without essential change from 5 ps to milliseconds, we conclude that the charge separated state is created at very early times and lives for milliseconds at 80K.

Definitive evidence for a complete electron transfer and a metastable charge separation is obtained from light induced electron spin resonance (LESR) experiments. If the charge separation is long-lived enough (to accumulate the number of spins detectable for the spin resonance within one modulation cycle) there will be two LESR signals observable originating from the positive polaron (donor cation radical) as well as from the anion of the

432

C_{60} (acceptor anion radical). The results of LESR experiments are displayed in Fig. 2 for two different conjugated polymer/C_{60} composites, comparatively.

The ultrafast photoinduced electron transfer from the conjugated polymers onto C_{60} and the following stabilization of the charge separated state should enhance the near steady state photoconductivity of the composites compared to the pure host polymer photoconductivity. This effect is observed to be fairly strong for P3OT/C_{60} as well as MEH-PPV/C_{60} composites [7].Addition of only a few percent of C_{60} into the conjugated polymer sensitizes its steady state photocurrent one order of magnitude over the entire spectral range from the near infrared to the ultraviolet [7].

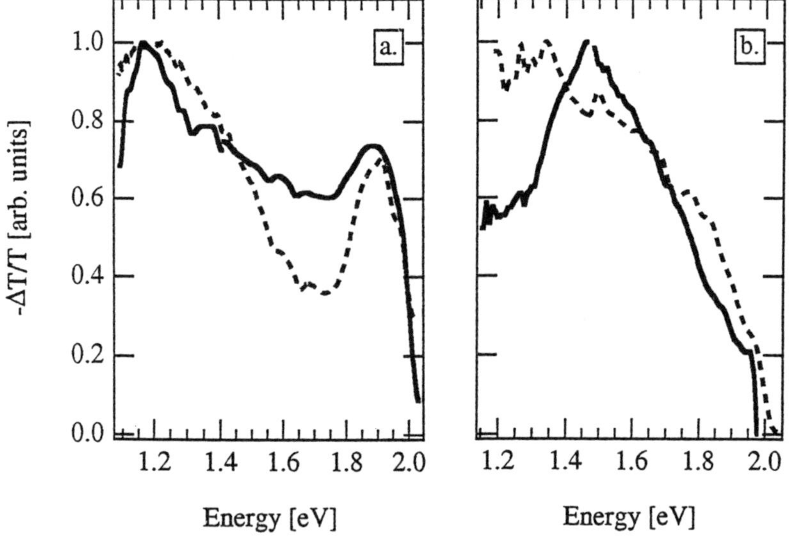

FIGURE 1   a.) Photoinduced changes in the absorption spectrum of P3OT in its pristine state (dashed line) and in composite with 50% C60 (solid line) for 700 fs time delay.

b.) Same for BCHA-PPV in its pristine state (dashed line) and in composite with 1% C_{60} (solid line) for 700 fs time delay.

Another implication of long-lived photoinduced charge separation is the enhancement of the photoinduced infrared vibrational absorption modes (IRAV). These modes originate from the activation of the totally symmetrical Raman modes in the infrared spectrum, due to the strong one dimensional electron phonon coupling in these systems which create local lattice distortions around injected (doping and/or photoexcited) charges.

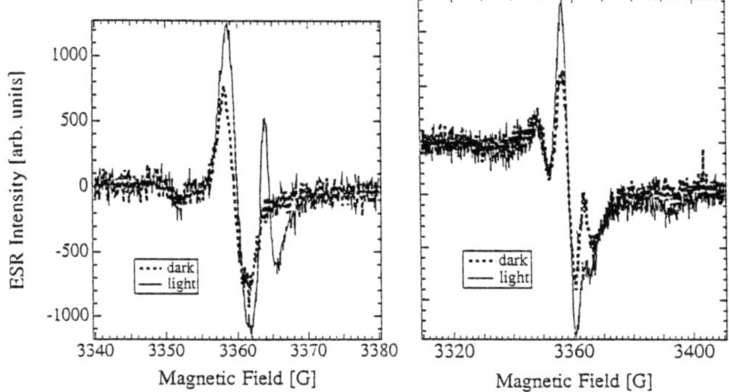

FIGURE 2   LESR spectra of P3OT/C_{60} (a.) and MEH-PPV/C_{60} (b.) composites excited with 514.5 nm Ar ion laser at 80K.

Since the photoinduced electron transfer is responsible for creating long-lived charge carriers (photodoping), this phenomenon is strongly enhanced in conjugated polymer/C_{60} composites [8].The conditions under which this photoinduced electron transfer is favored are not yet completely clear. The energetics of the donor-acceptor couple are certainly important, as described in Marcus theory [15]. The dielectric constant of the medium also plays an important role. In our comparative studies we observed that degenerate ground state conjugated polymers (PHDK) do not show a photoinduced electron transfer [5].

Recently, we looked at polydiacetylenes (PDA) which also do not show any photoinduced electron transfer onto C_{60}. Studies of picosecond transient photoconductivity, femtosecond photoinduced absorption as well as near steady state photoinduced absorption experiments clearly show that there is no **forward** electron transfer in PHDK/C_{60} and PDA/C_{60} composites, ruling out the possibility of a rapid forward transfer accompanied with a rapid back relaxation. Since the energetics of the PHDK as well as PDA are similar to P3OT (ionization potential, 5.5eV), the striking difference between PPV's (P3OT's) with an ultrafast photoinduced electron transfer, on one hand, and PHDK (PDA's) with complete inhibition of this process, on the other, must have its origin in the photophysics of the conjugated polymer donor itself. The solitonic stabilization deep in the gap for the degenerate ground state polymer is proposed to inhibit the electron transfer process [5]. This limits the forward electron transfer rate to the estimated soliton formation time ($\tau_{ET} > 10^{-13}$ s)[5]. For the case of polydiacetylenes we propose that strongly bound (0.5 eV) exciton formation in PDA's inhibits the electron transfer, which requires complete charge separation over the Coulomb barrier. This implies, however, that the Coulomb correlations in PPV's, P3OT's *etc.* are well screened, creating free electron-hole pairs upon photoinduction and no barrier for the rapid electron transfer onto C_{60}[16].

ACKNOWLEDGEMENTS

This work is supported by the Department of Energy (DOE # DEFG0393ER12138) and the Office of Naval Research (ONR # N00014-91-J-1235).

REFERENCES

1. See, Photoinduced Electron Transfer, edited by M. A. Fox and M. Chanon (Elsevier, Amsterdam, 1988)
2. M. A. Fox, J. Wayne, E. Jones and D. M. Watkins, Chemical & Engineering News, March 15, (1993)
3. N. S. Sariciftci, L. Smilowitz, A. J. Heeger, and F. Wudl, Science, 258, 1474 (1992).
4. N. S. Sariciftci, D. Braun, C. Zhang, V. I. Srdanov, A. J. Heeger, G. Stucky and F. Wudl, Appl. Phys. Lett., 62, 585 (1993)
5. L. Smilowitz, N. S. Sariciftci, R. Wu, C. Gettinger, A. J. Heeger, and F. Wudl, Phys. Rev. B, 47, 13835 (1993).
6. B. Kraabel, C. H. Lee, D. McBranch, D. Moses, N. S. Sariciftci, and A. J. Heeger, Chem. Phys. Lett., 213, 389 (1993)
7. C. H. Lee, G. Yu, D. Moses, K. Pakbaz, C. Zhang, N. S. Sariciftci, A. J. Heeger, and F. Wudl, Phys. Rev. B, 48, 15425 (1993)
8. K. H. Lee, R. A. J. Janssen, N. S. Sariciftci, and A. J. Heeger, Phys. Rev. B 49, No.8, (1994), in press
9. N. S. Sariciftci, L. Smilowitz, A. J. Heeger and F. Wudl, Synth. Metals, 59, 333 (1993).
10. N. S. Sariciftci and A. J. Heeger (Review), Int. J. of Mod. Phys. B, 8, No:3, 237, (1994).
11. K. Pakbaz, R. Wu, F. Wudl and A.J. Heeger, J. Chem. Phys., 99, 590 (1993)
12. P. M. Allemand, G. Srdanov, A. Koch, K. Khemani, F. Wudl, Y. Rubin, F. Diederich, M. M. Alvarez, S. J. Anz and R. L. Whetten, J. Am. Chem. Soc., 113, 2780 (1991).
13. K. Yoshino, Xiao Hong Yin, S. Morita, T. Kawai and A. A. Zakhidov, Sol. State Commun., 85, 85 (1993).
14. S. Morita, S. Kiyomatsu, Xiao Hong Yin, A. A. Zakhidov, T. Noguchi, T Ohnishi and K. Yoshino, J. Appl. Phys., 74, 2860 (1993).
15. R. A. Marcus, Rev. Mod. Phys., 65, 599 (1993)
16. N. S. Sariciftci, B. Kraabel, C. H. Lee, K. Pakbaz, A. J. Heeger and D. J. Sandman, Phys. Rev. B, submitted.

AN INVESTIGATION OF TIME RESOLVED
ELECTROLUMINESCENCE IN FULLERENE CRYSTALS

A. T. WERNER, H. J. BYRNE, D. O'BRIEN*, and S. ROTH.
*Max-Planck-Institut für Festkörperforschung, Heisenbergstrasse 1,
70569-Stuttgart, Germany.*

and

School of Science of Materials, Trinity College, Dublin 2, Ireland

ABSTRACT

Broadband electroluminescence emission from fullerene crystals is reported and described. The spectral distribution is comparable to that of the photoluminescence at high excitation densities. The emission intensity is nonlinearly dependent on the current. The response of the crystal to the application of an alternating current is investigated and the frequency dependence of the emission intensity is described. The observed high frequency cut-off behaviour cannot be mimicked by a simple equivalent circuit and so the system is modelled using a rate equation model to describe the state of the system. Fits of the model to the observed behaviour provide rate constants which compare favourably to those reported for excited state decay in fullerenes.

1. Introduction

At high excitation densities, a highly nonlinear luminescence process is affected in fullerenes[1] and transport studies support the association of this phenomenon with a transition from a localised molecular-like to a banded or extended state behaviour, i.e. an optically driven Mott-like transition[2]. The resultant nonlinear emission process is characterised by a cubic dependence of the output on the input intensity accompanied by a strong red shift of the emission spectrum. The unique nature of this process prompts an investigation of its applications potential, in particular in the form of an electroluminescent device. In this paper, the characteristics of the previously reported[3] electroluminescent output of fullerene crystals are briefly described. Furthermore the dynamic response of the emission is probed through the application of an ac voltage in an attempt to shed further light on the dynamics of the emission process.

2. Experimental

C_{60} crystals were generated by vacuum sublimation of fullerene powder as described previously[3]. The sample employed was of dimensions ~200x200x200μm^3 and was mounted using silver paste on a substrate with two gold strip lines. The crystal is therefore symmetrically contacted. All processing was performed under inert atmosphere, and the sample was mounted in a cryostat cooled to liquid nitrogen temperatures.

The sample was connected to a Keithley voltage source which provided a dc voltage. The light output was monitored using a photomultiplier. For the frequency dependent measurements, a Kontron function generator was connected in series and the transient response was monitored using a Hewlett-Packard 54111D digitising oscilloscope.

3. Results and Discussion

Employing the dc source only, the current, voltage and light output were simultaneously monitored. As is shown in figure 1, the current/voltage characteristics are approximately linear until a voltage of ~10V. Below this region, no light emission is detectable. Increasing the voltage above this point, however, results in the abrupt turn-on of light emission. The light emission increases highly nonlinearly with the current[3] above the threshold and the current-voltage characteristics become somewhat nonlinear. The emission has a broad distribution, as shown in figure 2, which is comparable to the photoluminescence observed from fullerene crystals at high excitation densities[1].

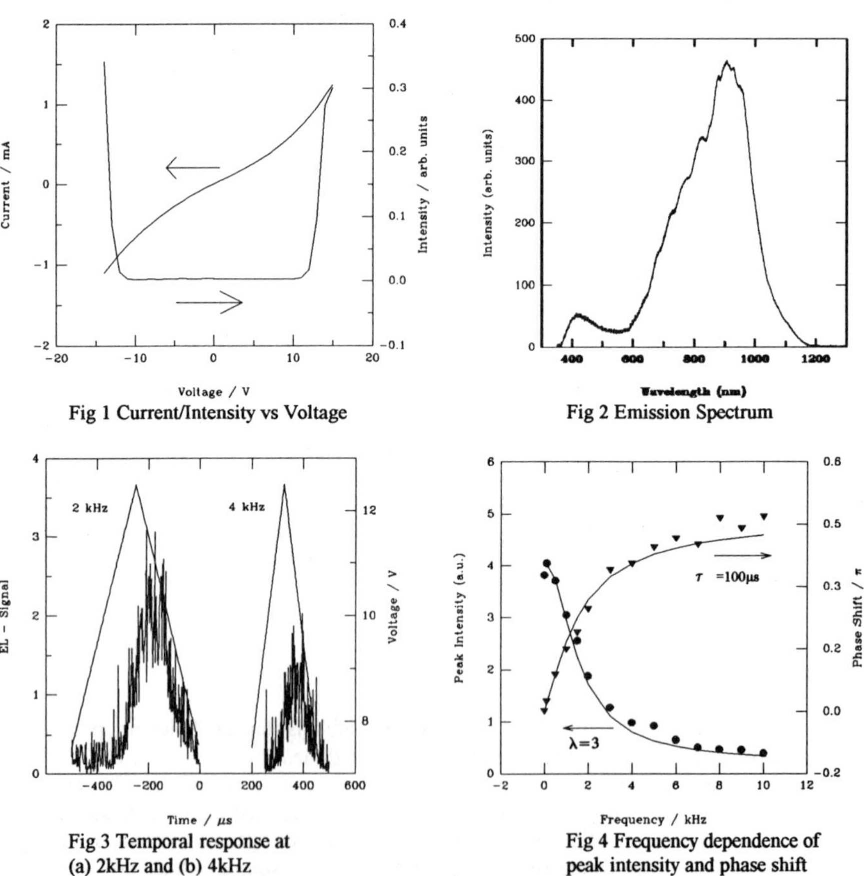

Fig 1 Current/Intensity vs Voltage

Fig 2 Emission Spectrum

Fig 3 Temporal response at
(a) 2kHz and (b) 4kHz

Fig 4 Frequency dependence of
peak intensity and phase shift

The origin of the nonlinearity of the optical and electronic properties of fullerenes at high excitation densities have been discussed in terms of an optically induced insulator to metal transition[4] dependent on a critical excited state density. The relationship of the electroluminescence emission to this behaviour has been discussed previously[3]. Of particular interest are the kinetics of the formation of this critical density in the case of electron injection and the dynamic response of the contacted crystal were probed.

The crystal was maintained just below the emission threshold by the application of 10V dc. The system was then driven periodically above threshold by the application of a triangular waveform of peak to peak voltage 5V. The net applied voltage was therefore driven between 7.5 and 12.5V. The driving frequency was varied in the range of 100Hz to 10kHz. At low frequencies, the time evolution of the light emission follows that of the driving waveform closely. However, with increasing frequencies, the rise of the emission ceases to follow that of the voltage, resulting in a phase shift as well as a reduction in the maximum of the light intensity. Figure 3 shows the temporal evolution of the luminescence compared to the driving voltage waveform at (a) 2kHz and (b) 4kHz. At frequencies >10kHz, negligible light emission is detectable. Figure 4 shows the frequency dependence of both the phase shift and the light emission. The phase shift, which is positive, asymptotically approaches $\pi/2$.

In considering an equivalent circuit for the system, a series RC combination is discarded on the basis of the requirement of a dc current. A parallel combination, can provide a similar frequency response with the exception that a frequency dependent phase shift is negative. It appears, therefore, that the origin of the bandwidth limitations have origins in the nature of the emitting state rather than the external circuitry.

The dynamics of light emission as a result of the application of a variable electrical current has been extensively explored for the case of laser diodes. The light output depends mainly on the density of excited states and the output intensity responds to an increase in current with a time constant determined by the recombination time t. So, to analyse the effect of a change in the current it is necessary to consider a rate equation for the excited state density n. This is given by

$$\frac{dn}{dt} = J - \frac{n}{\tau}$$

(1)

where J is defined as a particle current per unit volume and determines the injected carrier concentration. In a rough approximation, J is assumed to be a sum of the dc current i_{dc} and a sinusoidal contribution rather than the triangular waveform

$$J = \frac{1}{eV}(i_{dc} + i_0 \sin \omega\tau)$$

(2)

where V is the active volume of the crystal. The temporal dependence of n is given by (3) where the the phase shift φ fulfils the relationship in (4).

$$n = \frac{i_{dc}\tau}{eV} + n_0\sin(\omega\tau-\varphi)$$

(3)

$$\varphi = \arctan(\omega\tau)$$

(4)

which is in excellent agreement with the experimentally observed dependence as shown by the solid line in figure 4. The fit yields a value for the recombination lifetime of 100µs. This value is not inconsistent with the reported values of excited state lifetimes in fullerenes[5-7].

438

The peak population density is given by the equation,

$$n_o = \frac{i_o/eV}{\sqrt{\omega^2 + 1/\tau^2}}$$

(5)

The frequency dependence of the light output I, is limited by that of the population density, n. The nonlinear dependence of the photoluminescence on input intensity as well as the electroluminescence on the injection current indicate that the relationship between I and n is not linear and is best fitted with a relationship of the form,

$$I = K(n-n_{th})^\lambda$$

(6)

The data can be fitted well with a value of λ between 2 and 4. The solid line in figure 4 is of a power law of order 3.

On the basis of the fits, application of the model to the dynamic response of the electroluminescence from fullerenes appears justified. In particular, the time constant involved is within the range of those reported for fullerenes. The model is however, primitive. It assumes the conductivity and time constant of the crystal to be largely independent of excitation density, an assumption which seems inappropriate in the light of the nonlinear photoconductivity[2,7]. Further elucidation of the physics of the process may be afforded by short pulse injection, an examination of which is currently underway.

4. Conclusions

The electroluminescent emission from fullerene crystals is broadband with a strongly nonlinear dependence of light output on current. These characteristics strongly associate it with the nonlinear luminescence and photoconductive response of fullerenes above a critical excitation density. Examination of the dynamic response indicates that it is determined and limited by the material response time which is of order $100\mu s$.

5. References

1. H.J. Byrne, W.K. Maser, W.W. Rühle, A. Mittelbach and S. Roth, *Appl. Phys.* **A56**, (1993) 235
2. H.J. Byrne, W.K. Maser, M. Kaiser, L. Akselrod, J. Anders, W.W. Rühle, X.-Q. Zhou, A. Mittelbach and S. Roth, *Appl. Phys.* **A57**, (1993) 81
3. A.T. Werner, J. Anders, H.J. Byrne, W.K. Maser, M. Kaiser, A. Mittelbach and S. Roth, *Appl. Phys.* **A57** (1993) 157
4. H.J. Byrne, W.K. Maser, M. Kaiser, W.W. Rühle, L. Akselrod, A.T. Werner, J. Anders, X.-Q. Zhou, G. Mahler, T. Kuhn, A. Mittelbach and S. Roth, *Appl. Phys.* **A57**, (1993) 303
5. S.P. Sibley, S.M. Argentine and A.H. Francis, *Chem. Phys. Lett.*, **188**, (1992) 187
6. T.W. Ebbesen, *private commun.*
7. H.J. Byrne, A.T. Werner, D. O'Brien, W.K. Maser, M. Kaiser, L. Akselrod, W.W. Rühle and S. Roth, *these proceedings*

The nonlinear optical properties of C_{60}: z-scans and switching measurements

D. Wernham and F. Placido
Department of Physics, University of Paisley, High St., Paisley PA1 2BE, Renfrewshire, Scotland.

B. Stewart
Department of Chemistry and Chemical Engineering, University of Paisley, High St., Paisley PA1 2BE, Renfrewshire, Scotland.

A. Christie and S. McGeoch
Pilkington Optronics - Barr & Stroud Ltd., Linthouse Rd, Glasgow, Scotland.

The nonlinear absorption and the nonlinear refraction of solutions of purified C_{60} in toluene have been investigated using z-scan and switching measurements at 532nm (frequency-doubled Nd:YAG laser). From these experiments we show that C_{60} has a negative nonlinear refractive index at this wavelength and calculate the imaginary part of the third-order susceptibility.

1. Introduction

The nonlinear optical properties of C_{60} have been studied by a large number of groups [1-4], showing C_{60} to be a reverse saturable absorber at a number of wavelengths. The extent of the optical limiting behaviour, as quantified by the imaginary part of the third order susceptibility has been calculated using a variety of methods. These include degenerate four-wave mixing, third harmonic generation and switching measurements and give values of $\text{Im}\chi^{(3)}$ between 10^{-16} and 10^{-20} m^2V^{-2}, depending on the method used and whether or not the C_{60} is in solution or the solid state. There are a number of techniques that have been utilised to investigate nonlinear refraction including elipse rotation, beam distortion measurements, nonlinear interferometry and wave mixing techniques. The z-scan [5] technique offers comparable sensitivity but has the added advantage of a fairly simple experimental setup.

2. Theory

The z-scan technique allows us to investigate both the nonlinear refraction and the nonlinear absorption effects. The experimental setup is shown in figure 1. The transmittance of the medium is dependent upon the position of the sample with respect to the focus of the laser beam, measured via z-position. In the case of nonlinear absorption, we would expect the maximum effect to occur at the point where the largest absorbing volume is encompassed by the tightly focused beam, ie. at the centre of the sample, at the point z=0. Also, the z-scan should be symmetrical with respect to this point. If an aperture is included, the transmittance is dependent upon the vergence of the beam.

A nonlinear refractive material will exhibit self-lensing when subject to a large applied field. When the sample is far from the focus, the irradiance is low and the transmittance should be that of the linear case. Taking for example, a material which has a positive nonlinear refractive index; as the sample is brought into the proximity of the focus from the z domain, self focusing prior to the focus

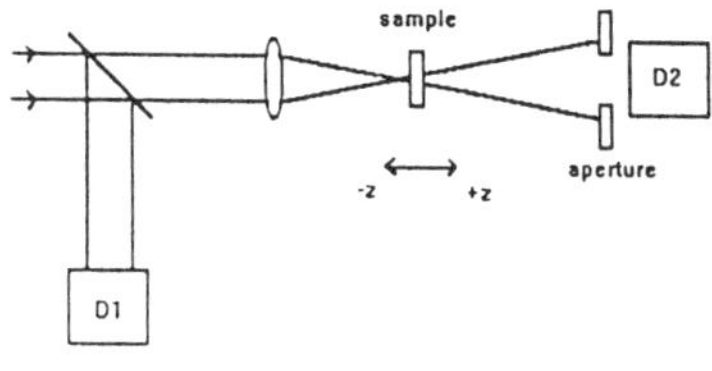

Figure 1. z-scan experimental setup

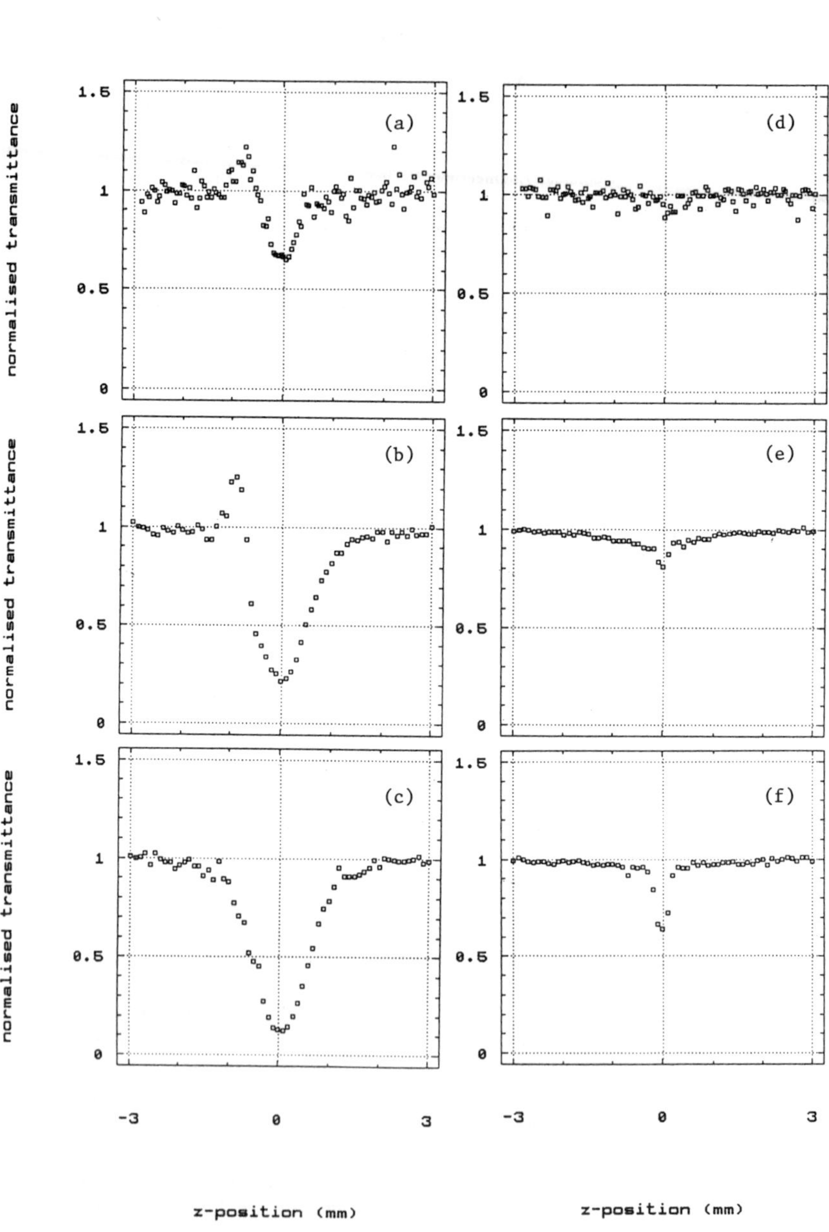

Figure 2. conventional and open aperture z-scans of C_{60} in toluene at 6, 11 and 33µJ

will result in beam divergence in the far field and a decrease in the measured transmittance. At the point z=0, when the focus is in the centre of the sample, there is a minimal change in the far field pattern of the beam and the measured transmittance should be that of the linear case. As the sample is moved into the +z domain, self focusing subsequent to the focus position will collimate the beam resulting in an increase in the measured transmittance. A material which exhibits negative nonlinear refraction (self-defocusing) will show the opposite configuration. ie. Increased transmittance in the -z domain and decreased transmittance in the +z domain. When an aperture is included, the z-scan is sensitive to both nonlinear refraction and nonlinear absorption effects. When the aperture is removed and all the transmitted light is collected, then the z-scan is sensitive only to nonlinear absorption.

3. Experimental

Fullerene-rich soot was produced by arc-ablation of graphite in an inert atmosphere of helium (10 torr), using a full-wave rectified a.c. source which delivered an average current of 80A at 15V. The crude soot was sonicated in toluene and the fullerenes C_{60} and C_{70} were separated from the extract by flash chromatography using activated carbon as the stationary phase and toluene as the eluent. The fullerenes were crystallised out of cyclohexane solutions and subsequently characterised by HPLC and UV/vis spectroscopy yielding C_{60} >99% purity. Solutions of C_{60} in toluene were made up and these were filtered to <0.2μm in order to minimise scattering effects. The laser used to perform all of the measurements was a Q-switched Nd:YAG, frequency-doubled to 532nm and with a pulse width of 15ns (FWHM). The light energy incident upon the sample was varied by means of calibrated, neutral density filters which gave a working range of 6-33μJ.

Figures 2(a), (b) and (c) show the conventional (incorporating aperture) z-scan curves for a 1.2 x 10^{-4} M solution of C_{60} in toluene at incident energies of 6, 11 and 33μJ respectively. The aperture was set up beforehand to allow a 40% transmittance of the incident beam energy. Figures 2(d), (e) and (f) show the equivalent scans with the aperture removed. The peak - valley configuration of 2(a) is indicative of the negative (defocusing) nature of the nonlinear refractive index. To ensure that the nonlinearity was due to the C_{60}, a z-scan of the neat toluene was performed and showed that no significant nonlinear refraction or absorption was due to the solvent at this energy. The degree of asymmetry in the conventional z-scans is seen to increase with increasing energy and at 33μJ, the peak has completely disappeared. This can be partially accounted for by the increase in the nonlinear absorption with increasing energy, evident in the open aperture z-scans. Also, the toluene starts to exhibit nonlinear (also defocusing) behaviour at the highest energy level. However, even when the effects due to the solvent and the absorption have been extracted from the conventional z-scan data we find that, somewhat reduced, asymmetry between the peak and valley portions persists. This indicates that there is attenuation of the beam by some other mechanism, possibly nonlinear scatter.

Once an open aperture z-scan has been performed and the maximum of the nonlinear absorption has been found, the effect can be further investigated by measuring the transmittance as a function of incident energy. Such a graph is shown in figure 3. This clearly shows a decrease in transmittance with increasing energy. This is indicative of efficient transfer of ground state population to an excited state trapping level and subsequent excited state absorption thereof. In the case of C_{60}, the laser pulse excites population from the ground state S_0 to the Frank-Condon singlet. There is a rapid relaxation to the lowest energy of the first excited singlet manifold.

442

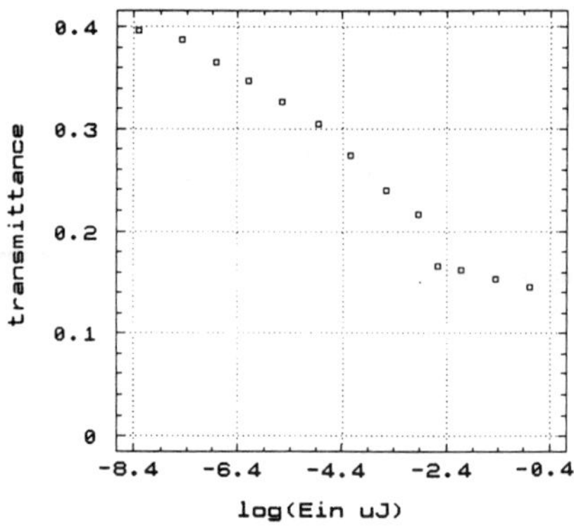

Figure 3. C_{60} switching

From here, there are three competing processes, namely, excited state absorption from the singlet, singlet relaxation to the ground state and formation of triplet states via inter-system crossing. The latter has been found to be the dominant effect with an ISC quantum efficiency in excess of 80% [6]. Once the triplet states are formed relaxation back to the ground state singlet is spin forbidden and hence slow. This means that the triplet acts as a trapping level.

Furthermore, triplet-triplet absorption can occur. The ground state absorption is governed by the absorption cross-section per molecule σ_{gr}, and the excited state absorption is governed σ_{ex}. The criteria for reverse saturable absorption is that $\sigma_{ex} > \sigma_{gr}$. We can quantify the ratio of the ground and excited state absorption cross sections from the transmittance graph.. Assuming a simple model, where there is complete conversion from the ground state singlet to the lowest lying triplet we find that $\sigma_{ex}:\sigma_{gr} \sim 2.6$. The nonlinear susceptibility $Im\chi^{(3)}$ is related to the change in the absorption $\Delta\alpha$ via the equation,

$$Im\chi^{(3)} = \frac{\varepsilon_0 cn\lambda\Delta\alpha}{4\pi I}$$

where n is the linear refractive index, c is the velocity of light in free space, λ is the wavelength of the laser pulse and I is the intensity. The change in the absorption coefficient $\Delta\alpha$ is dependent upon the relative populations of the ground and excited states as well as the absorption cross-section per molecule in these states. Where $I=I_{sat}$, the saturation intensity, the change in absorption is given by $\frac{1}{2}N(\sigma_{ex}-\sigma_{gr})$ where N is the total population density of the molecules. This then yields a value for the imaginary part of the third order susceptibility of $Im\chi^{(3)} = 3.6 \times 10^{-17} m^2V^{-2}$.

In conclusion, we have shown that C_{60} exhibits a negative nonlinear refractive index and nonlinear absorption by performing z-scans and switching measurements. From the latter we have estimated the imaginary part of the third order susceptibility $Im\chi^{(3)}$ using a simple model . We are also considering nonlinear scatter as a further attenuation process present when toluene solutions of C_{60} are subject to pulsed, high energy irradiation.

References

1. L.W. Tutt and A. Kost, Nature **356**, 225 - 226 (1992)

2. F. Henari, J Callaghan, H. Stiel, W. Blau, and D.J. Cardin, Chem. Phys. Lett. **199**, no. 1,2, pp144 - 148 (1992).

3. F. Kajzar, C. Taliani, R. Danieli, S. Rossini, and R. Zamboni, Chem. Phys. Lett. **217**, no. 4, pp418 - 422 (1994)

4. Q. Gong, Y. Sun, Z, Xia, Y.H. Zou, Z. Gu, X. Zhou, and D. Qiang, J. Appl. Phys. **71**, no. 6, pp 3025 - 3026 (1992) .

5. M. Sheik-Bahae, A.A. Said, T.H. Wei, D.J. Hagan and E.W. van Stryland, IEEE J. Quant. Elect. **26**, no. 4, pp760 - 769 (1990)

6. R.V. Bensasson, T. Hill, C. Lambert, E.J. Land, S. Leach, and T.G. Truscott, Chem. Phys. Lett. **201**, no. 1,2,3,4 pp326 - 335 (1993)

COMPLEX DIELECTRIC CONSTANT SPECTRA MEASUREMENTS BY MEANS OF SURFACE WAVES INTERFEROMETRY. PRINCIPLES, REALIZATIONS, EXAMPLES FOR HTSC-MATERIALS.

G. N. ZHIZHIN and V. A. YAKOVLEV

*Institute for Spectroscopy of Russian Academy of Sciences,
142092, Troitzk, Moscow region, Russia.*

ABSTRACT.

The surface electromagnetic waves (SEW) applications for optical constants in broad IR spectral range determination is reviewed. The SEW configuration with the maximum field on the surfaces of metals (surface plasmons) and dielectrics (surface polaritons) provides one order of magnitude gain in sensitivity for imaginary part determination. The real part can be interferometrically measured in the sample inclusive double beam interferometer with SEW excited by IR laser, free electron laser or IR synchrotron beam. The CO_2 laser radiation was used for 1-2-3 ceramic, single crystal and film on strontium titanate complex dielectric function measurements as well as fullerene film on quartz.

The investigation of the surface states of solids is very fast developing field of condensed matter physics. The understanding of the surface and thin film properties including transition layers, oxide films, physi- and chemisorbed monomolecular films, their transformations at the phase transitions and their chemical reaction are of great importance for studying many practical problems. The electron energy loss spectroscopy (EELS), the neutron, atom and molecular beams scattering, reflection-absorption IR spectroscopy, surface enhanced Raman scattering (SERS) methods are used for studying the vibrational spectra of the surfaces. The last two are nondestructive techniques. Last time the surface polaritons (plasmons) (SP) or surface electromagnetic waves (SEW) spectroscopy was developed [1-5].

The amplitude of the field of surface electromagnetic waves (SEW) is maximal at the interface between two media and falls off away from this boundary. The distribution of SEWs along the interface is described in terms of the effective refractive index of SEW in x direction [1]:

$$\chi_x = n'_e + in''_e = k_x c / w, \tag{1}$$

where k_x is the wave vector of SEW, ω is the angular frequency. At the interface between a vacuum and a medium with dielectric constant ε the effective refractive index is given by [1]:

$$\chi_x = n'_e + i * n''_e = \sqrt{\frac{\varepsilon}{\varepsilon + 1}} \tag{2}$$

The imaginary part n''_e represents the spatial attenuation of a SEW and can be determined by measuring how the intensity varies with the distance travelled by this wave (energy loss spectroscopy). The real part n'_e (phase spectroscopy of SEW) can be determined by the interference measurements [6,7]. Figure 1 outlines the idea of such an experiment for a surface wave launching in the gap between the specimen and the screen. The other part of the wave remains partially transformed into bulk radiation. The remaining portion of the surface wave also becomes bulk radiation at the far edge of the specimen. These two bulk radiation beams interfere with one another, and if there is an interference maximum or minimum at a point z [7]:

$$a Re\chi_x + \sqrt{b^2 + z^2} - \sqrt{(a+b)^2 + z^2} = (m + \Delta)/\nu \tag{3}$$

where a is the distance from screen to specimen edge, b is the distance from specimen edge to observation plane, m is an integer for maxima, or half-integer for minima, $\nu = \omega/2\pi c$ is the wavenumber. The quantity Δ has been introduced to allow for a possible phase shift between a bulk wave and the SEW.

Figure 2 shows the intensity plots as functions of z obtained in moving pyroelectric detector at a constant speed. The position of these interference may by used to determine the real part of the effective refractive index either graphically [7,8] or by least squares (by Eq. 3) on a computer. If the absorption coefficient of the surface wave is also known (from measurements by energy loss SEW spectroscopy on the instrument described in [9], then the dielectric constant of the specimen can be determined. The equation of dispersion of SEWs (Eq. 2) at a metallic surface becomes more complicated when this surface is covered by thin films for χ_x depends on the properties and thickness of the film. Solving the inverse problem one may determine the parameters of the film by the measured values of the real and imaginary parts of the SEW effective refractive index. Fig.1 shows the scheme of the experimental set-up. The laser beam was focused on the aperture gap between the sample surface and a razor blade. The gap width was about 10 μm. The interference field distribution was registered by the detector which was moving along the circle arc with the center on the sample edge. To increase the accuracy of determination of the refractive index and to obtain the SP attenuation constant, the interference distributions with various distances a were detected.

The specimens of high-temperature superconductors $YBa_2Cu_3O_{7-x}$ were investigated in the form of ceramic [10], single crystals [11] and films deposited on strontium titanate [12]. For single crystals in the 10 μm range, we obtained the complex constant ε=-45+i90. For 1000 Åcrystalline film with the c axis perpendicular to the plane of the substrate ([100] plane of strontium titanate single crystal) dielectric constant ε=-37+i61 in the temperature interval 10–300K (the superconducting transition temperature was 91K with the width 0.5K).

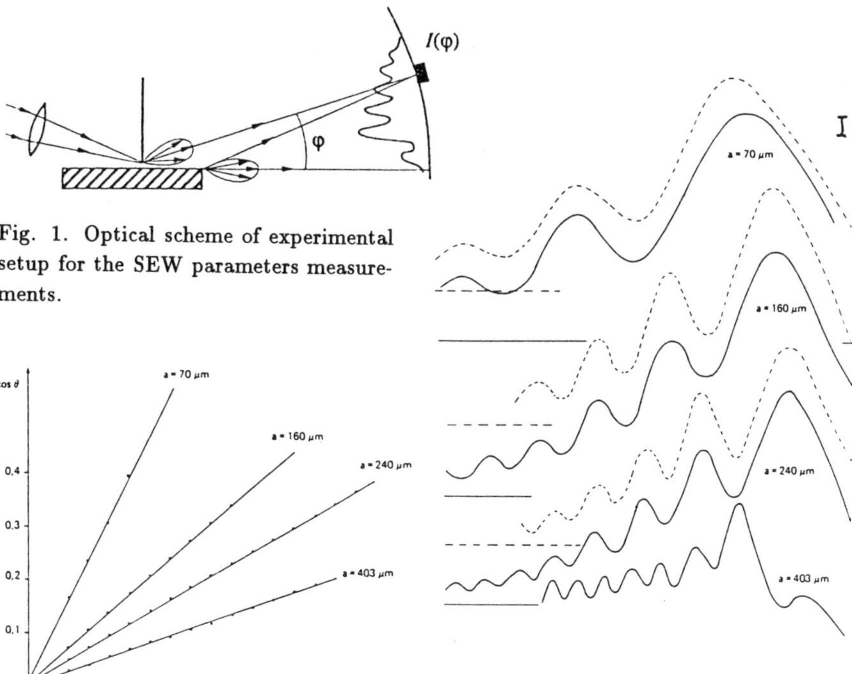

Fig. 1. Optical scheme of experimental setup for the SEW parameters measurements.

Fig. 3. Dependence of the value 1-cos θ_m on the interference order $m/2$ for the experimental interferograms shown in fig. 2.

Fig. 2. Experimental (solid) and calculated (dashed) interferograms at different distances a.

The thin (0.141 μm) fullerene film on quartz was studied by the same techniques. The real part of SEW's wave vector for quartz with C_{60} film was determined at wavenumber 1080cm^{-1}: $n'_e - 1 = (4.35 \pm 0.05) \times 10^{-2}$. From these data the dielectric constant of C_{60} film was determined: ε=3.5+i0.2, in a good agreement with that one obtained from the reflection studies [13].

The above mentioned results were obtained by means of CO_2 laser tunable in the range 930-1080 cm^{-1}. It was also used a Fourier spectrometer to perform interference measurements for silver films in a wide spectral range (700-2500 cm^{-1}) [14,15]. In this case the interference pattern in the spectrum can be recorded. The problem is to draw the SEW parameters from these spectra. Previously the absorption SEW spectra of the thermally oxidized copper surface [16], copper and aluminum surfaces with natural oxide films [1] were studied. The copper oxide film has not a selective SEW absorption in the spectral range 920-1090 cm^{-1}. So the presence of the oxide leads to the shift of the real and imaginary parts of the effective refractive index proportional to the square of wavenumber if Drude model for dielectric function of metal is valid. The spectrum of the aluminum oxide film possesses the selective absorption band. In this case anomalous dispersion in the effective refractive index was observed [8].

The study of the surface polaritons (SP) propagation on the dielectrics met with difficulties because of the large values of the SP attenuation constant. The SP attenuation on dielectrics is 10^2–10^3 times larger than the SEW attenuation on metals. However, the solving of the problem of SP propagation on dielectrics would allow us to use the SEW spectroscopy technique developed for metals for the study of dielectric surfaces. Moreover, it would be helpful for the investigation of the SEW and SP propagation in visible spectral range.

The first attempt to solve this problem was made using gratings on the sample [17]. The SP propagation on the α-quartz crystal surface was obtained. The wedge technique for the measurement of the SP large attenuation was used. The SP effective refractive index dispersion on the quartz crystal was studied for different orientations of the optical axis. The SP refractive index dispersion for three directions of SP wavevector relative to quartz optical axis was obtained. The effective refractive index was calculated from the angle maxima locations.

These curves show the SP propagation anisotropy which corresponds to the dielectric function anisotropy of α-quartz crystal [17]. In case of attenuation constants measurement [18] the SP propagation parallel and perpendicular direction of SP wavevector relative to optical axis was obtained.

Acknowledgements

The authors acknowledge the support of the State program "High Temperature superconductivity" (Grant No. 93209).

References

1. Zhizhin G.N., Moskalova M.A., Shomina E.V., Yakovlev V.A. in Surface Polaritons (eds. V.M.Agranovich and D.L.Mills), North-Holland Publishing Co., 1982, p.93.

2. V.M.Agranovich, Usp. Fiz. Nauk. **115** (1975) 199; (Sov.Phys. Usp.**18** (1975) 99).

3. M.A.Chesters, S.F.Parker and V.A.Yakovlev, Optics Comm. **55** (1985) 17.

4. G.N.Zhizhin, M.A.Moskalova, A.A.Sigarev and V.A.Yakovlev, Optika i Spec-

troscopiya (Sov) **52** (1982) 395.

5. Y.L.Chabal, A.J.Sievers, Phys.Rev.Lett. **44** (1980) 944.

6. V.A.Yakovlev, V.A.Sychugov and A.A.Khakimov. Kvant. Elektron. **10** (1983) 611.

7. G.N.Zhizhin, S.A.Kiselev, M.A.Moskalova, V.I.Silin and Y.A.Yakovlev. Zh. Techn. Phys. **29** (1984) 582.

8. V.I.Silin, S.A.Voronov, V.A.Yakovlev, G.N.Zhizhin, International Journal of Infrared and Millimeter Waves, **10** (1989) 101.

9. G.N.Zhizhin, A.I.Kuryatnikov, G.Yu.Lopatovsky, M.A.Moskalova, V.I.Silin, V.V.Tkachev, V.A.Yakovlev, Computer Enhanced Spectroscopy **2** (1984) 107.

10. G.N.Zhizhin, K.V.Kraiskaya, L.A.Kuzik, F.A.Uvarov and V.A.Yakovlev. Fiz. Tverd. Tela **30** (1988) 929.

11. A.F.Goncharov, G.N.Zhizhin, S.A.Kiselev, L.A.Kuzik and V.A.Yakovlev. Phys. Lett. B. **133** (1988) 163.

12. E.V.Alieva, G.N.Zhizhin, L.A.Kuzik, E.V.Pechen', E.I.Firsov, V.A.Yakovlev, Sov.Phys.JETP **70** (1990) 315.

13. Alieva E.V., Mattei G., Petrov Y.Y., Rossi G., Sukhorukov A.V., Yakovlev V.A., Zhizhin G.N., Proceedings of 9th International Conference on Fourier Transform Spectroscopy (Aug.1993, Calgary, Canada, eds. J.E.Bertie, H.Vieser), SPIE Proceedings series **2089** (1993) 402.

14. M.A.Chesters, S.A.Parker, V.A.Yakovlev. Opt. Commun. **55** (1985) 17.

15. Zhizhin G.N., Yakovlev V.A., Physics Reports **194** (1990) 281.

16. D.A.Bryan, D.L.Begley, K.Bhasin, R.W.Alexander, R.J.Bell, R.Gerson, Surf.Sci. **57** (1976) 53.

17. F.Gervais, P.Piriou, J.Phys.C: Sol.St.Phys. **7** (1974) 2374.

18. G.N.Zhizhin, V.I.Silin, V.A.Sychugov, V.A.Yakovlev, Solid State Commun. **51** (1984) 613.

Part Seven

MAGNETIC RESONANCE

PRESSURE EFFECTS ON THERMODYNAMIC AND ELECTRONIC PROPERTIES OF C_{60} AND K_3C_{60} STUDIED BY ^{13}C-NMR.

P. AUBAN-SENZIER, R. KERKOUD, J. GODARD, D. JEROME,
Laboratoire de Physique des Solides (associé au CNRS), Université Paris-Sud,
91405 Orsay, France

J.M. LAMBERT, A. ZAHAB, F. RACHDI and P. BERNIER
Groupe de Dynamique des Phases Condensées, Université des Sciences et Techniques du Languedoc, 34060 Montpellier, France

ABSTRACT

We have performed ^{13}C-NMR on C_{60} single crystals grown by sublimation and 8% enriched in ^{13}C, in the pressure range 0-5kbar and at different temperatures between 200 and 343K. The thermodynamic data, enthalpy and volume of activation, are deduced from the temperature and pressure dependences of the spin lattice relaxation time, T_1, in both stuctural phases, and compared to results in other plastic crystals.

In the superconducting fullerene K_3C_{60}, ^{13}C-NMR properties have been investigated under pressure up to 12kbar, at room temperature. Comparing pressure dependences of the hyperfine spin lattice relaxation rate and of the resonance frequency provides the first determination of the ^{13}C-Knight shift (61±4 ppm) and of the chemical shift of the C_{60}^{3-} molecule (125ppm/TMS).

1. Introduction:

High pressure is an interesting variable in molecular solids such as fullerenes as it provides a continuous tuning of intermolecular distances. Particularly, in neutral C_{60}, molecular motions and the order-disorder transition temperature T_c, are strongly affected by an applied pressure. This is evidenced by ^{13}C-NMR data, spin-lattice relaxation time, T_1, and spectral shapes, which are solely governed by molecular motions [1]. On the other side, in the superconducting fullerenes A_3C_{60}, NMR data, T_1^{-1} and the Knight shift (K), are correlated to the electronic properties, namely the density of states at the Fermi level, $N(E_F)$, and the superconducting transition which are both very sensitive to pressure [2,3]. We have used the proportionality between T_1^{-1} and K and their pressure variation to determine the actual value of K. ^{13}C-NMR experiments were performed in a magnetic field B_0=9.3T (100MHz). The pressure transmitted by isopentane, was applied directly on the sample space in the 5kbar range for the C_{60} experiment, while a double stage equipment (in-situ intensifier ratio x 4) was used for the K_3C_{60} experiment up to 12kbar.

2. Dynamics of C_{60}:

Our NMR study was performed on C_{60} single crystals (8% enriched in ^{13}C) grown by sublimation. One crystal ($\approx$0.7mg) was used at ambient pressure while several crystals ($\approx$7mg) were required for the experiments under pressure. At ambient pressure, the temperature dependence of T_1 defines a sharp transition at T_C=262K. This transition temperature, slightly higher than in powdered samples [1,4] and non-sublimed samples [5] was also confirmed by differential scanning calorimetry [6]. The first effect of pressure is to increase T_C with a slope, +16K/kbar, significantly higher than what is reported when Helium gas is used (+11.7K/kbar [7]) instead of isopentane as pressure transmitting fluid. This difference has already been reported [8] and attributed to the intercalation of He molecules in the interstitial sites of the C_{60} structure. T_C separates regions with two kinds of molecular motions: isotropic rotations in the faces centered cubic (f.c.c.) phase at high temperature and uniaxial jumps reorientations in the simple cubic (s.c.) phase, below T_C [9].

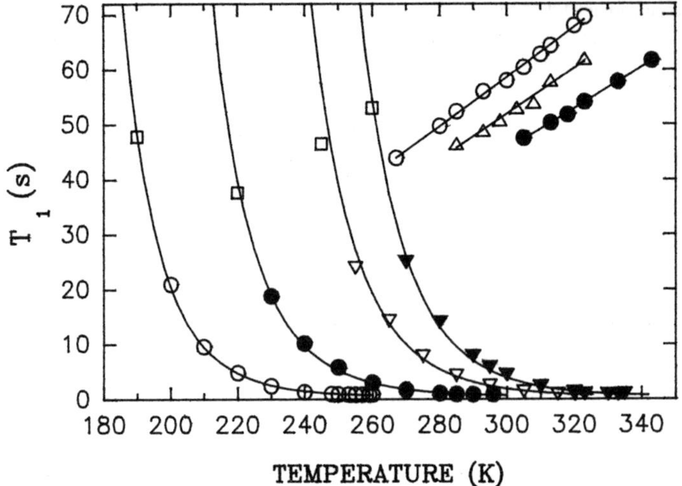

Figure 1: Temperature dependence of the ^{13}C spin-latice relaxation time for C_{60} single crystals, at different pressures: P = 1bar (o), 1kbar (Δ), 2kbar (o), 4kbar (∇) and 5kbar (▼). Solid lines are a fit of experimental points to the general law for molecular motions in the high and low temperature regions. ( ) points are deduced from biexponential relaxation.

Figure 1 shows the evolution of T_1 versus temperature at different pressures up to 5kbar, in both structural phases. The long T_1's at low temperature, confirm the high purity of the crystal and assure a good accuracy for the determination of the dynamics parameters. For high magnetic fields, the chemical shift anisotropy is the dominant mechanism for the relaxation. The correlation time for molecular motions τ, is thermally activated: $\tau=\tau_0 \exp(\Delta F/RT)$, where $\Delta F=\Delta H-T\Delta S$ is the free energy and enters the relaxation rate as follows [10]:

$$(T_1)^{-1} = \gamma^2 \, \delta B^2 \frac{2\tau}{(1+\omega^2\tau^2)}$$

where δB is the amplitude of the local field proportional to B_0 and fluctuating at the Larmor frequency ω. This relation yields

$$T_1 = C_1 . \tau \quad \text{for } \omega\tau \gg 1 \text{ (slow reorientations)}$$

$$\text{and} \quad T_1 = C_2 / \tau \quad \text{for } \omega\tau \ll 1 \text{ (rapid reorientations)}$$

Experimental points are fitted to these two laws respectively below and above the structural transition. Assuming that τ_0, C_1 and C_2 are independent of pressure and temperature in our range of measurements, we can deduce the enthalpies of activation ΔH, from the temperature dependence of T_1 at constant pressure results and the volumes of activation ΔV, from the pressure dependence of T_1 at constant temperature:

$$\Delta H = \pm R\left[\frac{\partial(\ln(T_1))}{\partial(1/T)} \right]_P \qquad\qquad \Delta V = RT\left[\frac{\partial(\ln(T_1))}{\partial P} \right]_T$$

s.c. phase $T < T_c$			f.c.c. phase $T > T_c$		
P (kbar)	$\Delta H/R$(K)	ΔH(kJ/mol)	P (kbar)	$\Delta H/R$(K)	ΔH(kJ/mol)
0.001	3300	27.44	0.001	696	5.79
2	3770	31.34	1	719	5.98
4	4490	37.33	2	728	6.05
5	5120	42.57			

s.c. phase $T < T_c$		f.c.c. phase $T > T_c$	
T (K)	ΔV(cm^3/mol)	T (K)	ΔV(cm^3/mol)
230	19.69	285	2.81
240	20,34	305	3.05
250	19.67	313	3.22
260	19.83	323	3.38
280	19.52		
295	20.38		
305	18.28		

Table 1: Enthalpies of activation ΔH and volumes of activation ΔV in C_{60}.

These values, listed in table 1, for both phases, compare well with data obtained for plastic crystals such as adamantane [11] which presents only rotational motions. As the enthalpies of activation are concerned, the ambient pressure values are consistent with previous results [1,4,5]. Then ΔH increases with pressure which acts as an additional constraint to molecular motion and this more efficiently in the s.c. phase. The volumes of activation, which represent the additional volume necessary for the reorientation of the molecule, are much smaller in the f.c.c. phase (≈ 3cm^3/mol) than in the s.c. phase (≈ 20cm^3/mol) reflecting the two different natures of motions.

454

3. ^{13}C-Knight shift on K_3C_{60}:

^{13}C-NMR under pressure was performed on 10mg of K_3C_{60} powder, isotopically ^{13}C enriched up to 10 at% and containing around 50 mol% of neutral C_{60} used as a magnetic field marker. The position of the NMR line in a conductor is determined by a chemical shift and a Knight shift (K). The chemical shift not only depends on the C_{60} molecule but also on the electronic charge it holds. As a consequence, the origin of K in K_3C_{60} is unknown. However, K is proportional to $N(E_F)$ and is the only part of the frequency shift to be affected by pressure. On the other side, the electronic component of the relaxation rate T_1^{-1} is of dipolar origin but remains proportional to $(N(E_F))^2$ [13].

The frequency shift δ, of the C_{60}^{3-} line relative to the C_{60}^0 one taken as a reference is measured at room temperature as a function of pressure, up to 8kbar (inset of Figure 2), from the relative position of the maxima of the two peaks. A linear pressure dependence is observed with a slope $d\delta/dP \approx -0.73$ppm/kbar and K is related to δ by $K(P) = \delta(P) - \delta_0$.

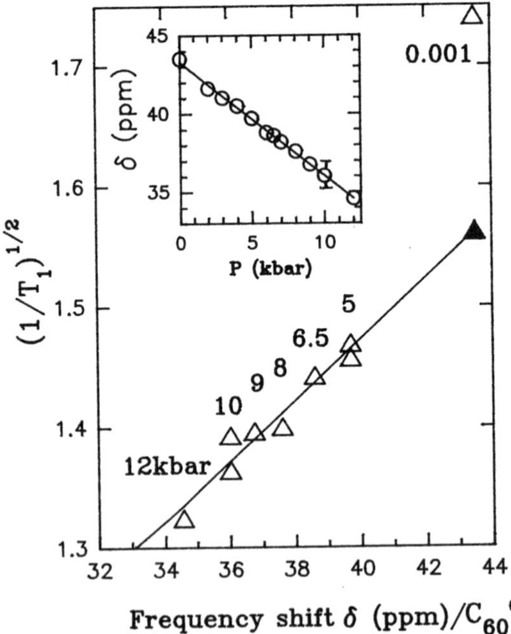

Figure 2: Relation between the C_{60}^{3-} electronic relaxation rate and the resonance frequency shift δ relative to C_{60}^0, measured at different pressures, plotted as $(T_1)^{-1/2}$ versus δ, yielding the actual Knight shift K.

Inset: Room temperature pressure dependence of the resonance frequency shift δ of the ^{13}C-C_{60}^{3-} line relative to the C_{60}^0 one in the C_{60}-riched K_3C_{60} sample. A typical error bar (± 0.7ppm) is shown.

In order to determine δ_0 we have used the pressure dependence of T_1^{-1} and more precisely its electronic component. Extrinsic relaxation channels are actually provided by molecular reorientations between 0 and 5kbar, as it was observed at ambient pressure around 230K [14]. Moreover above 5kbar spin diffusion between the two ^{13}C spin species is important when a large amount of C_{60} is present, due to the progressive overlap of the spectra. This is the reason why we have used T_1^{-1} values (between 5 and 12kbar) previously obtained on a K_3C_{60} sample containing only $\approx 5\%$ of unreacted C_{60} [12]. Figure 2 shows the proportionality relation between δ and $(T_1)^{-1/2}$ when the pressure is varied up to 12kbar. The extrapolation of the line position corresponding to zero relaxation rate gives $\delta_0 = -18 \pm 4$ppm, from which is deduced the actual Knight shift at P=1bar, K= 61 ± 4ppm. It decreases by ≈ 7ppm i.e.$11 \pm 1.5\%$ under 10kbar while at the same time, the electronic contribution to T_1^{-1} drops by $21.5 \pm 3\%$. These results compare very well with the extended Hückel band structure calculations [1,15] leading to a 11% reduction of $N(E_F)$ taking a lattice parameter reduction of 1.2%. NMR data under pressure can be therefore understood within a weak Coulomb interaction model. This result is also compatible with the ^{13}C-C_{60}^{3-} resonance in Rb_3C_{60} at 196ppm/TMS, obtained by MAS-NMR technique at 5T [16]. The difference of 10ppm between the two compounds may be ascribed only to an increase of the Knight shift as the chemical shift is fixed by the charge density on the C_{60} molecule. Finally, this chemical shift of C_{60}^{3-} (125ppm/TMS) deduced from K shows that the C_{60}^{3-} species is 18ppm more diamagnetic than the neutral molecule which could be due to anomalous paramagnetic ring currents in neutral C_{60} [17].

4. References:

1   R. Tycko et al., *Phys. Rev. Lett.* **67** (1991) 1886.

2   R. Fleming et al., *Nature* **352** (1991) 787.

3   K. Tanigaki ct al., *Nature* **356** (1992) 419.

4   R.D. Johnson et al., *Science* **255** (1992) 1235.

5   Y. Maniwa et al., *Synth. Metals* **56** (1993) 3057.

6   R. Kerkoud et al., *C.R. Acad. Sci., Paris,* **318** (1994) 159.

7   G. Kriza et al., *J. Phys. I France* **1** (1991) 1361.

8   G. A. Samara et al., *Phys. Rev.* **B47** (1993) 4756.

9   W.I.F. David et al., *Europhys. Lett.* **18** (1992) 219

10 C.P. Slichter, Principles of Magnetic Resonance (Springer New York, 1990).

11 N.I. Liu and J. Jonas, *Chem. Phys. Lett.* **14** (1972) 555.

12 G. Quirion et al., *Europhys. Lett.* **21** (1993) 233.

13 V.P. Antropov et al., *Phys. Rev.* **B47** (1993) 12373.

14 Y. Yoshinari et al., *Phys. Rev. Lett.* **71** (1993) 2413.

15 S. Saito and A. Oshiyama, *Phys. Rev. Lett.* **66** (1991) 2637.

16 G. Zimmer et al., Europhys. Lett. **24** (1993) 59.

17 A. Pasquarello, M. Schlüter and R.C. Haddon, *Phys. Rev.* **A47** (1993) 1783.

^{13}C and ^{87}Rb NMR in K_4C_{60} and Rb_3C_{60}

G. Zimmer, M. Helmle, and M. Mehring

2. Physikalisches Institut, Universität Stuttgart,

70550 Stuttgart, Fed. Rep. of Germany

J. Reichenbach and F. Rachdi

Groupe de Dynamique des Phases Condensées, USTL

4060 Montpellier, France

J. E. Fischer

Department of Materials Science and Engineering, University of Pennsylvania

Philadelphia, Pennsylvania 19104

Abstract: NMR methods were applied in order to unravel electronic as well as dynamic properties of the insulating K_4C_{60} and superconducting Rb_3C_{60} compounds. Two reorientational modes of the C_{60} molecule were observed in K_4C_{60} by relaxation measurements which are also found in K_3C_{60}. The T_1^{-1} relaxation rate shows in addition an activated electronic contribution. The observed line shift as a function of temperature indicates the importance of hyperfine interactions between paramagnetic electronspins and nuclei. In ^{87}Rb T_1 measurements a peak in the relaxation rate was found around the superconducting transition temperature.

1 Introduction

Both stoichiometric phases A_3C_{60} and A_4C_{60} show interesting physical properties. A_3C_{60} compounds are known to be superconductors, whereas the A_4C_{60} compounds are insulators. In this communication we report on ^{13}C NMR line shift and relaxation data in K_4C_{60} and on ^{87}Rb NMR spin lattice relaxation at the transition from the conducting to the superconducting regime.

The samples with natural abundance ^{13}C were prepared following the standard preparation method described elsewhere. The NMR measurements were performed with homebuilt spectrometers at different magnetic fields. The spectra are recorded by a Hahn-echo sequence. The inversion recovery technique was used for T_1 relaxation measurements in most cases. At low temperatures the saturation recovery sequence was applied in the ^{13}C T_1 measurements K_4C_{60}. We found no magnetic field dependence of T_1 for K_4C_{60} in the temperature range from 150 K - 460 K and also for Rb_3C_{60} around T_c. The T_2 relaxation times were measured by a Hahn echo sequence.

2   Results

^{13}C NMR in K$_4$C$_{60}$

The temperature dependence of the ^{13}C spectra and of the T_1^{-1}, T_2^{-1} relaxation rates is shown in Figs.2,2.

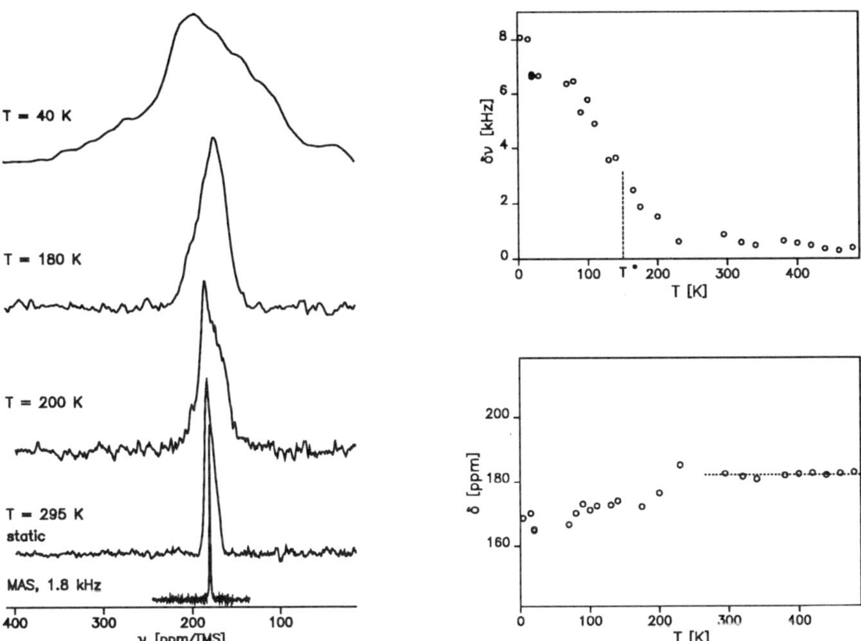

Figure 1: Left: ^{13}C spectra of K_4C_{60}. Right: Line shift δ as well as the square root of the second moment which is a measure of the line width.

The line shift of the insulating K$_4$C$_{60}$ is constant at 182 ppm above about 200 K and drops only slightly down to 169 ppm below. The amount of the shift in the high temperature region is about the same as observed in the conducting K$_3$C$_{60}$ compound. This leads us to the conclusion that a considerable part of the line shift is due to coupling of ^{13}C nuclei to paramagnetic electrons. Different from the observations in K$_3$C$_{60}$ we find a residual anisotropy at room temperature which is proved by comparing the static spectrum with the MAS spectrum. In the temperature range down to about 200 K, the spectra can essentially be described by an almost axially symmetric shift tensor.
On further cooling, the slowing down of C$_{60}^{4-}$ motion on the NMR time scale is reflected in the line width as shown in Fig.2. At the temperature $T^* \simeq 150$ K, the line width reached about half its maximum value. In addition, a peak occurs in T_2^{-1} which can be understood by the slowing down of reorientational motion:

$$\frac{1}{T_2} = \frac{1}{T_{20}} + \frac{(\Delta\omega)^2 \tau_{cS}}{1 + (\Delta\omega)^2 \tau_{cS}^2}, \qquad \tau_{cS} = \tau_{cS0} exp(\frac{\Delta E_S}{kT})$$

By subtracting a temperature independent $T_{20} = 12\,\mathrm{ms}$ which is probably due to the dipolar couplings of ^{13}C spins and setting $\Delta\omega = 2\pi \cdot 4.2\,\mathrm{kHz}$, we obtained an activation energy of $327\,\mathrm{meV}$ and $\tau_{cS0} = 4 \cdot 10^{-16}\,\mathrm{s}$ for the C_{60}^{4-} reorientational mode as observed by T_2 relaxation. These results are comparable to an analogous observation in K_3C_{60} [2], where $\tau_{cS0} = 7 \cdot 10^{-16}\,\mathrm{s}$ and $\Delta E = 586\,\mathrm{meV}$ ($6800\,\mathrm{K}$) was found. The difference in ΔE is evident already from the different characteristic temperatures T^* where the T_2 rates peak and is due probably to the different lattice parameters of the two compounds.

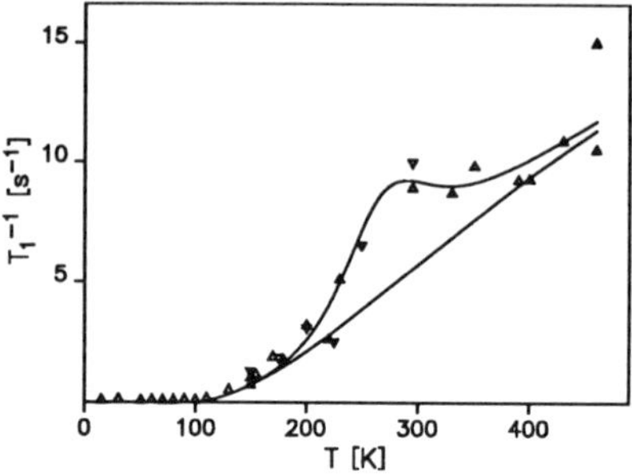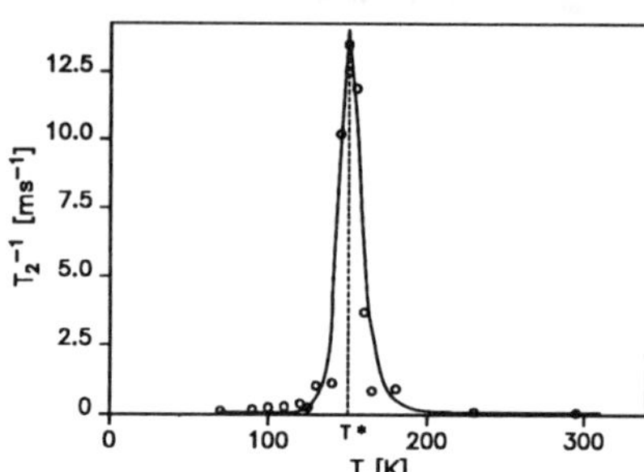

Figure 2: ^{13}C T_1 relaxation (left) and T_2 relaxation (right). The solid lines are simulations of the models described in the text.

On first sight, the T_1^{-1} rate exhibits a strange behaviour as a function of temperature. The measured rate can be decomposed into different contributions by the following model:

$$\frac{1}{T_1} = \frac{1}{T_{10}} + \frac{2\Delta\omega^2\tau_{cL}}{1+\omega_0^2\tau_{cL}^2} + C \cdot exp(\frac{-\Delta E_e}{kT}), \qquad \tau_{cL} = \tau_{cL0} \cdot exp(\frac{\Delta E_L}{kT}).$$

Here, T_{10} is a temperature independent background rate. The second term describes the relaxation caused by a fluctuating shift tensor.

The electronic part (third term) is characterized by an activation energy of $\Delta E_e = 50\,\mathrm{meV}$. This process has to be distinguished from the electronic energy gap which amounts to 300 meV as determined by μSR [3]. On the other hand, a similar activated behaviour with an activation energy of about $40\,\mathrm{meV}$ was found in Raman scattering [4]. Note, that the lifting of the t_{1u} degeneracy also amounts to about $40\,\mathrm{meV}$. Together with the observed line shift this leads to the conclusion that the insulating K_4C_{60} is paramagnetic with a Pauli like susceptibility above $200\,\mathrm{K}$, which is confirmed by ESR measurements [5].

Apart from the temperature independent as well as the electronic part a broad peak in the relaxation rate between 150 K - 400 K is observed due to the reorientational motion of the C_{60}^{4-} molecules. Using $\Delta\omega = 2\pi \cdot 7.5\,\mathrm{kHz}$ (spectral width observed at $4\,\mathrm{K}$), the parameters relevant for this C_{60}^{4-} reorientational motion, τ_{cL0} and ΔE_L, amount to $9 \cdot 10^{-13}\,\mathrm{s}$ and $2100\,\mathrm{K}$ ($180\,\mathrm{meV}$), similar to those obtained for K_3C_{60} ($\tau_{cL0} = 2 \cdot 10^{-14}\,\mathrm{s}$, $\Delta E_L = 2900\,\mathrm{K}$).

This finding leads to the conclusion that the reorientational modes of C_{60} molecules are similar in both compounds. The mode observed in T_2^{-1} shows a more pronounced dependence on the lattice structure as that observed in T_1 measurements. By considering the spectral shapes, we attribute the motion observed by T_2 measurements to an axial rotation mode, the one observed in T_1 to large angle reorientations.

^{87}Rb in Rb$_3$C$_{60}$ in the superconducting state

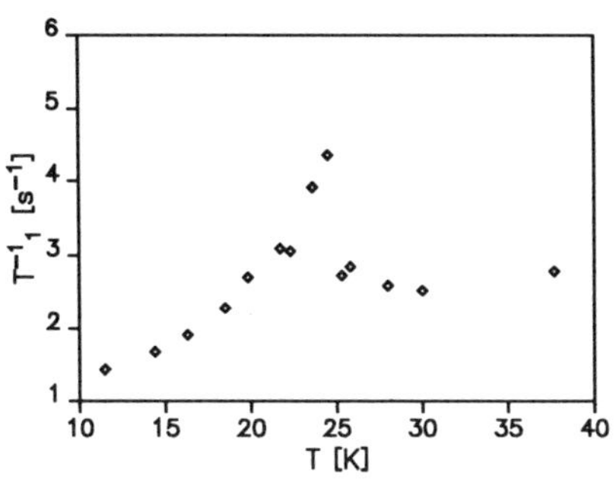

Figure 3: ^{87}Rb T_1^{-1} relaxation rates, measured at 7.8 T with a inversion recovery technique.

As shown in fig.2 for the tetrahedral line, the ^{87}Rb T_1^{-1} relaxation rate measured at 8 T with an inversion recovery method shows a peak around $T_c = 25$ K for all three lines. A similar peak was found in μSR relaxation, but not in the ^{13}C T_1^{-1} rate [6]. Although it is tempting to interpret this peak as a Hebel-Slichter peak, we are reluctant to do so because of the following observations: The relaxation rate increases slightly already above T_c with different enhancements for the three sites, which do not scale with the square of the hyperfine interactions. Furthermore a Hebel-Slichter peak should exhibit a field dependence, which is not observed in measurements at different fields. More details will be published elsewhere.

References

[1] D. W. Murphy, M. J. Rosseinsky, R. M. Fleming, R. Tycko, A. P. Ramirez, R. C. Haddon, T. Siegrist, G. Dabbagh, J. C.Tully, and R. E. Walstedt, J. Phys. Chem. Solids, **53**, 1321 (1992)

[2] Y. Yoshinari, H. Alloul, G. Kriza, and K. Holczer, Phys. Rev. Lett., **71**, 2413, (1993)

[3] R. F. Kiefl,T. L. Duty, J. W. Schneider, W. A. MacFarlane, K. H. Chow, J. W. Elzey, P. Mendels, G. D. Morris, J. H. Brewer, E. J. Ansaldo, C. Niedermeyer, D. R. Noakes, C. E. Stronach, B. Hitti, and J. E. Fischer, Phys. Rev. Lett., **69**, 2005 (1992)

[4] C. Taliani, private communication, unpublished results

[5] P. Küster, private communication

[6] R. Tycko, G. Dabbagh, M. J. Rosseinsky, D. W. Murphy, A. P. Ramirez, and R. M. Fleming, Phys. Rev. Lett. **68**, 1912 (1992)

[7] C. S. Yannoni, R. D. Johnson, G. Meijer, D. S. Bethune, and J. R. Salem, J. Phys. Chem. **95**, 9 (1991)

[8] R. F. Kiefl, W. A. MacFarlane, K. H. Chow, S.Dunsinger, T. L. Duty, T. M. S. Johnston, J. W. Schneider, J. Sonier, L. Brard, R. M. Strongin, J. E. Fischer, and A. B.Smith (III), Phys. Rev. Lett., **70**, 3987 (1993)

HIGH-RESOLUTION ^{13}C NMR IN ALKALI INTERCALATED C_{60}

C. Goze, F. Rachdi, J. Reichenbach, I. Luk'yanchuk*, G. Zimmer+, and M. Mehring+
Groupe de Dynamique des Phases Condensées, Université de Montpellier II,
Place Eugène Bataillon, F-34095 Montpellier Cedex 05, France
+ 2. Physikalisches Institut, Universität Stuttgart, Pfaffenwaldring 57,
D-70569 Stuttgart, Germany
*on leave from the L.D. Landau Institute for Theoretical Physics, Kosygina 2,
Moscow, Russia

ABSTRACT

We present the results of ^{13}C NMR measurements on potassium and rubidium doped C_{60} at room temperature. By using high-resolution MAS NMR spectroscopy we were able to identify and characterise the various thermodynamically stable phases. Based on an analysis of the obtained NMR spectra we discuss the effects of the intercalation on the molecular dynamics as well as on the electronic properties in these compounds.

The discovery of distinct phases with unusual properties in alkali fullerides has raised fundamental questions about their physical and chemical structure [1,2]. The fcc structure of pristine C_{60} contains two tetrahedral and one octahedral interstitial sites per molecule which are large enough to accommodate different atoms as intercalants. Spectroscopic studies such as photoemission, Raman, X-ray absorption and electron-energy-loss-spectroscopy have clearly confirmed that a nearly complete charge transfer takes place from the alkali atoms to the C_{60} molecules [3-7]. For the alkali metals K, Rb, and Cs thermodynamically stable phases $A_x C_{60}$ with x=3, 4, and 6 exhibiting different crystal structures have been well established through structural investigations [8,9]. Recently, the existence of a fourth stable phase in alkali fullerides with stoichiometry $A_1 C_{60}$ has been demonstrated [10]. Nuclear magnetic resonance is a powerful tool for studying the alkali fullerides because it supplies a microscopic probe which allows for studying the electronic structure as well as the molecular dynamics of the different phases. Although many NMR investigations in pristine and doped fullerene have been performed [11-15], the number of high-resolution studies in the solid state, especially for the doped compounds, has been rather limited [16-18]. This is mainly due to the fact that these samples are very air-sensitive; measurements require a special NMR probe head and spinner design, not available with standard commercial MAS probe heads. In this paper we report the results of ^{13}C nuclear magnetic resonance on the different phases in alkali fullerides $K_x C_{60}$ and $Rb_x C_{60}$ by using magic angle spinning at room temperature, and analyse and discuss their distinctive features in terms of chemical and/or Knight shifts.

Powder samples of $K_x C_{60}$ and $Rb_x C_{60}$ with x=3, 4, and 6 were synthesised by preparing in a first step the compound $A_6 C_{60}$ using a vapour transport method similar to that described in the literature [2]. In briefly, an excess amount of potassium or rubidium and pure C_{60} powder were sealed in a Pyrex tube under vacuum. Within the tube both materials were spatially separated and isolated from each other by a frit. The tube was put in a furnace and heated at 250 °C for several days. A temperature gradient of about 5 to 10 °C was maintained along the tube preventing condensation of the alkali-metal onto the C_{60} powder, which was deposited at the hotter part of the tube. Part of the A_6C_{60} compound was then transferred to an NMR tube and sealed under vacuum, while the rest of the sample was used to prepare A_3C_{60} and A_4C_{60} in a similar way by direct reaction of A_6C_{60} with additional stoichiometric amounts of pure C_{60}, respectively. The sample structures were checked by X-ray diffraction and in the case of the superconducting phases by non-resonant microwave absorption and Meissner shielding. The onset T_c value of the superconducting transition was found to be 27 K for Rb_3C_{60}.

NMR spectra were recorded at a ^{13}C frequency of 50.3 MHz (4.7 T field) on a Bruker ASX 200 spectrometer. All chemical shifts were referenced to tetramethylsilane (TMS). We applied magic angle spinning (MAS) by using a home built NMR probe head and a specially designed spinner which allows spinning of the air-sensitive samples in placed NMR tubes with frequencies up to 3 kHz. Approximately 30 mg of material was filled in the NMR tubes. T_1 values were determined with the inversion-recovery technique. The undoped starting material C_{60} was also checked by using ^{13}C MAS NMR and showed the usual resonance at 143.6 ppm (relative to TMS).

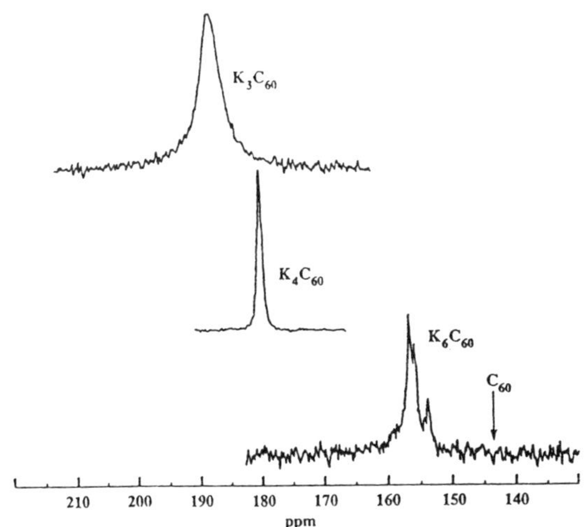

Fig. 1: High-resolution ^{13}C MAS NMR room temperature spectra of the different phases in potassium-doped fullerene K_xC_{60} (x=3, 4, 6) showing the different isotropic lines. The arrow marks the resonance position of pure C_{60}. Frequencies are in ppm with respect to tetramethylsilane.

Figure 1 shows the isotropic parts of the high-resolution ^{13}C MAS NMR spectra for the different phases of potassium-doped C_{60} at ambient temperature, which were all obtained with the same spinning frequency of about 2.5 kHz. As can be clearly seen, each phase is easily discernible by its characteristic resonance with the following values for the shifts δ (relative to TMS): K_3C_{60} (156 ppm), K_4C_{60} (180 ppm) and K_3C_{60} (187 ppm). The arrow marks the resonance of undoped C_{60} (143.6 ppm). For the used spinning frequency, K_3C_{60} and K_4C_{60} yield a single line shifted 44 ppm and 37 ppm downfield from undoped C_{60}, respectively, whereas the spectrum of K_6C_{60} shows three peaks, indicating the existence of three magnetically non-equivalent carbon sites in this compound [16]. Note that the full width at half maximum (FWHM) of the resonance in K_3C_{60} is distinctly larger ($\Delta\sim4$ ppm) as compared to the FWHM of the lines in K_4C_{60} and K_6C_{60} ($\Delta\sim1$ ppm).

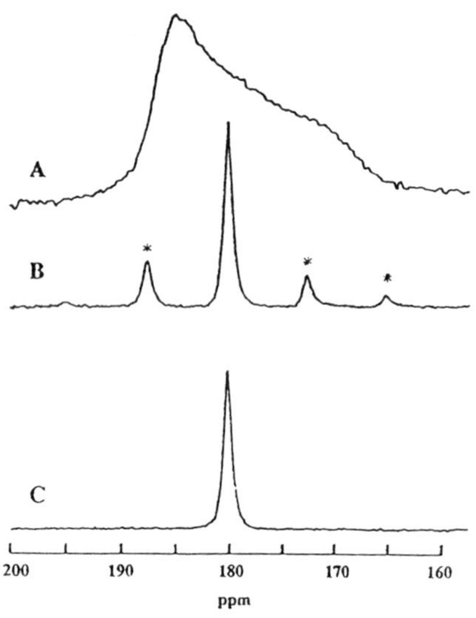

Fig. 2: (a) ^{13}C NMR spectrum of K_4C_{60} at room temperature. (b) ^{13}C MAS NMR spectrum with a spinning frequency of 370 Hz. (c) Same as (b) with spinning frequency of 3.1 kHz .

We present in Figure 2 the ^{13}C NMR spectra for K_4C_{60} at different spinning rates. In Fig. 2a we have plotted the static spectrum which shows a drastically reduced spectral spread of about 25 ppm compared to the roughly 300 ppm in A_6C_{60} [16]. This indicates the presence of molecular motion of the buckyballs, averaging to a high degree the anisotropies already present at room temperature. Spinning the sample at slow speed leads to a typical sideband pattern, which mimics the form of the static spectrum. At higher spinning rates (Fig. 2c) we are able to remove completely the anisotropies, thereby observing a single very sharp isotropic line at 180 ppm with a FWHM of about 1 ppm. The results for Rb_4C_{60} are very similar. The isotropic line at 180 ppm with a FWHM of about 1 ppm. The results for Rb_4C_{60} are very similar. The isotropic line of Rb_4C_{60} appears at 182 ppm. The most interesting fact is the very large spin-lattice relaxation (SLR) rate in these compounds. The T_1 values at ambient are 110 ms and 80 ms for K_4C_{60} and Rb_4C_{60}, respectively, and are even shorter than the ones in the conducting A_3C_{60} phases. Furthermore, detailed measurements on the temperature dependence of the ^{13}C SLR rate reveal a strong deviation from a Korringa-like behaviour [19]. With decreasing temperature T_1 increases to several seconds at 100 K. This can consistently be explained by the freezing in of different types of motion at low temperatures. There is no sign of a metallic state. Simple rigid-band models or band structure calculations, however, predict that the A_4C_{60} compounds should be metallic with four electrons occupying the three-fold degenerate LUMO band [20,21] which, on the other hand, is in striking contrast to experimental results. Recent low-temperature ultraviolet photoemission spectroscopy with high energy-resolution has revealed that K_4C_{60} is a semiconductor with an energy gap of more than 200 meV [22]. Furthermore, the results of preliminary measurements of the susceptibility χ by ESR reveal the absence of a Pauli-like behaviour and show a strong decrease of χ with decreasing temperature accompanied by a narrowing of the resonance line [19]. Although the low temperature ^{13}C relaxation can be explained by molecular motion, the high temperature behaviour points to a spin excitation which might be due to electron-electron correlations

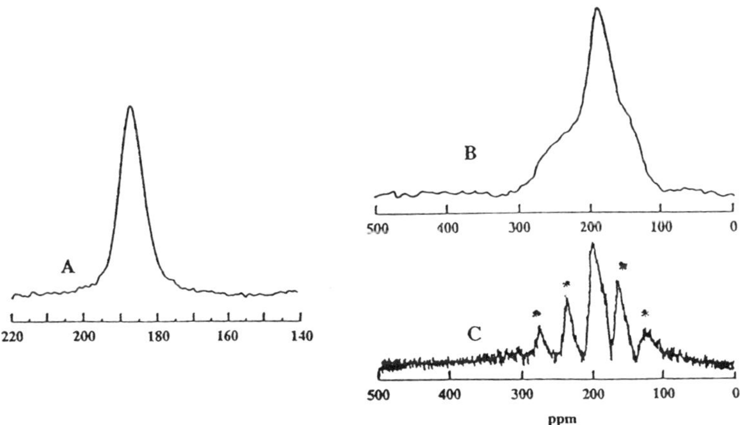

Fig. 3: (a) Static ^{13}C NMR of K_3C_{60} at room temperature. The line is narrow and symmetric due to motional averaging. (b) Static ^{13}C NMR of K_3C_{60} at room temperature. Note the inhomogeneously broadened line with an extent of roughly 200 ppm. (c) ^{13}C MAS NMR spectrum of Rb_3C_{60} with a spinning frequency of 1.8 kHz.

Finally, we turn to the A_3C_{60} phases. As can be seen in Fig. 3a, the static ^{13}C NMR spectrum for K_3C_{60} is narrow and symmetric (FWHM~7.5 ppm) due to fast re-orientations of the C_{60}^{3-} ions. Spinning of the sample at 2.5 kHz (see Fig. 1) yields a narrowing of the resonance by roughly a factor of two. In Fig. 3b we have plotted the static spectrum for Rb_3C_{60}. The line is inhomogeneously broadened due to the distribution of NMR shifts of about 200 ppm, which are incompletely averaged by the slowed-down motion of the molecules. At elevated temperatures (T=400 K) the resonance becomes motionally narrowed and symmetric. The measured SLR rates at room temperatures are 2.8 s^{-1} for K_3C_{60} and 4.5 s^{-1} for Rb_3C_{60}, respectively. The dominant spin-lattice relaxation mechanism in these conducting compounds is due to the hyperfine coupling to conduction electrons rather than by coupling to molecular reorientations through the chemical shift anisotropy as in pure C_{60} and in the A_6C_{60} compounds.

Recently, Yannoni et al. [16] reported high-resolution ^{13}C NMR spectra of Rb_3C_{60}. At ambient temperature the authors observed a single narrow line shifted downfield from undoped C_{60} (186 ppm relative to TMS) as well as a resonance at 157 ppm, which was assigned to an impurity. There exists a disagreement between their results and the ones presented in this paper. In the light of our study we assign the resonance at 157 ppm to the phase Rb_6C_{60}, whereas the narrow single line at 186 ppm is probably due to Rb_4C_{60}. As mentioned Rb_3C_{60} shows a broad line around 195 ppm (Fig. 3c), which is actually more shifted downfield by 24 ppm than the value reported and attributed to this phase by Tycko et al. [13]. We have observed the quoted line shifts in different samples and are quite confident that the values are correct.

In summary, high resolution ^{13}C NMR measurements revealed finer details about the structural and electronic properties and molecular dynamics of the different thermodynamically stable phases in K and Rb intercalated C_{60}. In the saturated phase, we have been able to distinguish the crystallographically non-equivalent carbon sites in the C_{60} molecule which confirms the orientationally ordered structure in this phase. The A_3C_{60} phase, which has been extensively studied is clearly identified as a conducting phase. We suggest here a qualitative explanation for the observed paramagnetic shift, based on the admixture of the atomic 2s-orbital to the p_z-orbital due to core polarisation and a possible contribution of the curvature of the C_{60} molecule. Concerning the A_4C_{60} phase we have found considerable anisotropic motion which gradually freezes in when reducing the temperature. The ^{13}C spin lattice relaxation and the shift give no indication for a metallic behaviour. However, electron-electron correlation might contribute.

Acknowledgement

We are grateful to N. Kirova for valuable discussions and to J.E. Fischer for providing the Rb_3C_{60} sample.

References

1. M.S. Dresselhaus, G. Dresselhaus, P.C. Eklund, *J. Mater. Res.* **8** (1993) 2054

2. A.F. Hebard, M.J. Rosseinsky, R.C. Haddon, D.W. Murphy, S.H. Glarum, T.T.M. Palstra, A.P. Ramirez, A.R. Kortan, *Nature* **350** (1991) 600

3. P.J. Benning, J.L. Martins, J.H. Weaver, L.P.F. Chibante, R.E. Smalley, *Science* **252** (1991) 1417

4. G.K. Wertheim, J.E. Rowe, D.N.E. Buchanan, E.E. Chaban, A.F. Hebard, A.R. Kortan, A.V. Makhija, R.C. Haddon, *Science* **252** (1991) 1419

5. T. Takahashi, S. Suzuki, T. Morikawa, H. Katayama-Yoshida, S. Hasegawa, H. Inoguchi, K. Kikuchi, S. Suzuki, K. Ikemoto, Y. Achiba, *Phys. Rev. Lett.* **68** (1992) 1232

6. E. Sohmen, J. Fink, W. Krätschmer, *Europhys. Lett.* **17**, (1992) 51

7. R.C. Haddon, A.F. Hebard, M.J. Rosseinsky, D.W. Murphy, S.J. Duclos, K.B. Lyons, B. Miller, J.M. Rosamilia, R.M. Fleming, A.R. Kortan, S.H. Glarum, A.V. Makhija, A.J. Miller, R.H. Eick, S.M. Zahurak, R. Tycko, G. Dabbagh, F.A. Thiel, *Nature* **350** (1991) 320

8. O. Zhou, J.E. Fischer, N. Coustel, S. Kycia, Q. Zhu, A.R. McGhie, W.J. Romanow, J.P. McCauley, A.B. Smith, D.E. Cox, *Nature* **351** (1991) 462

9. P.W. Stephens, L. Mihaly, P.L. Lee, R.L. Whetten, S.M. Huang, R. Kaner, F. Deiderich, K. Holczer, *Nature* **351** (1991) 632

10. Q. Zhu, O. Zhou, N. Bykovetz, J.E. Fischer, A.R. McGhie, W.J. Romanow, C.L. Lin, R.M. Strongin, M.A. Cichy, A.B. Smith III, *Phys. Rev. B* **47** (1993) 13948

11. C.S. Yannoni, R.D. Johnson, G. Meijer, D.S. Bethune, J.R. Salem, *J. Chem. Phys.* **95** (1991) 9

12. R. Tycko, G. Dabbagh, M.J. Rosseinsky, D.W. Murphy, R.M. Fleming, A.P. Ramirez, J.C. Tully, *Science* **253** (1991) 884

13. R. Tycko, G. Dabbagh, M.J. Rosseinsky, D.W. Murphy, A.P. Ramirez, R.M. Fleming, *Phys. Rev. Lett.* **68** (1992) 1912

14. D.W. Murphy, M.J. Rosseinsky, R.M. Fleming, R. Tycko, A.P. Ramirez, R.C. Haddon, T. Siegrist, G. Dabbagh, J.C. Tully, R.E. Walstedt, *J. Phys. Chem. Solids* **53** (1992) 1321

15 G. Quirion, C. Bourbonnais, E. Barthel, P. Auban, D. Jerome, J.M. Lambert, A. Zahab, P. Bernier, C. Fabre, A. Rassat, *Europhys. Lett.* **21** (1993) 233

16. F. Rachdi, J. Reichenbach, L. Firlej, P. Bernier, M. Ribet, R. Aznar, G. Zimmer, M. Helmle, M. Mehring, *Solid State Commun.* **87** (1993) 547

17. C.S. Yannoni, H.R. Wendt, M.S. de Vries, R.L. Siemens, J.R. Salem, J.L. Yerla, R.D. Johnson, M. Hoinkis, M.S. Crowder, C.A. Brown, D.S. Bethune, L. Taylor, D. Nguyen, P. Jedrzejewski, H.C. Dorn, *Synthetic Metals* **59** (1993) 279

18. G. Zimmer, M. Helmle, M. Mehring, F. Rachdi, J. Reichenbach, L. Firlej, P. Bernier, *Europhys. Lett.* **24** (1993) 59

19. G. Zimmer, M. Helmle, M. Mehring, F. Rachdi, *Europhys. Lett.*, submitted

20. S. Saito, A. Oshiyama, *Phys. Rev. B* **44** (1991) 11536

21. S.C. Erwin, W.E. Pickett, *Science* **254** (1991) 842

22. T. Morikawa, T. Takahashi, *Solid State Commun.* **87** (1993) 1017

^{13}C NMR IN Rb$_3$C$_{60}$ SAMPLES–TEMPERATURE AND PRESSURE DEPENDENCE OF RELAXATION

St. Kluthe, J. Roos, M. Mali, D. Brinkmann
Physik–Institut, Universität Zürich, 8057 Zürich, Switzerland
H.P. Lang, V. Thommen–Geiser, H.-J. Güntherodt
Institut für Physik, Universität Basel, 4056 Basel, Switzerland

Abstract

Preliminary studies of the temperature (T) and pressure (p) dependence of the ^{13}C NMR spin–lattice relaxation time, T_1, in several Rb$_x$C$_{60}$ samples are reported. We have measured $T_1(T)$ for the Rb$_3$C$_{60}$ phase obtaining a product $T_1 T$ which is independent of temperature in the range 80 – 300 K. At 295 K, $T_1(p)$ of ^{13}C in Rb$_4$C$_{60}$ was measured. Between normal pressure and 610 MPa, $1/T_1$ decreases by about 32%. This behavior resembles that of metallic K$_3$C$_{60}$. Pressurized helium induces an irreversible degradation of Rb$_3$C$_{60}$ but not of Rb$_4$C$_{60}$.

1. Introduction

The great scientific interest in alkali metal fullerides A$_x$C$_{60}$ is not only due to the discovery of superconductivity in some of these compounds with x = 3 but also to the interesting solid state properties of the different A$_x$C$_{60}$ phases. Nuclear magnetic resonance (NMR) is a well established method for studying such properties on the atomic scale. NMR measurements yield valuable insight in the electronic structure of metallic phases by probing the interactions between nuclear magnetic moments and conduction electrons and in the molecular motions, both influencing the NMR relaxation times. Since pressure as well as temperature is one of the most important thermodynamic variables in solid state physics, high pressure studies can provide a wealth of additional information.

In this study we investigated the ^{13}C spin–lattice relaxation time in the metallic Rb$_3$C$_{60}$ phase and the suppression of molecular motion by applying hydrostatic pressure in the insulating Rb$_4$C$_{60}$ phase.

2. Experimental

The two powder samples (A, B) were prepared from (99.9%, 99.5%) pure C$_{60}$ with naturally abundant ^{13}C. After annealing the C$_{60}$ powder for several hours in vacuum at $\sim$ 470 K, it was mixed in a glass vessel with metallic rubidium (molar ratio 1:3) under high purity argon gas atmosphere. The glass vessel was evacuated, sealed and heated for 80 hours at $\sim$ 670 K (A) and 60 hours at $\sim$ 725 K (B), respectively. Only sample A was treated additionally with intermediate homogenizing each 12 hours.

According to our ^{13}C NMR analysis, sample A contains $(75 \pm 5)\%$ Rb$_3$C$_{60}$ and $(25 \pm 5)\%$ unreacted C$_{60}$. Within experimental accuracy ($\sim 5\%$) we did not detect any Rb$_4$C$_{60}$. These findings are supported qualitatively by X–ray analysis. Sample B, according to ^{13}C magic-angle spinning (MAS) NMR, has a composition of $\sim 20\%$ Rb$_4$C$_{60}$ and $\sim 80\%$ Rb$_3$C$_{60}$, but no unreacted C$_{60}$. The variation in composition is attributed to the different sample preparation.

All ^{13}C NMR measurements were performed at a field of 4.7 T in a home made CuBe pressure cell similar to that described by Huber et $al.$[1]. The cell fits into a cylindrical commercial cryostat for temperature control. Inserting the sample tubes into the pressure cell was performed under helium gas. Because the pressure cell is gas tight against 700 MPa helium, no air can enter the cell from outside.

The T$_1$ data were taken using a saturation pulse sequence. The magnetization recovery was determined from the magnitude of the Fourier transform (FT) either using the free induction decay (FID) [sample B] or the spin echo [sample A].

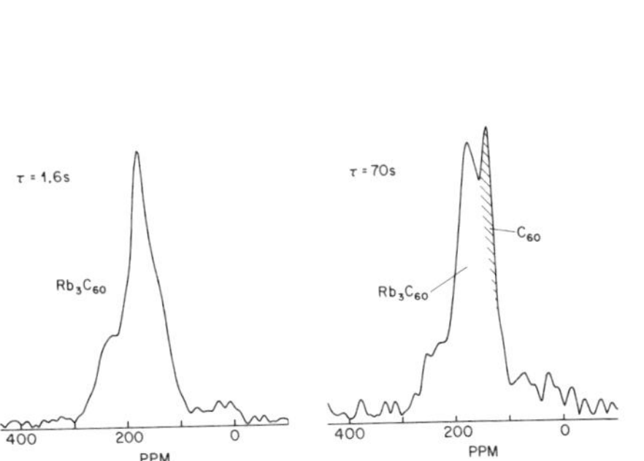

Fig. 1: Typical ^{13}C spectra of sample A at 294 K and ambient pressure for different recovery times τ (ppm relative to TMS).

Fig. 2: Magnetization recovery of ^{13}C in sample A at 294 K (integrated intensity) $\bullet$: corresponding spectra are shown in fig. 1.

3. Discussion

3.1 ^{13}C Spin–Lattice Relaxation Time of Rb$_3$C$_{60}$ in Sample A

Fig. 1 displays room temperature ^{13}C spectra for two different recovery times τ. For $\tau = 1.6$ s the spectrum is very similar to that observed by Murphy et $al.$[2] in Rb$_3$C$_{60}$, while for $\tau = 70$ s the pure C$_{60}$ signal is clearly detectable.

The magnetization recovery at 294 K as measured by the integrated intensity over the total ^{13}C spectrum (Fig. 2) reveals two relaxation processes, a fast one attributed to the Rb$_3$C$_{60}$ phase (see insert) and a very slow one due to unreacted C$_{60}$. Therefore, we fitted the data to the sum of two exponentials resulting in a perfect fit for the whole range from 10 ms to 100 s. The fit yields a Rb_3C_{60} content of 75%. Below ~ 120 K slight deviations in the fast relaxing part from a simple exponential behavior were detected.

Fig. 3 shows that $(T_1T)^{-1}$ for Rb_3C_{60} is temperature independent within experimental errors with a mean value of 0.0157 ± 0.008 $(sK)^{-1}$. The constant nature of $(T_1T)^{-1}$ seems to reflect the well known Korringa relation for a metal: $T_1TK^2 = h/(8\pi^2 k_B) \cdot (\gamma_e/\gamma_n)$, where γ_e, γ_n are the gyromagnetic ratios of electrons and nuclei, respectively, and K stands for the paramagnetic (Knight) shift of the resonance frequency. However, our T_1T value would yield K $\sim$ 260 ppm relative to insulating C_{60}^0. Since the experimental chemical and Knight shift values are in the range of 20 ppm [3] and 50 ppm [4], respectively, it is most probably not the Korringa relation which is responsible for the temperature independence of T_1T as already pointed out by Holczer et al.[5] for the case of K_3C_{60}.

Our data deviate slightly from earlier measurements in Rb_3C_{60} by Tycko et al.[6], which exhibit a nearly constant $(T_1T)^{-1}$ for the normal phase above 140 K followed by a gradual decrease down to T_c. T_1 was determined by fitting the peak height of the spectrum to a stretched exponential. Applying a similar procedure to our data, we obtain a result that, although qualitatively agreeing with that of Tycko et al., is not reliable because of the influence of the 25% C_{60} phase, in particular at lower temperatures.

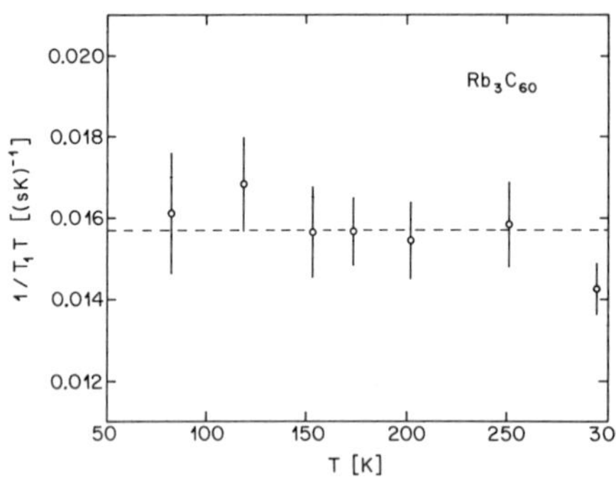

Fig. 3: $(T_1T)^{-1}$ of ^{13}C in the Rb_3C_{60} phase as a function of temperature (integrated intensity).

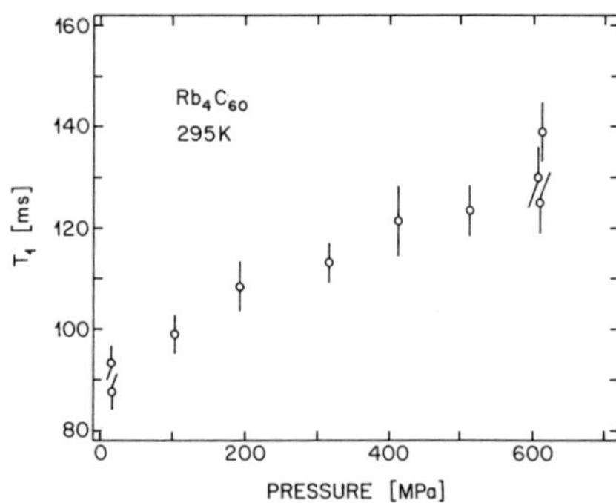

Fig. 4: T_1 of ^{13}C in the Rb_4C_{60} phase as a function of pressure at 294 K (peak intensity).

3.2 Sample B : Pressure Dependence of Relaxation in Rb_4C_{60}

Unfortunately, a pressure investigation of T_1 in sample A failed since it turned out that pressurized helium gas induces an irreversible degradation of Rb_3C_{60}. Possible explanations are an instability of the Rb_3C_{60} phase under pressure, the build up of other Rb_xC_{60} phases (e.g. x = 4) or a destruction due to helium entering the structure. We thus studied pressure effects in sample B only.

The ^{13}C NMR spectrum of sample B at 295 K consists of the typical broad feature belonging to Rb_3C_{60} and a narrower spectral component from Rb_4C_{60}. Although the peaks of these components could not be resolved, the magnetization recovery (at 295 K and 16 MPa) reveals the existence of two different phases, as also seen by MAS–NMR, with a fast relaxing peak ($T_1 = 90$ ms) and slower relaxing wings with T_1 values that agree with those of the the Rb_3C_{60} phase in sample A. The fast relaxing component is attributed to the Rb_4C_{60} phase because its T_1 agrees with MAS–NMR data of a *pure* Rb_4C_{60} sample and for intensity reasons.

Fig. 4 shows the pressure dependence of T_1 for Rb_4C_{60} as determined by single exponential analysis of the peak intensity. The data were taken in several pressure runs and were always reproducible. Within experimental error the frequency of the peak is not affected by pressure.

The increase of T_1 with pressure is in qualitative agreement with the behavior of metallic K_3C_{60} [7,3] below 500 MPa. It was argued[3] that the T_1 increase is dominated by the suppression of the molecular reorientations contributing to the relaxation at ambient pressure. Hence, this change in T_1 is not connected to the metallic properties of the material. Therefore our similar findings for Rb_4C_{60} probably are due to a pressure induced hindrance of the molecular motion.

We would like to thank M. Kanowski (Freie Universität, Berlin) for the discussion of his MAS–NMR results prior to publication. The partial support of our work by the Schweizerischer Nationalfonds is gratefully acknowledged.

References

1. H. Huber, M. Mali, J. Roos and D. Brinkmann, *Rev. Sci. Instrum.* **55** (1984) 1325.

2. D. W. Murphy, M. Rosseinsky, R. M. Fleming, R. Tycko, A. P. Ramirez, R. C. Haddon, T. Siegrist, G. Dabbagh, J. C. Tully and R. E. Walstedt, *J. Phys. Chem. Solids* **53** (1992) 1321.

3. R. Kerkoud, P. Auban–Senzier, D. Jérome, J. M. Lambert, A. Zahab and P. Bernier, *Europhys. Lett.* **25** (1994) 379.

4. G. Zimmer, M. Helmle, M. Mehring, F. Rachdi, J. Reichenbach, L. Firlej and P. Bernier, *Europhys. Lett.* **24** (1993) 59.

5. K. Holczer, O. Klein, H. Alloul, Y. Yoshinari, F. Hippert, S. M. Huang, R. B. Kaner and R. L. Whetten, *Europhys. Lett.* **23** (1993) 63.

6. R. Tycko, G. Dabbagh, M. Rosseinsky, D. W. Murphy, A. P. Ramirez and R. M. Fleming, *Phys. Rev. Lett.* **68** (1992) 1912.

7. G. Quirion, C. Bourbonnais, E. Barthel, P. Auban, D. Jérome, J. M. Lambert, A. Zahab, P. Bernier, C. Fabre and A. Rassat, *Europhys. Lett.* **21** (1993) 233.

Br NMR STUDY OF $C_{60}Br_{24} \cdot Br_2$

F. MILIA, G. PAPAVASSILIOU, A. LEVENTIS, G. PAPANTOPOULOS

NCSR "Demokritos", Institute of Materials Science
153 10 Ag. Paraskevi Attiki, Athens, Greece

and

P. VENTURINI, D. MIHAILOVIC, J. PIRNAT, AND R. BLINC

J. Stefan Institute, University of Ljubljana, Ljubljana, Slovenia

ABSTRACT

Here we report solid state [81]Br NMR measurements of powder halogeno derivatives of C_{60}. A second order quadrupole broadening 200 KHz wide Br NMR central $1/2 \rightarrow -1/2$ transition has been observed for a powdered sample. The lineshape and spin-lattice relaxation time T_1 vs. temperature can be assigned only to the $C_{60}Br_{24}$ bromine.

1. Introduction

The high reactivity of fullerenes C_{60} allowed recently the composition of halogenoderivatives $C_{60}X_n$ (X = Cl, Br, I and n = 6, 8, 24) (ref. 1,2). The halogen doped fullerenes have attracted a great deal of attention, since halogenated C_{60} species are important synthetic intermediates the study of which will help in the development of the chemistry and physics of fullerenes.

2. Experimental Procedure

The $C_{60}Br_{24} \cdot Br_2$ has been synthesized according to the procedure described in (ref. 3). C_{60} and liquid bromine formed by mixing a crystalline compound isolated as a bromine solvate. The composition of the crystal determined by elemental analysis showed one Br_2 molecule for every $C_{60}Br_{24}$ molecule.

Previous work, by means of X-ray, Raman, and IR techniques, indicates that $C_{60}Br_{24}$ displays T_h symmetry slightly reduced from the I_h symmetry of the parent molecule (ref. 1,3). This is so as sp^3 carbons at sites of bromination distort the surface. Additional ^{13}C NMR measurements, support the evidence for the existence of a highly symmetric structure (ref. 4).

2.1 Lineshape measurements

Figure 1. shows the lineshape vs. temperature in a 4.7 T superconductive magnet.

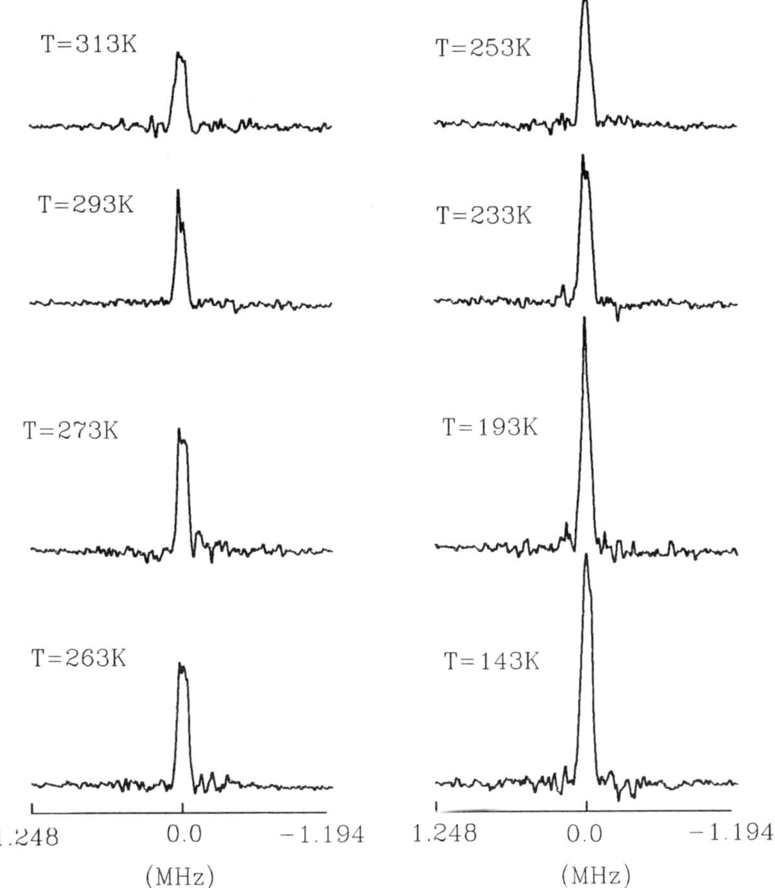

FIGURE 1: ^{81}Br NMR lineshape vs. temperature.

We observed the existence of an asymmetric line with approximately 200 KHz width at room temperature, due to a second order quadrupolar broadening. The quadrupolar coupling constants for both Br_2 and Br_{24} molecules are known to be (ref. 5),

$$\frac{e^2qQ}{h}(Br_2) \approx 350 \text{ MHz}$$

$$\frac{e^2qQ}{h}(C\text{-}Br_2) \approx 100\text{-}150 \text{ MHz} \tag{1}$$

The second order quadrupolar shift of the Zeeman Br NMR line gives a frequency distribution due to powder average

$$f\left[v^{(2)} - v_0\right] = \frac{v_Q}{v_L}\langle f(\theta,\phi,n)\rangle \qquad (2)$$

where $v_Q = \dfrac{e^2 qQ}{2h}\sqrt{1 + \dfrac{n^2}{3}}$, v_0 equals to the unperturbed Br Larmor frequency v_L, and n is the asymmetry parameter. By lowering the temperature the line becomes narrower and more intense till $T \approx 230$ K. From there the line is almost stable.

In Figure 2. the intensity and the total signal area are compared, as expected the line increase its intensity and becomes narrower down to $T \approx 230$ K (see also Figure 3. where the width is shown vs. temperature). From these measurements we see the possible existence of a first order phase transition at $T \approx 230$ K.

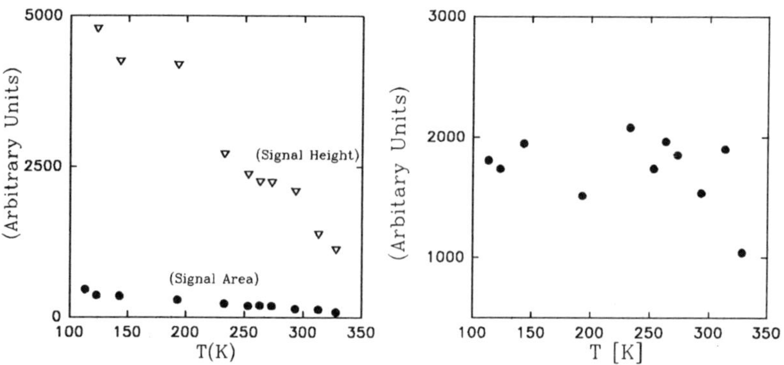

FIGURE 2: Intensity and total signal area vs. temperature.

FIGURE 3: Full width at half intensity vs. temperature.

2.2 Spin lattice relaxation time (T_1) measurements

For T_1 measurements the $90° - \tau - 90°$ pulse sequence was used and the data analysis was performed on the total signal area by performing a multi exponential fit. The best results as one can see from Figure 4. were observed for a two exponential fit. The analysis exhibits already two components that change with temperature with an average ratio $M_{0a}/M_{0b} \approx 2$. In Figure 5. we also observe the possible phase transition at $T \approx 230$ K.

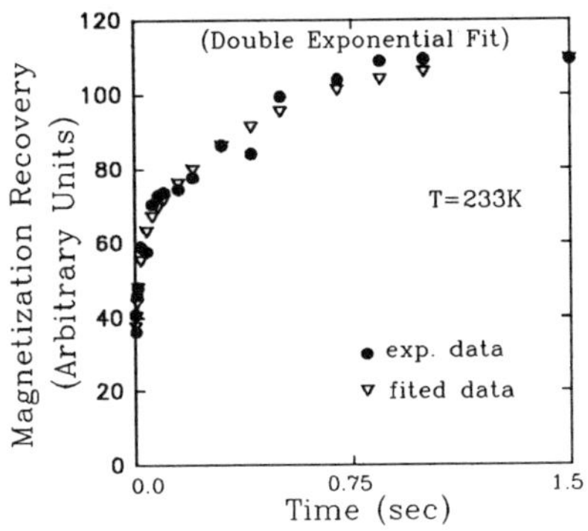

FIGURE 4: Double exponential T_1 fit.

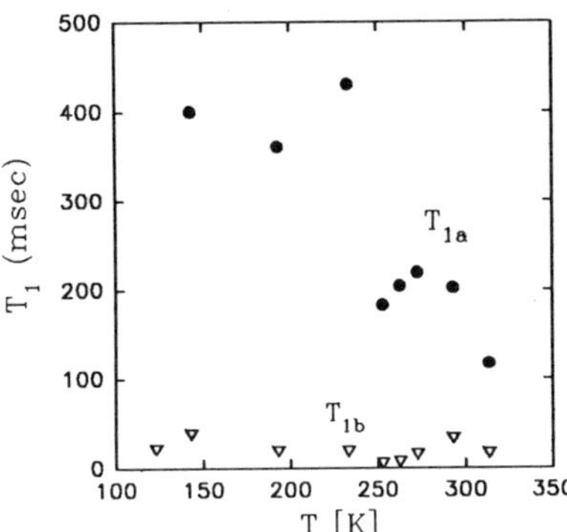

FIGURE 5: T_1 components vs. temperature.

3. Discussion

The above results, exclude the case that the line belongs to the dynamically disordered (over fourteen sites) Br_2 molecules. In such case the quadrupolar broadening should increase with lowering the temperature leading to a smaller (lower intensity) NMR signal. The increase in the molecular rotational correlation time with decreasing temperature would broaden the line into a powder pattern shape (ref. 4.). The signals from Br_{24} are much stronger than the signals of Br_2 mainly for two reasons.

(i) Smaller frequency spread because of the smaller ν_Q (Eqs. 1,2).

(ii) We have 12 times larger abundance.

As a consequence the signals from Br_2 would be smeared out in a powder sample over a large frequency range and lost in the noise background. We therefore can conclude that the observed NMR "line" is a singularity of the frequency distribution resulting from the powder averaging in an external magnetic field.
In addition, the temperature dependence of the spectrum means that the $C_{60}Br_{24}$ groups are rotating in the temperature interval study leading to a motional averaging.
The work is still in progress.

3. References

1. Paul R. Brikett, et al., *Nature*, **357**, (1992) 479.
2. Paul R. Brikett, et al., *Chem. Phys. Let.*, **205**, (1993) 399.
3. F.N. Teppe, et al., *Science*, **256**, (1992) 822.
4. C. S. Yannoni, et al., *J. Phys. Chem.*, **95**, (1991) 9;
   R. Tycko, et al., *J. Phys. Chem.*, **95**, (1991) 518.
5. A. Abragam, *The Principles of Nuclear Magnetism*, (Oxford, 1973) 232.

PARAMAGNETIC STATES IN PRISTINE FULLERENES

A. Bartl, L. Dunsch, J. Fröhner, U. Kirbach and B. Schandert
Institut für Festkörperforschung im
Institut für Festkörper- und Werkstofforschung e.V. Dresden,
Helmholtzstr. 20, D-01069 Dresden, Germany

ABSTRACT

In pristine fullerenes a very weak ESR signal is often observed. The ESR spectra of the solid soot extract and of C-60 and C-70 fractions show the same linewidths of 0.1 mT but the spin concentrations are quite different. The lowest spin concentration is found in the soot extract. In pure C-60/C-70 material it is about 10^{17} spins/g, but in all samples is this only one unpaired electron per 1000 to 10000 fullerene molecules.

Alternations of the ESR line shape can be observed with increasing temperature in pristine fullerene. Up to 300° C the lineshape of the signal changes and therefore it is concluded that the structure of the paramagnetic centres changes also. Above 400° C a symmetric single ESR line is found in vacuum and the ESR linewidth decreases with increasing temperature. A more dramatically change is demonstrated for the concentration of the paramagnetic species during the temperature treatment. In the region between 300° C and 600° C the spin concentration increases linearily by three orders of magnitude.

1. Introduction

Since 1985 it has been known that besides the two classical carbon modifications diamond and graphite there exist fullerenes as a third modification[1]. These compounds are regular carbon clusters differing in the number of carbon atoms, e. g. 60, 70, 82 etc. Since 1990 methods exist to produce these substances in fairly large quantities[2]. Today - by electrical burning of graphite electrodes in a helium atmosphere - fullerene material as a mixture of C-60 and C-70 extracted from soot is available in many laboratories[3].

There are also many different and controversial results concerning the ESR data. The reasons of these differences are mainly differences in the experimental conditions, e.g. sample preparation , sample properties, measuring conditions etc.

From this point of view the initial fullerene material has to be characterized by ESR with respect to spin concentration and linewidth of the spectra. For further investigations e. g. on chemical and electrochemical doping oxygen is to be removed by heating the fullerene samples in vacuum to obtain defined starting conditions.

The aim of this paper is to obtain from the ESR experiments information on the paramagnetic species in pristine fullerenes.

2. Experimental

Fullerene samples were prepared by evaporating graphite electrodes in 100 Torr Helium atmosphere described in[3]. For separation of C-60, C-70 and higher fullerenes a preparative HPLC column was used.

Mass spectra were recorded using a MAT 95 (FINNIGAN) spectrometer which allows the detection of fullerenes by electron impact, fast atom bombardment and chemical ionization. For all samples chemical ionization was used.

Electron spin resonance spectra were recorded by an ERS-300 X-band spectrometer (ZWG Berlin) with 100 kHz modulation, 9.5 GHz microwave frequency and 340 mT magnetic field strength. The glass sample tubes have a diameter of 4 mm.

3. Results and discussion

Generally in each fullerene sample a single line ESR spectrum can be found. The spin concentrations of these samples and the linewidths of the ESR signals measured on air and at room temperature are shown in Table 1.

Table 1: ESR properties of starting materials and different C-60 samples

materials	$\Delta B/mT$	$[R°]/10^{17}g^{-1}$
spectrum coal	1.82	2.7
soot	0.23	14.0
soot extract	0.11	0.3
C-60 (up to 50°C heated)	0.11	1.2
C-60 L (with solvent)	0.11	14.0
C-60 (150°C/220°C in vacuum)	0.13	2.0
C-70	0.13	4.0
C-60/C-70 (Hoechst AG)	-	< 0.0001

From the different ESR linewidths follows that the paramagnetic species in spectrum coal and also in soot are quite different from that in pure fullerenes. It is assumed that the paramagnetic centres in coal and soot are defects in the structure of the graphitic particles.

The ESR signals of soot extract and of the C-60 and C-70 fractions show the same linewidths but the spin concentrations are quite different. The lowest spin concentration is found in the soot extract. After different temperature treatments the spin concentration increases. In pure C-60/C-70 material the spin concentration is about 10^{17} spins/g, but this is only one unpaired electron per thousand fullerene molecules.

By temperature treatment between 20° C and 600° C under vacuum conditions better than 10^{-3} Pa in sealed ESR sample tubes remarkable changes of the ESR lines can be observed with increasing temperature. Up to 300° C the lineshape of the signal changes and therefore it is concluded that the structure of the paramagnetic centres changes also. Above 400° C a symmetric single line in vacuum is found and the ESR linewidth decreases drastically with increasing temperature.

A strong change is also seen in the concentration of the paramagnetic species during the temperature treatment (Fig. 1). Up to 300° C practically the same spin concentration is measured neglecting a small decrease in spin concentration. But above 300° C a strong increase can be observed. In the region between 300° C and 600° C the spin concentration increases by three orders of magnitude.

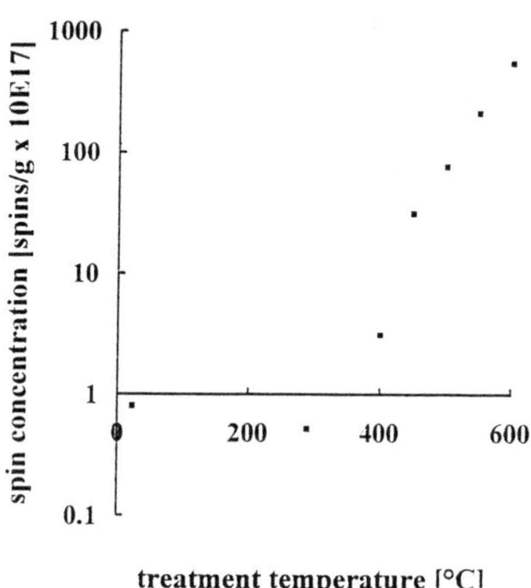

Fig. 1: Increase of spin concentration by heat treatment in C-60 powder in vacuum

An analogues behaviour shows the ESR linewidth as a function of the treatment temperature (Fig. 2). Starting at 20° C the linewidth increases from 0.1 mT to more than 0.6 mT at 400° C measured at room temperature. In the region of 400° C to 600° C the ESR linewidth decreases down to 0.05 mT.

This very small linewidth is comparable with that of high conjugated structures e. g. in conjugated polymers like polyacetylene with $\Delta B = 0.05$ mT or two dimensional polypyrrole with $\Delta B = 0.02$ mT measured in vacuum[5].

From all these results it is concluded that in temperature treated C-60 the paramagnetic centres are unpaired electrons trapped in an extented π-electron system. This π-system is more perfect after heat treatment probably because oxygen is removed and defects remain. The observed ESR line is narrow, that means that the electrons are delocalized over several carbon atoms of fullerene molecules and the ESR line is dynamically averaged on these molecules. Before heat treatment mass spectrometric measurements show fullerene molecules which might be obviously an epoxy structure.

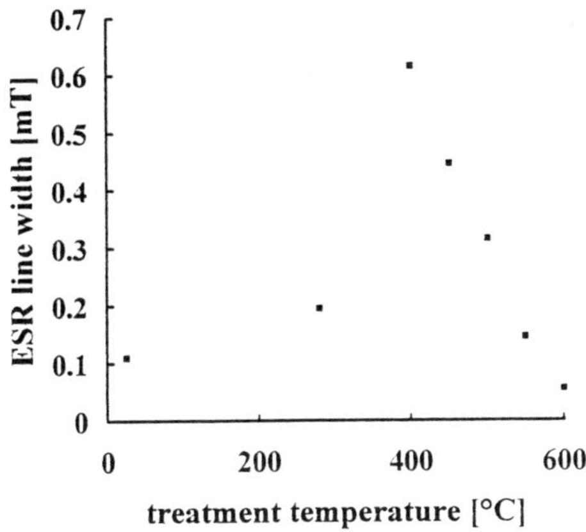

Fig. 2: ESR linewidth of C-60 powder as a function of temperature treatment in vacuum

In this respect it can also be explained that in samples without heat treatment ^{13}C satellit lines can be observed and in heat treated samples no ^{13}C satellit lines were seen. The ^{13}C hyperfine coupling constant was measured to be 0.06 G by time-resolved ESR in partial and full ^{13}C enriched C-60. Because of this splitting constant is so small, no distinct satellite line can be observed with ^{13}C in natural abundance at 20° C. In oxygen contaminated fullerenes the unpaired electrons are strongly localized and therefore the ^{13}C coupling constant is so large that these lines as satellite lines can be found. In our case the ^{13}C coupling constant is 1.2 mT. It means, that in this case oxygen is incorporated in the sample.

4. Acknowledgement

Financial support of this work by the **Sächsisches Ministerium für Wissenschaft und Kunst** is gratefully acknowledged.

5. References

1. H. W. Kroto, J. R. Heath, S. C. O'Brien, R. F. Curt and R. E. Smalley; Nature **318** (1985) 162.
2. Krätschmer, L. D. Lamb, K. Fostiropoulos and D. R. Huffman; Nature **347** (1990) 354.
3. L. Dunsch, F. Ziegs, J. Fröhner, U. Kirbach, K. Klostermann, A. Bartl and U. Feist; in Electronic Properties of Fullerenes (Springer Series in Solid State Sciences 117, 1993) p. 39.
4. D. Zhang, J. R. Norris, P. J. Krusic, E. Wassermann, C. C. Chen and Ch. M. Lieber; J. Phys. Chem. **97** (1993) 5886
5. A. Bartl, L. Dunsch, H. Naarmann, D. Schmeißer and W. Göpel, Synthetic Metals **61** (1993) 167.

Magnetic properties of Na_2C_{60} and K_4C_{60}

P. Petit[1], J. Robert[1], J.-J. André[1], T.Yildirim[2] and J. E. Fischer[3]

[1] Institut Charles Sadron, CNRS, 6, rue Boussingault, 67000 Strasbourg, France.
[2] Department of Physics and Laboratory for Research on the Structure of Matter University of Pennsylvania, Philadelphia, PA 19104-6396.
[3] Materials Science Department and Laboratory for Research on the Structure of Matter, University of Pennsylvania, Philadelphia, PA 19104-6272.

Abstract: Every alkaly doped C_{60} phases present evidences of phase transitions other than superconducting transitions, or phase separations. Here we focusse on ESR studies of the high temperature transitions due to changes in tC60 dynamics. These are responsible for strong change in spin - phonon interactions, leading to low spin - high spin equilibria. The activation energies of which are shown to be strongly dependent on the intraball phonon mode populations.

1. Introduction. Since fullerenes became available in macroscopic amounts [1], numerous intercalation compounds have been obtained, some leading to superconducting states [2-4]. However, very few ESR experiments as a function of temperature and on a wide temperature range have been reported on these paramagnetic systems, although this technique is a very powerfull probe of spin interactions [5, 6]. In pure C_{60}, the temperature dependence of ESR parameters reflects the passage from an orientationally ordered phase at low temperature to a disordered phase at high temperature in which molecules are freely rotating. The transition temperatures are strongly dependent on the alkali counter-ions in the intercalation compounds. Once doped, phase transitions in these paramagnetic compounds may be studied by ESR. Investigations of these transitions using ESR on Na_2C_{60} and K_4C_{60} are reported in the present paper.

2. Experimental. The samples have been prepared as previously reported [4, 7] and were sealed under helium atmosphere in 3 mm diameter quartz tubes and characterized as pure phase by x-ray diffraction.

Continuous wave (CW) ESR experiments have been performed in X band (10GHz), using a standard Bruker ESP 300 spectrometer equipped with a powermeter, a NMR gaussmeter and a frequencymeter. A standard Oxford He gas flow system was inserted into the cavity allowing to change only the sample temperature, while the measuring parameters, such as the quality factor (Q) of the resonator, remained unchanged. The absolute values of the susceptibility have been determined by comparing the integrated signal intensity to that of a known quantity of a coal reference sample. These measurements provide the susceptibility in terms of N_{eff}, the number of effective free spins (giving ESR signal), assuming that the susceptibility at any temperature can be expressed as $\chi = g^2\mu_B^2 N_{eff}(T)/(3k_BT)$.

Pulsed ESR experiments have been performed on a Bruker ESP 380 FT spectrometer equipped with an Oxford CF 935 cryostat and a 1kW TWT amplifier. Pulse lengths of 24 ns and 48ns have been used for the $\pi/2$ and π pulses respectively. Spin-spin (T_2) and spin-lattice (T_1) relaxation times have been measured respectively by fitting the free induction decays (FID) following $\pi/2$ pulses and the inversion-recovery (IR) curves of the magnetization obtained by the classical π–τ-$\pi/2$ pulse sequence. The equilibrium magnetizations (at zero time) have been determined by extrapolating the curve (FID and IR) fits at the middle of the first pulse. The dead time of the spectrometer has been measured for each temperature, the Q factor of the cavity slightly deacreasing with increasing temperature.

3. Results and discussion

$\underline{Na_2C_{60}}$.

The ESR susceptibility of Na_2C_{60} as a function of temperature is shown on figure 1.a. The low temperature regime is characteristic of a Curie-Weiss law between 200K and 10K. Below 10K, a drop of the susceptibility is observed, both by CW and pulsed ESR (figure 1.b). Above 200K, the susceptibility increases up to 360K and then slowly decreases. The transition between both two regimes is broad (about 150 degrees) and shows hysteresis of 15K (figure 1, inset).

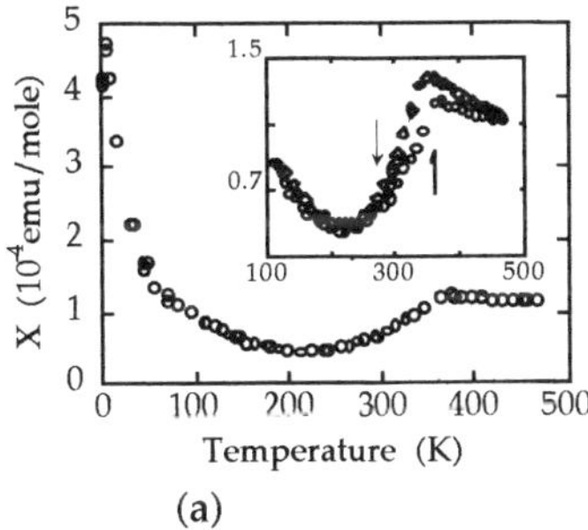
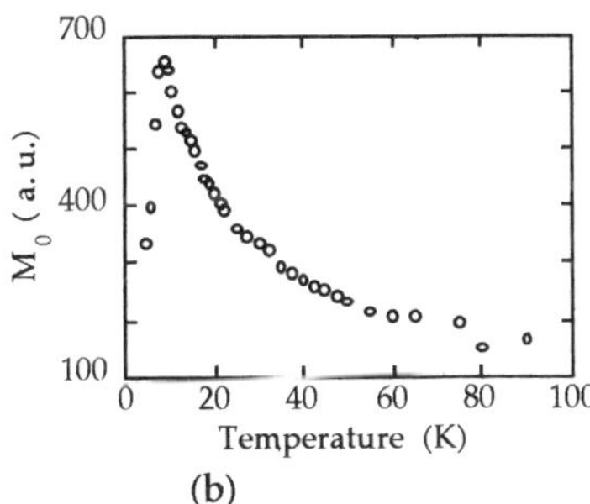

(a) (b)

<u>Figure 1</u>: a) Variation of the susceptibility as a function of the temperature. Inset: Hysteresis of the susceptibility at the transition (circles: heating, diamonds: cooling).

b) Variation of the equilibrium magnetization in the low temperature range.

The temperature dependence of the ESR linewidth also reflects this transition (figure 2). Above 400K, the peak-to-peak linewidth (of about 15 Gauss) increases with temperature. Passing through the transition on cooling, the peak-to-peak linewidth drops from 14 Gauss at 400K to 3 Gauss at 300K. The same hysteresis as seen in the susceptibility is observed in the linewidth. Below 300K, the linewidth decreases continuously, down to 0.5 Gauss at 4K.

480

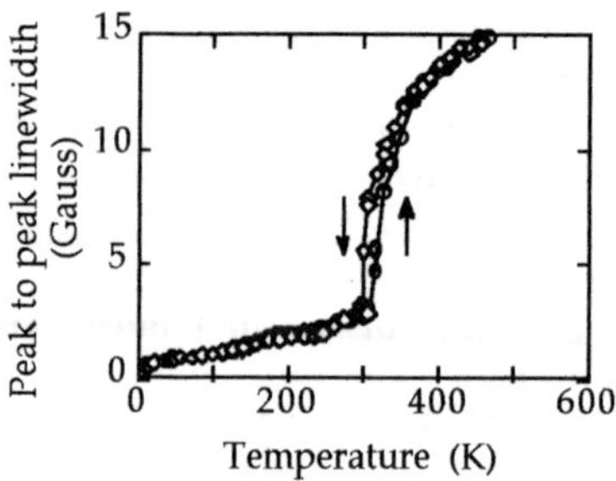

Figure 2: Variation of the peak-to-peak linewidth as a function of temperature (circles: heating, diamonds: cooling).

The low temperature Curie-Weiss dependence of the susceptibility may then be attributed to an other phase Na_xC_{60} ($x<1$). This assumption is supported by the high value of the absolute susceptibility, much higher than the one characteristic of defects in pure C_{60}. The fit of the low temperature regime gives an antiferromagnetic Weiss temperature of about 12K.

The increase of susceptibility from 250K to 470K can be understood as a singlet-triplet equilibrium (S-T) of the spin state of C_{60}^{2-} dianions. Fitting the susceptibility on the whole temperature range in terms of Curie-Weiss and singlet-triplet equilibrium gives an S-T activation energy of about 1000K. However, the fit is poor in the high temperature range (figure 3). The low and high temperature regimes have different space groups (orientationally ordered and disordered phases respectivelly), differentiated by the free rotation of the balls at high temperature ($T>T_m$) [8]. Thus we can suppose that the intraball phonon modes (populated above T_m) are responsible for the S-T equilibrium and that its activation energy decreases as the high frequency phonon modes become populated. We then introduce an activation energy of the form $\varepsilon(T)=\varepsilon_0\exp(E/kT)$, proportionnal to $1/<n>$, where $<n>$ is given by the Bose-Einstein statistics. This new fit of the spin susceptibility vs temperature (figure 3) gives a value of E of about 1000K, indicating that intraball phonons modes are responsible for the S-T equilibrium. Detailed analysis of the fit parameters will be given elsewhere [9].

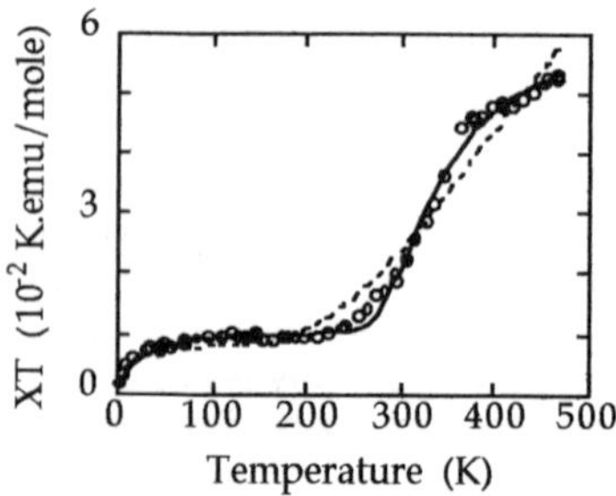

Figure 3: χT versus T. Dashed line: fit $\alpha/(1+\beta/T)+3\gamma/[3+\exp(\varepsilon/T)]$

Solid line: fit $\alpha/(1+\beta/T)+3\gamma/[3+\exp(\varepsilon_0/Txexp(E/T))]$

<u>K$_4$C$_{60}$</u>.

The general behaviour of the spin susceptibility of K$_4$C$_{60}$ is comparable to that observed on Na$_2$C$_{60}$ (figure 4). A broad transition takes place over a wide temperature range (200-350K) in which the susceptibility increases with increasing T.

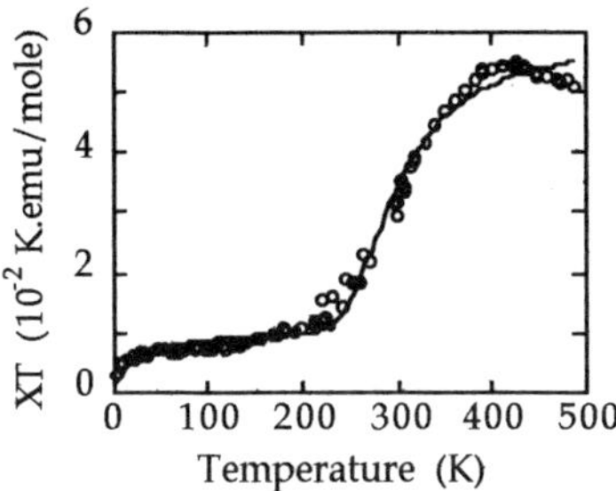

<u>Figure 4</u>: Variation of the susceptibility of K$_4$C$_{60}$ as a function of temperature. Solid line: same as for figure 3.

Using the same model as above, the value of E obtained is closed to the one for Na$_2$C$_{60}$. However, the reported results on K$_4$C$_{60}$ are preliminary ones and a possible equilibrium toward a highest spin state (s=2) may have to be taken into account.

Conclusion. Up to now, no transition of both studied compounds at high temperature have been observed. A possible transition from an insulating state at low temperature to a metallic one at high temperature can be ruled out. Indeed, such a transtion would lead to a Pauli susceptibility at high temperature, but the absolute value of this Pauli susceptibility could not be higher than the one of the insulating phase. Thus, the increase of susceptibility of these systems with increasing T reflects an intrinsic feature of both materials, which we interpret as a low spin - high spin equilibrium. The activation energies of these equilibria have been shown to be linked to the phonon modes arising above the orientational disordered - ordered phase transition.

Aknowledgements: The Conseil Regional is thanked for financial support. One of us (J. R.) benefits of a MRE Grant. Work at Penn supported by the NFS Materials Research Laboratory Program DMR91-20668.

References   [1] W. Krätschmer et al., Nature **347**, 354 (1990).

[2] A. F. Hebard et al., Nature **350**, 600 (1991).

[3] K. Holczer et al., Science **252**, 1154 (1991).

[4] K. Tanigakiet al., Nature **356**, 419 (1992).
[5] W. H. Wong et al., Europhys. Lett. **18**, 79 (1992).
[6] A. Jánossy et al., Phys. Rev. Lett. **71**, 1091 (1993).
[7] T. Yildirim et al., Nature **360**, 568 (1992).
[8] T. Yildirim et al., Phys. Rev. Lett. **71**, 1383 (1993).
[9] P. Petit et al., to be published.

in situ ESR-SPECTROSCOPIC STUDIES OF CHARGE TRANSFER AT FULLERENES

L. Dunsch
*Institut für Festkörperforschung, IFW Dresden e. V.,
Helmholtzstr. 20, D-01069 Dresden*

ABSTRACT

Electron spin resonance spectroscopy is used to characterize the electronic state of fullerenes in charge transfer reactions. This work is centered to C_{60} as the model for simple fullerenes. Studies of the formation of different anionic spin states and their equilibria by the in situ electrochemical ESR technique are presented.

The influence of the electrode surface in the electron transfer reactions of fullerenes and the formation of fullerene surface layers on the working electrode demonstrated. The mono- and dianion of the fullerene formed at the electrode are found by in situ ESR-spectroscopic measurements and their kinetic behaviour is discussed. Furthermore it is shown that functionalization of the fullerene molecule does not suppress the spin states of the anions and well established paramagnetic mono- and dianions are formed which still have the spin density localized at the C_{60} molecule.

1. Introduction

There are few ESR-spectroscopic studies[1-5] of the spin states of fullerenes produced by electrochemical reductions[6,7] while a first indication of the reversible oxidation of C_{60} was not successful in the detection of a paramagnetic species[8]. It has been demonstrated by several authors that electrochemical reduction forms different spin states of the fullerenes and the free electron at the fullerene molecule is delocalized resulting in a single line ESR-spectrum.

Up till now no detailed in situ ESR-spectroscopic study of fullerenes at electrodes was made taking into account the special situation at the electrode interface. Furthermore the reaction of the fullerene anions formed electrochemically was not followed in situ by ESR-spectroscopy. The role of substituents at the carbon cage of the fullerene in the formation of spin states and their reactivity was not studied in more detail elsewhere[12]. Therefore the in situ characterization of the electrochemical reactions of fullerenes by ESR-spectroscopy was expected to result in more insight in the structure of fullerene spin states and their kinetic behaviour.

2. Experimental

The in situ ESR-method at electrochemical systems and the equipment used for this study is described elsewhere[9]. Electron spin resonance spectroscopy was done with the ERS 221 X-band cw-spectrometer (ZWG Berlin) with 100 kHz field modulation and 1-15 mW microwave power at room temperature. The electrochemical equipment consisted of a computer-driven potentiostat PG 285 (HEKA, Lambrecht).

The quartz ESR-cell for in situ measurements of electrochemical reactions of fullerenes in solution under stable inert conditions is shown in Fig. 1. The cell is a modification of the flow-through type described earlier[9]. This cell is constructed in such a manner that it can be filled in a glove box under conditions of very low water and oxygen content, closed inside the box, transferred into the ESR-cavity and used in the in situ measurements. The material of the working electrode was either platinum or a carbon fibre bundle. As a reference electrode a silver wire coated with silverchloride was used. All electrode potentials are referred to this potential. The counter electrode was in each case a platinum wire. Methylenechloride was used as solvent and tetrabutyl-ammoniumtetrafluoroborate as supporting electrolyte. The UV-irradiation of the electrode inside the in situ ESR electrochemical cell was done with UV-lamp HBO 200 (Wetron).

The fullerene soot was produced according to the method of Krätzschmar et al.[11] by a dc arc vaporization using the apparatus developped by our group[10]. This apparatus is characterized by a strong water cooling of the burning rods and the metallic mantle of the burning chamber. The purification of the fullerene samples by HPLC and the characterization of the samples by mass spectroscopy is also described in ref.[10].

3. Results

Longterm experiments in the ESR spectroscopic studies at solid electrodes in solutions of fullerenes (C_{60}) have shown that there are stable paramagnetic centres already at off-potential which seems to be free spins in a surface layer of C_{60} at the

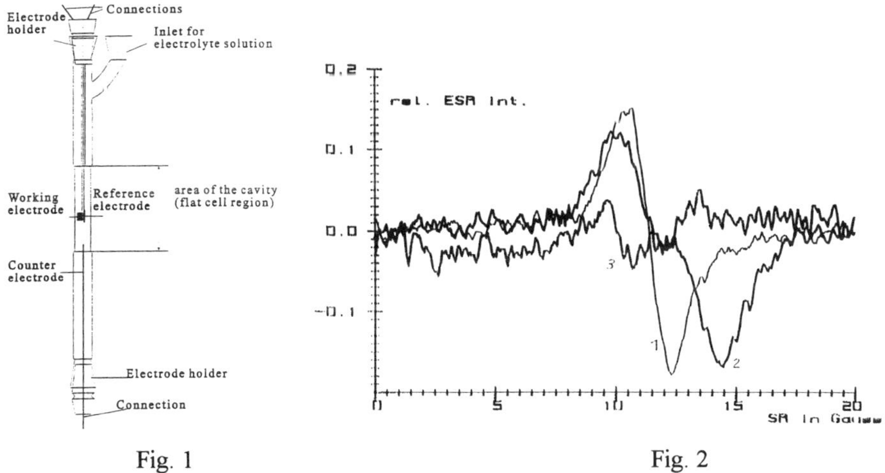

Fig. 1: ESR-electrochemical cell for in situ measurements of fullerenes

Fig. 2: ESR spectrum of a C_{60} film at the platinum electrode in $MeCl_2$ $TBAPF_6$ solution
       1- Off-potential situation
       2- Reduction at -500 mV
       3- Oxidation of the film at +100 mV

electrode. By visuell checking of the electrode during the experiments it is seen that a brown surface layer is formed at the electrode. The ESR-signal found at the off-potential (Fig. 2, curve 1) is changed by electrochemical reduction at -500 mV in linewidth and shape indicating the formation of a new paramagnetic species at the electrode (Fig. 2, curve 2). Taking into account that at this potential of the first voltammetric peak the monoanion is formed giving its own ESR-signal (see below) the paramagnetic center inside the surface layer is different from that of C_{60}. Thus chemical changes like a polymerization of the C_{60} are likely to occur in the electrochemical experiments especially at more negative potentials. The resonance of the surface film is diminished at anodic potential of +100 mV (Fig. 2, curve 3) which is more positive than the re-oxidation peak of the C_{60}-anion. This is an additional hint for the presence of a new chemical structure of the surface layer and its paramagnetic species.

If the electrochemical reduction of C_{60} at the platinum electrode in methylene-chloride solution is done under UV-irradiation the electrochemical behaviour is completely changed. The voltammetric peak of the first electron transfer reaction is deformed to a voltammetric step of higher current density indicating that the electron transfer to the C_{60} is enlarged by UV-irradiation (Fig. 3, curve 2). Furthermore a new voltammetric step at about -180 mV is found. The nature of the electron transfer at this potential is not yet clear, but the enlarged charge injection is accompanied by the formation of a surface layer with a new chemical structure being obviously different from that of the layer formed in long term electrolysis. There is no detailed structural information of the layer although an oligomeric is to be assumed.

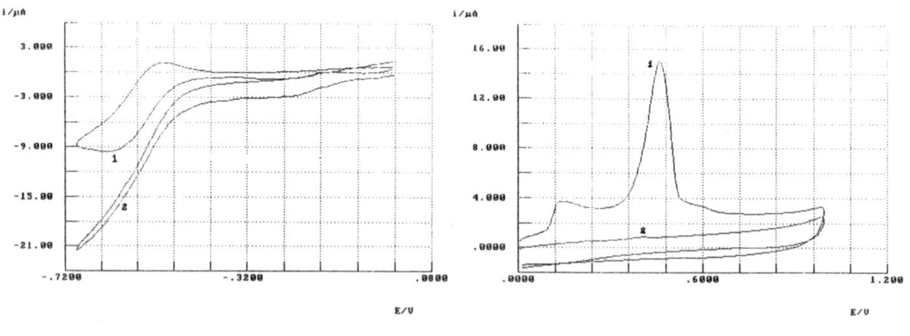

Fig. 3   Fig. 4

Fig. 3: Cyclic voltammogram of C_{60} in methylenechloride solution under UV-irradiation at the platinum electrode of the in situ ESR-cell in the cathodic range
> Curve 1- no irradiation
> Curve 2- under irradiation

Fig. 4: Cyclic voltammogram of a surface layer at the platinum electrode within the in situ ESR-cell in UV-irradiated C_{60}/methylenechloride solution in the anodic range
> Curve 1- first cycle
> Curve 2- second cycle

Looking for the anodic reaction of this surface layer it is seen that there exist two anodic peaks of different height attributed to two distinct (charged) states inside the film (Fig. 4, curve 1). The peaks are totally irreversible which seems to be due to a bond breaking. In the second cycle no further electrochemical reaction is to be seen (Fig. 4, curve 2).

It is important to point out that all in situ ESR-measurements of the C_{60}-solution under UV-irradiation did not result in the detection of any ESR-signal. Therefore the spin states of the fullerene and the fullerene surface layer are activated under UV-irradiation for chemical reactions causing irreversible behaviour.

To follow in situ the electrochemical formation of the C_{60} monoanion and dianion it is important to use clean electrode surfaces and no prolonged electrochemical experiments. In that way the ESR-signal of both the C_{60} monoanion and dianion can be detected by in situ measurements at room temperature (Fig. 5). The electrochemical formation of the monoanion is followed by comparing the spin concentration with the charge injected. The ratio found after subtraction of the charging current is of about 3. The linewidth of the ESR-signal is 1.02 G which is quite different from results of the study at frozen solutions where very broad lines (40-50 G) were found. In all our in situ experiments no broad ESR-signal could be detected.

Increasing the cathodic potential causes the formation of the dianion the linewidth of which is to some extent higher than that of the monoanion. Furthermore the spin concentration of the dianion is not doubled as expected. The factor for the enlargement of the spins is higher than 2. This behavior is explained by an spin pairing of the monoanion in solution at room temperature while the dianion exits in the non-aggregated form. Furthermore no remarkable difference in the g-factor could be seen in comparing the ESR-spectra of both the mono- and the dianion.

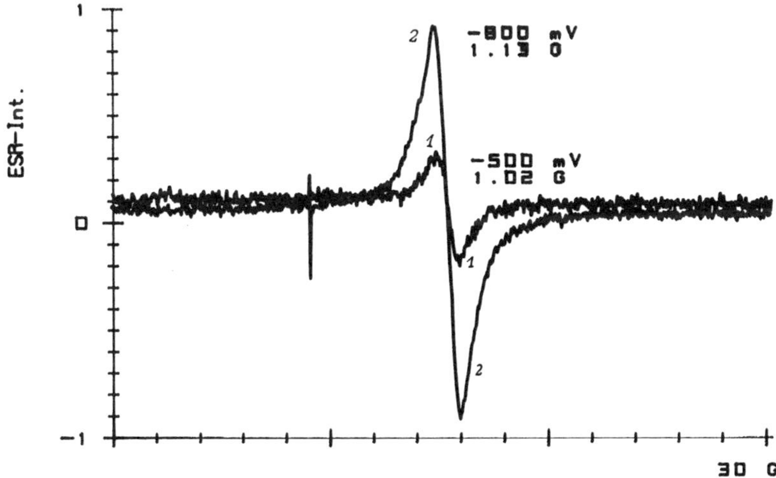

Fig. 5: ESR spectrum of C_{60} anion (curve 1) and dianion (curve 2) in methylenechloride solution formed electrochemically at -500 mV and -800 mV resp.

The kinetic behaviour of the monoanion and dianion formation at a platinum electrode is shown in Fig. 6. During the linear potential scan between +50 mV and -950 mV (Fig. 6, curve 1) the two current steps of monoanion and dianion formation can be detected. The ESR-intensity measured simultaneously to the current response clearly demonstrates that the spin concentration in the first step is much lower than that of the second. While the increase in the ESR-intensity is going parallel to the current increase in the forward scan the electrochemical back reaction is not characterized by distinct steps in the ESR-intensity which is due to the synproportionation of the dianion with the pristine C_{60} and the diffusion of the paramagnetic centers away from the electrode into the elctrolyte solution. It is obvious that homogeneous electron transfer reactions of fullerenes especially of different redox state take place.

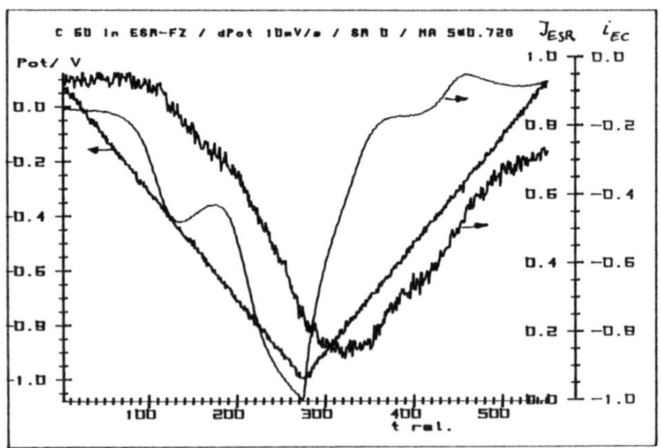

Fig. 6: Potential variation, current response and relative ESR-intensity at an in situ ESR electroche-mical cell filled with C_{60}/methylenechloride solution at a platinum electrode

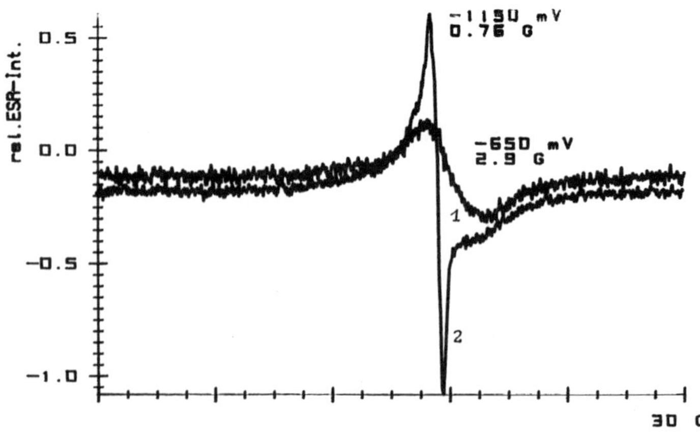

Fig. 7: ESR spectrum of {4-benzo[15]crown-5} phenylmethanofullerene monoanion (curve 1) and dianion (curve 2) in methylene chloride solution formed electrochemically at the platinum electrode

An example for this result is also given by the in situ ESR-spectroscopic measurement of {4-benzo[15]crown-5} phenylmethanofullerene(C_{60})[13] reduction at the platinum electrode. In the first reduction step at -650 mV the monoanion is formed resulting in an ESR-signal with a linewidth of 2.9 G. The reduction at the second voltammetric step (-1150 mV) gives an superimposed ESR-spectrum of the monoanion and the dianion signal with a linewidth of 0.75 G. This can be explained by the synproportionation reaction of the dianion with pristine C_{60}. Furthermore it is demonstrated that the C_{60} derivative gives no rise of a dramatic change in the electronic situation in the fullerene shell but the electron is delocalized at C_{60}.

4. Acknowledgement

This work was supported in part by a grant of the **Sächsisches Ministerium für Wissenschaft und Kunst** which is gratefully acknowledged. Furthermore we thank Prof. Fritz Vögtle und DC Jens Osterodt (University of Bonn) for the gift of the crown-fullerene, for discussions concerning ESR-measurements Dr. Anton Bartl and Dr. Andreas Petr and for preparative work on the fullerene samples Dr. Jürgen Fröhner and Uwe Kirbach. The technical assistence of Ulrike Feist and Frank Ziegs are duly acknowledged.

5. References

1. D. Dubois, M. T. Jones and K. M. Kadish,
   J. Am. Chem. Soc. 114 (1992) 6446.
2. D. Dubois, K. M. Kadish, S. Flanagan, R. E. Haufler, L. P. F. Chibante and
   L. J. Wilson, J. Am. Chem. Soc. 113 (1991) 4364.
3. T. Kato, T. Kodama, M. Oyama, S. Okasaki, T. Shida, T. Nakagawa, Y.
   Matsui, S. Susuki, K. Yamaguchi and Y. Achiba, Chem. Phys. Lett.
   186(1991) 35.
4. S. C. Kukolich and D. F. Huffman
   Chem. Phys. Lett. 182(1991) 263.
5. P. M. Allemand, G. Srdanov, A. Koch, K. Khemani and F. Wudl
   J. Am Chem. Soc. 113 (1991) 2785.
6. L. J. Wilson, S. Flanagan, L. P. F. Chibante and J. M. Alford, in:
   Buckminsterfullerenes (W.E.Billups and M.A.Ciufolini; Eds.), New York
   and Weinheim, VCH 1993, p.285.
7. D. Koruga, S. Hameroff, J. Withers, R. Loutfy and M. Sundareshan:
   Fullerene C_{60}: History, Physics, Nanobiology and Nanotechnology.
   Amsterdam, North Holland 1993; p. 305.
8. Q. Xie, F. Arias and L. Echegoyen
   J. Am. Chem. Soc. 115 (1993) 9818.
9. L. Dunsch and A. Petr,
   Ber. Bunsenges. Phys. Chem. 97 (1993) 436.
10. L. Dunsch, F. Ziegs, J. Fröhner, U. Kirbach, K. Klostermann, A. Bartl and U. Feist, in:
    H. Kuzmany, J. Fink, M. Mehring and S. Roth (Eds) "Electronic Properties of Fullerenes";
    Springer Ser. Solid-State Sci. 117 (1993) p. 39.
11. W. Krätschmer, L. D. Lamb, K. Fostiropoulos and D. R. Huffmann,
    Nature 347 (1990) 354.
12. T. Susuki, Y. Maruyama, T. Akasaka, W. Ando, K. Kobayashi and
    S. Nagase, J. Am. Chem. Soc. 116(1994) 1359.
13. J. Osterodt, M. Nieger, P.-M. Windscheif and F. Vögtle,
    Chem. Ber. 126 (1993) 2331

FT-EPR INVESTIGATIONS OF FULLERENES AND FULLERIDES

K.-P. DINSE
Phys. Chem. III, TH Darmstadt, Petersenstrasse 20, D64287-Darmstadt, Germany
and
D.M. WANG, R. BRAMLEY
Res. School of Chem., Austr. Nat. University, Canberra, ACT 0200, Australia
and
C.A. STEREN, H. VAN WILLIGEN
Chem. Dept., University of Massachusetts, Boston, MA 02125, USA

ABSTRACT

Pulsed EPR techniques have been utilized for an investigation of the photoexcited triplet state of C_{60} to clarify its rotational dynamics. The measured spin-lattice and spin-spin relaxation rates could be analysed invoking modulation of the dipolar interaction by molecular tumbling, spin-rotational interaction and by fast interconversion between Jahn-Teller states. Localized spins in superconducting Rb_3C_{60} could selectively be observed by applying standard pulse sequences. The observed line broadening in the superconducting state was attributed to formation of the vortex state, and its value utilized to derive the London penetration length $\lambda(0) = 480(40)$ nm.

1. Introduction

Photoexcitation of C_{60} results in the formation of rather long-lived electronically excited states. The intrinsic lifetime of the isolated triplet state $^3C_{60}$ was shown to be of the order of 200 µs, rendering this paramagnetic state accessible to magnetic resonance techniques. One of the intriguing features of $^3C_{60}$ was its narrow, single-line EPR spectrum, which was detected by several groups[1-3]. It was soon realized that the spin relaxation parameters T_1, T_2 could be used for an investigation of rotational dynamics in solution and for an investigation of electronical dynamics originating from the high orbital degeneracy of the molecule. The narrow-line resonance of $^3C_{60}$ ($\Delta\nu$(FWHM) = 750 kHz at 300K in toluene) can easily be investigated with Fourier-transform EPR (FT-EPR), probing not only its relaxation times T_1 and T_2 directly by multi-pulse techniques but also by detecting the amount of optically-generated electron spin polarization (OEP).

FT-EPR can also be utilized for a selective detection of localized paramagnetic spins in superconducting fullerides. In contrast to itinerant spins recently seen in cw-EPR, localized spins can be used as microscopic probes to measure local fields in the sample. The local field variance generated in the vortex state of the superconductor can be detected as a line-broadening of the EPR of the localized spins within the super-conducting domains. Under conditions of dispersion of localized spins on the length scale of the vortex lattice (amounting to 70 nm at 0.3 T), the resulting variance can be used for a determination of the London penetration length λ and of its temperature dependence.

2. Experimental

Solutions of C_{60} in toluene and methylcyclohexane were degassed on a high vacuum line by several freeze-pump-thaw cycles. Superconducting Rb_3C_{60} samples were prepared according to standard procedures. The pulsed EPR experiments were performed with home-built spectrometers of slightly varying design with respect to the resonator. Whereas experiments requiring optical access utilized a modified Varian TM_{102} cavity or a bimodal cavity, the experiments with the superconductor were performed with a loop-gap resonator optimized for short dead time over the full temperature range. C_{60} was excited either with the frequency-doubled output of a Nd:YAG laser with approximately 16 mJ pulse energy or with an excimer-pumped dye laser at 580 nm (3 mJ pulse energy).

3. Results

3.1. Spin dynamics of $^3C_{60}$ in solution

Fig. 1 shows the signal intensity of the EPR signal of $3C_{60}$ obtained after Fourier transformation of the FID as a function of delay time between laser and microwave pulse. Although the generation of $^3C_{60}$ via Inter System Crossing (ISC) should be instantaneous

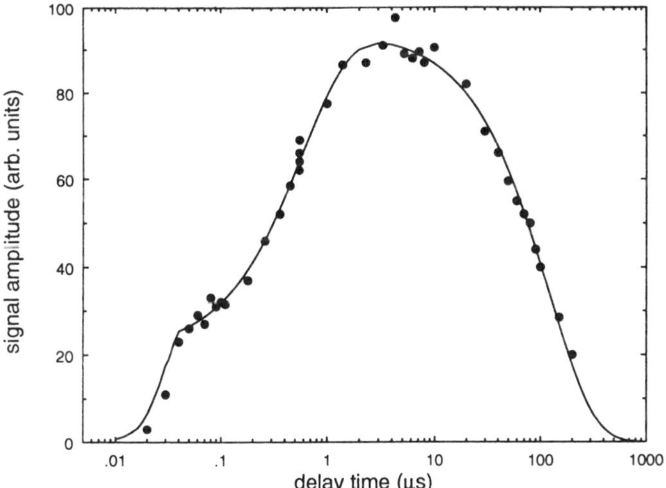

Fig. 1   FT-EPR signal intensity of $^3C_{60}$ as function of delay time at 190 K in toluene

considering our time resolution of 10 ns, the EPR signal initially increases and reaches a maximum at approximately 1 μs. This initial signal build-up can be ascribed to spin-lattice relaxation of the (nearly) unpolarized $^3C_{60}$ into its Boltzmann polarization, as was shown previously[4]. The signal later decreases by non-radiative deactivation (first-order kinetics) and triplet-triplet-annihilation (second-order kinetics), the spin system being in thermal equilibrium, however, as was measured by a standard π-$\pi/2$ sequence probing T_1. This

490

technique was used to verify that T_1 determined from the initial signal rise was identical to the value determined with the pulse sequence for later delay times. Similarly, T_2 could be determined either by simply relating it to the width Δv of the Lorentzian line ($T_2 = 1/(\pi \Delta v)$) or by measuring it from the echo decay performed in an inhomogeneous field.

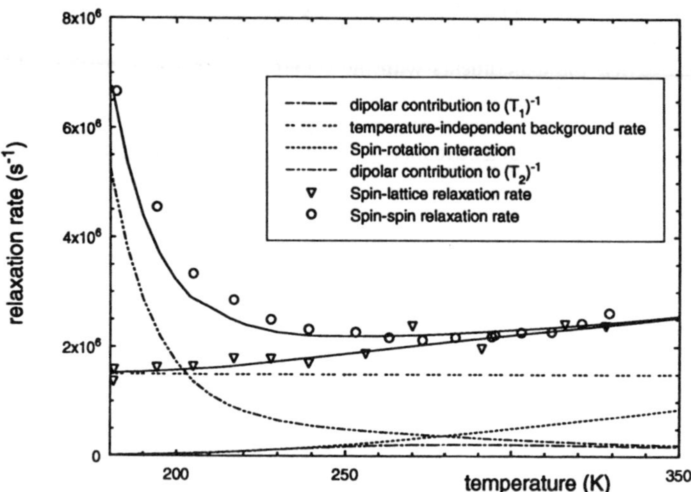

Fig. 2 Experimental T_1 and T_2 values as function of temperature together with their deconvolution in terms of a theoretical model

The resulting T_1 and T_2 relaxation rates are depicted in Fig. 2 in combination with a fit of the data. For this fit we had to assume that not only modulation of the electron spin dipolar interaction (DD) by molecular tumbling and spin-rotational interaction (SR) contribute to both relaxation rates as was suggested by Grupp et al.[5], but that there is also a noticeable temperature independent contribution to T_1^{-1} and T_2^{-1}, which actually gives a lower limit to T_1^{-1} at low temperatures and to T_2^{-1} at room temperature. A quantitative

$$T_2^{-1} = \frac{1}{15}(D^*)^2\,\tau_r\left\{3 + \frac{5}{1+\omega^2\tau_r^2} + \frac{2}{1+4\omega^2\tau_r^2}\right\} \tag{1}$$

$$OEP = \frac{2}{15}\frac{kT}{\hbar}(D^*)\tau^2 K\left\{\frac{1}{1+\omega^2\tau_r^2} + \frac{4}{1+4\omega^2\tau_r^2}\right\} \tag{2}$$

deconvolution of these contributions was possible because DD- and SR-derived rates dominate at low and high temperatures, respectively, and because the temperature dependence of τ_r can be related to the temperature dependence of the viscosity of the solvent. According to Eq. 1, T_2^{-1}(DD) is a function of the product of $(D^*)^2$, $D^* = (D^2+3E^2)^{1/2}$ and τ_r, and both factors can only be determined independently, if the experimental value for the initial Zeeman polarization is evaluated, too. As is seen from Eq. 2, this initial value generated by ISC is also depending on D^* and τ_r, although in a different analytical form. The dependence of OEP on τ_r describes the fact that the initial spin alignment is converted to spin polarisation in its maximum limiting value only if the spin system can adjust adiabatically to the fluctuating eigenfrequencies, i.e., that the Larmor frequency is fast on the time scale of the reciprocal rotational correlation time τ_r. $^3C_{60}$ with its exceptionally short τ_r does not allow adiabatic tracking of the electron spin precession axis to the molecular axes system, and with its room temperature value of $\tau_r = 16$ ps easily violates the adiabatic limit. As is seen from Fig. 3, the experimental OEP cannot only be used to determine τ_r, but furthermore the principal element D* of the ZFS tensor is also fixed when additionally using the ZFS-related contribution to T_2^{-1}. As a result we find that D* is reduced by an order-of-magnitude as compared to values for $^3C_{60}$ in solid matrices. This reduction of D* in solution in combination with the temperature-independent contribution to the relaxation rates can be taken as evidence for a fast interconversion between Jahn-Teller states derived from the orbitally degenerate T_{2g} level of $^3C_{60}$. Assigning different ZFS tensor elements to the D_{3d}, D_{5d}, and D_{2h} substates, an interconversion with a correlation time $\tau_c \leq 10^{-12}$ s would be consistent with our observation.

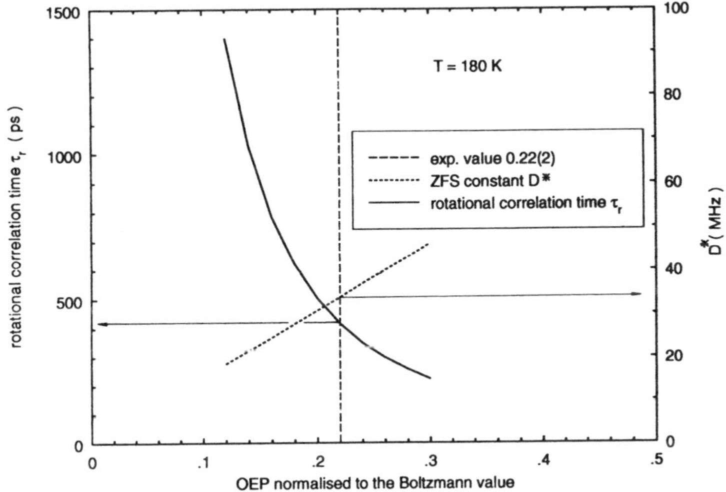

Fig. 3   D* and τ_r as determined from the initial electron polarisation

3.2. Spin dynamics in Rb_3C_{60}

The superconductivity of the sample investigated was probed by detecting microwave absorption at very low external fields and by measuring the microwave Meissner effect. Microwave field exclusion leads to an effective reduction of the resonator volume and as a result its eigenfrequency is increased below T_c. Fig. 4 shows the frequency shift as function of temperature. The absolute shift can be used for a determination of the superconducting volume of the sample which was estimated as 20%.

A 2-pulse ($\pi/2$-π) sequence was used to detect that amount of spin susceptibility that has a constant Larmor frequency on the time scale of the pulse sequence. The integrated echo intensity was used to determine the corresponding susceptibility as

492

function of the external field, thus mapping the absorption line (echo-induced EPR). Below T_c, a broadening of the absorption line with approximately Gaussian shape was observed, which after deconvolution exhibited a linear dependence when plotting it vs. l-$(T/T_c)^4$. This characteristic temperature dependence was taken as evidence that the observed broadening can be attributed to vortex-induced local field variations[6]. Extrapolated the vortex-induced variance for $T \rightarrow 0$, its value can be converted into $\lambda(0)$ = 480(40) nm, which compares reasonably well with the value $\lambda_\mu(0)$ = 370(20) nm, determined recently by μ^+SR[7]. From $\lambda(0)$ and the coherence length ξ = 2.9 nm, the effective carrier mass m* = 9 m_e was estimated.

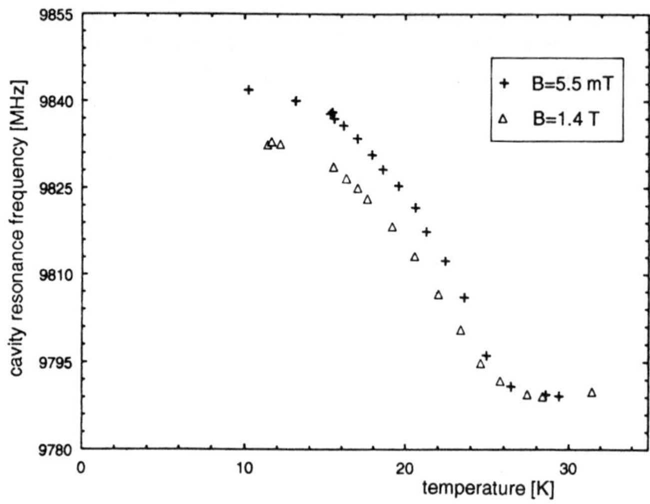

Fig. 4   Resonator detuning caused by the microwave Meissner effect of a Rb_3C_{60} sample

4. References

1. G.L. Closs, P. Gautam, D. Zhang, P.J. Krusic, S.A. Hill, E. Wasserman, *J. Chem. Phys.* **96** (1992) 5228

2. M. Rübsam, K.-P. Dinse, M. Plüschau, J. Fink, W. Krätschmer, K. Fostiropoulos, C. Taliani, *J. Amer. Chem. Soc.* **114** (1992) 10059

3. H. Levanon, V. Meiklyar, A. Michaeli, S. Michaeli, A.J. Regev, *J. Chem. Phys.* **96** *(1992) 6128*

4. C.A. Steren, P.R. Levstein, H. van Willigen, H. Linschitz, L. Biczok, *Chem. Phys. Lett.* **204** (1993) 23

5. A. Grupp, M. Bennati, M. Mehring, in: *Proc. IWEPS 93*, J. Fink, H. Kuzmani, M. Mehring, Eds., Springer Berlin (1994)

6. K.-P. Dinse, D.M. Wang, R. Bramley, J.W. White, *Physica* **C218** (1993) 341

7. R.F. Kiefl, W:A: MacFarlane, K.H. Chow, S. Dunsiger, T.L. Duty, T.M.S. Johnston, J.W. Schneider, J. Sonier, L. Brard, R.M Strongin, J.E. Fischer, A.B. Smith III, *Phys. Rev. Lett.* **70** (1993) 3987

Part Eight

ELECTRICAL TRANSPORT AND SUPERCONDUCTIVITY

STRUCTURE AND DYNAMICS OF SUPERCONDUCTING FULLERIDES

KOSMAS PRASSIDES
*School of Chemistry and Molecular Sciences, University of Sussex,
Brighton BN1 9QJ, United Kingdom*

and

KATSUMI TANIGAKI, JUN'ICHIRO MIZUKI, ICHIRO HIROSAWA
*Fundamental Research Laboratories, NEC Corporation,
34 Miyukigaoka, Tsukuba 305, Japan*

ABSTRACT

The structure and dynamics of fullerides with stoichiometry $A_2A'C_{60}$ are briefly reviewed. Three structural families have been at present characterised, each distinguished by the orientational state of the fulleride ions in the crystal. Superconducting K_3C_{60} and Rb_3C_{60} adopt a merohedrally disordered $Fm\overline{3}m$ structure in which the C_{60}^{3-} units perform small-amplitude librational motion up to 650 K. Superconducting Na_2CsC_{60} and Na_2RbC_{60} are isostructural with pristine C_{60} and show order→disorder ($Pa3 \rightarrow Fm\overline{3}m$) phase transitions near room temperature. The orientational potential is considerably softer than that in K_3C_{60} and the librations collapse to a single quasielastic line above the transition temperature. Non-superconducting Li_2CsC_{60} is fcc ($Fm\overline{3}m$) and contains quasi-spherical C_{60}^{3-} ions which show excess carbon density along the cube diagonals, signature of strong Li^+–C interactions. The structural details, the intercalate-carbon interactions and the orientational state of the fullerenes evidently affect their electronic and conducting properties very sensitively.

1. Introduction

The appearance of superconductivity in alkali fullerides with stoichiometry[1,2] $A_2A'C_{60}$ led to considerable efforts in attempting to understand their electronic, structural, and dynamic properties and to elucidate the origin of their high T_c's and the mechanism responsible for the pair formation. A growing series of reviews, covering various aspects of alkali fulleride chemistry and physics already exists[3-7]. Theoretical treatments of superconductivity in the fullerides have considered purely electronic[8] as well as phonon-induced pairing, involving either low-energy intermolecular[9] or high-energy intramolecular modes[10]. Inelastic neutron scattering measurements of the phonon spectra of fullerides[11] have provided evidence for a phonon-mediated mechanism of superconductivity but with an electron-phonon coupling strength distributed over a wide range of energies (33-195 meV) as a result of the finite curvature of the fullerene spherical cage. Additional supporting evidence has also come from Raman experiments[12]. Until now, the broad premise within which their structural properties have been discussed, consists of a merohedrally-disordered[13] face-centred-cubic (space group $Fm\overline{3}m$) array of essentially ionic C_{60}^{3-} units with the alkali ions occupying all available tetrahedral and octahedral interstices. However, recent structural and dynamical work on ternary fullerides intercalated at the tetrahedral sites with the small alkali ions Na^+ and Li^+ has revealed the existence of a much richer structural chemistry with profound consequences for the electronic and conducting properties of the materials. It is the purpose of this brief review to summarise the recent advances in our understanding of the properties of these fullerides.

2. Merohedrally-Disordered Fullerides, K_3C_{60} and Rb_3C_{60}

The icosahedral symmetry of C_{60} is incompatible with periodic translational symmetry and a cubic crystalline field will reduce the molecular symmetry to at most $m\bar{3}$. However, the crystal symmetry may be further raised through disorder: equal fractional occupancy of the two standard orientations of C_{60} (rotational 2-fold symmetry axes coincident with [100] axes) will result in $m\bar{3}m$ symmetry. Such an orientational state for C_{60}^{3-} has been found to exist in K_3C_{60}[13]. The two standard orientations are related either by 90° rotations about [100] or by ~44.5° rotations about [111] axes. The neutron diffraction profile at 12 K[14] shows pronounced diffuse scattering present even at this temperature; Fig. 1 shows that its calculated intensity within the merohedral disorder model and including a rms amplitude of libration of 1.1° is in good agreement with experiment. The alkali ions are located in the available octahedral and tetrahedral interstices and face either four hexagons (c.n. 24) (Fig. 2a) or six hexagon-hexagon (6:6) fusions (c.n. 12) of neighbouring fullerenes, respectively. Geometrical considerations of ccp fullerene spheres reveals that while the size of the octahedral holes (2.06 Å) is large enough to accommodate any alkali ion, the radius of the tetrahedral hole (1.12 Å) is smaller than the ionic radii of K^+, Rb^+ and Cs^+. As a result, rotational motion of the fulleride ions is restricted and the A^+-C_{60}^{3-} orientational potential is dominated by strong repulsive interactions. Optimisation occurs when C_{60}^{3-} exposes that part of its surface with the maximum possible area (hexagons), leading to maximisation of the A^+-C distances.

The harder intermolecular potential is manifested by the large librational energies (3.59(4) meV at 300 K)[14] measured by inelastic neutron scattering in K_3C_{60} in a wide temperature range (10-650 K). Assuming a simple sinusoidal hindrance potential for the librational motion of C_{60} and for small librational amplitudes, we can estimate the rotational barrier, $E_{rot} \sim 520$ meV in K_3C_{60} from the expression: $E_{rot} = \left(E_{lib}^2/B\right)\left(\theta/2\pi\right)^2$, where $B = 0.364$ μeV is the rotational constant and $\theta \sim 44.5°$ is the hopping angle between neighbouring potential minima. The librations soften and their amplitudes increase as the

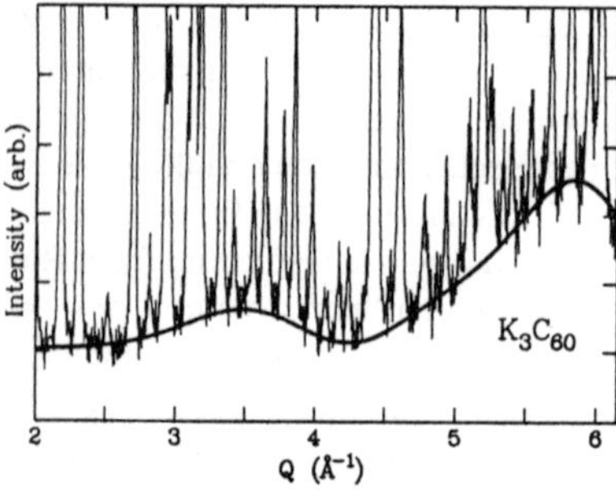

Fig. 1 Diffraction profile of K_3C_{60} at 12 K. The diffuse scattering is modeled using a merohedral disorder model and a rms librational amplitude of 1.1° for both orientations.

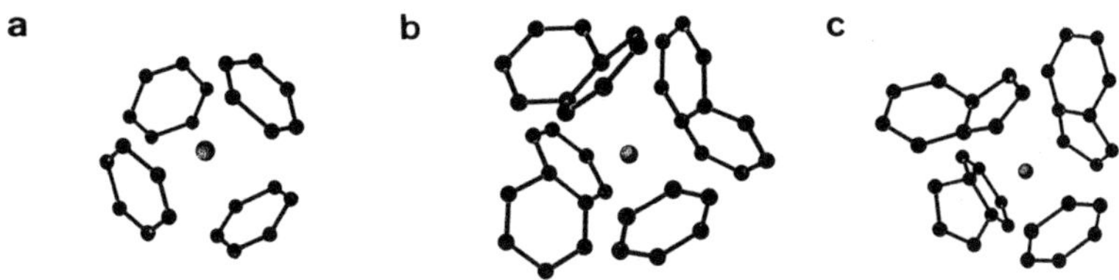

Fig. 2 Coordination environment of the alkali ions in the tetrahedral holes of: (a) the merohedrally-disordered fcc fulleride structures, (b) the major and (c) the minor orientation in primitive cubic fullerides.

temperature increases but there are no discontinuities up to 650 K. In addition, no anomalous stiffening or sharpening of the librations is observed below T_c, thus discounting any significant contribution to the total electron-phonon coupling strength from intermolecular librational modes ($\lambda_{el-lib} \leq 0.08$).

3. Primitive Cubic Fullerides, $Na_2A'C_{60}$ (A'= Cs, Rb)

The presence of Na^+ in the tetrahedral holes of a ccp array of C_{60}^{3-} ions has important structural consequences. Since its ionic radius (0.95 Å) is smaller than the size of the tetrahedral interstices, the C_{60}^{3-} ions are no longer necessarily confined to their two standard orientations found in K_3C_{60}. Evidently there is enough space for the C_{60}^{3-} ions to rotate in such a way as to optimise both the attractive Na^+-C_{60}^{3-} interactions and the C_{60}-C_{60} contacts. The resulting structure is experimentally found to be primitive cubic (space group $Pa\bar{3}$) for both Na_2CsC_{60}[15] and Na_2RbC_{60}[16]. Quite remarkably, optimisation occurs for the same orientational state found in solid C_{60}, namely an anticlockwise rotation about <111> of ~98°. For this setting, the C_{60}-C_{60} contacts are optimised as 6:6 fusions "nest" over pentagonal faces of neighbouring ions and at the same time, the Na^+-C_{60}^{3-} coordination is highly optimal, as each ion presents two hexagonal faces and six 6:6 fusions to its eight neighbouring Na^+ ions, reducing their coordination number to 12 (Fig. 2b). Optimal coordination is also preserved for the ions (Cs^+ or Rb^+) occupying the octahedral holes, as they now coordinate to six hexagon-pentagon (6:5) fusions of neighbouring fulleride ions. In a similar fashion to pristine C_{60}, orientational disorder is also present in these fullerides. 11.7(8)% of the ions are in a different orientation, characterised by a rotational angle of ~38°, a C_{60}-C_{60} contact of 6:6 fusions "nesting" over hexagons of neighbouring ions and an optimal Na^+-C_{60}^{3-} coordination to one hexagonal face and three 6:5 fusions (Fig. 2c). On heating, DSC measurements[17] show the existence of a first-order phase transition (at 299 K for Na_2CsC_{60} and 304.5 K for Na_2RbC_{60}); neutron diffraction experiments have identified this transition as an order-disorder transition to a fcc phase, best modeled as containing quasi-spherical C_{60}^{3-} ions[15]. These results have also important ramifications for the dynamic behaviour of the fulleride ions. Inelastic neutron scattering experiments[18] on Na_2RbC_{60} reveal much softer librations (2.83(17) meV at 50 K) than the ones reported in K_3C_{60}, also implying a softer

498

intermolecular potential (~300 meV). Their behaviour as a function of temperature mimics that found in solid C_{60}, in that they soften and broaden with increasing T until they collapse into a quasielastic line above the phase transition.

The neutron diffraction[15] results on Na_2CsC_{60} are of sufficiently high quality to allow the experimental determination of the C–C bond lengths in C_{60}^{3-}. Compared to C_{60}, in which the 6:6 and 6:5 bonds are of length 1.40(1) and 1.45(1) Å, respectively, we find that, within experimental error, both types of bonds are essentially equal at 1.43(1) Å. This may be compared to *ab initio* molecular dynamics calculations[19] which, while they correctly predict that the 6:6 bonds get longer and the 6:5 ones shorter on reduction, they somewhat underestimate the magnitude of this effect for C_{60}^{3-}.

In both superconductivity models for the fullerides[8,10], pair binding is assumed to be dominated by intramolecular properties. Moreover, the observed monotonic scaling of the superconducting transition temperatures with the unit cell size[20] in fcc fullerides had been rationalised as arising, to first-order, from the modulation of the electronic density-of-states by the interfullerene spacing (Fig. 3). The modified structure and intermolecular potential in the primitive cubic systems provide an additional dimension and evidently sensitively affect their electronic and conducting properties. The changed orientational state of the ions thus affects electron hopping between neighbouring ions and as a consequence, leads to a modified rate of change of T_c with interfullerene spacing in the $Pa\overline{3}$ structures. The seemingly "anomalous" behaviour of Na_2RbC_{60} (T_c= 3.5 K) can then be easily understood (Fig. 3). Further support for such an interpretation has now come from high-pressure measurements[21] of the superconducting behaviour of Na_2CsC_{60} which shows a faster depression of T_c with pressure, $\left(dT_c/dP\right) = -12.5(2)\,K/GPa$, than than that in the fcc fullerides[22].

4. Spherically Disordered Fullerides, Li_2CsC_{60}

High-resolution X-ray diffraction[23] has been used to show the existence of a third possible structural modification of alkali fullerides with stoichiometry $A_2A'C_{60}$, realised for the smallest (r=0.60 Å) and least electropositive alkali metal ion dopant A= Li^+ cation. We found no evidence in Li_2CsC_{60} for a transition to a superconducting state down to 50 mK (Fig. 4), while DSC measurements showed no indication of a structural phase transition between 100 and 450 K. We then probed its structural properties between 13 and 300 K by both laboratory and high-resolution synchrotron X-ray diffraction.

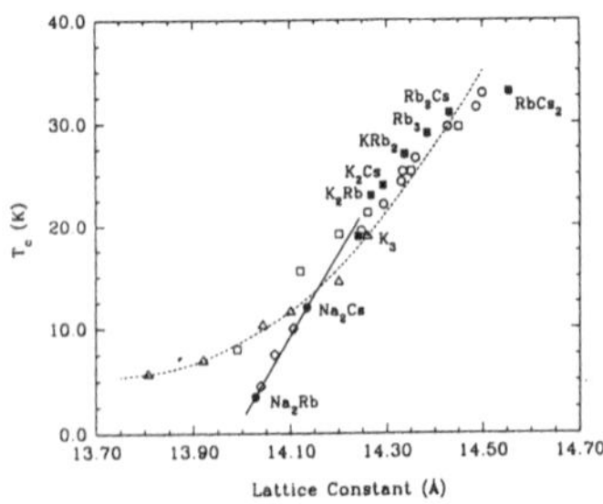

Fig. 3 Evolution of the critical temperature of superconducting fullerides with the cubic lattice constant. ☐, △ and ◊ symbols are high-pressure data[21,22]; all others are from ambient pressure measurements. The dotted and solid lines distinguish the behaviour of fcc and primitive cubic systems.

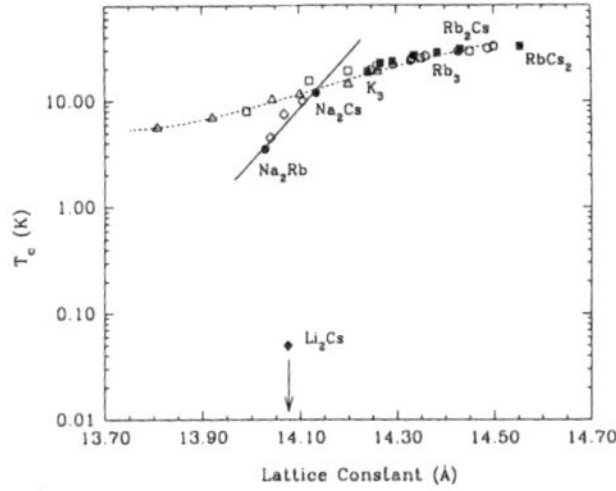

Fig. 4 Evolution of the critical temperature (logarithmic scale) of superconducting fullerides with the cubic lattice constant. ◆ refers to Li_2CsC_{60}, other symbols and meaning as in Fig. 3.

Unexpectedly, we found that its crystal structure did not belong to any of the two fulleride structural families known at present; it was fcc ($Fm\bar{3}m$) but with a novel orientational state for the C_{60}^{3-} ions which were best modeled as quasispherical units of radius 3.556(4) Å. Deviations from perfect sphericity were more pronounced than those encountered in the disordered phase of pristine C_{60}[24]. A symmetry-adapted spherical-harmonics (SASH) analysis of the orientational distribution function revealed a substantial excess of C atom density in the <111> directions facing the strongly polarising Li^+ ions and a deficit along <100> facing the Cs^+ ions, along with additional details (Fig. 5). These results may be understood if a strong stabilising Li^+–C interaction with some covalent character is present; this is unprecedented in fullerene chemistry, but reminiscent of many similar bonding situations, encountered in the widespread organometallic chemistry of Li. Steric factors also contribute to the stability of this novel orientational state of the fullerene units which usually tended in the solid state to align their hexagonal faces (and molecular 3-fold symmetry axes) along the <111> crystal directions. An orientational distribution function similar to pristine C_{60} would have led to a large coordination number for the tetrahedral ions and steric crowding for the small Li^+ ions. Our present results are thus consistent with the trend of decreasing coordination number of the tetrahedral ions in fulleride solids with decreasing ionic size[15].

C_{60} adopts in fcc Li_2CsC_{60} a different orientational state from that in superconducting fcc K_3C_{60} (merohedral disorder) and primitive cubic Na_2CsC_{60}

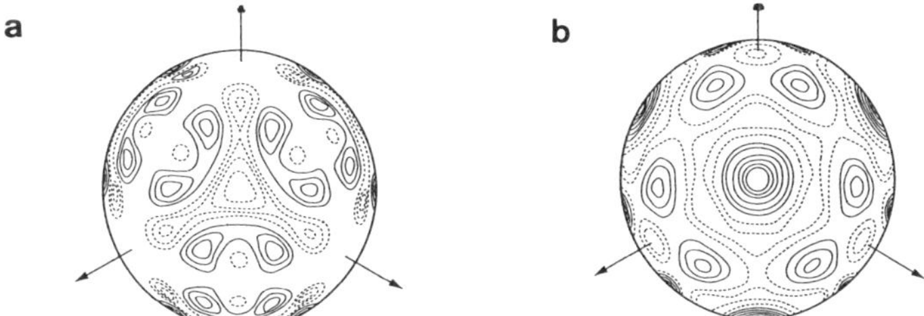

Fig. 5 The orientation distribution function for C_{60} in (a) C_{60} and (b) Li_2CsC_{60} at room temperature viewed down one of the <111> directions. The zero-order term is omitted to emphasise differences from spherical symmetry. Solid (dashed) contours indicate excess (deficit) carbon density with respect to the zero-order term. Contour intervals are in (a) ±0.024 and in (b) ±0.128.

500

(orientational order) with dramatic effects on the macroscopic properties. This further amplifies the important role of orientational disorder on the superconducting properties of the fullerides. Li_2CsC_{60} is not superconducting, even though the transition temperatures expected for the value of its lattice constant should be of the order of ~5-10 K (Fig. 4). The detrimental effect on superconductivity evidently comes from the strong lithium-carbon interaction which leads to reduced ionicity of the fulleride ions through hybridisation of carbon p_z and metal s orbitals and modifies the electronic structure at the Fermi level. In addition, electron hopping between neighbouring molecules may be adversely affected by the quasispherical disordered state of the fulleride ions.

5. References

1. A.F. Hebard, M.J. Rosseinsky, R.C. Haddon, D.W. Murphy, S.H. Glarum, T.T.M. Palstra, A.P. Ramirez and A.R. Kortan, *Nature* **318** (1991) 600.
2. K. Holczer, O. Klein, S.-M. Huang, R.B. Kaner, K.-J. Fu, R.L. Whetten and F. Diederich, *Science* **252** (1991) 1154.
3. K. Prassides, H.W. Kroto, R. Taylor, D.R.M. Walton, W.I.F. David, J. Tomkinson, M.J. Rosseinsky, D.W. Murphy and R.C. Haddon, *Carbon* **30** (1992) 1277.
4. K. Holczer and R.L. Whetten, *Carbon* **30** (1992) 1261.
5. J.E. Fischer, P.A. Heiney and A.B. Smith, *Acc. Chem. Res.* **25** (1992) 112.
6. D.W. Murphy, M.J. Rosseinsky, R.M. Fleming, R. Tycko, A.P. Ramirez, R.C. Haddon, T. Siegrist, G. Dabbagh, J.C. Tully and R.E. Walstedt, *J. Phys. Chem. Solids* **53** (1992) 1321.
7. J.H. Weaver, *J. Phys. Chem. Solids* **53** (1992) 1433.
8. S. Chakravarty, M.P. Gelfand and S. Kivelson, *Science* **254** (1991) 970.
9. O.V. Dolgov and I.I. Mazin, *Solid State Commun.* **81** (1992) 935.
10. C.M. Varma, J. Zaanen and K. Raghavachari, *Science* **254** (1991) 989; M. Schlüter, M. Lannoo, M. Needels, G.A. Baraff and D. Tománek, *Phys. Rev. Lett.* **68** (1992) 526.
11. K. Prassides, J. Tomkinson, C. Christides, M.J. Rosseinsky, D.W. Murphy and R.C. Haddon, *Nature* **354** (1991) 462.
12. M.G. Mitch, S.J. Chase and J.S. Lannin, *Phys. Rev. Lett.* **68** (1992) 883.
13. P.W. Stephens, L. Mihaly, P.L. Lee, R.L. Whetten, S.-M. Huang, R. Kaner, F. Diederich and K. Holczer, *Nature* **351** (1991) 632.
14. C. Christides, D.A. Neumann, K. Prassides, J.R.D. Copley, J.J. Rush, M.J. Rosseinsky, D.W. Murphy and R.C. Haddon, *Phys. Rev. B* **46** (1992) 12088.
15. K. Prassides, C. Christides, I.M. Thomas, J. Mizuki, K. Tanigaki, I. Hirosawa and T.W. Ebbesen, *Science* **263** (1994) 950.
16. K. Kniaz, J.E. Fischer, Q. Zhu, M.J. Rosseinsky, D.W. Murphy and O. Zhou, *Solid State Commun.* **88** (1993) 47.
17. K. Tanigaki, I. Hirosawa, T. Manako, J.S. Tsai, J. Mizuki and T.W. Ebbesen, *Phys. Rev. B*, in press.
18. C. Christides, K. Prassides, D.A. Neumann, J.R.D. Copley, J. Mizuki, K. Tanigaki, I. Hirosawa and T.W. Ebbesen, *Europhys. Lett.* **24** (1993) 755.
19. W. Andreoni, in *Physics and Chemistry of the Fullerenes*, ed. K. Prassides (Kluwer, Dordrecht, 1994) p. 169.
20. R.M. Fleming, A.P. ramirez, M.J. Rosseinsky, D.W. Murphy, R.C. Haddon, S.M. Zahurak and A.V. Makhija, *Nature* **352** (1991) 787.
21. J. Mizuki, M. Takai, H. Takahashi, N. Môri, K. Tanigaki, I. Hirosawa and K. Prassides, *Phys. Rev. B*, submitted.
22. G. Sparn, J.D. Thompson, R.L. Whetten, S.-M. Huang, R.B. Kaner, F. Diederich, G. Grüner and K. Holczer, *Phys. Rev. Lett.* **68** (1992) 1228.
23. I. Hirosawa, K. Prassides, J. Mizuki, K. Tanigaki, M. Gevaert, A. Lappas and J.K. Cockcroft, *Science*, in press.
24. P.C. Chow, X. Jiang, G. Reiter, P. Wochner, S.C. Moss, J.D. Axe, J.C. Hanson, R.K. McMullan, R.L. Meng and C.W. Chu, *Phys. Rev. Lett.* **69** (1992) 2943.

ALLEN'S FORMULA AND RAMAN SCATTERING FROM THE METALLIC FULLERIDES

M. J. Rice and P. Gomes da Costa

Xerox Webster Research Center, Webster, NY 14580, U.S.A.

ABSTRACT

The use of Allen's formula in analysing Raman scattering from the metallic fullerides (A_3C_{60}) is critically assessed for the crystalline and orientationally disordered forms of the materials.

For a conventional metal in which the phonon frequency $\Omega_{\vec{q}}$ is negligible by comparison to the Fermi energy ϵ_F Allen's formula [1],

$$\frac{1}{N} \sum_{\vec{q}} \frac{\Gamma_{\vec{q}}}{\Omega_{\vec{q}}^2} \;=\; \pi \, N_F \, \lambda_{SC} \tag{1}$$

relates the intrinsic phonon linewidth $\Gamma_{\vec{q}}$ to the dimensionless electron-phonon coupling constant λ_{SC} defined in the theory of superconductivity. In Eq.(1) N_F denotes the total density of states per atom at ϵ_F, and $\Gamma_{\vec{q}}$ is defined as the *full* width at half maximum. N denotes the number of atoms in the crystal and $\vec{q}$ collectively the wavevector and polarization of the phonon. This formula, originally derived by Allen[1] for inelastic neutron scattering, has been applied by several authors[2] to deduce λ_{SC} from the observed widths of the Raman active H_g vibrational modes of the C_{60}^{3-} molecular ions of the metallic fullerides. The pairing interaction responsible for superconductivity in these compounds is believed[3,4] to be mediated by the phonons derived from these vibrations. There are eight, five-fold degenerate, H_g modes and we denote their frequencies, before electron-phonon renormalization, by ω_α, where α is the mode index. The magnitude of these phonons extend from 272 to 1525 cm^{-1}. In this paper, which will be mainly tutorial in nature, we will discuss the validity of this application of Allen's formula.

Let us consider first the case of the crystal for which Raman spectroscopy probes only $\vec{q} = 0$ phonons. It is immediately obvious in this case that the measured linewidth Γ_α can never be related to the right hand side of Allen's formula, Eq(1), since λ_{SC} involves, of course, phonons of all wavevectors[1]. The correct answer for Γ_α, however, is an interesting one which we shall now derive.

Let us assume *weak* electron-phonon (e-p) coupling and neglect explicit effects of electron-electron interactions. The underlying e-p coupling constant[3] is $g^\alpha_{n\prime n}(\vec{k}\prime, \vec{k})$, which is the matrix element for the process in which an electron in the Bloch state with wavevector $\vec{k}$ and band index n is scattered into another state with wavevector $\vec{k}\prime$ and band index $n\prime$ by the dispersionless Hg-derived phonon of frequency ω_α. For the crystalline fulleride there are three narrow overlaping bands (n= 1, 2, 3) derived from intermolecular overlap of the molecular orbitals of molecular C_{60}[5]. In the limit $\vec{q} \longrightarrow 0$, second-order perturbation theory gives the following expressions for Γ_α and the shift in frequency $\Delta\omega_\alpha$ caused by the e-p interaction:

$$\frac{\Delta\omega_\alpha}{\omega_\alpha} = -\frac{1}{2}\,\Pi_R \tag{2}$$

$$\frac{\Gamma_\alpha}{\omega_\alpha^2} = \frac{\Pi_I}{\omega_\alpha} \tag{3}$$

where Π_R and Π_I are the real and imaginary parts of the phonon self energy

$$\Pi_{\vec{q}} = \frac{1}{3N}\sum_{\vec{k},n\prime,n}\frac{2\,|\,g^\alpha_{n\prime n}(\vec{k}+\vec{q},\vec{k})\,|^2}{\omega_\alpha}\frac{(f_{\vec{k}n}-f_{\vec{k}+\vec{q},n\prime})}{\epsilon_{\vec{k}+\vec{q},n\prime}-\epsilon_{\vec{k},n}-(\omega_\alpha+i\delta)} \tag{4}$$

taken in the limit $\vec{q} \longrightarrow 0$. In (4), $f_{\vec{k}n}$ denotes the Fermi function for the Bloch state of energy $\epsilon_{\vec{k}n}$ and the summation over $\vec{k}$ also implies a summation over spin polarization. The interesting feature of Eq (4) is that in the $\vec{q} \longrightarrow 0$ limit the *intra*band contributions ($n\prime = n$) to $\Pi_{\vec{q}}$ vanish. This is because for Einstein phonons $\vec{q}v_F/\omega_\alpha \longrightarrow 0$ as $\vec{q} \longrightarrow 0$, where v_F is the mean value of the Fermi velocity. For a conventional metal one has acoustic phonons for which $\vec{q}v_F/\omega_q = v_F/s \gg 1$ in the $\vec{q} \longrightarrow 0$ limit so that the usual intraband terms proportional to N_F survive in Eq.(4). Thus, the organic metal is quite different from the conventional metal in that the longwavelength

renormalization of its important phonons and is determined entirely by *inter*band e-p interaction[6].
Thus, it follows from Eqs. (4) and (3) that

$$\frac{\Gamma_\alpha}{\omega_\alpha^2} = \frac{\pi}{3N} {\sum_{\vec{k},n\prime,n}}' \frac{2 \mid g_{n\prime n}^\alpha(\vec{k},\vec{k}) \mid^2}{\omega_\alpha} \delta(\epsilon_{k\;n\prime} - \epsilon_{kn} - \omega_\alpha) \tag{5}$$

where the prime over the summation signs excludes terms for which $n\prime = n$. This result is of the form
$\Gamma_\alpha/\omega_\alpha^2 \approx \pi(2g_\alpha^2(ib)/\omega_\alpha)N_{ib}(\omega_\alpha)$, where $g_\alpha(ib)$ is an averaged interband electron-phonon coupling
constant and $N_{ib}(\omega)$ is the density of interband transitions (joint density of states). Similarly, the
result for $\Delta\omega_\alpha$ is of the form $\Delta\omega_\alpha/\omega_\alpha \approx (1/2)(2g_\alpha^2(ib)/\omega_\alpha)\chi_R(\omega_\alpha)$ where $\chi_R(\omega)$ is essentially a
Lorentz oscillator describing the continuum of interband transitions. Since $N_{ib}(\omega)$ is large, while
$\chi_R(\omega)$ is small at typical electronic interband frequencies, $\omega_{ib} \approx 0.1 - 0.2eV$, Γ_α and $\Delta\omega_\alpha$ should
be large and small, respectively, for the Hg phonons whose frequencies overlap ω_{ib}. This appears
to be consistent with experiment.[2]

Although intraband e-p coupling does not contribute to Γ_α for the crystal we note that
it still, of course, contributes to the phonon-mediated pairing potential $V(\vec{k}n,\vec{k}\prime n\prime)$ which in the
weak-coupling limit is

$$V(\vec{k}n,\vec{k}\prime n\prime) = -\sum_{n\prime n}\sum_\alpha \frac{(2 \mid g_{n\prime n}^\alpha(\vec{k}\prime,\vec{k}) \mid^2}{\omega_\alpha}$$

and is of the order of 50-100 meV[3,4].

We now consider the case of the orientationally disordered crystal[7], again assuming weak
e-p coupling. We may follow a familiar argument in metal physics that since there is *no momentum
conservation* in a disordered crystal, to calculate Γ_α we may average $\Gamma_\alpha(\vec{q})$ over all wavevectors $\vec{q}$.
For the fullerides this will be tantamount to assuming that disorder is so strong that a given Hg
vibrational mode on one C_{60} ion is completely uncorrelated with the same mode on any other C_{60}
ion in the crystal. In this case we have

$$\Gamma_\alpha = \frac{1}{N}\sum_{\vec{q}} \frac{\Gamma_\alpha(\vec{q})}{\omega_\alpha^2} = \frac{\pi}{3N^2} \sum_{\vec{k}\prime,n\prime}\sum_{\vec{k},n} \frac{2 \mid g_{n\prime n}^\alpha(\vec{k}\prime,\vec{k}) \mid^2}{\omega_\alpha} \frac{(f_{\vec{k}n} - f_{\vec{k}\prime n\prime})}{\omega_\alpha} \delta(\epsilon_{k\prime n\prime} - \epsilon_{kn} - \omega_\alpha). \tag{6}$$

504

Although now similar in form to Allen's formula, Eq.(6) does not in general reduce to Eq.(1) since the phonon frequencies cannot be neglected by comparison to ϵ_F. However, we may obtain an approximate form of Allen's formula by assuming that (a) the coupling constants $g^{\alpha}_{n'n}(\vec{k}', \vec{k}) = g_{\alpha}$, a constant independent of the band labelling and (b) a constant density of states (per spin per molecule) $N_0(\epsilon) = 3/W$, for $0 \leq \epsilon \leq W$, and $N_0(\epsilon) = 0$ otherwise. This simplification yields the analytical results

$$\frac{\Gamma_\alpha}{\omega_\alpha^2} = \frac{1}{3} \left(\frac{2g_\alpha^2}{\omega_\alpha}\right)\left(\frac{\chi_I(\omega_\alpha)}{\omega_\alpha}\right) \tag{7}$$

$$\frac{\Delta\omega_\alpha}{\omega_\alpha} = \frac{1}{6}\left(\frac{2g_\alpha^2}{\omega_\alpha}\right)\chi_R(\omega_\alpha) \tag{8}$$

where

$$\chi_I(\omega) = 2\pi N_0^2\omega \qquad \text{for } 0 \leq \omega \leq W/2 \tag{9}$$

$$= 2\pi N_0^2(W - \omega) \qquad \text{for } W/2 < \omega \leq W \tag{10}$$

and $\chi_R(\omega) = \chi_1(\omega) + \chi_1(-\omega)$, where

$$\chi_1(\omega) = 2N_0^2W\{(\frac{\omega}{W})ln \mid \frac{W}{\omega} \mid +(1 - \frac{\omega}{W})ln \mid 1 - \frac{\omega}{W} \mid$$
$$- (\frac{\epsilon_F - \omega}{W})ln \mid \frac{\epsilon_F - \omega}{W} \mid -(\frac{W - \epsilon_F - \omega}{W})ln \mid \frac{W - \epsilon_F - \omega}{W} \mid\} \quad . \tag{11}$$

The Fermi energy $\epsilon_F = W/2$. $\chi_R(\omega)$ and $\chi_I(\omega)$ are plotted in Fig.s 1 and 2.

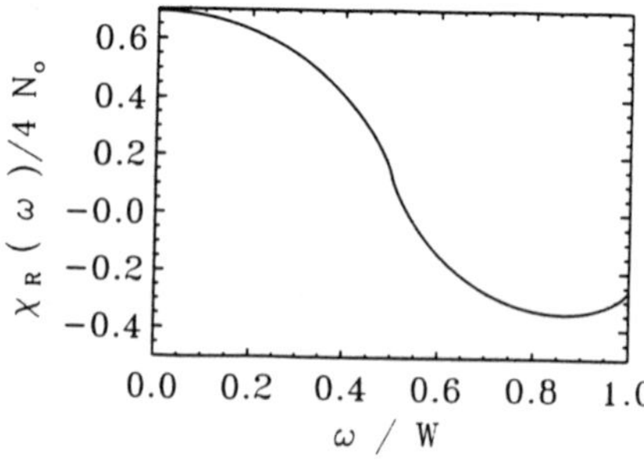

Fig. 1 . The function $\chi_R(\omega)$ vs. ω/W.

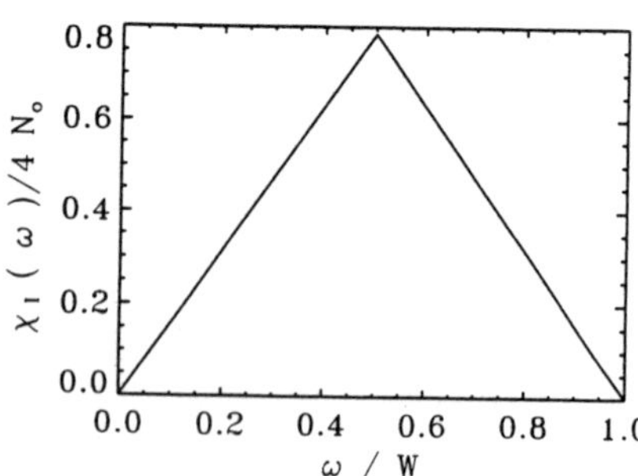

Fig. 2 . The function $\chi_I(\omega)$ vs. ω/W.

In view of eq.s (9) and (7) we obtain

$$\frac{\Gamma_\alpha}{\omega_\alpha^2} = \pi\lambda_\alpha N_F \qquad\qquad \text{for } \omega_\alpha < (\frac{W}{2}) \qquad\qquad (12)$$

$$= \pi\lambda_\alpha N_F(\frac{W}{\omega_\alpha} - 1) \qquad\qquad \text{for } \omega_\alpha > (\frac{W}{2}) \qquad\qquad (13)$$

where $\lambda_\alpha = (N_0/3)(2g_\alpha^2/\omega_\alpha)$ is the "one-band" weak coupling BCS coupling constant for the phonon mode α. Eq.(12) is Allen's formula, while Eq.(13) reflects the shrinking phase space for energy conservation for $\omega_\alpha > \epsilon_F$. Since $\chi_R(\omega) \longrightarrow 0$ as $\omega \longrightarrow \epsilon_F$, Eq.(8) shows that the effect of the e-p interaction to shift the frequencies of phonons with energies $\omega_\alpha \leq \epsilon_F$ is strongly weakened.

In conclusion we have pointed out that Allen's formula is not applicable to Raman spectroscopy for the case of a crystalline fulleride. Its applicability is, however, approximately justified for the case of the strongly disordered fulleride ($l_{MFP} \sim a$) with electron-phonon coupling sufficiently weak for perturbation theory to be valid.

References

[1]P. B. Allen, Phys. Rev. **B 6**, 2577 (1972).

[2]D. M. G. Mitch et al., Phys. Rev. Lett. **68**, 883 (1992); P. Zhou et al., Phys. Rev. **B 46**, 2595 (1992).

[3]C. M. Varma et al., Science **254**, 898 (1991).

[4]M. Schluter et al., Phys. Rev. Lett. **68**, 526 (1992).

[5]S. Saito and A. Oshiyama, Phys. Rev. Lett. **66**, 2647 (1991).

[6]For further discussion see, e.g., M. J. Rice et al., Phys. Rev. **B 49**, 3687 (1994).

[7]P. Stevens et al., Nature **351**, 632 (1991); M. P. Gelfand and J. P. Lu, Phys. Rev. Lett.**68**, 1050 (1992).

MICROWAVE CONDUCTIVITY AND CONDUCTION ELECTRON SPIN RELAXATION OF Rb$_3$C$_{60}$

A. Janossy, O. Legeza, A. Beleznay, R. Gaal and G. Mihaly
Research Institute for Solid State Physics
Budapest H-1525 POB 49, Hungary
and Institute of Physics of the Technical University
Budapest H-1521, Hungary

and

O. Chauvet, L. Forro
Laboratoire de Physique des Solides Semicristallins, IGA
Department de Physique, École Polytechnique Federal de Lausanne,
1015 Lausanne, Switzerland

The 9 GHz conductivity, $\sigma(T)$ and the conduction electron spin relaxation time $T_1(T)$ of the superconducting alkali fullerene compound, Rb$_3$C$_{60}$ are presented as a function of temperature, T. In contrast to normal metals, above 200 K both $\sigma(T)$ and $T_1(T)$ increase with increasing T. The short T_1 at low T is indicative of a molecular disorder. Below the superconducting transition at $T_c = 29$ K a small peak appears in the real part of the conductivity suggestive of a Hebel-Slichter peak. The relative amplitude of the conductivity peak, $\sigma_{max}/\sigma(30\ K)=1.13$ is small and resembles the peak in the muon spin relaxation rate observed by Kiefl et al. (Phys. Rev. Lett. 70, 3987 (1993).

1. Introduction

The absorption of electromagnetic radiation at microwave frequencies in the alkali fullerenes A$_3$C$_{60}$ is of interest both in the superconducting and the normal state. Below T_c coherence effects lead to a maximum in both the real part of the conductivity and the nuclear spin relaxation rate. This effect depends on the strength of the electron phonon coupling, the anistropy of the gap function and in extreme cases on the electron momentum lifetime. A small peak has been observed recently in the muonium spin relaxation[1] of Rb$_3$C$_{60}$ which was attributed to the coherence effect. The effect, however, has not been observed in the ^{13}C nuclear relaxation rate of A$_3$C$_{60}$ compounds. In this paper we report on a coherence peak in the real part of the conductivity at 9 GHz which is similar to that observed by muonium spin relaxation[1].

The 9 GHz electric conductivity in the normal state of Rb$_3$C$_{60}$ is almost certainly the same as at zero frequency. The technique has the advantage of relative ease since no electric contacts are required and powder samples with well defined stoichiometry can be used. Our results for the conductivity $\sigma(T)$ of Rb$_3$C$_{60}$ in the normal state reported in this paper differ qualitatively from reports on thin films[2] and single crystals[3,4] using dc technics and measurements at microwave frequencies on compressed pellet[5]. These previous measurements suggested that in spite of the very large residual resistivity at low T, the resistivity increases linearly with temperature like in pure normal metals. Such a behaviour is not expected since it implies a mean free path which is much shorter than the intermolecular spacing[2]. Our new measurements show a rather complicated temperature dependence of the resistivity. In contrast to previous reports[2-5] we find that above 200 K the resistivity, $1/\sigma$ decreases with increasing T. The new results are in agreement with the temperature dependence of the conduction electron spin relaxation rate $1/T_1$ which in normal metals is proportional to the momentum scattering rate $1/\tau$ measured by the resistivity. We also report on a new measurement of $1/T_1$ in Rb_3C_{60} which confirms earlier findings[6] of an anomalously fast spin relaxation.

2. Samples

4 samples with nominal compositions of $Rb:C_{60}$, $2Rb:C_{60}$, $3Rb:C_{60}$ and $4Rb:C_{60}$, (Samples 1 , 2, 3 and 4 respectively) were prepared by heating at 260°C for several weeks the appropriate amounts of Rb and purified C_{60}.

The ESR and microwave conductivity of Samples 2 and 3 were measured as a function of temperature. Sample 1 (with nominal composition $Rb:C60$) was studied to ascertain that the microwave losses arise from the superconducting Rb_3C_{60} compound and not from Rb_1C_{60} which recently has been shown[6,7] to be a metal. The relative amount of the two conducting phases, Rb_3C_{60} and Rb_1C_{60} was estimated from the ESR intensities of the appropriate lines. Sample 1 had a strong narrow ESR line arising from Rb_1C_{60}, the broad Rb_3C_{60} line was not observed and microwave losses were small. Sample 2 contained comparable amounts of both phases while Sample 3 contained only a small amount (about 15%) of Rb_1C_{60}. In Sample 4 the ESR was almost entirely due to the CESR (conduction electron spin resonance) of the Rb_3C_{60} phase.

ESR and microwave conductivity were measured on the same samples. The special sample holder allowed the subtraction of background magnetic losses (for the ESR) and electric losses (for microwave conductivity) with the quartz sample holder left in place but the sample moved out of the microwave field.

2.1. Relation between microwave absorption and real part of conductivity

The microwave absorption P and frequency shift S were measured at a frequency $\omega/2\pi$ = 9.3 GHz by placing the powder form sample in the center of a high Q (Q=25000 at 300 K) TE101 cavity. In this geometry and for crystallite sizes small with respect to the penetration depth of the microwave field the absorption P is proportional to the real part of the conductivity :

$$P = K \, \sigma_1 \, \langle E^2 \rangle \qquad (1)$$

where $\langle E^2 \rangle$ is an average over the square of the microwave electric field, K is a filling factor proportional to the volume V of the sample. $\langle E^2 \rangle$ is proportional to $\langle r^2 \rangle$ for a small spherical sample with radius r. The relative absorption P/V is much larger for powders with larger crystallites. Depolarization effects are not important in the geometry of the experiment because the sample is at the maximum microwave magnetic field and currents are flowing inhibited around the particles.

508

The absorption measures directly the real part of the conductivity for sufficiently small sample dimensions when the penetration of microwave fields is not significantly screened by eddy or superconducting currents. In the normal state this corresponds to:

$$\delta > d \qquad (2)$$

where $\delta = (\mu_0 \sigma \omega / 2)^{-1/2}$ is the skin depth. For our samples this condition was fulfilled in the normal state at all temperatures. We estimated $d=3$ μm for the typical largest linear dimension of single crystallites of the Rb_3C_{60} powder. Various estimates of the conductivity[2,4,8] of Rb_3C_{60} show that in the normal state σ is low. For example from an extrapolation of the infrared reflectivity[8] $1/\sigma=0.4$ mΩcm was found corresponding to a skin depth of $\delta \cong 10\mu$m at a frequency $\omega/2\pi = 9$GHz. The undistorted Lorentzian lineshape of the conduction electron spin resonance of the Rb_3C_{60} phase is a proof that the condition $\delta > d$ is fulfilled. In a narrow temperature range of a few degrees below T_c the condition of small sample $d > \lambda_L$ is still fulfilled, here λ_L is the London penetration depth.

3. Meissner effect

Figure 1a shows the change of the resonance frequency below the superconducting transition of Sample 2. The shifts were compared to the shift of fine copper powder with particle size much larger than

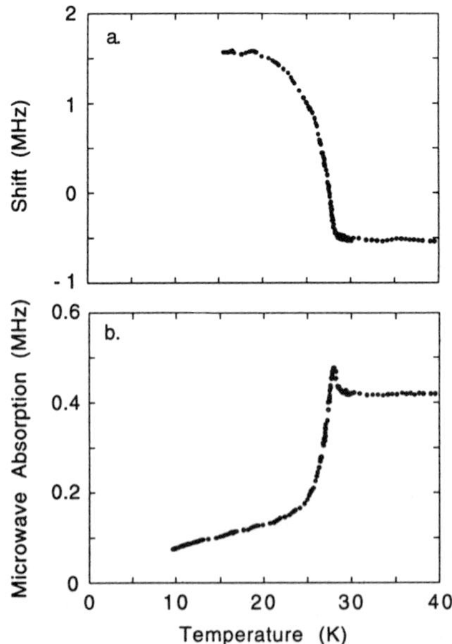

Figure 1a. Shift of the resonance frequency vs temperature for Rb_3C_{60}. In the superconducting state the shift is due to the exclusion of field by the Meissner effect.

Figure 1b. Peak in the real part of the conductivity σ_1 of Rb_3C_{60} below T_c at 9GHz. Below the peak the microwave penetration is limited by the Meissner effect and the absorption is no more proportional to σ_1.

the skin depth of Cu. For Sample 3 the Meissner field exclusion is 35% at low T. This figure is characteristic of relatively high quality Rb_3C_{60} samples, the incomplete shielding is due to the finite size of particles with respect to λ_L and the presence of other phases. Sample 2 contains less of the Rb_3C_{60} phase and the Meissner field exclusion was only 12%. The large frequency shift S between the normal and superconducting states is the best evidence that condition (2) holds in the normal state.

4. Coherence peak in the real part of the conductivity of Rb_3C_{60}

Figure 1b shows the microwave absorption data below 40 K. The most remarkable feature is the narrow peak within a few degrees below T_c which arises from a maximum in the real part of the conductivity. The ratio of the peak conductivity to the normal state conductivity at 30 K is $\sigma_{max}/\sigma_0=1.13$. In this temperature range the shift is still small thus the imaginary part of the conductivity is small and the field penetrates almost completely. A peak with the same relative amplitude was observed in Samples 2 and 3. A plausible interpretation is that it is due to coherence effects of the electronic wavefunctions below T_c which gives rise to an anomalous increase of the nuclear magnetic spin relaxation rate $1/T_1$ (usually denoted as a Hebel-Slichter peak[9]) and a similar increase in the electric conductivity. The conductivity peak below T_c has been clearly demonstrated in the conventional superconductor Pb[10]. In the perovskite high temperature superconductors $Bi_2Sr_2CaCu_2O_8$[11] and $YBa_2Cu_3O_7$ [12] a peak in the real part of the conductivity has also been attributed to coherence effects. No coherence peak has been found in the NMR spin relaxation rate and other interpretations of the conductivity peak in the perovskites have also been suggested. For example Nuss et al.[13] argued that below T_c the decrease of inelastic scattering of quasi particles may be observed in the conductivity but not in $1/T_1$.

The observation[1] of a similar peak in the muon spin relaxation rate of endohedral muonium in Rb_3C_{60} as reported here for the conductivity gives strong support for attributing it to coherence effects. The peak has not been observed in the ^{13}C nuclear magnetic spin relaxation in K_3C_{60}.

The conductivity and the muon spin relaxation peaks are small and narrow and are difficult to understand in terms of a weak coupling BCS case II limit. As pointed out by Kiefl et al.[1], smearing of the density of states by a very short momentum lifetime τ or special band structure effects[14] are possible reasons for the smallness of the peak. As shown below, the conduction electron spin relaxation indicates a short value of τ.

5. Temperature dependence of the conductivity of Rb_3C_{60} in the normal state

Figure 2 displays the temperature dependence of the absorption at 9 GHz for Sample 3. The normal state behaviour is rather complicated and is very different from the $1/T$ dependence of σ reported[2-5] by several authors. We find that the conductivity is constant from 29 K to about 75 K, it has a shallow minimum at about 180 K and increases at higher temperatures.

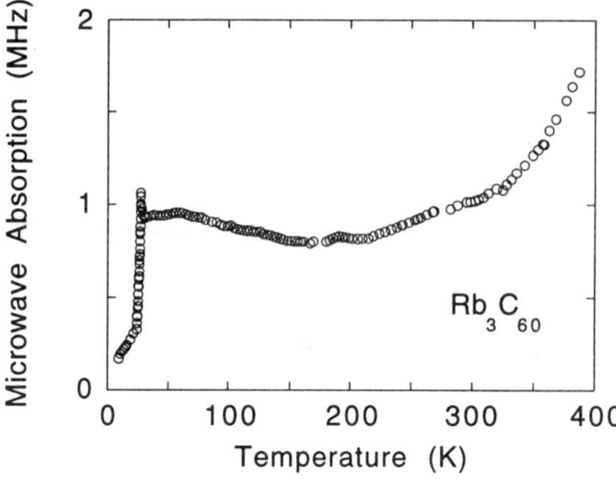

Figure 2.. Microwave absorption vs temperature of Rb_3C_{60}. The absorption is proportional to the conductivity.

The data resemble somewhat the raw data of the microwave absorption by Gruner[5] but we used a different sample geometry and the temperature dependence of the conductivity σ found by us differs qualitatively from theirs. In ref. 5 the sample consisted of a compressed pellet placed at the bottom of a cavity. It was assumed that the sample dimensions were large compared to the penetration depth at all temperatures. In this interpretation P is proportional to $\sigma^{-1/2}$ in contrast to our case where in the normal state P is proportional to σ.

The temperature dependence of the absorption is similar for Samples 2 and 3 with nominal compositions of $2Rb:C_{60}$ and $3Rb:C_{60}$ respectively. For Sample 2 P/V is 1/3 than for sample 3 which has a much larger Rb_3C_{60} phase content. The absorption of Sample 1 which contains no superconducting phase was only 2% of that of Sample 3. This supports our view that the absorption in Samples 2 and 3 is due to the superconducting Rb_3C_{60} phase only. (It is likely that in Samples 1 and 2 which contain large amounts of the conducting Rb_1C_{60} phase the smaller particle dimensions and the anisotropy of the conductivity reduce the absorption of this phase.)

6. Conduction electron spin resonance of Rb_3C_{60}

The Rb_3C_{60} conduction electron spin resonance spectra of Sample 4 with a nominal $4Rb:C_{60}$ composition is shown on figure 3. In the Rb rich sample the Rb_1C_{60} content is negligible and the spectrum is composed almost entirely of the broad Rb_3C_{60} resonance. The ESR spectra of fig. 3 are not the usual magnetic field derivative spectra but the absorption spectra proportional to the imaginary part of the dynamic susceptibility, χ''. We modified the Bruker 300E spectrometer to obtain χ'' directly. In the modified setup the microwave power is chopped by a PIN modulator instead of

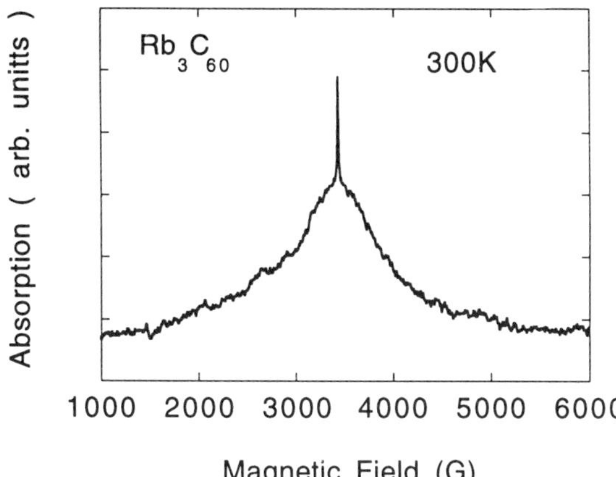

Figure 3. The conduction electron spin resonance absorption spectrum in Rb$_3$C$_{60}$. The linewidth is ~900G .The narrow peak at the central field is from a minority phase.

modulating the magnetic field[15]. The ESR linewidth of this sample is the same as for a 3Rb:C$_{60}$ nominal composition sample reported in ref. 6. We have found the same Rb$_3$C$_{60}$ ESR linewidth in several samples with various nominal compositions and conclude that the unusually large width is intrinsic. The narrow ESR lines reported by several authors is usually due to the Rb$_1$C$_{60}$ phase although defects in the insulating phases may also give rise to a narrow ESR line.

The large linewidth, ΔH, of the conduction electron spin resonance in Rb$_3$C$_{60}$ is another indication[6] of an anomalously short τ. In normal metals the conduction electron spin lifetime T_1 is determined by the Elliott-Yafet mechanism related to static or dynamic fluctuations of the spin orbit coupling caused by impurities or phonons[16]. According to this theory $T_1 = C\tau$ where C is proportional to the square of the (average) spin orbit coupling constant λ. C is independent of temperature and is the same for scattering by phonons, impurities or dislocations[17,18]. In the Elliott-Yafet process the Zeeman levels are homogeneously broadened and except for extreme cases, $\Delta H = 2/(\gamma T_1)$ is a good approximation. Δg, the shift of the g factor from the free electron value of 2.0023 is a measure of λ and empirically the relation

$$\tau = \alpha(\Delta g)^2 T_1 \qquad (1)$$

is fulfilled in normal metals with α varying less than an order of magnitude. The linewidth of Rb$_3$C$_{60}$ is too broad to measure Δg accurately. The g shift is, however, not very different from that of Rb$_1$C$_{60}$ for which in the fcc (high temperature) phase we found g=1.9993 and in the orthorhombic phase at 300K g=2.0007. $\tau = 1.8\ 10^{-15}$ sec is found[6] at 30 K from T_1 and relation (1) by using $\Delta g = 2 \times 10^{-3}$ and $\alpha = 4.8$, the value appropriate for Na which is a simple metal with a similar g shift as Rb$_3$C$_{60}$.

512

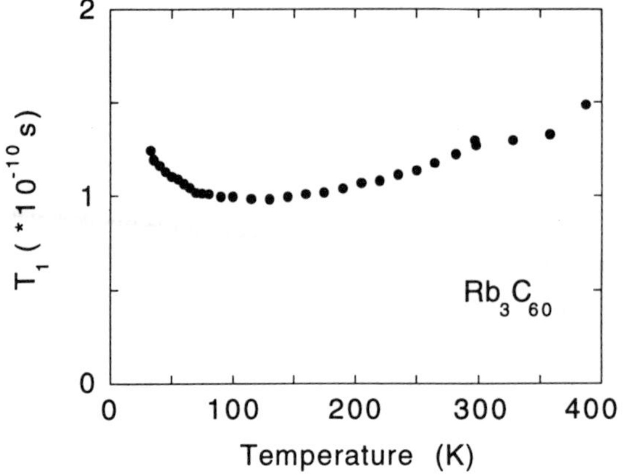

Figure 4. The conduction electron spin-lattice relaxation time T_1 vs temperature of Rb$_3$C$_{60}$. T_1 was deduced from the ESR linewidth.

Figure 4 shows the temperature dependence of T_1 deduced from the ESR linewidth for Sample 3 for which the microwave absorption was measured. Although the precision of the ESR data is not very high, the similarity of the temperature dependences of T_1 and σ is well established. Unlike normal metals, in Rb$_3$C$_{60}$ above 200 K T_1 increases with increasing temperature. As expected from relation (1), the anomalous temperature dependence of σ is accompanied by a similar anomaly in T_1.

7. Conclusions.

We found that the temperature dependence of the conductivity is qualitatively different than that reported previously by several authors. The peak in the real part of the conductivity of Rb$_3$C$_{60}$ below T_c is indicative of a Hebel-Slichter coherence peak. It gives strong support to the interpretation of Kiefl et al[1] of a similar peak in the spin relaxation rate of muonium. The smallness of the peak may be due to the short momentum lifetime indicated also by the fast CESR relaxation. In the normal state the temperature dependence of the conductivity is complicated. It is similar to the temperature dependence of the conduction electron spin relaxation rate.

Acknowledgements. Work at Budapest was supported by the grant OTKA 2932 and 2979. Work at Lausanne was supported by the grants of the Swiss Natinal Foundation for Scientfic Research, grant No. 2100-037318 and 7UNPJ0384426.

8. References.

1. R.F. Kiefl et al. *Phys. Rev. Lett.* 70 (1993) 3987.
2. A. F. Hebbard, T. T. M. Palstra, R. C. Haddon and R. M. Flemming, *Phys. Rev.* B48 (1993) 9945
3. Y. Maruyama, T. Inabe, H. Ogata, Y. Achiba, S. Suzuki, K. Kikuchi, and I. Ikemoto. *Chem. Lett.* 10, (1991) 1849
4. X. -D. Xiang, J. G. Hou, G. Breceno, W.A Bareka, R. Mostovoy, A. Zettl, V. H. Crespi, and M.L. Cohen, *Science* 256, (1992) 1190 .

5. O. Klein, G. Gruner, S.-M. Huang, J. B. Wiley, and R. B. Kaner, *Phys. Rev.* B46 (1992) 11247.
6. A. Janossy, O. Chauvet, S. Pekker, J. R. Cooper, and L. Forro, *Phys. Rev. Lett.* 71 (1993) 1091 .
7. O. Chauvet, G. Oszlanyi, L. Forro, P.W. Stephens, M. Tegze, G. Faigel, and A. Janossy, *Phys. Rev. Lett.*, in press.
8. L.D. Rotter et al., *Nature* (London) 352, (1992) 532,.
9. L. C. Hebel and C. P. Slichter, *Phys. Rev.* 113, (1959) 1504.
10. K. Holczer, O. Klein, and G. Gruner, *Solid State Commun.* 78, (1991) 875 .
11. K. Holczer, L. Forro, L. Mihaly, and G. Gruner , *Phys. Rev. Lett.* 67 (1991) 152.
12. O. Klein, K. Holczer, and G. Gruner, *Phys. Rev. Lett.*, 68, (1992) 2407.
13. M. C. Nuss, P. M. Mankiewich, M. L. O'Malley, E. H. Westerwick, and P. B. Littlewood, *Phys. Rev. Lett.* 66, (1991) 3305.
14. E. J. Mele, and S. C. Erwin, *Phys. Rev.* B47, (1993) 2948.
15. A. Janossy and R. Chicault, *Physica* C 192 (1992) 399
16. Y. Yafet, in *Solid State Physics,* edited by F. Seitz and D. Turnbull (Academic, New York, 1963). Vol. 14, p.1.
17. P. Monod and F. Beneu, *Phys. Rev.* B19 (1979) 911.
18. F. Beneu and P. Monod *Phys. Rev.* B 13, (1976) 3424.

NEW FULLERENE CHARGE TRANSFER MATERIALS:
A SEARCH FOR NEW MAGNETIC MATERIALS AND SUPERCONDUCTORS BY ESR

P.Venturini, V.Kraševec, D.Mihailovič
J.Stefan Institute, University of Ljubljana, Slovenia

M.Eiermann, G.Srdanov, N.S.Sariciftci, C.Li, F.Wudl
Institute for Polymers and Organic Solids, University of California Santa Barbara, USA

ABSTRACT

A number of very different compounds were synthesized with aim of finding either ferromagnetic or superconducting organic materials based on the fullerene C_{60}. In this work we present fullerene C_{60}, dinitro-spiromethanofullerene ($C_{61}"NO_2"$), p-dimethylamino-diphenyl-methanofulleren (C61(PhNMe2)2) and methanofullerene ($C_{61}H_2$) doped with electron donors like tetrakis(dimethylamino)ethylene (TDAE), cobaltocene (Cp_2Co) and the conducting polymer poly(dipropargylhexylamine) (PDPHA). ESR data shows a range of properties, from ferro-, antiferro- to paramagnetic-like properties but so far no superconductivity.

1. Introduction

Very interesting solid state properties of doped fullerene C_{60} have so far been discovered: superconductivity in alkali metal doped C_{60}, M_3C_{60} (M=K,Rb,RbCs,...)[1] and magnetic ordering in the complex of C_{60} with strong organic electron donor tetrakis(dimethylamino)ethylene (TDAE)[2]. In order to find fullerene materials with higher superconducting or magnetic ordering transition temperatures one can try variety of electron donor substances, take higher fullerenes or functionalize C_{60}. Here we present measurements on some of the novel charge transfer complexes synthesized from the fullerene derivatives.

2. Experiments

Among various methods for functualization of C_{60} the cyclo addition of diazo compounds producing methanofullerene is by far the simplest and electronic properties of methanofullerenes were found to be similar to parent molecule[3]. Methano bridging in C_{60} occurs at the 6-6- and the 6-5- ring junctions.

The methano fullerenes were synthesized as described before[3,4,5].

For doping of methano fullerenes with TDAE in glove box access TDAE (97% Aldrich) was added to about 10 mg of methanofullerenes and stirred overnight. Unreacted TDAE was washed out with hexane and the resulting complex was dried in vacuum.

Cobaltocene (Cp_2Co) doping was performed so that Cp_2Co and about 10 mg of methanofullerenes were separately dissolved in toluene. Cp_2Co was drop wise added to the methanofullerene solution in molar ratio 1:1. The resulting precipitate was washed with hexane and dried in vacuum.

The synthesis of polymer ploy(dipropargylhexylamine) doped C_{60}, PDPHA-C_{60} was performed similarly: polymer PDPHA was dissolved in toluene and drop wise added to a solution of C_{60} in toluene to form a 1:1 ratio monomer to C_{60}. Solutions of C_{60} as well as PDPHA in toluene are purple. After mixing color changes to red-brown. Also UV-visible absorption of the obtained product showed new absorptions at 720 and 1200 nm.

For ESR experiments we used a BRUKER ESP 300 spectrometer equipped with an Oxford Instruments ESR 900 cryostat and Oxford Instruments ITC temperature controller. The temperature dependence of ESR signal was measured for all samples synthesized.

3. Results and discussion

ESR spectral shapes of cobaltocene doped dinitro-spiromethanofullerene are from 3.3 K to RT of simple Lorenzian form without hyperfine structure. The ESR line width was broadened from 1.0 Gauss at RT to 7.0 Gauss at 3.3 K. The shape of χ^*T, where χ is spin susceptibility, shows a typical paramagnetic type behavior above 8 K (Figure 1.) whereas bellow 8 K magnetic ordering with a ferromagnetic-like or spin glass-like contribution to spin susceptibility is observed. Similar to TDAE-C_{60}, which shows spin glass transition at about 16 K, cobaltocene doped dinitro spiromethanofullerene is first of the derivatives showing magnetic ordering behavior.

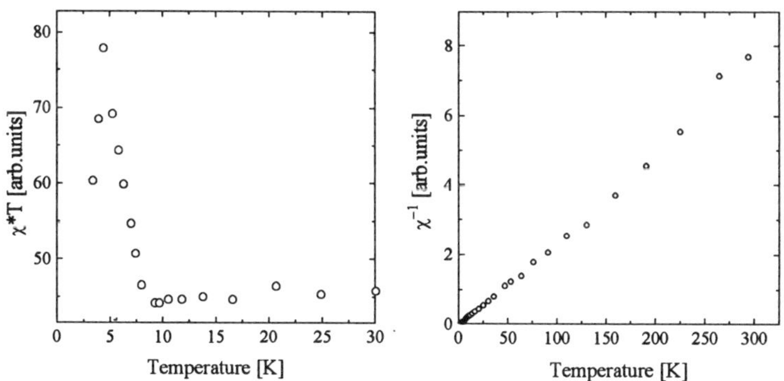

Figure 1. Temperature dependence of χ^*T and χ^{-1} for dinitro-spiromethano-fullerene doped with cobaltocene (Cp_2Co-C_{61}"NO_2").

Doping of dinitro-spiromethanofullerene with TDAE gives a charge transfer material showing significantly different behavior. The ESR line shapes are like cobaltocene doped material symmetrical in the whole temperature range and the temperature dependence of χ^{-1} ($\propto$1/ESR signal intensity) is Curie-like above 20 K (Figure 2.). Below 20 K is observed an antiferromagnetic contribution to the spin susceptibility instead of ferromagnetic or glassy behavior. For both dinitro-spiromethanofullerene doped materials the g-value was 2.0028 at room temperature which could be taken as evidence that spins mostly reside on the methanofullerenes.

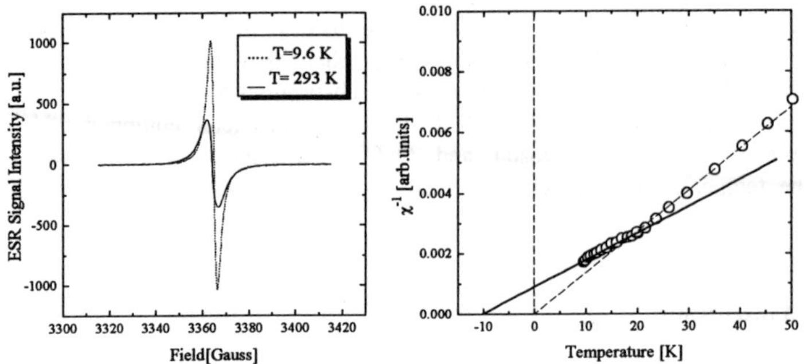

Figure 2. ESR line shapes and χ^{-1} versus temperature of TDAE doped dinitro_ spiromethanofullerene (TDAE-C_{61}"NO_2").

Charge transfer complexes of p-dimethylamino-diphenylmethanofullerene ($C_{61}(PhNMe_2)_2$) with cobaltocene and TDAE again show different behavior. The g-values are 2.0004 and 2.0009 respectively at room temperature. ESR measurements suggest Curie-like localized spins in the whole temperature range for the TDAE doped material, whereas the cobaltocene doped compound shows deviation from Curie-like behavior indicating antiferromagnetic spin correlations in the sample (Figure 3.).

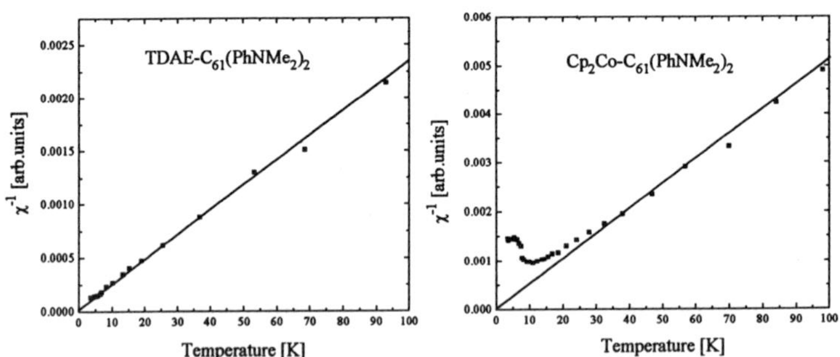

Figure 3. Temperature dependence of χ^{-1} for TDAE and Cp_2Co doped $C_{61}(PhNMe_2)_2$.

In contrast to other two C_{60} derivatives presented, [6,5] $C_{61}H_2$ has open ring structure and has 60 π electrons just like parent C_{60} molecule. The complex of $C_{61}H_2$ with TDAE shows antiferromagnetic properties. From temperature dependence of χ^{-1}(Figure 4.), the Neel temperature, T_N, was determined to be 17 K.

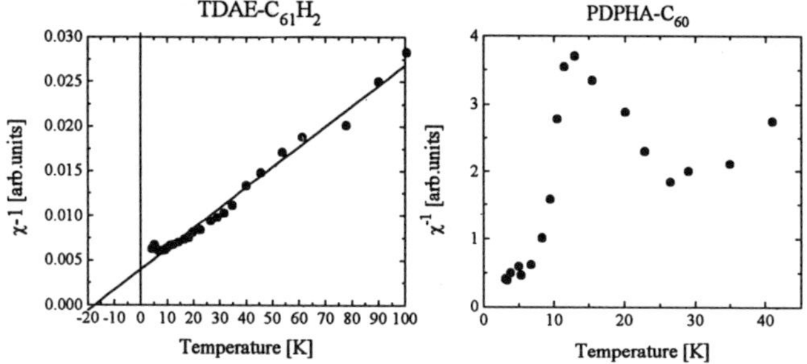

Figure 4. Temperature dependence of χ^{-1} for TDAE-C$_{61}$H$_2$ and PDPHA-C$_{60}$.

The ESR line shape of polymer PDPHA doped C$_{60}$ is asymmetrical at high temperatures and is simple Lorenzian at 3.4 K. The g-value of PDPHA-C$_{60}$ is 2.0004 at room temperature, in good agreement with 2.0003-2.0014 of TDAE-C$_{60}$[6,7,8]. The temperature dependence of χ^{-1} shows paramagnetic behavior at high temperatures, whereas there is anomaly at low temperatures from 25 to 10 K. Below 10 K the ordering "collapses" and the Curie-like behavior recovers at lower temperatures.

4. Conclusion

The ESR measurements on some novel fullerene charge transfer complexes show that the properties are dramatically changed in doped C$_{60}$ derivatives compared to doped C$_{60}$. We have found materials that show paramagnetic behavior (e.g. TDAE-C$_{61}$(PhNMe$_2$)$_2$) as well as materials that show ferromagnetic or spin glass type (e.g. Cp$_2$Co-C$_{61}$"NO$_2$") or antiferromagnetic (e.g. TDAE-C$_{61}$H$_2$) correlations of spins.

5.References

[1] A.F.Hebard et al., *Nature* **350**(1991)600.

[2] P.-M.Allemand et al., *Science* **253**(1991)301.

[3] F.Wudl et al., *Acc.Chem.Res.***25**(1992)157.

[4] T Suzuki et al., *Science* **254**(1991)1186.

[5] T.Suzuki et al., *J.Am.Chem.Soc.* **114**(1992)7301.

[6] K.Tanaka et al., *Phys.Lett. A* 164 (1992) 221.

[7] P.W.Stephens et al., *Nature* 355,(1992)331.

[8] D.Mihailovič et al., *Sol.Stat.Comm.* **89**(1994)209.

LOW-FREQUENCY PHONONS, SUPERCONDUCTIVITY AND RESISTIVITY OF FULLERIDES

S.-L. Drechsler*

Institut für Festkörperforschung am IFW Dresden e.V., D-01171 Dresden, Germany

O.V. Dolgov[†] and U. Nitzsche

Max-Planck-AG "Elektronensysteme" a. d. Techn. Universität Dresden, Germany

S. Shulga (a) and T. Galbaatar (b)

(a) Inst. of Spectroscopy, Troitsk; (b) Joint Inst. f. Nuclear Research, Dubna, Russia

ABSTRACT

Sizeable linear and/or quadratic coupling of electrons to low-frequency phonons of fullerides are discussed within the standard Eliashberg theory. The quadratic T-dependence of the resistivity for $T < 250$ K excludes strong linear coupling of electrons to harmonic phonons. It requires a substantial quadratic el-phonon interaction. Coupling to strong anharmonic vibrations of alkaline ions in octahedral positions cannot be ruled out. Madelung energy studies of C_{60}^{-3}-ions with homogeneous and inhomogeneous onball-charge distributions show that the quadratic electron-libron matrix element can be enhanced by an order of magnitude for strongly localized cyclic polarons.

1. Introduction

Our present understanding of the superconductivity in fullerides has not been reached a quantitative level so far, although commonly various high-energy onball phonon mechanisms are considered as the predominant ones (see e.g. [1]). In the present paper we show that in standard strong coupling theory several superconducting thermodynamic and normal state properties might be sizeable affected by low-frequency interball-phonons. Among them, librons coupled quadratically to electrons and strong anharmonic alkaline octahedral vibrations are of special interest.

2. Eliashberg-theory analysis of superconductivity in fullerides

We model the spectral density $\alpha^2 F(\omega)$ by several peaks centered at ω_i. Their partial coupling strengths are given by λ_i. In addition to moderate coupling strength $\lambda_{opt} \approx 0.5$ at high energies $\hbar\omega_{opt} \sim 50-100$ meV giving the most important contribution $\geq 60\%$ to the critical temperature T_c, a sizeable coupling $\lambda_{1,2} \sim 1$ to appropriate soft modes, e.g. $\hbar\omega_O = 3 \div 4$ meV and $\hbar\omega_q = 8$ meV (see below), is suggested by the large $2\Delta(0)/T_c$ ratios (given in parentheses) derived from tunneling (5.2) [2],

---

*Supported by the Deusche Forschungsgemeinschaft

[†]Permanent address P.N. Lebedev Physical Institute, Moscow, Russia

photoemission (4.1) [3], and Raman (6.8) [4] data. Further support of this picture follows from the density of states dependence of T_c. Assuming constant frequencies and $\lambda_i \propto N(0)$, we obtained for η given by $\frac{dT_c}{T_c} = \eta \frac{dN(0)}{N(0)}$ the following relation

$$\frac{1}{\eta} = 0.83 = \lambda_{opt} - \mu^* + 2\sum_i \lambda_i \left(\frac{\tau_i}{1+\tau_i^2}\right)^2 = \lambda_{opt} - \mu^* + 0.1\lambda_O + 0.48\lambda_q, \qquad (1)$$

where $\tau_i = \hbar\omega_i/2\pi T_c$ has been introduced. The lhs-value follows from susceptibility data [5] and the numerical factors of the rhs correspond to soft modes considered below. Disregarding any influence of soft modes, one gets a large effective coupling constant for T_c in contradiction to $T_c \leq 35K$.

Furthermore, the shape of $\Delta(T)$ and Holstein-like peaks near 3 and 4.5 meV visible in IR-data [6] cannot be explained within a weak coupling picture. Some experimental data as relaxational rates (NMR and μSR) yielding a BCS-like Arrhenius laws, a small jump of the specific heat [5], and the penetration depth $\lambda^{-2} \propto [1 - (T/T_c)^4]$ (Dinse et al. (94)) are *not* conclusive for a purely weak coupling scenario in contrast to common believe. Thus, the multi-peak model with strongly coupled soft-modes mentioned above explains the absence of a Hebel-Slichter peak and yields *two* Arrhenius-like regions below T_c. Within the upper region $0.2T_c < T < 0.9T_c$ an effective Arrhenius energy $E_A \sim \sqrt{\Omega_{low}\Delta(0)}$ accidentally near the BCS-value $1.76k_BT_c$ can be obtained. The true gap appears in the low $T < 0.1T_c$ region, only.

From the other hand, small $2\Delta(0)/T_c$-ratio derived from IR-data [6] contradict our simple isotropic picture, but they could be explained by anisotropy or multiband effects. The occurence of a Hebel-Slichter like peak in μSR (Kiefl et al. (93,94)), too small changes of the libronic linewidth below T_c, and the strange huge isotope effect for mixtures of ^{12}C-balls and ^{13}C-ones require more sophisticated studies.

3. Semi-microscopic estimate of electron-soft mode coupling constants

There are two potentially important soft-modes: librons and anharmonic octahedral K-vibrations near 3 meV. The latter are suggested by the observation of an unexpected increase of spectral weight in K_3C_{60} [7,8] below 25 K. The corresponding coupling constant λ_O can be estimated as

$$\lambda_O = N(0)\mid I \mid_{FS}^2 d_0^2 \frac{\tanh(\hbar\omega_O/2k_BT)}{\hbar\omega_O} \approx 0.03N(0)[\text{states/spin/C}_{60}], \qquad (2)$$

where $2d_0 \sim 0.25$ Å is the double-well distance and the value of $I^2 \approx 6.5\ 10^{-3}$ eV^2/Å^2 was taken from LDA-calculations [1]. With typical values $N(0) \sim 10 \div 15$, we get a non-negligible coupling constant λ_O without contradiction to the nearly zero-Rb(K) isotope effect for T_c caused by the low frequency and the strong anharmonicity. On the contrary, a sizeable alkaline isotope effect for the gap is expected.

To estimate the magnitude of the quadratic el-lib coupling constant λ_q

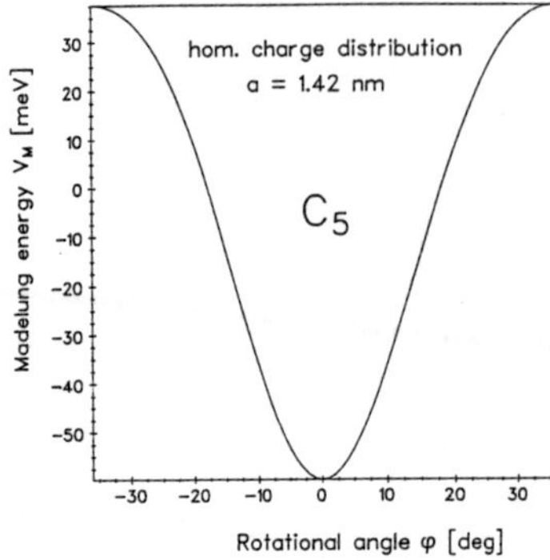

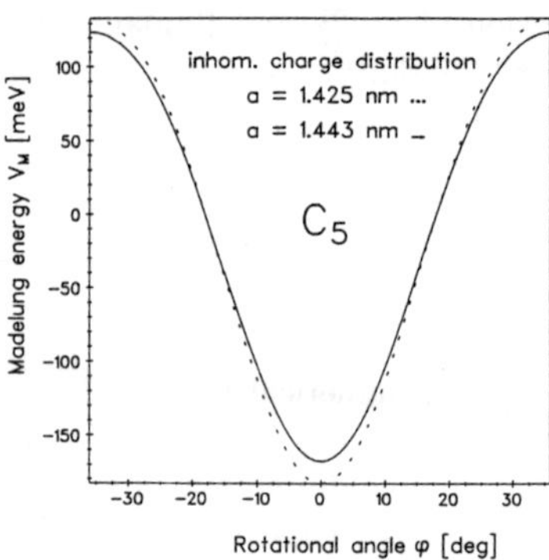

Fig. 1. Libronic Madelung potential for rotation around a five-fold axis.

Fig. 2. The same as in Fig. 3 for an inhomogeneous local charge distribution

$$\lambda_q(T) = \frac{N(0)\,|\,V(\varphi)''\,|^2_{FS}\,\hbar}{I^2\omega_{lib}^3}\,\coth(\hbar\omega_{lib}/2k_BT) \tag{3}$$

we evaluated the change of the Madelung energy under rotation around a local two- or five-fold axis of a C_{60}^{-3} anion with a rigid local charge distribution. For the sake of simplicity we approximated all neighbouring balls by point charges. Two limiting cases for the on-ball charge distribution are considered: homogeneous and inhomogeneous ones. In the latter case we adopt a charge distribution like in cyclic polarons found for isolated ions. The calculated potentials shown in Figs. 1 and 2 can be quite well approximated by two cosine terms:

$$V(\varphi) = V_0 - V_1\cos(n\varphi) - V_2\cos(2n\varphi) + ..., \quad \text{with} \quad V_2 \approx \frac{1}{8}V_1 \quad n = 2,5. \tag{4}$$

Notice the large effect of the charge inhomogeneity, enhancing V_1 by a factor of three compared with the homogenous case. With $V_1 \approx 150$ meV from Eq. 4 one easily estimates $V'' \approx 5 \div 6$eV and arrives under favourable circumstances at

$$\lambda_q(0) \sim N(0)\hbar\frac{\omega_{lib}}{3} \approx 0.001N(0)[\text{states /eV/spin/C}_{60}].$$

Using again $N(0) \sim 10 \div 15$, at $T \leq T_c$ only a small coupling constant follows. However, a further increase until $\lambda_q \approx 0.5$ might be obtained within a more realistic picture of self-consistent 3D-extended carbon charges on the ball and a discrete distribution on nearest neighbour balls as well as some modulation of the interball transfer integrals, provided the cyclic tripolarons survive the solid state effect.

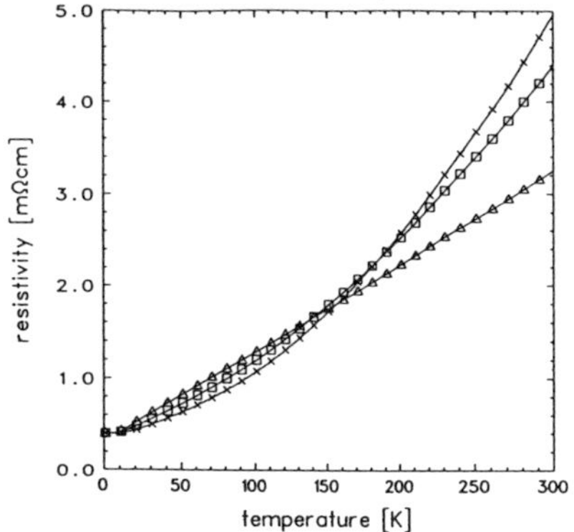

Fig. 3. Resistivity vs. T within the three peak model of Sec. 2.
$\triangle$ – linear coupling:
$\lambda_{lib} = 2.75$, $\lambda_{opt} = 0.53$;
$\square$ – linear + quadratic coupling:
$\lambda_{lib} = 1.4$, $\lambda_q = 0.75$, $\lambda_{opt} = 0.5$;
$\times$ – $\rho_0 + AT^2$.

4. Resistivity and normal state properties

Adopting $\alpha^2 F(\omega)_{sc} \approx \alpha^2 F(\omega)_{tr}$, $\hbar\omega_{pl} = 1.2$ eV, and a residual resistivity $\rho_0 = 0.5\text{m}\Omega\text{cm}$, the resistivity ρ has been calculated using the Ziman-formula (see Fig. 3)

$$\rho(T) = \frac{4\pi^2\hbar}{T\omega_{pl}^2} \int \frac{\omega\alpha^2 F(\omega)}{\sinh^2(\hbar\omega/k_B T)})d\omega \tag{5}$$

For linear coupling to harmonic soft modes one obtains from Eq. 5 a linear term for $4k_B T > \hbar\omega_i$ discriminated, however, by the experiment. Unfortunately, no information about possible linear coupling of anharmonic alkaline modes can be deduced from $\rho(T)$ because the corresponding term $\propto 1/[T \sinh(\hbar\omega_i/k_B T)]$, remains nearly constant at $T > 30$K. By contrast, quadratic coupling to harmonic soft modes yields a desired quadratic contribution to $\rho(T)$ caused by the coth-factor in Eq. 3. The observed linear behaviour for $T \geq 300 \div 500$ K might be a consequence of corrections to Eq. 3 due to the short mean free path and anharmonic effects. Finally, the T-dependence of the thermopower $S(T)$ at $T_c < T < 150$ K can be explained either by renormalization due to strong coupling of soft modes or by phonon drag effects.

References

1. V.P. Antropov et al. , Phys. Rev. B, **48**, 7651 (1993).
2. Z. Zhang et al., Science **254**, 1619 (1991).
3. F. Gu et al., High-T_c-Update, **8**, (1994).
4. G. Els et al., Verhandlungen der DPG **5**, 1599 (1994).
5. A.P. Ramirez et al., Phys. Rev. Lett. **69**, 1687 (1992).
6. L. Degeorgi et al., Phys. Rev. Lett. **69**, 2987 (1992).
7. B. Renker et. al., Zs. Phys. B, **92**, 451 (1993).
8. S.-L. Drechsler et al., these Proceedings (1994).

MULTI-BAND EFFECTS ON THE SUPERCONDUCTIVITY IN ORIENTATIONALLY ORDERED DOPED C_{60}

Han-Yong Choi [a]

Department of Physics, Sung Kyun Kwan University, Suwon 440-746, Korea

and

M. J. Rice

Xerox Webster Research Center, 0114-39D, Webster, NY 14580, USA

ABSTRACT

We consider multi-band effects on the superconductivity in the alkali-doped fullerenes. It is shown that the multi-band superconducting gap is reduced to the well known Bardeen-Cooper-Schrieffer expression due to a cyclic symmetry in orientationally ordered phases. This justifies many previous works assuming single-band s-wave pairing without considering several conduction bands explicitly.

The recently found Buckminsterfullerene has lots of interesting and novel physical properties, and has spurred an intense search for various physical properties of this class of material.[1] Perhaps the most interesting is the occurrence of superconductivity at relatively high temperatures.[1,2] The coupling of the conduction electrons to the phonons derived from the Raman active H_g and A_g vibrations of the C_{60} molecular ion may be relevant to the remarkable occurrence of superconductivity in these compounds.[1] As well known[3-5], multi-band effects should be considered explicitly in the theory of superconductivity for materials having more than one conduction band. The so called multi-band superconductivity (MBS) is a logical consequence of the existence of several conduction bands. We therefore explicitly consider the several conduction bands derived from the three-fold degenerate t_{1u} molecular orbitals (MO) in this paper. For an orientationally disordered phase, the Anderson theory of dirty superconductivity[6] is applicable. For an orientationally ordered phase, on the other hand, it will be shown that the multi-band superconductivity is again reduced to the Bardeen-Cooper-Schrieffer (BCS) theory[7] of single-band s-wave pairing. This will justify many previous works on the superconductivity in doped fullerenes assuming single-band s-wave theory without considering MBS theory explicitly.[8,9] This is done within the *weak-*

coupling scheme on the assumption that superconductivity in A_3C_{60} (A = K, Rb, and/or Cs) is due to electron-intramolecular-vibration (EMV) coupling.[1]

We describe the conduction electron states and their symmetry allowed local coupling to the molecular C_{60} H_g and A_g vibrational modes by the following tight-binding Hamiltonian.

$$H = - \sum_{j,\delta,\nu,\nu'} t^\delta_{\nu'\nu} a^+_{j+\delta,\nu'} a_{j,\nu} + \sum_{j,\alpha\lambda} \left\{ \left(b^+_{j,\alpha\lambda} b_{j,\alpha\lambda} + \frac{1}{2} \right) \omega_\alpha + \sum_{\nu,\nu'} g^{\alpha\lambda}_{\nu\nu'} a^+_{j,\nu} a_{j,\nu} \left(b^+_{j,\alpha\lambda} + b_{j,\alpha\lambda} \right) \right\}. \tag{1}$$

The first term describes the intermolecular hopping of electrons between the t_{1u} orbitals, the second term the H_g and A_g vibrational excitations, and the third term the EMV coupling at each molecular site. One obtains the following equation from the first term of Eq. (1).

$$\sum_{\nu'=1}^{3} H_{\nu\nu'}(k) A_{n\nu'}(k) = \varepsilon_{k,n} A_{n\nu}(k), \quad \text{with} \quad \sum_{\nu=1}^{3} A_{n\nu}(k)^2 = 1, \tag{2.a}$$

where

$$H_{\nu\nu'}(k) = \begin{pmatrix} t_1 c_y c_z + t_2 c_z c_x + t_3 c_x c_y & t' s_x s_y & t' s_z s_x \\ t' s_x s_y & t_3 c_y c_z + t_1 c_z c_x + t_2 c_x c_y & t' s_y s_z \\ t' s_z s_x & t' s_y s_z & t_2 c_y c_z + t_3 c_z c_x + t_1 c_x c_y \end{pmatrix}, \tag{2.b}$$

with $c_x = \cos(k_x a/2)$, $s_x = \sin(k_x a/2)$, and similar expressions for y and z components. It is important to point out that Eq. (2) possesses a cyclic symmetry: Under a simultaneous transformation

$$(k_x, k_y, k_z) \rightarrow (k_y, k_z, k_x) \text{ and } (A_{n1}, A_{n2}, A_{n3}) \rightarrow (A_{n2}, A_{n3}, A_{n1}), \tag{3}$$

Eq. (2.a) remains unchanged. It is precisely due to this symmetry that the MBS theory reduces to one-band theory as will be shown later.

In icosahedral symmetry, only the A_g and H_g phonons can couple to two t_{1u} electrons. Taking advantage of the symmetry of the coupling constants, the H_g and A_g phonon mediated pairing potential is obtained as[5]

$$V(n,k;n',k') = V_0 + V_1 \sum_{v=1}^{3} \{A_{nv}(k) A_{n'v}(k')\}^2, \tag{4.a}$$

where

$$V_0 = 3V_H/4, \quad V_1 = V_A + V_H/4, \quad V_H = \sum_{\alpha \in H_g} 2g^2_\alpha/\omega_\alpha \text{ and } V_A = \sum_{\alpha \in A_g} 2g^2_\alpha/\omega_\alpha. \tag{4.b}$$

In MBS theory the standard BCS gap equation is extended as follows to include band indices:[3]

$$\Delta_{k,n} = \sum_{k',n'} V(k,n;k',n') \frac{\tanh(\frac{1}{2}\beta E_{k',n'})}{2E_{k',n'}} \Delta_{k',n'}, \tag{5}$$

where $E_{k,n} = \sqrt{(\varepsilon_{k,n}-\mu)^2+\Delta_{k,n}^2}$, μ the chemical potential, $\Delta_{k,n}$ the superconducting gap, and $V(k,n;k',n')$ is the superconducting pairing potential. Inspecting this equation, we can immediately see that $\Delta_{k,n}$ must have the following form.

$$\Delta_{k,n} = \sum_{i=1}^{3} \Delta_i \left[A_{ni}(k)\right]^2, \tag{6}$$

where Δ_i's are constants which depend on temperature alone. Substituting Eq. (6) into the gap equation, we obtain

$$\Delta_i = \sum_j S_{ij}\Delta_j, \quad S_{ij} \equiv V_0 F_j + V_1 F_{ij}, \tag{7}$$

where $F_j = f[A_{nj}]$, $F_{ij} = f[A_{ni}A_{nj}]$, $F = f[1]$, where $f[g]$ is defined as

$$f[g] \equiv \sum_{k,n} \frac{\tanh(\frac{1}{2}\beta E_{k,n})}{2E_{k,n}} [g(k)]^2. \tag{8}$$

Due to the cyclic symmetry of Eq. (4) and the normalization condition of the Bloch amplitude A_{nj}, there exist the following relations between the elements of F_j and F_{ij}.

$$F_i = \Sigma_j F_{ij} = (1/3)F, \quad F_{12} = F_{23} = F_{31}, \quad F_{11} = F_{22} = F_{33}. \tag{9}$$

The above relations render the matrix S_{ij} of Eq. (7) to the form of

$$S_{ij} = \begin{pmatrix} a & b & b \\ b & a & b \\ b & b & a \end{pmatrix}. \tag{10}$$

The pairing field having the highest transition temperature is the one with the largest eigenvalue of S_{ij}, which in this case is $\Delta_i = \Delta$ with the corresponding eigenvalue of $a + 2b$. We therefore have $\Delta_{k,n} = \Delta$ from Eq. (6), where Δ depends on the temperature alone. The gap equation of Eq. (5) is then reduced to

$$1 = V F, \quad V = V_0 + V_1/3 = (5/6)V_H + V_A/3. \tag{11}$$

This implies that V_H modes are more important than V_A modes in determining one-band superconductivity as mentioned before. Eq. (11) is just the self-consistency equation of the BCS theory, which determines the transition temperature. It should be pointed out that other previous works[8,9] simply averaged the pairing potential S_{ij} as $(1/3)^2\Sigma_{ij}S_{ij}$ instead of finding its largest eigenvalue as described. This procedure, for S_{ij} of the form of Eq. (10), is equal to taking the largest eigenvalue, and is now justified. This shows how the multi-band gap in the orientationally ordered A_3C_{60} is reduced to the single-band s-wave gap, provided that the superconductivity in doped fullerenes is due to the EMV coupling.

Experimentally, the fullerene molecules in A_3C_{60} are quenched into an orientationally disordered state.[10] The cyclic symmetry will be lifted in this case, which will cause some deviations from the uniform s-wave gap. The degree of this gap anisotropy depends on the degree of disorder, and for completely disordered phase the gap will be isotropic again. Experimental situations will correspond to somewhere between the orientationally ordered and completely disordered phases, which may then show some degree of gap anisotropy.

In summary, we have explicitly considered the effects of several conduction bands on the superconductivity in A_3C_{60} by solving the multi-band superconductivity gap equation and have shown that the the multi-band theory is reduced to the simple one-band s-wave theory due to a symmetry in orientationally ordered phases, assuming that superconductivity in A_3C_{60} is due to electron-intramolecular-vibration coupling.

One of us (HYC) would like to acknowledge the support from Korean Science and Engineering Foundation (KOSEF) through Grant No. 931-0200-003-2 and through Center for Theoretical Physics, Seoul National University.

(a) E-mail: hychoi@yurim.skku.ac.kr

References

1. See, for example, A. F. Hebard, *Physics Today* **45**, No. 11, 26, (1992), and references therein.
2. A. F. Hebard *et al.*, *Nature* **350**, 600 (1991); M. J. Rosseinsky *et al.*, *Phys. Rev. Lett.* **66**, 2830 (1991); K. Holczer *et al.*, *Science* **252**, 1154 (1991).
3. H. Suhl, B. T. Matthias, and L. R. Walker, *Phys. Rev. Lett.* **3**, 552 (1959).
4. M. J. Rice, H.-Y. Choi, and Y. R. Wang, *Phys. Rev. B* **44**, 10414 (1991); E. J. Mele and S. Erwin, *ibid* **47**, 2948 (1993).
5. H.-Y. Choi and M. J. Rice, to be published in *Phys. Rev. B* **49** (1994).
6. P. W. Anderson, *J. Phys. Chem. Solids* **11**, 26 (1959).
7. J. Bardeen, L. N. Cooper, and J. R. Schrieffer, *Phys. Rev.* **106**, 162 (1957).
8. C. M. Varma, J. Zaanen, and K. Raghavachari, *Science* **254**, 989 (1991).
9. M. Schluter *et al.*, *Phys. Rev. Lett.* **68**, 526 (1992).
10. P. W. Stephens *et al.*, *Nature* **351**, 632 (1991).

POLARONIC SIGNATURES
IN CUPRATE AND FULLEREN SUPERCONDUCTORS

J. RANNINGER

Centre de Recherches sur les Très Basses Températures,
laboratoire associé à l'Université Joseph Fourier,
C.N.R.S., BP 166, 38042 Grenoble-Cédex 9, France

ABSTRACT

Localized polaronic charge carriers are a common feature of all high T_c superconductors in their insulating phase. For understanding the mechanism of superconductivity in these materials it is of vital importance to establish whether these polaronic charge carriers also exist in the metallic phase and in particular if they exist in form of coherent Bloch-like band states. We propose a number of spectroscopic measurements which test the internal electric and vibrational structure of small polarons and which could provide clues to distinguish itinerant from localized polaronic states.

1. Introduction

Over the past five years it has become increasingly clear that polaronic features are common to all the high T_c superconductors—cuprates, bismutates and fullerens included. Contrary to amorphous semiconductors and transition metal oxides where polarons exist as well localized states, in the high T_c superconductors (HT_cSC) they can exist in form of fluctuating valence states. This is made possible i) by an overlap of the polaronic energies with the energies of itinerant electrons in these systems and ii) by the existence of polaronic centers in form of particular molecular clusters capable of internal deformation fluctuations which drive fluctuations between different valence states. As examples for the latter, we note : in $YBa_2Cu_3O_{6+x}$ the clusters $2O^{2-}(1)$-$2O^{2-}(4)$-$Cu^{2+}(1)$ can change into effective clusters $2O^{2-}(4)$-$Cu^+(1)$ upon changing the Cu(1)-O(4) distance from 1.94 Å to 1.8 Å ; in $BaBi_{1-x}Pb_xO_3$, the six fold octahedral environment $6O^{2-}$-Bi^{5+} can change into an effective five fold distorted pyramidal one $5O^{2-}$-Bi^{3+} by a concomitant moving up of the Bi ions toward the apical oxygen and a moving down of the four oxygens of the basal plane ; in the fullerens we observe a $C - C$ bonding configuration which lies between the flat one characterizing graphite (being of sp^2-type and having a bond angle of 120°) and the diamond sp^3 configuration (having a bond angle of 107°). The tangential A_g mode of 1469 cm^{-1} in C_{60} permits to fluctuate between the two extreme positions of graphite and diamond. All these dynamical fluctuations of atomic clusters are combined with sizeable changes in the bond length and hence provide the necessary conditions for self-trapping of small polarons.

In the low doped HT_cSC, polarons exist as localized entities in form of randomly distributed and non-overlapping polaronic centers. The system then is insulating because of a generally filled valence band and empty conduction band. With increased doping these polaronic centers reorganize themselves leading to strings of such entities[1] which result in local metallic pockets due to an emptying out of the states on top of the valence band and the onset of charge fluctuations between the polaronic centers and the itinerant electron subsystems[2]. The transition from insulating to metallic behaviour occurs gradually with increasing the doping and results in fully metallic samples only after the formation of a macroscopic network of linked together and overlapping polaronic units[3]. While the polaronic nature of the localized polarons in the low doped semiconducting samples of HT_cSC has been established beyond doubt[4] it remains to be verified for the metallic samples. Indications that even in the metallic regime of HT_cSC part of the charge carriers have polaronic features comes from indirect evidence such as the comparable modulation of local lattice modes due to either the presence of photo-induced or of chemically induced charge carriers[5].

The verification of polaronic charge carriers in the metallic phase of HT_cSC is of prime interest for any attempt to understand the basic mechanism driving superconductivity in these materials. In the following section 2 we shall briefly review the major experimental evidence for polaron formation in HT_cSC. The basic theoretical concepts of small polarons will be discussed in section 3 and experimental tests studying the nature of small polarons in HT_cSC will be proposed in Section 4.

2. Experimental Evidence for Small Polarons in HT_cSC : Cuprates, Bismutates and Fullerens

As we shall see in more details in the next section, a small polaron is an electron which is self-trapped inside a strong local lattice deformation. It is a consequence of strong electron-lattice coupling and all the tests concerning small polarons consist in changing the electronic state (either by stripping off the charge carrier from the trapping lattice deformation or by externally populating an electronically excited state) and examining the resulting relaxation of the lattice deformation caused by such a perturbation.

The results of such tests are the following :
a) Chemically induced doping leads to sizeable ($\sim$ 10 %) shifts of the frequencies of certain local modes which is a consequence of the change of local deformations around the self-trapped charge carriers. In $YBa_2Cu_3O_{6+x}$ these modes correspond to the c-axis polarized vibrations of the O(4)-Cu(1)-O(4) dumbbells, changing from 475 cm^{-1} to 505 cm^{-1} upon changing $Cu^+(1)$ to $Cu^{2+}(1)$ by extraction an electron upon doping[4]. In $BaBiO_3$ [6] (the parent compound of $BaBi_{1-x}Pb_xO_3$) they correspond to modes describing deformations leading from Bi^{3+} stable complexes to Bi^{5+}

stable complexes (discussed above) and are characterized by a 250 cm^{-1} mode which splits into two upon photoexcitations. In $K_x C_{60}$ these modes correspond to tangential A_g breathing modes of the C_{60} on pentagon rings with a change in frequency from 1468 cm$^{-1}(x = 0)$ to 1430 cm$^{-1}(x = 6)$[7].

b) In both, chemically doped metallic compounds and photodoped insulating compounds, the infrared excited Raman scattering is characterized by peculiar resonance conditions. The strong resonant enhanced scattering intensities show a series of well defined overtones of the intrinsic harmonic vibrational frequency associated with particular polaronic centers. This is due to strongly enhanced amplitudes of the molecular vibrations[8] when the overlap of the intramolecular deformation in the excited state is compatible with the shape of the vibrational deformations.

c) In metallic samples of HT_cSC a broad mid-infrared band is observed in the optical absorption spectra. With increasing temperature the intensity of this peak diminishes and its spectral weight is transfered to the Drude component of itinerant carriers[9]. The temperature dependence of this MIR band can adequately be described in terms of small polarons[10]. Its dependence on doping tracks the variation of the critical temperature T_c with doping which is a strong evidence that superconductivity is determined by the charge carriers in polaronic states.

3. Basic Notions of Small Polarons

We shall here briefly review, on a qualitative level, the basic underlying physics of small polarons. Let us for that purpose consider a one-dimensional chain of diatomic molecules and electrons coupling to the intramolecular deformations given by the Hamiltonian :

$$H = t \sum_{<i \neq j> \sigma} c_{i\sigma}^{+} c_{j\sigma} - g \sum_{i\sigma} n_{i\sigma} X_i + \sum_i \left(\frac{M}{2} \dot{X}_i^2 + \frac{M\omega^2}{2} X_i^2 \right) \tag{1}$$

The first term describes hopping of the electrons between adjacent molecules, the second term a local coupling of the electron density $n_i = n_{i\uparrow} + n_{i\downarrow}$ to the intramolecular deformations and the third term represents the harmonic oscillators of the molecules, neglecting any elastic coupling between them. The electron-lattice coupling proportional to g can be eliminated by a shift of the equilibrium positions of the atoms in the diatomic molecules

$$\tilde{X}_i = X_i - \frac{g n_i}{M\omega^2} \tag{2}$$

leading to

$$H = t \sum_{<i \neq j> \sigma} c_{i\sigma}^{+} c_{j\sigma} + \sum_i \left(\frac{M}{2} \dot{\tilde{X}}_i^2 + \frac{M}{2} \omega^2 \tilde{X}_i^2 \right) - \frac{g^2 n_i}{2M\omega^2} - g^2 \frac{n_{i\uparrow} n_{i\downarrow}}{M\omega^2}$$
$$+ g \frac{\dot{\tilde{X}}_i \dot{n}_i}{\omega^2} + \frac{M}{2} g^2 (\dot{n})^2 / (M\omega^2)^2 \tag{3}$$

From this form of the Hamiltonian we can read off immediately all the essential physics contained in the problem of strong electron-lattice coupled systems.

Let us first consider the case of an electron localized on a molecular site, i.e. $t \equiv 0$ and consequently $\dot{n}_i \equiv 0$. Expression (3) then reduces to three terms :

– a diagonal term in the electron number operator, measuring the polaronic level shift $\epsilon_p = g^2 n_i / 2M\omega^2$,

– a local polaron induced attraction between electrons $v_{ii} = -2\epsilon_p$,

– an harmonic oscillator for a diatomic molecular whose equilibrium positions are shifted by $X_0 = gn_i/M\omega^2$.

An electron, held fixed on a given site, hence gains energy ϵ_p via a relaxation of the lattice around it. Such an induced lattice relaxation then in turn re-acts back onto the electron and in this way leads to self-trapping. A localized polaronic state is hence described by :

$$\tilde{c}_{i\sigma}^+ |0> = c_{i\sigma}^+ |0> |\phi(x - X_0)>_i \qquad (4)$$

where $|\phi(x-X_0)>$ describes an oscillator state whose equilibrium positions are shifted by X_0, and where $\alpha = (\epsilon_p/\omega)^{1/2} = X_0(M\omega/2\hbar)^{1/2}$ denotes an effective dimensionless electron-lattice coupling constant.

Let us next consider the case of a delocalized polaronic state. The electron then will have to choose between either becoming an almost localized state by gaining the energy ϵ_p or remaining a quasi free electron and gaining the kinetic energy zt, z being the coordination number. In the strong coupling limit ($\alpha \gg 1$) and for $\omega \sim t$ this energy balance can be assessed easily, since in that case the hopping term of the electrons can be treated as a perturbation with respect to the rest of the Hamiltonian which has been paraphrased in terms of localized polaronic states, given by the third, fourth and fifth term in expression (3). The effective hopping matrix element for polarons is then given by $t^* = t \exp(-\alpha^2)$ which in general will be small compared to t.

In the strong coupling limit ($\alpha \gg 1$), the many electron state can hence adequately be described in terms of quasi free small polarons

$$\prod_k \tilde{c}_k^+ |0> \equiv \prod_k \frac{1}{\sqrt{N}} \sum_i e^{ikr_i} c_i^+ |0> |\phi(x - X_0)>_i \prod_{j \neq i} |\phi(x)>_j \qquad (5)$$

One of the key experiments to be carried out on HT_cSC is the verification of such itinerant bipolaronic or polaronic states in the metallic phase of these compounds which is the pre-requisite of superfluidity[11] of bipolarons or enhanced T_c BCS type superconductivity of small polarons[12].

4. Signatures of Itinerant Polaronic States

The characteristic feature of a small polaron is the self-trapping of the electron by which the electron localizes on a given site during a characteristic time $\sim \hbar/t^*$. Such

530

partial localization of the charge carrier will involve the ensemble of Bloch states of the electron, characterized by the wave vectors $\{k\}$. If we assume that band-like polaronic states exist (which might be possible for a Coulomb repulsion roughly compensating the polaron-polaron attraction) then we can label them with wave vector $\mathbf{k}$ such that :

$$\tilde{c}_{\mathbf{k}}^{+}|0> = \frac{1}{\sqrt{N}} \sum_{\mathbf{q}} c_{\mathbf{k+q}}^{+}|0> |\phi(x - X_0)>_{\mathbf{q}} \tag{6}$$

From expression (6) we see that all the electron states $\mathbf{k}$ in the Brillouin zone participate in constructing such a polaronic state and that at the same time the local lattice deformation $|\phi(x - X_0)>_i$ must be in a coherent Bloch-like state :

$$|\phi(x - X_0)>_{\mathbf{q}} = \frac{1}{\sqrt{N}} \sum_i e^{i\mathbf{q}\mathbf{r}_i}|\phi(x - X_0)>_i \; . \tag{7}$$

Then the vibrational degrees of freedom are irrevocably tied together with the charge degrees of freedom and hence any spectroscopical measurement coupling to either the lattice or the charge must necessarily sense both, electronic as well as vibronic excitations.

In particular, inelastic neutron scattering in a system containing a single polaron in the ground state (ie $\mathbf{k} = 0$) and given by the cross-section

$$S(\mathbf{q},\omega) \propto FT < T\{X_{\mathbf{q}}(0)X_{-\mathbf{q}}(t)\} >\propto \delta(\omega - \epsilon_{\mathbf{q}}) \tag{8}$$

measures directly the full dispersion of the small polaron[13] $\epsilon_{\mathbf{q}} = t^* \left(1 - \frac{1}{z}\sum_a \cos \mathbf{q} \cdot \mathbf{a}\right)$ just as angle resolved photoemission (ARPES) would do it via the correlation function[13]

$$I(\mathbf{q},\omega) \propto FT < c_{\mathbf{q}}^{+}(t)c_{\mathbf{q}}(0) >\propto \delta(\omega - \epsilon_{\mathbf{q}}) \tag{9}$$

The fact that in any single polaron the entire ensemble of wave vectors of electronic states is involved leads to kinematical interactions between polarons in a many polaron system which is manifest in a strongly modified spectrum for inelastic neutron scattering

$$S(\mathbf{q},\omega) \propto \begin{cases} \frac{\omega}{|q|} & \text{for } \omega < 2|q|a\sqrt{2\mu t^*}, q \to 0 \\[2ex] \rho\left(\frac{\omega}{2}\right) \cdot \theta(\omega - 2\mu)\theta(W - \omega), q = q_B \end{cases} \tag{10}$$

exhibiting a broad spectrum of scattered intensity and a gap-like structure for $q \sim q_B$[13]. Inelastic neutron scattering might then be an ameanable tool for examining the possibly itinerant nature of small polarons which for the time being is not feasable by the more direct means of ARPES because of limitations due to resolution. ARPES however is a useful probe for the spectral properties of small polarons as far as the incoherent part of those electronic excitations is concerned and given by a distribution of overtones of some basic oscillator mode from which the electron-phonon coupling constant α can in principle be determined[14].

Small cluster systems with polarons such as C_{60}^- could possibly serve as testing materials for improving our understanding of small polaron dynamics and to check if in principle they can exist as itinerant states. The comparison of photoemission of C_{60}^- in the gaseous phase[15] with that on metallic $Rb_3\,C_{60}$ [16] clearly shows to what extent itinerant electrons in the latter are participating in polaronic states on individual C_{60} molecular as can be infered from the huge incoherent part in the photoemission spectrum.

5. References

1. A. Bianconi *et al.*, in *Lattice Effects in High T_c Superconductors*, ed. Y. Bar-Yam, T. Egami, J. Mustre de Leon and R.A. Bishop, *World Scientific*, (Singapore) 1992 p. 65; S.J.L. Billinge and T. Egami, ibid p. 93.

2. J. Ranninger, in *Superconductivity and Strongly Correlated Electron Systems World Scientific*, (Singapore) 1994, to be published.

3. A. Zibold *et al.*, *PhysicaC* **212** (1993) 365.

4. C. Taliani *et al.*, in *Electronic Properties of High T_c Superconductors and Related Compounds*, ed. H. Kuzmany, M. Mehring and J. Fink, Springer Series of *Solid State Science* (Berlin) **99** (1990) 280 ; C.M. Foster *et al.*, *Synthetic Materials* **33** (1989) 171.

5. D. Mihailovic *et al.*, *Phys. Rev.* **B 42** (1990) 7989.

6. Ref. 4 and P. Ruani, private communication.

7. E. Denisov and B.N. Mavrin, *Sov. Phys. JETP* **75** (1992) 158.

8. R. Zamboni *et al.*, *Solid State Commun.* **70** (1989) 813.

9. J.P. Falck *et al.*, *Phys. Rev.* **B 48** (1993) 4043.

10. Xiang-Xin Bi and P. Ecklund, *Phys. Rev. Lett.* **70** (1993) 2625.

11. A.S. Alexandrov and J. Ranninger, *Phys. Rev.* **B 23** (1981) 1796.

12. A.S. Alexandrov, *Phys. Rev.* **B 38** (1988) 925.

13. J. Ranninger, *Solid State Commun.* **85** (1993) 929.

14. J. Ranninger, *Phys. Rev.* **B 48** (1993) 13166.

15. W. Eberhardth, see these proceedings.

16. M. Knupfer *et al.*, *Phys. Rev.* **B 47** (1993) 13944.

Part Nine

SUMMARY

ARE THERE APPLICATIONS IN FULLERENE RESEARCH?

H.U. ter Meer
Hoechst AG Frankfurt, Germany

Since the epoch making publication on a process for producing fullerenes by Krätschmer and Huffman [1] in 1990 the number of scientific publications on this subject as well as the number of groups active in the field has steadily increased (see Table 1). As Wasserman said [2]: "In contrast (to the development of superconductors after the first rush) fullerenes after two years of effort are continuing to attract more practitioners to the field. Related areas of research are appearing such as giant fullerenes and nested fullerenes as well as buckytubes. Such growth and branching is a sign of a developing discipline."

Table 1: Explosion of fullerene literature

Year	Number of Publications
1990	35
1991 (I+II)	93
1991 (III+IV)	242
1992 (I+II)	426
1992 (III+IV)	572
1993 (I+II)	680
1993 (III+IV)	847

The topic of fullerenes rose to a broader awareness when superconductivity was detected in alkali metal intercalates by the research group at AT&T [3]. This was but the first discovery of many unexpected properties of this new form of carbon. Investigation of the solid state behaviour revealed many properties that may be useful. Examples are photoconductivity [5], non-linear absorption [7], electroluminescence [8], magnetic behaviour [9], semiconduction [10] to name but a few.

If any of those properties will lead to an application is an open question. The answer depends not only on the usefulness of the properties, i.e. its capability of solving a problem or need, but also on the relative performance. On competing with already existing solutions it must be better, cheaper or more acceptable to the user. If offering a solution to an unsolved problem, its applicability has to be measured against the cost/benefit for the user. Most of the research is still based on understanding the basic properties of the material. It is difficult, hence, to evaluate the likelihood of success for any of the proposed applications. If one looks, however, to the areas of scien-

536

tific research (Table 2) it appears, that most of the research done thus far is dealing with the physical properties, the chemistry on the other hand is just beginning to emerge. It is likely, that many new areas of application may arise from there in the future.

Table 2: Publications by scientific area

Physical Science	64 %
Chemistry	22 %
Material Science	8 %
Production	2 %
Separation	2 %
Polymers	1 %
Others	1 %

In the following, some scientific results from the last year showing some promise for an application will be high-lighted.

The single most remarkable event was certainly the finding of bucky-tubes and the investigations of its properties [4]. The most basic of these properties are their predicted high tensional strength [11] that will make them superior to any known carbon fiber. Another exciting property is their transition from metal to insulator, depending on their diameter. Should it become possible to produce those fibers cheaply in bulk will these advantages make them good candidates for a novel form of carbon that went into application.

Fullerene on its own possesses a unique property that allows its use in analytical chemistry. The work of Kiyokatsu et al. [12] showed that C_{60} can be used as a stationary phase for the separation of polycyclic aromatic hydrocarbons (PAH's). This class of chemicals is of growing ecological concern and the need for specific characterisation of this material is increasing.

Another area of research that currently is emerging is the use of fullerenes as a photoconductor [6]. Here we see, that the combination of a polymer (PMPS) with the C_{60} is giving properties that under certain conditions can be superior to existing systems.

The charge generation in those systems can well be described by the Onsager theory [25], using two parameters. One parameter (r_0) describes the efficiency at low field strength. The other (ϕ_0) is proportional to the maximally achievable efficiency. The C_{60}/PMPS system (r_0=27 Å, ϕ_0=0.85) shows a significantly larger saturation efficiency, compared to the performance of the best known system of an organic photoconductor (thiapyrilium/polycarbonate r_0=44 Å, ϕ_0=0.58).

The combination of fullerenes and polymers is not only limited to the simple mixing, but various methods to attach fullerenes chemically to polymers have been reported [13-17]. Thus, fullerenes have been bound to the surface of polyethylene, glass, polyanilines and other polymers. Samulski et al. grafted polystyrene onto C_{60} [18].

This allows the formation of nanophase separated polymeric materials. And, last but not least, Loy and Assink [17] co-polymerised C_{60} with para-cyclophane, thus creating a high temperature stable polymer.

None of those polymers have yet shown any technical application. To transform the soft crystals of C_{60} into a cohesive material, however, is a prerequisite for any application of fullerenes.

Another, potentially very important finding was the discovery, that some fullerene derivatives show biological activity. The first of such finding was an interdisciplinary effort of several Californian universities [19], showing an inhibition of the HIV-protease. Here the shape and hydrophobic character of C_{60} is used.

Another area of biological activity was reported by Tokuyama et al. [20] who made use of the photophysical properties of C_{60}. On irradiating a solution of C_{60} and DNA he was able to selectively fragment of the DNA-string at the G-base position using the efficient singlet oxygen generation by C_{60}.

With these short outlooks into promising field of fullerene science I would like to turn now to the area of patents on fullerenes. Currently we can see a strong upturn on the number of patents and patent applications. As is inherent in the patenting process, there is a lag of two to three years between scientific publications and patent disclosures. If the explosive growth of scientific papers is anything to go by, the big surge of patents is still to come. As expected, the majority of applications is in the field of production and separation methods. They span a wide range of conceivable methods. To name but a few: "Fullerene production by evaporating a carbon source in a plasma" [21], "Fullerene synthesis in flames" [22], "Refining method for C_{60}" [23] or "Recognition selectors for carbon cluster compounds" [24]. The next biggest field of applications is material science. Some examples are the use as superconductors [26], as optical limiter [27], as toners for photocopiers [28] or in electrophotography [29]. The field of chemical applications is also beginning to show up in the patent literature. Two examples are the use as bromating agent of fullerene bromides [30] or the method of producing polysubstituted fullerenes [31].

To summarise this review I would like to quote again Wasserman's speech [2]: "As a field of basic research in industry fullerenes are attractive as there are a number of possible applications. Some of these could provide new business opportunities. In the short term extension of existing business and variations of existing products are likely to be the focus. It is difficult to discuss specific areas here as present programmes are still primarily concerned with fundamental understanding. Applications will have to wait on that understanding"

References

1. W. Krätschmer, L.D. Lamb, K. Fostiropulous, D. Huffman, *Nature* **347** (1990) 354
2. E. Wasserman, *Phil. Trans. Royal Soc. London A* **343** (1993) 129
3. A.F. Hebbard, M.J. Rosseinsky, R.C. Haddon, D.W. Murphy, S.H. Glarum, T.T.M. Palstra, A.P. Ramirez, A.R. Kortan, *Nature* **350** (1991) 600

4. T.W. Ebbesen, P.M. Ajayan, *Nature* **358** (1992) 220
5. a) J. Mort et al., *Chem. Phys. Lett.* **186** (1991) 281
    b) K. Pichler et al., *J. Phys.: Cond. Matter* 3 (1991) 9259
6. Y. Wan et al., *J. Am. Chem. Soc.* **115** (1993) 3844
7. W. Tutt, A. Kost, *Nature* **356** (1992) 225
8. a) M. Uchida et al., *Jap. J. Appl. Phys. II* 30 (1991) L2104
    b) A.T. Werner et al., *Appl. Phys. A* **57** (1993) 157
9. P.M. Allemand et al., *Science* **253** (1991) 301
10. a) T. Enoki, *Kagatem* **46** (1991) 850)
    b) J. Mort et al., *Chem. Phys. Lett.* **186** (1991) 284
11. Overney et al., *Z. Phys. D: AI, Mol. Clusters* **27** (1993) 93
12. Chromatographic Material: J. Kiyokatsu et al., *J. Microcol. Sep.* **4** (1992) 187
13. A.O. Patil et al., *Polym. Bull.* **30** (1993) 187
14. D.E. Bergbreiter, H. Gay, *J. Chem. Soc. Chem. Commun.* **645** (1993) 645
15. K. Geckeler, A. Hirsch, *J. Am. Chem. Soc.* **115** (1993) 385
16. J. Chuper et al., *J. Am. Chem. Soc.* **115** (1993) 4383
17. D.A. Loy, R. Assink, *J. Am. Chem. Soc.* **114** (1993) 3977
18. Samulski et al., *Chem. Mat.* **4** (1992) 1153
19. a) R.F. Schinazi et al., *Antimic. Agents Chenoth.* **37** (1993) 1707
    b) S.H. Friedman et al., *J. Am. Chem. Soc.* **115** (1993) 6506
20. H. Okuyama et al., *J. Am. Chem. Soc.* **115** (1993) 7918
21. US 5227038
22. WO 9220622
23. JP 05085711
24. WO 9322272
25. a) A. Onsager, *Phys. Rev.* **54** (1938) 554
    b) A. Mozamder, *J. Chem. Phys.* **60** (1974) 4300
26. US 523479
27. US 5172278
28. US 5188918
29. US 5215841
30. GB 2265613
31. US 5177248

AUTHORS INDEX